TELECOMMUNICATION SYSTEMS ENGINEERING

PRENTICE-HALL INFORMATION AND SYSTEM SCIENCES SERIES

Thomas Kailath, *Editor*

BERGER	*Rate Distortion Theory: A Mathematical Basis for Data Compression*
DI FRANCO & RUBIN	*Radar Detection*
DOWNING	*Modulation Systems and Noise*
DUBES	*The Theory of Applied Probability*
FRANKS	*Signal Theory*
GOLOMB, ET AL.	*Digital Communications with Space Applications*
LINDSEY	*Synchronization Systems in Communication and Control*
LINDSEY & SIMON	*Telecommunication Systems Engineering*
MELSA & SAGE	*An Introduction to Probability and Stochastic Processes*
PATRICK	*Fundamentals of Pattern Recognition*
RAEMER	*Statistical Communication Theory and Applications*
STIFFLER	*Theory of Synchronous Communications*
VAN DER ZIEL	*Noise: Sources, Characterization, Measurement*

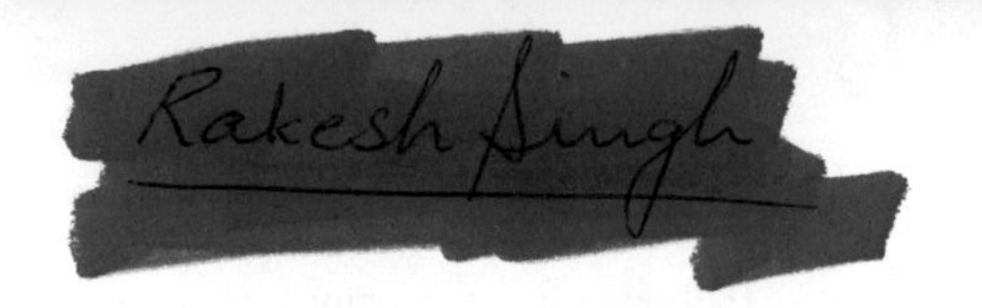

TELECOMMUNICATION SYSTEMS ENGINEERING

William C. Lindsey
University of Southern California

Marvin K. Simon
Jet Propulsion Laboratory

PRENTICE-HALL, INC.
Englewood Cliffs, New Jersey

Library of Congress Cataloging in Publication Data

(Prentice-Hall series in information and system sciences)
Included bibliographical references.
1. Signal theory (Telecommunication) 2. Artificial satellites in telecommunication. #. Data transmission systems. I. Simon, Marvin Kenneth, 1939- joint author. II. Title.
TK 5102.5.L57 621.38'042 73-9630
ISBN 0-13-902429-8

10 9 8 7 6 5 4 3 2 1

Printed in the United States of America

PRENTICE-HALL INTERNATIONAL, INC., *London*
PRENTICE-HALL OF AUSTRALIA, PTY. LTD., *Sydney*
PRENTICE-HALL OF JAPAN, INC., *Tokyo*
PRENTICE-HALL OF CANADA, LTD., *Toronto*
PRENTICE-HALL OF INDIA PRIVATE LTD., *New Delhi*

To

DOROTHY ANITA
JOHN BRETTE

AND OUR PARENTS

CONTENTS

PREFACE

During recent years a vast body of knowledge central to problems arising in the design and planning of telecommunication systems for space applications has accumulated in various professional journals and other publications. The words telecommunication or telecommunication system imply different things to different people. In telephony it implies simultaneous transmission of thousands of voice signals, teletype signals, facsimile and television via cables or microwave links. In cinema, it is used in determining and assessing the social, political and cultural impact which the news and television media have on the public, while in computer communication networks it takes on yet another meaning.

In this book the term telecommunication systems engineering is used to imply the design and planning of systems which must perform the functions of digital data transmission, tracking, telemetry, and command via radio links. Present technology, equipment constraints, propagation factors, and bandwidth are forcing the design engineer to favor the use of microwaves, e.g., the L, S, C, X and K frequency bands. At present, the technology is less developed in the C, X and K bands.

Telecommunication systems engineering, in the context addressed in this book, began at the Jet Propulsion Laboratory in the late 1940s when engineers were faced with the problem of providing telemetry and radio guidance for short-range ballistic missiles, and later on the problem of producing reliable communications under conditions of heavy interference. Solutions to these problems led to an intensive study and expansion of the earlier work of C. Shannon, N. Wiener and the staffs of MIT Radiation and Lincoln Laboratories. A very rapid development of the theory and operation of servomechanisms functioning in the presence of noise followed. Applications of this theory in the early fifties by E. Rechtin and R. Jaffe led to the development of phase-locked receivers, automatic gain control systems, coherent two-way velocity and range measuring systems, etc. These techniques led to the implementation of the Microlock system under Army auspices. The advent of Sputnik I in 1957 launched the USA into the space business and shortly thereafter the National Aeronautics and Space Administration was created. The telecommunication technique just mentioned along with considerable amounts of the hardware were immediately applicable to new problems of telemetry, tracking, and command. Indeed, the telemetry techniques developed for the Microlock system were used for the first American satellites, the Explorer series.

Space exploration and the use of satellites for commercial and military purposes posed new problems in utilization of radio signals near threshold, use of real time telemetry and commands, and the design and operation of small and large precision antenna structures. Over the past ten years, ground antennas have grown from the relatively small helical type used for the early Microlock system to the giant 64 meter type parabolic structure used in the Deep Space Network. Ground receiver noise temperatures have dropped from over 1000°K to less than 25°K. Data processors have evolved from hand reduction based on strip-chart records to large interconnected computer systems. Data rates for deep space applications have increased from a few hundred bits per second to several kilobits per second at communication distances which presently reach to the outer planets. Current satellite technology is capable of supporting data rates of the order of one to two gigabits per second. Organizations too numerous to mention have contributed to this base of technology and knowledge.

This text concentrates on the presentation and development of a theory that can be used in the design and planning of telecommunication systems which must operate with either small or large performance margins. Excess margins on the one had imply excess power, weight and cost; on the other hand, negative margins imply unacceptable performance. In order to design an efficient system, i.e. one that operates reliably with a small performance margin, it is necessary to have an accurate theory which accounts for the deleterious effects that affect this margin. For those designs wherein margin is plentiful and of no concern, then seat-of-the-pants engineering can be used to carry out the design.

Several features make this text ideally suited for practicing engineers as well as for graduate level students in communication systems courses.

1. For the first time in one text a theory for use in the design of one-way and two-way phase-coherent tracking and communication systems is presented.
2. Various methods of carrier (subcarrier) and suppressed carrier (subcarrier) synchronization techniques, e.g., generalized Costas loops, *N*th power loops, decision-directed loops, delay-locked loops, hybrid and data-aided loops, etc. are presented, analyzed and compared.
3. An improved treatment of the band-pass limiter theory is given and applied toward evaluating the performance of phase-coherent tracking

and communication systems which employ band-pass limiters in the early stages of the receiving system.

4. The subject of phase-coherent detection with perfect and noisy synchronization reference signals is unified. Results are obtained and graphically illustrated for orthogonal, bi-orthogonal, transorthogonal, polyphase, and L-orthogonal transmitted signal sets. Tabulations of error probability vs. signal-to-noise ratio for thse signal sets are provided.
5. The problem of ambiguity resolution (associated with suppressed carrier transmission and receiving systems) by differential encoding of the data is treated. Both coherent and differentially coherent detection of the differentially encoded signals are covered extensively. A comparison is drawn among the performances of systems employing these various signaling and detection techniques, e.g., biphase, quadriphase, octaphase.
6. A number of results pertaining to the design and performance of communication systems employing convolutional codes are presented and summarized.
7. Various methods of obtaining symbol (bit, clock) synchronization from the data-bearing signal itself as well as via separate channel techniques are discussed, analyzed and compared.
8. The problem of noncoherent detection of M-ary signals is treated, and special approaches to data detection by means of the Fast Fourier Transform (FFT) and Finite Time Autocorrelation are discussed. The evaluation of performance in the presence of receiver frequency and time uncertainties is included along with methods for achieving time and frequency synchronization.
9. A comprehensive set of homework problems which demonstrate the application of the theory developed is given at the end of the chapters. Solutions to many of these problems are available from the authors upon request.

The authors are grateful for the help of many colleagues during the preparation of this manuscript, and are particularly, indebted to Mr. James C. Springett of the Jet Propulsion Laboratory and Dr. Martin Nesenbergs of the Institute of Telecommunications Sciences, Department of Commerce, for their comments and criticisms of this volume. In addition, the authors appreciate and acknowledge the technical support provided by Professors Robert Gagliardi, Robert L. Scholtz, Lloyd Welch and Charles L. Weber of the University of Southern California and Professor Thomas L. Kailath of Stanford University. Much credit is also due to Miss Corinne J. Leslie who contributed a great deal in preparing the typescript.

An important factor in the completion of this book was the sustained financial support provided by the Office of Naval Research's Statistics and

Probability Program under the direction of Dr. Bruce J. McDonald. Finally, we wish to acknowledge the financial sponsorship provided by Mr. Leslie Klein and Dr. Sherman Karp of the Transportation Systems Center, Department of Transportation.

Pasadena, California

WILLIAM C. LINDSEY
MARVIN K. SIMON

Chapter 1

TELECOMMUNICATION NETWORK CONCEPTS

1-1. INTRODUCTION

This chapter examines the various factors that affect the design of a modern telecommunication network, specifies their relative significance, and introduces the reader to some basic terminology as it as used in this text. The word *telecommunication* by itself is literally translated as "communication at a distance." A *network*, simply defined, is an interconnected or interrelated chain of systems, each of which performs a particular function. A *telecommunication network*, then, is a group or collection of systems that combine through various links to provide a closed information loop—that is, one that provides for inquiry or command, response, and interpretation of response in relation to inquiry or command.

The particular type of telecommunication network of interest in this text is a *tracking and data acquisition network* with emphasis being placed on the *ground station to vehicle link*. Most of the tracking and data acquisition networks in operation today are designed to support many flight missions simultaneously; multimission support is the key to economical operation. For example, the United States National Aeronautics and Space Administration (NASA) has three tracking and data acquisition networks for supporting space exploration. The Space Tracking and Data Acquisition Network (STADAN)[1] is designed to handle automated unmanned satellites in Earth orbits with either high or low inclinations and with various orbit heights ranging from the minimum altitudes up to those in support of the manned space explorations. The Manned Space Flight Network (MSFN) is designed specifically to support the manned space explorations.[1] The NASA Deep

Space Network (DSN) provides tracking and data acquisition support to lunar, planetary, and deep-space missions.[1,2]

The DSN, as well as the MSFN, employs the *unified S-band* concept. The Unified S-Band System was originally used on the Mariner Mars 1964 mission and the Lunar Orbiter to provide reliable tracking and communications. It was later adapted to the Apollo manned spacecraft program.[1,2] In principle, the Unified S-Band System employs a single radio frequency (RF) carrier in each direction for the transmission of all tracking and communications data between a space vehicle—for example, a missile, satellite, or spacecraft—and the ground. Any *uplink* or *downlink* data to the space vehicle or to the ground is usually modulated onto *subcarriers*, which are in turn modulated onto the RF carrier. Such network design is discussed in detail in several of the chapters that follow.

The most attractive concept for the next generation tracking and data acquisition network is the TDRSN, which consists of synchronous-orbit *tracking* and *data relay satellites* for covering launches and low-orbit Earth satellites plus a few selected stations for supporting spacecraft in high Earth orbit and lunar orbit.[3] Figure 1-1 illustrates the TDRSN concept. Applications of a modified version include *air traffic control* and *navigation*,[4] relay of data collected via *earth resource satellites*[5] and support of *earth/space/base* stations.[6] The unique characteristic of a satellite is its ability to view a complete geographic area. This permits an entirely new approach to domestic,

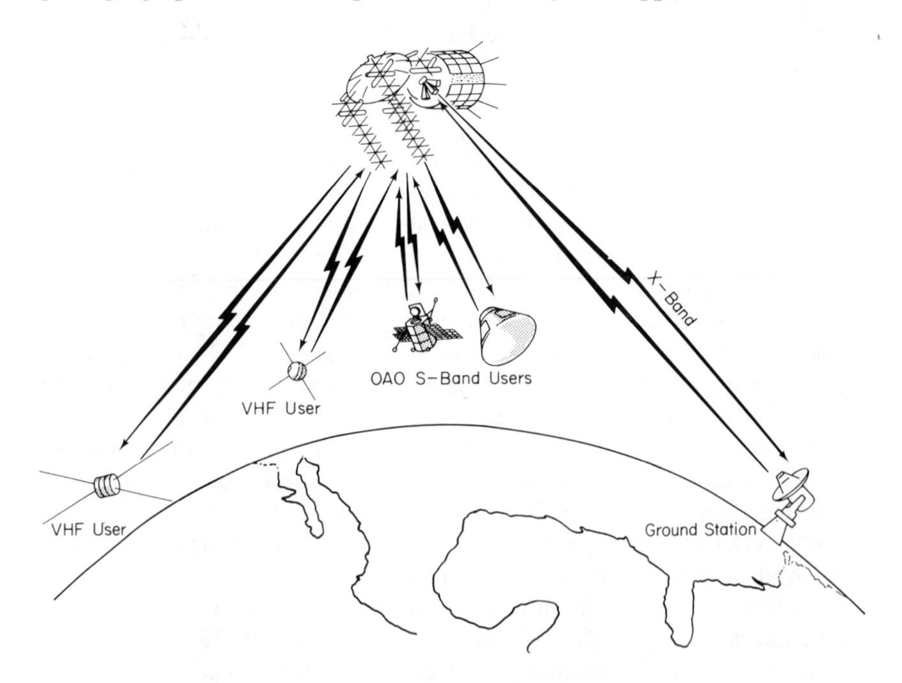

Fig. 1-1 Tracking and data relay satellite concept

commercial, and military communications. For example, under existing conditions a call from Los Angeles to New York via exclusively terrestrial means normally would pass through approximately ten distinct switching offices. By contrast, a call via a satellite is capable of going directly from the originating end office to the satellite and down to the end office of the party called. By this means only one switching level is required at each end, thereby eliminating eight switching centers and the necessary expense attendant to their use. Newer space vehicles are incorporating computer-type systems in the space vehicle that may require programming and memory load through the ground command link.

An example of a tracking and data acquisition network designed for the purposes of navigation and traffic control is NAVSTAR, studied by Thompson-Ramo-Wooldridge (TRW) Systems, Inc. for NASA (Fig. 1-2).[4] This navigation and traffic control network was intended to provide: (1) Atlantic coverage with the capability of eventual expansion to a worldwide network, (2) a navigation technique accommodating passive users, and (3) service for the broadest possible range of users, for example, large aircraft such as the SST, DC-10, and Boeing 747, general aviation, and marine craft. Conceivably, such a network could also be used to monitor meteorological conditions throughout the world.

Military applications frequently call for the design of a network that integrates the functions of communication, navigation, and identification —that is, the ICNI network concept.[7] Such a network requires (1) long-range

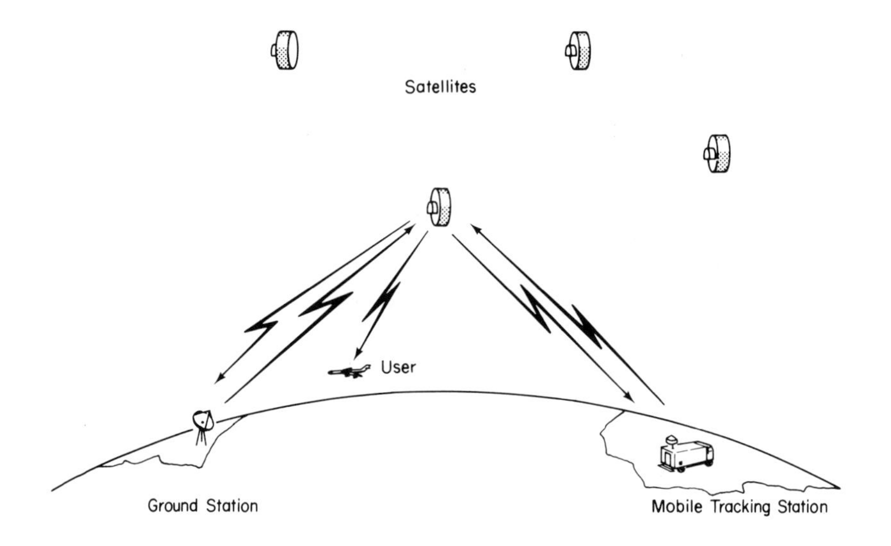

Fig. 1-2 NAVSTAR information network. From Technical Report TRW No. 08710–6012–R000, December, 1967.

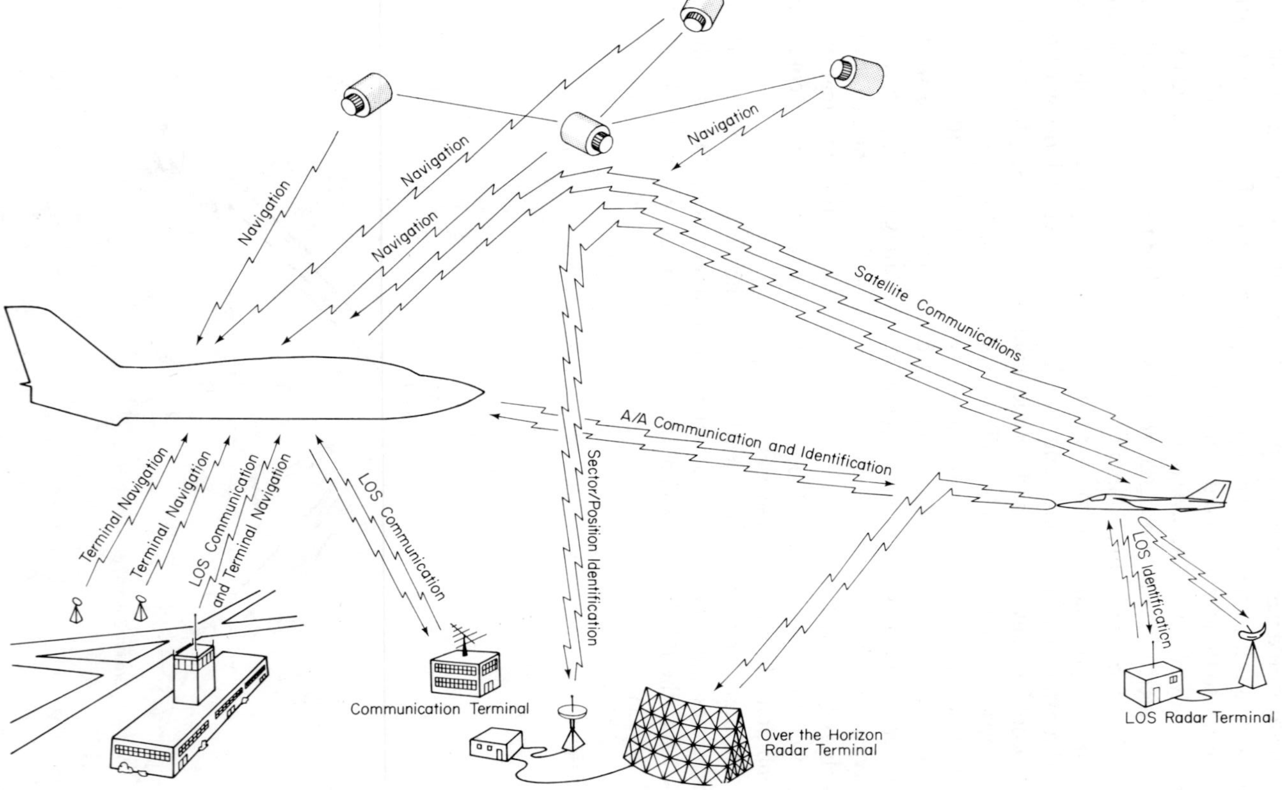

Fig. 1-3 ICNI network concept

communications, (2) short-range or line-of-sight (LOS) communications, (3) global navigation, (4) an LOS identification capability, and (5) a long-range identification capability. Figure 1-3 illustrates the appropriate network model along with the functional relationships and links between the various system terminals.

A typical element of a telecommunication network includes two distinct systems of operation: *data processing* and *telecommunication*. Our interest lies only in the development of a theory that may be used to design, evaluate, and test a telecommunication system.

1-2. TELECOMMUNICATION SYSTEM FUNCTIONS IN A TRACKING AND DATA ACQUISITION NETWORK

The telecommunication system must perform the functions of *signal acquisition and tracking*, *telemetry or data acquisition*, *command*, and *synchronization*. Here we define and qualitatively discuss these functions, reserving strict quantitative measures for later chapters.

The Signal Acquisition and Tracking Function

The signal acquisition function must be performed in many dimensions since it involves several different aspects of the general problem of acquiring the space vehicle. If the acquisition signal is being received from a transmitter aboard an Earth satellite, then the receiving antenna must first be pointed in a direction corresponding to the vehicle's position in the sky. This procedure involves acquisition in two spherical coordinates. In addition to angular acquisition, the receiver must be tuned to the proper frequency for reception of the tracking information. This is accomplished by a one-dimensional search through an appropriate portion of the frequency spectrum. Finally, for phase-coherent reception and for certain ranging purposes, a phase acquisition is also required. Hence, signal acquisition poses a four- or more- dimensional problem.

Although the term *tracking* is sometimes used to mean all communications with the space vehicle, it is more correctly restricted to the problem of obtaining data that can be used to determine the vehicle's position and velocity relative to the sun and planets. Three types of data are used for this purpose: (1) *angle data*, the angle at which the radiated signal arrives with respect to certain Earth reference points; (2) *Doppler data*, the change in the carrier RF due to the space vehicle's velocity with respect to the Earth station; and (3) *range data*, the distance between the point of signal reception and the vehicle. Since all tracking data to and from the space vehicle is processed in ground installations, the flow of data through the network is a pattern of unsteady arrivals and retransmissions at different points in time.

Although angle data can easily be measured to an accuracy of 0.01 to 0.02 deg, such data are insufficient for orbit determinations except during the first few hours after launch. During this time, however, angle measurements are important in determining the space vehicle's position with respect to the earth, and the resulting data are used in the first orbital calculations. Thereafter, orbit determinations are made from Doppler data because of their extreme accuracies.

Very accurate space vehicle velocities can be obtained from Doppler measurements made with a phase-coherent tracking system. Such a system utilizes a *space vehicle transponder*, which coherently transmits to Earth an RF carrier whose frequency is at a fixed ratio relative to that received from Earth. For example, in the Unified S-Band System, the frequency transmitted from Earth to the spacecraft (in the 2110- to 2120-MHz band) is tracked by means of a *phase-locked loop* (PLL), multiplied by the ratio 240/221 and retransmitted to Earth. There the observed carrier component is tracked by means of another phase-locked loop, multiplied by the inverse ratio, or 221/240, and the product is subtracted from a sample of the Earth-to-space vehicle transmitted frequency. Such a technique is referred to as *two-way* tracking. The frequency difference is equal to twice the *one-way* Doppler frequency, and since a very stable atomic standard is used as a frequency source, the accuracy of the Doppler data is limited basically by the phase instabilities of the transmitters and exciters. Velocity accuracies of 0.1 to 0.01 mm/sec are obtainable with correlation periods of one minute. By integrating the Doppler frequencies, accurate range calculations can be made, and the movement of the spacecraft can be fitted to an orbital trajectory. With such data accuracies, spacecraft position is easily computed to within 10 km at the moon and 200 km at the near planets.

Range data are obtained by modulating the Earth-to-space vehicle RF carrier with an appropriately designed ranging signal. The space vehicle retransmits this ranging signal to the ground, reconstructing it when necessary before retransmittal. There it is correlated with the transmitted signal to measure the time delay due to round-trip transmission time. Since Earth's ionospheric and tropospheric transmission delays are known (and are relatively very small) and the carrier velocity in space is equal to the velocity of light, the range to the space vehicle can be easily calculated. Range measurement accuracies of 15 m at all ranges can be achieved with range resolutions of about 1 to 2 m. Range measurements provide a separate and distinct method of determining the location of a space vehicle since they depend on a measurement of the carrier group (photon) velocity, whereas Doppler data depend on the phase velocity. Range data are particularly useful for quickly determining the position of a space vehicle when it is in orbit around a planet, while Doppler data are most valuable in measuring changing velocities (as during maneuvers) and in establishing accurate orbits where precise curve fitting is possible.

The Telemetry or Data Acquisition Function

The *telemetering* of data from a space vehicle to Earth, commonly called data acquisition, can be accomplished by several methods. For example, one or more *subcarrier frequencies* can be modulated with data received from space vehicle sensors and transducers. In turn, the RF carrier is then phase modulated with the subcarrier(s). The ground receiver uses the *phase-locked loop principle* to track the RF carrier component. The output of the loop's phase detector can be used to extract the *modulated data subcarriers*. These, in turn, are applied to the *data demodulator* to obtain the original modulation signals. More will be said later in the chapter about this *modulation-demodulation* process.

Consider the deep-space communication problem where the modulating signals can be separated into engineering and science signals. The former are the so-called "housekeeping" signals, such as temperatures, pressures of gas and fuel containers, currents and voltages of power units, solar cells, and instrument packages, which measure the status of space vehicle operating units. The science signals represent the transmission of pictures and the measurements made by the scientific instruments of phenomena such as electromagnetic, cosmic, and nuclear radiation; micrometeorites; and magnetic fields.

The telemetry data conveyed by the telemetry signals flow continuously from the space vehicle and are received and processed on the ground via the downlink. In some applications, telemetry data flow to the space vehicle and are retransmitted to a distant ground station.

The Command Function

The operational control of a space vehicle is accomplished by commands transmitted via the uplink. These signals cause certain functions to be performed such as ignition of rocket motors, turn-on of electrical power to high-power transmitters or various scientific sensors, or start of a magnetic tape recorder. Some of these operations are so time-dependent on previous space operational events, or happen so rapidly, that they have to be done automatically. It is also desirable to initiate some operations automatically because it may not be possible to do so with real-time commands.

Deep space exploratory space vehicles, for example, use automatic controllers, which either have the commands permanently stored in their memories or have memories that can be programmed from the ground.

The stored commands can be either the primary commands or an alternative to real-time commands. In addition to the stored commands, provision is made for receiving commands from the ground that are to be acted on immediately. These are called real-time commands, even though it may take several minutes for the transmission from the ground to reach the space vehicle. If the timing of the command is critical, the transmission time is taken into account in establishing the moment when the command is to be sent.

Most commands, stored program changes or real-time ones, are sent in two-way communication with the space vehicle so that confirmation of the receipt and acceptance of the command by the space vehicle can be relayed back to the ground. In emergencies, commands can be sent in the blind —that is, without confirming knowledge that uplink lock has been achieved. For this purpose, most deep-space vehicles have a receiver permanently connected to a broad beam antenna so that commands can be sent and received regardless of the space vehicle's attitude. Hence, the flow of command data via the uplink is unsteady. Because of the importance of commands, the design of the uplink is generally such that error probabilities will not be greater than one in 10^5 or 10^6. This is contrasted with acceptable telemetry error probabilities (on the downlink) of one in 10^3, or even 10^2 when some *a priori* knowledge of the data is available.

The Synchronization Function

Basically, the synchronization function consists of estimating two parameters in the received signal: frequency and time relative to a given origin. In what follows, we must distinguish between at least three types or levels of synchronization. First, carrier and subcarrier reference synchronization is required for coherent reception of digital modulation. Second, symbol synchronization is needed for purposes of estimating the instants of time when the modulation may change state. Third, word synchronization determines the end of one code word and the beginning of another. Other types, such as frame synchronization, can be classified with word synchronization. In this text, we are concerned primarily with the first two types of synchronization.

1-3. TELECOMMUNICATION SYSTEM DESIGN FOR SPACE APPLICATIONS

The design of any telecommunication system makes use of the so-called *communications equation*, which contains all of the basic elements inherent in the communication system and makes possible the specification of the received signal-to-noise spectral density ratio with a high degree of certainty:

$$P_tG_t\left(\frac{\lambda}{4\pi d}\right)^2 L_\theta L_p G_r \frac{1}{kT^\circ} = \frac{P_{\text{rec}}}{N_0}$$

$$\frac{S}{N_0} = (1 - L_m)\frac{P_{\text{rec}}}{N_0}$$

$$\frac{P_c}{N_0} = L_m \frac{P_{\text{rec}}}{N_0} \tag{1-1}$$

Each element of these "power budget" equations is defined and discussed here as to its effect on the system design. The parameter P_t represents the *total power transmitted:* power that is measurable with a wideband wattmeter at the transmitter antenna input. The parameter L_m represents the *transmitter modulation loss.* Essentially it specifies the power division between the carrier and information sidebands resulting from a particular *modulation process.* If the basic information carrier is a microwave, continuous-wave (CW) sinusoidal voltage, the modulation can be impressed on the transmitted carrier by *angle modulation* [phase (PM) or frequency (FM)], *amplitude modulation* (AM), or a simultaneous combination of the two. Analytically speaking, the instantaneous value of a modulated RF carrier can be expressed by

$$s(t) = \sqrt{2}\,Am(t)\cos\left[\omega_0 t + \theta(t)\right] \tag{1-2}$$

where $m(t)$ characterizes the transmitter amplitude modulation, $\theta(t)$ characterizes the transmitter angle modulation, ω_0 is the carrier radian frequency, and $\sqrt{2}\,A$ denotes the *peak carrier amplitude* when $m(t)$ is unity. The modulation signals—the information to be transmitted—can be classified as digital, analog, or a combination of the two.

Phase modulation appears to be the most attractive for tracking and communication purposes. This is because:

1. A carrier component is available for Doppler recovery (not true in FM).
2. Large amounts of carrier power can be converted to sideband power.
3. Nonlinear amplification does not distort the modulation waveform (not true in AM).
4. The ratio of peak power to average power is unity, providing a more efficient system (not true in AM).
5. The modulation mechanism is simple and reliable (not true in AM).

The sideband spectra resulting from various types of modulation are well documented in standard texts on communications engineering.[8-11] A typical spectrum resulting from the modulation of a subcarrier by binary data and the subsequent modulation of this waveform onto an RF carrier are shown in Fig. 1-4. Several representations for the binary waveform ap-

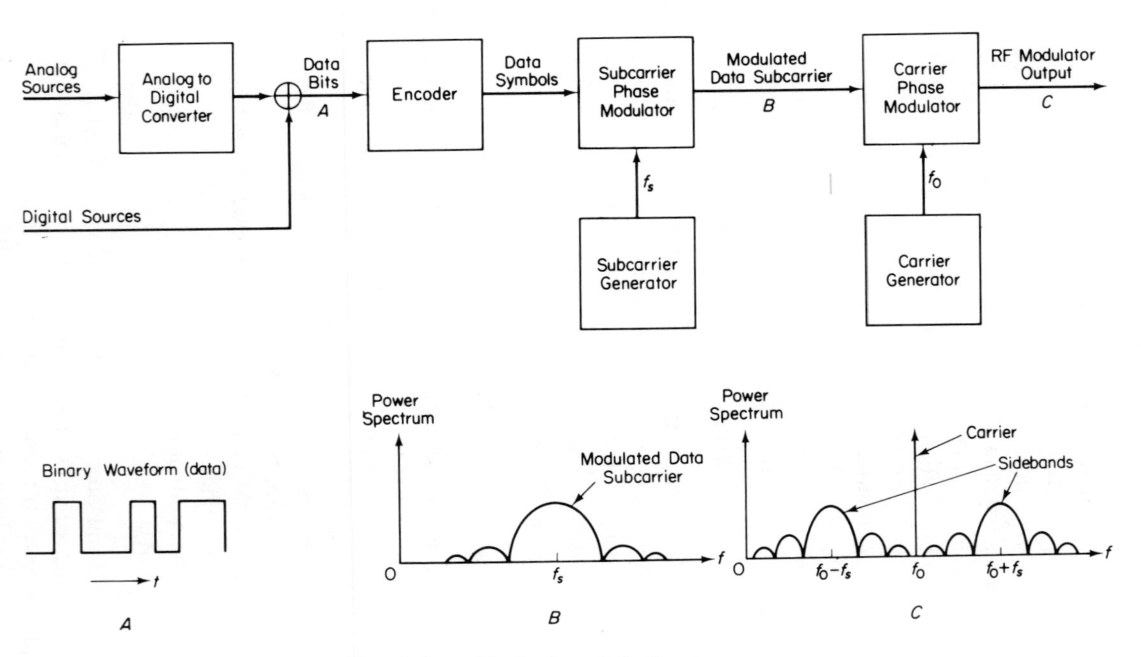

Fig. 1-4 Typical modulation spectrum

pearing at point A in Fig. 1-4 are permissible. Typical graphic and verbal descriptions of several of these conventions are illustrated in Fig. 1-5 for various *pulse code modulation* (PCM) formats.[12]

The parameter G_t in Eq. (1-1) represents the *transmitter antenna gain.* This parameter is defined as

$$G_t = \frac{4\pi A_t}{\lambda^2} \tag{1-3}$$

where $\lambda = 2\pi c/\omega_0$ is the transmitting wavelength with $c = 3 \times 10^8$m/sec the propagation velocity of the transmission medium, and A_t is the antenna *capture area;* G_t is assumed to include the aperture efficiency.

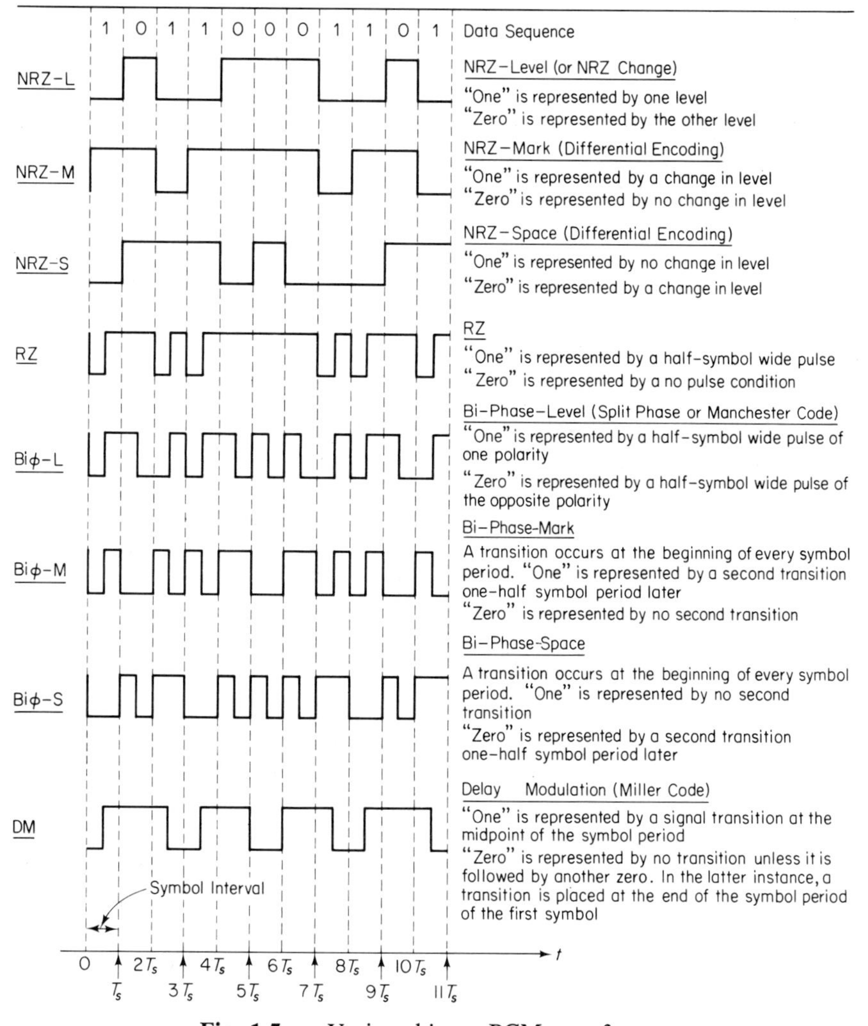

Fig. 1-5 Various binary PCM waveforms

The gain of any antenna is referenced to an *isotropic* (equally radiative in all directions) *antenna* operating at the same frequency. The aperture or capture area of an isotropic antenna (gain = 1) is defined as being equal to $\lambda^2/4\pi$. In this definition, the gain of an isotropic antenna is referenced to a physically realizable antenna, such as a dipole, whose resonant size, and hence, capture area, is a function of frequency.

The gain of an antenna is related to its directivity, or the increase in sensitivity in a particular direction that results from decreasing the sensitivity in other directions. The transmitted effective radiated power (over that available through an isotropic antenna) in a particular direction will be increased by the gain of the antenna. The obvious penalty for increased gain is reduced beamwidth. As a typical example, the one-half-power beamwidth in degrees as determined from the pattern of a large circular aperture is approximately $180/\sqrt{G_t}$. The pattern of a large aperture usually has a $(\sin x)/x$ type of angular distribution of radiated power.

The factor $(\lambda/4\pi d)^2$ represents the so-called *space loss*. The power P_r available at the aperture of a receiving antenna from an isotropic transmitting antenna is equal to the ratio of the area of the receiving antenna aperture A_r to the area of a sphere whose radius is the distance d between the two antennas times the power output of the transmitting antenna P_{to}. Thus,

$$P_r = \left(\frac{A_r}{4\pi d^2}\right)P_{to} \tag{1-4}$$

Substituting for A_r the expression equivalent to Eq. (1-3) in terms of the *receiving antenna gain* G_r, we obtain

$$\frac{P_r}{P_{to}} = G_r\left(\frac{\lambda}{4\pi d}\right)^2 \tag{1-5}$$

and, for the general equation, where the transmitting antenna has a gain G_t,

$$\frac{P_r}{P_t} = G_t G_r\left(\frac{\lambda}{4\pi d}\right)^2 \tag{1-6}$$

The parameter L_θ represents pointing loss of either the *transmitting* or *receiving antenna*. This factor is most important in communications involving a space vehicle omnidirectional antenna. It is possible to design a semi-omnidirectional antenna that functions well in one hemisphere, although nulls of 5 to 40 dB would occur in the opposite hemisphere. One of the design problems is to arrange the antenna orientation so that, over the normal pointing history expected during a flight, none of these nulls is encountered. Where this cannot be done, as for instance during a midcourse maneuver, the appropriate selection of the pitch, roll, and yaw sequence will usually prevent any deep nulls from being encountered. This procedure requires that complete spherical patterns be available for comparison and analysis when the maneuver is planned so as to preclude communication outages.

The pointing loss factor is also important when the space vehicle attitude-stablization accuracy approaches the beamwidth of the receiving or transmitting antenna.

The parameter L_p represents the *polarization loss factor*. Polarization loss results when the received polarization (usually a function of the design of the spacecraft transmitting antenna) does not match the polarization characteristics for which the ground receiving antenna was designed. There is a net loss of signal because the antenna cannot extract all the energy from the incoming wave. Theoretically, the polarization loss factor can vary from zero to infinity. Practically, it varies from a few tenths of a dB to 20 or 30 dB for cross-polarized signals (linear vertical to linear horizontal or right-hand circular to left-hand circular).

Polarization loss becomes significant in communications with omnidirectional antennas because of the difficulty of maintaining the design polarization over large look angles. As a typical example, the Mariner 4 omnidirectional antenna was essentially circularly polarized over most of the forward hemisphere, but at the 120° points, it was nearly linearly polarized.

There are other polarization loss factors, such as Faraday rotation through the ionosphere; but these losses do not generally disturb signals received with a circularly polarized antenna.

The parameter G_r represents *receiving antenna gain*. Since this factor is generally defined with respect to isotropic antenna gain at a particular frequency, a constant-area antenna will show an increase in gain with an increase in frequency. Actually, the power intercepted by a fixed-area receiving antenna is constant, and the apparent gain increase results from the capture area of an isotropic antenna being an inverse function of frequency squared. Omnidirectional antenna gain follows this inverse square law.

The parameter $N_0 = kT^\circ$ represents the normalized noise *power spectral density*, in watts per Hertz (w/Hz). This factor is usually referenced to the input stage of the receiver, since the signal-to-noise ratio is established at that point. Here, T° denotes the system noise temperature in degrees Kelvin (°K) and $k = 1.38 \times 10^{-23}$ w/Hz-°K is *Boltzmann's constant*. In a well-designed *diplexed, S-band, angle-tracking feed system*, the factors that affect N_0 are as given below, along with typical values in °K:

Galactic and radio source noise	4°
Atmospheric noise	2°
Earth temperature noise	1°
Antenna and transmission loss temperature noise	20°
Microwave amplifier noise	10°
Total noise temperature, T°	37°

The standard technique employed to measure the noise temperature of a system is to compare, in a fixed bandwidth B, a measurement of system *noise power* N with the noise power from a resistor at a known temperature.

Since N and B are related by $N = kT^\circ B$, the unknown system noise temperature can be determined quite simply. The quantity B is defined as the *single-sided noise bandwidth*, and noise power is always measured with respect to it.

Throughout this book we shall assume that the channel noise is white and Gaussian with a single-sided spectral density of $N_0 = kT^\circ$ w/Hz. A broad class of important communication channels are disturbed by time-varying multipath, non-Gaussian noise, atmospheric interference, etc; however, a thorough treatment of this subject requires documentation in a separate text.

The ratio P_{rec}/N_0 represents the *total received signal-to-noise spectral density ratio*. The ratio S/N_0 represents the *received signal-to-noise spectral density ratio in the data channel.* This ratio is generally established by the system objectives and type of data to be transmitted. Finally, P_c/N_0 is the *carrier signal-to-noise spectral density ratio* and is completely specified once the ratios P_{rec}/N_0 and S/N_0 have been selected.

Table 1-1 presents a set of basic communication parameters typical of deep-space communication. In this table, the notation dBm stands for decibels relative to one milliwatt of power.

TABLE 1-1 Typical Deep-Space Communication System Parameters

Parameter	*Value*
Total transmitter power (10 w), P_t	+40.0 dBm
Transmitter modulation loss, L_m	−4.1 dB
Transmitting-antenna gain, G_t	+23.5 dB
Space loss, $(\lambda/4\pi d)^2$ (at 2295 MHz, $d = 2.46 \times 10^8$ km)	−267.5 dB
Transmitting-antenna pointing loss, L_θ	−1.1 dB
Polarization loss, L_p	Included in transmitting-antenna pointing loss
Receiving-antenna gain, G_r	+ 53.0 dB
Receiving-attenna pointing loss, L_θ	0 dB
Total received power, P_{rec}	−152.1 dBm
Received data power, S	−154.2 dBm
Received carrier power, P_c	−156.2 dBm
Receiver noise spectral density, $N_0 = kT^\circ$ ($T^\circ = 55 \pm 1^\circ$K)	−181.2 dBm/Hz

1-4. SHANNON'S THEOREM AND COMMUNICATION SYSTEM EFFICIENCY

It is important to note here the promise theorized by Shannon[13–15] with regard to the efficiency of a communication system. Shannon shows that for the additive white Gaussian noise channel, coding schemes exist for which it is possible to communicate data *error free* over the channel provided the system *data rate* $\mathcal{R}_b$ (in bits per second) is less than or equal to what he called the *channel capacity* C. If $\mathcal{R}_b > C$, then error-free transmission is not possible. The channel capacity C, in bits per second, is defined by

$$C = B \log_2 \left(1 + \frac{S}{N}\right) \tag{1-7}$$

where S represents the *received signal data power* in watts, and B and N have previously been defined. Now S/N can be written as

$$\frac{S}{N} = \frac{ST_b\mathcal{R}_b}{N_0 B} = \frac{ST_b/N_0}{B/\mathcal{R}_b} = \delta_b R_b \tag{1-8}$$

where T_b is the time duration of a data bit in seconds—that is, the inverse of the data rate, $\delta_b = \mathcal{R}_b/B$, is the ratio of the data rate to the single-sided channel noise bandwidth in bits/Hz and $R_b = ST_b/N_0$ is the energy per bit-to-noise spectral density ratio. Frequently, the parameter R_b is referred to as the *relative efficiency* or *communication efficiency* of a particular communication scheme.

Substituting Eq. (1-8) and $B = \mathcal{R}_b/\delta_b$ into (1-7) and assuming communication at channel capacity ($\mathcal{R}_b = C$), we get

$$\delta_b = \log_2 (1 + \delta_b R_b) \tag{1-9}$$

Equation (1-9) is plotted in Fig. 1-6 as a function of $1/\delta_b$ and R_b. Notice that as the value of R_b decreases, the transmission bandwidth necessary

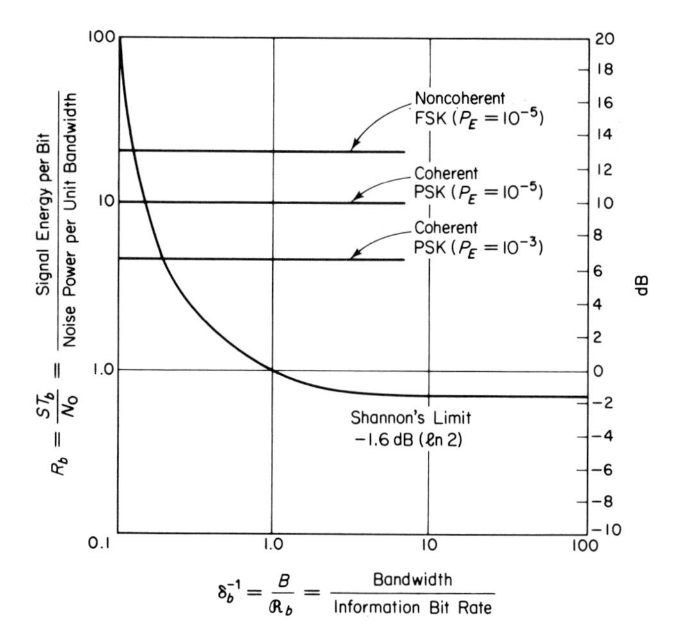

Fig. 1-6 Relative efficiency versus the ratio of bandwidth to information bit rate.

to achieve 100 percent efficiency—operation at *C*—*increases*. It is therefore instructive to allow B to approach infinity in Eq. (1-7) or equivalently let δ_b go to zero in (1-9). Taking the appropriate limit then gives

$$R_b = \ln 2 = -1.6\ \text{dB} \tag{1-10}$$

Consequently, coding schemes exist for which error-free transmission is possible in the *unconstrained*, additive, white Gaussian noise channel provided that $ST_b/N_0 \geq -1.6$ dB. Unfortunately, Shannon's theorem does not specify the data encoding method to accomplish this task.

The relative efficiency of any practical communication system can be compared with the most efficient on the basis of how closely it approaches the *Shannon limit*, Eq. (1-9), for a given error probability. In the following chapters we consider various modulation, demodulation, coding, decoding, and data-detection methods that attempt to approach the limit expressed by Shannon's theorem. Typical conclusions taken from later studies of various methods are compared with the Shannon limit in Figure 1-6.

1-5. SPECTRAL OCCUPANCY AND BANDWIDTH CONSIDERATIONS

As suggested by Shannon's theorem discussed in the previous section, two factors that are important in evaluating the efficiency of any communication system are the bandwidth required to transmit the chosen signaling or modulation technique and the energy per bit-to-noise spectral density ratio. Whereas the latter factor is a well-defined quantity, the former can be defined in many ways.[16] For example, a common definition of bandwidth is the width, in Hz, of an ideal rectangular filter that is necessary to pass a pre-specified percentage of the total signal energy. If one is referring to baseband signaling—that is, prior to any modulation of a carrier or subcarrier—then the ideal filter is a low-pass filter. On the other hand, if one desires to specify the bandwidth requirements at the carrier level, then this ideal filter becomes a band-pass filter. In either case, the required transmission bandwidth is directly related to the *power spectral density* of the signaling scheme.[16]

Aside from bandwidth considerations, a number of other criteria exist which must be considered when attempting to select a particular baseband signaling technique. The following four criteria are used most often to compare various methods:

(1) signal synchronization capabilities,
(2) signal error-detecting capabilities,
(3) signal interference and noise immunity properties, and
(4) cost and complexity of implementation of the transmitter and data detector.

In what follows, we discuss only the spectral occupancy of the signaling scheme. In particular, we present a method whereby the power spectral density (or autocorrelation function) of a broad class of random sequences of N-ary waveforms can be evaluated.[17] The remaining criteria of selection are considered in Chap. 9.

Power Spectral Density of a Random Data Sequence Generated by a Markov Source

Consider a random N-ary source which every $T_b = T_s$ seconds emits an elementary signal from the set $\{s_i(t); i = 1, 2, \ldots, N\}$ with probability p_i. If the source is Markov, then the sequence of waveforms so generated is characterized by the set of probabilities $\{p_i; i = 1, 2, \ldots, N\}$ (often called the *stationary probabilities*) and the set of *transition probabilities* $\{p_{ik}; i = 1, 2, \ldots, N\}$. Transition probability p_{ik} is the probability that signal $s_k(t)$ is transmitted in any given transmission interval after the occurrence of the signal $s_i(t)$ in the previous transmission interval. These transition probabilities are conveniently arranged in a *transition matrix* **P** defined by

$$\mathbf{P} \triangleq \begin{pmatrix} p_{11} & p_{12} & \cdots & p_{1N} \\ p_{21} & p_{22} & \cdots & p_{2N} \\ \cdot & \cdot & \cdot & \cdot \\ \cdot & \cdot & \cdot & \cdot \\ \cdot & \cdot & \cdot & \cdot \\ p_{N1} & p_{N2} & \cdots & p_{NN} \end{pmatrix} \tag{1-11}$$

From this statistical description of the source, the power spectral density of a data sequence generated by this source is given by[17]

$$\begin{aligned} S(f) = {} & \frac{1}{T_s^2} \sum_{n=-\infty}^{\infty} \left| \sum_{i=1}^{N} p_i S_i\left(\frac{n}{T_s}\right) \right|^2 \delta\left(f - \frac{n}{T_s}\right) \\ & + \frac{1}{T_s} \sum_{i=1}^{N} p_i |S_i(f)|^2 \\ & + \frac{2}{T_s} Re\left[\sum_{i=1}^{N} \sum_{k=1}^{N} p_i S_i^*(f) S_k(f) p_{ik}(e^{-j2\pi f T_s})\right] \end{aligned} \tag{1-12}$$

where

$\delta(f)$ is the Dirac delta function

$S_i(f) \triangleq \int_0^{T_s} s_i(t) e^{-j2\pi f t}\, dt$ is the Fourier transform of the ith elementary signal

and

$$p_{ik}(z) \triangleq \sum_{n=1}^{\infty} p_{ik}^{(n)} z^n \tag{1-13}$$

In Eq. (1-12), the asterisk denotes complex conjugate. The quantity $p_{ik}^{(n)}$ is defined as the probability that the elementary signal $s_k(t)$ is transmitted n

signaling intervals after the occurrence of $s_i(t)$. Hence, from the properties of Markov sequences, $p_{ik}^{(n)}$ is the ikth element of the matrix $\mathbf{P}^n$. Also by definition, $p_{ik}^{(1)} \triangleq p_{ik}$.

Notice that the first term [hereafter called the *line* (*spike*) *spectrum*] of Eq. (1-12) vanishes when

$$\sum_{i=1}^{N} p_i S_i\left(\frac{n}{T_s}\right) = 0 \tag{1-14}$$

which implies that a necessary and sufficient condition for the absence of a line spectrum is that

$$\sum_{i=1}^{N} p_i s_i(t) = 0 \tag{1-15}$$

Many special classes of signals exist for which the general power spectral density result of (1-12) can be simplified. The first class is generated by the source defined by the conditions:

(1) for each signal waveform $s_i(t)$ of the set of elementary signals, the negative $-s_i(t)$ is also in the set,
(2) the stationary probabilities on $s_i(t)$ and $-s_i(t)$ are equal, and
(3) the transition probability $p_{ik} = p_{rs}$ whenever $s_j(t) = \pm s_r(t)$ and $s_k(t) = \pm s_s(t)$.

Such a signaling source is said to be *negative equally probable* (NEP). Its overall spectrum is characterized by the absence of a line spectrum and furthermore is independent of the transition probabilities themselves. For this special case, Eq. (1-12) reduces to

$$S(f) = \frac{1}{T_s} \sum_{i=1}^{N} p_i \,|\, S_i(f) \,|^2 \tag{1-16}$$

which is the weighted sum of the energy spectrum of the elementary signaling set.

Another class of signals of interest is those generated by a purely random source, i.e., one that emits an elementary signal in a given signaling interval independent of those emitted in previous signaling intervals. Such a source can be modeled as a degenerate case of a Markov source (i.e., $\mathbf{P}^{-1}$ does not exist) whose transition matrix is given by

$$\mathbf{P} = \begin{pmatrix} p_1 & p_2 & \cdots & p_N \\ p_1 & p_2 & \cdots & p_N \\ \cdot & \cdot & \cdot & \cdot \\ \cdot & \cdot & \cdot & \cdot \\ \cdot & \cdot & \cdot & \cdot \\ p_1 & p_2 & \cdots & p_N \end{pmatrix} \tag{1-17}$$

and has the property that $\mathbf{P}^n = \mathbf{P}$ for all $n \geq 1$. In this case, the power spectral density of Eq. (1-12) simplifies to

$$\begin{aligned} S(f) = \frac{1}{T_s^2} \sum_{n=-\infty}^{\infty} \left| \sum_{i=1}^{N} p_i S_i\left(\frac{n}{T_s}\right) \right|^2 \delta\left(f - \frac{n}{T_s}\right) \\ + \frac{1}{T_s} \sum_{i=1}^{N} p_i(1 - p_i) |S_i(f)|^2 \\ - \frac{2}{T_s} \sum_{\substack{i=1 \\ i \neq k \\ i<k}}^{N} \sum_{k=1}^{N} p_i p_k Re[S_i(f) S_k^*(f)] \end{aligned} \tag{1-18}$$

When $N = 2$ and $p_1 = p$, $p_2 = 1 - p$, Eq. (1-18) reduces to

$$\begin{aligned} S(f) = \frac{1}{T_s^2} \sum_{n=-\infty}^{\infty} \left| p S_1\left(\frac{n}{T_s}\right) + (1 - p) S_2\left(\frac{n}{T_s}\right) \right|^2 \delta\left(f - \frac{n}{T_s}\right) \\ + \frac{1}{T_s} p(1 - p) \left| S_1(f) - S_2(f) \right|^2 \end{aligned} \tag{1-19}$$

Obviously, if $s_1(t) = -s_2(t)$ and $p = \frac{1}{2}$, the signaling format becomes NEP and (1-19) reduces to (1-16).

At this point, we turn to an evaluation of the power spectral densities of the various PCM signaling formats illustrated in Fig. 1–5.

NRZ Baseband Signaling. Referring to Fig. 1-5 and assuming that the binary waveform levels are $\pm A$, the NRZ signaling format falls into the class of signals whose spectrum is given by (1-19). Since the elementary signal is a rectangular pulse of width T_s, its Fourier transform is

$$S_1(f) = -S_2(f) = AT_s \exp(-j\pi f T_s) \frac{\sin(\pi f T_s)}{\pi f T_s} \tag{1-20}$$

Substituting (1-20) into (1-19) and letting $E_s = A^2 T_s$, we get

$$\frac{S(f)}{E_s} = \frac{1}{T_s}(1 - 2p)^2 \delta(f) + 4p(1 - p)\left[\frac{\sin^2(\pi f T_s)}{(\pi f T_s)^2}\right]$$

When $p = \frac{1}{2}$, the dc spike at the origin disappears and the NRZ signaling format falls into the NEP class with

$$\frac{S(f)}{E_s} = \frac{\sin^2(\pi f T_s)}{(\pi f T_s)^2} \tag{1-21}$$

RZ Baseband Signaling. Here we have a situation where $S_1(f) = 0$ and $S_2(f)$ corresponds to the Fourier transform of a half-symbol wide pulse, i.e.,

$$S_2(f) = \frac{AT_s}{2} \exp\left(-\frac{j\pi f T_s}{2}\right)\left[\frac{\sin(\pi f T_s/2)}{\pi f T_s/2}\right] \tag{1-22}$$

Since the source is again purely random, substituting (1-22) into (1-19) gives

$$\begin{aligned} \frac{S(f)}{E_s} = \frac{1}{4T_s}(1 - p)^2 \delta(f) + \frac{1}{4T_s}(1 - p)^2 \sum_{\substack{n=-\infty \\ n \neq 0}}^{\infty} \left(\frac{2}{n\pi}\right)^2 \delta\left(f - \frac{n}{T_s}\right) \\ + \frac{1}{4} p(1 - p) \frac{\sin^2(\pi f T_s/2)}{(\pi f T_s/2)^2} \end{aligned} \tag{1-23}$$

*Bi-Phase Baseband Signaling.** Here,

$$s_1(t) = A \text{ for } 0 \le t \le T_s/2$$
$$s_1(t) = -A \text{ for } T_s/2 \le t \le T_s \quad \text{and} \tag{1-24}$$
$$s_1(t) = -s_2(t)$$

Substituting the Fourier transform of (1-24) into (1-19) gives

$$\frac{S(f)}{E_s} = \frac{1}{T_s}(1-2p)^2 \sum_{\substack{n=-\infty \\ n\neq 0}}^{\infty} \left(\frac{2}{n\pi}\right)^2 \delta\left(f - \frac{n}{T_s}\right) + 4p(1-p)\left[\frac{\sin^4(\pi f T_s/2)}{(\pi f T_s/2)^2}\right]$$

For $p = \frac{1}{2}$ the line spectrum disappears and

$$\frac{S(f)}{E_s} = \frac{\sin^4(\pi f T_s/2)}{(\pi f T_s/2)^2} \tag{1-25}$$

Delay Modulation or Miller Coding. As indicated by Hecht and Guida[18], the Miller coding scheme can be modeled as a Markov source with four states whose stationary probabilities are all equal to $\frac{1}{4}$ and whose transition matrix is given by

$$\mathbf{P} = \begin{pmatrix} 0 & \frac{1}{2} & 0 & \frac{1}{2} \\ 0 & 0 & \frac{1}{2} & \frac{1}{2} \\ \frac{1}{2} & \frac{1}{2} & 0 & 0 \\ \frac{1}{2} & 0 & \frac{1}{2} & 0 \end{pmatrix} \tag{1-26}$$

Another property of the Miller code is that it satisfies the recursion relation

$$\mathbf{P}^{4+l}\mathbf{S} = -\tfrac{1}{4}\mathbf{P}^{l}\mathbf{S} \qquad l \ge 0 \tag{1-27}$$

where S is the signal correlation matrix whose *ik*th element is defined by

$$s_{ik} \triangleq \frac{1}{T_s}\int_0^{T_s} s_i(t)s_k(t)\,dt \qquad i, k = 1, 2, 3, 4 \tag{1-28}$$

For the Miller code, the four elementary signals are defined by

$$s_1(t) = -s_4(t) = A \qquad \text{for } 0 \le t \le T_s$$
$$s_2(t) = -s_3(t) = \begin{cases} A & \text{for } 0 \le t < T_s/2 \\ -A & \text{for } T_s/2 \le t \le T_s \end{cases} \tag{1-29}$$

Substituting (1-29) into (1-28) and arranging the results in the form of a matrix, we get

$$\mathbf{S} = \begin{pmatrix} 1 & 0 & 0 & -1 \\ 0 & 1 & -1 & 0 \\ 0 & -1 & 1 & 0 \\ -1 & 0 & 0 & 1 \end{pmatrix} \tag{1-30}$$

* The term *bi-phase* as used here is not to be confused with the similar parlance used later in the text for phase-shift keying of a carrier or subcarrier.

Finally, using (1-26), (1-27), and (1-30) in the general power spectral density result of (1-12) yields the result for the Miller code:

$$\frac{S(f)}{E_s} = \frac{1}{2\theta^2(17 + 8\cos 8\theta)}(23 - 2\cos\theta - 22\cos 2\theta - 12\cos 3\theta + 5\cos 4\theta + 12\cos 5\theta + 2\cos 6\theta - 8\cos 7\theta + 2\cos 8\theta) \tag{1-31}$$

where $\theta \triangleq \pi f T_s$.

The power spectral densities of the NRZ, bi-phase, and delay modulation signaling schemes as given by (1-21), (1-25), and (1-31), respectively, are plotted in Fig. 1-7.

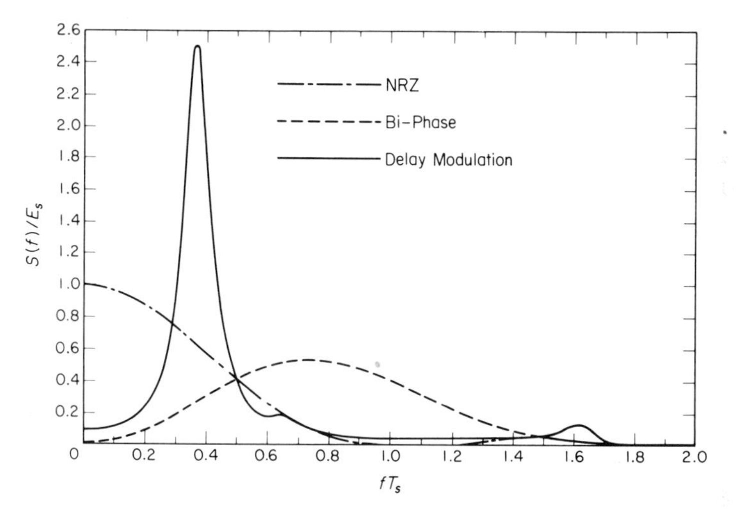

Fig. 1-7 Two-sided spectral densities of NRZ, bi-phase, and delay modulation waveforms.

Spectral properties of the Miller code that make it attractive for magnetic tape recording as well as phase-shift-keyed signaling include:

(1) The majority of the signaling energy lies in frequencies less than one-half the symbol rate, $\mathcal{R}_s = 1/T_s$.

(2) The spectrum is small in the vicinity of $f = 0$. This spectral minimum facilitates carrier tracking, which because of property (1) can also be more efficiently achieved than Manchester coding. A reduced spectral density in the vicinity of $f = 0$ is an important consideration in tape recording primarily because of the poor dc response.

(3) As a result of the second property, a lower magnetic tape recording speed (higher packing density) can be used.

(4) The Miller code is insensitive to the 180° phase ambiguity common to NRZ–L and Manchester coding.

(5) Bandwidth requirements are approximately $\frac{1}{2}$ those needed by Manchester coding.

Other well-known techniques for reducing spectral bandwidth rely on ternary, rather than binary, alphabets;[12] however, this is achieved at the expense of an increase in error probability performance and a complication of implementation.

1-6. FURTHER STUDIES

It is clear that in the short space of a chapter, one cannot hope to suggest all possible applications and system configurations that utilize satellite and space probes nor how they interconnect with ground terminals in a telecommunication network. However, what we have attempted to accomplish is to motivate the reader with several applications and network configurations which have been suggested in the literature and direct him to a set of additional references which summarize others. In this connection, some of the early qualitative work in telemetry for aerospace applications is documented in Ref. 19. The impact of the space program realized in the early 1960's brought about new techniques in telemetry, command, and tracking systems—in particular, those developed in conjunction with NASA's program of space probes and satellites planned for the 1970's and beyond. Some of these contributions are documented in Ref. 20. The technological and system aspects of domestic, military, and foreign satellite communication systems planned for service in the 1970's and beyond are presented in Refs. 21–24.

PROBLEMS

1-1. Find the antenna capture areas of the transmitting and receiving antennas in a communication system whose transmitter antenna gain is 26.4 dB (4-ft parabolic) and receiving antenna gain is 51.0 dB (85-ft parabolic). Assume that the system is designed to operate at either of the two carrier frequencies $f_0 = 2300$ MHz (S-band) or $f_0 = 9200$ MHz (X-band).

1-2. Determine the space loss at distances of 22,000 miles (typical of a satellite in synchronous orbit about Earth) and 1,000,000 miles (approximately four lunar distances) when $f_0 = 150$ MHz, 960 MHz, 2300 MHz, and 9200 MHz; that is, VHF, L-band, S-band, and X-band.

1-3. Using the parameters in Table 1-1:

a) Compute the ratio P_r/P_t.

b) Evaluate the received signal-to-noise spectral density ratios P_{rec}/N_0, S/N_0, P_c/N_0.

c) Find R_b for a data rate of 10 bits per second (typical of command system applications).

1-4. A spacecraft transmission system designed to operate at the distance $d = 2.2 \times 10^8$ km (Earth to Mars) and a transmitting carrier frequency $f_0 = 2300$ MHz has the following parameters: $P_t = 44$ dBm, $L_m = -3$ dB, $G_t = 26.4$ dB (4-ft parabolic), $L_\theta = -2$ dB, $L_p = 0$ dB, $G_r = 51$ dB (85-ft parabolic), and $T° = 40°$K. Find:

a) The space loss.
b) P_r/P_t.
c) The receiver noise spectral density N_0.
d) P_{rec}/N_0, S/N_0, P_c/N_0.

1-5. If the spacecraft system of Problem 1-4 is operating at 230 bits per second, find R_b. Reduce the rate to 3.0 bits per second and determine the required R_b.

1-6. Assume that the data detector in a receiving system is to operate at a bit error probability of $P_E = 10^{-3}$ and that this requires $R_b = 6.8$ dB. Find the maximum data rate at which the system can operate if $S/N_0 = 1000$. If the system performance requirements are changed such that $R_b = 10$ dB, find the new operating data rate.

1-7. A satellite-to-ground communication link and ground-to-satellite communication link operate with wavelengths of 0.48 ft and 2.1 ft, respectively.

a) Find the operating carrier frequencies.
b) If $d = 1.35 \times 10^8$ ft, find the space losses for each frequency found in part (a).
c) Determine the noise spectral density N_0 of the ground receiver if $T° = 80°$K.

1-8. System parameters in a ground-to-satellite communication link are given as follows: $P_t = 13$ dB, $G_t = 43.3$ dB, $\lambda = 2.1$ ft, $d = 1.35 \times 10^8$ ft, $G_r = 2$ dB, $L_p = L_\theta = 0$ dB, and $T° = 1000°$K. If $L_m = -2$ dB, find S/N_0 and the data rate which the link can support if $R_b = 10$ dB.

1-9. Compute the path attenuation (space·loss) in the Nimbus satellite communication system if the assigned transmitter frequency is 136.5 MHz and the horizon distance is 3700 km. Find N_0 if the system noise temperature is 1000°K.

1-10. Evaluate the wavelengths and space losses in the Pioneer V communication system at a distance of one million miles if the ground transmitting system operates at an assigned frequency of 378 MHz and the ground receiving system operates at 400 MHz.

1-11. Starting with Eq. (1-12) as the general expression for the power spectral density of a random data sequence generated by a Markov source:

a) Derive Eq. (1-18) for the special case of a purely random source.
b) Show that (1-18) simplifies to (1-19) in the binary case.

1-12. Evaluate the power spectral density for the frequency-shift-keyed (FSK) signaling technique where the elementary signals are given by:

$$s_1(t) = \sin \omega_1 t$$
$$s_2(t) = \sin \omega_2 t$$

with $\omega_1 = n_1\pi/T_s$, $\omega_2 = n_2\pi/T_s$ and where n_1 and n_2 are arbitrary nonzero integers. Assume that $p_1 = p_2 = \frac{1}{2}$ and sketch your result.

1-13. Repeat Problem 1-12 for the multiple frequency-shift-keyed (MFSK) signaling technique where the elementary signals are defined by $s_i(t) = \sin \omega_i t$ with $\omega_i = n_i\pi/T_s$, n_i an arbitrary nonzero integer, and $\omega_i - \omega_j = 2\pi k(i - j)/T_s$ for all $i \neq j$. Assume that $p_i = 1/N$ for all $i = 1, 2, \ldots, N$ and sketch your result.

1-14. Evaluate the power spectral density for the polyphase (MPSK) signaling scheme where the equiprobable elementary signals are given by

$$s_i(t) = \sqrt{2}A \sin(\omega_k t + \theta_i); \qquad i = 1, 2, \ldots, N$$

with $\omega_k = n_k\pi/T_s$, n_k an arbitrary nonzero integer, and $\theta_i \triangleq 2(i - 1)\pi/N$. Sketch your result for the special case $N = 2$—that is, phase-shift keying (PSK).

1-15. Repeat Problem 1-14 with θ_i considered to be a random variable uniformly distributed over the range $(0, 2\pi)$.

1-16. Starting with Eq. (1-12) and the properties of the Miller code as given in (1-26), (1-27), and (1-30), derive its power spectral density—that is, Eq. (1-31).

REFERENCES

1. COATES, R. J., "Tracking and Data Acquisition for Space Exploration," Goddard Space Flight Center, Greenbelt, Md., NASA No. N69–35443 (May, 1969).
2. HALL, J. R., LINNES, K. W., MUDGWAY, D. J., SIEGMETH, A. J., and THATCHER, J. W., "The General Problem of Data Return From Deep Space," *Space Science Reviews*, 8 (1968) 595–664.
3. STAMPFL, R. A., and JONES, A. E., "Tracking and Data Relay Satellites," *IEEE Transactions on Aerospace and Electronic Systems*, Vol. AES-6, No. 3 (May, 1970) 276–289.
4. OTTEN, D. D., "Study of Navigation and Traffic Control Technique Employing Satellites," TRW Systems, Inc., Redondo Beach, Calif., Final Report, Vol. 1, Summary, TRW No. 08710–6012–R000, NASA No. N68–31277 (December, 1969).
5. WELTI, G. R., and DURRANI, S. H., "Communication System Configuration for the Earth Resources Satellite" (see Ref. 23, pp. 289–314).

6. DURRANI, S. H., and LIPKE, D. W., "Satellite Communications for Manned Spacecraft" (see Ref. 24, pp. 128–139).

7. DIAMOND, P. M., "Satellite Systems for Integrated Communications, Navigation, and Identification" (see Ref. 23, pp. 437–456).

8. BLACK, H. S., *Modulation Theory*. D. Van Nostrand Co., Inc. Princeton, N. J., 1953.

9. MIDDLETON, D., *An Introduction to Statistical Communication Theory*. McGraw-Hill Book Company, New York, N.Y., 1960.

10. SCHWARTZ, M., *Information Transmission, Modulation, and Noise*, 2nd ed. McGraw-Hill Book Company, New York, N.Y., 1970.

11. TAUB, H., and SCHILLING, D. L., *Principles of Communication Systems*. McGraw-Hill Book Company, New York, N.Y., 1971.

12. DEFFEBACH, H. L., and FROST, W. O., "A Survey of Digital Baseband Signaling Techniques," NASA No. N71–37703 (June, 1971).

13. SHANNON, C. E., and WEAVER, W., *The Mathematical Theory of Communication*. University of Illinois Press, Urbana, Ill., 1959.

14. GALLAGER, R. G., *Information Theory and Reliable Communication*. John Wiley & Sons, Inc., New York, N.Y., 1968.

15. WOZENCRAFT, J. M., and JACOBS, I. M., *Principles of Communication Engineering*. John Wiley & Sons, Inc., New York, N.Y., 1965.

16. SCHOLTZ, R. A., "How Do You Define Bandwidth?", *Proceedings of the International Telemetering Conference*, Los Angeles, Calif. (October, 1972) 281–288.

17. TITSWORTH, R. C., and WELCH, L. R., "Power Spectra of Signals Modulated by Random and Pseudorandom Sequences," Jet Propulsion Laboratory, Pasadena, Calif., Technical Report No. 32–140 (October, 1961).

18. HECHT, M., and GUIDA, A., "Delay Modulation," *Proceedings of the IEEE*, Vol. 57, No. 7 (July, 1969) 1314–1316.

19. STILTZ, H. L., ed., *Aerospace Telemetry*, Vols. I, II, Prentice-Hall, Inc., Englewood Cliffs, N.J., 1961.

20. BALAKRISHNAN, A. V., ed., *Advances in Communications Systems*, Vol. 1, Academic Press Inc., New York, N.Y., 1965.

21. MARSTEN, R. B., ed., *Communication Satellite Systems Technology*, Progress in Astronautics and Aeronautics series, Vol. 19, MIT Press, Cambridge, Mass., 1966.

22. FELDMAN, N. E., and KELLY, C. M., eds., *Communication Satellites for the 70's: Technology*, Progress in Astronautics and Aeronautics series, Vol. 25, MIT Press, Cambridge, Mass., 1971.

23. FELDMAN, N. E., and KELLY, C. M., eds., *Communication Satellites for the 70's: Systems*, Progress in Astronautics and Aeronautics series, Vol. 26, MIT Press, Cambridge, Mass., 1971.

24. MORROW, W. E., Jr., ed., *Proceedings of the IEEE*, Special Issue on Satellite Communications (February, 1971).

Chapter 2

CARRIER-TRACKING LOOPS EMPLOYING THE PHASE-LOCK PRINCIPLE

2-1. INTRODUCTION

Successful transmission of information through a phase-coherent communication system requires, by definition, a receiver capable of determining or estimating the phase and frequency of the received signal with as little error as possible. Quite often the data-bearing signal is modulated on an RF carrier in such a way that a residual component at the RF exists in the overall signal power spectrum. This component could be tracked with a narrowband phase-locked loop (PLL) and used to provide the desired reference signal. On the other hand, the power contained in this residual component does not convey any information other than the phase and frequency of the carrier. Thus, it represents power not available for the transmission of data, and, in practice, techniques that conserve power are always of interest. More specifically, the data are often angle modulated on a secondary carrier—a subcarrier—which, in the absence of a dc component in the data signal spectrum, requires some type of suppressed carrier-tracking loop for establishing the coherent subcarrier reference. In what follows, we shall not distinguish between carrier and subcarrier (they will both be called carriers) since the methods developed apply to either situation.

Certain concepts pertinent to the behavior of carrier-tracking loops (PLLs and others using the phase-lock principle) are summarized in this chapter. We concern ourselves primarily with the tracking behavior of such loops, and furthermore only with those characteristics which are needed for the material that follows in subsequent chapters. Reference 1 provides a complete exposition on the theory, design, and performance of PLLs, including various aspects of signal acquisition and tracking: acquisition time and pull-in range, loop stability, phase error variance, moments of the first slip time, average number of slips per second, and the like.

2-2. PHASE-LOCKED LOOP OPERATION

The theory of operation of the second-order PLL is most easily explained using the simple functional block diagram in Fig. 2-1. The harmonic oscillations $s(t) = \sqrt{2P_c} \sin [\omega_0 t + \theta(t)]$ of the signal generator and $r(t) = \sqrt{2}K_1 \times \cos [\omega_0 t + \hat{\theta}(t)]$ of the voltage control oscillator (VCO) output act on the

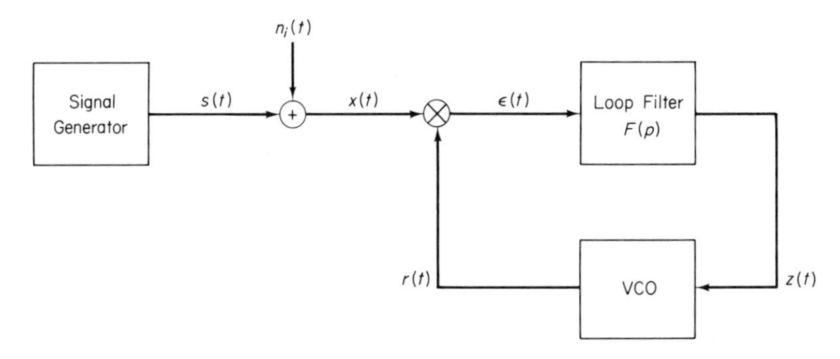

Fig. 2-1 A simple communication system with a phase-locked loop

multiplier (phase detector) to produce a voltage $\epsilon(t)$. This voltage $\epsilon(t)$ depends on the difference between the phase and frequency of the signals $r(t)$ and $s(t)$, and on the additive noise $n_i(t)$. The resulting signal $\epsilon(t)$, after being filtered by the loop filter, changes the output frequency and phase of the VCO, causing them to coincide (in the absence of noise) with the frequency and phase of the signal generator. Even in the absence of external noise, the practical operation of the PLL is considerably complicated by the presence of noise fluctuations due to frequency instabilities in the VCO. In the signal-to-noise ratio region where PLLs are expected to operate, the effect of these internal fluctuations are usually small in comparison with those produced by the additive noise $n_i(t)$. Hence, in what follows, we shall neglect any instabilities in the VCO.[1-4] By a slight modification in the PLL theory which follows, the VCO instabilities may be included.[1]

Linear PLL Theory

The received waveform $x(t)$ is defined by

$$x(t) = \sqrt{2P_c} \sin [\omega_0 t + \theta(t)] + n_i(t) \tag{2-1}$$

where P_c denotes the total transmitted carrier power in watts and

$$n_i(t) = \sqrt{2}[n_c(t) \cos \omega_0 t - n_s(t) \sin \omega_0 t] \tag{2-2}$$

represents the narrowband additive noise process. Both $n_c(t)$ and $n_s(t)$ are assumed to be statistically independent, stationary, white Gaussian noise processes of single-sided spectral density N_0 w/Hz, and single-sided bandwidth $W_i/2$ Hz less than $\omega_0/2\pi$. Using simple trigonometric identities, Eq. (2-2) can also be written as

$$n_i(t) = \sqrt{2}\{N_c(t) \cos [\omega_0 t + \theta(t)] - N_s(t) \sin [\omega_0 t + \theta(t)]\} \tag{2-3}$$

where

$$\begin{aligned} N_c(t) &= n_c(t) \cos \theta(t) + n_s(t) \sin \theta(t) \\ N_s(t) &= -n_c(t) \sin \theta(t) + n_s(t) \cos \theta(t) \end{aligned} \tag{2-4}$$

We shall have occasion throughout the text to use both (2-2) and (2-3) as representative noise models. We shall also assume throughout this chapter that the angle modulation is characterized by $\theta(t) = \Omega_0 t + \theta_0$ where Ω_0 represents initial *loop detuning* or *frequency offset* and θ_0 is a constant, independent of time. Analyses for more general angle modulations are given in Ref. 1.

The reference signal $r(t)$ at the output of the VCO is assumed to be a sinusoid of power K_1^2 whose instantaneous phase is related to the input control voltage $z(t)$ through the relationship

$$\omega_0 t + \hat{\theta}(t) = \omega_0 t + K_V \int^t z(\gamma)\, d\gamma \tag{2-5}$$

where K_V is the VCO gain in radians/volt-sec.* Noting that $\epsilon(t) = K_m x(t) \times r(t)$ and $z(t) = F(p)\epsilon(t)$, the stochastic integro-differential equation that relates the phase error $\varphi(t)$ to the input process $\theta(t)$ is given (in operator form) by

$$\begin{aligned} \varphi(t) &\triangleq \theta(t) - \hat{\theta}(t) \\ &= \theta(t) - \frac{KF(p)}{p}\{\sqrt{P_c} \sin \varphi(t) + N[t, \varphi(t)]\} \end{aligned} \tag{2-6}$$

where $K \triangleq K_1 K_m K_V$ with K_m the multiplier gain and

$$N[t, \varphi(t)] \triangleq N_c(t) \cos \varphi(t) - N_s(t) \sin \varphi(t) \tag{2-7}$$

In the above, all double-frequency terms have been neglected because neither the loop filter nor the VCO will respond significantly to them. Further, under certain physically justifiable assumptions, it may be shown[1] that

* In what follows, we shall write differential equations in compact form by introducing the Heaviside operator $p \triangleq d/dt$. In general, if $I(t)$ represents the input to a linear filter of transfer function $F(s)$, where s is the usual Laplace transform variable, then the output $O(t)$ can be written in differential form as $O(t) = F(p)I(t)$.

$N[t, \varphi(t)]$ is approximated by a low-pass white Gaussian noise process with the same spectral density as that of the original additive process $n_i(t)$. Thus, in a study of receiver structure between input and output, (2-6) and (2-7) provide the pertinent quantities, and the PLL may be conveniently represented by the block diagram of Fig. 2-2.

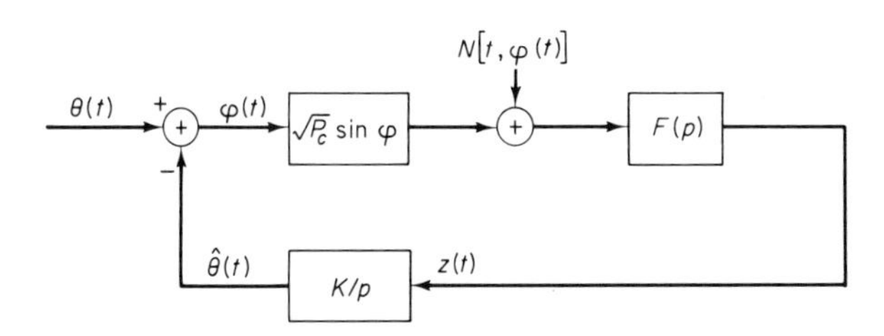

Fig. 2-2 Equivalent model of a phase-locked loop

If the loop is now linearized (assume that $\sin \varphi \cong \varphi$), then the closed-loop transfer function $H(s)$ of the PLL relative to the input signal $\theta(t)$ may be written by inspection from Fig. 2-2 as*

$$H(s) \triangleq \frac{\hat{\theta}(s)}{\theta(s)} = \frac{\sqrt{P_c}KF(s)}{s + \sqrt{P_c}KF(s)} \tag{2-8}$$

Therefore, the linearized closed-loop transfer function is specified once the loop filter is given. In practice, the carrier-tracking loop filter of greatest interest is given by

$$F(s) = \frac{1 + \tau_2 s}{1 + \tau_1 s} \tag{2-9}$$

Substituting Eq. (2-9) into (2-8), we have

$$H(s) = \frac{1 + \tau_2 s}{1 + [\tau_2 + (1/\sqrt{P_c}K)]s + (\tau_1/\sqrt{P_c}K)s^2} \tag{2-10}$$

In practice, the variance of the phase error is an important parameter. For the simple linear model, the variance of the phase error due to the additive noise is given by[1]

$$\sigma_\varphi^2 = \frac{N_0 W_L}{2P_c} = \frac{N_0}{2P_c}\left\{\frac{1 + (\sqrt{P_c}K\tau_2^2)/\tau_1}{2\tau_2[1 + (1/\sqrt{P_c}K\tau_2)]}\right\} \tag{2-11}$$

where

$$W_L \triangleq \frac{1}{2\pi j}\int_{-j\infty}^{j\infty} |H(s)|^2\, ds \tag{2-12}$$

is the two-sided loop bandwidth (in Hz) defined by the linear PLL theory.

* From this point on it will be convenient to use the Heaviside operator p and the Laplace transform variable s interchangeably. It should be clear from the context in which they are used which one is actually meant.[1]

If we define the *loop damping parameter* r by

$$r \triangleq \frac{\sqrt{P_c}K\tau_2^2}{\tau_1} = 4\zeta^2 \tag{2-13}$$

where ζ is the *loop damping* and assume $r\tau_1 \gg \tau_2$ (the case of greatest interest in practice), then from (2-11) we have

$$W_L \cong \frac{r+1}{2\tau_2} \tag{2-14}$$

In terms of r and W_L, Eq. (2-10) becomes

$$H(s) = \frac{1 + [(r+1)/2W_L]s}{1 + [(r+1)/2W_L]s + (1/r)[(r+1)/2W_L]^2 s^2} \tag{2-15}$$

which is the form of greatest interest in the exposition that follows.

Nonlinear PLL Theory

The most promising analytical approach in developing an exact nonlinear theory of PLLs is based on the Fokker-Planck theory.[1] This particular method leads to an expression for the probability density function (p.d.f.) of the nonstationary phase error, which in the steady-state gives an unbounded variance. This behavior of the variance is a result of the cycle slipping phenomenon associated with PLLs. Viterbi[5] and Tikhonov[3,4] were successful in applying this method to the analysis of the first-order loop by recognizing that the phase error reduced modulo 2π is stationary and possesses a bounded variance. A first attempt at extending the Fokker-Planck method to a second-order PLL, where the loop filter is of the proportional-plus-integral control type [Eq. (2-9)], is given in Ref. 7. The theory developed there suggests the possibility of approximating the second-order loop phase error p.d.f. by the more tractable results obtained from analysis of the first-order loop. Use of the expression for the first-order loop phase error p.d.f. in describing the statistics of the phase error in a second-order PLL was certainly a step in the right direction in view of the results that followed.

The most recent contributions relative to the statistical dynamics of the phase error process are given in Refs. 1 and 6, where the Fokker-Planck equation is solved in a more exact fashion than had previously been considered. The intent here is simply to summarize these results and at the same time provide physical insight into the problem.

The actual phase error process $\varphi(t)$ in a PLL undergoes diffusion much like a particle in Brownian motion; hence, the variance of the phase error becomes infinite in the steady-state. Previous work[3-5] on determining the p.d.f. of the phase error in the steady-state of a first-order loop was accomplished by reducing the phase error modulo 2π to a process that we shall denote by $\phi(t)$. For finding telemetry error probabilities, for example, this reduction gives sufficient information; whereas for estimating tracking accuracy, the statistical dynamics of the $\varphi(t)$ process itself must be studied.

To completely describe the $\phi(t)$ process, one must account for the component of its variance that results from diffusion—that is, cycle slipping. The steady-state effect of cycle slipping is perhaps best described by evaluating the diffusion coefficient—the rate at which the variance of the phase error is approaching infinity.[1,6] This quantity must then be combined with the variance of the phase error reduced modulo 2π to reflect the overall performance of the PLL.

For a second-order PLL, an approximate solution for the steady-state distribution of the phase error reduced modulo 2π is given below. The regions of validity for this solution are established by comparing it to experimental results.[1,7] In addition, approximate formulas are given for: (1) the diffusion coefficient of the phase error process, (2) the expected values of the time intervals between cycle slipping events, (3) the expected number of cycles slipped per unit of time, (4) the expected number of cycles slipped "to the right" and "to the left," (5) the expected value of the phase error rate in the steady-state, and (6) the mean time to the first slip. In the limit as the system damping approaches infinity, the results of these formulas are valid for the first-order loop.

Loop Model and Phase Error P.D.F. Reduced Modulo 2π. For a PLL with loop filter as given by (2-9), it has been shown[1] that the steady-state p.d.f. of the modulo 2π reduced phase error is given to a good approximation by

$$p(\phi) = \frac{\exp(\beta\phi + \alpha\cos\phi)}{4\pi^2 \exp(-\pi\beta)\,|I_{j\beta}(\alpha)|^2} \int_{\phi}^{\phi+2\pi} \exp(-\beta x - \alpha\cos x)\,dx \tag{2-16}$$

where $I_{\upsilon}(x)$ is the modified Bessel function of order υ and argument x. The domain of definition for ϕ in Eq. (2-16) is any interval of width 2π centered about any lock point $2n\pi$, with n an arbitrary integer. The parameters α and β that characterize (2-16) are related to the various system parameters through

$$\left.\begin{aligned} F_1 &= \frac{\tau_2}{\tau_1} \\ \alpha &= \left(\frac{r+1}{r}\right)\rho - \frac{1-F_1}{r\sigma_G^2} \\ \beta &= \left(\frac{r+1}{r}\right)^2 \frac{\rho}{2W_L}[\Omega_0 - \sqrt{P_c}K(1-F_1)\overline{\sin\phi}] + \alpha\,\overline{\sin\phi} \\ \rho &= \frac{2P_c}{N_0 W_L} \end{aligned}\right\} \tag{2-17}$$

where $G \triangleq \sin\phi - \overline{\sin\phi}$, the superbar denotes statistical average, and ρ is the signal-to-noise ratio in the loop bandwidth. The parameter $\sqrt{P_c}K$ represents the open-loop gain in sec^{-1}.

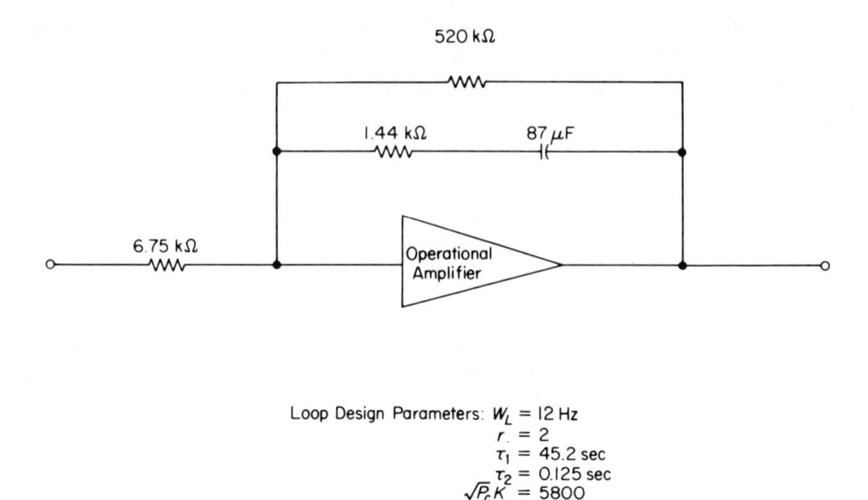

Fig. 2-3 A particular mechanization of a filter for an imperfect second-order loop

A particular mechanization of the loop filter of Eq. (2-9) is illustrated in Fig. 2-3.* The validity of using (2-16) to represent the statistical properties of the phase error in a second-order loop can be verified by comparing the equation graphically with an experimentally derived p.d.f. using the filter of Fig. 2-3 and assuming zero detuning. The closeness of the equation's approximation to the statistical properties is clearly evident from Fig. 2-4, where the analytical p.d.f. (solid curve) is superimposed over the experimental one (point plot) for various values of ρ in the range $0 < \rho < 6.5$ dB. The cumulative distributions of the measured phase error are shown in Fig. 2-5 for the same values of ρ. As shown, the probability of the loop-losing lock is extremely small when $\rho > 6.41$ dB. It is also evident from the cumulative distributions that the phase error density tends to become uniform for low signal-to-noise ratios. In Fig. 2-6, the variances of the phase error as computed from Eq. (2-11)—the linear model—and from (2-16) are compared with the variance of the measured phase error over the range of signal-to-noise ratio $0 < \rho < 6.5$ dB. In subsequent chapters, these results will be used to characterize the nonlinear behavior and performance of one-way and two-way phase-coherent communication systems. The expected value of the phase error can be found from (2-16) and the well-known Bessel

* We note that the dc gain of this filter has not been adjusted to unity; thus Eq. (2-9) applies only to within a multiplicative constant.

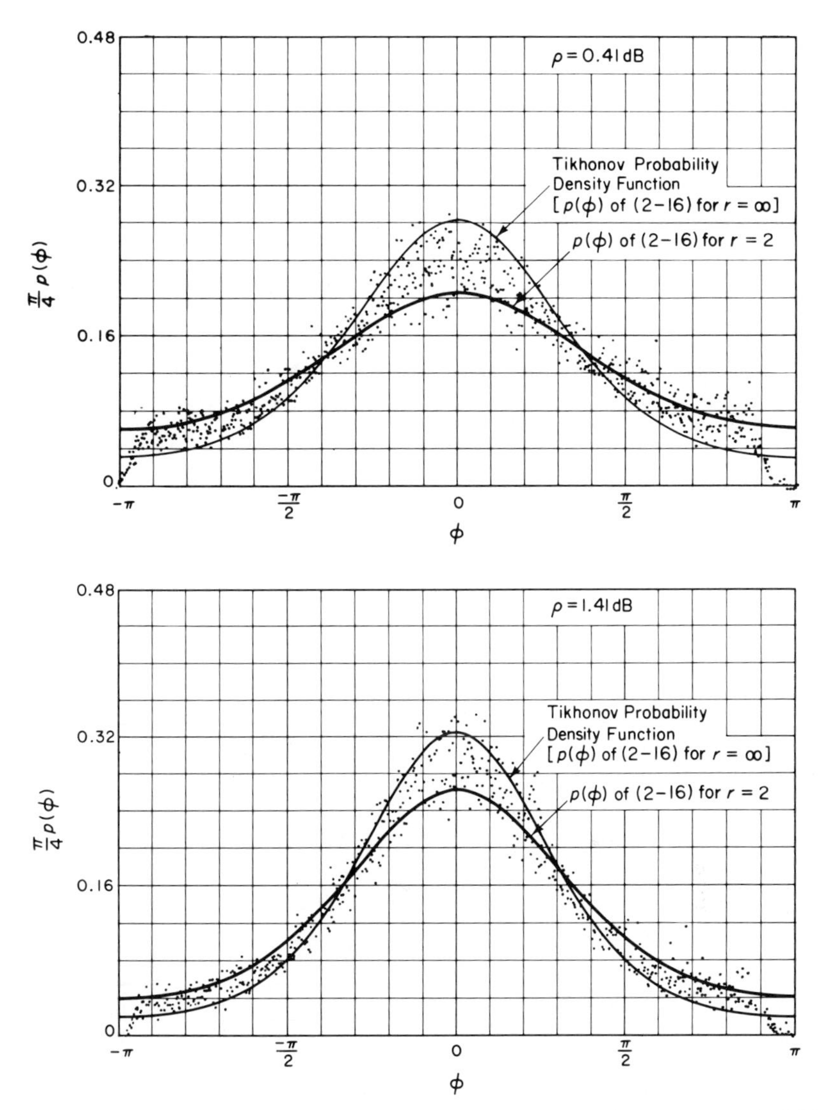

Fig. 2-4 A comparison of analytical and experimental phase error probability density functions for several values of loop signal-to-noise ratio

function expansions of exp $(\pm x \cos \phi)$,[1,6]

$$\begin{aligned}\bar{\phi} &= \int_{-\pi}^{\pi} \phi p(\phi)\, d\phi \\ &= \frac{2 \sinh \pi\beta}{\pi |I_{j\beta}(\alpha)|^2} \sum_{m=1}^{\infty} \frac{mI_m(\alpha)}{m^2 + \beta^2}\left[\frac{I_0(\alpha)}{m} + \frac{I_m(\alpha)}{4m} + \sum_{\substack{k=1\\k\neq m}}^{\infty} \frac{2m(-1)^k I_k(\alpha)}{m^2 - k^2}\right] \qquad \text{(2-18)}\end{aligned}$$

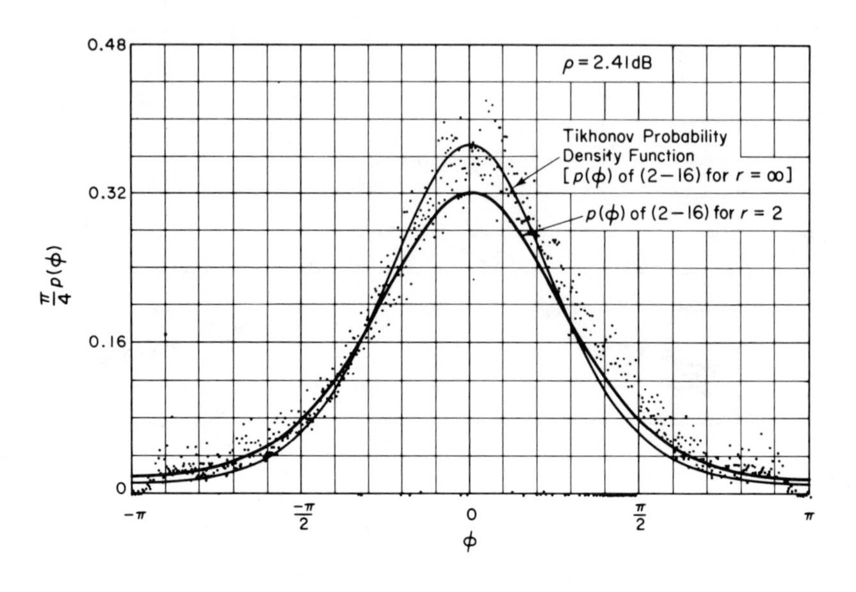
0.48
0.32
0.16
0
$\frac{\pi}{4}p(\phi)$
$\rho = 2.41$ dB
Tikhonov Probability
Density Function
[$p(\phi)$ of (2–16) for $r = \infty$]
$p(\phi)$ of (2–16) for $r = 2$
$-\pi$
$\frac{-\pi}{2}$
0
$\frac{\pi}{2}$
π
ϕ

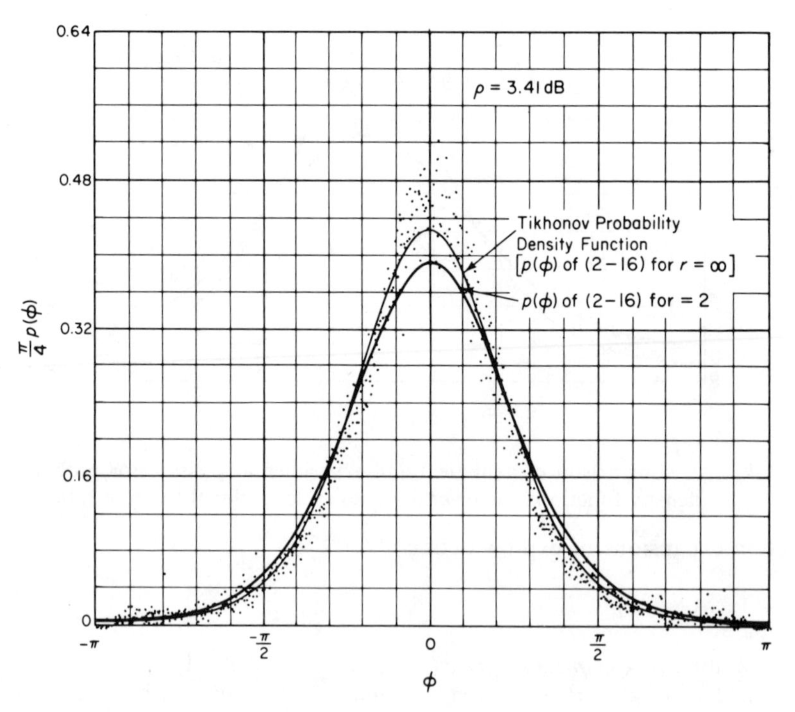
0.64
0.48
0.32
0.16
0
$\frac{\pi}{4}p(\phi)$
$\rho = 3.41$ dB
Tikhonov Probability
Density Function
[$p(\phi)$ of (2–16) for $r = \infty$]
$p(\phi)$ of (2–16) for = 2
$-\pi$
$-\frac{\pi}{2}$
0
$\frac{\pi}{2}$
π
ϕ

Fig. 2-4 (Cont.)

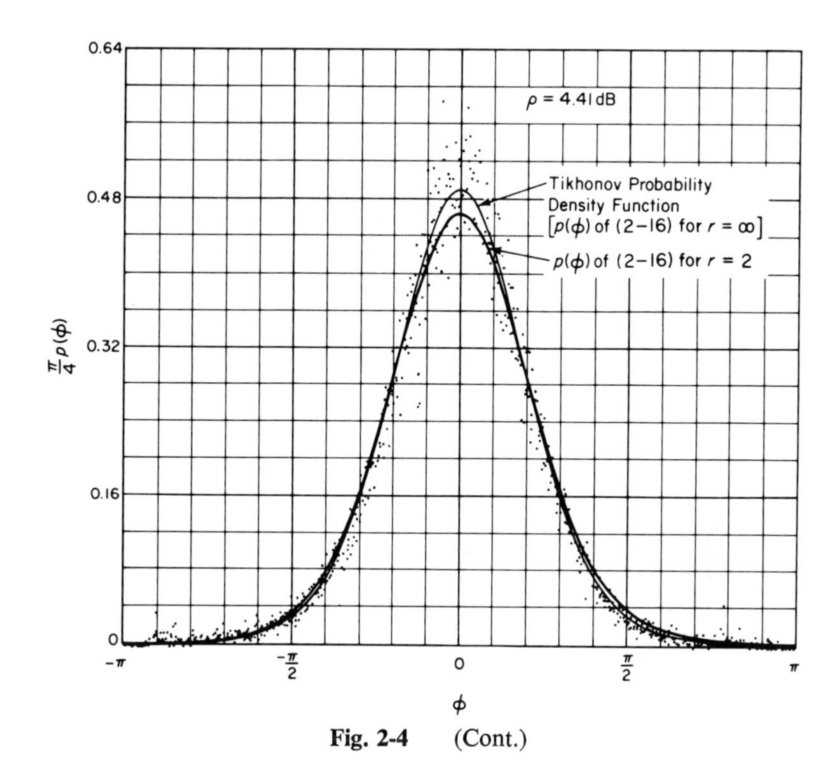

Fig. 2-4 (Cont.)

It is clear from (2-18) that with $\beta = 0$, $\bar{\phi} = 0$, and hence the phase error p.d.f. is symmetric in ϕ about the origin. Furthermore, $\overline{\phi^2}$ is given by[1,6]

$$\begin{aligned}\overline{\phi^2} &= \int_{-\pi}^{\pi} \phi^2 p(\phi)\, d\phi \\ &= \frac{\sinh \pi\beta}{\pi\,|I_{j\beta}(\alpha)|^2}\left\{\frac{I_0(\alpha)}{\beta}\left[\frac{\pi^2 I_0(\alpha)}{3} + 4\sum_{k=1}^{\infty}\frac{(-1)^k I_k(\alpha)}{k^2}\right] + 2\beta I_0(\alpha)\sum_{k=1}^{\infty}\frac{I_k(\alpha)}{k^2(\beta^2+k^2)}\right. \\ &\quad \left. + 2\beta\sum_{k=1}^{\infty}\frac{(-1)^k I_k(\alpha)}{\beta^2+k^2}\left[\left(\frac{\pi^2}{3}+\frac{1}{2k^2}\right)I_k(\alpha) + 4\sum_{\substack{m=1\\ m\neq k}}^{\infty}\frac{(-1)^m(k^2+m^2)I_m(\alpha)}{(k^2-m^2)^2}\right]\right\}\end{aligned} \tag{2-19}$$

The variance $\sigma_\phi^2 = \overline{\phi^2} - (\bar{\phi})^2$ is minimized when the loop is operating such that $\beta = 0$ and α is maximized.[1,6] For this case we have, from Eqs. (2-18) and (2-19),

$$\sigma_\phi^2 = \frac{\pi^2}{3} + \frac{4}{I_0(\alpha)}\sum_{k=1}^{\infty}\frac{(-1)^k}{k^2} I_k(\alpha) \tag{2-20}$$

Figure 2-7 illustrates the behavior of the variance of the phase error σ_ϕ^2 as a function of r with $\beta = 0$. The limiting case of $r = \infty$, $F_1 = 1$, and

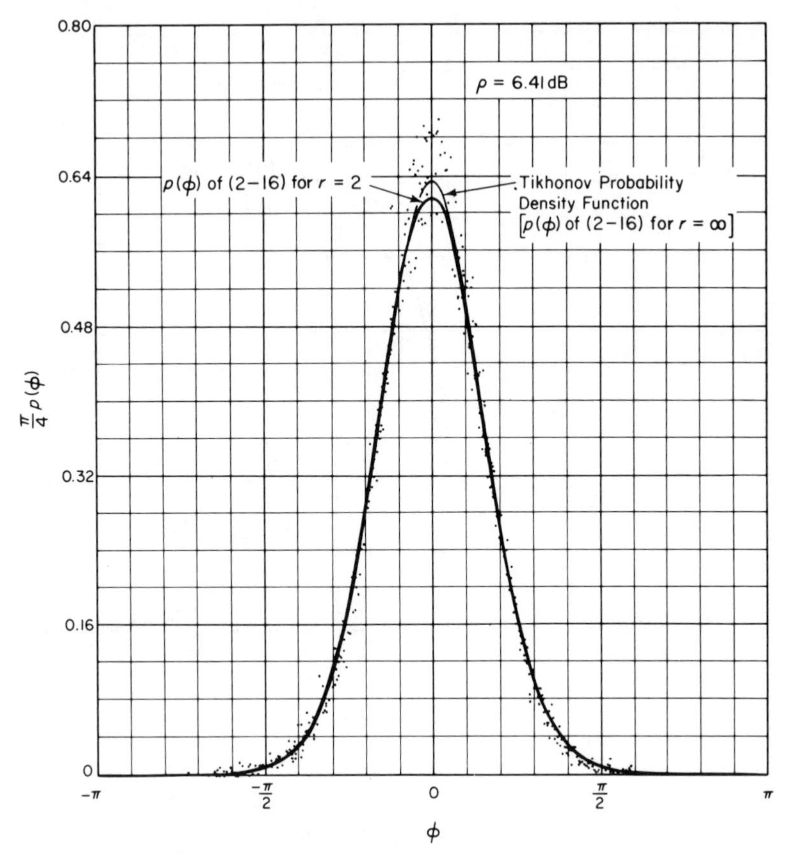

Fig. 2-4 (Cont.)

$\beta = 0$ gives the results for the first-order loop, with zero detuning, namely,

$$p(\phi) = \frac{\exp(\rho \cos \phi)}{2\pi I_0(\rho)} \qquad |\phi| \leq \pi \tag{2-21}$$

and σ_ϕ^2 is determined from (2-20) with ρ substituted for α. In the definition of ρ as given by (2-17), W_L for the first-order loop is equal to $\sqrt{P_c}K/2$.

Certain other steady-state statistics of the phase error process will be of interest when the problem of carrier suppression in two-way systems is considered later in the text. In particular, the *circular moments* of $p(\phi)$ are given by[1]

$$\overline{\sin n\phi} = \text{Im}\left[\frac{I_{n-j\beta}(\alpha)}{I_{-j\beta}(\alpha)}\right] \qquad \overline{\cos n\phi} = \text{Re}\left[\frac{I_{n-j\beta}(\alpha)}{I_{-j\beta}(\alpha)}\right] \tag{2-22}$$

from which

$$\overline{\cos^2 \phi} = \frac{1 + \overline{\cos 2\phi}}{2} \qquad \overline{\sin^2 \phi} = \frac{1 - \overline{\cos 2\phi}}{2} \tag{2-23}$$

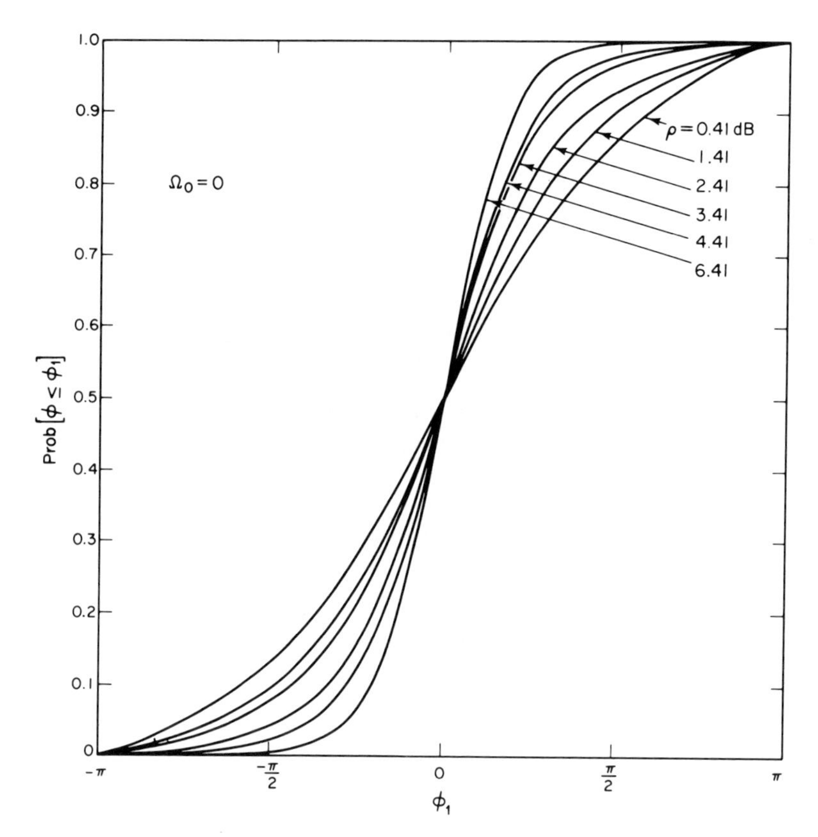

Fig. 2-5 The cumulative distribution function of the measured phase error for several values of loop signal-to-noise ratio

and

$$\sigma^2_{\cos\phi} = \overline{\cos^2\phi} - (\overline{\cos\phi})^2 \qquad \sigma^2_{\sin\phi} = \overline{\sin^2\phi} - (\overline{\sin\phi})^2 \tag{2-24}$$

As we shall see, the circular moments have application in specifying the carrier-suppression effects in two-way systems that use phase-coherent transponders. In Eq. (2-22), Re and Im denote the real and the imaginary parts, respectively, of the quantity in brackets. Plots of $\overline{\sin\phi}$, $\overline{\cos\phi}$, $\sigma^2_{\sin\phi}$, and $\sigma^2_{\cos\phi}$ versus $|\beta|/\alpha$ with α as a parameter are given in Figs. 2-8 through 2-11.

Finally, the expected value of the phase error rate $\dot{\phi}$ in a first-order loop is given by[6]

$$\overline{\dot{\phi}} = \Omega_0 - \sqrt{P_c}K\,\overline{\sin\phi} \tag{2-25}$$

where $\overline{\sin\phi}$ is given in (2-22) with $n = 1$.

As has been seen, the phase error p.d.f. and its moments depend

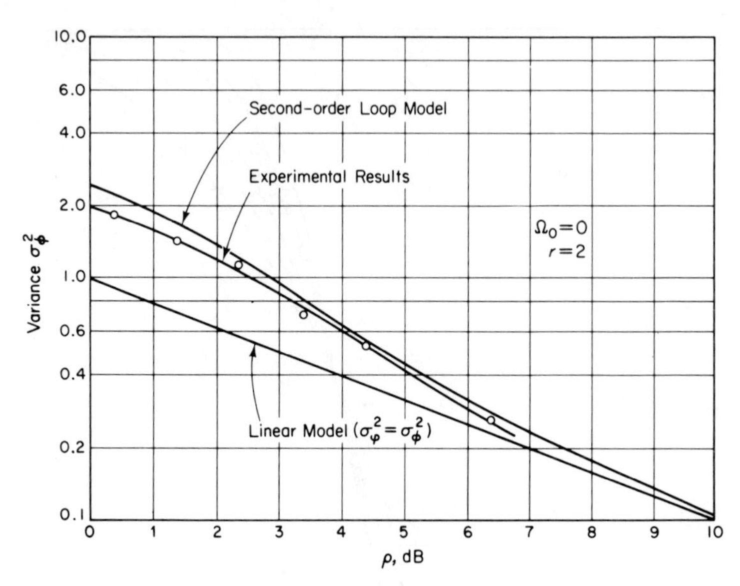

Fig. 2-6 Phase error variance vs loop signal-to-noise ratio for second-order phase-locked loop (a comparison of linear and nonlinear theories with experimental results)

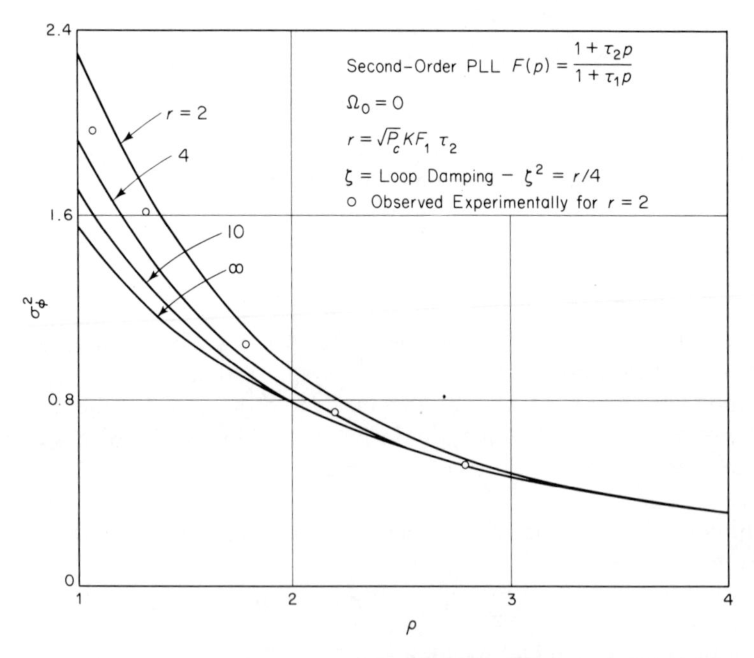

Fig. 2-7 Phase error variance of a second-order phase-locked loop vs loop signal-to-noise ratio for several values of r

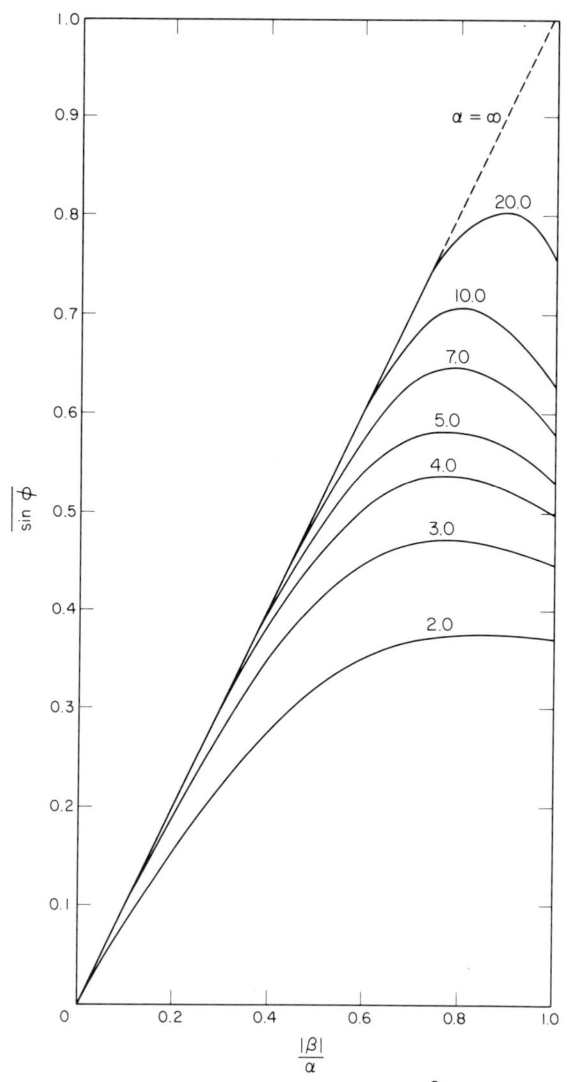

Fig. 2-8 Mean of $\sin \phi$ vs $\dfrac{|\beta|}{\alpha}$

solely on two parameters: α and β. On the other hand, it is desirable to evaluate the performance of a digital communication system employing a second-order PLL as a function of the design parameters normalized to loop detuning, $\Lambda \triangleq \Omega_0/\sqrt{P_c}K$ and loop signal-to-noise ratio ρ. Since the solution of (2-16) combined with (2-17) requires, in any given situation, a functional iteration on a digital computer, it is desirable from the standpoint

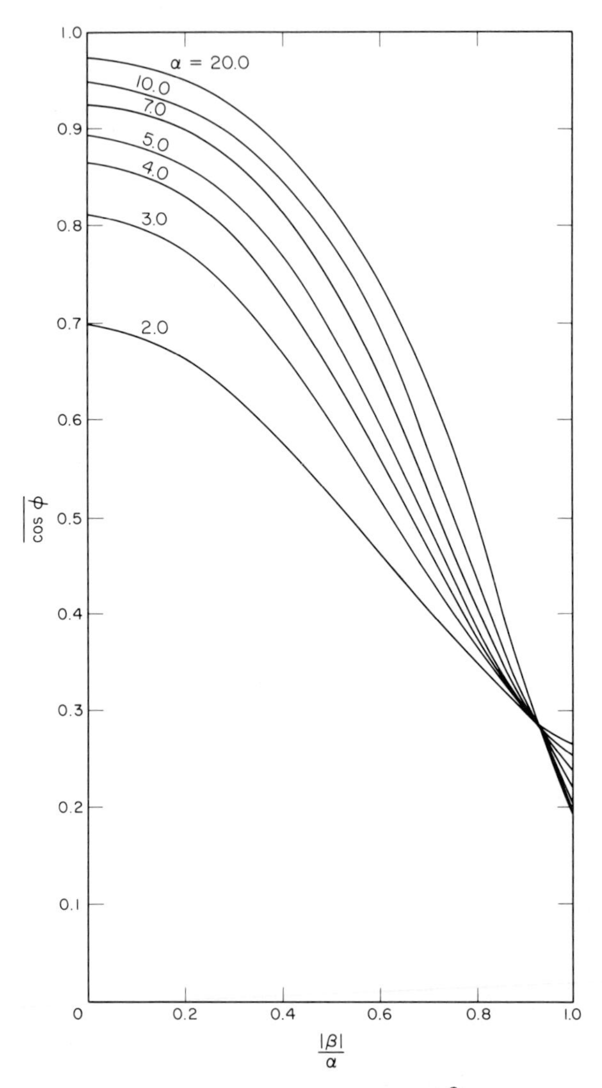

Fig. 2-9 Mean of $\cos\phi$ vs $\frac{|\beta|}{\alpha}$

of the design engineer to have available several grids relating ρ and Λ to α and β for fixed values of the loop damping parameter r. Figures 2-12 and 2-13 plot ρ versus $|\Lambda|$ with α and $|\beta|/\alpha$ as parameters, $F_1 = 0.002$ and $r = 2, 4$ corresponding respectively to the cases of 0.707 and critical damping. It is interesting to observe that for $\rho > 3$ dB, $|\beta|/\alpha \cong |\Lambda|$.

The Phase Error Diffusion Coefficient. One method of accounting for the

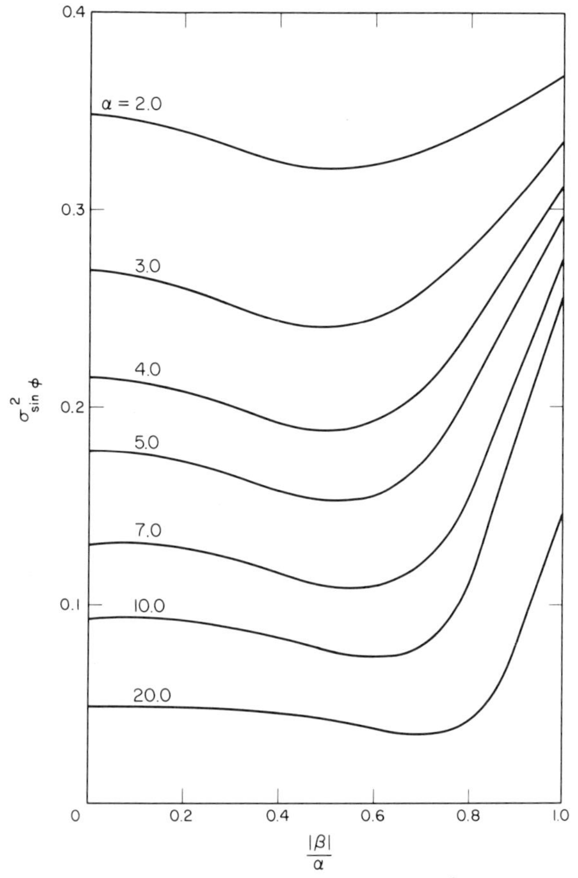

Fig. 2-10 Variance of $\sin \phi$ vs $\frac{|\beta|}{\alpha}$

fact that the loop skips cycles is to evaluate the rate at which the phase error process undergoes diffusion. This parameter, the so-called *phase error diffusion coefficient* D, is $(2\pi)^2$ times, $\bar{S}$, the total average number of phase jumps per unit of time. For a first-order loop, it is given by[1]

$$D \triangleq (2\pi)^2 \bar{S} = \frac{4W_L \cosh(\pi\rho\Lambda)}{\rho\,|\,\mathrm{I}_{j\rho\Lambda}(\rho)\,|^2} \tag{2-26}$$

The mean time interval between successive phase jumps ΔT is then equal to the reciprocal of $\bar{S}$. We now investigate the manner in which the quantity D contributes to the variance of the actual phase error process $\varphi(t)$ at time t, assuming zero error at time $t = 0$.

Recalling that $\phi(t)$ is the modulo 2π reduced version of $\varphi(t)$, it is con-

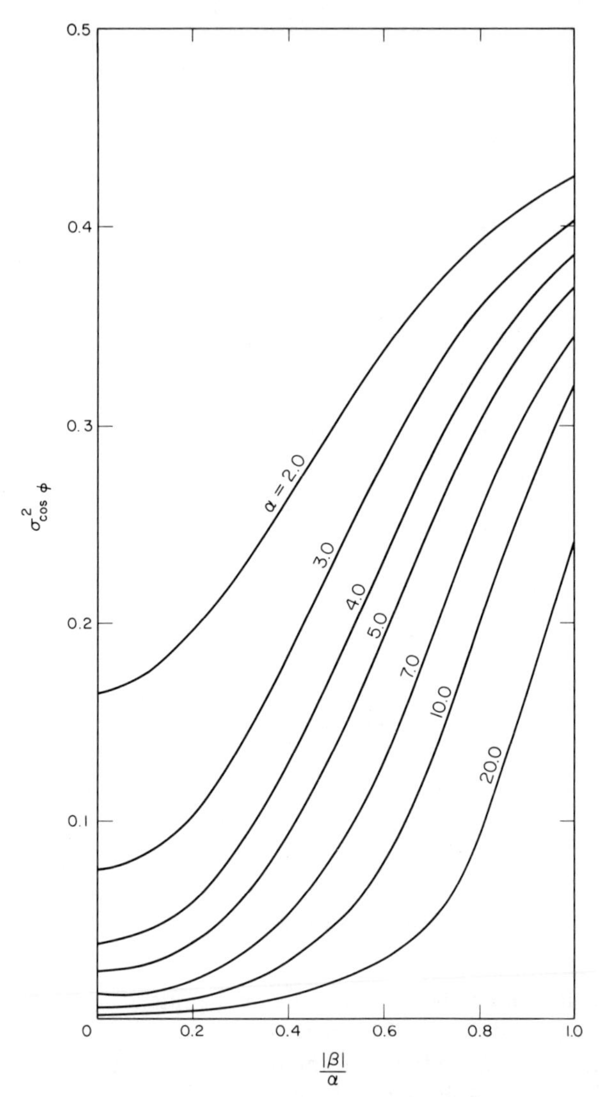

Fig. 2-11 Variance of $\cos\phi$ vs $\frac{|\beta|}{\alpha}$

venient to write

$$\varphi(t) = \phi(t) + 2\pi J(t) \tag{2-27}$$

where at any instant t, $J(t)$ is an integer-valued random function of t. Thus, J is a discrete random variable that takes on integer values at random points

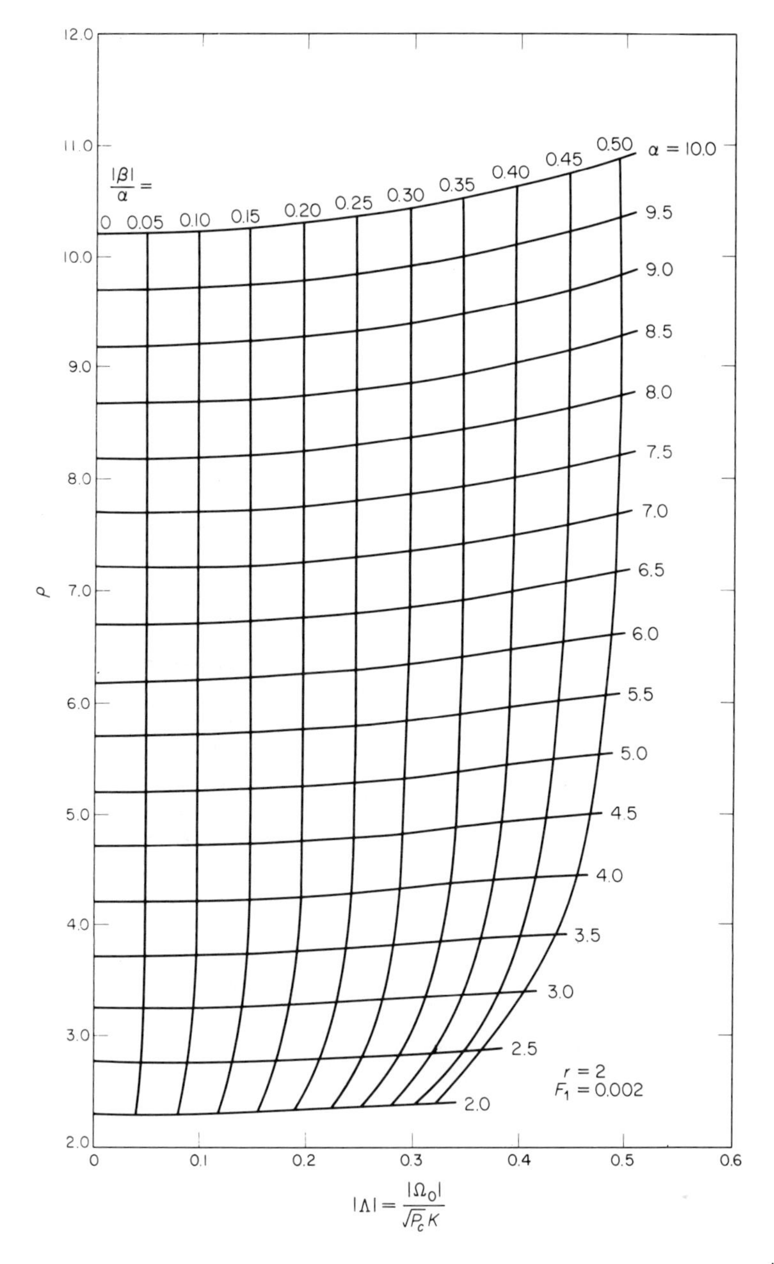

Fig. 2-12 Loop signal-to-noise ratio vs normalized detuning with α and $\frac{|\beta|}{\alpha}$ as parameters: $r = 2$, $F_1 = .002$

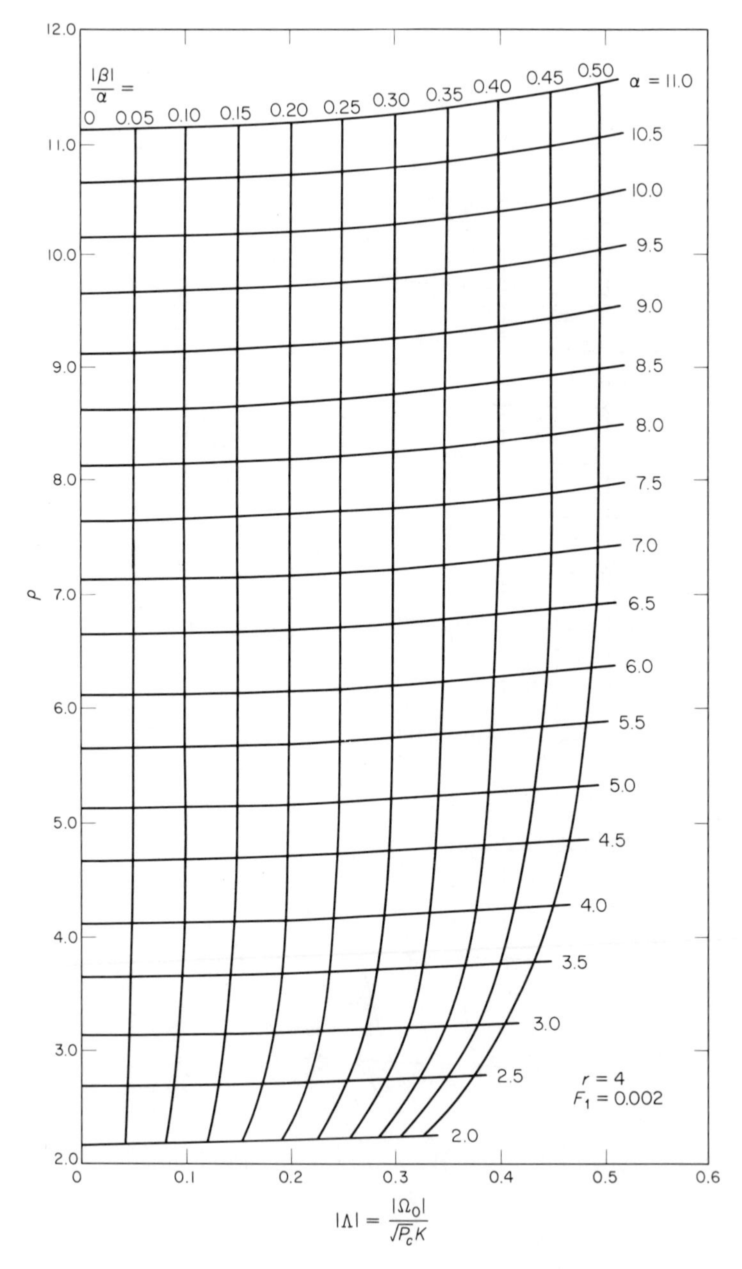

Fig. 2-13 Loop signal-to-noise ratio vs. normalized detuning with α and $\frac{|\beta|}{\alpha}$ as parameters: $r = 4$, $F_1 = .002$

in time. The variance of the $\varphi(t)$ process is given by

$$\sigma_\varphi^2(t) = \sigma_\phi^2(t) + (2\pi)^2\,\sigma_J^2 + 4\pi\,(\overline{J\phi} - \bar{J}\bar{\phi}) \tag{2-28}$$

We denote by $\mathcal{S}$ the event that j phase jumps occur in t seconds. If we assume that $\mathcal{S}$ is a Poisson-type process, then the quantity $\bar{S}$, representing the total average number of phase jumps per unit time, can be used to produce an approximate probabilistic model for the phase-jumping process that causes diffusion of the phase error process $\varphi(t)$, namely,

$$P(\mathcal{S}) \triangleq \text{prob}\,\{J = j\} \cong \frac{(\bar{S}t)^j \exp(-\bar{S}t)}{j!} \tag{2-29}$$

with mean $\bar{J} = \bar{S}t$ and variance $\sigma_J^2 = \bar{S}t$. Experimental evidence supporting the Poisson assumption is given in Refs. 7 through 9. Substituting $\sigma_J^2 = \bar{S}t$ in (2-28) gives

$$\sigma_\varphi^2(t) = \sigma_\phi^2(t) + (2\pi)^2\,\bar{S}t + 4\pi\,(\overline{J\phi} - \bar{J}\bar{\phi}) \tag{2-30}$$

Recalling Eq. (2-26) and making the reasonable assumption that J and ϕ are independent in the steady-state, we can write

$$\lim_{\substack{t\to\infty \\ \Delta T\to 0}} \frac{\sigma_\varphi^2(t+\Delta T) - \sigma_\varphi^2(t)}{\Delta T} = D \tag{2-31}$$

It should now be clear that D, the phase error diffusion coefficient of the $\varphi(t)$ process, is a measure of the rate at which the variance is approaching infinity. Equation (2-30) can be written in an alternate form for the steady-state case:

$$\lim_{t\to\infty} \sigma_\varphi^2(t) = \overline{\phi^2} - (\bar{\phi})^2 + \lim_{t\to\infty} Dt \tag{2-32}$$

where $\overline{\phi^2}$ is given by Eq. (2-19) and $\bar{\phi}$ by (2-18). Finally, the probability of losing phase lock in t seconds—that is, the probability of slipping one or more cycles—is given by

$$P(J \geq 1) \cong 1 - \exp(-\bar{S}t) \tag{2-33}$$

Figure 2-14 illustrates a plot of $\bar{S}/W_L$ vs ρ for a first-order loop. For a second-order loop it appears[1] that $\bar{S}$ is more difficult to evaluate; however, computer simulation techniques have yielded the results given in Fig. 2-14 for a second-order loop.

For the case where $\Omega_0 \neq 0$ the average number of cycles slipped per second "to the right" and "to the left" are of interest. Denoting those to the right per second by N_+ and those to the left per second by N_-, it has been shown[1] that

$$N_+ = \frac{\mathcal{J} \exp(\pi\rho\Lambda)}{2 \sinh(\pi\rho\Lambda)} \tag{2-34}$$

$$N_- = \frac{\mathcal{J} \exp(-\pi\rho\Lambda)}{2 \sinh(\pi\rho\Lambda)} \tag{2-35}$$

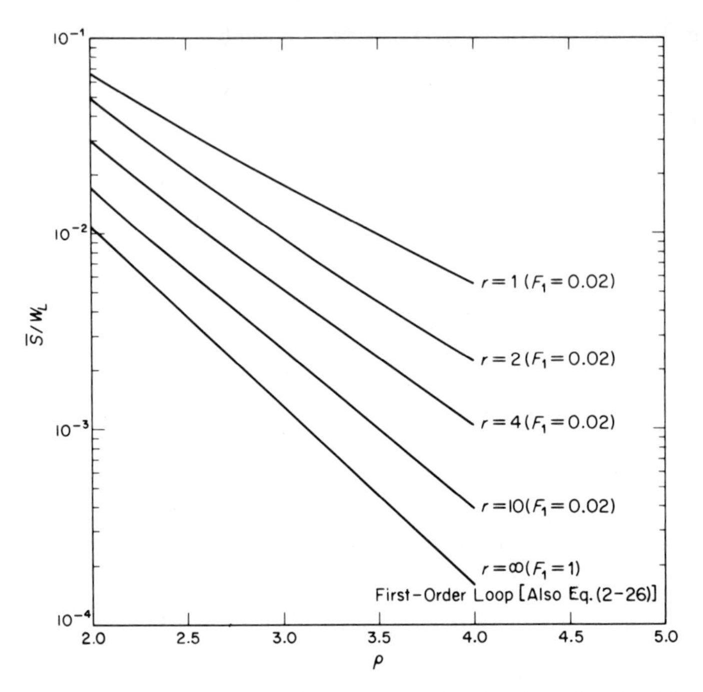

Fig. 2-14 Average rate of cycle slipping normalized to loop bandwidth vs loop signal-to-noise ratio for several values of r

where $\mathcal{J}$ is the (net) average number of cycle slips per second, that is,

$$\mathcal{J} = N_+ - N_- = \frac{W_L \sinh(\pi\rho\Lambda)}{\pi^2 \rho \,|\, \mathrm{I}_{j\rho\Lambda}(\rho)\,|^2} \tag{2-36}$$

For the unstressed loop with $\Omega_0 = 0$, $N_+ = N_-$ and $\mathcal{J} = 0$. The quantity $\mathcal{J}$ is related to $\bar{S} = N_+ + N_-$ through the equation

$$\mathcal{J} = \tanh(\pi\rho\Lambda)\bar{S} \tag{2-37}$$

Mean Time to First Slip or First Loss of Phase Synchronization. We have discussed the steady-state effect of the cycle slipping phenomenon in terms of the diffusion coefficient or equivalently the total average rate of cycle slips $\bar{S}$. Of perhaps less importance to the design engineer is the transient behavior of the phase error process, which can in part be described by its mean time to first slip from a given initial condition of the VCO. A study of this quantity is directly related to the first passage time problem for a continuous Markov process.[1,10] In fact, the mean time to first slip is defined equal to the first moment of the time for the phase error to cross either the plus or minus 2π boundaries from a given initial phase position. An approximate expression for this first moment normalized to the loop bandwidth is given by[1]

$$W_L\tau(2\pi\,|\,\varphi_0) \cong \left(\frac{r+1}{r}\right)^2 \frac{\rho}{2} \int_{-2\pi}^{2\pi} \int_{-2\pi}^{\varphi} [C_0(0) - u(x-\varphi_0)]$$
$$\exp\left[\rho\left(\frac{r+1}{r}\right)\{\cos\varphi - \cos x\} + \frac{\rho}{2r}(\varphi^2 - x^2) + \rho\Lambda(\varphi - x)\right] dx\, d\varphi \tag{2-38}$$

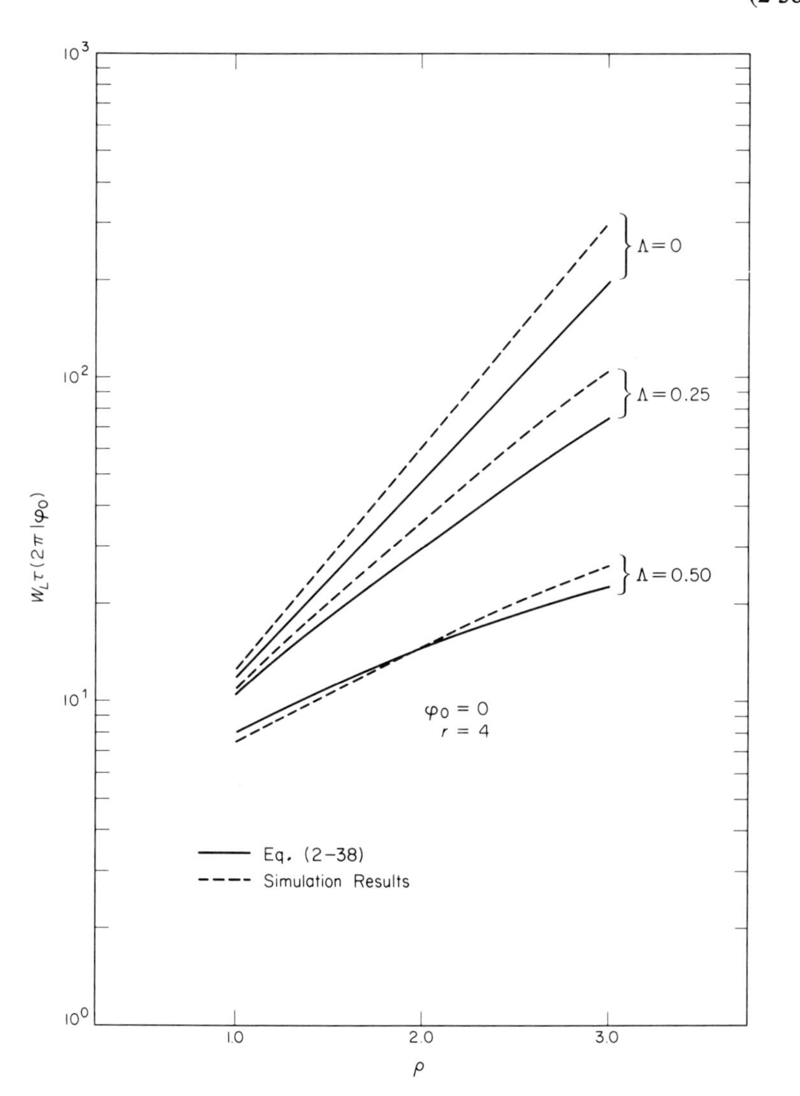

Fig. 2-15 Normalized mean first slip time vs loop signal-to-noise ratio: $r = 4$

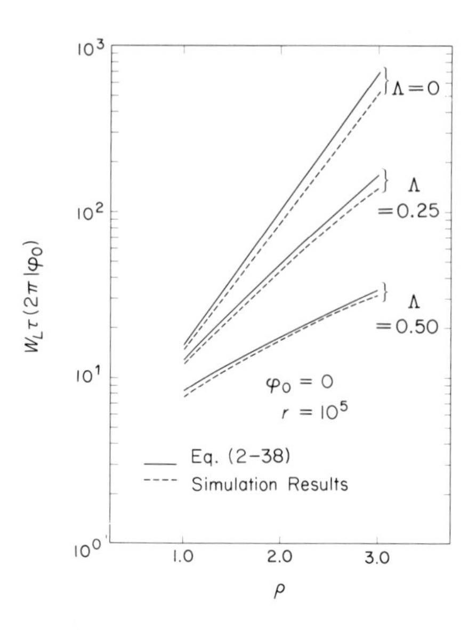

Fig. 2-16 Normalized mean first slip time vs loop signal-to-noise ratio: $r = 10^5$

where

$$C_0(0) \triangleq \frac{\int_{\varphi_0}^{2\pi} \exp\{-[(r+1)/r]\rho\cos x - (\rho/2r)x^2 - \rho\Lambda x\}dx}{\int_{-2\pi}^{2\pi} \exp\{-[(r+1)/r]\rho\cos x - (\rho/2r)x^2 - \rho\Lambda x\}dx} \tag{2-39}$$

$u(x)$ is the unit step function, and φ_0 is the initial condition on φ. This result represents a generalization of a previous result due to Tausworthe.[11,12] In fact, for $\Omega_0 = 0$, $\varphi_0 = 0$, and $C_0(0) = 1/2$, Eq. (2-38) reduces to his result. The product $W_L\tau(2\pi|\varphi_0)$ of (2-38) is plotted versus ρ in Figs. 2-15 and 2-16 for various values of r at fixed Λ with $\varphi_0 = 0$. Superimposed on these curves are experimental results obtained via computer simulation.[8,13]

An alternate approximation has been derived for the mean time to first slip based on the use of the orthogonality principle to evaluate certain conditional expectations.[1,6] The expression has the advantage of being in the same canonical form as that for a first-order loop and is given by

$$W_L\tau(2\pi|\varphi_0) \cong \left(\frac{r+1}{r}\right)^2 \frac{\rho}{2} \int_{-2\pi}^{2\pi} \int_{-2\pi}^{\varphi} [C_0(0) - u(x-\varphi_0)] \exp[\alpha(\cos\varphi - \cos x) + \beta(\varphi - x)]\, dx\, d\varphi \tag{2-40}$$

where α and β are defined in (2-17). However, experimental evidence suggests that this approximation may be less accurate than the former, particularly in the region of low signal-to-noise ratio. The validity of either of these expressions in the high signal-to-noise ratio region must be argued strictly on a theoretical basis since experimental evidence is difficult to achieve.

2-3. THE SECOND-ORDER PHASE-LOCKED LOOP PRECEDED BY A BAND-PASS LIMITER

In the practical operation of a coherent receiver employing a PLL, the loop is often preceded by a band-pass limiter (BPL) because the loop bandwidth is a function of the signal power P_c, and any fluctuation in P_c (owing to a change in range, for instance) causes the loop bandwidth to fluctuate similarly. BPLs are used also to protect various loop components, the multiplier in particular, where signal and noise levels can vary over several orders of magnitude and exceed the dynamic range of these components. In this section, the theoretical results needed to explain the behavior of a PLL when preceded by a BPL are discussed. The detailed theory and operation of the BPL are covered more completely in Refs. 1 and 14 through 18.

Linear PLL Theory and the Effects of Band-Pass Limiting

The mechanization of a BPL is illustrated in Fig. 2-17; the limiter incorporated in the PLL system is shown in Fig. 2-18. Assuming the output of the

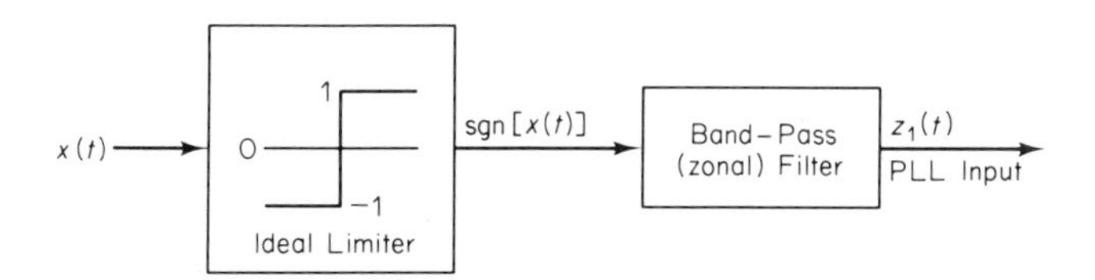

Fig. 2-17 A mechanization of a band-pass limiter

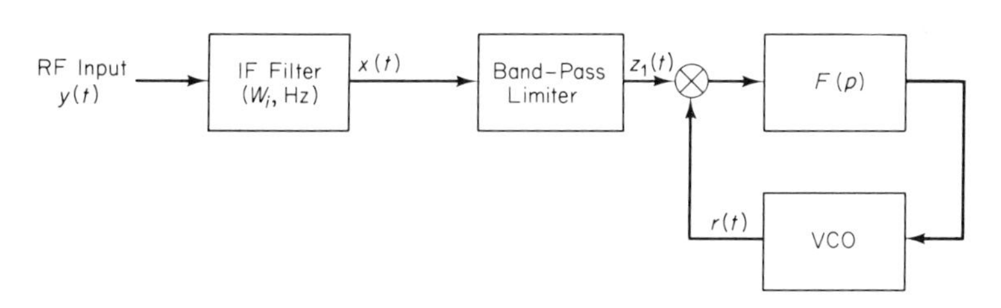

Fig. 2-18 A phase-locked receiver incorporating a band-pass limiter

IF filter is given by Eqs. (2-1) and (2-3) where the noise spectrum is flat over a bandwidth $W_i/2$ centered about the carrier frequency, then the first zone limiter output $z_1(t)$ can be expressed in the form,

$$z_1(t) = \sqrt{2\tilde{\alpha}_1^2 P_1} \sin [\omega_0 t + \theta(t)] + \sqrt{2}\{N_A(t) \cos [\omega_0 t + \theta(t)] - N_B(t) \sin [\omega_0 t + \theta(t)]\} \qquad (2\text{-}41)$$

In practice, $\tilde{\alpha}_1$ is usually referred to as the *signal amplitude suppression factor*, and is given by[15]

$$\tilde{\alpha}_1 = \sqrt{\frac{\pi}{2}}\left(\frac{\rho_i}{2}\right)^{1/2}\exp\left(-\frac{\rho_i}{2}\right)\left[I_0\left(\frac{\rho_i}{2}\right) + I_1\left(\frac{\rho_i}{2}\right)\right] \tag{2-42}$$

where $\rho_i = 2P_c/N_0W_i$ is the signal-to-noise ratio input to the limiter.* The parameter $P_1 = 8/\pi^2$ represents the fraction of the signal plus noise power that falls in the first zone. The noise processes $N_A(t)$ and $N_B(t)$ are zero mean and uncorrelated, but most important they are not, in general, Gaussian nor do they possess equal noise bandwidths or spectra. In fact $N_A(t)$ and $N_B(t)$ can be related back to the equivalent in-phase and quadrature components of the input noise $N_c(t)$ and $N_s(t)$ by

$$\begin{aligned} N_A(t) &= \frac{2\sqrt{2}}{\pi}\left\{\frac{N_c(t)}{\sqrt{[\sqrt{P_c} - N_s(t)]^2 + N_c^2(t)}}\right\} \\ N_B(t) &= \frac{2\sqrt{2}}{\pi}\left\{\tilde{\alpha}_1 - \frac{\sqrt{P_c} - N_s(t)}{\sqrt{[\sqrt{P_c} - N_s(t)]^2 + N_c^2(t)}}\right\} \end{aligned} \tag{2-43}$$

The p.d.f.'s of $N_A(t)$ and $N_B(t)$ have been studied in Ref. 18 with the results

$$\begin{aligned} p_A(N) &= \frac{1}{\pi\sqrt{1 - N^2}}\exp[-\rho_i]\{1 + \sqrt{\pi\rho_i}(1 - N^2) \\ &\quad \exp[\rho_i(1 - N^2)]\operatorname{erf}[\sqrt{\rho_i(1 - N^2)}]\} \qquad -1 \le N \le 1 \\ p_B(N) &= \frac{1}{\pi\sqrt{1 - (\tilde{\alpha}_1 - N)^2}}\exp[-\rho_i]\{1 + \sqrt{\pi\rho_i}(\tilde{\alpha}_1 - N) \\ &\quad \exp[\rho_i(\tilde{\alpha}_1 - N)^2][1 + \operatorname{erf}(\sqrt{\rho_i}(\tilde{\alpha}_1 - N))]\} \qquad \tilde{\alpha}_1 - 1 \le N \le \tilde{\alpha}_1 + 1 \end{aligned} \tag{2-44}$$

where $p_A(N)$ and $p_B(N)$ represent the p.d.f.'s of the normalized noise processes $[\pi/(2\sqrt{2})]N_A(t)$ and $[\pi/(2\sqrt{2})]N_B(t)$, respectively.** As will be seen, only $N_A(t)$ is of interest in studying the performance of a PLL preceded by a BPL. On the other hand, after coherent demodulation by an in-phase reference signal (as in matched filter detection), the $N_B(t)$ process is the one which affects system performance.

Figures 2-19 and 2-20 illustrate the behavior of the p.d.f.'s $p_A(N)$ and $p_B(N)$ as a function of N with ρ_i, the input signal-to-noise ratio, as a parameter. It is interesting to note that for large ρ_i, $p_A(N)$ becomes Gaus-

---

* A rational function approximation to $\tilde{\alpha}_1$, which is simpler to use in system design, is suggested in Ref. 2 and is given by

$$\tilde{\alpha}_1 = \sqrt{\frac{.7854\rho_i + .4768\rho_i^2}{1 + 1.024\rho_i + .4768\rho_i^2}}$$

** The function erf (x) as used in this text is defined as

$$\operatorname{erf}(x) \triangleq \frac{2}{\sqrt{\pi}}\int_0^x \exp(-t^2)dt$$

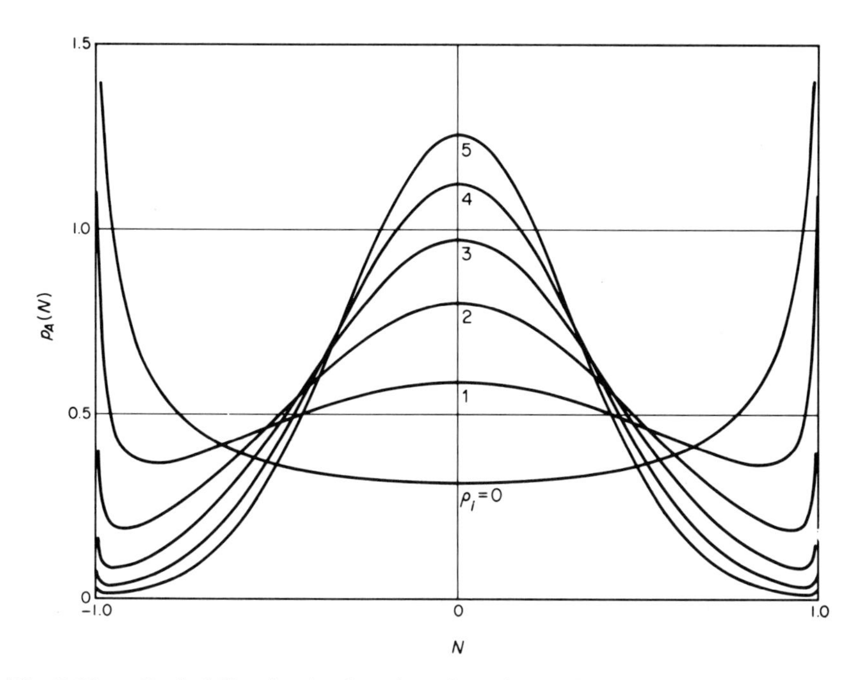

Fig. 2-19 Probability density function of $N_A(t)$ as a function of input signal-to-noise ratio

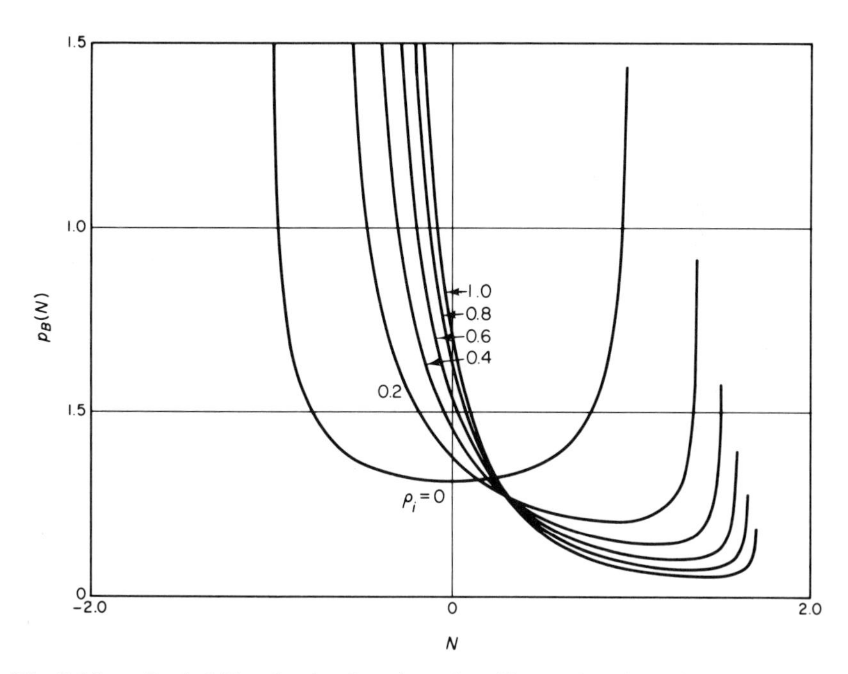

Fig. 2-20 Probability density function of $N_B(t)$ as a function of input signal-to-noise ratio

sian, whereas in the limit as $\rho_i \longrightarrow 0$, both $p_A(N)$ and $p_B(N)$ behave like $1/\pi\sqrt{1 - N^2}$.

Returning now to the PLL application, the output of the multiplier (phase detector) in Fig. 2-18 is (neglecting double-frequency terms)

$$\epsilon(t) = \sqrt{\tilde{\alpha}_1^2 P_1} \sin \varphi(t) + N_A(t) \cos \varphi(t) - N_B(t) \sin \varphi(t) \qquad (2\text{-}45)$$

For steady-state operation in the linear region (small φ), we define an equivalent signal-to-noise ratio at the output of the phase detector by

$$\rho_o \triangleq \frac{2P_e}{N_{0e} W_e} = \frac{\tilde{\alpha}_1^2 P_1}{\sigma_A^2} \qquad (2\text{-}46)$$

where P_e, N_{0e}, and W_e are respectively the equivalent power, noise spectral density, and bandwidth at the phase detector output, and σ_A^2 denotes the variance of $N_A(t)$. The concept of equivalent noise bandwidth is more formally defined in Ref. 1.

Using Eq. (2-42) and the p.d.f. of $\pi/(2\sqrt{2})N_A(t)$ as given in (2-44), ρ_o can be related to the input signal-to-noise ratio ρ_i by

$$\frac{\rho_0}{\rho_i} = \frac{\tilde{\alpha}_1^2}{(1 - e^{-\rho_i})/2} = \frac{\pi(\rho_i/2) \exp(-\rho_i)[I_0(\rho_i/2) + I_1(\rho_i/2)]^2}{1 - e^{-\rho_i}} \qquad (2\text{-}47)$$

which is only a function of ρ_i.

We now define a factor Γ_p called the *limiter performance factor* by the ratio of the input signal-to-noise spectral density to that at the output:

$$\Gamma_p = \frac{P_c/N_0}{P_e/N_{0e}} = \frac{1}{(\rho_0/\rho_i)(W_e/W_i)} \qquad (2\text{-}48)$$

Thus, the reciprocal of Γ_p is the product of the input/output signal-to-noise ratio as given by Eq. (2-47) and the ratio of the equivalent noise bandwidth at the phase detector output to the IF bandwidth. For any value of ρ_i, it is clear that $W_e/W_i \geq 1/2$ since passing noise of known bandwidth through a zero-memory nonlinear device can only increase the equivalent noise bandwidth. The factor of 1/2 is due to the band-pass to low-pass bandwidth transformation. Hence, an upper bound on Γ_p is simply given by two times the reciprocal of (ρ_0/ρ_i) as defined in (2-47). Furthermore, an exact analytical expression for Γ_p (or equivalently W_e/W_i) is quite difficult to develop and in general depends on the autocorrelation function of the input component noise processes $N_c(t)$ and $N_s(t)$. For the case of a rectangular noise spectrum (as has been heretofore considered), a good approximation to Γ_p is given by[18]

$$\Gamma_p = \frac{1 - e^{-\rho_i}}{(\pi\rho_i/4)e^{-\rho_i}[I_0(\rho_i/2) + I_1(\rho_i/2)]^2[1 + (4/\Gamma_0\pi - 1)e^{-\rho_i(1-\pi/4)}]} \qquad (2\text{-}49)$$

where Γ_0 is the value of Γ_p at $\rho_i = 0$, which for an ideal band-pass filter is approximately equal to 1/0.862.* The function $1/\Gamma_p$ with Γ_p of (2-49) is plotted versus ρ_i in Fig. 2-21 along with its lower bound $\rho_o/2\rho_i$. For an arbitrary input noise correlation function, some additional bounds on Γ_p at $\rho_i = 0$

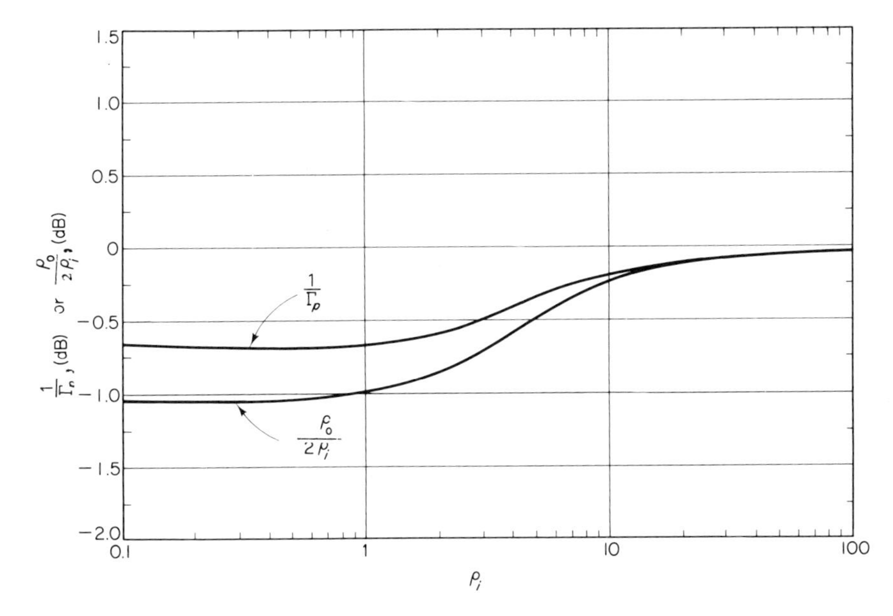

Fig. 2-21 The reciprocal of the limiter performance factor vs input signal-to-noise ratio

and $\rho_i = \infty$ can be found by noting that[18]

$$\left.\frac{W_e}{W_i}\right|_{\rho_i=0} \leq \frac{2}{\pi} \qquad \left.\frac{W_e}{W_i}\right|_{\rho_i=\infty} = \frac{1}{2} \tag{2-50}$$

Therefore,

$$1 \leq \Gamma_p|_{\rho_i=0} \leq \frac{4}{\pi} \qquad \Gamma_p|_{\rho_i=\infty} = 1 \tag{2-51}$$

As before, the variance of the phase error is an important parameter in specifying the loop response to a sine wave plus noise. Letting W_L also denote the loop bandwidth in the presence of the BPL (to be defined shortly), the

* A rational function approximation to Γ_p, which is simpler to use in system design, is suggested in Ref. 2 (later corrected in Ref. 17) and is given by

$$\Gamma_p = \frac{1 + \rho_i}{0.862 + \rho_i}$$

phase error variance in the linear region is given by

$$\sigma_\varphi^2 = \frac{1}{(P_e/N_{0e})(2/W_L)} = \frac{\Gamma_p}{(P_c/N_0)(2/W_L)} = \left(\frac{N_0 W_L}{2P_c}\right)\Gamma_p$$

We now seek to investigate the manner in which the BPL affects the closed-loop transfer function $H(s)$ and loop bandwidth W_L of the PLL that follows it. Since the effective signal power at the input to the PLL is now $\tilde{\alpha}_1^2 P_1$, the transfer function for a PLL preceded by a BPL is [from Eq. (2-10)]

$$H(s) = \frac{1 + \tau_2 s}{1 + [\tau_2 + (1/\tilde{\alpha}_1\sqrt{P_1}K)]s + (\tau_1/\tilde{\alpha}_1\sqrt{P_1}K)s^2} \tag{2-52}$$

and the corresponding W_L becomes

$$W_L = \frac{1 + (\tilde{\alpha}_1\sqrt{P_1}K\tau_2^2/\tau_1)}{2\tau_2[1 + (1/\tilde{\alpha}_1\sqrt{P_1}K\tau_2)]} = \frac{1 + r}{2\tau_2[1 + (\tau_2/r\tau_1)]} \tag{2-53}$$

where

$$r \triangleq \frac{\tilde{\alpha}_1\sqrt{P_1}K\tau_2^2}{\tau_1} \tag{2-54}$$

Notice that W_L is now a function of the limiter suppression factor $\tilde{\alpha}_1$, and hence the validity of the assumption $r\tau_1/\tau_2 \gg 1$ depends on the value of the input signal-to-noise ratio ρ_i.

Nonlinear PLL Theory and the Effects of Band-Pass Limiting

Results presented in the preceding section can be used in accounting for the effects of preceding the PLL by a BPL when the loop signal-to-noise ratio is sufficiently large to justify the use of linear PLL theory. In practice, the nonlinear behavior is of considerable interest in carrying out a particular design. As a rule, PLL parameters are usually specified at a *design point*, which in the past has arbitrarily been taken to be the condition where

$$2P_{c0} = N_0 W_{L0} \tag{2-55}$$

with

$$W_{L0} = \frac{1 + r_0}{2\tau_2(1 + \tau_2/r_0\tau_1)} \qquad r_0 = \frac{\tilde{\alpha}_{10}\sqrt{P_1}K\tau_2^2}{\tau_1} \tag{2-56}$$

The zero subscript on P, W_L, and r refers to their respective values at the design point. Let P_{c0} represent a signal power at which the linear PLL theory does not apply; hence, a suitable nonlinear model must be proposed from which one may predicate system performance in the actual region of operation. When the loop bandwidth W_L is designed to be small relative to the equivalent noise bandwidth at the phase detector output W_e, the component noise processes $N_A(t)$ and $N_B(t)$ are approximately independent of the $\varphi(t)$ process. Hence, from (2-45) total noise power at the phase detector output is approximately

$$\sigma_e^2 = \sigma_A^2 \cos^2 \varphi + \sigma_B^2 \sin^2 \varphi \tag{2-57}$$

where

$$\sigma_A^2 = \frac{8}{\pi^2}\left[\frac{1 - e^{-\rho_i}}{2\rho_i}\right]$$

$$\sigma_B^2 = \frac{8}{\pi^2}\left[1 - \frac{1 - e^{-\rho_i}}{2\rho_i} - \tilde{\alpha}_1^2\right] \tag{2-58}$$

and $\tilde{\alpha}_1$ is once again defined by Eq. (2-42). If, in addition, the input signal-to-noise ratio ρ_i is small (the usual case of interest for a PLL preceded by a BPL), then

$$\sigma_B^2 \cong \sigma_A^2 = \sigma_e^2 \tag{2-59}$$

It then follows that the effective noise in the loop is approximately Gaussian distributed with one-sided flat spectrum $2\sigma_e^2/W_L$. Thus, the effective signal-to-noise ratio in the loop bandwidth is given by

$$\rho \triangleq \frac{2P_c}{N_0 W_L}\left(\frac{1}{\Gamma_p}\right) = 1/\sigma_\varphi^2 \tag{2-60}$$

In the above, one should keep in mind that both W_L and Γ_p are dependent on the input signal-to-noise ratio through Eqs. (2-42), (2-49), and (2-53).

As we have already observed, the p.d.f. of the phase error for a second-order PLL with zero detuning is characterized by the parameter α defined by (2-17) in terms of loop signal-to-noise ratio ρ and system damping r. It is sufficient then in applying this result and those derived from it to the case of a PLL preceded by a BPL to use (2-60) [together with (2-49) and (2-53)] for ρ and (2-54) for r. We now consider the design of a PLL in terms of specifying system parameters at the design point.

If once again it is assumed that $r_0\tau_1/\tau_2 \gg 1$, then from (2-53) and (2-56) the actual operating condition loop bandwidth W_L can be related to the design point bandwidth W_{L0} by

$$W_L = \left(\frac{1 + r_0/\mu}{1 + r_0}\right) W_{L0} \tag{2-61}$$

where

$$\mu \triangleq \frac{\tilde{\alpha}_{10}}{\tilde{\alpha}_1} = \frac{r_0}{r} \tag{2-62}$$

and $\tilde{\alpha}_{10}$ is the value of $\tilde{\alpha}_1$ at the design point. Furthermore, the effective signal-to-noise ratio in the loop as defined in (2-60) can be written as

$$\rho = \frac{2P_c}{N_0 W_{L0}}\left(\frac{1}{\Gamma_p}\right)\left(\frac{1 + r_0}{1 + r_0/\mu}\right) \tag{2-63}$$

For purposes of plotting let $x = 2P_c/N_0 W_{L0}$ and $y = W_{L0}/W_i$. Note that $\rho_i = xy$, so that (2-63) can be rewritten as

$$\rho = x\left(\frac{1}{\Gamma_p}\right)\left(\frac{1 + r_0}{1 + r_0/\mu}\right) \tag{2-64}$$

Plotted in Fig. 2-22 is the variance of the phase error as determined from the linear theory ($\sigma_\varphi^2 = 1/\rho$) versus the signal-to-noise ratio in the design point loop bandwidth x, with $r_0 = 2$ and y as a parameter. Figure 2-23 illustrates the same plot as determined from the nonlinear theory.

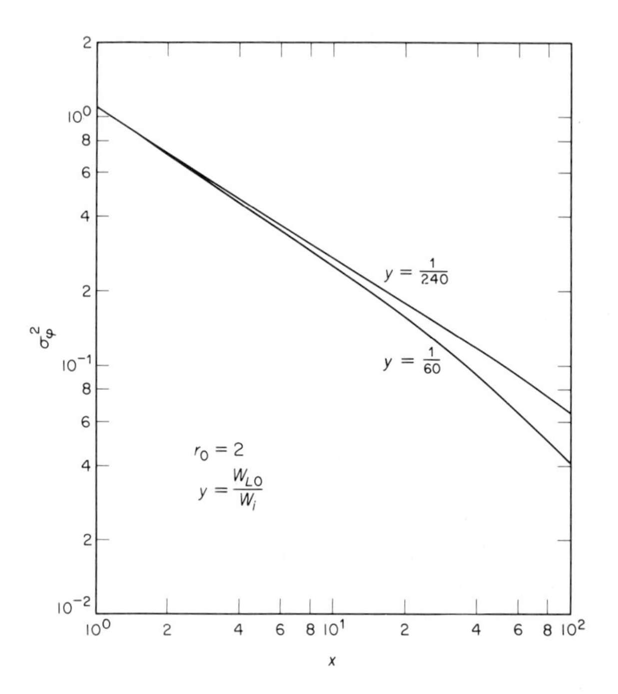

Fig. 2-22 Linear phase error variance vs signal-to-noise ratio in design point loop bandwidth with y as a parameter

2-4. SUPPRESSED CARRIER-TRACKING LOOPS

Various communication systems transmit information in the form $s(t) = \sqrt{2S}\, m(t) \sin [\omega_0 t + \theta(t)]$ where the linear modulation $m(t)$ possesses no dc component in its power spectrum. Since no residual spectral component exists at the frequency ω_0, it is not possible to use the conventional PLL discussed in the previous sections for establishing the required coherent reference.

A number of methods have been proposed for generating a reference carrier from the received waveform which contains a suppressed carrier signal component. Several of these approaches are discussed and compared here. The first of these, the *squaring loop method*, has been analyzed by a number of authors.[19-21] A second method, proposed by Costas,[22] is shown to be equivalent, theoretically, to the squaring loop. A third method in-

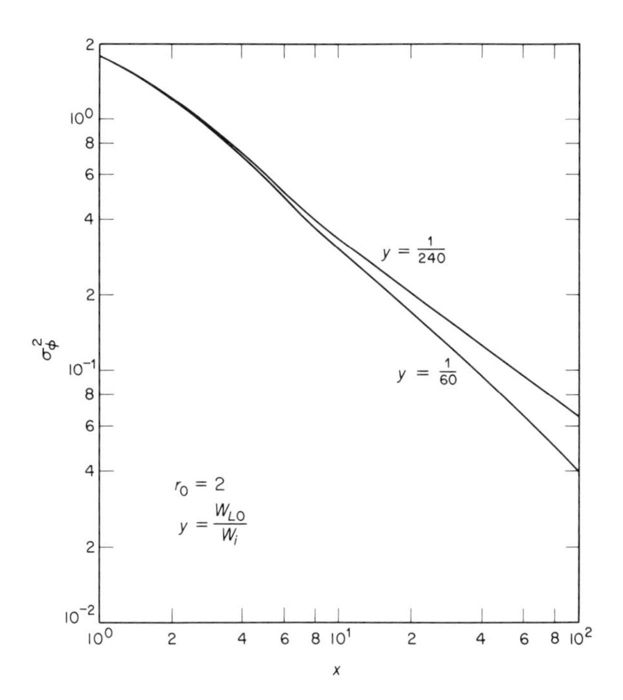

Fig. 2-23 Nonlinear phase error variance vs signal-to-noise ratio in design point loop bandwidth with y as a parameter

volves the principle of decision-directed feedback.[23] The basic procedure is as follows: first, the modulation itself is estimated, and an attempt is then made, by means of this estimate, to eliminate the modulation from the carrier, and to leave, as nearly as possible, an unmodulated sinusoid which can be tracked with a PLL.

The Squaring Loop Method

Our main concern here is to establish a coherent carrier reference for modulation extraction. The mechanization of a typical squaring loop is illustrated in Fig. 2-24. The received signal $y(t)$ is band-pass filtered and

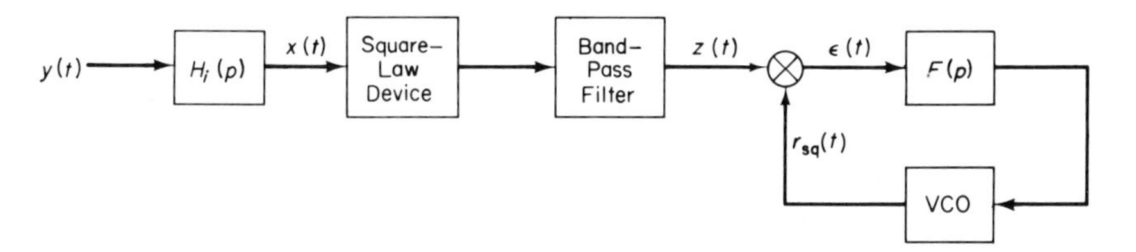

Fig. 2-24 The squaring loop

squared to remove the modulation $m(t)$, and after band-pass filtering the resultant double-frequency term is tracked by means of a PLL. When the output of the PLL is frequency divided by two, a coherent reference signal is available for demodulation and tracking purposes.

In deciding on a method of determining the performance of the squaring loop, a significant parameter is the bandwidth of the band-pass filter whose transfer function is denoted by $H_i(s)$. In fact, if the input is contaminated by white Gaussian noise, and if the bandwidth of the filter is so large that the *correlation time* τ_x of its output noise is much smaller than the time constant $1/W_L$ of the PLL, the squaring loop may be analyzed by using the mathematical techniques available from the theory of Markov processes. The correlation time of the zero mean random process $x(t)$ is defined by the relation

$$\tau_x = \frac{1}{R_x(0)} \int_0^\infty |R_x(\tau)|\, d\tau$$

where $R_x(\tau)$ is the correlation function of the process. The parameter τ_x gives some idea of the duration of the time interval over which correlation extends between values of the process $x(t)$.

Let the observed data $y(t)$ be given by

$$y(t) = \sqrt{2S}\, m(t) \sin \Phi(t) + n(t) = s(t) + n(t) \tag{2-65}$$

where $\Phi(t) = \omega_0 t + \theta(t)$, $m(t)$ is the signal modulation and $n(t)$ is the additive channel noise. The received signal is then band-pass filtered by $H_i(s)$ with the resulting output $x(t)$ in the form

$$x(t) = \sqrt{2S}\, m(t) \sin \Phi(t) + n_i(t) \tag{2-66}$$

where the two-sided bandwidth W_i of $H_i(s)$ has been assumed wide enough to pass the modulation $m(t)$ undistorted. For the case of binary digital data transmission, the modulation will be in the form of a ± 1 pulse train with bit interval T_b and the carrier is said to be *biphase modulated.* Furthermore, if $W_i < \omega_0/\pi$, then the noise $n_i(t)$ can be expressed in the form of a narrowband process about the actual frequency of the input observed data as in Eq. (2-3).

Assuming a perfect square-law device, the input to the PLL part of the overall squaring loop is (keeping only terms around $2\omega_0$)

$$\begin{aligned} z(t) = x^2(t) = {} & [-Sm^2(t) + N_c^2(t) - N_s^2(t) + 2\sqrt{S}\, m(t) N_s(t)] \cos 2\Phi(t) \\ & + [2\sqrt{S}\, m(t) N_c(t) - 2N_c(t) N_s(t)] \sin 2\Phi(t) \end{aligned} \tag{2-67}$$

The reference signal $r_{sq}(t)$, in Fig. 2-24 is conveniently represented by

$$r_{sq}(t) = 2K_1 \sin 2\hat{\Phi}(t) = 2K_1 \sin [2\omega_0 t + 2\hat{\theta}(t)] \tag{2-68}$$

The dynamic error signal then becomes (keeping only baseband frequency components)

$$\epsilon(t) = K_m z(t) r_{sq}(t) = K_m K_1 \{[Sm^2(t) - N_c^2(t) + N_s^2(t) \\ - 2\sqrt{S}\, m(t) N_s(t)] \sin 2\varphi(t) + [2\sqrt{S}\, m(t) N_c(t) - 2N_c(t) N_s(t)] \cos 2\varphi(t)\} \tag{2-69}$$

The instantaneous frequency at the VCO output is related to $\epsilon(t)$ by

$$\frac{2d\hat{\Phi}(t)}{dt} = K_V[F(p)\epsilon(t)] + 2\omega_0 \tag{2-70}$$

and hence the stochastic integro-differential equation of operation of Fig. 2-24 becomes

$$2\frac{d\varphi(t)}{dt} = 2\Omega_0 - KF(p)\{Sm^2(t) \sin 2\varphi(t) + v_2[t, 2\varphi(t)]\} \tag{2-71}$$

where $$v_2[t, 2\varphi(t)] = [-N_c^2(t) + N_s^2(t) - 2\sqrt{S}\, m(t) N_s(t)] \sin 2\varphi(t) \\ + [2\sqrt{S}\, m(t) N_c(t) - 2N_c(t) N_s(t)] \cos 2\varphi(t) \tag{2-72}$$

Note that $2\varphi(t)$ represents the actual phase error being tracked by the loop. In the case of digital modulation, $m(t) = \pm 1$ so that $m^2(t) = 1$. For the case where the message $m(t)$ is a zero mean random process that is high-pass relative to the equivalent loop bandwidth, then one can replace $m^2(t)$ in Eq. (2-71) by its mean value. This fact becomes obvious if we decompose $m^2(t) \sin 2\varphi(t)$ as

$$m^2(t) \sin 2\varphi(t) = \overline{m^2(t)} \sin 2\varphi(t) + [m^2(t) - \overline{m^2(t)}] \sin 2\varphi(t) \tag{2-73}$$

where the second term is a zero mean high-pass process and will be outside the bandwidth of the tracking loop. Furthermore, if $m(t)$ is not in itself high-pass in nature, we can make it so by modulating it onto a subcarrier. Assuming that the process is normalized such that $\overline{m^2(t)} = 1$, we have

$$2\dot{\varphi}(t) = 2\Omega_0 - KF(p)\{S \sin 2\varphi(t) + v_2[t, 2\varphi(t)]\} \tag{2-74}$$

To determine the steady-state p.d.f. of $2\varphi(t)$ we apply a procedure based on the Fokker-Planck equation—the fluctuation equation approach.[24] The method acknowledges the fact that under certain circumstances, a non-Markov process can be replaced or approximated by a Markov process.

Assuming a loop filter of the form given in Eq. (2-9), the p.d.f. of the modulo 2π reduced phase error $2\phi(t)$ can be approximated[21] in the form of Eq. (2-16) with ϕ replaced by 2ϕ and

$$\alpha \triangleq \frac{4}{N_{sq} F_1^2 K^2}\left[SKF_1 - \frac{KN_{sq}}{4S\sigma_G^2}\left(\frac{1 - F_1}{\tau_1}\right)\right]$$

$$\beta \triangleq \frac{4}{N_{sq} F_1^2 K^2}[2\Omega_0 - SK(1 - F_1)\overline{\sin 2\phi}] + \alpha \overline{\sin 2\phi}$$

$$\sigma_G^2 = \overline{\sin^2 2\phi} - (\overline{\sin 2\phi})^2$$

$$N_{sq} \triangleq 2\int_{-\infty}^{\infty} R_{v_2}(\tau)\, d\tau \tag{2-75}$$

where
$$R_{v_2}(\tau) \triangleq \overline{v_2(t, 2\phi)v_2(t+\tau, 2\phi)} \tag{2-76}$$

With $v_2(t, 2\phi)$ as functionally defined in Eq. (2-72), it is relatively straightforward to show (see Appendix A) that

$$R_{v_2}(\tau) = 4[SR_{N_c}(\tau) + R^2_{N_c}(\tau)] \tag{2-77}$$

where
$$R_{N_c}(\tau) = \overline{N_c(t)N_c(t+\tau)} = \frac{N_0}{2}\int_{-\infty}^{\infty} |H_\ell(j2\pi f)|^2 e^{j2\pi f\tau}\,df$$

and $H_\ell(j\omega)$ is the low-pass equivalent of $H_i(j\omega)$. Substitution of (2-77) into (2-75) gives

$$N_{\text{sq}} = 4SN_0\mathcal{S}_L^{-1} \tag{2-78}$$

where
$$\mathcal{S}_L \triangleq \left\{1 + \frac{2}{SN_0}\left[\int_{-\infty}^{\infty} R^2_{N_c}(\tau)\,d\tau\right]\right\}^{-1} \tag{2-79}$$

is defined to be the circuit *squaring loss*. It is assumed in the above that $H_\ell(j\omega)$ is normalized such that $H_\ell(0) = 1$. At this point, our results are general in that they hold for a broad class of prefilter characteristics $H_i(j\omega)$. We now consider several special cases of practical interest.

The quantities β and α that characterize the p.d.f. of 2ϕ can be related to an equivalent set of system parameters ρ' and r by

$$\alpha = \left(\frac{r+1}{r}\right)\rho' - \frac{1-F_1}{r\sigma_G^2}$$
$$\beta = \left(\frac{r+1}{r}\right)^2 \frac{\rho'}{2W_L}[2\Omega_0 - SK(1-F_1)\overline{\sin 2\phi}] + \alpha\,\overline{\sin 2\phi} \tag{2-80}$$

where

r = second-order loop damping parameter $= SKF_1\tau_2$
ρ' = effective signal-to-noise ratio in the loop bandwidth $= (\rho/4)\mathcal{S}_L = (\rho_i/2\gamma)\mathcal{S}_L$
ρ = equivalent signal-to-noise ratio in the loop bandwidth of second-order PLL $= 2S/N_0W_L$
ρ_i = input signal-to-noise ratio in input bandwidth $= 2S/N_0W_i$
γ = ratio of two-sided noise bandwidth of loop to that of $H_\ell(j\omega)$ $= 2W_L/W_i$ (2-81)

To specify the result any further it is necessary to consider various prefiltering characteristics from which the squaring loss coefficient may be evaluated.

As a first example, consider a band-pass filter with an RC transfer function. The equivalent low-pass spectrum for $N_c(t)$ or $N_s(t)$ has a correlation function:

$$R_{N_c}(\tau) = R_{N_s}(\tau) = \frac{N_0}{4}W_i \exp(-W_i|\tau|) \tag{2-82}$$

Thus, from Eqs. (2-79), (2-81), and (2-82),

$$\mathcal{S}_L = \frac{1}{1 + 1/2\rho\gamma} \tag{2-83}$$

For an ideal band-pass filter ($H_\ell(j\omega) = 1; |f| < W_i/4$, and zero otherwise),

$$\mathcal{S}_L = \frac{1}{1 + 1/\rho\gamma} \tag{2-84}$$

We note that these are the two limiting cases of the Butterworth spectra.

One obvious question is to ask for the presquaring filter which minimizes the squaring loss. This problem is quite difficult to solve. A more tractable one (but certainly not equivalent) is to find the presquaring filter that maximizes the signal-to-noise ratio at the input of the PLL. This problem has been solved, and for the case where the modulating spectrum is narrow with respect to the carrier frequency, the filter has been shown to be given by[25]

$$H_i(s) = k\left[\frac{S_s(s)}{S_s(s) + N_0/2}\right]^{1/2} \tag{2-85}$$

where k is an arbitrary positive constant chosen such that $H_i(\omega_0) = 1$ and $S_s(s)$ is the power spectral density of the modulated signal $s(t)$. For large input signal-to-noise conditions, the presquaring filter given by (2-85) becomes $H_i(s) = 1$ (i.e., an ideal band-pass filter), while for small signal-to-noise conditions the filter becomes

$$H_i(s) = k\left[\frac{2S_s(s)}{N_0}\right]^{1/2} \tag{2-86}$$

This says that for small signal-to-noise conditions, the filter is matched to the signaling spectrum. For an input modulation $m(t)$ characterized by a random ± 1 pulse train of period T_b, the corresponding squaring loss is given by

$$\mathcal{S}_L = \frac{1}{1 + 2/3\rho\gamma} \tag{2-87}$$

Thus, it is interesting to note that the filter of (2-86) gives a larger squaring loss than the RC band-pass filter. Various other examples of prefiltering characteristics along with their associated squaring loss are given in Table 2-1.

The squaring loop reference signal $r_{sq}(t)$ is frequency divided by two to provide a noisy reference for demodulation of the data off the carrier. The variance of the phase jitter asociated with this noisy reference is (for small ϕ)

$$\sigma_\phi^2 = \frac{1}{4}\sigma_{2\phi}^2 = \frac{\mathcal{S}_L^{-1}}{\rho} \tag{2-88}$$

and hence the penalty (relative to the linear performance of a PLL) is incor-

TABLE 2-1

Squaring Loss: $\mathcal{S}_L = \left[1 + \frac{K_L}{\rho\gamma}\right]^{-1}$

Prefilter Type	*Equivalent Low-Pass Transfer Characteristic* $\lvert H_\ell(j\omega)\rvert^2$	K_L
nth-order Butterworth	$\frac{1}{1+(\omega/\omega_i)^{2n}}$; $\omega_i = \frac{(2n-1)\pi W_i}{2\Gamma(1/2n)\Gamma[2-(1/2n)]}$	$\frac{2n-1}{2n}$
Gaussian	$\exp\left[-2\left(\frac{\omega}{\omega_i}\right)^2\right]$; $\omega_i = \sqrt{2\pi}W_i$	$\frac{1}{\sqrt{2}}$
Sinusoidal Rolloff $(0 \le \xi \le 1)$	$\frac{1}{4}\left[1 - \sin\frac{\pi}{2}\left(\frac{\omega-\omega_i}{\xi\omega_i}\right)\right]^2$ for $\lvert\omega-\omega_i\rvert \le \xi\omega_i$; 0 for $\omega-\omega_i \ge \xi\omega_i$; 1 for $-\omega_i \le \omega-\omega_i \le -\xi\omega_i$; $\omega_i = \frac{\pi W_i}{2[1-(1/4)\xi]}$	$\frac{1-(29/64)\xi}{1-(1/4)\xi}$

porated in the squaring loss. The nonlinear phase noise performance may be obtained by using Eqs. (2-18) and (2-19) to evaluate $\sigma_{2\phi}^2$ with α and β as defined in (2-80).

The Costas or I-Q Loop

In the Costas loop (Fig. 2-25), the phase of the data carrier is extracted from the suppressed carrier signal $s(t)$ plus noise $n_i(t)$ by multiplying the input voltages of the two phase detectors (multipliers) with that produced from the output of the VCO and a 90° phase shift of that voltage, filtering the results, and using this signal to control the phase and frequency of the loop's VCO output. When the filters in the in-phase and quadrature-phase arms of the Costas loop are mechanized with integrate-and-dump circuits, then the loop is referred to as an I-Q loop.

If we denote the output of the upper loop multiplier (gain $\sqrt{K_m}$) by $z_c(t)$ and the output of the lower loop multiplier (gain $\sqrt{K_m}$) by $z_s(t)$ (Fig. 2-25), then the output $z_c(t)$ is

$$z_c(t) = x(t)\,[2\sqrt{K_1K_m}\cos\hat{\Phi}(t)] \tag{2-89}$$

while the output of the low-pass filter becomes

$$\begin{aligned} y_c(t) &= \sqrt{2K_1K_m}\,[\sqrt{S}\,m(t) - N_s(t)]\sin\varphi(t) \\ &\quad + \sqrt{2K_1K_m}N_c(t)\cos\varphi(t) \end{aligned} \tag{2-90}$$

when Eq. (2-66) is substituted into (2-89) and all double-frequency terms

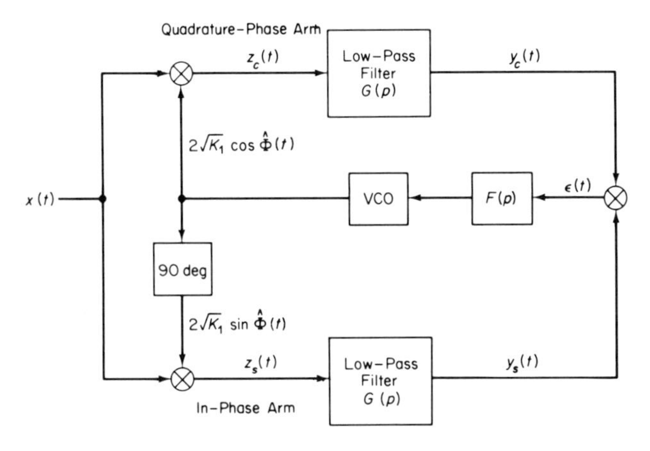

Fig. 2-25 The Costas loop

are neglected. Similarly, the output $y_s(t)$ is given by

$$y_s(t) = \sqrt{2K_1K_m}\,[\sqrt{S}m(t) - N_s(t)]\cos\varphi(t) - \sqrt{2K_1K_m}N_c(t)\sin\varphi(t) \tag{2-91}$$

We note in the above that the gains of the reference signal and multiplier are the square roots of their equivalents in the squaring loop. Multiplying the two low-pass filter outputs gives the dynamic error signal,

$$\epsilon(t) = K_1K_m\{[\sqrt{S}m(t) - N_s(t)]^2 - N_c^2(t)\}\sin 2\varphi(t) + 2K_1K_mN_c(t)[\sqrt{S}m(t) - N_s(t)]\cos 2\varphi(t) \tag{2-92}$$

Comparing the above with (2-69), it is clear that the two dynamic error signals are identical provided that the shaping of the noise spectrum produced by $G(s)$ of Fig. 2-25 is equivalent to that produced alone by the low-pass equivalent of $H_i(s)$ in Fig. 2-24. Furthermore, if the loop filter is identical to that used in the squaring loop, then the stochastic integro-differential equation describing the behavior of the Costas loop is also given by Eq. (2-74).* Hence, the solutions for the p.d.f. of the phase error 2ϕ are identical, and the tracking performance of the two circuits is the same. Note that the Costas loop can be implemented with integrate and dump circuits (matched filters) replacing the low-pass filters in the in-phase and quadrature-phase arms, thereby giving improved noise immunity. If this is to be accomplished, one must supply timing information to these circuits—that is, the instant at which to initiate and terminate the integration. On the other hand, if, for simplicity, one still desires to use low-pass filters, then they can be designed to approximate the above integrate and dump by making W_i

* Note that since the VCO in the Costas loop has a center frequency at ω_0, its gain in radians/volt-sec must be one-half of the VCO gain in the squaring loop.

$\simeq 2/T_b$ and hence $\gamma \simeq W_L T_b = 2/\delta$ where $\delta \triangleq 2/W_L T_b$. Both methods of carrier tracking exhibit the usual 180° phase ambiguity inherent in all systems that attempt to recover the carrier phase from a biphase modulated signal; that is, changing the sign of the received signal leaves the sign of the recovered carrier unaltered.

Methods often used in practice to distinguish between the true message and its inverse include: (1) tracking the suppressed carrier and using *differential encoding* of the data symbols—that is, the information to be transmitted is represented in terms of the changes between successive data symbols rather than the symbols themselves (see Chap. 5); (2) using the natural redundancy inserted in the data stream for frame synchronization purposes; (3) using the redundancy inserted in the data stream for error control; and (4) tracking the suppressed carrier and retaining a small carrier component for ambiguity resolution; (5) employing *Miller coding* (*delay modulation*).

Decision-Feedback Loop

A third technique for deriving a coherent reference from a suppressed carrier modulated signal makes use of the principle of decision-feedback and theoretically appears to give significant improvement in tracking performance over the devices discussed above. The configuration presented in Fig. 2-26 is one possible implementation of a decision-feedback loop.[23] The input $x(t)$ as defined in Eq. (2-66) is demodulated by a reference $r(t) = \sqrt{2}K_1 \cos \hat{\Phi}(t)$ to produce the dynamic error signal

$$\epsilon(t) = K_1 K_m \{\sqrt{S}m(t) \sin \varphi(t) + N[t, \varphi(t)]\} \tag{2-93}$$

This signal is delayed one bit interval and multiplied by an estimate of

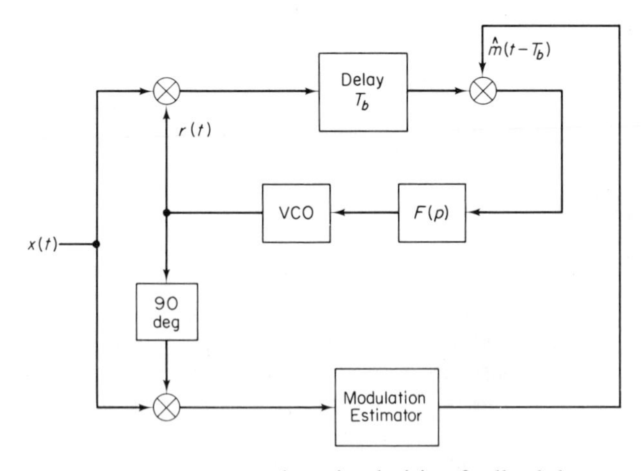

Fig. 2-26 A suppressed carrier decision-feedback loop

the data produced at the output of the data demodulator, $\hat{m}(t - T_b)$.* The data detector is nothing more than an integrate and dump circuit followed by a hard limiter where the reference signal used for demodulation of the data is derived from the decision-feedback loop itself. When the bandwidth of the loop is small relative to the signaling data rate, $\mathcal{R}_b = 1/T_b$; that is, the phase error is essentially constant for several time intervals (each of length T_b), then the product $m(t)\hat{m}(t)$ can be replaced by its average value: $E\{m(t)\hat{m}(t)\} = (+1) \times \text{Prob}\,\{m(t) = \hat{m}(t)\} + (-1) \times \text{Prob}\,\{m(t) \neq \hat{m}(t)\} = 1 - 2P_E[\varphi(t)]$ where $P_E[\varphi(t)]$ is the error probability of the data detector conditioned on the loop phase error. Under this assumption, the input to the VCO, $z(t)$, is given by

$$z(t) = K_1 K_m F(p)\{\exp[-pT_b][\sqrt{S}\{1-2P_E[\varphi(t)]\}\sin\varphi(t) + \hat{m}(t)N_\ell[t, \varphi(t)]]\} \tag{2-94}$$

where $N_\ell[t, \varphi(t)]$ can be modelled exactly the same way as $N[t, \varphi(t)]$ of (2-7).

It is clear that in general the transfer function factor $\exp(-pT_b)$ with $p = j\omega$ affects loop stability and reduces the signal acquisition or pull-in range.[1] As we are not concerned in this text with the theory pertaining to these problems, we shall make a simplifying assumption which enables us to neglect this exponential factor in regard to predicting steady-state performance. If the equivalent bandwidth of the loop (to be defined shortly) is narrow with respect to $\mathcal{R}_b$ (the usual case of interest), $\exp(-j\omega T_b)$ is approximately unity for all ω within this equivalent bandwidth. Hence, from a steady-state performance standpoint, this factor has negligible effect. Mechanistically, however, the delay T_b is important in assuring that $m(t)$ is multiplied by $\hat{m}(t)$ corresponding to the same bit interval.

Making the usual assumption that the instantaneous VCO phase, $\hat{\theta}(t)$ is related to its input $z(t)$ through

$$\hat{\theta}(t) = \frac{K_V}{p} z(t) \tag{2-95}$$

then the stochastic integro-differential equation that defines the loop operation is in the steady-state

$$\dot{\varphi}(t) = \dot{\theta}(t) - KF(p)\{\sqrt{S}\{1 - 2P_E[\varphi(t)]\}\sin\varphi(t) + \hat{m}(t)N_\ell[t, \varphi(t)]\} \tag{2-96}$$

One can readily see that the form of (2-96) is identical to that of a PLL where the equivalent nonlinearity is now $\sin\varphi[1 - 2P_E(\varphi)]$ and the equivalent additive noise is $\hat{m}N_\ell(t, \varphi)$.

As was true for the squaring and Costas loops, the decision-feedback loop also exhibits a 180° phase ambiguity. Assuming that $\varphi(t)$ varies slowly

---

* If the modulation $m(t)$ consists of data $d(t)$ on a square-wave subcarrier $\mathbf{S}(t)$, then prior to the delay in the loop, the subcarrier must be demodulated by a reference signal provided by a subcarrier tracking loop (see Chap. 11).

enough so it can be treated as though it were essentially constant over several bit intervals, then the conditional error probability $P_E[\varphi(t)]$ is given by (see Chap. 6)

$$P_E[\varphi(t)] \triangleq P_E(\varphi) = \frac{1}{2}\,\text{erfc}\,(\sqrt{R_b}\cos\varphi) \tag{2-97}$$

where
$$\text{erfc}\,(x) \triangleq 1 - \text{erf}(x) = \frac{2}{\sqrt{\pi}}\int_x^\infty \exp(-y^2)dy \tag{2-98}$$

and $R_b \triangleq ST_b/N_0$. Assuming that $P_E(\varphi)$ is characterized as in (2-97), then it can be shown[1] that the steady-state p.d.f. of the phase error reduced modulo 2π is given by

$$p(\phi) = C_0'\exp[-U_0(\phi)]\int_\phi^{\phi+2\pi}\exp[U_0(x)]\,dx \tag{2-99}$$

where C_0' is the normalization constant and the potential function $U_0(\phi)$ is given by

$$U_0(\phi) = -\beta\phi - \alpha\left(\frac{\cos\phi\,\text{erf}\,(\sqrt{R_b}\cos\phi) + (1/\sqrt{\pi R_b})\exp(-R_b\cos^2\phi)}{\text{erf}\,(\sqrt{R_b})}\right) \tag{2-100}$$

where α, β are defined by Eq. (2-17) with P_c replaced by $S\,\text{erf}^2\,(\sqrt{R_b})$, and $\sin\phi$ by $\{(\sin\phi)\,[\text{erf}\,(\sqrt{R_b}\cos\phi)/\text{erf}\,(\sqrt{R_b})]\}$. For a first-order loop the potential function can be simplified and expressed in terms of the system parameters by

$$U_0(\phi) = -\frac{4\Omega_0}{N_0K^2}\phi - R_b\delta_e\,\text{erf}\,(\sqrt{R_b})\left\{\cos\phi\,\text{erf}\,(\sqrt{R_b}\cos\phi) + \frac{1}{\sqrt{\pi R_b}}\exp(-R_b\cos^2\phi)\right\} \tag{2-101}$$

where
$$K = \frac{4}{\delta_e\sqrt{S\,T_b}\,\text{erf}\,(\sqrt{R_b})}$$
$$\delta_e = \frac{2}{W_{Le}T_b} \tag{2-102}$$

and W_{Le} is the effective two-sided bandwidth of the decision-feedback loop given by

$$W_{Le} = \frac{\sqrt{S}\,K\,\text{erf}\,(\sqrt{R_b})}{2} \tag{2-103}$$

Figure 2-27 illustrates the phase error density $p(\phi)$ for $\Omega_0 = 0$, $\delta_e = 10$, and R_b as a parameter. This figure also reiterates the fact that the decision-feedback loop exhibits the same 180° phase ambiguity as the squaring or Costas loop.

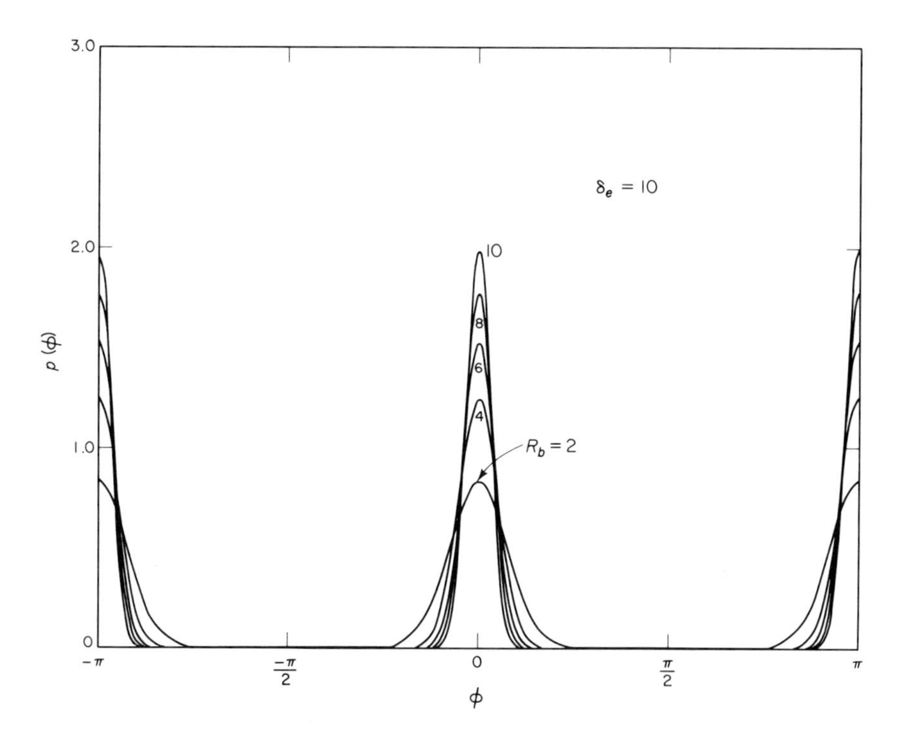

Fig. 2-27 Probability density function of the phase error in a decision-feedback loop with signal-to-noise ratio per bit as a parameter

We conclude this section with a comparison of the variance of the phase error for three configurations assuming $\Omega_0 = 0$ and for simplicity first-order loops.[26] An ideal band-pass filter precedes the squaring loop since this results in the maximum squaring loss and hence the most optimistic comparison. Figures 2-28 and 2-29 represent plots of the variance of the phase error versus R_b for $\delta_e = \delta = 10$ and 40 respectively. The performance of the decision-feedback loop is independent of $\gamma = W_L/2W_i$; however, the performance of the squaring loop is degraded with decreasing γ. In fact, for all cases examined it is found that the performance of the squaring loop is inferior to that of the decision-feedback loop. Figure 2-30 is a plot of the ratio in dB of the variance of the phase error in a squaring loop to that in the decision-feedback loop for various values of γ with $\delta = 10$. This figure is an alternate representation of the data presented in Fig. 2-29.

Other "hybrid"-type configurations can be proposed which in effect combine the advantages of the PLL with those of the suppressed carrier loops. This subject is treated in Chap. 11.

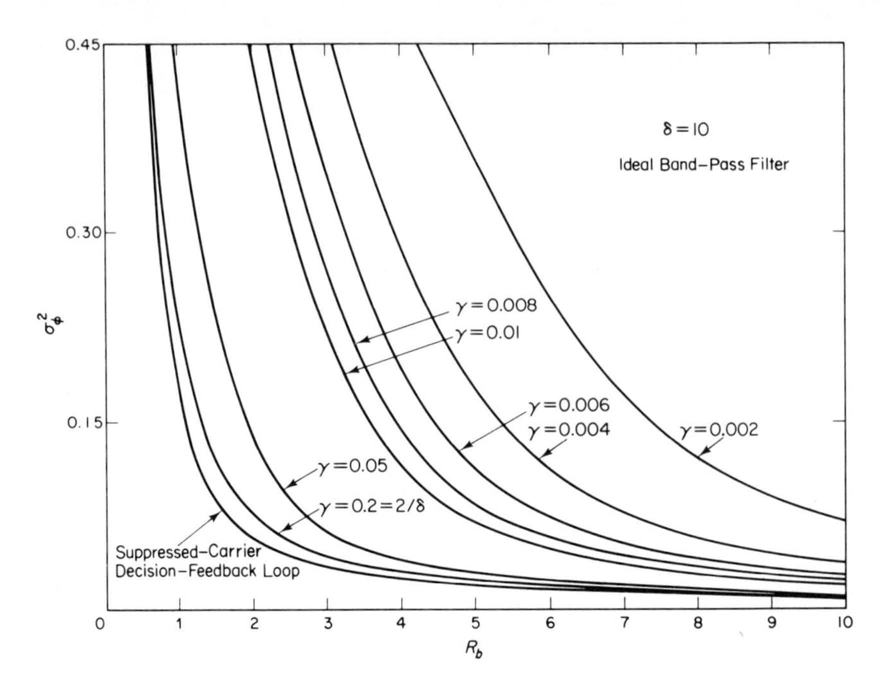

Fig. 2-28 Phase error variance in a decision-feedback loop vs signal-to-noise ratio per bit with γ as a parameter: $\delta = 10$

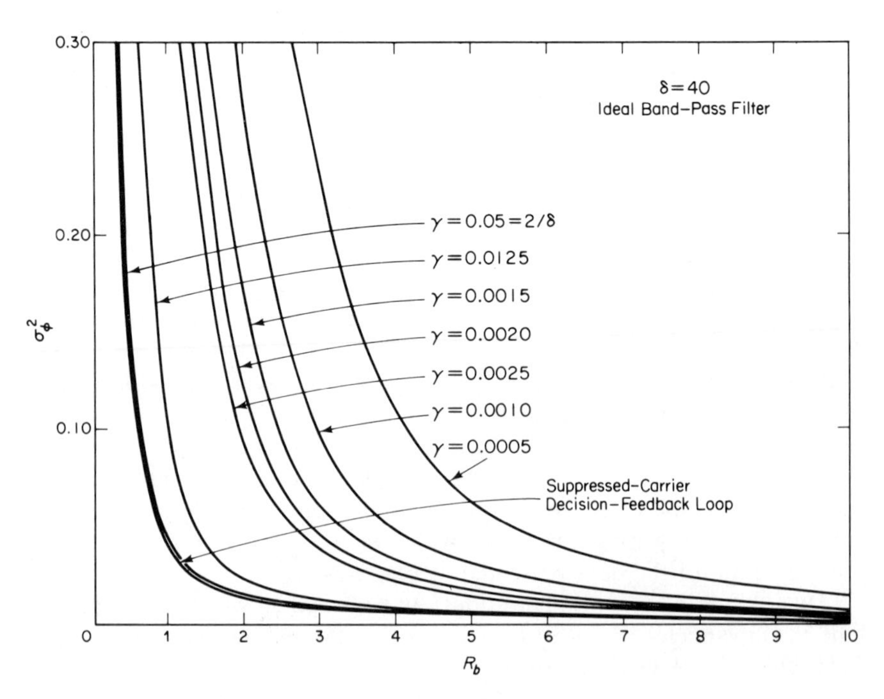

Fig. 2-29 Phase error variance in a decision-feedback loop vs signal-to-noise ratio per bit with γ as a parameter: $\delta = 40$

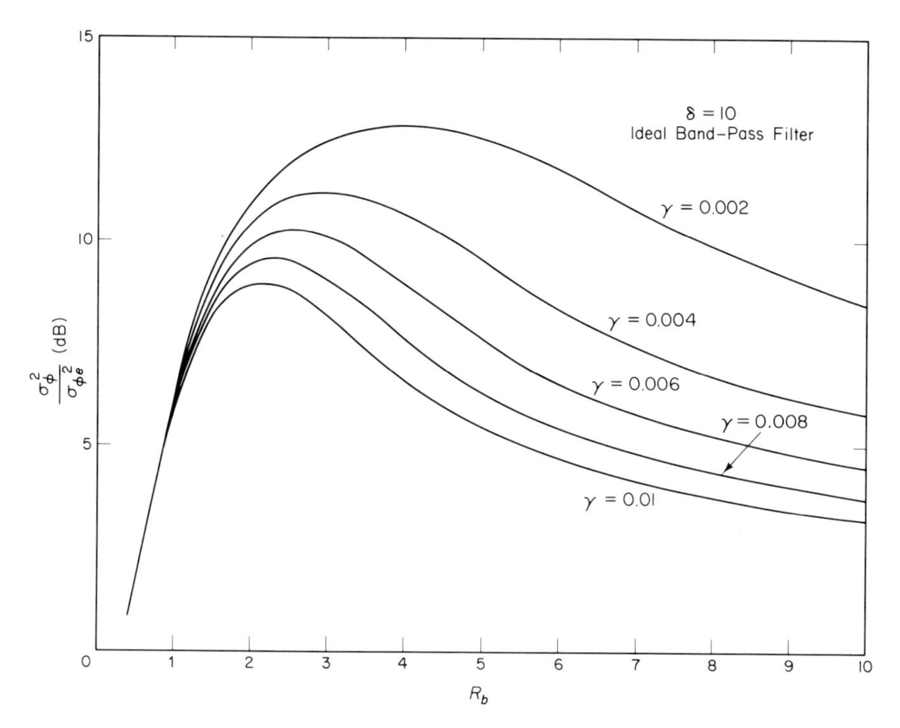

Fig. 2-30 The ratio of the variance of the phase error in a decision-feedback loop to that in a squaring loop vs signal-to-noise ratio per bit with γ as a parameter

2-5. CARRIER-TRACKING LOOPS FOR POLYPHASE SIGNALS

We have discussed carrier-tracking loops that can be used to establish a coherent reference signal from a *bi*phase-modulated input signal whose carrier component is either partially or totally suppressed. In Chap. 5 we introduce the concept of *poly*phase modulation wherein the observed data are of the form*

$$y(t) = \sqrt{2S}\sin\left(\Phi(t) + 2(k-1)\frac{\pi}{N}\right) + n(t) \qquad k = 1, 2, \ldots, N \tag{2-104}$$

For $N = 2$, Eq. (2-104) reduces to (2-65) if $m(t)$ represents digital modulation as previously discussed, while for $N = 4$ we get the familiar *quadriphase* signal representation.

---

* During a given transmission interval of $T = (\log_2 N)T_b$ seconds, the transmitted signal is characterized by the signal component of Eq. (2-104).

The reconstruction of a carrier for a polyphase signal can be accomplished in many ways; however, if the modulation scheme is to be successfully applied, one must provide an efficient and accurate method for establishing coherent reference signals in the receiver that are independent of the modulation. The significance of this statement is that the loop must track the random phase $\Phi(t)$ without concern for which of the N data signals phase-modulates the carrier.

The simplest class of loops that satisfy this requirement consists of the generalization of the suppressed carrier-tracking loops of Sec. 2-4 to N phases. The analysis of such N-phase loops parallels the development presented in the previous section for their biphase counterparts. Hence, wherever possible we shall draw upon these analogies.

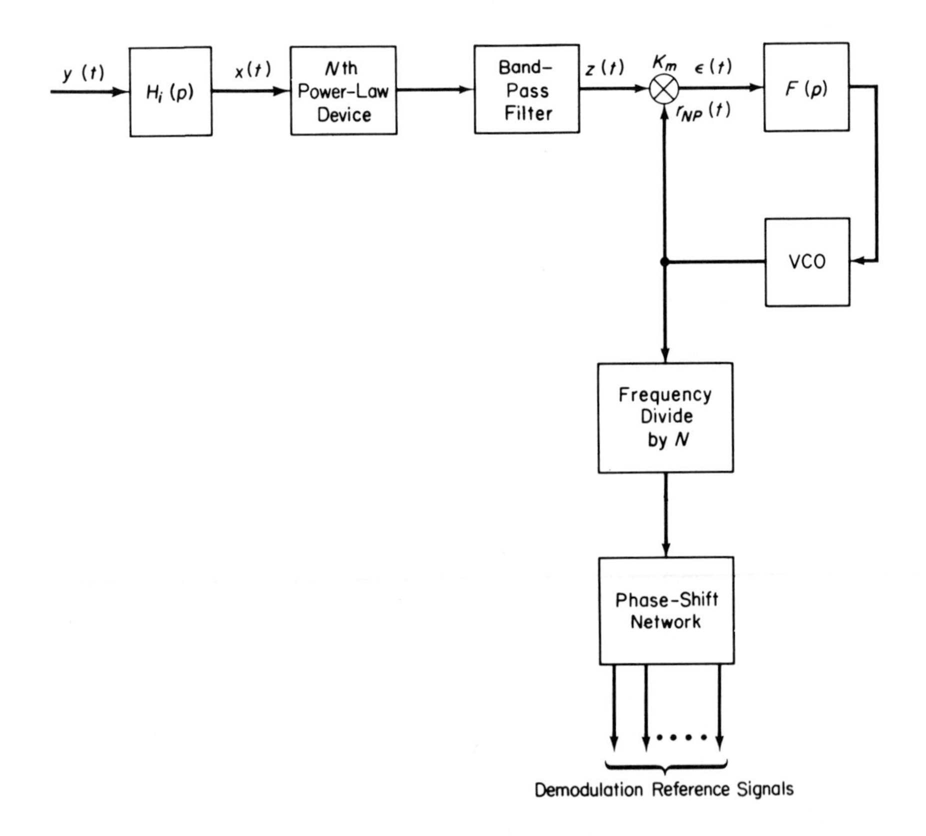

Fig. 2-31 The Nth power loop

The Nth Power Loop

Rather than seek totally new configurations, it is natural to ask whether the tracking loops already discussed can be generalized to accommodate a polyphase input signal. In this connection, consider the Nth power loop illustrated in Fig. 2-31, which is a generalization of the squaring loop of Fig. 2-24. Here the input after being band-pass filtered is raised to the Nth power, and the Nth harmonic of the carrier so produced is tracked by a PLL. The development of the stochastic integro-differential equation for the Nth power loop parallels that given in Sec. 2-4 for the squaring loop, hence we present only the key results.

The output of the Nth law device can be written in the form

$$\begin{aligned} z(t) = x^N(t) = \sum_{r=0}^{N} {}_N C_r \left[\sqrt{2S} \sin\left(\Phi(t) + 2(k-1)\frac{\pi}{N}\right)\right]^r \\ \times \left[\sqrt{2}\, N_i(t) \cos\left(\Phi(t) + \theta_i(t)\right)\right]^{N-r} \end{aligned} \quad (2\text{-}105)$$

where ${}_N C_r = N!/r!(N-r)!$ is the binomial coefficient and

$$\begin{aligned} N_i(t) &= \sqrt{N_c^2(t) + N_s^2(t)} \\ \theta_i(t) &= \tan^{-1}\frac{N_s(t)}{N_c(t)} \end{aligned} \quad (2\text{-}106)$$

Using well-known Fourier series expansions of $\sin^n x$ and $\cos^n x$ and keeping only terms around $N\omega_0$, Eq. (2-105) becomes

$$\begin{aligned} z(t) = \frac{\sin N\Phi(t)}{2^{(N/2)-1}} \Bigg\{ &\sum_{r=1,3,5\ldots}^{N} (-1)^{(r-1)/2} {}_N C_r \cos\left[\frac{r}{N} 2(k-1)\pi\right] S^{r/2} \\ &\times \sum_{\ell=0,2,4\ldots}^{N-r} (-1)^{\ell/2} {}_{N-r}C_\ell N_s^\ell(t) N_c^{N-r-\ell}(t) \\ &- \sum_{r=0,2,4\ldots}^{N} (-1)^{r/2} {}_N C_r \sin\left[\frac{r}{N} 2(k-1)\pi\right] S^{r/2} \\ &\times \sum_{\ell=0,2,4\ldots}^{N-r} (-1)^{\ell/2} {}_{N-r}C_\ell N_s^\ell(t) N_c^{N-r-\ell}(t) \\ &- \sum_{r=1,3,5\ldots}^{N} (-1)^{(r-1)/2} {}_N C_r \sin\left[\frac{r}{N} 2(k-1)\pi\right] S^{r/2} \\ &\times \sum_{\ell=1,3,5\ldots}^{N-r} (-1)^{(\ell-1)/2} {}_{N-r}C_\ell N_s^\ell(t) N_c^{N-r-\ell}(t) \\ &- \sum_{r=0,2,4\ldots}^{N} (-1)^{r/2} {}_N C_r \cos\left[\frac{r}{N} 2(k-1)\pi\right] S^{r/2} \\ &\times \sum_{\ell=1,3,5\ldots}^{N-r} (-1)^{(\ell-1)/2} {}_{N-r}C_\ell N_s^\ell(t) N_c^{N-r-\ell}(t) \Bigg\} \end{aligned}$$

$$+\frac{\cos N\Phi(t)}{2^{(N/2)-1}}\Big\{\sum_{r=0,2,4\dots}^{N}(-1)^{r/2}{}_NC_r\cos\left[\frac{r}{N}2(k-1)\pi\right]S^{r/2}$$
$$\times\sum_{\ell=0,2,4\dots}^{N-r}(-1)^{\ell/2}{}_{N-r}C_\ell N_s^\ell(t)N_c^{N-r-\ell}(t)$$
$$+\sum_{r=1,3,5\dots}^{N}(-1)^{(r-1)/2}{}_NC_r\sin\left[\frac{r}{N}2(k-1)\pi\right]S^{r/2}$$
$$\times\sum_{\ell=0,2,4\dots}^{N-r}(-1)^{\ell/2}{}_{N-r}C_\ell N_s^\ell(t)N_c^{N-r-\ell}(t)$$
$$-\sum_{r=0,2,4\dots}^{N}(-1)^{r/2}{}_NC_r\sin\left[\frac{r}{N}2(k-1)\pi\right]S^{r/2}$$
$$\times\sum_{\ell=1,3,5\dots}^{N-r}(-1)^{(\ell-1)/2}{}_{N-r}C_\ell N_s^\ell(t)N_c^{N-r-\ell}(t)$$
$$+\sum_{r=1,3,5\dots}^{N}(-1)^{(r-1)/2}{}_NC_r\cos\left[\frac{r}{N}2(k-1)\pi\right]S^{r/2}$$
$$\times\sum_{\ell=1,3,5\dots}^{N-r}(-1)^{(\ell-1)/2}{}_{N-r}C_\ell N_s^\ell(t)N_c^{N-r-\ell}(t)\Big\}$$

Separating the $r=0$, $l=0$ term in the fifth double summation, we can write $z(t)$ in the abbreviated form

$$z(t)=A_s(t)\frac{\sin N\Phi(t)}{2^{(N/2)-1}}+\left[A_c(t)-S^{N/2}\right]\frac{\cos N\Phi(t)}{2^{(N/2)-1}} \tag{2-107}$$

where $A_s(t)$ and $A_c(t)$ are zero mean noise processes.

If the reference signal $r_{NP}(t)$ in Fig. 2-31 is represented by

$$r_{NP}(t)=2^{N/2}K_1\sin N\hat{\Phi}(t) \tag{2-108}$$

then the dynamic error signal (keeping only baseband components) becomes

$$\begin{aligned}\epsilon(t)&=K_mz(t)r_{NP}(t)=K_mK_1\{-[A_c(t)-S^{N/2}]\sin N\varphi(t)+A_s(t)\cos N\varphi(t)\}\\&=K_mK_1S^{N/2}\sin N\varphi(t)+K_mK_1v_N[t,N\varphi(t)]\end{aligned} \tag{2-109}$$

where
$$v_N[t,N\varphi(t)]=A_s(t)\cos N\varphi(t)-A_c(t)\sin N\varphi(t) \tag{2-110}$$

The instantaneous frequency at the VCO output is related to $\epsilon(t)$ by

$$\frac{Nd\hat{\Phi}(t)}{dt}=K_V[F(p)\epsilon(t)]+N\omega_0 \tag{2-111}$$

hence the stochastic integro-differential equation of operation of the Nth power loop becomes

$$\frac{Nd\varphi(t)}{dt}=N\Omega_0-KF(p)\{S^{N/2}\sin N\varphi(t)+v_N[t,N\varphi(t)]\} \tag{2-112}$$

Assuming once again a loop filter of the form given in Eq. (2-9), the p.d.f. of the loop phase error, $N\phi(t)$, can be approximated in the form of (2-16) with

$$\alpha \triangleq \frac{4}{N_{NP}F_1^2K^2}\left[S^{N/2}KF_1 - \frac{KN_{NP}}{4S^{N/2}\sigma_G^2}\left(\frac{1-F_1}{\tau_1}\right)\right]$$

$$\beta \triangleq \frac{4}{N_{NP}F_1^2K^2}\left[N\Omega_0 - S^{N/2}K(1-F_1)\,\overline{\sin N\phi}\right] + \alpha\,\overline{\sin N\phi}$$

$$\sigma_G^2 \triangleq \overline{\sin^2 N\phi} - (\overline{\sin N\phi})^2$$

$$N_{NP} \triangleq 2\int_{-\infty}^{\infty} R_{v_N}(\tau)\,d\tau \tag{2-113}$$

with

$$R_{v_N}(\tau) \triangleq \overline{v_N(t, N\phi)v_N(t+\tau, N\phi)} \tag{2-114}$$

By an extension of the procedure outlined in Appendix A, one can derive the autocorrelation function of $v_N(t, N\phi)$ with the result[27]

$$R_{v_N}(\tau) = \sum_{r=0}^{N-1} ({}_NC_r)^2(N-r)!\,2^{N-r-1}S^rR_{Nc}^{N-r}(\tau) \tag{2-115}$$

Substitution of Eq. (2-115) into (2-113) gives

$$N_{NP} = N^2S^{N-1}N_0\mathcal{S}_L^{-1} \tag{2-116}$$

where the *N-phase loss* $\mathcal{S}_L$ is defined by

$$\mathcal{S}_L \triangleq \left\{1 + \sum_{r=0}^{N-2}\frac{({}_NC_r)^2}{N^2}\frac{(N-r)!\,2^{N-r}}{S^{N-1-r}N_0}\left[\int_{-\infty}^{\infty} R_{N_c}^{N-r}(\tau)\,d\tau\right]\right\}^{-1} \tag{2-117}$$

Unfortunately, it is not a simple matter to obtain closed form expressions for the N-phase loss for the class of filters considered in Table 2-1. However, the two extremes of the class of Butterworth filters—the single pole RC filter and the ideal band-pass filter—can be worked out in series form with the results, respectively

$$\mathcal{S}_L = \left\{1 + \sum_{r=0}^{N-2}\frac{({}_NC_r)^2}{N^2}\left[\frac{(N-r-1)!\,2^{N-r-1}}{(\rho\gamma)^{N-r-1}}\right]\right\}^{-1}$$

and

$$\mathcal{S}_L = \left\{1 + \sum_{r=0}^{N-2}\frac{({}_NC_r)^2}{N^2}\left[\frac{(N-r)\sum_{i=0,2,4\ldots}^{N-r-1}(-1)^{i/2}{}_{N-r}C_{i/2}(N-r-i)^{N-r-1}}{(\rho\gamma)^{N-r-1}}\right]\right\}^{-1} \tag{2-118}$$

Plotted in Fig. 2-32 are the reciprocals of the N-phase losses of Eq. (2-118) as a function of $\rho\gamma$ for values of $N = 2$, 4, and 8.

The coherent reference signal $r_{NP}(t)$ established by the N-phase loop is frequency divided by N to provide a noisy reference. By passing this reference through an appropriate phase shift network, the N references necessary for demodulating the N-phase data signal off the carrier can be obtained (see Fig. 2-31). The performance of the demodulator using the Nth power loop as a source of demodulating reference signal is treated in Chap. 6.

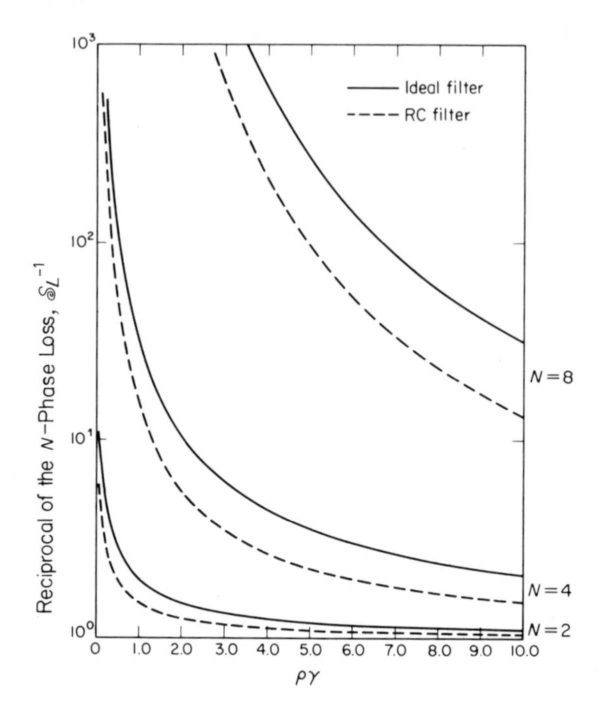

Fig. 2-32 Reciprocal of the N-phase loss as a function of N and input signal-to-noise ratio for RC and ideal band-pass filters

Meanwhile, the variance of the phase jitter associated with the noisy reference signal is, for small ϕ:

$$\sigma_\phi^2 = \frac{1}{N^2}\sigma_{N\phi}^2 = \frac{\mathcal{S}_L^{-1}}{\rho} \tag{2-119}$$

which is the identical result as Eq. (2-88) using, however, (2-117) as the definition of $\mathcal{S}_L$.

The N-Phase Costas (I-Q) Loop

It is straightforward to show that the N-phase Costas loop illustrated in Fig. 2-33 has the identical stochastic integro-differential equation as the Nth power loop of the previous subsection; hence, from a theoretical point of view the two are equivalent. This result should not be surprising in view of the equivalence already demonstrated for the biphase circuits. From a practical point of view, the N-phase Costas loop suffers for large N in that the amount of equipment needed for implementation becomes prohibitive.

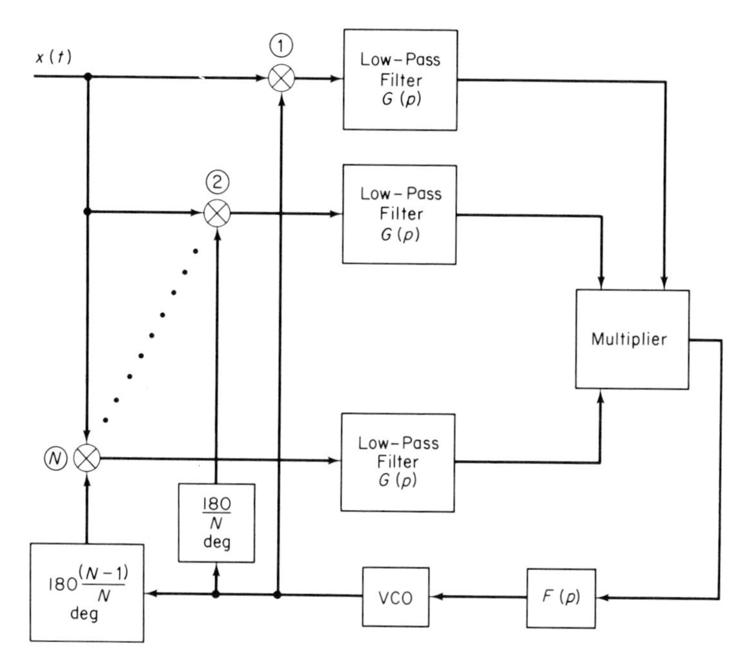

Fig. 2-33 The N-phase Costas loop

Other practical considerations regarding the use of one circuit over the other follow the comments made for the biphase circuits.

One final point is that the p.d.f. for either N-phase tracking loop exhibits N ambiguities in the interval $(0, 2\pi)$; hence, practical use of these circuits in combination with a data demodulator requires some means of resolving these ambiguities. Several techniques for doing such were discussed in Sec. 2-4.

N-Phase Decision-Feedback Loops

The suppressed carrier decision-feedback loop of Fig. 2-26 can easily be modified to accommodate a quadriphase input signal. Before illustrating the specific loop configuration, however, we first examine the polyphase signal of Eq. (2-104) for the case $N = 4$ and rewrite it in the form:

$$\begin{aligned} y(t) = {} & \sqrt{S}\, m_1(t) \sin [\Phi(t) - \pi/4] \\ & + \sqrt{S}\, m_2(t) \cos [\Phi(t) - \pi/4] + n(t) \end{aligned} \tag{2-120}$$

where $m_1(t)$ and $m_2(t)$ are two ± 1 digital signals whose transitions occur at T-second intervals, and whose amplitudes are related to the index k of

(2-104) by

$$\begin{aligned} m_1(t) &= 1 & m_2(t) &= 1 & &\text{when } k = 1 \\ m_1(t) &= -1 & m_2(t) &= 1 & &\text{when } k = 2 \\ m_1(t) &= -1 & m_2(t) &= -1 & &\text{when } k = 3 \\ m_1(t) &= 1 & m_2(t) &= -1 & &\text{when } k = 4 \end{aligned} \tag{2-121}$$

We see then from (2-120) that the quadriphase signal input can be modeled as the sum of two orthogonal equal power suppressed-carrier biphase signals. One might anticipate therefore that the necessary extension of Fig. 2-26 is the addition of a quadrature loop to the existing configuration. Indeed this is the case as illustrated in the quadriphase decision-feedback loop of Fig. 2-34.

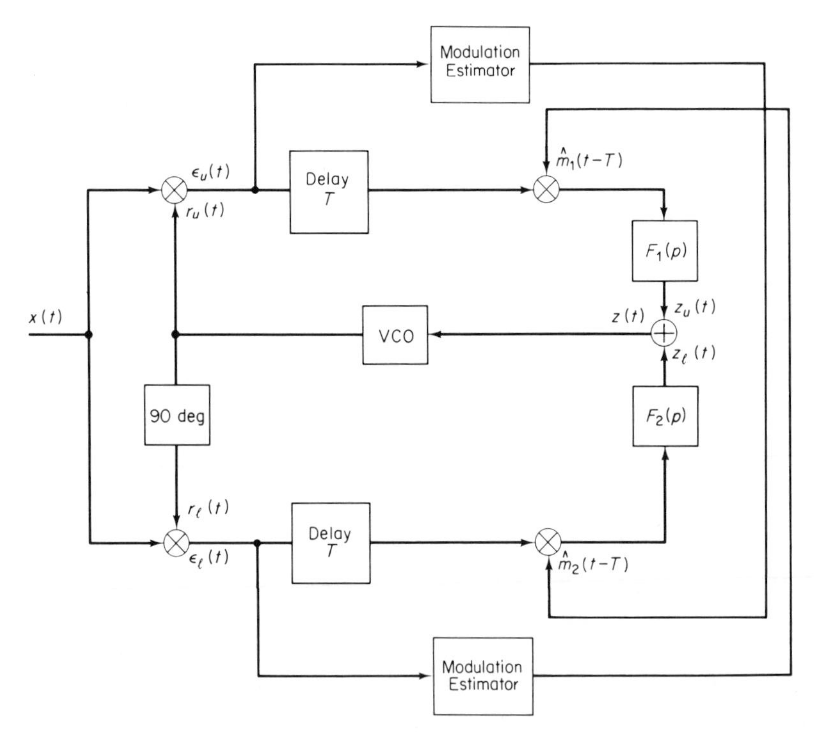

Fig. 2-34 A quadriphase decision-feedback loop

Letting

$$\begin{aligned} r_u(t) &= \sqrt{2}\,K_1 \cos\,[\hat{\Phi}(t) - \pi/4] \\ r_\ell(t) &= -\sqrt{2}\,K_1 \sin\,[\hat{\Phi}(t) - \pi/4] \end{aligned} \tag{2-122}$$

respectively, denote the upper and lower loop reference signals, then the upper and lower dynamic reference signals $\epsilon_u(t)$ and $\epsilon_\ell(t)$ are given by

$$\epsilon_u(t) = \frac{K_1 K_{1m}}{\sqrt{2}}\{\sqrt{S}m_1(t)\sin\varphi(t) + \sqrt{S}m_2(t)\cos\varphi(t) + \sqrt{2}N_u[t,\varphi(t)]\}$$

$$\epsilon_\ell(t) = \frac{K_1 K_{2m}}{\sqrt{2}}\{-\sqrt{S}m_1(t)\cos\varphi(t) + \sqrt{S}m_2(t)\sin\varphi(t) + \sqrt{2}N_\ell[t,\varphi(t)]\} \tag{2-123}$$

where $N_u[t, \varphi(t)]$ and $N_\ell[t, \varphi(t)]$ are uncorrelated noise processes, which are modeled exactly the same way as $N[t, \varphi(t)]$ of Eq. (2-7). Following the development given in the previous section for the biphase decision-feedback loop, the signals $z_u(t)$ and $z_\ell(t)$ are given by

$$z_u(t) = \frac{K_1 K_{1m}}{\sqrt{2}}F_1(p)\exp(-pT)\{\sqrt{S}m_1(t)\hat{m}_1(t)\sin\varphi(t) + \sqrt{S}m_2(t)\hat{m}_1(t)\cos\varphi(t) + \sqrt{2}\hat{m}_1(t)N_u[t,\varphi(t)]\}$$

$$z_\ell(t) = \frac{K_1 K_{2m}}{\sqrt{2}}F_2(p)\exp(-pT)\{-\sqrt{S}m_1(t)\hat{m}_2(t)\cos\varphi(t) + \sqrt{S}m_2(t)\hat{m}_2(t)\sin\varphi(t) + \sqrt{2}\hat{m}_2(t)N_\ell[t,\varphi(t)]\} \tag{2-124}$$

Following the same averaging argument as given in arriving at Eq. (2-94), and the fact that $E[m_i(t)\hat{m}_j(t)] = 0$ for $i \neq j$, Eq. (2-124) reduces to

$$z_u(t) = \frac{K_1 K_{2m}}{\sqrt{2}}F_1(p)\exp(-pT)\Big\{\sqrt{S}\Big[1 - 2P_{E1}[\varphi(t)]\Big]\sin\varphi(t) + \sqrt{2}\hat{m}_1(t)N_u[t,\varphi(t)]\Big\}$$

$$z_\ell(t) = \frac{K_1 K_{2m}}{\sqrt{2}}F_2(p)\exp(-pT)\Big\{\sqrt{S}\Big[1 - 2P_{E2}[\varphi(t)]\Big]\sin\varphi(t) + \sqrt{2}\hat{m}_2(t)N_\ell[t,\varphi(t)]\Big\} \tag{2-125}$$

Letting $z(t) = z_u(t) + z_\ell(t)$ and using Eq. (2-95), the steady-state stochastic integro-differential equation of the quadriphase decision-feedback loop becomes

$$\dot{\varphi}(t) = \dot{\theta}(t) - \frac{K_u}{\sqrt{2}}F_1(p)\Big\{\sqrt{S}\Big[1 - 2P_{E1}[\varphi(t)]\Big]\sin\varphi(t) + \sqrt{2}m_1(t)N_u[t,\varphi(t)]\Big\} - \frac{K_\ell}{\sqrt{2}}F_2(p)\Big\{\sqrt{S}\Big[1 - 2P_{E2}[\varphi(t)]\Big]\sin\varphi(t) + \sqrt{2}m_2(t)N_\ell[t,\varphi(t)]\Big\} \tag{2-126}$$

where $K_u \triangleq K_1 K_{1m} K_V$ and $K_\ell \triangleq K_1 K_{2m} K_V$. The above equation represents the generalization of Eq. (2-96) to a quadriphase system.

The quadriphase decision-feedback loop of Fig. 2-34 can, in principle, be extended to N phases. To see this, we first note that the received polyphase signal in (2-104) can also be written in the binary representation

$$y(t) = \left(\sin \frac{\pi}{N}\right)\sqrt{2S} \sum_{k=0}^{(N/2)-1} m_k(t) \sin\left(\Phi(t) + (k-1)\frac{\pi}{N}\right) + n(t) \quad (2\text{-}127)$$

where the set $\{m_k(t); k = 0, 1, 2, \ldots, (N/2) - 1\}$ are again ± 1 digital waveforms whose transitions occur at T-second intervals. Equation (2-127) is a generalization of the quadriphase signal representation given in (2-120). The N-phase decision-feedback loop then would have $N/2$ arms corresponding to the $N/2$ decisions $\hat{m}_k(t); k = 0, 1, 2, \ldots, (N/2) - 1$ on the binary signals $m_k(t); k = 0, 1, 2, \ldots, (N/2) - 1$. The output of the VCO would be phase-shifted by $(k-1)\pi/N$; $k = 0, 1, 2, \ldots, (N/2) - 1$ to provide $N/2$ coherent reference signals. As a result, the $N/2$ equivalent additive low-pass noise processes at the input of the summing junction are no longer mutually uncorrelated and, hence, for $N > 4$, the loop becomes less efficient.

Another N-phase tracking loop which makes use of the principle of decision-feedback and is equally efficient for all N is illustrated in Fig. 2-35.

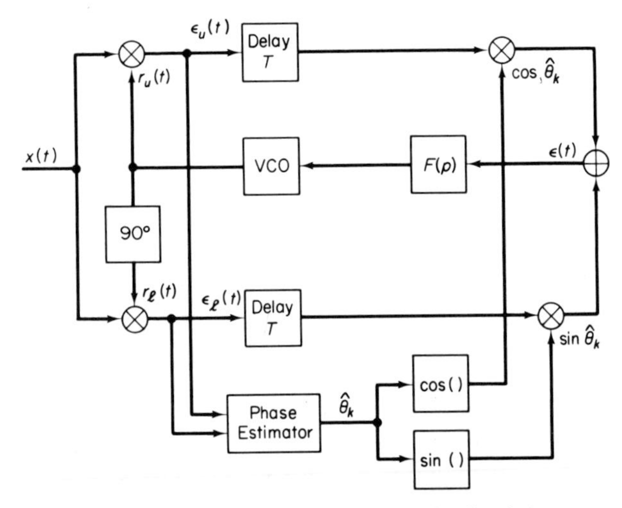

Fig. 2-35 An N-phase decision-feedback loop

Here, in each T-second interval, the decision $\hat{\theta}_k$ on the transmitted phase symbol $\theta_k = 2(k-1)\pi/N$ is used to produce the decision-feedback signals. Under the same assumptions given for the previous decision-feedback loop, the dynamic error signal becomes

$$\begin{aligned}\epsilon(t) = K_1K_m\{&\sqrt{S}\cos(\theta_k - \hat{\theta}_k)\sin\varphi(t) + \sqrt{S}\sin(\theta_k - \hat{\theta}_k)\cos\varphi(t) \\ &+ \cos\hat{\theta}_k N_u[t, \varphi(t)] + \sin\hat{\theta}_k N_\ell[t, \varphi(t)]\}\end{aligned} \quad (2\text{-}128)$$

Based upon the same averaging argument used previously, the output of the loop filter in the tracking mode can be expressed in terms of the circular moments of $\theta_k - \hat{\theta}_k$,

$$z(t) = K_1 K_m F(p)\{\sqrt{S}\,\overline{\cos(\theta_k - \hat{\theta}_k)} \sin \varphi(t) + \sqrt{S}\,\overline{\sin(\theta_k - \hat{\theta}_k)} \cos \varphi(t) + \cos \hat{\theta}_k N_u[t, \varphi(t)] + \sin \hat{\theta}_k N_\ell[t, \varphi(t)]\} \tag{2-129}$$

This discrete random variable $\theta_k - \hat{\theta}_k$ ranges over the set of allowable values $\{2j\pi/N; j = 0, \pm 1, \pm 2, \ldots, \pm(N/2) - 1, N/2\}$ with probabilities (see Chap. 6)

$$\begin{aligned} P_j(\varphi) &\triangleq \text{Prob}\{\theta_k - \hat{\theta}_k = 2j\pi/N\} \\ &= \frac{1}{\pi}\int_0^\infty \Big[\exp\{-(u - \sqrt{R_d}\cos\varphi)^2\} \\ &\quad \times \int_{u\tan[(2j-1)\pi/N]}^{u\tan[(2j+1)\pi/N]} \exp\{-(v - \sqrt{R_d}\sin\varphi)^2\}dv\Big]\,du \end{aligned} \tag{2-130}$$

where we have again assumed that the loop phase error $\varphi(t)$ is essentially constant over several signalling intervals. Also, from the law of total probability,

$$\sum_{j=-(N/2)+1}^{N/2} P_j(\varphi) = 1 \tag{2-131}$$

Thus, from (2-130) and (2-131), the circular moments of $\theta_k - \hat{\theta}_k$ can be expressed as

$$\begin{aligned} \overline{\sin(\theta_k - \hat{\theta}_k)} &\triangleq \sum_{j=-(N/2)+1}^{N/2} P_j(\varphi)\sin(2j\pi/N) = \sum_{j=-(N/2)+1}^{N/2}{}' P_j(\varphi)\sin(2j\pi/N) \\ \overline{\cos(\theta_k - \hat{\theta}_k)} &\triangleq \sum_{j=-(N/2)+1}^{N/2} P_j(\varphi)\cos(2j\pi/N) = P_0(\varphi) + \sum_{j=-(N/2)+1}^{N/2}{}' P_j(\varphi)\cos(2j\pi/N) \end{aligned} \tag{2-132}$$

where the prime on the summation denotes omission of the $j = 0$ term. Looking ahead to Chap. 6, we note that $P_0(\varphi)$ is the conditional probability that the decision on θ_k is correct, given the phase error φ. Thus, letting

$$P_E(\varphi) \triangleq 1 - P_0(\varphi) = \sum_{j=-(N/2)+1}^{N/2}{}' P_j(\varphi) \tag{2-133}$$

Eq. (2-132) may be expressed in the equivalent form

$$\begin{aligned} \overline{\sin(\theta_k - \hat{\theta}_k)} &= \sum_{j=-(N/2)+1}^{N/2}{}' P_j(\varphi)\sin(2j\pi/N) \\ \overline{\cos(\theta_k - \hat{\theta}_k)} &= 1 - 2P_E(\varphi) + \sum_{j=-(N/2)+1}^{N/2}{}' P_j(\varphi)[1 + \cos(2j\pi/N)] \end{aligned} \tag{2-134}$$

Substituting (2-134) into (2-129) and recalling (2-95), the stochastic integro-differential equation of operation for the N-phase decision-feedback loop of Fig. 2-35 becomes

$$\varphi(t) = \dot{\theta}(t) - KF(p)\{\sqrt{S}\{1 - 2P_E(\varphi)] \sin \varphi$$
$$+ \sqrt{S} \sum_{j=-(N/2)+1}^{N/2}{}' P_j(\varphi) \sin \varphi[1 + \cos(2j\pi/N)]$$
$$+ \sqrt{S} \sum_{j=-(N/2)+1}^{N/2}{}' P_j(\varphi) \cos \varphi \sin(2j\pi/N)$$
$$+ \cos \hat{\theta}_k N_u(t, \varphi) + \sin \hat{\theta}_k N_\ell(t, \varphi)\} \tag{2-135}$$

where $K \triangleq K_1 K_m K_V$. Recognizing from (2-130) that $P_j(\varphi) = P_{-j}(-\varphi)$, the second and third terms of (2-135) are odd functions of φ and as such contribute to the overall tracking error characteristic. When $N = 2$, both of these terms are zero and (2-135) reduces to the stochastic integro-differential equation of operation for the biphase decision-feedback loop of Fig. 2-26 as given by (2-96).

APPENDIX A
EVALUATION OF THE AUTOCORRELATION FUNCTION OF $v_2(t, 2\phi)$

The autocorrelation function of $v_2(t, 2\phi)$ is defined by

$$R_{v_2}(\tau) = E[v_2(t, 2\phi)v_2(t + \tau, 2\phi)] \tag{A-1}$$

with $v_2(t, 2\phi)$ given by

$$v_2(t, 2\phi) = [2\sqrt{S}m(t)N_c(t) - 2N_c(t)N_s(t)] \cos 2\phi$$
$$+ [-N_c^2(t) + N_s^2(t) - 2\sqrt{S}m(t)N_s(t)] \sin 2\phi \tag{A-2}$$

and $N_c(t)$ and $N_s(t)$ independent zero mean Gaussian random processes.

Since $T_b \gg \tau_n$, with τ_n the correlation time of the noise, it is valid to assume that $m(t) = m(t + \tau)$ for all τ of interest. Hence, substituting Eq. (A-2) in (A-1) gives, on simplification,

$$R_{v_2}(\tau) = [4SR_{N_c}(\tau) + 4R_{N_c}(\tau)R_{N_s}(\tau)] \cos^2 2\phi$$
$$+ [R_{N_c}^2(0) + 2R_{N_s}^2(\tau) + R_{N_s}^2(0)$$
$$+ 2R_{N_s}^2(\tau) - 2R_{N_c}(0)R_{N_s}(0)$$
$$+ 4SR_{N_s}(\tau)] \sin^2 2\phi \tag{A-3}$$

If $R_{N_c}(\tau) = R_{N_s}(\tau)$, then

$$R_{v_2}(\tau) = [4SR_{N_c}(\tau) + 4R_{N_c}^2(\tau)] \cos^2 2\phi$$
$$+ [4SR_{N_c}(\tau) + 4R_{N_c}^2(\tau)] \sin^2 2\phi$$
$$= 4[SR_{N_c}(\tau) + R_{N_c}^2(\tau)] \tag{A-4}$$

PROBLEMS

2-1. For the loop filter implemented in Fig. 2-3, verify that $\tau_1 = 45.2$ sec and $\tau_2 = 0.125$ sec.

2-2. An imperfect, second-order PLL is designed to track the carrier in a coherent communication system. The operating parameters are chosen such that $r = 2$, $F_1 = 0.002$, $\sqrt{P_c'K} = 6000$, and $\Omega_0 - 0$.

a) Find the loop filter time constants τ_1 and τ_2.
b) Determine the loop bandwidth W_L.
c) If $N_0 = -161$ dBm/Hz, find the received carrier power which yields $\sigma_\varphi^2 = 1$ on the basis of the linear theory.
d) What mean-squared phase error (in degrees²) does this imply if the nonlinear PLL theory is used?

2-3. A second-order PLL operates with $\rho = 3$ and $W_L = 18$ Hz.

a) Find the mean time to first slip if $\Lambda = 0, 0.25, 0.50$.
b) Determine the mean-squared value of the phase error (in degrees²) on the basis of the linear and nonlinear PLL theories (assume zero detuning).
c) Find the average number of cycle slips per second if $\Lambda = 0$.
d) Determine the mean and variance of the static phase error, $\sin \phi$.
e) Determine the mean and variance of the signal-to-noise ratio degradation due to noisy references; that is, $\cos \phi$.

2-4. A band-pass limiter precedes a second-order, narrowband PLL which operates with zero detuning. Assuming that the loop design point is chosen such that $r_0 = 2$ and $W_{L0} = 20$ Hz, the IF filter has a two-sided bandwidth $W_i = 4.8$ kHz, and $N_0 = -161$ dBm/Hz, determine

a) the signal amplitude suppression factor, $\tilde{\alpha}_1$, and the limiter performance factor, Γ_p, for values of received carrier power $P_c = -148$, -145, -141, and -131 dBm.
b) for the same values of P_c, the mean-squared value of the phase error (in degrees²). Use both the linear and nonlinear PLL theories and compare your answers.

2-5. Find the variances σ_A^2 and σ_B^2 of the zero mean noise processes $N_A(t)$ and $N_B(t)$, respectively, as defined in Eq. (2-43). Use the expression for σ_A^2 to verify (2-47).

2-6. A squaring loop is designed using the loop filter of Problem 2-2. The spectral density of the additive Gaussian noise is $N_0 = -161$ dBm/Hz and the received data power is $S = -131$ dBm. Furthermore, the operating parameters are chosen such that $W_L = 20$ Hz and $W_i = 2$ kHz.

a) Compute the squaring loss, $\mathcal{S}_L$ (in dB) for an ideal rectangular filter and a single pole RC filter.
b) Evaluate the mean-squared phase error (in degrees²) using the linear and nonlinear theories (assume zero detuning).
c) Repeat parts (a) and (b) for received data powers of $S = -141$ and -151 dBm.

2-7. For the parameters given in Problem 2-6, evaluate the mean-squared phase error (in degrees^2) if $\delta = 10$ and 100 in the decision-directed loop of Fig. 2-26. Compare your results with those obtained in Problem 2-6.

2-8. Starting with Eq. (2-117), derive the expressions given in (2-118) for the squaring loss of an Nth power loop with single pole RC and ideal band-pass filters.

2-9. Develop the stochastic integro-differential equation of operation for the N-phase Costas loop and show that it is identical to that given in Eq. (2-112) for the Nth power loop.

2-10. For the N-phase Costas or Nth power loop, the equivalent noise $v_N[t, N\varphi(t)]$ in the stochastic integro-differential equation of operation is given by Eq. (2-110). By an extension of the procedure given in Appendix A, derive the autocorrelation function of $v_N(t, N\phi)$ for the special case $N = 4$ (that is, quadriphase). Check that your result agrees with the general result for arbitrary N given in (2-115).

REFERENCES

1. Lindsey, W. C., *Synchronization Systems in Communication and Control*, Prentice-Hall, Inc., Englewood Cliffs, N.J., 1972.
2. Tausworthe, R. C., *Theory and Practical Design of Phase-Locked Receivers*, Jet Propulsion Laboratory, Pasadena, Calif., Technical Report 32–819, 1966.
3. Tikhonov, V. I., "The Effect of Noise on Phase-Locked Oscillator Operation," *Automation and Remote Control*, Vol. 20, (1959) 1160–1168. Translated from *Automatika i Telemekhaniki*, Akademya Nauk SSSR, Vol. 20, Sept., 1959.
4. Tikhonov, V. I., "Phase-Lock Automatic Frequency Control Operation in the Presence of Noise," *Automation and Remote Control*, Vol. 21, (1960) 209–214. Translated from *Automatika i Telemekhaniki*, Akademya Nauk SSSR, Vol. 21, Mar., 1960.
5. Viterbi, A. J., "Phase-Locked Loop Dynamics in the Presence of Noise by Fokker-Planck Techniques," *Proceedings of the IEEE*, Vol. 51, No. 12 (Dec., 1963) 1737–1753.
6. Lindsey, W. C., "Nonlinear Analysis and Synthesis of Generalized Tracking Systems," *Proceedings of the IEEE*, Vol. 57, No. 10 (Oct., 1969) 1705–1722.
7. Charles, F. J., and Lindsey, W. C., "Some Analytical and Experimental Phase-Locked Loop Results for Low Signal-to-Noise Ratios," *Proceedings of the IEEE*, Vol. 55, No. 9 (Sept., 1966) 1152–1166.
8. Tausworthe, R. C., and Sanger, D., "Experimental Study of the First-Slip Statistics of the Second-Order Phase-Locked Loop," Jet Propulsion Laboratory, Pasadena, Calif., SPS 37–43, Vol. 3 (1967) 76–80.
9. Sanneman, R. W., and Rowbotham, J. R., "Random Characteristics of the Type II Phase-Locked Loop, *IEEE Transactions on Aerospace and Electronic Systems*, Vol. AES-3, No. 4 (July, 1967) 604–612.

10. Darling, D. A. and Siegert, A. J. F., "The First Passage Time Problem for a Continuous Markov Process," *Annals of Mathematical Statistics* Vol. 24 (1953) 624–639.

11. Tausworthe, R. C., "Cycle Slipping in Phase-Locked Loops," *IEEE Transactions on Communication Technology*, Vol. COM-15, No. 3 (June, 1967) 417–421.

12. Tausworthe, R. C., "Simplified Formula for Mean Cycle-Slip Time of Phase-Locked Loops with Steady-State Phase Error," *IEEE Transactions on Communication Technology* Vol. COM-20, No. 2 (June, 1972) 331–337.

13. Holmes, J. K., "First Slip Time Versus Static Phase Error Offset for the First- and Passive Second-Order Phase-Locked Loop," *IEEE Transactions on Communication Technology*, Vol. COM-19, No. 2 (April, 1971) 234–235.

14. Springett, J. C., "Signal-to-Noise and Signal-to-Noise Spectral-Density Ratios at the Output of a Filter-Limiter Combination," Jet Propulsion Laboratory, Pasadena, Calif., SPS 37–36, Vol. 4 (1965) 241–248.

15. Davenport, W. B., "Signal-to-Noise Ratios in Band-Pass Limiters," *Journal of Applied Physics*, Vol. 24, No. 6 (June, 1953) 720–727.

16. Forney, G. D., "Coding in Coherent Deep-Space Telemetry," McDonnell Douglas Astronautics Corp., Santa Monica, Calif. (March 19, 1967).

17. Tausworthe, R. C., "Limiters in Phase-Locked Loops: A Correction to Previous Theory," Jet Propulsion Laboratory, Pasadena, Calif., SPS 37–54, Vol. 3 (1968) 201–204.

18. Springett, J. C., and Simon, M. K., "An Analysis of the Phase Coherent/Incoherent Output of the Bandpass Limiter," *IEEE Transactions on Communication Technology*, Vol. COM-19, No. 1 (Feb., 1971) 42–49.

19. Van Trees, H., "Optimum Power Division in Coherent Communication Systems," M.I.T. Lincoln Lab., Lexington, Mass., Technical Report 301, February, 1963.

20. Didday, R. L., and Lindsey, W. C., "Subcarrier Tracking Methods and Communication System Design," *IEEE Transactions on Communication Technology*, Vol. COM-16, No. 4 (Aug., 1968) 541–550.

21. Lindsey, W. C., and Simon, M. K., "Nonlinear Analysis of Suppressed Carrier Tracking Loops in the Presence of Frequency Detuning," *Proceedings of the IEEE*, Vol. 58, No. 9 (Sept., 1970) 1302–1321.

22. Costas, J. P., "Synchronous Communications," *Proceedings of the IRE*, Vol. 44 (Dec., 1956) 1713–1718.

23. Natali, F. D., and Walbesser, W. J., "Phase-Locked Loop Detection of Binary PSK Signals Utilizing Decision Feedback," *IEEE Transactions on Aerospace and Electronic Systems*, Vol. AES-5, No. 1 (Jan., 1969) 83–90.

24. Stratonovich, R. L., *Topics in the Theory of Random Noise*, Vol. 1, Gordon and Breach, New York, N.Y., 1963.

25. Layland, J. W., "An Optimum Squaring Loop Filter," *IEEE Transactions on Communication Technology*, Vol. COM-18, No. 5 (Oct., 1970) 695–697.

26. Lindsey, W. C., and Simon, M. K., "A Comparison of the Performance of Costas or Squaring Loops with Data-Aided Loops," Jet Propulsion Laboratory, Pasadena, Calif., SPS 37–60, Vol. III (1970) 56–58.

27. LINDSEY, W. C., and SIMON, M. K., "Carrier Synchronization and Detection of Polyphase Signals," *IEEE Transactions on Communication Technology*, Vol. COM-20, No. 3 (June, 1972) 441–454.

Chapter 3

PHASE AND DOPPLER MEASUREMENTS IN TWO-WAY PHASE-COHERENT TRACKING SYSTEMS

3-1. INTRODUCTION

In a phase-coherent communication system for space applications, the telemetry and command signals usually phase modulate the RF carrier with modulation indices sufficiently low to permit the spectrum of the transmitted waveform to retain a portion of the total transmitter power at the carrier frequency. The two reasons for this are most simply described in terms of a two-path (reference-vehicle-reference) transmission. First, if the telemetry and command data are to be detected coherently, the measurement or estimation of the phase of the observed carrier in noise is essential. Second, the vehicle must be tracked with extreme accuracy. As discussed in Chap. 1, tracking data consists of the vehicle's velocity, range, and angular position. Velocity of the vehicle is usually based on an estimate of the received signal's Doppler shift in frequency. This Doppler shift, which is proportional to the velocity of the vehicle along a line connecting the vehicle to the reference system, is sometimes referred to as the Doppler velocity, range rate, or slant velocity; in this discussion, the parameter will be referred to simply as Doppler. Doppler may be used to accurately specify the flight path or trajectory of the vehicle, and it is of fundamental importance in all uses of telemetry data.

There are several techniques for measuring Doppler in a phase-coherent communication system. The one considered in this chapter is the measurement of a two-way Doppler shift by transmitting a known signal to the vehicle and *coherently transponding* the observed signal back to the reference system; that is, the vehicle coherently demodulates the uplink carrier and retransmits on the downlink a frequency-translated version of it where the amount of frequency translation introduced is known *a priori* at the reference system (see Fig. 3-1). In the ground reference system, a *velocity extraction unit* compares a continuing sample of the uplink carrier frequency with the downlink carrier frequency and for a preset period of time counts the number of cycles in the difference between these two frequencies. After subtracting out the known frequency translation introduced in the vehicle transponder, the remaining number of cycles is a measure of Doppler shift—vehicle velocity.

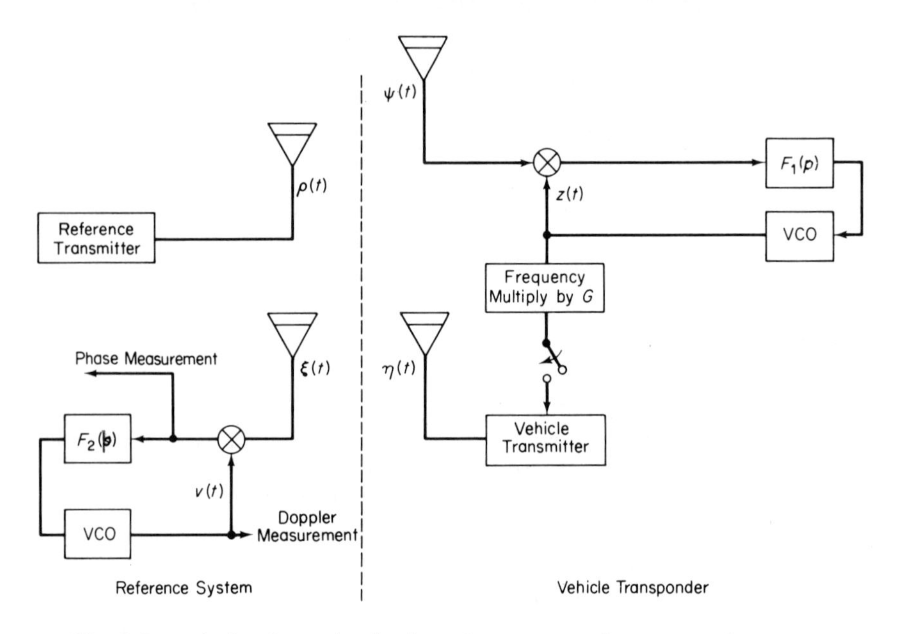

Fig. 3-1 A simple mechanization of a two-way phase measuring system

A one-way Doppler measurement is made by transmitting a signal from a free-running oscillator on board the vehicle and estimating the received Doppler shift. One-way Doppler is, of course, inferior to two-way Doppler because the frequency of the reference oscillator on board the vehicle is usually not known *a priori* at the reference system. The practical reason for this is that the vehicle frequency drifts because of variations in the physical environment. Thus, to obtain accurate Doppler measurements in a one-way Doppler system, separate telemetered information about the space vehicle's oscillator is required. Consequently, from a practical point of view, two-way Doppler is more attractive than one-way Doppler.

This chapter will develop a theory on which the steady-state performance of a two-way phase and Doppler measuring system may be designed and evaluated. The theory presented includes the effects of BPLs in both the vehicle and the reference systems, and the effects of nonlinear behavior of the PLLs on system design. The linear PLL theory with and without BPLs preceding the PLLs is also considered. The last part of the chapter generalizes the material on phase and Doppler measurements for use in the design of one- and two-way, n-step phase-coherent tracking systems.

There are several other factors affecting the design of a two-way tracking system, such as the time required to achieve signal acquisition, loop acquisition range, and stability of the various carrier-tracking loops. These subjects are discussed in detail in Ref. 1.

3-2. TWO-WAY PHASE MEASUREMENTS

Basic System Model

Figure 3-1 describes the simplest mechanization of a two-way phase measuring system—that is, a two-path (reference-vehicle-reference) transmission link. The reference system emits, on the uplink, the waveform

$$\rho(t) = \sqrt{2P_{c1}} \sin (\omega_0 t + \theta_0) \tag{3-1}$$

where θ_0 is assumed to be constant. After transmission, the channel introduces an arbitrary (but unknown) phase shift θ_1 in the transmitted waveform and further disturbs $\rho(t)$ with additive, narrowband white Gaussian noise $n_1(t)$ of single-sided spectral density N_{01} w/Hz. Thus, we observe in the vehicle the Doppler-shifted, phase-shifted, noise-corrupted waveform

$$\psi(t) = \sqrt{2P_{c1}} \sin (\omega_1 t + \theta_0 + \theta_1) + n_1(t) \tag{3-2}$$

where θ_1 is also assumed to be constant. The narrowband additive noise process $n_1(t)$ may be represented by

$$n_1(t) = \sqrt{2}[N_{1c}(t) \cos (\omega_1 t + \theta_0 + \theta_1) - N_{1s}(t) \sin (\omega_1 t + \theta_0 + \theta_1)] \tag{3-3}$$

where $N_{1c}(t)$ and $N_{1s}(t)$ are statistically independent, white Gaussian noise processes of single-sided spectral density N_{01} w/Hz and single-sided bandwidth, $W_{i1}/2$. The vehicle coherently tracks the carrier by means of a narrowband PLL, producing the estimate

$$z(t) = \sqrt{2} \cos [\omega_1 t + \hat{\theta}_1(t)] \tag{3-4}$$

where $\hat{\theta}_1(t)$ is the vehicle PLL estimate of $\theta_0 + \theta_1$.

The carrier for the downlink in a two-way phase measuring system is derived from the vehicle's carrier-tracking loop; that is, the switch in the vehicle transponder of Fig. 3-1 is closed. Thus, on the downlink, we transmit

$$\eta(t) = \sqrt{2P_{c2}} \sin [\omega_{10} t + \hat{\theta}_{10}(t)] \tag{3-5}$$

and observe, at the reference system, the Doppler-shifted, phase-shifted, noise-corrupted waveform

$$\xi(t) = \sqrt{2P_{c2}} \sin [\omega_2 t + \hat{\theta}_{10}(t) + \theta_2] + n_2(t) \tag{3-6}$$

where $n_2(t)$ is narrowband white Gaussian noise of single-sided spectral density N_{02} w/Hz, $\omega_{10} = G\omega_1$, $\hat{\theta}_{10}(t) = G\hat{\theta}_1(t)$, and θ_2 is assumed to be constant. Here G denotes the *static phase gain of the vehicle transponder*. A convenient representation for $n_2(t)$ is given by

$$n_2(t) = \sqrt{2}\{N_{2c}(t) \cos [\omega_2 t + \hat{\theta}_{10}(t) + \theta_2] - N_{2s}(t) \sin [\omega_2 t + \hat{\theta}_{10}(t) + \theta_2]\} \tag{3-7}$$

where $N_{2c}(t)$ and $N_{2s}(t)$ are statistically independent, white Gaussian noise processes of single-sided spectral density N_{02} w/Hz and single-sided bandwidth $W_{i2}/2$. The reference receiver tracks the carrier component in $\xi(t)$, which provides the receiver with the estimate

$$v(t) = \sqrt{2} \cos [\omega_2 t + \hat{\theta}_2(t)] \tag{3-8}$$

The quantity $\hat{\theta}_2(t)$ is the reference receiver PLL estimate of $\hat{\theta}_{10}(t) + \theta_2$.

In the design of such a system there are at least two significant errors: (1) the reference receiver's two-way tracking phase error $\varphi_{r2}(t) \triangleq \hat{\theta}_{10}(t) + \theta_2 - \hat{\theta}_2(t)$; and (2) the two-way Doppler phase error $\varphi_{d2}(t) \triangleq \hat{\theta}_2(t) - \theta_0$. In the next section, we shall consider the phase error $\varphi_{r2}(t)$ and its relationship to system parameters employing the linear PLL theory and in a later section remodel the problem so that the nonlinear effects of the PLLs are considered. Finally, we shall study the phase error when BPLs precede the vehicle and reference system PLLs.

Two-Way Tracking Phase Error

Linear PLL Theory. The important features of the system of Fig. 3-1 are illustrated in linearized form in Fig. 3-2. The filter functions $H_n(s)$, $n = 1, 2$,

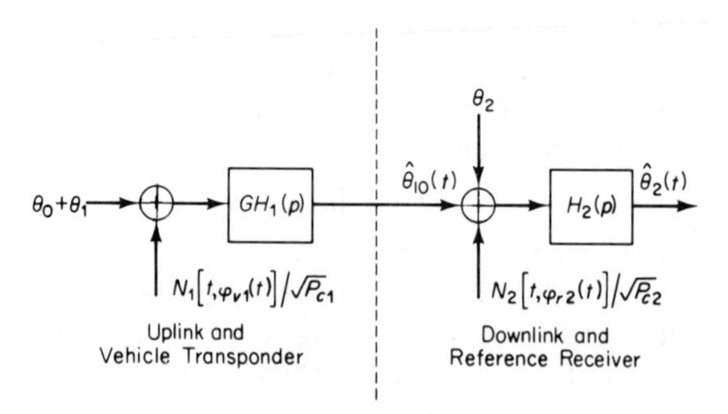

Fig. 3-2 An equivalent linear representation of a two-way phase measuring system

are the closed-loop transfer functions of the system's carrier-tracking loops; that is,

$$H_n(s) = \frac{1 + [(r_n + 1)/2W_{Ln}]s}{1 + [(r_n + 1)/2W_{Ln}]s + (1/r_n)[(r_n + 1)/2W_{Ln}]^2 s^2} \tag{3-9}$$

where the loop filters $F_n(s)$, $n = 1, 2$, are of the proportional-plus-integral control type:

$$F_n(s) = \frac{1 + \tau_{2n} s}{1 + \tau_{1n} s} \tag{3-10}$$

and

$$r_n = \frac{\sqrt{P_{cn}} K_n \tau_{2n}^2}{\tau_{1n}} \tag{3-11}$$

$$W_{Ln} = \frac{1}{2\pi j} \int_{-j\infty}^{j\infty} |H_n(s)|^2 \, ds \cong \frac{1 + r_n}{2\tau_{2n}} \tag{3-12}$$

These results follow directly from the material presented in Chap. 2.

The reference receiver's tracking phase error $\varphi_{r2}(t)$ is defined as the difference between the receiver output phase and the receiver input phase:

$$\varphi_{r2}(t) \triangleq \hat{\theta}_{10}(t) + \theta_2 - \hat{\theta}_2(t) \tag{3-13}$$

The noise that enters the vehicle transponder must be considered as part of the reference receiver's input signal $\hat{\theta}_{10}(t) + \theta_2$. Thus,

$$\varphi_{r2}(t) = \frac{GH_1(p)[1 - H_2(p)]N_1[t, \varphi_{v1}(t)]}{\sqrt{P_{c1}}} - \frac{H_2(p)N_2[t, \varphi_{r2}(t)]}{\sqrt{P_{c2}}} + GH_1(p)[1 - H_2(p)]\theta_1 + [1 - H_2(p)]\theta_2 \tag{3-14}$$

where $\varphi_{v1}(t) \triangleq \theta_0 + \theta_1 - \hat{\theta}_1(t)$ is the vehicle receiver's tracking phase error and $N_1[t, \varphi_{v1}(t)]$, $N_2[t, \varphi_{r2}(t)]$ are defined by equations similar to Eq. (2-7) in terms of the noise components $N_{1c}(t)$, $N_{1s}(t)$ and $N_{2c}(t)$, $N_{2s}(t)$, respectively. The mean-squared value of the reference receiver's tracking phase error due to noise only becomes

$$\sigma_{\varphi_{r2}}^2 = \frac{1}{2\pi j}\left[\frac{N_{01}G^2}{2P_{c1}} \int_{-j\infty}^{j\infty} |H_1(s)|^2 |1 - H_2(s)|^2 \, ds + \frac{N_{02}}{2P_{c2}} \int_{-j\infty}^{j\infty} |H_2(s)|^2 \, ds\right] \tag{3-15}$$

Substitution of (3-9) in (3-15) and integrating yields

$$\sigma_{\varphi_{r2}}^2 = \frac{1}{\alpha_1} + \frac{1}{\rho_2} \tag{3-16}$$

where

$$\alpha_1 \triangleq \frac{\rho_1}{G^2 K_R(r_1, r_2, \xi)} \tag{3-17}$$

$$\rho_n \triangleq \frac{2P_{cn}}{N_{02} W_{Ln}} \qquad n = 1, 2$$

and

$$K_R(r_1, r_2, \xi) \triangleq \frac{r_1\xi}{r_2(r_1+1)}\left[\frac{r_2 + r_1r_2\xi(1+\xi) + \xi^2(1+\xi) + r_1\xi^3}{r_2/r_1 + r_2\xi + \xi^2(r_1+r_2-2) + r_1\xi^3 + r_1\xi^4/r_2}\right] \tag{3-18}$$

with $\xi \triangleq \dfrac{W_{L1}(r_2+1)}{W_{L2}(r_1+1)}$

The function $K_R(r_1, r_2, \xi)$ has been computed, and the results are plotted in Fig. 3-3 for various values of $r_1 = r_2 = r$ and ξ. Clearly, the parameters r_1, r_2, and ξ are significant in the design of two-way phase measuring systems.

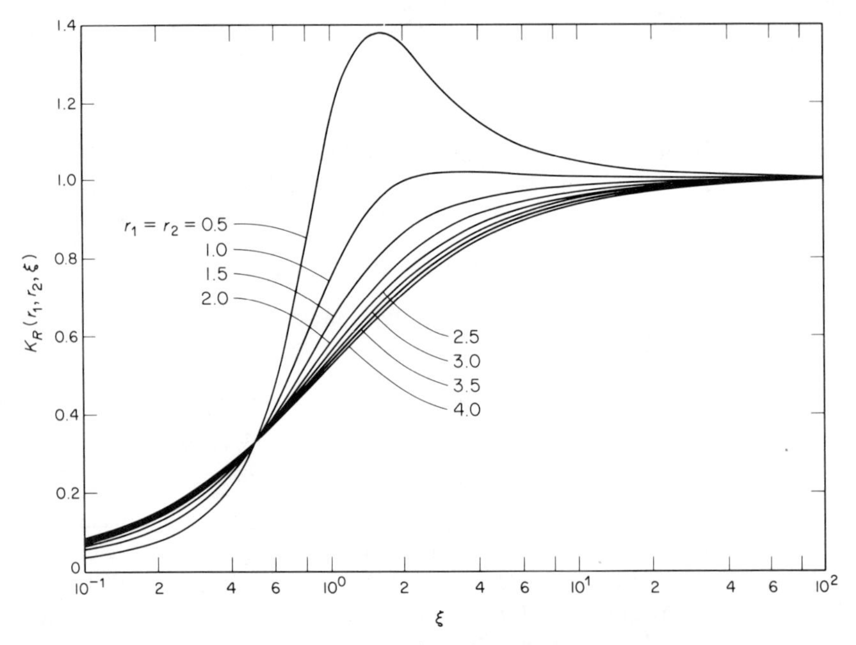

Fig. 3-3 A plot of $K_R(r_1, r_2, \xi)$ vs ξ for several values of $r_1 = r_2$

In a one-way phase measurement, the signal relayed to the reference system is derived from a free-running oscillator on board the vehicle; that is, the switch in the vehicle transponder of Fig. 3-1 is open. The variance of the one-way tracking phase error is obtained quite simply from Eq. (3-16) by letting $\alpha_1 \longrightarrow \infty$:

$$\sigma^2_{\varphi_{r1}} = \frac{N_{02}W_{L2}}{2P_{c2}} = \frac{1}{\rho_2} \tag{3-19}$$

which agrees with the linear PLL theory discussed in Chap. 2.

Nonlinear PLL Theory. The exact nonlinear PLL theory that pertains to the two-way phase measurement appears to be formidable; however, an approximate model for the p.d.f. of the two-way phase error φ_{r2} may be developed on the basis of the PLL measurements presented in Chap. 2. The ability of this model to predict the exact performance of a two-way phase measuring system must be checked against results obtained in the laboratory. Unfortunately, experimental measurements for the two-way system have not yet been completed; however, the approximate p.d.f. of the phase error proposed here checks at the extremes of high and low signal-to-noise ratios. Preliminary test results attest to its validity for a range of parameter values of interest in practice.

Based on the results contained in Sec. 2-2, it is suggested[1] that the p.d.f. of the two-way modulo 2π reduced phase error ϕ_{r2} be approximated by

$$p(\phi_{r2}) = \frac{I_0(|\alpha_1 + \rho_2 \exp(j\phi_{r2})|)}{2\pi I_0(\alpha_1)I_0(\rho_2)} \qquad |\phi_{r2}| \leq \pi \tag{3-20}$$

where α_1 and ρ_2 are defined in Eq. (3-17). This p.d.f. model is valid only if G is approximately equal to unity, which thus far in practice has been the case of greatest interest.* Also, the p.d.f. of (3-20) is of the same generic form as that derived later on in this chapter for an n-cascade of PLLs perturbed by statistically independent, additive, white Gaussian noise sources. The validity of using the above density function as a model for the p.d.f. of the two-way phase error may be checked by considering the limiting cases of high and low signal-to-noise ratios. For large signal-to-noise ratios in the uplink and downlink carrier-tracking loops, it can be shown that (3-20) is Gaussian with zero mean and variance $\alpha_1^{-1} + \rho_2^{-1}$:

$$p(\phi_{r2}) = \frac{\exp\{-[\phi_{r2}^2/2(\alpha_1^{-1} + \rho_2^{-1})]\}}{\sqrt{2\pi(\alpha_1^{-1} + \rho_2^{-1})}} \tag{3-21}$$

Thus, the p.d.f. checks with the linear PLL theory for large signal-to-noise ratios. In the limit as α_1 and ρ_2 approach infinity, we have

$$\lim_{\substack{\rho_2 \to \infty \\ \alpha_1 \to \infty}} p(\phi_{r2}) = \delta(\phi_{r2}) \tag{3-22}$$

which means that perfect phase measurement is obtained.

For weak signal conditions on either the uplink or the downlink or both, (3-20) becomes uniformly distributed; that is,

$$\lim_{\alpha_1 \text{ or } \rho_2 \to 0} p(\phi_{r2}) = \frac{1}{2\pi} \qquad |\phi_{r2}| \leq \pi \tag{3-23}$$

This is in direct agreement with one's intuition for weak signal strengths.

---

* As mentioned in Chap. 1. the unified S-band system employs an S-band uplink and an S-band downlink for which $G = 240/221 \cong 1$. More recently, attention has been given to extending the link capabilities to X-band as well as S-band. For S-band uplink and X-band downlink, the ratio $G = 880/221 \cong 4$.[2] A p.d.f. model which, under certain circumstances, appears to give improved results for such values of G has been suggested by Yuen[3] and is based upon his doctoral dissertation.[4]

Plotted in Fig. 3-4 is the p.d.f. of $\phi_{r2}(t)$ as given by Eq. (3-20) for various values of $\alpha_1 = \rho_2 = \alpha$.

The variance of ϕ_{r2} is important in practice; that is,

$$\sigma^2_{\phi_{r2}} = \int_{-\pi}^{\pi} \phi^2_{r2} p(\phi_{r2}) \, d\phi_{r2} \tag{3-24}$$

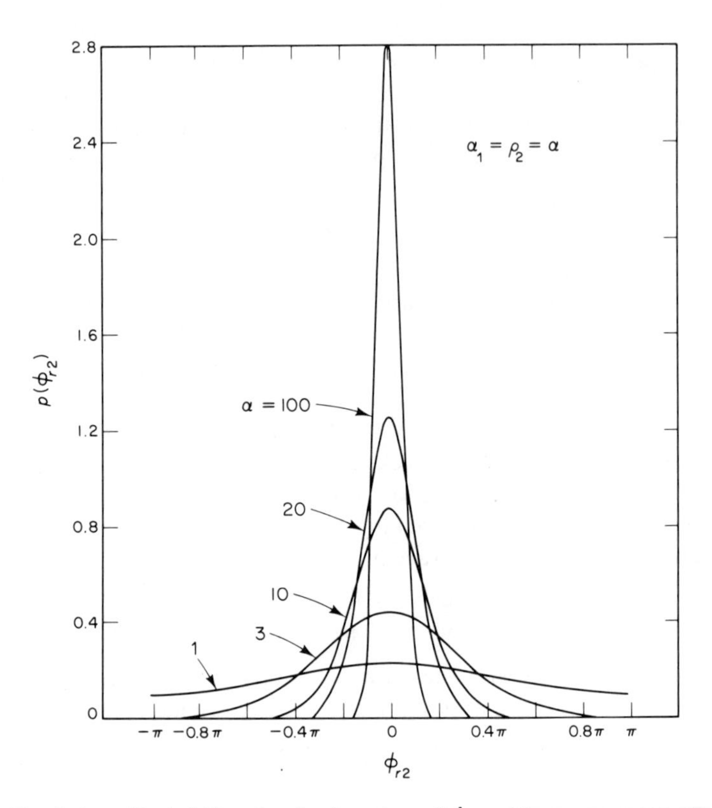

Fig. 3-4 Probability density function of ϕ_{r2} with α as a parameter

Substitution of (3-20) into (3-24) and the use of Neumann's addition theorem (see Problem 3-8) leads to

$$\sigma^2_{\phi_{r2}} = \frac{\pi^2}{3} + 4 \sum_{k=1}^{\infty} \frac{(-1)^k}{k^2} \frac{I_k(\alpha_1) I_k(\rho_2)}{I_0(\alpha_1) I_0(\rho_2)} \tag{3-25}$$

To illustrate the general behavior of this variance, $\sigma^2_{\phi_{r2}}$ as computed from (3-25) and $\sigma^2_{\varphi_{r2}}$ from (3-16) are plotted in Fig. 3-5 as a function of the reciprocal of $\alpha_1 = \rho_2 = \alpha$. For comparison purposes we illustrate results from the linear and nonlinear theory of one-way links; that is, $n = 1$.

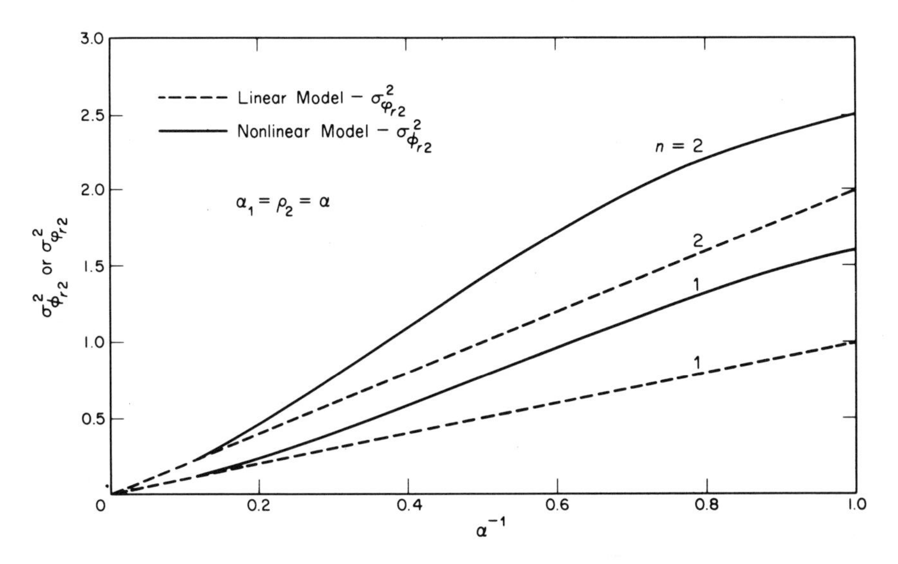

Fig. 3-5 A comparison of the variance of the reference receiver tracking error vs $1/\alpha$ as predicted by the linear and nonlinear theories

Two-Way Tracking Phase Error with Carrier-Tracking Loops Preceded by BPLs

Linear PLL Theory. For the reasons discussed in Sec. 2-3, it is customary in practice to precede carrier-tracking loops with BPLs. Fig. 3-6 illustrates a typical mechanization of a practical two-way phase measuring system.

Employing the linear PLL theory given in Chap. 2, we can show that the variance of the two-way phase measurement is given by

$$\sigma^2_{\varphi_{r2}} = \frac{1}{\alpha_{1\ell}} + \frac{1}{\rho_{2\ell}} \tag{3-26}$$

where

$$\alpha_{1\ell} \triangleq \frac{\rho_{1\ell}}{G^2 K_R(k_1, k_2, \xi)}$$

$$\rho_{n\ell} \triangleq \frac{2P_{cn}}{N_{0n}W_{0n}}\left(\frac{1 + r_{n0}}{1 + r_{n0}/\mu_n}\right)\frac{1}{\Gamma_{pn}} \qquad n = 1, 2 \tag{3-27}$$

and Γ_{pn}, $n = 1, 2$, are the limiter performance factors defined by

$$\Gamma_{pn} \triangleq \frac{1 - e^{-x_n y_n}}{(\pi x_n y_n/4)e^{-x_n y_n}[I_0(x_n y_n/2) + I_1(x_n y_n/2)]^2[1 + (4/\Gamma_0\pi - 1)e^{-x_n y_n(1-\pi/4)}]} \tag{3-28}$$

with

$$x_n \triangleq \frac{2P_{cn}}{N_{0n}W_{0n}} \qquad y_n \triangleq \frac{W_{0n}}{W_{in}} \tag{3-29}$$

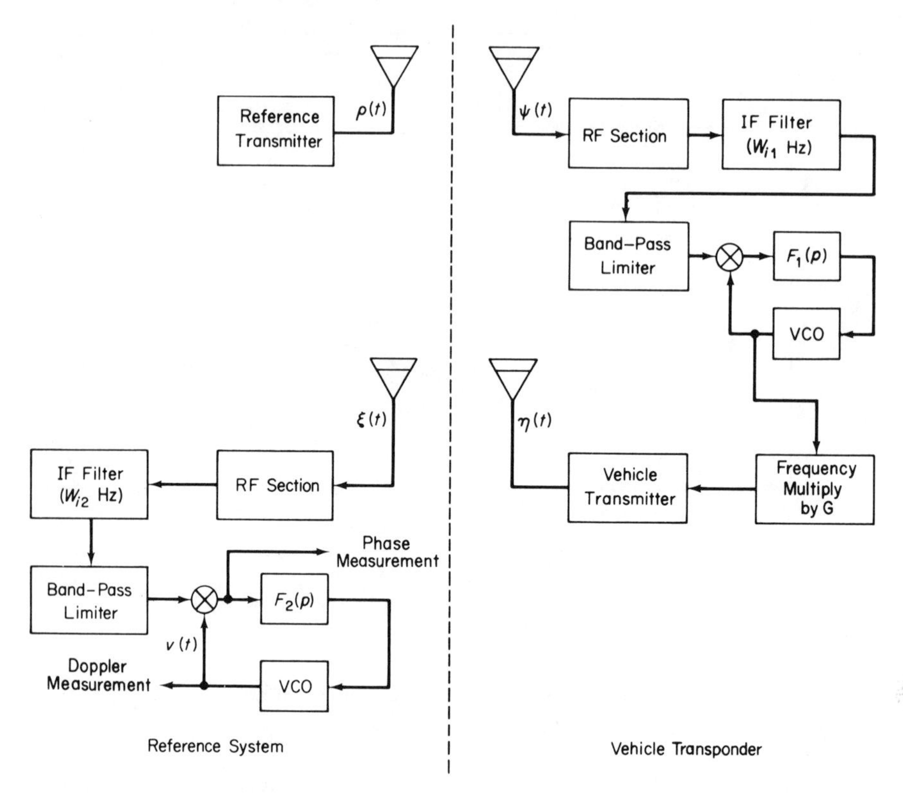

Fig. 3-6 A simple mechanization of a two-way phase measuring system employing band-pass limiters

The parameters W_{0n}, $n = 1, 2$, are the bandwidths of the carrier-tracking loops at the design point defined by Eq. (2-55), and W_{in}, $n = 1, 2$, are the bandwidths of the IF filters that precede the carrier-tracking loops. Further, the parameters r_{n0} and μ_n are given by

$$r_{n0} = \tilde{\alpha}_{0n}\sqrt{P_1}K_n\tau_{2n}^2/\tau_{1n} \qquad \mu_n \triangleq \frac{\tilde{\alpha}_{0n}}{\tilde{\alpha}_n} = \frac{r_{n0}}{r_n} \tag{3-30}$$

where the $\tilde{\alpha}_n$ are the limiter-suppression factors whose functional form is given in (2-42). Note in the above a slight change in notation to avoid triple subscripts and at the same time accommodate the two-way system model: the L subscript on W is omitted for the bandwidth of the carrier-tracking loop, and the $\tilde{\alpha}_n$ are assumed to correspond to the first zone. Finally, $K_R(k_1, k_2, \xi)$ is given by

$$K_R(k_1, k_2, \xi) \triangleq \frac{k_2\xi}{k_1(2/k_1 + 1)}\left[\frac{2k_1 + 4\xi(1 + \xi) + \xi^2(1 + \xi)k_1k_2 + 2k_2\xi^3}{k_1^2 + 2k_1\xi + 2\xi^2(k_1 + k_2 - k_1k_2) + 2\xi^3k_2 + k_2^2\xi^4}\right] \tag{3-31}$$

where $k_n \triangleq \dfrac{2\mu_n}{r_{n0}}$ and $\xi \triangleq \dfrac{W_{01}}{W_{02}} \dfrac{(r_{20}+1)}{(r_{10}+1)}$

The function $K_R(k_1, k_2, \xi)$ multiplied by $(1 + r_{10}/\mu_1)/(1 + r_{10})$ is plotted in Fig. 3-7 versus ξ for various values of the signal-to-noise ratio in the design

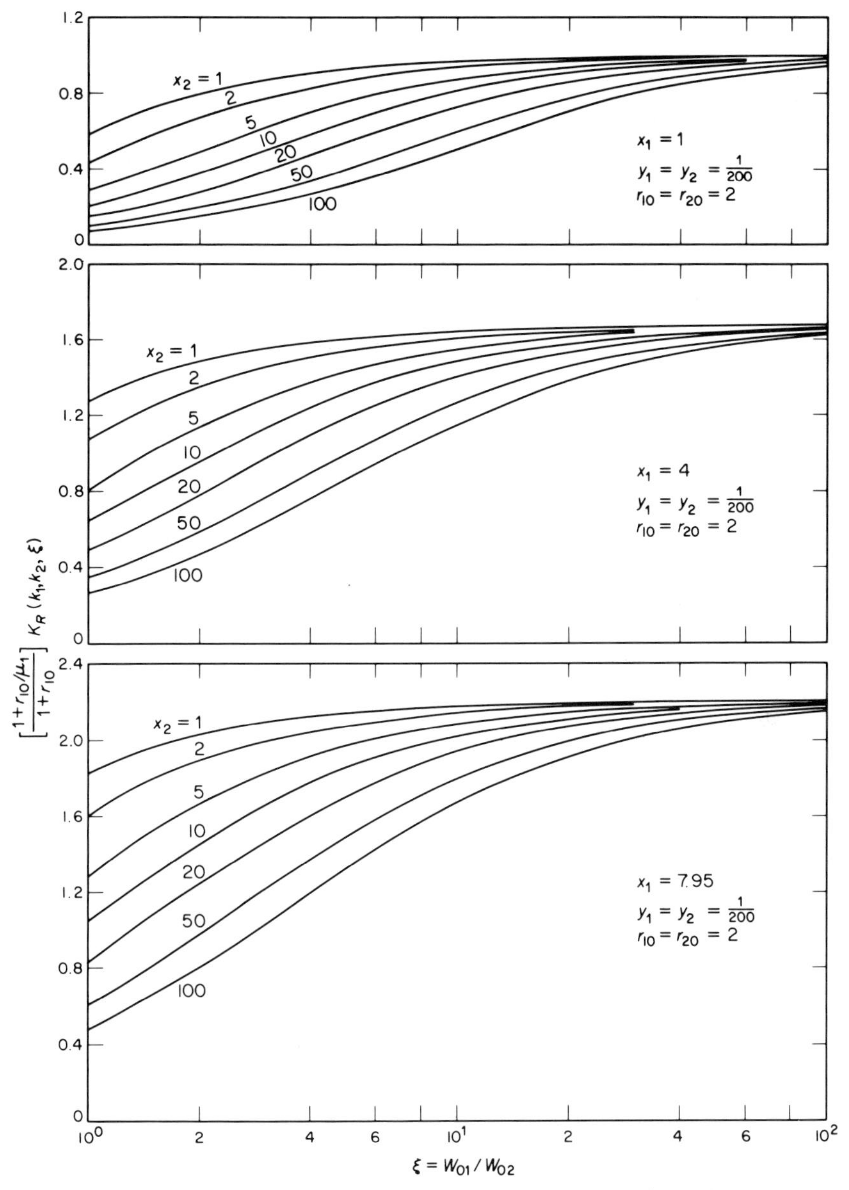

Fig. 3-7 Several plots of $[(1 + r_{10}/\mu_1)/(1 + r_{10})]K_R(k_1, k_2, \xi)$ vs ξ for various values of the parameters $x_1, x_2, y_1, y_2, r_{10}, r_{20}$

point loop bandwidths of the vehicle and ground receiver carrier-tracking loops; limiters are present in both receivers and $y_1 = y_2 = y$, $r_{10} = r_{20}$ are assumed. In Fig. 3-8, $\sigma^2_{\varphi_{r2}}$ as computed from Eq. (3-16) is plotted against various values of the signal-to-noise ratio in the design point loop bandwidths of the vehicle and ground receiver carrier-tracking loops. It may be

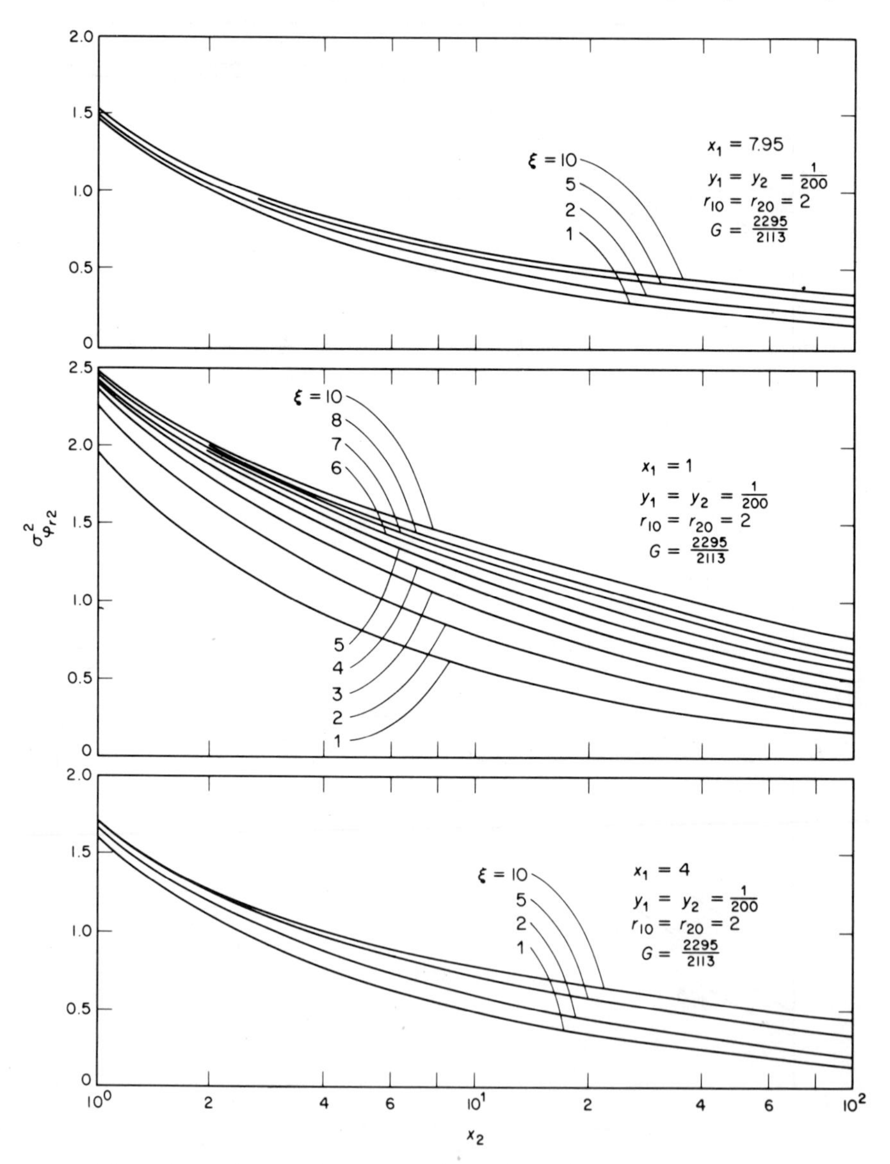

Fig. 3-8 Several plots of $\sigma^2_{\varphi_{r2}}$ vs x_2 for various values of the parameters x_1, y_1, y_2, G, ξ, $r_{10} = r_{20}$

shown that this plot is independent of y for the cases of greatest interest in practice—(small y). Linear PLL theory is assumed, and BPLs are present in the system. Again the performance of one-way phase measuring links may be obtained from the above equations by allowing $\alpha_{1\ell}$ to approach infinity.

Nonlinear PLL Theory. The p.d.f. of the phase error ϕ_{r2} may be modeled in a manner similar to that for the two-way phase error φ_{r2}. Thus, when BPLs precede the carrier-tracking loops, the p.d.f. of the phase error ϕ_{r2} is well approximated by

$$p(\phi_{r2}) = \frac{I_0(|\alpha_{1\ell} + \rho_{2\ell}\exp(j\phi_{r2})|)}{2\pi I_0(\alpha_{1\ell})I_0(\rho_{2\ell})} \qquad |\phi_{r2}| \leq \pi \tag{3-32}$$

where $\alpha_{1\ell}$ and $\rho_{2\ell}$ are defined in (3-27). The variance of ϕ_{r2},

$$\sigma^2_{\phi_{r2}} = \int_0^\pi \phi^2_{r2} \frac{I_0(\sqrt{\alpha^2_{1\ell} + \rho^2_{2\ell} + 2\alpha_{1\ell}\rho_{2\ell}\cos\phi_{r2}})}{\pi I_0(\alpha_{1\ell})I_0(\rho_{2\ell})}\, d\phi_{r2} \tag{3-33}$$

becomes, on carrying out the integration,

$$\sigma^2_{\phi_{r2}} = \frac{\pi^2}{3} + 4\sum_{k=1}^{\infty} \frac{(-1)^k}{k^2} \frac{I_k(\alpha_{1\ell})I_k(\rho_{2\ell})}{I_0(\alpha_{1\ell})I_0(\rho_{2\ell})} \tag{3-34}$$

Plotted in Fig. 3-9 is the variance of the tracking phase error as evaluated from Eq. (3-34) for various system mechanizations. The variance shown in this figure exists in the design point loop bandwidths of the vehicle and ground receiver carrier-tracking loops. Again, this plot is independent of y for the cases of greatest interest in practice. Nonlinear PLL theory is assumed, and BPLs are present in the system.

3-3. TWO-WAY DOPPLER MEASUREMENTS

In the design of a system such as that discussed in Sec. 3-2, the variance of the two-way Doppler error $\varphi_{d2}(t)$ is a significant parameter; namely,

$$\sigma^2_{\varphi_{d2}} \triangleq \overline{(\hat{\theta}_2 - \theta_0)^2} = \overline{\varphi^2_{d2}} \tag{3-35}$$

In this section, we consider the Doppler error $\varphi_{d2}(t)$ and its relationship to the various system parameters. The development parallels that of Sec. 3-2: the results are first derived employing the linear PLL theory followed by the appropriate remodeling to permit study of nonlinear PLL effects. Finally, the Doppler error when BPLs precede the vehicle and reference system carrier-tracking loops is considered.

Two-Way Doppler Error

Linear PLL Theory. The two-way Doppler phase error $\varphi_{d2}(t)$ is defined as the difference between the reference receiver output phase $\hat{\theta}_2(t)$ and the original phase of the transmitted carrier θ_0. Thus, the spectral density of the

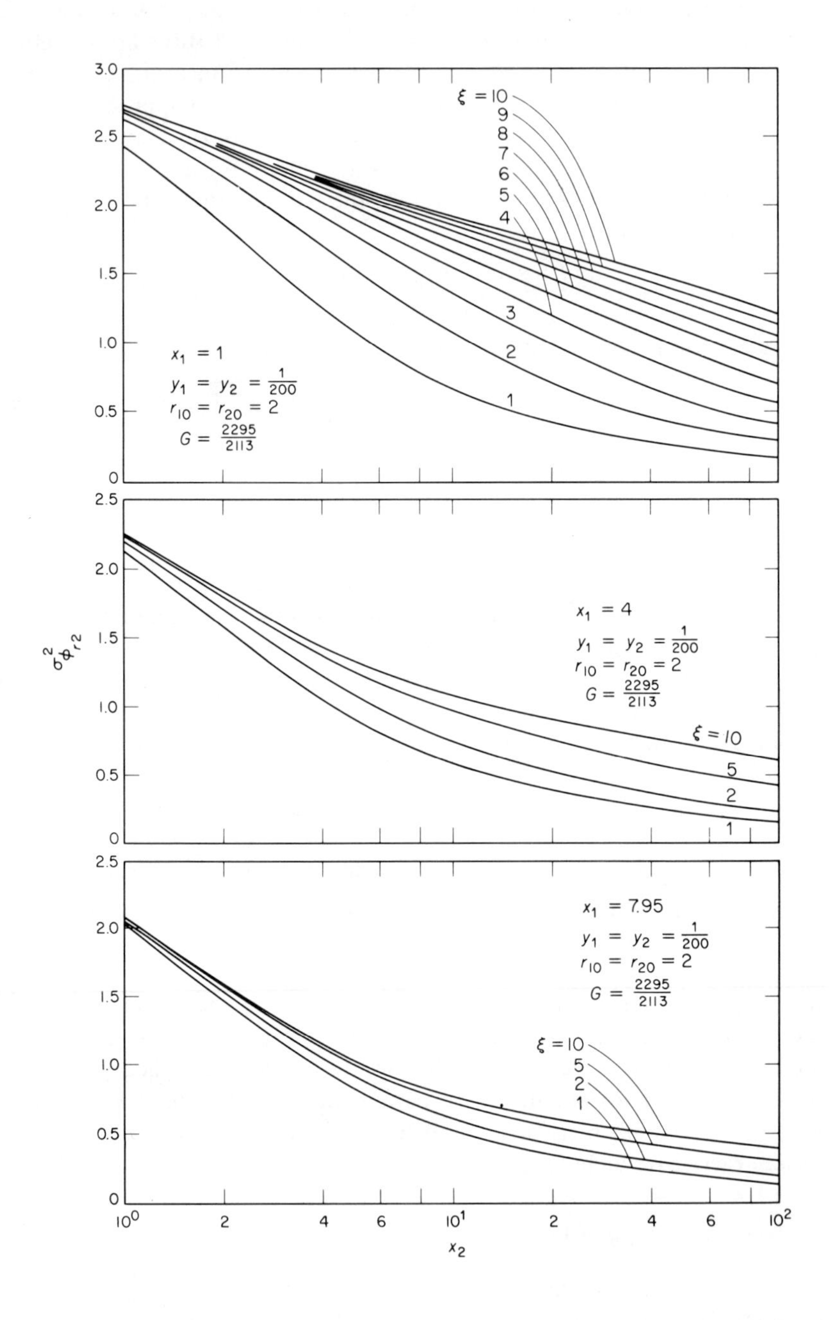

Fig. 3-9 Several plots of $\sigma^2_{\phi_{r2}}$ vs x_2 for various values of the parameters x_1, y_1, y_2, G, ξ, $r_{10} = r_{20}$

Doppler error due to noise alone is (in Laplace transform notation)

$$\varphi_{d2}(s)\varphi_{d2}(-s) = \frac{N_{01}G^2}{2P_{c1}}|H_1(s)H_2(s)|^2 + \frac{N_{02}}{2P_{c2}}|H_2(s)|^2 \tag{3-36}$$

and the variance of the Doppler error becomes

$$\sigma^2_{\varphi_{d2}} = \frac{1}{2\pi j}\left\{\frac{N_{01}G^2}{2P_{c1}}\int_{-j\infty}^{j\infty}|H_1(s)H_2(s)|^2\,ds + \frac{N_{02}}{2P_{c2}}\int_{-j\infty}^{j\infty}|H_2(s)|^2\,ds\right\} \tag{3-37}$$

Substitution of Eq. (3-9) into (3-37) and carrying out the integration gives

$$\sigma^2_{\varphi_{d2}} = \frac{1}{d_1} + \frac{1}{\rho_2} \tag{3-38}$$

where

$$d_1 \triangleq \frac{\rho_1}{G^2K_D(r_1, r_2, \xi)} \tag{3-39}$$

and $K_D(r_1, r_2, \xi)$

$$\triangleq \frac{1}{(r_1+1)}\left[\frac{r_2(r_1+1)+r_1(r_1+r_2+r_1r_2)(\xi+\xi^2)+r_1(r_1r_2+r_1)\xi^3/r_2}{r_2+r_1r_2\xi+r_1(r_1+r_2-2)\xi^2+r_1^2\xi^3+r_1^2\xi^4/r_2}\right] \tag{3-40}$$

The function $K_D(r_1, r_2, \xi)$ has been computed, and the results are plotted versus ξ in Fig. 3-10 for various values of $r_1 = r_2 = r$. For the limiting case of $r \longrightarrow \infty$—first-order PLLs,

$$\lim_{r\to\infty} K_D(r, r, \xi) = \lim_{r\to\infty}\frac{K_R(r, r, \xi)}{\xi} = \frac{1}{1+\xi} \tag{3-41}$$

Nonlinear PLL Theory. Following arguments similar to those made in Sec. 3-2, the p.d.f. of the two-way Doppler error may be approximated by[1]

$$p(\phi_{d2}) = \frac{I_0[|d_1 + \rho_2\exp(j\phi_{d2})|]}{2\pi I_0(d_1)I_0(\rho_2)} \qquad |\phi_{d2}| \le \pi \tag{3-42}$$

where d_1 is defined in Eq. (3-39). The validity of using (3-42) as a model for the two-way Doppler error p.d.f. may be checked again by considering the limiting cases of high and low signal-to-noise ratios. A comparison of (3-42) and (3-20) should be sufficient to justify for the Doppler case results similar to (3-21) through (3-23). Also, by direct analogy,

$$\sigma^2_{\phi_{d2}} = \frac{\pi^2}{3} + 4\sum_{k=1}^{\infty}\frac{(-1)^k}{k^2}\frac{I_k(d_1)I_k(\rho_2)}{I_0(d_1)I_0(\rho_2)} \tag{3-43}$$

Hence, the curves of Fig. 3-5 are directly applicable to the Doppler measurement if α is replaced by $d = d_1 = \rho_2$. It should be clear by now that for the Doppler measuring system, the form of the results and hence their derivation is identical to that of the phase measuring system previously described. Thus, in the next subsection we merely present the final results to avoid unnecessary repetition.

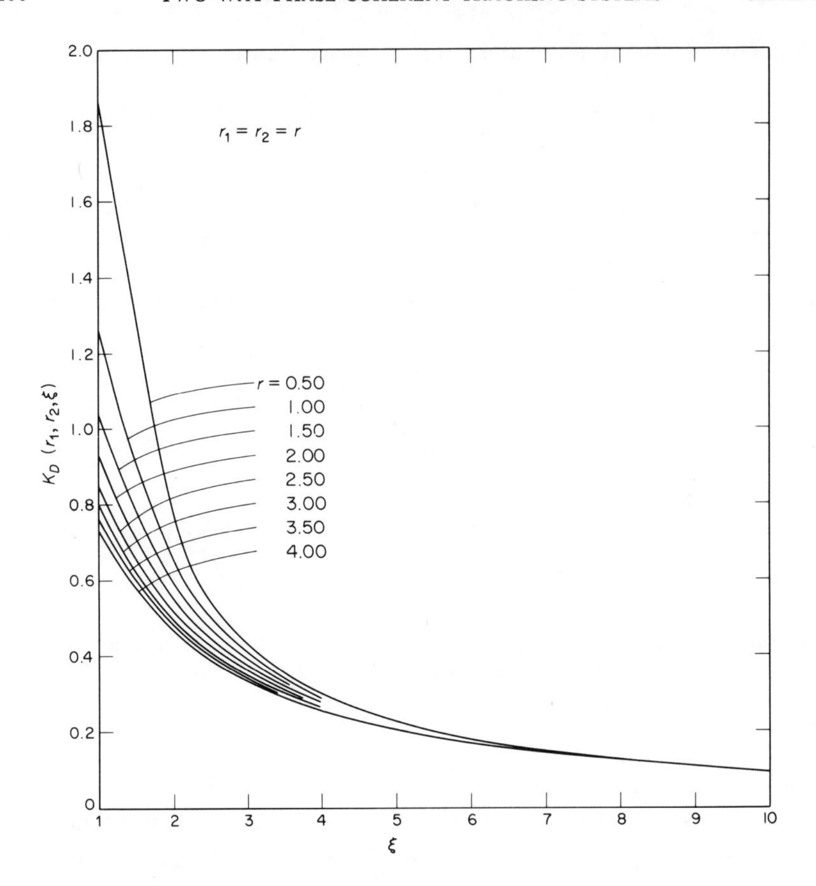

Fig. 3-10 A plot of $K_D(r_1, r_2, \xi)$ vs ξ for several values of $r_1 = r_2$

Two-Way Doppler Error with Carrier-Tracking Loops Preceded by BPLs

Linear PLL Theory

$$\sigma^2_{\varphi d2} = \frac{1}{d_{1\ell}} + \frac{1}{\rho_{2\ell}} \tag{3-44}$$

where

$$d_{1\ell} \triangleq \frac{\rho_{1\ell}}{G^2 K_D(k_1, k_2, \xi)} \tag{3-45}$$

$K_D(k_1, k_2, \xi)$

$$\triangleq \frac{1}{(2/k_1 + 1)}\left[\frac{k_1(2 + k_1) + 2(k_1 + k_2 + 2)(\xi + \xi^2) + k_2(2 + k_2)\xi^3}{k_1^2 + 2k_1\xi + 2(k_1 + k_2 - k_1k_2)\xi^2 + 2k_2\xi^3 + k_2^2\xi^4}\right] \tag{3-46}$$

and k_n, $n = 1, 2$ and ξ are defined following (3-31). The function $K_D(k_1, k_2, \xi)$

multiplied by $(1 + r_{10}/\mu_1)/(1 + r_{10})$ is plotted versus ξ in Fig. 3-11 for various values of the signal-to-noise ratios in the design point loop bandwidth of the vehicle and ground receiver carrier-tracking loops. The performance of a one-way Doppler measuring system may be obtained from Eqs. (3-44) and

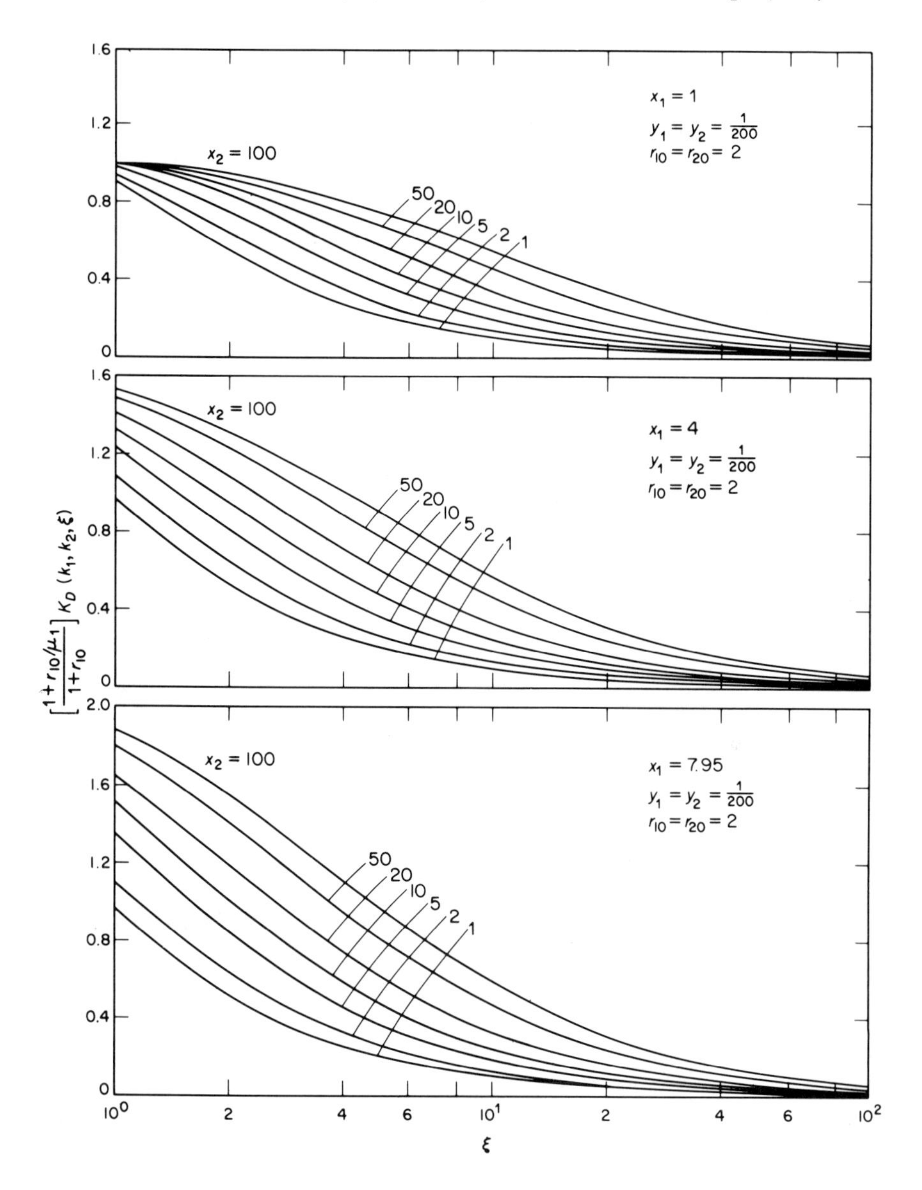

Fig. 3-11 Several plots of $[(1 + r_{10}/\mu_1)/(1 + r_{10})]K_D(k_1, k_2, \xi)$ vs ξ for various values of the parameters $x_1, x_2, y_1, y_2, r_{10} = r_{20}$

(3-45) by letting $d_{1\ell}$ approach infinity. The variance of $\varphi_{d2}(t)$ as determined from (3-44) is plotted in Fig. 3-12 for various system mechanizations.

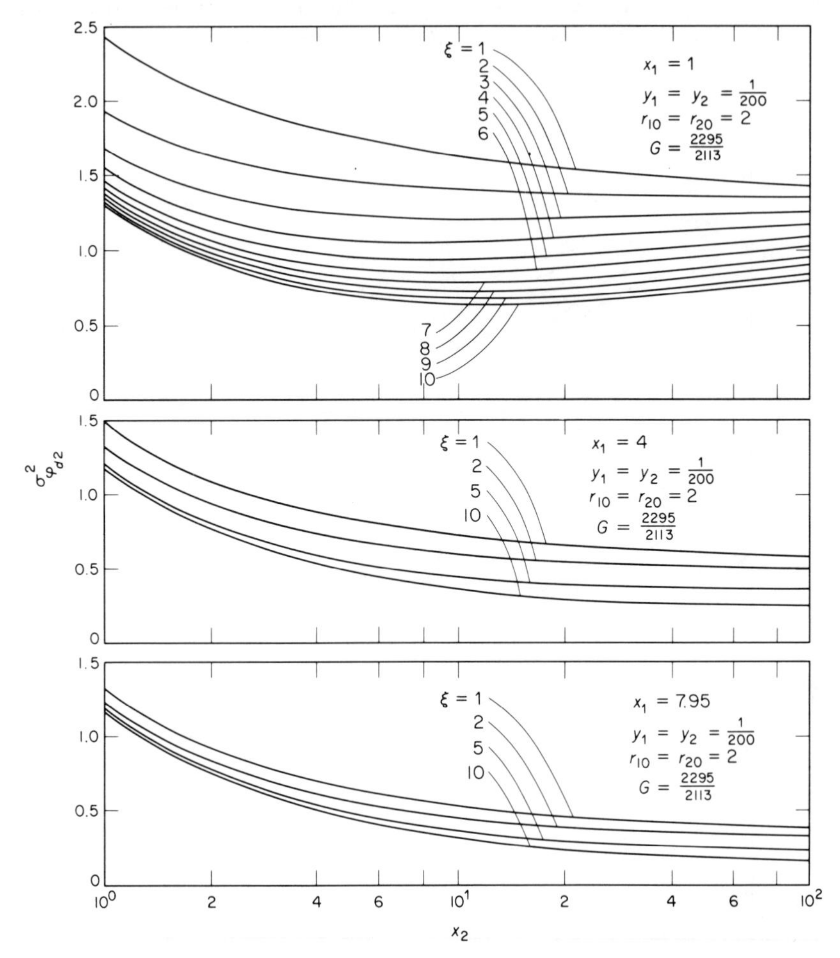

Fig. 3-12 Several plots of $\sigma^2_{\varphi_{d2}}$ vs x_2 for various values of the parameters x_1, $y_1, y_2, G, \xi, r_{10} = r_{20}$

Nonlinear PLL Theory

$$p(\phi_{d2}) = \frac{I_0[|\, d_{1\ell} + \rho_{2\ell} \exp(j\phi_{d2})\,|]}{2\pi I_0(d_{1\ell}) I_0(\rho_{2\ell})} \qquad |\phi_{d2}| \leq \pi \tag{3-47}$$

where $d_{1\ell}$ is defined in Eq. (3-45).

$$\sigma^2_{\phi_{d2}} = \frac{\pi^2}{3} + 4 \sum_{k=1}^{\infty} \frac{(-1)^k}{k^2} \frac{I_k(d_{1\ell}) I_k(\rho_{2\ell})}{I_0(d_{1\ell}) I_0(\rho_{2\ell})} \tag{3-48}$$

Plotted in Fig 3-13 is $\sigma^2_{\phi_{d2}}$ as determined by Eq. (3-48) for various system

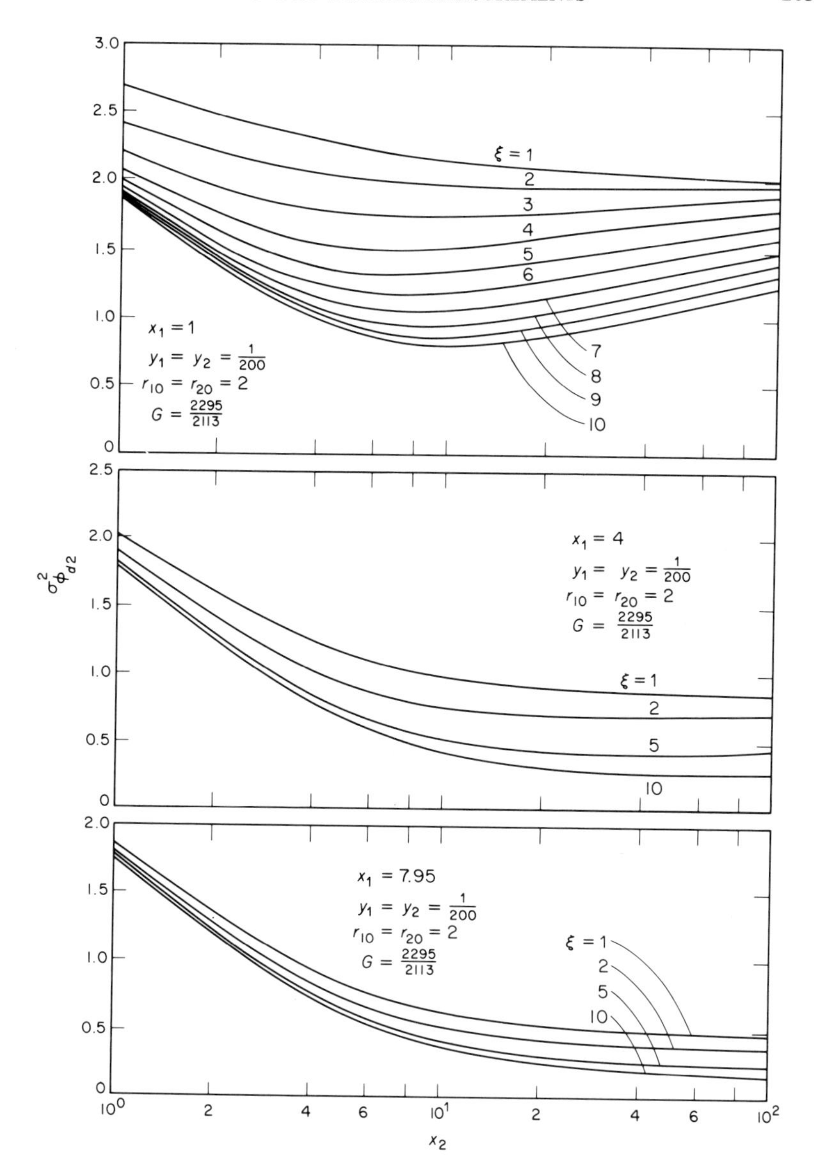

Fig. 3-13 Several plots of $\sigma^2_{\phi_{d2}}$ vs x_2 for various values of the parameters x_1, y_1, y_2, G, ξ, $r_{10} = r_{20}$

mechanizations. The behavior of the Doppler variance as a function of the signal-to-noise ratio in the design point loop bandwidth x_2 may be attributed to the fact that the bandwidth of the ground receiver's carrier-tracking loop increases as the downlink signal-to-noise ratio increases.

3-4. DOWNLINK CARRIER-SUPPRESSION EFFECTS DUE TO ADDITIVE NOISE ON THE UPLINK

In this section, we shall use the second moment theory of random processes to study the suppression effect produced on the downlink carrier by the random modulation that exists on the vehicle's VCO output. This random modulation, $\hat{\theta}_1(t)$, which is due to the additive noise on the uplink, reduces the power remaining in the downlink carrier component below that value obtained if the system transmitted a clean carrier back to the reference system. In practice, this suppression of the downlink carrier, which may be measured near the end of a mission, is important in the design of a two-way link.

Carrier Suppression

Linear PLL Theory. If the linear PLL theory is used, then $\hat{\theta}_{10}(t)$, the random phase modulation on the downlink due to the additive noise on the uplink, is well approximated by a stationary, Gaussian process with zero mean and variance $\sigma^2_{\hat{\theta}_{10}} = N_{01}W_{L1}G^2/2P_{c1}$. If we denote the covariance function of the phase modulation $\hat{\theta}_{10}(t)$ by $R_{\hat{\theta}_{10}}(\tau)$ with $R_{\hat{\theta}_{10}}(0) = 1$, then it may be shown[5] that the covariance function $R_\eta(\tau)$ of $\eta(t)$ as given by Eq. (3-5) is

$$R_\eta(\tau) = P_{c2} \exp\{-\sigma^2_{\hat{\theta}_{10}}[R_{\hat{\theta}_{10}}(0) - R_{\hat{\theta}_{10}}(\tau)]\} \cos \omega_{10}\tau \tag{3-49}$$

At $\tau = 0$, $R_\eta(0) = P_{c2}$; that is, as expected, the mean total power is P_{c2}, inasmuch as the modulation is entirely in the phase term. The intensity of the downlink carrier component is

$$I_c = R_\eta(\infty) = P_{c2} \exp\left(-\frac{N_{01}W_{L1}G^2}{2P_{c1}}\right) \tag{3-50}$$

For a Gaussian phase modulation, it is clear that $I_c > 0$ so that there is always a discrete carrier component, although it may represent a trivial fraction of the total power P_{c2} if $\sigma_{\hat{\theta}_{10}}$ is large.

The carrier-suppression factor $\tilde{S}$ is defined as the ratio of the power remaining in the carrier component when the uplink additive noise affects the downlink transmission to the power remaining in the carrier component when the downlink carrier is derived in the vehicle from a free-running oscillator. Thus, from (3-50) we have

$$\begin{aligned}\tilde{S} \triangleq \frac{I_c}{P_{c2}} &= \exp\left(-\frac{N_{01}W_{L1}G^2}{2P_{c1}}\right) \\ &= \exp(-\sigma^2_{\hat{\theta}_{10}})\end{aligned} \tag{3-51}$$

In Fig. 3-14, the carrier-suppression factor is plotted versus $1/\sigma^2_{\hat{\theta}_{10}} = \rho_1/G^2$ where ρ_1 is again the signal-to-noise ratio in the vehicle's carrier-tracking loop.

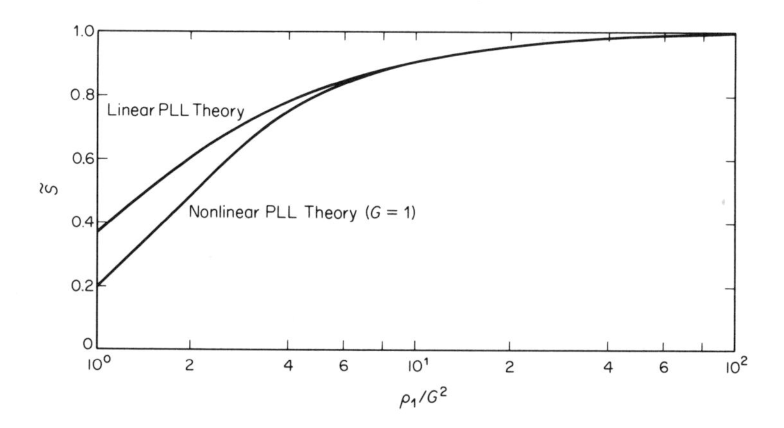

Fig. 3-14 Carrier-suppression factor vs ρ_1/G^2 (linear and nonlinear theories)

Nonlinear PLL Theory. If the phase modulation process is not Gaussian, the covariance function $R_\eta(\tau)$ is not mathematically tractable; however, the carrier-suppression factor may still be computed. Gagliardi[6] has shown that the carrier suppression of a carrier, angle modulated by an arbitrary first-order, stationary, stochastic process $\theta_m(t)$, depends only on the characteristic function of the modulation $\Psi_{\theta_m}(\omega)$, namely,

$$\tilde{S} = |\Psi_{\theta_m}(1)|^2 \tag{3-52}$$

where

$$\Psi_{\theta_m}(\omega) \triangleq \int_{-\infty}^{\infty} p(\theta_m) e^{j\omega\theta_m}\, d\theta_m \tag{3-53}$$

and $p(\theta_m)$ is the p.d.f. of the random modulation.

From the results of Chap. 2, the distribution of $\hat{\theta}_{10}(t)$ is well approximated by

$$p(\hat{\theta}_{10}) \triangleq p[G(\theta_0 + \theta_1 - \varphi_{v1})] = \frac{\exp[\rho_1 \cos(\theta_0 + \theta_1 - \hat{\theta}_{10}/G)]}{2\pi G I_0(\rho_1)}; \qquad \left|\theta_0 + \theta_1 - \frac{\hat{\theta}_{10}}{G}\right| \leq \pi \tag{3-54}$$

with ρ_1 defined in (3-17). Letting $\theta_m(t) = \hat{\theta}_{10}(t)$ in (3-53), we get for the characteristic function of $\hat{\theta}_{10}(t)$,

$$\Psi_{\hat{\theta}_{10}}(\omega) = e^{j\omega G(\theta_0+\theta_1)}\left[\frac{I_{G\omega}(\rho_1)}{I_0(\rho_1)}\right] \tag{3-55}$$

Letting $\omega = 1$ in (3-55) and substituting into (3-52) gives the desired expression for carrier suppression:

$$\tilde{S} = \left[\frac{I_G(\rho_1)}{I_0(\rho_1)}\right]^2 \tag{3-56}$$

In the more general case where frequency detuning exists between the uplink carrier and the vehicle local oscillator, the p.d.f. of $\hat{\theta}_{10}(t)$ should be characterized by an equation similar to (2-16). For this case, the carrier suppression becomes

$$\tilde{S} = \left[\frac{I_{G-j\beta}(\alpha)}{I_{-j\beta}(\alpha)}\right]^2 \tag{3-57}$$

where the real and imaginary parts of $I_{G-j\beta}(\alpha)$ are the circular moments of the phase error p.d.f. as defined in (2-22).

The factor $\tilde{S}$ as given by (3-56) with $G = 1$ is plotted in Fig. 3-14 vs. ρ_1 for comparison with the result obtained using the linear PLL theory. Note that as $\rho_1 = \sigma_{\hat{\theta}_{10}}^{-2}$ approaches zero, the downlink carrier component becomes completely suppressed in the vehicle.

Carrier Suppression with BPLs Preceding the Carrier-Tracking Loop

Linear PLL Theory. Again we assume that the bandwidth of the vehicle's carrier-tracking loop is small enough for the process $\hat{\theta}_{10}(t)$ to be Gaussian; that is, the linear theory may be used. Thus, the variance of the process $\hat{\theta}_{10}(t)$ adequately characterizes the carrier suppression.

From Sec. 3-2, the variance of $\hat{\theta}_{10}$ becomes

$$\sigma_{\hat{\theta}_{10}}^2 = \left(\frac{N_{01}W_{01}}{2P_{c1}}\right)(G^2\Gamma_{p1})\left(\frac{1 + r_{10}/\mu_1}{1 + r_{10}}\right) \tag{3-58}$$

and Γ_{p1} and μ_1 are defined in Eqs. (3-28) and (3-30), respectively. Thus, the downlink carrier suppression is given by (3-51) where $\sigma_{\hat{\theta}_{10}}^2$ is defined in (3-58). In Fig. 3-15, this suppression is plotted against x_1/G^2 for a range of system parameters.

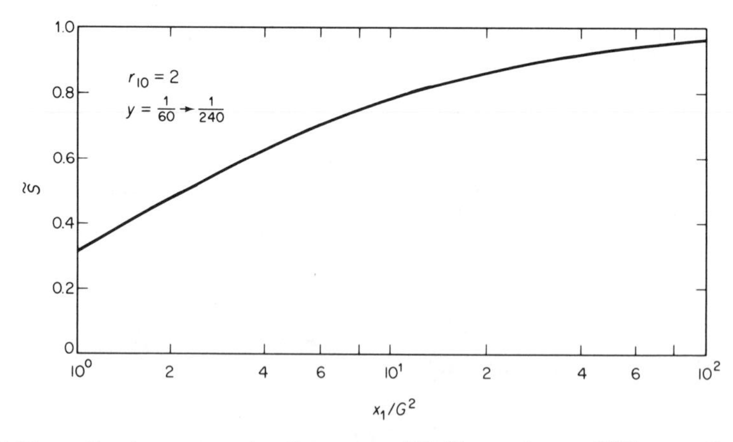

Fig. 3-15 Carrier-suppression factor vs x_1/G^2 (linear theory, BPL preceding the PLL)

Nonlinear PLL Theory. In this case, the covariance $R_\eta(\tau)$ again cannot be determined mathematically; however, on the basis of (2-63) and the results given above, the carrier-suppression factor is still given by (3-56) where ρ_1 is now defined by

$$\rho_1 = \frac{2P_{c1}}{N_{01}W_{01}}\left(\frac{1}{\Gamma_{p1}}\right)\left(\frac{1 + r_{10}}{1 + r_{10}/\mu_1}\right) \tag{3-59}$$

As before, we note that the downlink carrier is completely suppressed when the uplink signal-to-noise ratio approaches zero. This result further supports the uniformity of $p(\phi_{r2})$ of (3-20) as ρ_2 approaches zero, and supports the analytical model in this signal-to-noise ratio region. In Fig. 3-16, $\tilde{S}$ as determined from Eqs. (3-56) and (3-59) with $G = 1$ is plotted versus x_1 for a a range of system parameters.

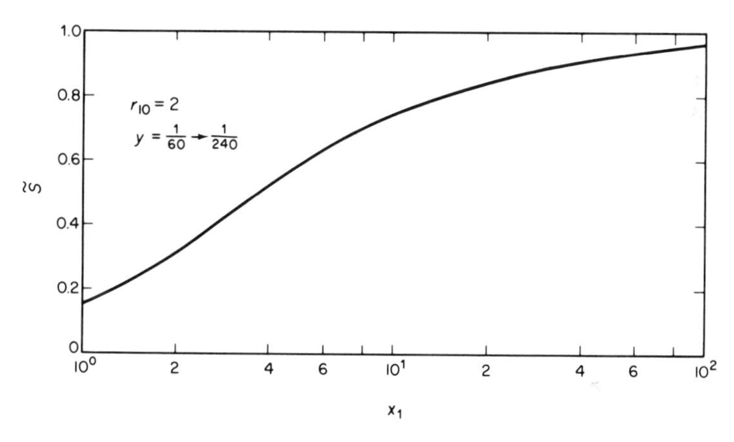

Fig. 3-16 Carrier-suppression factor vs x_1 (nonlinear theory, BPL preceding the PLL)

3-5. DIVERSITY COMBINING TO IMPROVE PHASE AND DOPPLER MEASUREMENTS

In the area of space communications, the interest in large antenna apertures arises from the need for greater signal capture area and gain, rather than angular resolution, the major interest being in the radioastronomy field. From the standpoint of communication theory, a second difference is that the large antenna performance characteristics can be traded off against against other communication link performance parameters. For example, the tradeoffs among data rate, Doppler and tracking accuracy, obtained with modifications in the transmitting and receiving system are related in Table 3-1. Thus, from this table, one qualitatively observes the tradeoffs between the transmitter and receiver parameters which the communications engineer may make in order to improve system performance.

In this section, we investigate briefly the improvement to be realized in the Doppler and tracking error by operating on the receiver parameters, in particular the receiving antenna gain $G_r(f)$ (see Table 3-1). Except for

TABLE 3-1 Factors Affecting Doppler System Design

System Performance Parameters		*Transmitter Parameters*		*Transmission Media*		*Receiver Parameters*
$\mathcal{R}$ Data Rate	$\propto$	Gain of Transmitting Antenna $M_T P_T G_T(f)$ Modulation Technique Transmitter Power	$\times$	$\frac{1}{f^2 D_T^2}$ Frequency Transmission Distance	$\times$	Gain of Receiving Antenna $\frac{G_r(f)}{T_r(f)}$ Receiver Noise Temperature
$\sigma^2_{1\text{-way}}$ Tracking or Doppler Accuracy	$\propto$	$\frac{1}{M_T P_T G_T(f)}$	$\times$	$\frac{1}{f^2 D_T^2}$	$\times$	Tracking Loop Bandwidth $\frac{T_r(f) W_L}{G_r(f)}$
$\sigma^2_{2\text{-way}}$ Tracking or Doppler Accuracy	$\propto$	$\frac{K_D(r_1, r_2, \xi)}{M_{T_1} P_{T_1} G_{T_1}(f_1)}$	$\times$	$\frac{1}{f_1^2 D_T^2}$ UPLINK CONTRIBUTION	$\times$	$\frac{T_{r_1}(f_1) W_{L1}}{G_{r_1}(f_1)}$ $+$
		$\frac{1}{M_{T_2} P_{T_2} G_{T_2}(f_2)}$	$\times$	$\frac{1}{f_2^2 D_T^2}$ DOWNLINK CONTRIBUTION	$\times$	$\frac{T_{r_2}(f_2) W_{L2}}{G_{r_2}(f_2)}$

very unusual situations, the space-to-Earth link is more critical; therefore, the receiving equipment is identified as being Earth-based. The method of improvement in tracking and Doppler accuracy to be minutely explored here is that of *diversity reception.*

Diversity is defined here as a general technique that utilizes two or more copies of a signal with varying degrees of disturbance to obtain, by a selection or combination scheme, a larger signal-to-noise ratio than is achievable from any one of the individual copies alone. Although diversity is commonly understood to be aimed at improving the reliability of reception of signals that are subject to fading in the presence of random noise, the significance of the term will be extended here to cover conceptually related techniques, particularly those intended to reduce the effective phase and Doppler error occurring in space communication systems.

The first problem in diversity reception is the procurement of the "diverse" copies of the disturbed signal, or, if only one copy is available, the operation on this copy to generate additional "diversified" copies. In space communications, the transmission medium can be tapped for an ever-available supply of diversified copies by employing more than one receiving antenna, and if physically separated by several wavelengths, the observed signal copies will be statistically independent. Herein lies the key to the improvements that are realizable through diversity reception: if in fact the copies are *not* statistically independent, the gain obtained will be due only to an increase in area.

The second problem in diversity is the question of how to utilize the available disturbed copies of the signal in order to achieve the least possible loss of information in extracting the desired message or parameter. The technique of particular interest here is to employ the method of combining, i.e., to *coherently* sum the observed copies at the observed frequencies. In practice, this is a difficult feat; nevertheless, for purposes of discussion, we assume that this can be accomplished with perfect precision.

The Signal-Combining Technique

Denote the input to the ℓth receiving antenna ($\ell = 1, 2, \ldots, L$) by

$$\xi_\ell(t) = \sqrt{2P_{c2}} \sin [\omega_{2\ell} t + \hat{\theta}_{1\ell}(t) + \theta_{2\ell}] + n_{2\ell}(t) \tag{3-60}$$

where $\omega_{2\ell}$ is the radian frequency of the carrier component, $\hat{\theta}_{1\ell}(t)$ is the phase jitter on the downlink carrier due to the additive noise on the uplink, $\theta_{2\ell}$ is the phase shift on the downlink, and $n_{2\ell}(t)$ is narrowband white Gaussian noise of single-sided spectral density N_{02} w/Hz. Thus, the observed waveform appearing at the output of the signal combiner (Fig. 3-17) is given by

$$\xi(t) = \sum_{\ell=1}^{L} \xi_\ell(t) \tag{3-61}$$

which becomes, upon using Eq. (3-60) with $\hat{\theta}_{1\ell}(t) = \hat{\theta}_1(t)$, $\theta_{2\ell} = \theta_2$, and $\omega_{2\ell} = \omega_2$ for all values of $\ell = 1, 2, \ldots, L$,

$$\xi(t) = \sqrt{2L^2 P_{c2}} \sin [\omega_2 t + \hat{\theta}_1(t) + \theta_2] + n_{2L}(t) \tag{3-62}$$

where $n_{2L}(t)$ is white Gaussian noise of single-sided spectral density LN_{02} w/Hz. Thus, combining has increased the signal power by a factor of L^2 while the noise power per unit bandwidth is increased only by the factor L.

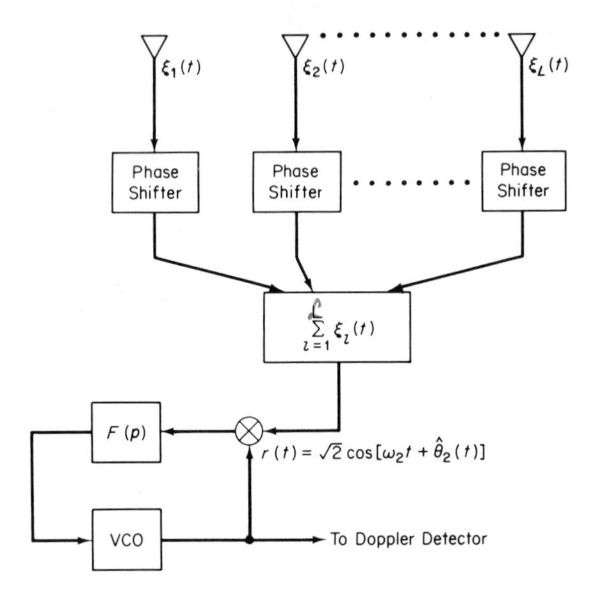

Fig. 3-17 Basic signal combining method for a two-way Doppler and RF phase measuring system employing antenna diversity

Improvements Realized in Phase and Doppler Measurements with the Use of Diversity Techniques

Comparing (3-62) and (3-6) it is a simple matter to generalize the results of Secs. 3-2 and 3-3 to a diversity combining system employing L antennas. In fact, one merely replaces ρ_2 by $L\rho_2$, and the argument of Γ_2, i.e. x_2y_2 by Lx_2y_2 in all equations in Secs. 3-2 and 3-3 to validate them for predicting L-fold diversity performance. As an example, we give the Doppler results based on nonlinear theory where the carrier tracking loops are preceded by BPLs. The p.d.f. and variance of ϕ_{d2} are then

$$p(\phi_{d2}) = \frac{I_0(|d_{1\ell} + \rho_{2\ell}\exp(j\phi_{d2})|)}{2\pi I_0(d_{1\ell})I_0(\rho_{2\ell})} \qquad |\phi_{d2}| \leq \pi \tag{3-63}$$

$$\sigma_{d2}^2 = \frac{\pi^2}{3} + 4\sum_{k=1}^{\infty}\frac{(-1)^k}{k^2}\frac{I_k(d_{1\ell})I_k(\rho_{2\ell})}{I_0(d_{1\ell})I_0(\rho_{2\ell})} \tag{3-64}$$

where

$$d_{1\ell} = \frac{2P_{c1}}{N_{01}W_{01}}\frac{1}{G^2\Gamma_{p1}K_D(k_1, k_2, \xi)}$$
$$\rho_{2\ell} = \frac{2LP_{c2}}{N_{02}W_{02}\Gamma_{p2}}\left[\frac{1 + r_{20}}{1 + r_{20}/\mu_2}\right] \tag{3-65}$$

and

$$\Gamma_{p2} = \frac{1 - e^{-Lx_2y_2}}{\begin{array}{r}(L\pi x_2 y_2/4)e^{-Lx_2y_2}[I_0(Lx_2y_2/2) + I_1(Lx_2y_2/2)]^2 \\ [1 + (4/\Gamma_0\pi - 1)e^{-Lx_2y_2(1-\pi/4)}]\end{array}} \tag{3-66}$$

All other parameters have been defined in Sec. 3-3.

3-6. GENERALIZATION OF TWO-WAY PHASE AND DOPPLER MEASUREMENTS TO AN *N*-STEP NETWORK

In this section, we present a probability model for the phase and Doppler errors associated with a set $l = \{l_1, l_2, \ldots, l_n\}$ of PLLs in cascade that are perturbed by statistically independent, additive, white Gaussian noise sources. Such an arrangement of PLLs occurs in communication networks required to track distant vehicles. In an Earth-to-planet-to-Earth type network, for example, the set $l = \{l_1, l_2, l_3\}$ of loops is required. For the special case of a two-link network, the results reduce to those obtained in the previous sections of this chapter.

Among other things, the probability model developed here is useful in later discussions where it is necessary to specify the performance of phase-coherent communication links having noisy reference signals, and for optimal allocation of the total transmitter power between the carrier and one or more of its sidebands.

The problem of Doppler and phase measurement may be abstracted into statistical communication theory as a problem of channel measurement. Knowledge of the channel may be conveniently divided into two types: *a priori*, usually based on a physical model of the channel; and *a posteriori*, based on channel soundings, that is, measurement of the channel characteristics by means of probing signals.

In this section, we assume a communication network consisting of n receivers and n transponders—which are moving with respect to the origin of some suitably defined coordinate system (Fig. 3-18). The n transponders are to be used to perform integrated Doppler and phase measurements. In particular, the purpose of the ith ($i = 0, 1, 2, \ldots, n - 1$) receiver is to provide the ith transponder with an *a posteriori* measurement of the phase and frequency of the observed data arriving from the $i - 1$st transponder. This measurement is to be used as a source of energy to phase modulate the carrier stored in the ith transponder. The resulting waveform is to be relayed to the $i + 1$st receiver in the network and an *a posteriori* measurement of the received data phase is then made by the $i + 1$st receiver. This measurement serves as a source of information for phase

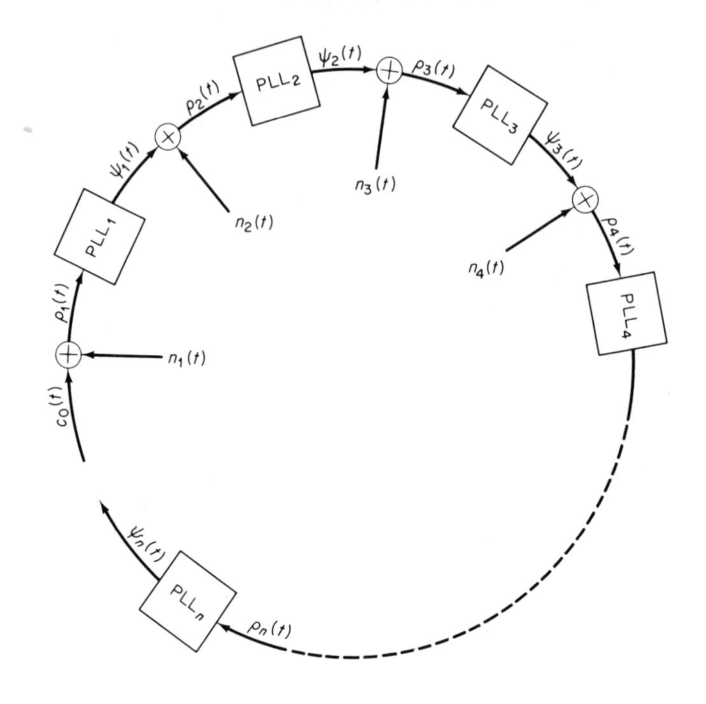

Fig. 3-18 An n-step communication network

modulating the $i + 1$st carrier stored in the $i + 1$st transponder. The process is repeated sequentially until the nth phase measurement is complete, and the result of the nth measurement is subsequently compared with the phase and frequency of a master reference carrier (usually located on the ground) for final processing and ultimate Doppler readout. Such Doppler-measuring networks have the distinct advantage of providing the reference receiver with a continuous and up-to-date Doppler reading, as well as compensating for any instabilities in equipment oscillators. An accompanying disadvantage, however, is that if any one phase or Doppler measurement is poor, the resulting measurement will be poor.

The network model may be briefly described as follows. Stored in the ith transponder is the carrier waveform, $c_i(t) = \sqrt{2P_{ci}} \cos \omega_i t$ possessing an average power of P_{ci} watts. The waveform $c_0(t)$ will be taken as the system reference. The phase measurement $\hat{\theta}_i(t)$ of $\theta_i(t)$ derived from the $i - 1$st piece of observed data is used to phase modulate $c_i(t)$, producing the emitted waveform

$$\rho_{i+1}(t) = \sqrt{2P_{c,i+1}} \cos [\omega_i t + \hat{\theta}_i(t)] \qquad i = 0, 1, \ldots, n - 1 \tag{3-67}$$

In passing through the ith channel, the waveform $\rho_{i+1}(t)$ is Doppler-shifted, phase modulated by the channel phase characteristic $\theta_{i+1}(t)$, and further perturbed by additive white Gaussian noise $n_{i+1}(t)$ of single-sided spectral density $N_{0,i+1}$ w/Hz.

The signal received at the $i+1$st observation point in the network may be written as

$$\psi_{i+1}(t) = \sqrt{2P_{c,i+1}} \cos [\omega_{i+1}t + \hat{\theta}_i(t) + \theta_{i+1}(t)] + n_{i+1}(t) \qquad i = 0, 1, \ldots, n-1 \tag{3-68}$$

where $\theta_{i+1}(t)$ is assumed to be a slowly varying random process. By "slowly-varying" we mean that the process $\theta_{i+1}(t)$ does not appreciably change during an observation interval of T sec, or more specifically, that the variations of $\theta_{i+1}(t)$ are slow when compared to the transient response of the measuring device. Because the measurement is made in the presence of noise, the *a posteriori* knowledge will not be exact; all we can hope for is an *a posteriori* p.d.f. of the characteristics of the medium, to be derived in the $i+1$st receiver by some operation on $\theta_{i+1}(t)$, assuming that the $i+1$st receiver knows ω_{i+1} exactly. For brevity, we denote this p.d.f. by

$$p(\hat{\theta}_{i+1} \mid \theta_{i+1}, \rho_{i+1}, \psi_{i+1}) = p_{\psi_{i+1}}(\hat{\theta}_{i+1} \mid \hat{\theta}_i) \tag{3-69}$$

where $\hat{\theta}_{i+1}$ is the *a posteriori* estimate of $\hat{\theta}_1 + \theta_{i+1}$. The *a posteriori* p.d.f. $p_{\psi_{i+1}}(\hat{\theta}_{i+1} \mid \hat{\theta}_i)$ is derived in the $i+1$st receiver by means of a PLL that tracks the incoming signal $\psi_{i+1}(t)$. In each case, the *a posteriori* statistics resulting from the ith measurement become the *a priori* statistics for use in measuring the phase shift at the $i+1$st phase-locked receiver. We assume that the *a posteriori* p.d.f. (for frequency lock conditions) of the $i+1$st measurement is given by

$$p_{\psi_{i+1}}(\hat{\theta}_{i+1} \mid \hat{\theta}_i) = \frac{\exp [u_{i+1} \cos (\hat{\theta}_i + \theta_{i+1} - \hat{\theta}_{i+1})]}{2\pi I_0(u_{i+1})} \tag{3-70}$$

where u_{i+1} is the parameter that characterizes the p.d.f. This assumption is predicated upon results obtained from the PLL measurements given in Chap. 2. For the present, we ignore the problem of relating the measurement parameter u_{i+1} to the parameters in the network; suffice it to say, however, that u_{i+1}, $i = 1, \ldots, n-1$, may be related to the signal powers $p_{c,i+1}$, the channel noises $N_{0,i+1}$, and the bandwidths of the carrier-tracking loops.

Looking now at an n-step network, the joint p.d.f. of the phase (or Doppler) measurements is

$$p_n(\hat{\theta}_1, \hat{\theta}_2, \ldots, \hat{\theta}_n) = \prod_{i=1}^{n-1} p_{\psi_{i+1}}(\hat{\theta}_{i+1} \mid \hat{\theta}_i) p_1(\hat{\theta}_1) \tag{3-71}$$

where $p(\hat{\theta}_1)$ is exactly given by

$$p(\hat{\theta}_1) \triangleq p_1(\hat{\theta}_1) = \frac{\exp [u_1 \cos (\theta_0 + \theta_1 - \hat{\theta}_1)]}{2\pi I_0(u_1)} \tag{3-72}$$

and $p_{\psi_{i+1}}(\hat{\theta}_{i+1} \mid \hat{\theta}_i)$ is defined in Eq. (3-70).

Substituting (3-70) and (3-72) into (3-71) gives

$$p_n(\hat{\theta}_1, \hat{\theta}_2, \ldots, \hat{\theta}_n) = \frac{\exp\left[\sum_{i=1}^{n} u_i \cos(\hat{\theta}_{i-1} + \theta_i - \hat{\theta}_i)\right]}{(2\pi)^n \prod_{i=1}^{n} I_0(u_i)} \tag{3-73}$$

where for simplicity of notation we set $\hat{\theta}_0 = \theta_0$. The p.d.f. for the round-trip measurement may be derived from (3-73) by integrating out the random variables $\hat{\theta}_1, \hat{\theta}_2, \ldots, \hat{\theta}_{n-1}$, that is,

$$p(\hat{\theta}_n) = \underbrace{\int_0^{2\pi}\int_0^{2\pi}\cdots\int_0^{2\pi}}_{(n-1)\text{-fold}} p_n(\hat{\theta}_1, \hat{\theta}_2, \ldots, \hat{\theta}_n)\, d\hat{\theta}_1 d\hat{\theta}_2 \cdots d\hat{\theta}_{n-1} \tag{3-74}$$

If we define the total error in the phase measurement by

$$\theta_E = \sum_{i=0}^{n} \theta_i - \hat{\theta}_n \tag{3-75}$$

and make use of the Jacobian of the transformation, then the p.d.f. of the total error $q(\theta_E)$ becomes

$$q(\theta_E) \triangleq p\left(\sum_{i=0}^{n} \theta_i - \theta_E\right) \tag{3-76}$$

where $p(x)$ is the result of the $(n-1)$-fold integration in (3-74). This $(n-1)$-fold integration on $p_n(\hat{\theta}_1, \hat{\theta}_2, \ldots, \hat{\theta}_n)$ as defined in (3-73) can be evaluated and a closed-form expression obtained for $q(\theta_E)$ by deriving the results for $n = 2, 3$, and then generalizing by mathematical induction to the case of arbitrary n. Thus, for $n = 2$,

$$p_2(\hat{\theta}_1, \hat{\theta}_2) = \frac{\exp[u_1 \cos(\theta_0 + \theta_1 - \hat{\theta}_1) + u_2 \cos(\hat{\theta}_1 + \theta_2 - \hat{\theta}_2)]}{(2\pi)^2 I_0(u_1) I_0(u_2)} \tag{3-77}$$

and

$$p(\hat{\theta}_2) = \int_0^{2\pi} p_2(\hat{\theta}_1, \hat{\theta}_2) d\hat{\theta}_1 = \frac{I_0[\sqrt{u_1^2 + u_2^2 + 2u_1u_2\cos(\theta_0 + \theta_1 + \theta_2 - \hat{\theta}_2)}]}{2\pi I_0(u_1) I_0(u_2)} \tag{3-78}$$

Hence, from (3-76),

$$q(\theta_E) = \frac{I_0[\sqrt{u_1^2 + u_2^2 + 2u_1u_2\cos\theta_E}]}{2\pi I_0(u_1) I_0(u_2)} \tag{3-79}$$

The first and second moments of the measurement are of interest. Using Neumann's addition theorem, it is easy to show that the mean of θ_E is zero while the variance is

$$\sigma_{\theta_E}^2 = \frac{\pi^2}{3} + 4\sum_{k=1}^{\infty} \frac{(-1)^k}{k^2} \frac{I_k(u_1) I_k(u_2)}{I_0(u_1) I_0(u_2)} \tag{3-80}$$

Note that the first term in this expression is the variance of a uniformly distributed random variable. The cumulative probability distribution of the absolute value of the total phase error

$$\text{Prob}[|\theta_E| < \theta] = \int_{-\theta}^{\theta} q(\theta_E)\, d\theta_E \tag{3-81}$$

is also of interest since it indicates the fraction of time during which the absolute value of the total phase error in the two-way measurement is less than a given value, say θ. Substituting and integrating gives

$$\text{Prob}[|\theta_E| < \theta] = \frac{\theta}{\pi}\left[1 + 2\sum_{k=1}^{\infty} \frac{I_k(u_1)I_k(u_2)}{I_0(u_1)I_0(u_2)}\left(\frac{\sin k\theta}{k\theta}\right)\right] \tag{3-82}$$

Proceeding now to the case $n = 3$,

$$\begin{aligned} p(\hat{\theta}_3) &= \int_0^{2\pi}\int_0^{2\pi} p_3(\hat{\theta}_1, \hat{\theta}_2, \hat{\theta}_3)\, d\hat{\theta}_1 d\hat{\theta}_2 = \int_0^{2\pi}\int_0^{2\pi} p_{\psi_3}(\hat{\theta}_3 | \hat{\theta}_2) p_2(\hat{\theta}_1, \hat{\theta}_2)\, d\hat{\theta}_1 d\hat{\theta}_2 \\ &= \int_0^{2\pi} p_{\psi_3}(\hat{\theta}_3 | \hat{\theta}_2) p(\hat{\theta}_2)\, d\hat{\theta}_2 \end{aligned} \tag{3-83}$$

where $p(\hat{\theta}_2)$ is given by (3-78) and $p_{\psi_3}(\hat{\theta}_3 | \hat{\theta}_2)$ by (3-70) with i set equal to 2. Substituting into Eq. (3-83) and integrating out the random variable $\hat{\theta}_2$ gives

$$\begin{aligned} p(\hat{\theta}_3) &= \int_0^{2\pi} \frac{I_0[\sqrt{u_1^2 + u_2^2 + 2u_1u_2 \cos(\theta_0 + \theta_1 + \theta_2 - \hat{\theta}_2)}]}{(2\pi)^2 I_0(u_1)I_0(u_2)I_0(u_3)} \exp[u_3 \cos(\hat{\theta}_2 + \theta_3 - \hat{\theta}_3)]\, d\hat{\theta}_2 \\ &= [2\pi I_0(u_1)I_0(u_2)I_0(u_3)]^{-1} \sum_{k=0}^{\infty} \epsilon_k I_k(u_1)I_k(u_2)I_k(u_3) \cos k(\theta_0 + \theta_1 + \theta_2 + \theta_3 - \hat{\theta}_3) \end{aligned} \tag{3-84}$$

where $\epsilon_k = 1$ if $k = 0$ and $\epsilon_k = 2$ if $k \neq 0$. It follows immediately from (3-84) that

$$q(\theta_E) = [2\pi I_0(u_1)I_0(u_2)I_0(u_3)]^{-1} \sum_{k=0}^{\infty} \epsilon_k I_k(u_1)I_k(u_2)I_k(u_3) \cos k\theta_E \tag{3-85}$$

The mean of the total phase error θ_E is zero while the variance of the measurement is

$$\sigma_{\theta_E}^2 = \frac{\pi^2}{3} + 4\sum_{k=1}^{\infty} \frac{(-1)^k}{k^2} \frac{I_k(u_1)I_k(u_2)I_k(u_3)}{I_0(u_1)I_0(u_2)I_0(u_3)} \tag{3-86}$$

If $u_1 = 0$, $u_2 = 0$, or $u_3 = 0$, then $\sigma_{\theta_E}^2 = \pi^2/3$, which is the variance of a uniform distribution. Thus the *a posteriori* p.d.f. of the phase error measurement becomes uniformly distributed if any one of the sublinks becomes excessively noisy. This corresponds to the physical situation where any one of the three measuring devices thresholds or loses lock.

Consider the situation where any one of the three sublinks becomes non-noisy as, for example, when we let N_{03} approach zero. Hence, $\hat{\theta}_3$

approaches $\hat{\theta}_2 + \theta_3$ and (3-85) becomes

$$q(\theta_E) = [2\pi I_0(u_1)I_0(u_2)]^{-1} \sum_{k=0}^{\infty} \epsilon_k I_k(u_1)I_k(u_2) \cos k\theta_E \tag{3-87}$$

where we have made use of the asymptotic formula $\lim_{x\to\infty} I_k(x) \sim \exp(x)/\sqrt{2\pi x}$. Using Neumann's addition theorem, (3-87) becomes

$$q(\theta_E) = \frac{I_0(\sqrt{u_1^2 + u_2^2 + 2u_1u_2 \cos \theta_E})}{2\pi I_0(u_1)I_0(u_2)} \tag{3-88}$$

which is the p.d.f. of the two-link measurement [see (3-79)]. This result says that any non-noisy link does not affect the total phase error.

The probability that the absolute value of the total three-way phase error $\theta_E = (\theta_0 + \theta_1 + \theta_2 + \theta_3) - \hat{\theta}_3$ remains less than θ rad is

$$\text{Prob}[|\theta_E| < \theta] = \frac{\theta}{\pi}\left[1 + 2\sum_{k=1}^{\infty} \frac{I_k(u_1)I_k(u_2)I_k(u_3)}{I_0(u_1)I_0(u_2)I_0(u_3)}\left(\frac{\sin k\theta}{k\theta}\right)\right] \qquad 0 \le \theta \le \pi \tag{3-89}$$

Finally, for the n-step network, the error statistics of the nth phase measurement θ_E become

$$q(\theta_E) = \left[2\pi \prod_{i=1}^{n} I_0(u_i)\right]^{-1} \sum_{k=0}^{\infty} \epsilon_k \prod_{i=1}^{n} I_k(u_i) \cos k\theta_E \tag{3-90}$$

$$\bar{\theta}_E = 0$$

$$\sigma_{\theta_E}^2 = \frac{\pi^2}{3} + 4\sum_{k=1}^{\infty} \frac{(-1)^k}{k^2} \frac{\prod_{i=1}^{n} I_k(u_i)}{\prod_{i=1}^{n} I_0(u_i)} \tag{3-91}$$

The probability that the absolute value of the total phase error of the n-step network remains less than θ rad is

$$\text{Prob}[|\theta_E| < \theta] = \frac{\theta}{\pi}\left[1 + 2\sum_{k=1}^{\infty} \frac{\prod_{i=1}^{n} I_k(u_i)}{\prod_{i=1}^{n} I_0(u_i)}\left(\frac{\sin k\theta}{k\theta}\right)\right] \tag{3-92}$$

The statistics of the total phase error are of interest when $n = \infty$. Since $I_k(x) < I_0(x)$ for all $k \ge 1$, we have

$$\prod_{i=1}^{n} \frac{I_k(u_i)}{I_0(u_i)} < 1 \qquad k \ge 1$$

Hence

$$\lim_{n\to\infty} \prod_{i=1}^{n} \frac{I_k(u_i)}{I_0(u_i)} = 0 \qquad \text{and} \qquad \lim_{n\to\infty} q(\theta_E) = \frac{1}{2\pi} \tag{3-93}$$

which says that the p.d.f. of the total phase error becomes uniformly distributed.

Several interesting and general conclusions may be made about the network from the error statistics of (3-90). If any one of the $u_i (i = 1, \ldots, n)$ becomes small at any time during the measurement procedure, then $q(\theta_E)$

becomes uniformly distributed and the resulting measurement is not a useful one. This corresponds to the physical situation where any one of the PLLs in the network thresholds and no phase information may be extracted from the measurement. Further, if any $(n - k)$ of the u_i becomes excessively large (large signal-to-noise ratios in the tracking loops), these $(n - k)$ links may be dropped from the network since they do not degrade the phase measurement at all. The disturbances on the remaining k links, however, determine the total phase error. To obtain the performance of an n-step system one must evaluate (3-90), (3-91), and (3-92) for various values of the measurement parameters $(u_1, u_2, \ldots, u_n)$.

The p.d.f. of θ_E as computed from (3-90) is plotted in Fig. 3-19 versus

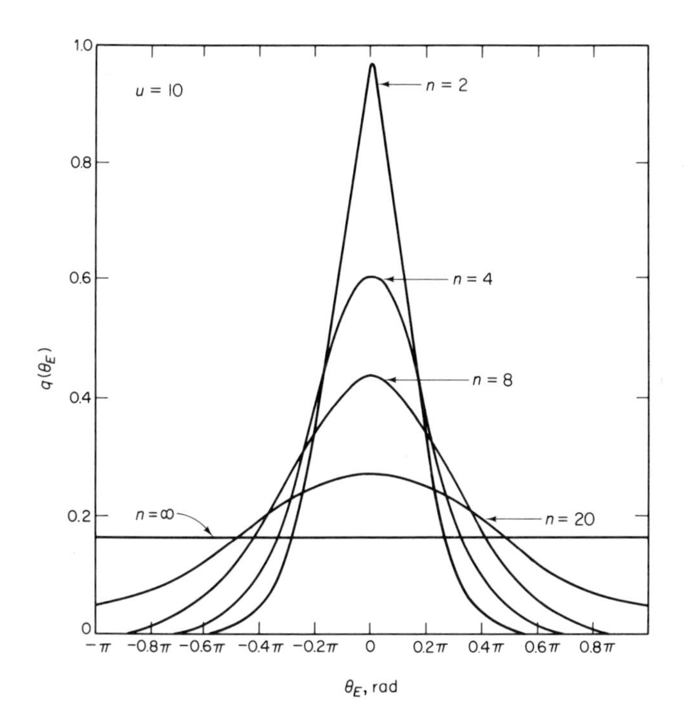

Fig. 3-19 Total phase error probability density function ($u = 10$)

the total phase error θ_E for various values of the parameter n and $u_1 = u_2 = \cdots = u_n = u = 10$. Notice how the distribution becomes more flat as n changes from 2 to infinity. This increasing "flatness" is, of course, due to a larger variance in the measurement.

Figure 3-20 illustrates the behavior of $\sigma^2_{\theta_E}$, as determined from Eq. (3-91), as a function of the reciprocal of the common measurement parameter u for various values of the parameter n. For comparison purposes, we superimpose the corresponding results that would be obtained if the linear model

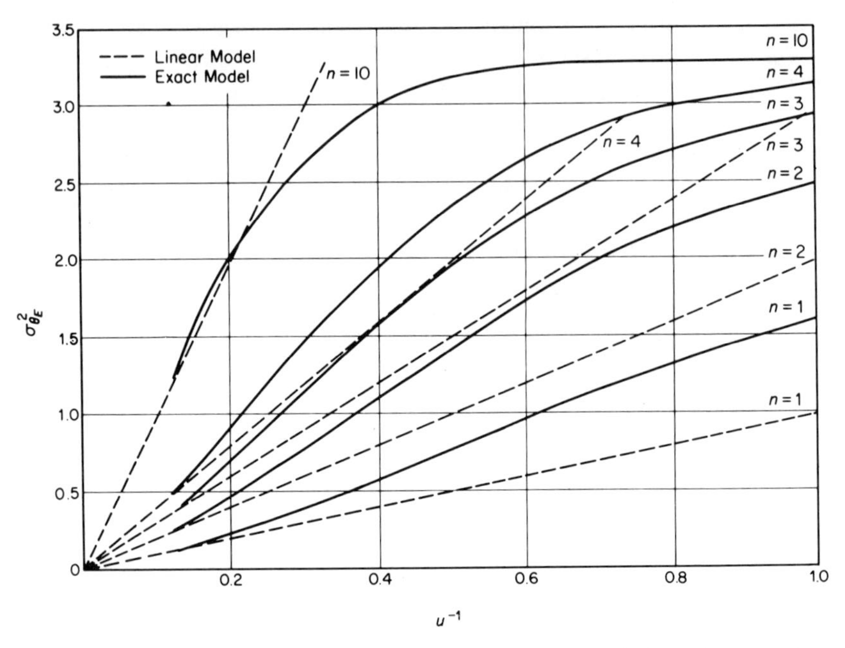

Fig. 3-20 Variance of the total phase error as a function of the reciprocal of the common measurement parameter

of the PLL were employed. Note that for values of the measurement parameter $u \geq 5$ the linear model is quite good. Figures 3-21, 3-22, and 3-23 are plots of the cumulative distribution of θ_E as given by Eq. (3-92) with u as a parameter.

In order to use the two-way n-step theory just developed in a Doppler or phase measurement system it is necessary to relate the measurement parameters u_i, $i = 1, 2, \ldots, n$, to the basic parameters associated with the carrier-tracking loops. For the case of Doppler measurements $u_i = d_i$, and using linear PLL theory,

$$\frac{1}{d_i} = \frac{1}{2\pi j} \int_{-j\infty}^{j\infty} \frac{N_{0i}}{2P_{ci}} \prod_{i=k}^{n} G_i^2 \, |H_i(s)|^2 \, ds \tag{3-94}$$

for $i = 1, 2, \ldots, n$. The parameters G_i, $i = 1, 2, \ldots, n$, are respectively the static phase gains of the n transponders. Therefore, the variance of the Doppler error as determined from the linear PLL theory is

$$\sigma^2_{\varphi_{d2}}(n) = \sum_{i=1}^{n} d_i^{-1} \tag{3-95}$$

The variance as determined from the nonlinear model is given by (3-91) with u_i replaced by d_i which are found from (3-94) with $G_i = 1, i = 1, 2, \ldots, n$.

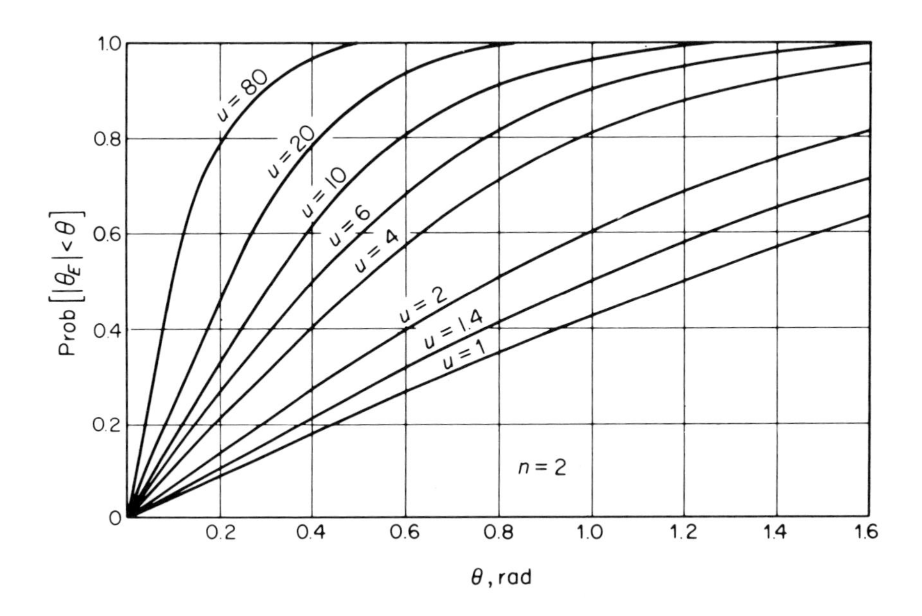

Fig. 3-21 Cumulative probability distribution function for θ_E; $n = 2$

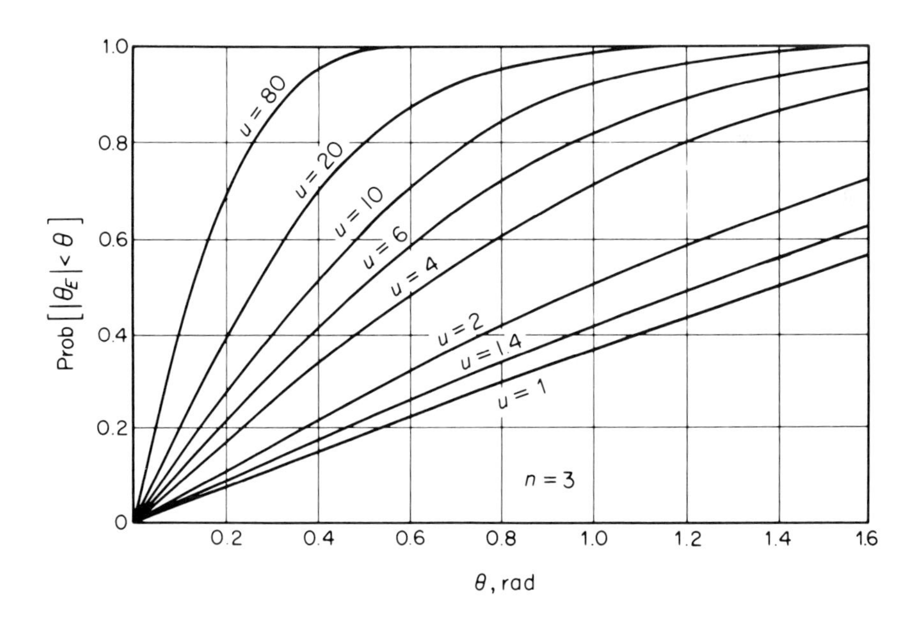

Fig. 3-22 Cumulative probability distribution function for θ_E; $n = 3$

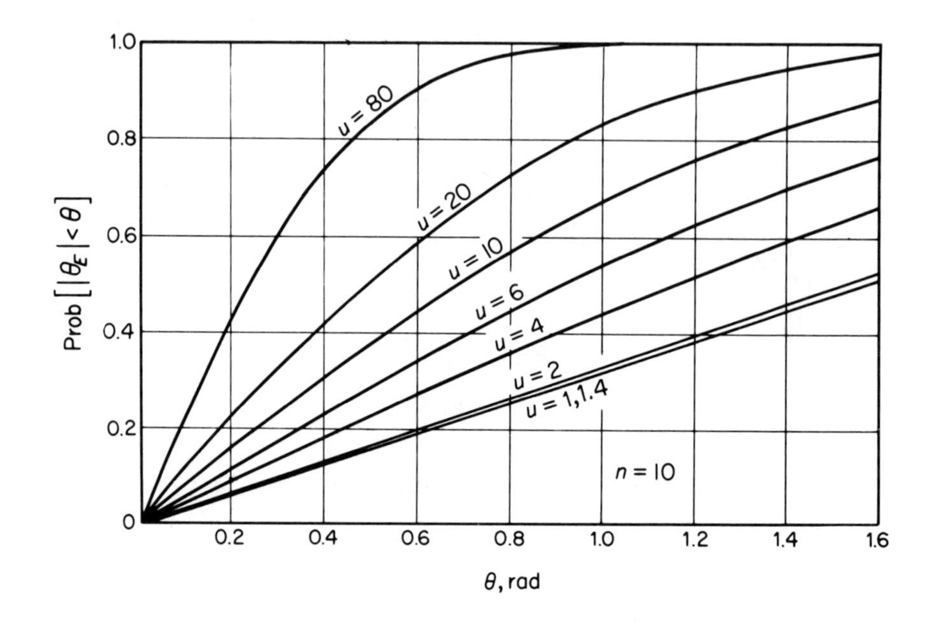

Fig. 3-23 Cumulative probability distribution function for θ_E; $n = 10$

For $n = 2$ and ρ_2 denoted by d_2, these results check with those given in Sec. 3-3.

For the case of phase measurements, the model p.d.f. for the phase error is obtained from Eq. (3-90) by replacing the u_i by α_i. In this case, we have

$$\alpha_i^{-1} = \frac{1}{2\pi j}\int_{-j\infty}^{j\infty} \frac{N_{0i}}{2P_{ci}} \prod_{\ell=1}^{n-1} |G_\ell H_\ell(s)|^2 \, |1 - H_n(s)|^2 \, ds \qquad i = 1, 2, \ldots, n-1 \tag{3-96}$$

$$\alpha_n^{-1} = \frac{1}{2\pi j}\int_{-j\infty}^{j\infty} \frac{N_{0n}}{2P_{cn}} |H_n(s)|^2 \, ds \tag{3-97}$$

and from the linear theory, the variance of the phase error is

$$\sigma^2_{\varphi_{r2}}(n) = \sum_{i=1}^{n} \alpha_i^{-1} \tag{3-98}$$

The specification of the distribution parameters for phase and Doppler measurement systems in which the carrier-tracking loops are preceded by BPLs follows directly from the results of Secs. 3-2 and 3-3.

Finally, one can compute the carrier-suppression factor for an n-step network by an extension of the results of Sec. 3-4. For example, using the nonlinear theory and no BPLs preceding the carrier-tracking loops, the suppression factor is given by

$$\tilde{S} = \prod_{i=1}^{n} \frac{I_1(\rho_i)}{I_0(\rho_i)} \tag{3-99}$$

where $\rho_i = 2P_{ci}/N_{0i}W_{Li}$. In arriving at (3-99), we have used (3-90) and (3-75) to characterize the probability density of $\hat{\theta}_n$ and have assumed that $G_i = 1$, $i = 1, 2, \ldots, n$.

PROBLEMS

3-1. A coherent, two-way tracking system operates with the following parameters:

Vehicle System	*Ground System*
$W_{L1} = 18$ Hz	$W_{L2} = 12$ Hz
$\sqrt{P_{c_1}}K_1 = 39{,}780\ \text{sec}^{-1}$	$\sqrt{P_{c_2}}K_2 = 9{,}600\ \text{sec}^{-1}$
$\tau_{21} = 0.0833$ sec	$\tau_{22} = 0.1388$ sec
$r_1 = 2$	$r_2 = 2$

a) Find the vehicle's loop filter time constant τ_{11}, and the ground receiver's loop filter time constant, τ_{12}.
b) Evaluate $K_R(r_1, r_2, \xi)$ and $K_D(r_1, r_2, \xi)$.
c) If the static phase gain of the coherent transponder is $G = 1.1$ and $\rho_1 = \rho_2 = 6$, find the mean-squared value of the ground receiver's tracking error and the mean-squared value of the Doppler error. Use linear theory. Repeat with $\rho_1 = 20$ and $\rho_2 = 10$.

3-2. Repeat part (c) of Problem 3-1 using the nonlinear theory and compare your answers.

3-3. Using the closed-loop transfer functions of the ground and vehicle receivers' carrier-tracking loops as defined in Eq. (3-9) and performing the complex integrations as required in (3-15) and (3-37), develop the expressions for $K_R(r_1, r_2, \xi)$ and $K_D(r_1, r_2, \xi)$ as given by (3-18) and (3-40), respectively.

3-4. Show that in the limit as $r_1 = r_2 = r$ approaches infinity corresponding to a two-way system in which both the vehicle and ground receivers employ first-order tracking loops,

$$\lim_{r\to\infty} K_D(r, r, \xi) = \lim_{r\to\infty} \frac{K_R(r, r, \xi)}{\xi} = \frac{1}{1+\xi}$$

3-5. If the vehicle and ground receivers' carrier-tracking loops are preceded by BPLs, then using the theory developed in Sec. 3 of Chap. 2 and the definitions of k_n and ξ which follow Eq. (3-31), verify, without resorting to complex integration, the expressions for $K_R(k_1, k_2, \xi)$ and $K_D(k_1, k_2, \xi)$ as given, respectively, by (3-31) and (3-46).

3-6. A coherent, two-way tracking system with BPLs preceding the carrier-tracking loops operates with the following parameters:

Vehicle System		*Ground System*	
W_{01}	$= 20$ Hz	W_{02}	$= 20$ Hz
W_{i1}	$= 4.8$ kHz	W_{i2}	$= 4.8$ kHz
P_{c1}/N_{01}	$= 13$ dB	P_{c2}/N_{02}	$= 20$ dB
r_{10}	$= 2$	r_{20}	$= 2$

a) Recalling that $x_1 = x_2 = 1$ at the design point, find μ_1, μ_2, k_1, k_2, and ξ.

b) Evaluate $K_R(k_1, k_2, \xi)$ and $K_D(k_1, k_2, \xi)$.

c) If the static phase gain of the coherent transponder is $G = 1.1$, find the mean-squared value of the ground receiver's tracking error and the mean-squared value of the Doppler error. Use linear theory.

3-7. Repeat part (c) of Problem 3-6 using the nonlinear theory and compare your answers.

3-8. Applying Neumann's addition theorem (Watson, G. N., *Theory of Bessel Functions*, 2nd ed., Cambridge University Press, 1962, p. 358):

$$I_0(\sqrt{x^2 + y^2 + 2xy\cos\theta}) = \sum_{k=0}^{\infty} \epsilon_k I_k(x) I_k(y) \cos k\theta \qquad \begin{array}{l}\epsilon_0 = 1\\ \epsilon_k = 2;\ k \geq 1\end{array}$$

to the p.d.f.'s of the two-way modulo 2π reduced phase errors ϕ_{r2} and ϕ_{d2}, develop the expressions for $\sigma^2_{\phi_{r2}}$ and $\sigma^2_{\phi_{d2}}$ as given by (3-25) and (3-43), respectively.

3-9. In a coherent, two-way tracking system, assume that the vehicle and ground receivers' tracking errors θ_1 and θ_2, respectively, have first-order and conditional p.d.f.'s which can be modeled as

$$p(\theta_1) = \frac{\exp(x\cos\theta_1)}{2\pi I_0(x)} \qquad |\theta_1| \leq \pi$$

and

$$p(\theta_2 | \theta_1) = \frac{\exp[y\cos(\theta_2 - \theta_1)]}{2\pi I_0(y)}; \qquad |\theta_2| \leq \pi$$

a) Using the Jacobi-Anger formula

$$\exp(z\cos\theta) = \sum_{k=0}^{\infty} \epsilon_k I_k(z)\cos(k\theta) \qquad \begin{array}{l}\epsilon_0 = 1\\ \epsilon_k = 2;\ k \geq 1\end{array}$$

find the p.d.f. $p(\theta_2)$ in series form.

b) Place your answer in closed form by applying Neumann's addition theorem given in Problem 3-8.

c) Using $p(\theta_2)$ as found in part (a), derive an expression for the variance of θ_2; that is, σ_2^2.

3-10. Find the carrier suppression on the downlink carrier of a two-way system if:

a) The transponder static phase gain $G = 1.1$ and $\rho_1 = 6$. Use both the linear and nonlinear PLL theories and compare answers.
b) Repeat part (a) for $\rho_1 = 10$.
c) Repeat parts (a) and (b) for $G = 4$ (for example, S-band uplink and X-band downlink).

3-11. As an alternate to coherent translation, spacecraft transponders often use simple turnaround retransmission methods coupled with some form of limiting. For example, the uplink carrier signal and additive narrowband Gaussian noise are band-pass limited, amplified, and directly retransmitted to the ground receiver. Assuming that the uplink carrier has a constant phase angle, the phase angle of the retransmitted carrier has the first-order p.d.f.

$$p(\theta_m) = \begin{cases} \dfrac{1}{2\pi}\displaystyle\int_0^{\infty} u \exp\left[-\left(\frac{u^2}{2} + \rho_i\right)\right] \exp\left[-\sqrt{2\rho_i}u \cos\theta_m\right] du & |\theta_m| \leq \pi \\ 0 & |\theta_m| > \pi \end{cases}$$

where ρ_i is the uplink carrier-to-noise ratio in the band-pass filter bandwidth. Using this p.d.f. in (3-53),

a) Find the carrier suppression on the downlink carrier.
b) Compare this expression with the square of the signal amplitude suppression factor as given by (2-42).

3-12. Turnaround transponding can also be achieved by phase demodulating the uplink carrier, limiting and filtering, and remodulating on a downlink carrier. If the system is modeled as an ideal limiter followed by a single-pole RC low-pass filter, and if the phase demodulator output is modeled as white noise passed through a single-pole RC low-pass filter, the phase angle on the downlink carrier has a first-order p.d.f. given approximately by

$$p(\theta_m) = \begin{cases} \dfrac{(1 - \theta_m^2)^{k-1}}{B(k, 1/2)} & |\theta_m| \leq 1 \\ 0 & |\theta_m| > 1 \end{cases}$$

where $B(x, y)$ is the beta function and k is a positive parameter depending upon the ratio of the limiter output filter 3 dB bandwidth to the bandwidth of the demodulated process. Find the carrier suppression on the downlink carrier.

3-13. Starting with Eq. (3-77) for the joint p.d.f. of $\hat{\theta}_1$ and $\hat{\theta}_2$ and using the Jacobi-Anger formula given in Problem 3-9 and Neumann's addition theorem given in Problem 3-8, derive $p(\hat{\theta}_2)$ as given by Eq. (3-78).

3-14. Using (3-79) to characterize the first-order p.d.f. of the total phase error θ_E, show that the cumulative probability distribution of a two-way system is given by Eq. (3-82). Extend your result to a three-way system; that is, develop Eq. (3-89).

3-15. Letting (3-90) characterize the p.d.f. of the total phase error θ_E and using (3-75) to define the phase estimate $\hat{\theta}_n$, then letting $\theta_m = \hat{\theta}_n$ in (3-53), derive the carrier suppression for an n-step network as given by (3-99). Assume that all transponder static phase gains have unit value.

REFERENCES

1. LINDSEY, W. C., "Optimal Design of One-Way and Two-Way Phase-Coherent Communication Systems," *IEEE Transactions on Communication Technology*, Vol. COM-14, No. 4 (August, 1966) 418–431.

2. EDELSON, R. E., et al, "Telecommunications System Design Techniques Handbook," Jet Propulsion Laboratories, Pasadena, Calif., Technical Memorandum No. 33–571 (July 15, 1972) 370.

3. LINDSEY, W. C., and YUEN, J. H., "Two-Way Coherent Tracking Systems," *Proceedings of the International Telemetering Conference*, Washington, D.C. (Sept., 1971) 209–217.

4. YUEN, J. H., "Theory of Cascaded and Parallel Tracking Systems with Applications," Doctoral Dissertation, University of Southern California, Los Angeles, Calif., June, 1971.

5. MIDDLETON, D., *An Introduction to Statistical Communication Theory*, McGraw-Hill Book Company, New York, N.Y., 1960.

6. GAGLIARDI, R., "Carrier Suppression in Angle Modulated Telemetry," *IEEE Transactions on Communication Technology*, Vol. COM-19, No. 2 (April, 1971) 236–238.

Chapter 4

RANGE MEASUREMENTS BY PHASE-COHERENT TECHNIQUES

4-1. INTRODUCTION

The problem of measuring the *range* to a distant cooperative vehicle arises in a variety of applications, e.g., tracking a deep space vehicle, missile tracking, trajectory determination of artificial satellites, navigation, surveillance, and time service systems, etc.[1,2,3] For these applications the problem of determining range is equivalent to that of measuring the distance between two points; however, it should be noted that the way in which range data is obtained and used differs from application to application.

A *ranging system* determines range by measuring the time required for a signal to propagate to an object and return. The transmitter, see Fig. 4-1, is modulated with an appropriately designed periodic waveform—the *ranging signal.* The receiver recovers a noise-corrupted, time-delayed, version of the transmitted ranging signal and from this extracts the relative time delay between the transmitted and received signals. This time delay is a measure of the round-trip propagation time and is therefore proportional to range.

Before we present various range measurement techniques it is worthwhile to consider two concrete examples of systems in which ranging is a requirement; namely, multiple *satellite navigation* and multiple *satellite surveillance* systems.

In a multiple satellite navigation system, the ranging problem manifests itself in the determination by the user of his position relative to a satellite constellation whose ephemeris is known (see Fig. 4-2). Since the user makes the range measurement for the purpose of navigating his desired course, he is considered a *passive user*. On the other hand, in a satellite surveillance

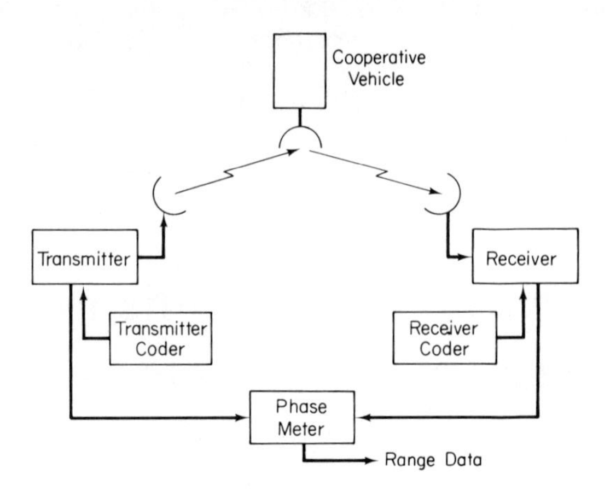

Fig. 4-1 Basic ranging system

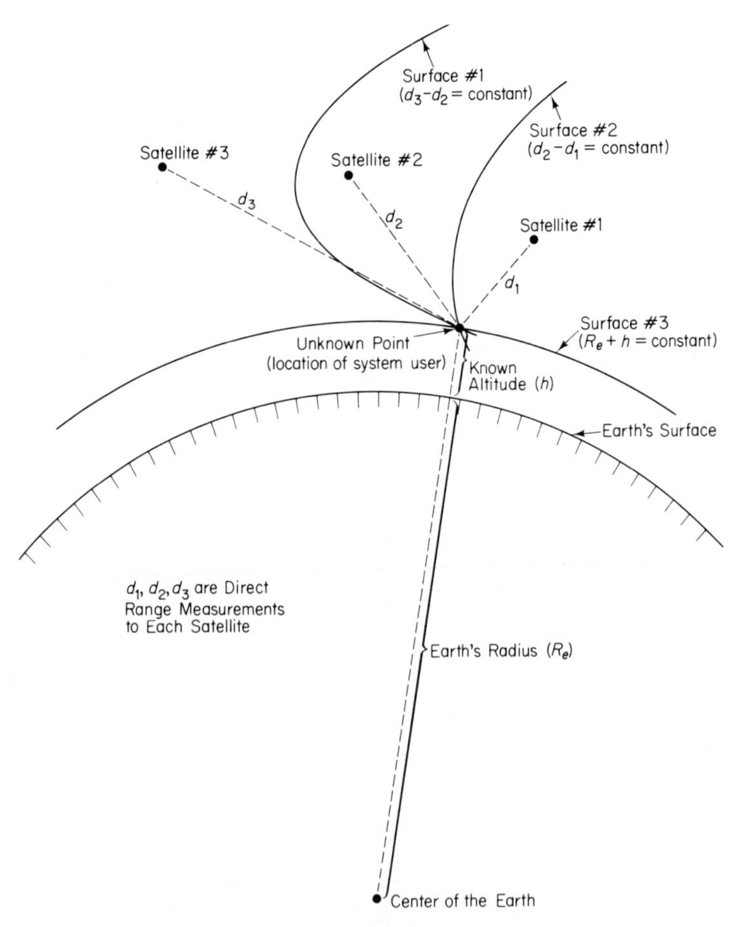

Fig. 4-2 A range difference measurement using three satellites (User's altitude assumed known)

system, the user is considered to be *active* as he himself does not make the range measurement. Instead, the user at his unknown position receives a ranging signal which has been transmitted from a ground station and relayed to him by an appropriate satellite. He then transponds this ranging signal back through the satellite to the ground station from which a round-trip range measurement determines his position. The object of making the range measurement is to determine and monitor the position of various land-based or mobile vehicles, e.g., ship, aircraft, etc. by a central control agency for purposes of traffic (maritime or air) control.

In either system application, the crux of the problem is to determine the position of the cooperating vehicle in a convenient set of coordinates. Position can be described by specifying *altitude*, *latitude*, and *longitude*. The altitude is measured in terms of distance from the center of the Earth. Latitude and longitude are specified as angles which, with the altitude, form a three-dimensional polar coordinate system with coordinate points (d, θ, ϕ). There are several methods in which range measurements can be used in combination with the ephemeris of the satellite constellation to evaluate the position (d, θ, ϕ). These include: (1) *range difference* measurements, (2) *range sum* measurements, and (3) *direct range* measurements.

Range difference measurements from an unknown point (the location of the system user) to a pair of satellites defines a surface in which $\Delta d_1 = d_2 - d_1$ is a constant (see Fig. 4-2). Here d_1 and d_2 respectively denote the distances from the user to each of the two satellites. Since a hyperbola is defined as the locus of a point that moves so that the difference of its distance from two fixed points (foci) is equal to a constant, the surface indicated is a hyperboloid and the satellites are located at its foci. If a third satellite is (see Fig. 4-2) introduced, a range difference measurement, $\Delta d_2 = d_3 - d_2$, involving the unknown point, the third satellite, and either of the other two satellites defines a second hyperboloid. Finally, if the altitude of the system user is known—for example, the radius of the Earth if the user is land based—then a geocentric sphere whose radius equals the user's altitude defines a third surface. The intersection of the two hyperboloids with the geocentric sphere determines a point (d, θ, ϕ) whose coordinates give the location of the system user (Fig. 4-2). If the user's altitude is unknown, then a fourth satellite is required to form a third independent hyperboloid (Fig. 4-3) which determines $\Delta d_3 = d_4 - d_3$. The user's location is now found from the intersection of three hyperbolic surfaces. The method of using *range sum* measurements follows the same lines as that just described for the range difference approach except that the conic surfaces of revolution are now ellipsoids.

The method of *direct range* measurement has the advantage over that of range difference or sums in that one less satellite is required. Each satellite by itself defines a spherical surface, with itself at the center and a radius equal to the range measurement; that is, the distance to the user. The intersection of these three spheres defines a point whose coordinates identify the position

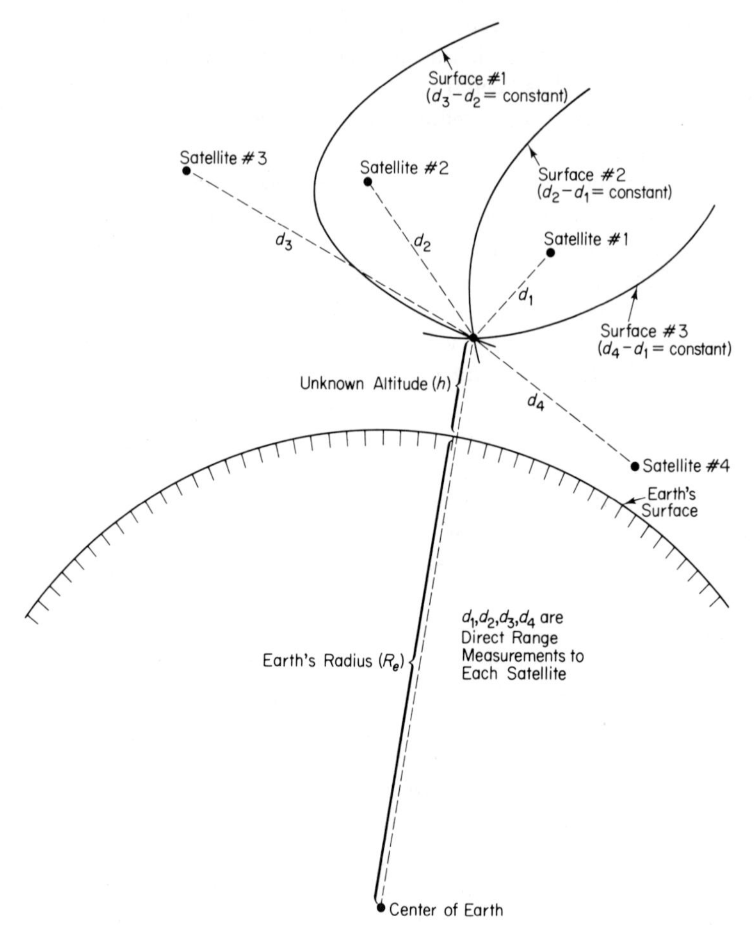

Fig. 4-3 A range difference measurement using four satellites (User's altitude unknown)

of the system user. Use of this method requires time synchronization of clocks located in the satellites. In this chapter we will concentrate on the design of ranging signals and range-tracking receivers without further emphasizing the applications.

4-2. RANGE ESTIMATION

Of basic importance to a discussion of ranging techniques is a theory describing how to estimate range optimally from a noise-corrupted version of the transmitted *ranging signal*, say $s(t)$, that has been transmitted through a medium with *range delay* τ_d. The time delay between the transmitted signal and its return from a cooperating vehicle, at distance d, is related to the range through $\tau_d = 2d/c$. Here c is the velocity of propagation. In what follows, τ_d is assumed to be a random variable characterized by a p.d.f. $p(\tau_d)$ that is uniformly distributed over the interval $(0, \tau_{\max} = 2d_{\max}/c)$. Here, $d_{\max}$ represents the maximum distance at which a range measurement is to be made and $\tau_{\max}$ is the corresponding maximum round-trip propagation time.

To produce an optimum estimate, say $\hat{\tau}_d$, of τ_d one has at his disposal the noise-corrupted received waveform $y(t)$, $0 \leq t \leq T_0$, and a stored replica of the periodic ranging signal $s(t)$ over the same time interval. Selection of the observation interval T_0 is determined by two considerations: one having to do with range accuracy and the other having to do with range ambiguity. To make full use of the distribution of energy designed into the ranging signal, one should observe the received signal over a time interval equal to at least one period T of the transmitted ranging signal. As we shall see shortly, the larger we choose the value of T_0, the more *accurate* the range measurement becomes. Range ambiguity has to do with the problem of distinguishing among intervals of length T in the returned signal. Since we desire an *unambiguous* range measurement, the ranging signal period T must be selected larger than the round-trip propagation time $\tau_{\max}$. Thus, for unambiguous range measurement, observation time T_0 will be greater than $\tau_{\max}$.

The measure of *accuracy* between the true value of delay, say τ_s, and the estimated value $\hat{\tau}_d$ depends on the criterion of optimization chosen. The criterion of optimization to be used here in defining an optimum range estimator is maximization of the *a posteriori* probability $\{p[\tau_d | y(t)], 0 \leq t \leq T_0\}$ of τ_d given $\{y(t), 0 \leq t \leq T_0\}$. The theory, generally referred to as MAP estimation theory, is well documented.[4] Since our interest here does not center on reproducing the theory, rather its applications, we merely extract the significant results. In effect, the optimum estimator is a device which produces, at its output, the function $\{p[\tau_d | y(t)], 0 \leq t \leq T_0\}$ when $y(t)$ is the input.

If the transmitted signal $s(t)$ is corrupted by additive white Gaussian noise $n(t)$, of single-sided spectral density N_0, it has been shown that the evaluation of $p[\tau_d | y(t)], 0 \leq t \leq T_0$ is equivalent to determining the *cross-correlation function*

$$q(\tau_d) = \int_0^{T_0} y(t)s(t - \tau_d)dt \qquad 0 \leq \tau_d \leq \tau_{\max} \tag{4-1}$$

The optimum estimator then selects that value, $\hat{\tau}_d$, which gives rise to the maximum value of $q(\tau_d)$. From (4-1), it is clear that the optimum estimator requires a continuum of cross-correlators.

In the absence of noise, the function $q(\tau_d)$ peaks at a value of $\tau_d = \tau_s$; thus, an exact measurement of true delay τ_s is obtained. When noise is present, the location $\hat{\tau}_d$ of the peak of $q(\tau_d)$ will be displaced from its noise-free value, introducing an error in the measurement of τ_s. Thus, the accuracy of the range estimate can be characterized by the moments of the random variable $\tau_e \triangleq \tau_s - \hat{\tau}_d$. We shall call the quantity τ_e the *range delay error*.

To compute the moments of the cross-correlation $q(\tau_d)$, we assume that a particular observation of the received signal can be expressed in the form

$$y(t) = s(t - \tau_s) + n(t) \qquad 0 \le t \le T_0 \tag{4-2}$$

Substituting this particular observation into (4-1), the function $q(\tau_d)$ can be written in terms of the autocorrelation function $R_s(\tau)$ of the signal and the cross-correlation function $R_{ns}(\tau)$ of the signal and noise. The result is

$$q(\tau_d) = T_0[R_s(\tau_s - \tau_d) + R_{ns}(\tau_d)] \tag{4-3}$$

where

$$R_s(\tau_s - \tau_d) \triangleq \frac{1}{T_0}\int_0^{T_0} s(t - \tau_s)s(t - \tau_d)\,dt$$

$$R_{ns}(\tau_d) \triangleq \frac{1}{T_0}\int_0^{T_0} n(t)s(t - \tau_d)\,dt \tag{4-4}$$

Defining the ratio of the signal energy in the observation interval to the noise spectral density by

$$R_d \triangleq \frac{ST_0}{N_0} = \frac{1}{N_0}\int_0^{T_0} s^2(t)\,dt \tag{4-5}$$

then, from the definition of the range delay error and the assumption that R_d is large, the first two moments of τ_e are approximately given by (see Prob. 4-1)

$$\bar{\tau}_e \cong 0$$

$$\sigma_{\tau_e}^2 \cong \frac{1}{2\beta_s^2 R_d} \qquad R_d \gg 1 \tag{4-6}$$

where

$$\beta_s^2 \triangleq \frac{\left(\frac{1}{2\pi}\right)^2\left\{\int_{-\infty}^{\infty}\int_{-\infty}^{\infty} v^2 S(u)S^*(v)\left\{\frac{\sin[(u - v)T_0/2]}{(u - v)T_0/2}\right\}du\,dv\right\}^2}{S\int_{-\infty}^{\infty}\int_{-\infty}^{\infty} uvS(u)S^*(v)\left\{\frac{\sin[(u - v)T_0/2]}{(u - v)T_0/2}\right\}du\,dv} \tag{4-7}$$

and

$$S(\omega) \triangleq \lim_{T\to\infty}\int_{-T/2}^{T/2} s(t)e^{-j\omega t}\,dt \tag{4-8}$$

The value of β_s is characteristic of the ranging signal chosen and represents a measure of its *frequency spread*. Clearly, from (4-6), one desirable property of a ranging signal is that it should be designed such that β_s is as large as practical considerations will allow.

From our discussions thus far we can suggest other desirable properties of the ranging signal $s(t)$ now. Since $\hat{\tau}_d$ is determined by selecting the value of τ_d

for which $q(\tau_d)$ is a maximum and since $R_{ns}(\tau_d)$ is, in general, independent of τ_d, it is desirable to design the ranging signal such that its autocorrelation function is sharply peaked in the vicinity of $\tau_d = \tau_s$, that is, $R_s(0) \gg R_s(\tau_s - \tau_d)$; $\tau_d \neq \tau_s$. Another desirable property of the autocorrelation function is that its value away from the true delay τ_s should be indicative of which direction one should shift τ_d in order to approach the peak located at τ_s, and perhaps indicate also the magnitude of this required shift. This property aids considerably in acquiring range; that is, in bringing the locally generated signal $s(t - \tau_d)$ into time alignment with the received signal component $s(t - \tau_s)$. Ranging signals whose autocorrelation functions have the latter property are said to be *rapidly acquirable*. Examples of such ranging signals are in the next section.

4-3. RANGING TECHNIQUES

In discussing various ranging techniques that make use of the theory developed, herein, it is often convenient to categorize the technique in terms of the characteristics of the signal. There are several ways to form this classification. In one instance, one might think of dividing ranging techniques into those using pulse-type signals and those of the CW type. Pulse-type signals are used in systems commonly found in radar applications. Ranging systems that employ CW signals have application in navigation and surveillance systems, deep-space missions, and satellite-tracking networks. These signals can be further divided into *fixed-* or *swept-tone*, and digitally coded such as *pseudonoise* (PN) and *rapid acquisition* (BINOR) signals. An alternate method for classifying ranging techniques is to form a dichotomy between ranging systems that transmit *harmonic* signals and those that employ *nonharmonic* signals. In this structure, fixed-tone and swept-tone techniques would fall under the harmonic classification and the others under the nonharmonic classification. Next we discuss ranging techniques in terms of how each satisfies the desirable properties of the signal autocorrelation function that were suggested in Sec. 4-2.

The Fixed-Tone Ranging Technique

First, consider the suitability of using the sinusoid $s(t) = \sqrt{2S} \cos \omega t$ as a ranging signal. The finite time autocorrelation function of $s(t)$ is $R_s(\tau) = S \cos \omega\tau$. As frequency ω is increased, the sharpness of the peaks in $R_s(\tau)$ increases, thereby improving range accuracy; however, increasing the frequency of $s(t)$ also reduces the distance between successive peaks. Thus, for a given value of $\tau_{\max}$, multiple indistinguishable peaks will occur in $R_s(\tau)$ which give rise to the problem of *range ambiguity;* that is, which peak should be selected. Because of this effect, we conclude that a single frequency sinusoid is not sufficient to meet the requirements of a good ranging signal.

To overcome this problem, one could transmit a high-frequency sinusoid to achieve the desired accuracy along with a set of lower frequency sinusoids to determine the correct peak in $R_s(\tau)$. Such an approach is taken in a ranging system that uses the *fixed-tone* technique. In a fixed-tone ranging system, one selects as a ranging signal a properly weighted combination of N sinusoids of different frequencies (called tones) such that the lowest frequency, say $f_1 = \omega_1/2\pi$, satisfies the range ambiguity requirement, i.e., $\tau_{\max} = 2\pi/\omega_1 = 2d_{\max}/c$, whereas the highest frequency, say $f_N = \omega_N/2\pi$, controls the desired accuracy. If $R_s(\tau)$ is to be periodic with period $\tau_{\max}$, then the frequencies of the sinusoidal components must be harmonically related and properly phased. A signal that possesses the above properties is given by

$$s(t) = \sum_{n=1}^{N} a_n \cos \omega_n t \tag{4-9}$$

where $\omega_{n+1}/\omega_n = M$, M is any integer greater than 1, and the a_n's are chosen so as to apportion the total transmitted power among the tones. The ranging signal of (4-9) is called a *fixed-tone ranging signal.* The finite time autocorrelation function of (4-9) is given by

$$\begin{aligned} R_s(\tau) &\triangleq \frac{1}{\tau_{\max}} \int_0^{\tau_{\max}} s(t)s(t+\tau)\,dt \\ &= \sum_{n=1}^{N} S_{tn} \cos \omega_n \tau \end{aligned} \tag{4-10}$$

where $S_{tn} \triangleq a_n^2/2$ is the average power in the nth sidetone, and we have set $T_0 = \tau_{\max}$.

To compute the accuracy of the range measurement when the fixed-tone ranging signal of (4-9) is transmitted, one must first evaluate β_s^2 of (4-7) for this signal. Taking the Fourier transform of (4-9) and substituting it into (4-7) yields (see Prob. 4-2)

$$\beta_s^2 = \frac{\sum_{n=1}^{N} \omega_n^2 S_{tn}}{\sum_{n=1}^{N} S_{tn}} = \frac{\sum_{n=1}^{N} \omega_n^2 S_{tn}}{S} \tag{4-11}$$

Then, from (4-6) the variance of the range delay error becomes

$$\sigma_{\tau_e}^2 = \frac{N_0}{2\tau_{\max}} \left[\frac{1}{\sum_{n=1}^{N} \omega_n^2 S_{tn}} \right] = \frac{N_0}{4\pi} \left[\frac{\omega_1}{\sum_{n=1}^{N} \omega_n^2 S_{tn}} \right] \tag{4-12}$$

The corresponding variance of the error in distance is given by

$$\sigma_d^2 = \left(\frac{c\sigma_{\tau_e}}{2} \right)^2 = \frac{cd_{\max}}{16\pi^2} \left[\frac{\omega_1^2}{\sum_{n=1}^{N} (S_{tn}/N_0)\omega_n^2} \right] \tag{4-13}$$

If $S_{tn} = S_t$ for all n, then (4-13) simplifies to

$$\sigma_d^2 = \frac{cd_{\max}}{16\pi^2(S_t/N_0)}\left[\frac{M^2 - 1}{M^{2N} - 1}\right] \tag{4-14}$$

Either (4-13) or (4-14) characterizes the performance of the MAP estimator of range.

Since the MAP estimator of range is impossible to build, we consider the performance of a suboptimum range estimator. In particular, we postulate a ranging receiver that isolates each tone and tracks its phase and frequency by means of a narrowband PLL (see Fig. 4-4). The output $r_n(t) = \sqrt{2}K_1$

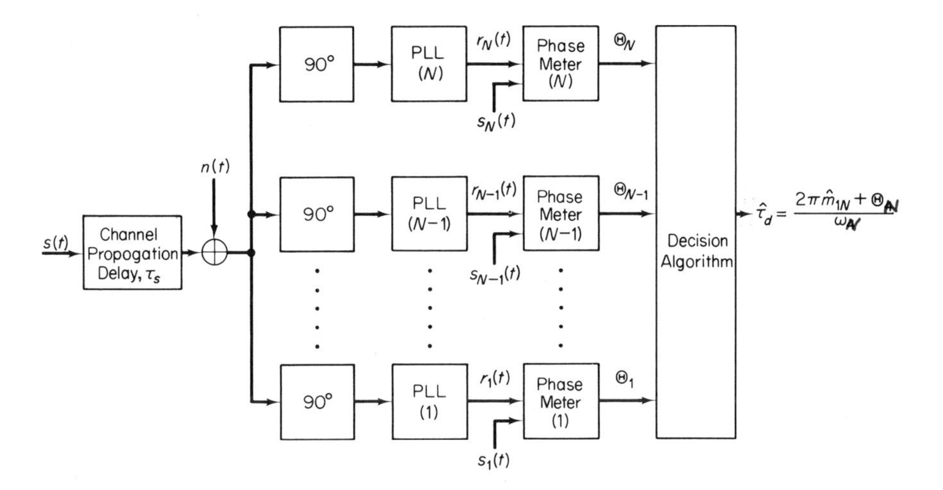

Fig. 4-4 A suboptimum receiver for estimation of range delay

$\times \cos[\omega_n(t - \tau_s) - \varphi_n]$ of each PLL, along with component tones $s_n(t) = a_n \cos \omega_n t$, are applied to N phase meters. Each phase meter reads the phase difference Θ_n between $s_n(t)$ and $r_n(t)$ i.e.

$$\Theta_n \triangleq \{[\omega_n \tau_s + \varphi_n + \pi] \text{ modulo } 2\pi\} - \pi \qquad n = 1, 2, \ldots, N$$
$$-\pi \leq \Theta_n \leq \pi$$

These phase measurements are then combined according to a prescribed decision algorithm (to be given shortly) to produce the estimate $\hat{\tau}_d$ of range. This approach to ranging is analogous to that employed in the Goddard Range and Range Rate system.[5]

In the suboptimum range estimation procedure just described, phase measurement Θ_N on the highest frequency tone determines the accuracy of the estimate, whereas the remaining $N - 1$ tones are used to resolve the range ambiguity, that is introduced by the fixed-tone ranging signal. In fact, if the variances $\sigma_{\varphi_n}^2$; $n = 1, 2, \ldots, N - 1$ are not too large, range ambiguity is correctly reduced by a factor of M each time a phase measurement is made on the next lower frequency tone. Finally, a point is reached where the range becomes unambiguous because the period of the lowest frequency tone has been chosen equal to $\tau_{\max}$.

The accuracy of the range measurement can be established by noting that in the linear region of loop operation, the variance of the phase error φ_n in tracking the nth tone is given by (see Chap. 2)

$$\sigma_{\varphi_n}^2 = \frac{1}{\rho_n} \tag{4-15}$$

where

$$\rho_n = \frac{2S_{tn}}{N_0 W_{Ln}} \tag{4-16}$$

is the signal-to-noise ratio in the loop bandwidth W_{Ln}, $n = 1, 2, \ldots, N$. Since, in this case, the highest frequency tone alone determines range accuracy, the variance of the error in distance measurement is given by

$$\sigma_d^2 = \left(\frac{c}{2}\right)^2 \frac{1}{\omega_N^2 \rho_N} = \frac{c d_{\max}}{16\pi^2 (S_{tN}/N_0)} \left[\frac{M^2}{M^{2N}}\right] \frac{2\pi W_{LN}}{\omega_1} \tag{4-17}$$

Comparing (4-14) and (4-17), we see that for $M > 1$, $S_{tn} = S_t$ and a loop bandwidth of the order of the lowest frequency tone, the variance of the error in range measurement is approximately equivalent to that produced by the MAP estimate (see Prob. 4-4).

The way that the suboptimum receiver extracts range can be made more explicit by considering a concrete example. Letting the frequency ratio $M = 2$ and the number of tones $N = 4$, Fig. 4-5 illustrates the component tones $s_n(t)$, $n = 1, 2, 3, 4$. Beginning with the highest frequency tone $s_4(t)$ (since this establishes the accuracy of the range measurement), a PLL whose VCO has a nominal frequency $\omega_4 = 8\omega_1$ tracks the frequency and phase of this tone and establishes a coherent reference signal $r_4(t) = \sqrt{2}\,K_1 \cos[\omega_4(t - \tau_s) - \varphi_4]$. Since the positive peaks of $r_4(t)$ occur at the time instants $t_4 = [\omega_4\tau_s + \varphi_4 + 2k\pi]/\omega_4$, with k any integer, then in the interval $(0, \tau_{\max})$, $r_4(t)$ will have $2^{N-1} = 8$ positive peaks which can be related to the phase meter measurements by

$$t_4 = \frac{\omega_4\tau_s + \varphi_4 + 2k\pi}{\omega_4} = \frac{2m_4\pi + \Theta_4}{\omega_4} \qquad m_4 = (0), 1, 2, \ldots, (8)$$

These time instants are indicated in Fig. 4-5 as vertical arrows above $s_4(t - \tau_s)$. The parentheses around the integers $m_4 = 0$ and 8 denote the fact that *either* 0 or 8 is correct with 0 being used when Θ_4 is positive, and 8 being used when Θ_4 is negative. One of these time instants will be chosen ultimately as our estimate of the true delay. Thus, the remaining phase measurements to be made on the $N - 1 = 3$ lower frequency tones will be used merely to resolve the ambiguity: that is, choose the correct peak or equivalently the

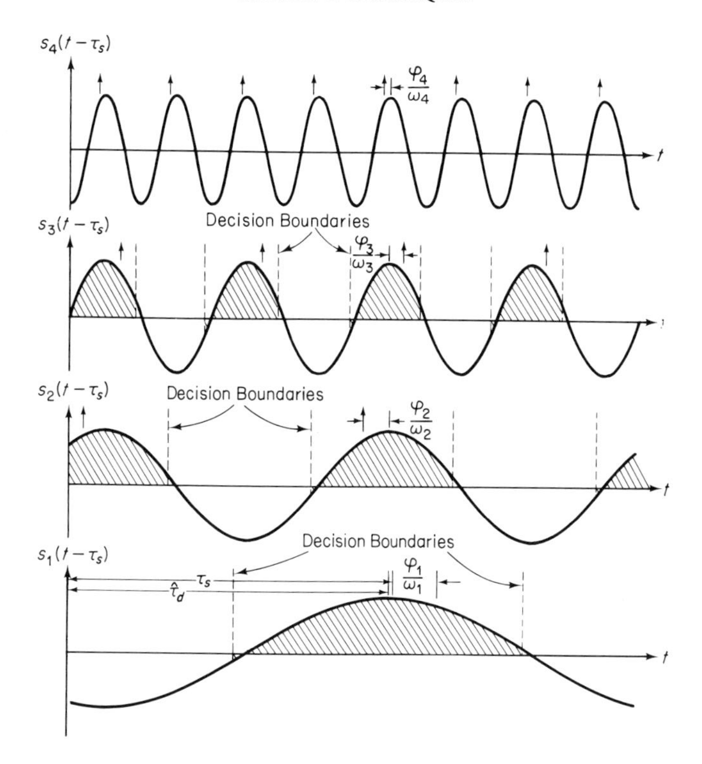

Fig. 4-5 Sinusoidal tone range ambiguity resolution

value of m_4. The step-by-step procedure for doing this is described as follows. The next lowest frequency tone is tracked by another PLL whose VCO has a nominal frequency $\omega_3 = 4\omega_1$. The reference signal $r_3(t)$ produced by this loop is used to reduce the number of ambiguous peaks by a factor of two; that is, on $(0, \tau_{\max})$ $r_3(t)$ has only four positive peaks which are displaced from those of $s_3(t - \tau_s)$ by $\tau_3 \triangleq \varphi_3/\omega_3$ sec where φ_3 is the loop phase error in the third PLL. Again these peaks are indicated in Fig. 4-5 by vertical arrows above $s_3(t - \tau_s)$. Whether or not the number of ambiguous peaks is correctly reduced by a factor of two—that is, the four *positive* peaks of $r_3(t)$ which occur in the interval $(0, \tau_{\max})$ are chosen rather than the four *negative* ones—depends on the value of φ_3 relative to φ_4. The rule for deciding this follows: Decision boundaries are drawn midway between the peaks of $r_4(t)$; that is, at the time instants $t_{\pm} = [\omega_4\tau_s + \varphi_4 + (2k \pm 1)\pi]/\omega_4$. Recognizing that the four positive peaks of $r_3(t)$ on $(0, \tau_{\max})$ occur at time instants

$$t_3 = \frac{\omega_3\tau_s + \varphi_3 + 2k\pi}{\omega_3} = \frac{2\pi m_3 + \Theta_3}{\omega_3} \qquad m_3 = (0), 1, 2, 3, (4)$$

then these peaks are chosen if the value of φ_3 is such that they lie in the shaded zones (see Fig. 4-5). In mathematical terms, the number of ambiguous peaks will be correctly reduced by a factor of two if

$$\tau_s + \frac{\varphi_4}{\omega_4} - \frac{\pi}{\omega_4} < \tau_s + \frac{\varphi_3}{\omega_3} < \tau_s + \frac{\varphi_4}{\omega_4} + \frac{\pi}{\omega_4}$$

or equivalently

$$\left| \frac{\varphi_3}{\omega_3} - \frac{\varphi_4}{\omega_4} \right| < \frac{\pi}{\omega_4}$$

Using the relationships between phase measurements Θ_3 and Θ_4, and the time instants at which the PLL reference signals $r_3(t)$ and $r_4(t)$ peak, the preceding inequality can be rewritten as

$$\frac{\Theta_4}{2} < \Theta_3 + \left(2m_3 - m_4 + \frac{1}{2}\right)\pi < \frac{\Theta_4}{4} + \pi$$

Considering first values of $m_4 = 2m_3$, the above inequality simplifies to

$$-\frac{\pi}{2} + \frac{\Theta_4}{2} < \Theta_3 < \frac{\Theta_4}{2} + \frac{\pi}{2}$$

or equivalently

$$0 < |\Theta_{34}| < \frac{\pi}{2}$$

where Θ_{34} is defined by

$$\Theta_{34} \triangleq \left\{ \left[\Theta_3 - \frac{\Theta_4}{2} + \pi \right] \text{modulo } 2\pi \right\} - \pi$$

For values of $m_4 = 2m_3 \pm 1$, we get

$$\frac{\pi}{2} < |\Theta_{34}| < \pi$$

Thus, in terms of the phase measurements Θ_3 and Θ_4, the decision rule for reducing the number of ambiguous peaks by a factor of two is:

choose $\hat{m}_{34} = (0), 2, 4, 6, (8)$ when $0 < |\Theta_{34}| < \frac{\pi}{2}$

and choose $\hat{m}_{34} = 1, 3, 5, 7$ when $\frac{\pi}{2} < |\Theta_{34}| < \pi$

where $\hat{m}_{34}$ denotes the estimate of m_4 based upon these measurements.

Continuing this procedure, a phase measurement between the next lowest tone $s_2(t)$ and $r_2(t)$ can be used to reduce the number of ambiguous peaks by another factor of two given that the previous reduction was correct. Here the decision boundaries are drawn midway between *alternate* peaks of $r_4(t)$ and a correct decision is made—that is, the two positive peaks of $r_2(t)$ on $(0, \tau_{\max})$ are chosen rather than the two negative ones—if these positive peaks lie within the shaded zones in Fig. 4-5, i.e.,

$$\left| \frac{\varphi_2}{\omega_2} - \frac{\varphi_4}{\omega_4} \right| < \frac{2\pi}{\omega_4}$$

Since the time instants at which $r_2(t)$ has its positive peaks are given by

$$t_2 = \frac{\omega_2 \tau_s + \varphi_2 + 2k\pi}{\omega_2} = \frac{2\pi m_2 + \Theta_2}{\omega_2} \qquad m_2 = (0), 1, (2)$$

then using this expression together with the similar relation between t_4 and Θ_4 allows us to write the above inequality as

$$\frac{\Theta_4}{4} < \Theta_2 + \left(\frac{4m_2 - m_4 + 1}{2}\right)\pi < \frac{\Theta_4}{4} + \pi$$

Separating the even values of m_4 into two sets, each containing alternating even integers, then the above inequality reduces to

$$0 < |\Theta_{24}| < \frac{\pi}{2} \qquad \text{when } m_4 = 4m_3$$

and

$$\frac{\pi}{2} < |\Theta_{24}| < \pi \qquad \text{when } m_4 = 4m_3 \pm 2$$

where

$$\Theta_{24} \triangleq \left\{\left[\Theta_2 - \frac{\Theta_4}{4} + \pi\right] \text{modulo } 2\pi\right\} - \pi$$

A similar subdivision of the odd values of m_4 gives

$$-\pi < \Theta_{24} < 0 \qquad \text{when } m_4 = 4m_3 - 1$$

and

$$0 < \Theta_{24} < \pi \qquad \text{when } m_4 = 4m_3 + 1 \text{ and } m_4 = 4m_3 - 3.$$

Thus, in terms of phase measurements Θ_2 and Θ_4, the decision rule is described as follows.

If the previous decision selected the values $\hat{m}_{34} = (0), 2, 4, 6, (8)$, then

$$\text{choose } \hat{m}_{24} = (0), 4, 8 \qquad \text{when } 0 < |\Theta_{24}| < \frac{\pi}{2}$$

and

$$\text{choose } \hat{m}_{24} = 2, 6 \qquad \text{when } \frac{\pi}{2} < |\Theta_{24}| < \pi.$$

If the previous decision selected the values $\hat{m}_{34} = 1, 3, 5, 7$, then

$$\text{choose } \hat{m}_{24} = 1, 5 \qquad \text{if } 0 < \Theta_{24} < \pi$$

and

$$\text{choose } \hat{m}_{24} = 3, 7 \qquad \text{if } -\pi < \Theta_{24} < 0.$$

In this, $\hat{m}_{24}$ is the conditional estimate of m_4, based upon the phase measurements Θ_2 and Θ_4 and the previous estimates $\hat{m}_{34}$. Using similar arguments to those given, we arrive at a decision rule for the final step in the reduction of the number of ambiguous peaks.

Defining the phase angle Θ_{14} by

$$\Theta_{14} \triangleq \left\{\left[\Theta_1 - \frac{\Theta_4}{8} + \pi\right] \text{modulo } 2\pi\right\} - \pi$$

then this decision rule is described as follows.

If the previous decision selected $\hat{m}_{24} = (0), 4, (8)$, then

$$\text{choose } \hat{m}_{14} = (0), (8) \qquad \text{when } 0 < |\Theta_{14}| < \frac{\pi}{2}$$

and

$$\text{choose } \hat{m}_{14} = 4 \qquad \text{when } \frac{\pi}{2} < |\Theta_{14}| < \pi.$$

If the previous decision selected $\hat{m}_{24} = 2, 6$, then

$$\text{choose } \hat{m}_{14} = 2 \qquad \text{when } 0 < \Theta_{14} < \pi$$

and

$$\text{choose } \hat{m}_{14} = 6 \qquad \text{when } -\pi < \Theta_{14} < 0.$$

If the previous decision selected $\hat{m}_{24} = 1, 5$, then

$$\text{choose } \hat{m}_{14} = 1 \qquad \text{when } -\frac{\pi}{4} < \Theta_{14} < \frac{3\pi}{4}$$

and

$$\text{choose } \hat{m}_{14} = 5 \qquad \text{when} \begin{Bmatrix} \frac{3\pi}{4} < \Theta_{14} < \pi \\ -\pi < \Theta_{14} < -\frac{\pi}{4} \end{Bmatrix}$$

If the previous decision selected $\hat{m}_{14} = 3, 7$, then

$$\text{choose } \hat{m}_{14} = 3 \qquad \text{when} \begin{Bmatrix} \frac{\pi}{4} < \Theta_{14} < \pi \\ -\pi < \Theta_{14} < -\frac{3\pi}{4} \end{Bmatrix}$$

and

$$\text{choose } \hat{m}_{14} = 7 \qquad \text{when } -\frac{3\pi}{4} < \Theta_{14} < \frac{\pi}{4}.$$

We note that this final step in the reduction of ambiguous peaks provides us with a single value of m_4, i.e., $\hat{m}_{14}$. Thus, the estimate of true delay based upon this suboptimum procedure is given by

$$\hat{\tau}_d = (2\pi\hat{m}_{14} + \Theta_4)/\omega_4$$

The preceding algorithm can be generalized to the case where N and M are arbitrary.

We now evaluate the probability of incorrectly resolving the ambiguity by proper choice of peaks at each step in the range measurement process. This probability can be determined using the fact that for large signal-to-noise ratio ρ_n, the p.d.f. of the random variable $\varphi_n \cong \phi_n$ is approximately Gaussian with zero mean and variance $1/\rho_n$. Defining $\tau_n \triangleq \varphi_n/\omega_n$ and generalizing the inequalities to arbitrary M and N, a phase measurement on the $n - 1$st tone will correctly reduce the ambiguity by a factor of M if

$$\left|\tau_{n-1} - \tau_N\right| < \frac{\pi M^{N-n}}{\omega_N} \qquad n = 2, 3, \ldots, N \tag{4-18}$$

or in terms of the frequency ratio M

$$\left| M\varphi_{n-1} - \frac{\varphi_N}{M^{N-n}} \right| < \pi \tag{4-19}$$

Assuming that the random variables φ_n, $n = 1, 2, \ldots, N$ are statistically independent, then the probability of incorrectly reducing the number of ambiguous peaks by a factor of M from a phase measurement on the $n - 1$st tone is given approximately by

$$P_E(n-1) = \text{Prob}\left\{\left| M\varphi_{n-1} - \frac{\varphi_N}{M^{N-n}} \right| \geq \pi\right\} = 1 - \text{erf}\left[\frac{\pi}{\sqrt{2}\,\sigma_{n-1}}\right] \tag{4-20}$$

where
$$\text{erf}(x) \triangleq \frac{2}{\sqrt{\pi}} \int_0^x \exp(-t^2)\,dt$$

and
$$\sigma_{n-1} = \sqrt{\frac{M^2}{\rho_{n-1}} + \frac{1}{M^{2(N-n)}\rho_N}} \tag{4-21}$$

Since ρ_N is determined by the required range accuracy as specified by the variance of the range error of (4-17) and $\omega_N = M^{N-1}\omega_1$, Eq. (4-21) can be rewritten as

$$\sigma_{n-1} = \sqrt{\frac{M^2}{\rho_{n-1}} + \left(\frac{2}{c}\right)^2 \sigma_d^2 \omega_1^2 M^{2(n-1)}} \tag{4-22}$$

Practical considerations associated with generating and processing low frequency tones provide a lower bound on the lowest frequency in the tone set (usually around 10 Hz). Other considerations in the choice of system parameters are a specification on the highest frequency (ordinarily limited by the allowable RF bandwidth occupancy) and a limitation on the total power output of the transmitter. It is further desirable to keep the number of tones N to a minimum because of the equal number of required PLLs and other auxiliary circuits.

Thus far in our discussion of the fixed-tone technique we have not addressed the problem of *acquisition time*. For second-order PLLs implemented with imperfect integrating loop filters, it is known that when a frequency offset of Δf Hz serves as the loop input, then, in the absence of noise, the time T_f' required to achieve frequency lock is approximated by

$$T_f' \cong \tau_2 \left\{ \frac{(2\pi\Delta f \tau_2/r)^2 - \frac{1}{2}\ln(4\pi\Delta f \tau_2/r)}{1 - [F_1/(2 - F_1)][2\pi\Delta f \tau_2/r]^2} \right\} \tag{4-23}$$

where $1 < (2\pi\Delta f \tau_2/r)^2 < 2/F_1 - 1$. The time required to achieve phase lock once frequency lock has been obtained is usually on the order of $5(r + 1)/rW_L$. The *acquisition range* can be found by noting the condition that makes T_f' approach infinity. Since the frequency uncertainty due to Doppler shift is largest in the highest frequency tone, we would expect that acquisition of this tone would take the longest. Also, assuming that the tones are acquired in parallel, the acquisition time of the highest frequency tone would then dominate.

More recently, there has been increasing interest in ranging systems that transmit and acquire tones sequentially. The advantage of such systems is that the full power capability of the transmitter can be used for each tone independently in succession. A single PLL can be used by changing its nominal VCO frequency to match the tone it is trying to acquire and track. It is necessary, of course, to incorporate a *lock detector* into the system, which by an out-of-lock indication will inform the PLL when the tone frequency changes. We also note that sequential acquisition of tones necessarily increases the total acquisition time, thus making such a system undesirable for certain applications.

Another possibility is the use of square wave "tones" rather than sinusoids. A sequential ranging system[6] based on this technique has been employed by the Jet Propulsion Laboratories on the Mariner-Mars 1969 and 1971 missions (see Prob. 4-5).

The Swept-Tone Ranging Technique

A swept-tone system differs from a fixed-tone system in that the group of tones previously used for reducing the number of ambiguities in the phase measurement on the highest frequency tone is replaced by a single tone, which is swept over a frequency range sufficiently wide to perform the same function. The sweep procedure provides a coarse measurement of range, whereas the highest frequency tone remaining at the end of the sweep is still used for fine range measurement, i.e., to determine range accuracy.

In practice, the presence of Doppler shifts requires that the system be configured with two channels. One of the channels is used to measure the Doppler while the other is used as a calibration channel.* An example of such a two-channel system which employs the swept-tone ranging technique is the so-called Mistram system. In what follows, we shall discuss the way in which the swept-tone ranging signal transmitted through the calibration channel is processed by the ranging receiver to determine the true range delay τ_s.

The transmitted ranging signal has the form

$$s(t) = \sqrt{2S}\cos[\omega(t)t] \tag{4-24}$$

where $\omega(t)$ is a periodic function which in any given period, e.g., $0 \leq t \leq T$

* As in the previous sections, we shall consider, for analysis purposes, the case where the channel Doppler is zero. The analysis is somewhat complicated when a Doppler offset is present (see Prob. 4–6).

is characterized by

$$\omega(t) \triangleq \begin{cases} \omega_i & 0 \le t \le T_i \\ \omega_i + (\omega_f - \omega_i)m_{SW}(t - T_i) & T_i \le t \le T_i + T_{SW} \\ \omega_f & T_i + T_{SW} \le t \le T_i + T_{SW} + T_f \\ \omega_f - (\omega_f - \omega_i)m_r(t - T_i + T_{SW} + T_f) & \\ \qquad T_i + T_{SW} + T_f \le t \le T & \end{cases} \tag{4-25}$$

In the above, ω_i and ω_f are the initial and final radian sweep frequencies, respectively, $m_{SW}(t)$ is the sweep waveform which monotonically increases from zero to one over the interval $0 \le t \le T_{SW}$ and $m_r(t)$ is the retrace waveform which monotonically increases from zero to one over the interval $0 \le t \le T - T_i - T_{SW} - T_f$. The received ranging signal can be described by

$$y(t) = \sqrt{2S} \cos \{\omega(t - \tau_s)[t - \tau_s]\} + n(t) \tag{4-26}$$

A PLL whose VCO center radian frequency is ω_0 tracks the frequency and phase of a 90-degree phase shift of $y(t)$ and produces the reference signal $r(t) = \sqrt{2}K_1 \cos \{\omega(t - \tau_s)[t - \tau_s] + \varphi(t)\}$. A phase meter then reads the difference in phase between $r(t)$ and a replica of the transmitted signal $s(t)$, i.e.,

$$\Theta(t) \triangleq \{[\omega(t - \tau_s)\tau_s + [\omega(t) - \omega(t - \tau_s)]t + \varphi(t) + \pi] \text{ modulo } 2\pi\} - \pi$$
$$-\pi \le \Theta(t) \le \pi \tag{4-27}$$

If the duration T_i of transmission of the initial frequency is chosen to be greater than the maximum range delay $\tau_{\max}$, then for any τ_s and $\tau_s \le t \le T_i$, the phase meter reads

$$\Theta_i = \{[\omega_i\tau_s + \varphi_i + \pi] \text{ modulo } 2\pi\} - \pi \tag{4-28}$$

where φ_i denotes the PLL phase error during the time interval the loop is phase locked to the initial frequency ω_i. When the transmitter is swept from ω_i to ω_f the output of the phase meter will pass through many 2π cycles. This number of cycles N_{SW} can be recorded by a counter which updates its count by one every time the phase meter output reaches π radians. Finally, if the duration T_f of transmission of the final frequency is also chosen to be greater than $\tau_{\max}$, then for $T_i + T_{SW} + \tau_s \le t \le T_i + T_{SW} + T_f$, the phase meter will read

$$\Theta_f = \{[\omega_f\tau_s + \varphi_f + \pi] \text{ modulo } 2\pi\} - \pi \tag{4-29}$$

where φ_f is the loop phase error corresponding to phase lock at ω_f. From (4-28) and (4-29), we see that θ_i, θ_f, and N_{SW} satisfy the relation

$$\omega_f\tau_s + \varphi_f - \omega_i\tau_s - \varphi_i = \Theta_f - \Theta_i + 2\pi N_{SW} \tag{4-30}$$

Defining the quantities

$$\Delta\omega \triangleq \omega_f - \omega_i$$
$$\Delta\varphi \triangleq \varphi_f - \varphi_i \tag{4-31}$$

Eq. (4-30) takes on the simplified form

$$\Delta\omega\tau_s + \Delta\varphi = \Theta_f - \Theta_i + 2\pi N_{SW} \tag{4-32}$$

Thus, a coarse estimate $\hat{\tau}_{d_c}$ of τ_s can be obtained from the relation

$$\hat{\tau}_{d_c} \triangleq \frac{\Theta_f - \Theta_i + 2\pi N_{SW}}{\Delta\omega} \tag{4-33}$$

Fine range measurement is made at the end of the sweep interval T_{SW} and depends only on the phase measurement θ_f. Following the notation introduced in our discussion of the fixed-tone system, $r(t)$ has positive peaks which occur at the time instants $t_f = [\omega_f\tau_s + \varphi_f + 2k\pi]/\omega_f$ with k any integer. Thus, in a time interval of duration $\tau_{\max}$, there are $N_p = [\omega_f\tau_{\max}/2\pi]$ positive peaks which can be related to the phase meter measurement θ_f by

$$t_f = \frac{\omega_f\tau_s + \varphi_f + 2k\pi}{\omega_f} = \frac{2m_f\pi + \Theta_f}{\omega_f} \qquad m_f = (0), 1, 2, \ldots, (N_p) \tag{4-34}$$

The square brackets around $\omega_f\tau_{\max}/2\pi$ denote the integer part of this quantity, and the parentheses around integers $m_f = 0$ and $m_f = N_p$ have the same significance as in the fixed-tone case. Since one of these time instants is to be chosen ultimately as our estimate of the true delay, the following rule is used to make this decision.

Choose $\hat{\tau}_d$ equal to that value of t_f which is closest in absolute value to $\hat{\tau}_{d_c}$. In mathematical terms,

$$\hat{\tau}_d = t'_f \tag{4-35}$$

where t'_f is the value of t_f corresponding to

$$\min_{t_f} |\hat{\tau}_{d_c} - t_f|$$

Using this rule, then by analogy with the discussion of the fixed-tone system, our decision will be correct when $\hat{\tau}_{d_c}$ satisfies the inequality

$$\tau_s + \frac{\varphi_f}{\omega_f} - \frac{\pi}{\omega_f} < \hat{\tau}_{d_c} < \tau_s + \frac{\varphi_f}{\omega_f} + \frac{\pi}{\omega_f} \tag{4-36}$$

or equivalently from (4-31) and (4-32),

$$\frac{\varphi_f}{\omega_f} - \frac{\pi}{\omega_f} < \frac{\Delta\varphi}{\Delta\omega} < \frac{\varphi_f}{\omega_f} + \frac{\pi}{\omega_f} \tag{4-37}$$

Rewriting (4-36) in simpler form we have

$$\left|\frac{\Delta\varphi}{\Delta\omega} - \frac{\varphi_f}{\omega_f}\right| < \frac{\pi}{\omega_f} \tag{4-38}$$

We note from this result that the ambiguous peaks present in the high frequency measurement are removed here in one step by the coarse estimate of delay. Thus, in this sense alone, the swept-tone system is similar to a fixed-tone system that employs only two tones.

The probability of incorrectly resolving the ambiguity due to multiple peaks associated with the phase measurement on the highest frequency tone, or equivalently, the error probability, of the range measurement is given by

$$P_E = \text{Prob}\left\{\left|\left(\frac{\omega_f}{\Delta\omega}\right)\Delta\varphi - \varphi_f\right| > \pi\right\} = \text{Prob}\left\{\left|\left(\frac{\omega_i}{\Delta\omega}\right)\varphi_f - \left(\frac{\omega_f}{\Delta\omega}\right)\varphi_i\right| > \pi\right\} \tag{4-39}$$

To evaluate (4-39) one must characterize the p.d.f.'s of the random variables φ_i and φ_f. These p.d.f.'s depend among other things on the detuning of the initial and final VCO center randian frequencies, ω_{0i} and ω_{0f}, respectively, relative to the initial and final transmitted frequencies. For a large loop signal-to-noise ratio ρ, the p.d.f.'s of φ_i and φ_f are approximately Gaussian with variance $1/\rho$ and means $(\omega_{0i} - \omega_i)/\sqrt{SK}$ and $(\omega_{0f} - \omega_f)/\sqrt{SK}$ respectively. Assuming further that φ_i and φ_f are statistically independent, then

$$P_E = 1 - \frac{1}{2}\,\text{erf}\left[\frac{\pi - \dfrac{(\omega_i/\omega_f)\omega_{0f} - \omega_{0i}}{\sqrt{SK}(1 - \omega_i/\omega_f)}}{\sqrt{2}\,\sigma}\right] - \frac{1}{2}\,\text{erf}\left[\frac{\pi + \dfrac{(\omega_i/\omega_f)\omega_{0f} - \omega_{0i}}{\sqrt{SK}(1 - \omega_i/\omega_f)}}{\sqrt{2}\,\sigma}\right] \tag{4-40}$$

where

$$\sigma = \sqrt{\frac{1 + (\omega_i/\omega_f)^2}{\rho(1 - \omega_i/\omega_f)^2}} \tag{4-41}$$

If the VCO is tuned so that $\omega_{0i} = \omega_i$ and $\omega_{0f} = \omega_f$, then (4-40) simplifies to

$$P_E = 1 - \text{erf}\left\{\pi\sqrt{\frac{\rho(1 - \omega_i/\omega_f)^2}{2[1 + (\omega_i/\omega_f)^2]}}\right\} \tag{4-42}$$

The results presented thus far appear to be independent of the sweep waveform $m_{SW}(t)$ and the duration of the sweep T_{SW}. Actually, we have assumed these parameters out of the problem by concerning ourselves only with measurements made at the beginning and end of the sweep interval. In practice, the choice of a sweep waveform and a sweep interval are governed by the limitation on the rate at which the VCO can be swept in frequency without having the PLL lose lock. These considerations affect the design of the loop filter.

Pseudonoise Ranging Techniques

As one attempts to measure larger and larger ranges, as in deep space applications, a point is reached where the generation of a low frequency tone whose period is compatible with the round-trip transit time becomes impractical (the frequency becomes too low). In situations such as these, one often turns to the transmission of a binary sequence of digits (*ranging code*) as a signal whose period for certain classes of codes can be designed to be arbitrarily long; hence, range ambiguity presents no problem. On the other hand, the use of a digital ranging signal brings with it the problem of range resolution in that the range measurement on the returned signal using correlation techniques can be resolved only to within one code-digit time interval. Clearly, a tradeoff exists here, since the finer the range resolution (that is, the shorter the duration of a code-digit) the longer the code sequence must be to avoid range ambiguity. If the range measurement is performed by sliding (in one code-digit increments) a replica of the transmitted code sequence past the received signal and searching for the maximum correlation, then long code sequences necessarily imply large acquisition times. In fact, since one usually has no *a priori* information about the range to be determined, cross-correlating the received signal with the transmitted signal replica shifted one digit at a time would in cases of distant range lead to unacceptable acquisition performance. The way in which this problem is solved depends upon a knowledge of the correlation properties of certain basic code sequences, which we shall now discuss.

Correlation Properties of Periodic Binary Sequences. We first define the correlation operation for a pair of sequences of discrete digits. It is convenient to assign a position to each code digit in the sequence so that an n-digit shift of the sequence corresponds to a translation (right or left) through n positions. Hence, the notation d_n refers to the code digit in the nth position of the defining sequence. We are interested here only in so-called "periodic sequences" defined by $d_{p+k} = d_k$, $k = 1, 2, \ldots$, where p is the number of code positions in one complete cycle. Thus, a positive shift of n digits of a periodic sequence $d_1, d_2, \ldots, d_p$ produces the sequence $d_{n+1}d_{n+2}\cdots d_p d_1 d_2 \cdots d_n$. Sometimes, this type of shift is referred to as a "circular" or "cyclic" shift because of the way in which the sequence closes on itself after one period.

One can define a correlation function for two sequences only if both contain the same total number of digits. For periodic sequences, this number of code-digits is chosen to be the minimum length for which both sequences contain an integral number of periods. Thus, for two periodic sequences with periods p_1 and p_2, respectively, the minimum length sequence over which the correlation can be defined is the least common multiple (l.c.m) between p_1 and p_2. If p_1 and p_2 are relatively prime, then this minimum length is the product $p_1 p_2$, whereas if $p_1 = p_2 = p$, we need correlate only over a sequence of length p. With this in mind we define the cross-correlation between two periodic sequences with fundamental patterns $c_1 c_2 \cdots c_{p_1}$ and $d_1 d_2 \cdots d_{p_2}$ by

$$R_{cd}(k) \triangleq \frac{\sum_{n=1}^{\text{l.c.m}(p_1 p_2)} c_n d_{n+k}}{\text{l.c.m}(p_1, p_2)} \tag{4-43}$$

where k represents the phase position of the d sequence relative to the c sequence and p_1, p_2 are the periods of the c and d sequences, respectively. When the c and d sequences are identical, we call $R_{cc}(k)$ the autocorrelation function of the c sequence. If for simplicity of notation we assume that p_1 is the smaller of the two periods, then a p_1-digit shift of one sequence with respect to the other does not alter the value of the sum in (4-43). Hence, $R_{cd}(k)$ is a periodic function with period p_1. Clearly, if $p_1 = p_2 = p$, then $R_{cd}(k)$ has period p.

If the code digits c_i, d_i are allowed to take on plus and minus values, then the multiplication and addition operations in (4-43) will both be interpreted in the algebraic sense. If corresponding positions of the sequences $\{c_n\}$ and $\{d_{n+k}\}$ have the same polarity—both are positive or both are negative—then their product is positive. If, on the other hand, corresponding positions have opposite polarity, then their product is negative. Since these are the only possibilities for binary-valued sequences, (4-43) can be rephrased as

$$R_{cd}(k) = \left[\frac{A(k) - D(k)}{A(k) + D(k)}\right] cd \tag{4-44}$$

where $A(k)$ is the number of code positions in which the two sequences agree in polarity, $D(k)$ is the number of code positions in which they disagree, and c, d are the magnitudes of the digits in the sequences $\{c_n\}$, $\{d_{n+k}\}$, respectively. When the allowable binary values assigned to the code digits are 0 and 1, then the multiplication operation is interpreted as a *logical* or *Boolean addition* modulo 2. If zeros are replaced by ones and ones are replaced by minus ones in the resulting sequence, then the summation operation in (4-43) may still be interpreted algebraically. Alternately, allowing the resulting sequence to maintain its one–zero identification, the autocorrelation function may be expressed as

$$R_{cd}(k) = \frac{Z(k) - O(k)}{Z(k) + O(k)} \tag{4-45}$$

where $O(k)$ and $Z(k)$ are the number of ones and zeros, respectively, in the sequence formed.

The Generation of Pseudonoise (PN) Sequences. In Sec. 4-2, we discussed several desirable properties for the correlation function of a signal suitable for ranging purposes. When the signal is a digital sequence, as here, it is possible to rephrase these properties in terms of the basic characteristics of the ranging code itself and any of the above definitions of correlation. In summary, we desire a ranging code having the following four characteristics:

1. To avoid ambiguity in range measurement, the length of one complete code cycle (measured in code digits per period times the duration of one code symbol) should be greater than the maximum anticipated round-trip transit time.

2. The code symbol repetition rate must be sufficiently high to meet the required specification on the resolution or accuracy of the range measurement.
3. The autocorrelation function of the code should be of the two-level type: the sequence should have a maximum correlation when compared with itself and a uniform degree of mismatch when compared with its k-digit shifts. In terms of the autocorrelation function definition of (4-44), the maximum difference between agreements and disagreements should occur for $k = 0$ (all agreements), while for all other values of k within the code period, this difference should be uniformly low. Actually, this property of maximum separation between the autocorrelation peak and *all* its other values is desirable only from the standpoint of detectability of measurement—that is, maximizing the probability of correctly identifying the location of the autocorrelation peak in its noisy environment. For purposes of acquisition, it is highly undesirable to have a uniformly low correlation away from the peak, since it provides no measure of the direction or extent to which the locally generated code should be shifted to rapidly achieve the range measurement. Nevertheless, an understanding of the characteristics of two-level autocorrelation sequences is basic to the consideration of the more sophisticated type of rapid acquisition code.
4. To improve efficiency in transmission, it is desirable to provide balanced use of power in the carrier sidebands by requiring the ranging code to have nearly the same number of ones and zeros within one complete period of the sequence pattern.

Ranging codes possessing the above desirable properties are commonly formed from a class of binary sequences known as *pseudonoise* (PN) sequences.[7,8] The word *pseudo* is used to denote the resemblance between the autocorrelation function of these sequences and that of band-limited white random noise with uniform spectral density over a wide range of frequencies.

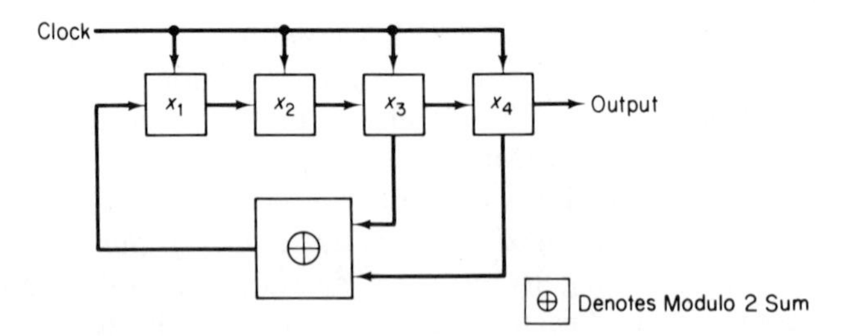

Fig. 4-6 A four-stage binary shift register with a feedback connection that results in a PN sequence.

The generation of periodic sequences of the PN type is accomplished by means of a shift register with prescribed feedback connections. Consider the four-stage shift register in Fig. 4-6, where a feedback path consisting of a modulo 2 sum of the third- and fourth-stage outputs is applied to the input of the first stage. In any type of feedback shift register device, the structure of the recurring output sequence is in general a function of the initial contents of the register stages. However, by choosing the proper feedback connections, as in the figure one can make the output sequence independent of the initial contents of the register, except for a possible shift in phase. The exception to this statement is the *all zero* initial condition where, regardless of the feedback connections, the output sequence will be all zeros. Table 4-1 illustrates for two different initial conditions the step-by-step contents

TABLE 4-1 Generation of a PN Sequence

Step	*Register Stage*	*Step*	*Register Stage*
	1 2 3 4		1 2 3 4
0 (initial contents)	0 0 1 0	0 (initial contents)	1 0 1 0
1	1 0 0 1	1	1 1 0 1
2	1 1 0 0	2	1 1 1 0
3	0 1 1 0	3	1 1 1 1
4	1 0 1 1	4	0 1 1 1
5	0 1 0 1	5	0 0 1 1
6	1 0 1 0	6	0 0 0 1
7	1 1 0 1	7	1 0 0 0
8	1 1 1 0	8	0 1 0 0
9	1 1 1 1	9	0 0 1 0
10	0 1 1 1	10	0 0 0 1
11	0 0 1 1	11	1 1 0 0
12	0 0 0 1	12	0 1 1 0
13	1 0 0 0	13	1 0 1 1
14	0 1 0 0	14	0 1 0 1
15	0 0 1 0	15	1 0 1 0
.	. . . .	.	. . . .
.	. . . .	.	. . . .
.	. . . .	.	. . . .

(Stage 4 column in each half marked "output sequence ↓")

of the shift register after each clock pulse excitation. The function of the clock is to advance the state of the register by shifting each digit one stage to the right. The output sequence is obtained by sequentially examining the digits stored in register 4 as one progresses step by step through the various register stages. It is interesting to observe from Table 4-1 how the structure of the code has been achieved independent of the initial condition. We notice that in forming one complete cycle of the output sequence, the shift register has been cycled through 15 different four-digit words, which are in fact all the possible nontrivial four-digit sequences that can be formed from the (0, 1) binary alphabet. The trivial sequence is by definition the *all-zero* sequence. Thus, regardless of where in the complete cycle we start the register—so long as the feedback connections are such that the register contents pass only one time through each and every nontrivial four-digit binary word—the output sequence is unique except for a possible cyclic shift. In fact, because we have exhausted all possible four-digit sequences in forming one cycle of the output code, we readily observe that a 15-digit output sequence is the longest sequence one can generate with a four-stage linear feedback shift register of the type considered above. The term *linear* refers to the modulo 2 sum operation in the feedback path. Sequences generated as above are called *maximal-length shift register sequences* of which the PN sequences are a special case. In general, then, an N-stage shift register with appropriate feedback connections is capable of generating an output sequence whose maximum periodic length is $2^N - 1$ binary digits.

Correlation Properties of PN Sequences. We begin our discussion of PN correlation properties by investigating the composition of maximal-length shift register sequences in terms of the frequency of occurrence of zeros and ones in a single sequence. In considering the 2^N possible sequences formed from N digits, the last digit in each sequence is zero as often as it is one. Since the all-zero sequence is not allowable as a possible shift register state, then for the remaining $2^N - 1$ sequences, the last digit takes on the value unity one more time than it does zero. Hence, in one complete cycle of the output sequence, the number of ones exceeds the number of zeros by unity. Letting $p = 2^N - 1$, there are $(p + 1)/2$ ones and $(p - 1)/2$ zeros in a single fundamental period of a PN sequence.

It is clear from the definition of autocorrelation in (4–45) that the autocorrelation function $R_{\text{PN}}(k)$ of a PN sequence has unity value at the origin; that is,

$$R_{\text{PN}}(0) = 1 \tag{4-46}$$

The above follows from the fact that all code positions are in agreement when the sequence is compared to itself. To evaluate $R_{\text{PN}}(k)$ at values of k other than zero, one must investigate the "shift and add" property of PN sequences. Since the digits in a PN sequence obey a recursion relationship based on the feedback connections in their associated register generator, it can be shown quite easily that the modulo 2 sum of a given PN sequence and a cyclic shift of itself produces in general another cyclic shift of the original sequence—that is, another PN sequence. In mathematical terms, letting $\{d_k\}$ denote a given PN sequence and $\{d_{k+m}\}$ an m-digit shift of this sequence, then

$$\{d_k\} \oplus \{d_{k+m}\} = \{d_{k+q}\} \tag{4-47}$$

where $\oplus$ denotes modulo 2 addition and q is in general unequal to m. Since the result of a shift and add is a PN sequence, the number of ones in the resulting sequence exceeds the number of zeros by unity. Recalling the definition of correlation as given by (4-45), we conclude that

$$R_{PN}(k) = -\frac{1}{p} \qquad k = \pm 1, \pm 2, \ldots, \pm(p-1) \tag{4-48}$$

where p is the period of the sequence and hence also the period of the autocorrelation function. Thus, *the autocorrelation function of a PN sequence is a two-level discrete function of the discrete shift variable* k, *possessing one value for* k $= 0$ *and another for all other* k $\neq 0$.

Corresponding to every PN sequence, one can associate a *binary waveform*, which is constructed by assigning an equal interval of time called the *digit period*, denoted by t_0, to each code digit in the sequence. The level of the binary waveform during each digit period is determined by assigning a unit positive level to each *one* in the corresponding sequence and a unit negative level to each *zero*. Hence, the autocorrelation function of the binary waveform, $s_{PN}(t)$, generated by a PN sequence is a periodic continuous function of a continuous time shift parameter τ, whose values at the sample points, $kt_0, k = 0, \pm 1, \pm 2, \ldots, \pm(p-1)$, agree with those of the discrete function described in (4–46) and (4-48). More specifically,

$$R_{PN}(\tau) = \begin{cases} 1 - \dfrac{1 + 1/p}{t_0}|\tau| & |\tau| \leq t_0 \\ -\dfrac{1}{p} & t_0 \leq |\tau| \leq pt_0 \end{cases} \tag{4-49}$$

The power spectrum of a PN binary waveform whose autocorrelation is given above is[7]

$$S_{PN}(\omega) = \left(\frac{p+1}{p^2}\right)\left[\frac{\sin(\omega t_0/2)}{\omega t_0/2}\right]^2 \sum_{\substack{n=-\infty \\ n\neq 0}}^{\infty} \delta\left(\omega - \frac{2\pi n}{pt_0}\right) + \frac{1}{p^2}\delta(\omega) \tag{4-50}$$

that is, a line spectrum. On examination of (4–50) it is clear that contributions to the line spectrum occur at frequencies that are multiples of the fundamental frequency of the autocorrelation function: $1/pt_0$. Also, since the binary waveform is a square wave whose amplitude alternates between plus and minus one in accordance with the code digits in the PN sequence, its average power is constant irrespective of the digit period t_0. Thus, for a fixed value of p, increasing the digit period makes the spectral lines more dense and reduces their respective amplitudes proportionally. The envelope of the spectrum for large p is essentially dependent only upon t_0. Hence, increasing the sequence period by increasing p at fixed t_0 leaves the envelope—and also the bandwidth required to transmit the waveform—unchanged. Also, the spectrum has periodic nulls at integral multiples of the digit repetition frequency, $1/t_0$.

Other properties of PN sequences have been investigated.[7, 8] For our purposes, however, the above discussion is sufficient for an understanding of how PN sequences form the basis of acceptable digital ranging codes.

The Use of PN Sequences in Forming Ranging Codes. A PN sequence with a long period p could by itself be used to provide unambiguous and accurate range data. As previously noted, the unambiguous ranging capability of a coded sequence increases in direct proportion to its length, whereas precision is determined by the width of each pulse—that is, the digit period in the transmitted code. Thus, the maximum unambiguous range capable of being measured is $d_{max} = (c/2)pt_0$. Assuming further that the range delay is equally likely to occur anywhere within the digit interval, an rms resolution error of $ct_0/4\sqrt{3}$ in range can always be expected irrespective of the phase jitter due to thermal noise.

Another important practical consideration affecting the use of PN sequences is the time required to acquire the code. For example, if a ranging code was derived from a sequence of million digit length and there was no *a priori* information about the phase of the returned code, a half-million trial correlations would be required, on the average, to find the correct code phase. Further, a PN type of autocorrelation function has the disadvantage of not providing any indication during the correlation search process of which way or how far to shift the local replica of the delayed event to achieve the desired time coincidence. It is impossible to learn anything from each trial except that exact correlation has or has not been achieved.

In practice, this number of trial cross-correlations can be significantly reduced by forming the required long ranging code as a combination of a number of shorter PN codes. If several, say l, continuously repeating PN sequences with relatively prime periods, say $p_1, p_2, \ldots, p_l$, are combined digit by digit according to the correct *Boolean function*,* the period of the resulting combined sequence is equal to the product of the periods of the component sequences—that is, $p = p_1 p_2 \cdots p_l$.[9] Furthermore, if the correct Boolean function is chosen for combining the sequences, the cross-correlation function of the code derived from the combined sequences with any of the component codes is periodic, with the period being that of the component. It has a peak when the component code is in phase with the corresponding component of the combined code i.e., $k = 0$ and has a uniformly low value when they are out of phase by more than one digit period.[9] If such a combined code is used for ranging, the phase of each component code can be determined by cross-correlating it with the combined code. Since the phase of the combined code is uniquely determined by the phases of the components, a number of trial correlations (binary decisions) equal to the sum $p_1 + p_2 + \cdots + p_l$ of the lengths of the component sequences must be performed to

---

* A function with N binary inputs and one binary output is called a Boolean function of N variables. There are 2^{2^N} different Boolean functions for a given N.

select the correct phase from among a total number equal to the product of the lengths of the component sequences, $p = p_1 p_2 \cdots p_l$. From an information theoretic point of view, only

$$\log_2 p = \sum_{i=1}^{l} \log_2 p_i \tag{4-51}$$

binary decisions should be needed for a given value of p. The large number of excess decisions

$$\sum_{i=1}^{l} (p_i - \log_2 p_i) \tag{4-52}$$

is primarily attributed to the fact that the component sequence periods p_i are constrained to be relatively prime.

The *acquisition time* of such a ranging code can be determined as follows. A single component sequence—for example, one having period p_k—together with the $p_k - 1$ other sequences obtained by shifting the original sequence through all of its $p_k - 1$ code phases approximately form an orthogonal code set with respect to the received signal sequence. Although this term will be more formally defined in Chap. 5, we use it here in the sense that one member of the code set is highly correlated with the received sequence while all others are essentially uncorrelated with it. Although the cross-correlation between the kth component sequence and the received sequence is uniformly nonzero when the two are not in phase, this small degree of correlation can be assumed to be negligible when p_k is large. Thus, we shall assume that the cross-correlation function between the unit power replica of the kth component sequence and the total ranging signal of average power S is approximated by $\lambda_k^2 S$, $(0 < \lambda_k < 1)$, when the two are in phase and by zero elsewhere. The value of λ_k is a function of the length of the kth component sequence p_k and the Boolean function used to combine the components. Under these assumptions, one may determine the acquisition error probability for the kth component in the same manner as one finds the word error probability for an orthogonally coded phase-coherent system. Using a result derived in Chap. 5, the probability of incorrectly acquiring the phase of the kth component code is approximated by

$$P_{E_k} = 1 - \int_{-\infty}^{\infty} \frac{\exp(-z^2/2)}{\sqrt{2\pi}} \left[\operatorname{erf}\left(z + \sqrt{\frac{2\lambda_k^2 S T_{c_k}}{N_0}}\right)\right]^{p_k} dz \tag{4-53}$$

where T_{c_k} is the correlation time for each of the p_k correlations of the kth component code with the received sequence. Hence, for a given P_{E_k}, which is usually chosen to be independent of k, and fixed S/N_0, T_{c_k} can be found from an inverse solution to Eq. (4–53). This is usually done from a set of curves of P_E versus ST/N_0 for orthogonally coded phase-coherent systems (see Fig. 5-10). The total acquisition time for the ranging code T_{acq}, exclusive of the time required to acquire the clock component, is given in terms of the above solution for T_{c_k} by

$$T_{\mathrm{acq}} = \sum_{k=1}^{l} p_k T_{c_k} \tag{4-54}$$

The final requirement of a ranging code from a practical point of view is that it be capable of being tracked by the ranging receiver. This requirement is usually met by including a component, called the "clock," in the ranging code. The clock is merely a sequence of alternating pulses—that is, a square wave—whose frequency is normally chosen such that one complete square wave cycle corresponds to the duration t_0 of one PN symbol. The ± 1 PN sequence is biphase modulated onto the clock to form the composite digital ranging signal.

Rapid Acquisition Sequences (BINOR Codes)

One technique for transmitting square wave "tones" rather than sinusoids as a ranging signal—the *binary optimum ranging* (*BINOR*) *code*—is based on the parallel transmission of square wave tones with geometrically related frequencies.[10] If one were to sum up a group of N plus/minus-one coherent square waves, the resulting signal would be a digital waveform with, in general, $N + 1$ possible levels. By passing this waveform through a hard limiter, we generate a binary waveform whose pulses alternate at a maximum rate equal to that of the highest frequency square wave of the group. By choosing the component square wave frequencies ω_i, $i = 1, 2, \ldots, N$ to be geometrically related, the period of the hard-limited binary waveform will be identical to that of the lowest frequency square wave ω_1, and its length (in sequence symbols) will be equal to 2^N. Such is the generation of a BINOR code, which is illustrated in Fig. 4–7 for $N = 5$ and the ratio of adjacent square wave frequencies $M = 2$. It is convenient to have N odd so that the zero level never occurs in the summed waveform; hence, there is no ambiguity in applying the hard limiting device, although as we shall see this is not a necessity. Note from Fig. 4-7 that the BINOR code has mirror symmetry about its half-period point.

The motivation for forming such a binary ranging signal stems from the following considerations. It is desired to acquire the phase of a sequence of period Lt_0, where t_0 is the duration of a single pulse (symbol) and $L = 2^N$ is the number of symbols in one complete cycle of the sequence. We seek a method that accomplishes this using the minimum number of binary decisions—N—and hence is optimum from the standpoint of minimum time to acquisition. If N distinct channels were available, one could transmit a square wave of frequency $1/(2^{N-k+1} t_0)$ over the kth channel ($k = 1, 2, \ldots, N$), which when cross-correlated with a local replica of itself at the receiver could be used to reduce the number of phase ambiguities in the $k + 1$st channel by two. Hence, using this technique a number of binary de-

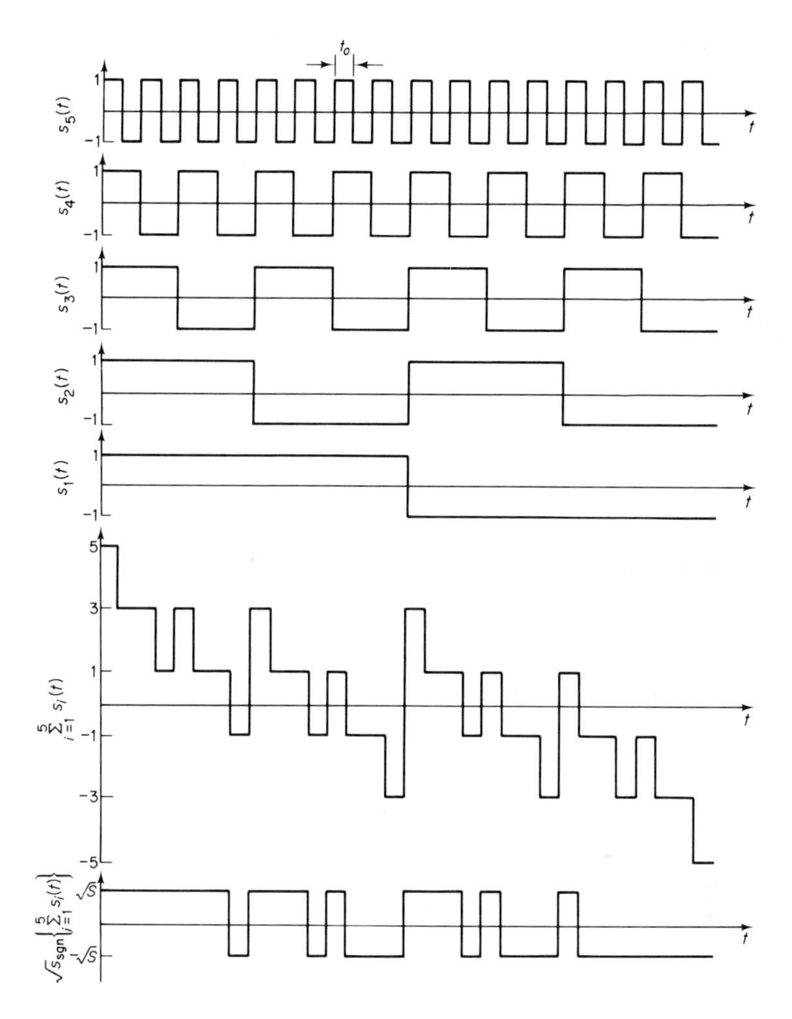

Fig. 4-7 Generation of a BINOR code

cisions equal to N would be sufficient to correctly acquire the sequence phase from a total of 2^N positions.

Although the previous method is optimum from an acquisition time standpoint, it has the decided disadvantage of requiring N channels. One can get around this limitation and still have an optimum ranging code for acquisition purposes by recognizing that the set of square waves transmitted separately over each channel are mutually orthogonal over a time interval equal to the period of the lowest frequency waveform Lt_0. Hence, an alternate procedure requiring only one channel would be to algebraically sum up the component ± 1 square waves and transmit this waveform as a composite ranging signal. Of course, in doing the above we have created a signal that in general has $N+1$ levels and thus cannot be transmitted over a binary channel. It is at this point that one must deviate from an optimum acquisition ranging signal. If, however, our objective is still to keep the degradation in acquisition time performance to a minimum, then the BINOR code method suggests hard limiting the $N+1$ level signal, thereby creating once again a binary signal.

Letting $d'_{k1}, d'_{k2}, \ldots . d'_{kL}$ denote the ± 1 symbols in one complete cycle of a square wave with frequency $1/(2^{N-k+1} t_0)$ then the jth symbol of the BINOR code d_j is described by

$$d_j = \sqrt{S} \operatorname{sgn} \left(\sum_{k=1}^{N} d'_{kj} \right) \tag{4-55}$$

where for our purposes here sgn (x) is defined by

$$\operatorname{sgn}(x) = \begin{cases} 1 & x \geq 0 \\ -1 & x < 0 \end{cases} \tag{4-56}$$

The value of sgn(x) at $x = 0$ has been arbitrarily chosen equal to $+1$ and is only of interest when the number of square wave tones N is even. The cross-correlation $R_{dd'_k}(0)$ between the BINOR code and the kth square wave symbol sequence is of interest in evaluating the performance of the range measurement at the receiver. Since each one of these cross-correlation operations is intended to reduce by a factor of two the number of phase ambiguities of the previous measurement—that is, provide one bit of information—it is desirable to design the ranging code so that $R_{dd'_k}(0)$ is independent of k, thereby achieving a uniform power distribution. To show that the BINOR code possesses this behavior, we consider

$$R_{dd'_k}(0) - R_{dd'_i}(0) = \frac{1}{L} \sum_{j=1}^{L} (d'_{kj} - d'_{ij}) d_j \tag{4-57}$$

There are three cases of interest in evaluating the sum in the above equation. First, those phase positions for which $d'_{kj} = d'_{ij}$ offer zero contribution to the sum irrespective of the corresponding value of d_j. Thus, we are left with two possibilities: either $d'_{kj} = -\sqrt{S}$ and $d'_{ij} = +\sqrt{S}$, or $d'_{kj} = +\sqrt{S}$ and $d'_{ij} = -\sqrt{S}$. However, the set of sequences of length N described by $\{d'_{1l}, d'_{2l}, \ldots, d'_{Nl}\}$, $l = 1, 2, \ldots, L$, contains once only each of the binary representations (in ± 1 arithmetic) of the integers 0 through L. Thus, it is always possible to find two phase positions p and q for which $d'_{kp} = -\sqrt{S}$, $d'_{ip} = +\sqrt{S}$, and $d'_{kq} = +\sqrt{S}$, $d'_{iq} = -\sqrt{S}$. But the remainder of the symbol pairs (d'_{np}, d'_{nq}), $n \neq p \neq q$, have matching digits; hence from (4–55) $d_p = d_q$. By arranging the sum in (4–57) in such pairs of phase positions, we force its value to zero—$R_{dd'_k}(0) = R_{dd'_i}(0) \triangleq R_{dd'}(0)$, for all k and i. To evaluate $R_{dd'}(0)$ as a function of N, we write

$$R_{dd'}(0) = \frac{1}{N} \sum_{k=1}^{N} R_{dd'_k}(0) = \frac{\sqrt{S}}{LN} \sum_{k=1}^{N} \sum_{l=1}^{L} d'_{kl} \operatorname{sgn} \sum_{n=1}^{N} d'_{nl} \tag{4-58}$$

which on interchanging orders of the first two sums and noting the independence of the third sum on k becomes

$$R_{dd'}(0) = \frac{\sqrt{S}}{LN} \sum_{l=1}^{L} \left| \sum_{k=1}^{N} d'_{kl} \right| \tag{4-59}$$

The preceding expression has been evaluated by Stiffler[10] with the result

$$R_{dd'}(0) = \begin{cases} \dfrac{\sqrt{S}}{2^{N-1}} \, {}_{N-1}C_{(N-1)/2} & N \text{ odd} \\ \dfrac{\sqrt{S}}{2^N} \, {}_{N}C_{N/2} & N \text{ even} \end{cases} \tag{4-60}$$

where ${}_NC_k$ is the binomial coefficient of N things taken k at a time. For large N, Stirling's formula can be used to give an asymptotic value for $R_{dd'}(0)$:

$$R_{dd'}(0) \simeq \sqrt{\frac{2S}{\pi N}} \tag{4-61}$$

Comparing this result with that obtained from the optimum separate channel method, where each square wave is assumed to have amplitude $\sqrt{S/N}$ ($1/N$th of the total power), we see that the BINOR code possesses an effective degradation of $\sqrt{2/\pi}$ in signal amplitude relative to the optimum system. Furthermore, this penalty is the maximum and decreases as N is reduced. Another interesting fact is that the choice of binary code digits by any rule other than (4-55) results in an average correlation less than $R_{dd'}(0)$ as specified by (4-60). Furthermore, the minimum correlation of that set cannot be greater than $R_{dd'}(0)$ of (4-60). Thus, the BINOR code results in the maximum average code correlation.

If it is assumed that a separate clock accompanies the transmission of the BINOR code, then a ranging receiver that tracks this clock would produce—on 100 percent acquisition of the code—the same mean-squared phase measurement error as would be if the clock were modulated by a PN code. Thus, with regard to accuracy of range measurement, the BINOR and PN codes would yield identical performance. Continuing under this assumption, the acquisition time of the BINOR code may be obtained by considering the amount of correlation time required to yield a fixed probability of correct decision at any stage of the search procedure. Since at each of the N stages a binary decision is being made on the basis of whether the correlation output is positive or negative—that is, the local replica of the ith square wave is in phase or 180 degrees out of phase with the received sequence—the probability of correct decision for the ith stage is given by

$$P_c(i) = \frac{1}{2}\left[1 + \operatorname{erf}\left(\frac{\mu}{\sqrt{2}\sigma}\right)\right] \tag{4-62}$$

where $\mu = R_{dd'}(0)T_c$, $\sigma = \sqrt{(N_0/2)T_c}$, and T_c, the length of the correlation interval, is an integral multiple of the sequence period Lt_0. Since N identical independent binary decisions are required (one for each square wave) the probability of correctly acquiring the BINOR code is given by

$$P_c = \prod_{i=1}^{N} P_c(i) = \left\{\frac{1}{2}\left[1 + \text{erf}\left(R_{dd'}(0)\sqrt{\frac{T_c}{N_0}}\right)\right]\right\}^N$$
$$= \left\{1 - \frac{1}{2}\left[1 - \text{erf}\left(R_{dd'}(0)\sqrt{\frac{T_c}{N_0}}\right)\right]\right\}^N \qquad (4\text{-}63)$$

If the acquisition error probability $P_E = 1 - P_c$ is to be small and N is assumed large, then using (4-61) for $R_{dd'}(0)$ and applying the binomial expansion to (4-62) gives

$$P_E \cong \frac{N}{2}\,\text{erfc}\left[\sqrt{\frac{2}{\pi N}\left(\frac{ST_c}{N_0}\right)}\right] \qquad (4\text{-}64)$$

For a given small acquisition error probability and fixed values of N and S/N_0, the solution of (4-64) for T_c, when multiplied by the number of successive binary decisions required for acquisition, gives the total acquisition time exclusive of the time required to track the clock:

$$T_{\text{acq}} = NT_c = \frac{\pi N^2}{2(S/N_0)}\left[\text{erfc}^{-1}\left(\frac{2P_E}{N}\right)\right]^2 \qquad (4\text{-}65)$$

Figure 4-8 plots T_{acq} versus N for $P_E = 0.001$ and several values of S/N_0 in dB.

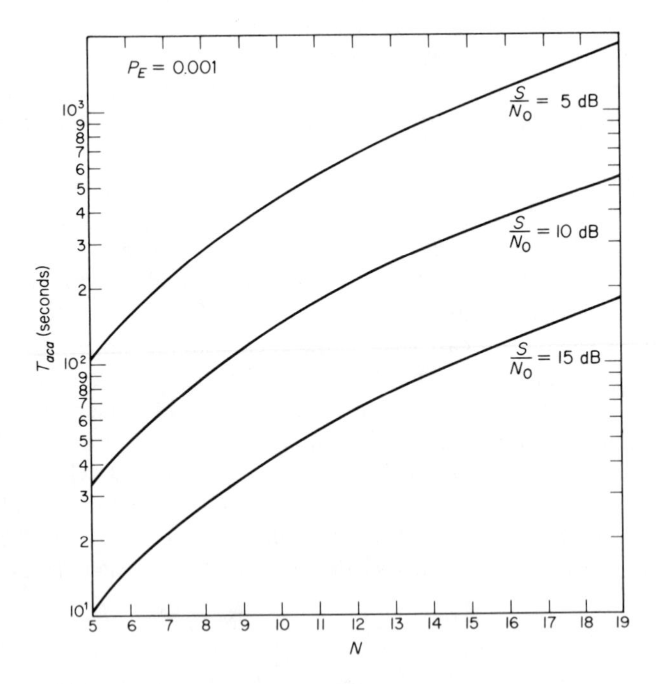

Fig. 4-8 Acquisition time of a BINOR code versus number of square-wave components

In practice, the highest frequency square wave in the sequence often is used directly as the clock, which then is acquired by the clock-tracking loop, and $N - 1$ correlations are required on the remaining $N - 1$ square waves. The total acquisition time would now be the sum of the clock loop acquisition time plus $(N - 1)T_c$. Also, since one is no longer only performing an in- or out-of-phase measurement on the highest frequency square wave but instead is obtaining an estimate of its actual phase, it might be desirable to allocate more power to this component.

Pulse Signal Ranging

All the ranging techniques considered thus far are of the CW type in which the total signal energy is spread out over the entire transmission time. Furthermore, they all have one feature in common: a phase-locked receiver is required to provide the fine accuracy range measurement. In contrast to CW systems, a pulse-type ranging system concentrates the total signal energy in a single short-duration pulse and the receiver is noncoherent, being interested only in detecting the presence or absence of a pulse. Since achieving the range measurement depends entirely on the detection of a single pulse (or possibly two successive pulses), we see that a primary advantage of the pulse signal ranging system is the reduction in acquisition time that it affords. In fact, since the pulse repetition frequency f_p is related to the maximum unambiguous range by $d_{\max} = c/2f_p$, the largest acquisition time would merely be the time $1/f_p$ between adjacent pulses. The major drawback of such a system lies in the high peak-power requirements placed upon the transmitter. Because of this problem, it is common in peak-power-limited systems to transmit long pulses and use a form of *pulse compression* at the receiver to achieve the desired range accuracy. Spreading a fixed energy pulse in time reduces its average power, while compressing it increases its effective bandwidth and hence reduces the mean-squared error in measuring its time of arrival.

The above statements may be put in mathematical terms by recalling Eqs. (4-6) and (4-7). Since for a pulse-type ranging signal, the observation interval T_0 is much larger than the pulse duration, we may let T_0 approach infinity in which case (4-7) simplifies to

$$\beta_s^2 = \frac{(1/2\pi)\int_{-\infty}^{\infty} \omega^2 |S(\omega)|^2 d\omega}{(1/2\pi)\int_{-\infty}^{\infty} |S(\omega)|^2 d\omega}$$

Then, for a rectangular pulse of width T_r passed through an ideal filter extending in frequency from $-W/2$ to $W/2$ $(W \gg 1/T_r)$,

$$\beta_s^2 \cong \frac{2W}{T_r} \tag{4-66}$$

Substituting (4-66) in (4-6) gives an expression for the variance of the time delay error:

$$\sigma_{\tau_e}^2 \cong \frac{T_r}{4WR_d} \tag{4-67}$$

where R_d is interpreted as the ratio of the pulse energy to the noise spectral density. The above result can be obtained by a completely different method of attack—namely, a threshold-crossing criterion is used to determine the time of arrival. If we assume that the pulse signal-to-noise ratio is large, then the effect of the noise is merely to shift the location of the threshold crossing in time without disturbing the slope of the waveform as it crosses the threshold. Under these conditions, a combination of independent threshold-crossing measurements made on the leading and falling edges of the pulse gives a time delay error variance

$$\sigma_{\tau_e}^2 \cong \frac{t_r T_r}{4R_d} \tag{4-68}$$

where t_r is the pulse rise time (assumed equal to the fall time) and T_r is now measured at the threshold level. Since the pulse rise time is limited by the system bandwidth W, then by making the assumption $t_r \cong 1/W$, we reduce (4-68) to (4-67).

To consider the effect of pulse compression on range measurement accuracy, we consider a particular type of pulse compression system, namely, one using frequency modulation at the transmitter. A carrier of amplitude $\sqrt{S}$ is linearly increased in frequency from f_1 to f_2 over the rectangular pulse duration T_r. The returned frequency modulated signal at the receiver is passed through a *pulse compression filter* whose time delay is a function of frequency in the same manner as is the modulation waveform. The output of this filter is approximately given by $\sqrt{SWT} \sin \pi Wt/\pi Wt$, where $W = f_2 - f_1$. Since the duration of the filter output waveform is approximately $1/W$, the compression of the transmitted pulse duration by a factor WT_r is transformed into an effective increase in instantaneous peak power by the same factor. The factor WT_r is suitably called the *pulse compression ratio*. For large WT_r, the spectrum of the pulse compression filter output waveform approaches a rectangular spectrum with $\beta_s^2 = \pi^2 W^2/3$.[11] Hence, from (4-6) we have

$$\sigma_{\tau_e}^2 \cong \frac{3}{2\pi^2 W^2 R_d} \tag{4-69}$$

Comparing this result with that obtained for the unmodulated case [see Eq. (4-67)], we find that pulse compression affords an effective improvement in $\sigma_{\tau_e}^2$ by a factor of $(\pi^2/6)\ WT_r$.

Pulse compression systems such as the above are readily found in the radar literature,[11–13] particularly in regard to the study of "chirp" radars.[14] The original U. S. patents on the technique are credited to Dicke[15] in 1953 and Darlington[16] in 1954. There are several notable works on the subject of pulse radar to which the interested reader is referred.[17–19]

4-4. PN RANGE TRACKING RECEIVERS

As noted in previous sections, the ranging problem is composed of two stages: the ranging signal is first acquired in phase and once this has taken place, the ranging code must be tracked to maintain a continuous range measurement. Analog and digital correlation techniques for handling the acquisition aspect of the ranging problem have been discussed. If the target range did not change with time, then it would be sufficient to perform the range measurements once, and the received signal and the locally generated replica of the transmitted signal would remain forever in phase. However, if the range changes as a function of time, then either the acquisition measurements must be repeated frequently or, more practically, the range must be continuously monitored with the use of a tracking device. The range acquisition and tracking equipment and the carrier-tracking system together constitute the *ranging receiver*. In this section, we are concerned primarily with the range tracking function and specific devices for accomplishing this task.

There are primarily two philosophies with regard to tracking a digital ranging signal. The so-called *direct* tracking approach makes direct use of the correlation properties of the code itself to provide an error signal when the received code tends to drift from the delayed locally generated replica of the transmitted code. The second method includes in the ranging signal format a clock component that the ranging receiver locks onto. The clock estimate formed in the receiver tracking loop drives a code generator which produces a local replica of the code. Both these philosophies are covered in the examples of range tracking receivers in this section.

The Delay-Locked Loop

To begin our discussion of the delay-locked loop (DLL), we consider the correlation function formed by delaying and advancing by one code digit the PN correlation function and then subtracting the results. From (4-49),

$$R_{\text{DLL}}(\tau) \triangleq R_{\text{PN}}(\tau - t_0) - R_{\text{PN}}(\tau + t_0)$$

$$= \begin{cases} \left(\dfrac{p+1}{p}\right)\dfrac{\tau}{t_0} & |\tau| \leq t_0 \\ -\left(\dfrac{p+1}{p}\right)\dfrac{\tau}{t_0}\left[1 - \dfrac{2t_0}{|\tau|}\right] & t_0 \leq |\tau| \leq 2t_0 \\ 0 & 2t_0 \leq |\tau| \leq (p-2)t_0 \end{cases} \tag{4-70}$$

$$R_{\text{DLL}}(\tau) = R_{\text{DLL}}(|\tau| - pt_0) \qquad (p-2)t_0 \leq |\tau| \leq (p+2)t_0$$

For comparison purposes, $R_{PN}(\tau)$ and $R_{DLL}(\tau)$ are illustrated in Fig. 4-9. We shall refer to $R_{DLL}(\tau)$ as the *phase detector characteristic* corresponding to the autocorrelation of a maximal-length binary shift-register sequence.* As shown, $R_{DLL}(\tau)$ varies linearly with τ for $|\tau| \leq t_0$ and hence provides a good tracking region for a loop that implements $R_{DLL}(\tau)$ as its equivalent error curve. In the ranging application, the parameter τ would correspond to $\tau_e = \tau_s - \hat{\tau}_d$—the delay error between the actual propagation delay of the ranging signal and the estimate of it made at the receiver.

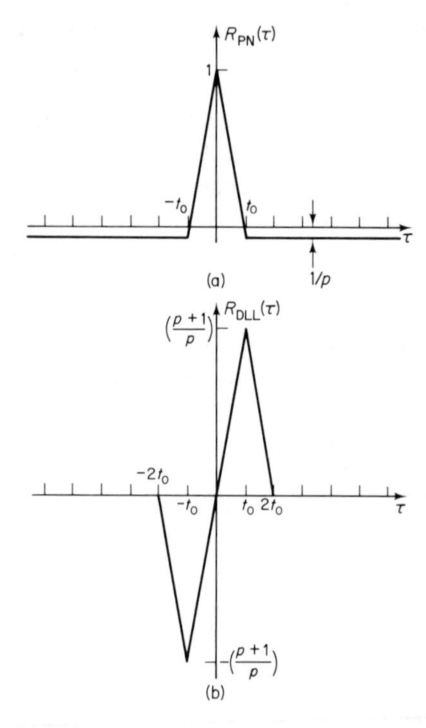

Fig. 4-9 Autocorrelation function, $R_{PN}(\tau)$, of a PN waveform and equivalent phase detector characteristic, $R_{DLL}(\tau)$, for a delay-locked loop

The DLL[20,21] achieves such an error curve by cross-correlating the received ranging signal sequence with a delayed and advanced version of the locally generated code replica (Fig. 4-10). The local code replica is generated in the feedback path with an N-stage shift register identical to that used in the transmitter for generating the code initially. The frequency of the clock that drives this sequence generator is controlled by a

* Frequently, the application of *dither* is used to alter the slope of $R_{DLL}(\tau)$ which provides "vibrational smoothing" of the tracking characteristics of the loop.

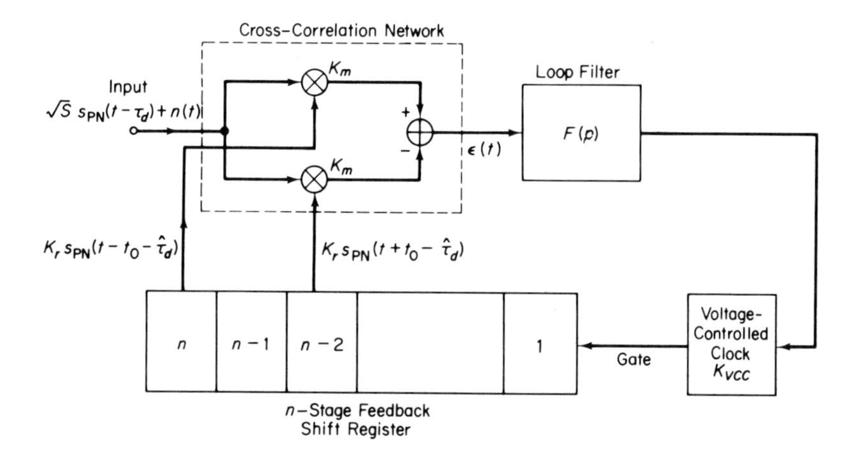

Fig. 4-10 The delay-locked loop

filtered version of the loop error signal at the output of the cross-correlation network.

Letting $s_{PN}(t)$ be a unit power signal corresponding to a PN sequence of period $p = 2^N - 1$, then $\epsilon(t)$, the loop error signal at the cross-correlation network output, is given by

$$\epsilon(t) = \sqrt{S} K_r K_m \delta s_{PN}(t - \hat{\tau}_d)[s_{PN}(t - \tau_s) + n(t)] \tag{4-71}$$

where $\delta s_{PN}(t)$ denotes the signal difference $s_{PN}(t - t_0) - s_{PN}(t + t_0)$. According to Spilker[20,21], the signal component of (4–71) can be written in the form

$$\delta s_{PN}(t - \hat{\tau}_d) s_{PN}(t - \tau_s) = R_{DLL}(\tau_e) + n_s(t, \tau_e) \tag{4-72}$$

where $n_s(t, \tau_e)$ is a zero mean process referred to as the *self noise* of the correlator output and, in effect, is the remainder of the signal component of $\epsilon(t)$ after the mean has been subtracted out. Since $s_{PN}(t)$ is periodic with period pt_0, then both $R_{DLL}(\tau_e)$ and $n_s(t, \tau_e)$ are also periodic with the same period. To assess the effect of $n_s(t, \tau_e)$ on system performance, one must evaluate its power spectrum. Based on the cycle and add property of PN codes, $n_s(t, \tau_e = mt_0)$ may be simplified to

$$n_s(t, \tau_e = mt_0) = \begin{cases} s_{PN}(t - \tau_s + jt_0) - s_{PN}(t - \tau_s + (j-1)t_0) & m = 0 \\ -m[s_{PN}(t - \tau_s + nt_0)] + \dfrac{1}{p} & m = \pm 1 \\ s_{PN}(t - \tau_s + kt_0) - s_{PN}(t - \tau_s + lt_0) & k \neq l \\ & m = \pm 2, \pm 3, \ldots, \pm(p-2) \end{cases} \tag{4-73}$$

where j, k, l, and n are integers dependent on m. Hence, the spectrum of the self noise can be determined from a knowledge of the spectrum of a PN sequence as given by (4-50). For $m = 0$, the correlation function of the self noise $R_{n_s}(\tau, m)$ is given by

$$R_{n_s}(\tau, 0) = 2R_{\text{PN}}(\tau) - R_{\text{PN}}(\tau - t_0) - R_{\text{PN}}(\tau + t_0) \tag{4-74}$$

with a corresponding spectrum

$$S_{n_s}(\omega, 0) = 2S_{\text{PN}}(\omega)(1 - \cos \omega t_0) \tag{4-75}$$

For $m = \pm 1$, the self-noise spectrum is identical to that of a PN sequence with its dc value removed:

$$S_{n_s}(\omega, \pm 1) = S_{\text{PN}}(\omega) - \frac{1}{p^2}\delta(\omega) \tag{4-76}$$

Finally, for the other values of m,

$$S_{n_s}(\omega, m) = 2S_{\text{PN}}(\omega)[1 - \cos \omega(k - l)t_0)] \tag{4-77}$$

The normalized envelope of the spectrum as defined by (4-75) is illustrated in Fig. 4–11. The null of this spectrum at the origin is quite desirable, since it permits removal of the self noise by a narrowband loop filter.

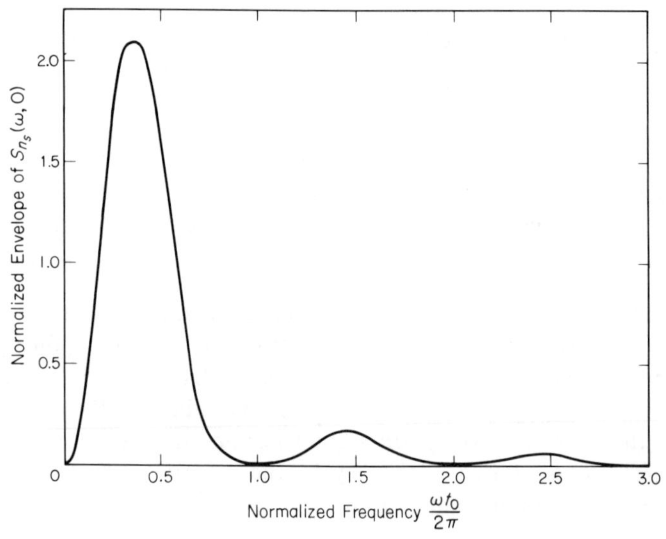

Fig. 4-11 Normalized envelope of the self-noise spectral density evaluated at $\tau_e = 0$

The additive input noise $n(t)$, when multiplied by the equivalent normalized feedback reference signal $\delta s_{\text{PN}}(t - \hat{\tau}_d)$, produces an essentially white noise process $n_n(t)$ at the cross-correlator output whose spectral density is approximately given by

$$S_{n_n}(\omega) = \left[2\left(\frac{p+1}{p}\right)\right]\left(\frac{N_0}{2}\right) \tag{4-78}$$

where $2(p + 1)/p$ is the average power of $\delta s_{\text{PN}}(t - \hat{\tau}_d)$. Thus, the total output of the cross-correlation network can be approximated by

$$\epsilon(t) = \sqrt{S}K_r K_m\left[R_{\text{DLL}}(\tau_e) + n_s(t, \tau_e) + \frac{n_n(t)}{\sqrt{S}}\right] \tag{4-79}$$

Since the output of the loop filter approximately represents the derivative of the delay estimate $\hat{\tau}_d$, then the stochastic integro-differential equation describing the operation of the loop is

$$\frac{d\tau_e}{dt} \cong \frac{d\tau_s}{dt} - \sqrt{S}Kt_0F(p)\left[R_{\text{DLL}}(\tau_e) + n_s(t, \tau_e) + \frac{n_n(t)}{\sqrt{S}}\right] \tag{4-80}$$

where $K = K_r K_m K_{\text{VCC}}$ and K_{VCC} is the gain of the voltage-controlled clock in Hz/v.

If one assumes that the loop filter is designed sufficiently narrow to remove the self-noise term in the above, then the steady-state tracking performance of the DLL can be obtained by application of the Fokker-Planck technique to (4-80). From the theory developed in Chap. 14 of Ref. 22, the p.d.f. of the delay error is approximately given by an equation similar to (2-99); that is,

$$p(\tau_e) = C_0' \exp\left[-U_0(\tau_e)\right]\int_{\tau_e}^{\tau_e + pt_0} \exp\left[U_0(x)\right] dx \qquad |\tau_e| \leq \frac{pt_0}{2} \tag{4-81}$$

where C_0' is the normalization constant and the potential function $U_0(\tau_e)$ is given by

$$U_0(\tau_e) \triangleq -\beta\tau_e + \alpha\int_{t_0}^{\tau_e} R_{\text{DLL}}(\tau)d\tau$$

$$= \begin{cases} -\beta\tau_e + \alpha\left[\dfrac{\tau_e^2 - t_0^2}{2}\right] & |\tau_e| \leq t_0 \\ -\beta\tau_e + \alpha\left[\dfrac{t_0^2 - (|\tau_e| - 2t_0)^2}{2}\right] & t_0 \leq |\tau_e| \leq 2t_0 \\ -\beta\tau_e + \alpha\left[\dfrac{t_0^2}{2}\right] & 2t_0 \leq |\tau_e| \leq \dfrac{pt_0}{2} \end{cases} \tag{4-82}$$

The parameters α and β are those defined in (2-17) with the following replacements:

$$\begin{aligned} P_c &\longrightarrow S\left(\frac{p+1}{p}\right)^2 \\ \sin\phi &\longrightarrow \left(\frac{p}{p+1}\right)t_0 R_{\text{DLL}}(\tau_e) \\ \Omega_0 &\longrightarrow \frac{d\tau_s}{dt} \\ N_0 &\longrightarrow 2N_0\left(\frac{p+1}{p}\right)t_0^2 \end{aligned} \tag{4-83}$$

For a first-order loop with τ_s essentially constant, Nishimura[23] has evaluated the normalization constant C_0' and the variance of the delay error, $\sigma_{\tau_e}^2$. The results are as follows:

$$\frac{1}{C_0'} = \sqrt{\frac{2\pi}{\rho}} \exp\left[\frac{\rho t_0^2}{2}\right]\left\{\operatorname{erf}\left(\sqrt{\frac{\rho t_0^2}{2}}\right) + \exp\left[-\rho t_0^2\right] h\left(\sqrt{\frac{\rho t_0^2}{2}}\right)\right\}$$
$$+ (p-4)t_0 \exp\left[-\frac{\rho t_0^2}{2}\right] \tag{4-84}$$

where
$$h(x) = \frac{2}{\sqrt{\pi}} \int_0^x \exp(t^2)\,dt = \frac{2}{\sqrt{\pi}} \sum_{n=0}^{\infty} \frac{x^{2n+1}}{n!(2n+1)} \tag{4-85}$$

and from (4–83),

$$\rho = \frac{4\sqrt{S}[(p+1)/p]^2}{2N_0[(p+1)/p]t_0^2 K} = \frac{1}{2t_0^2}\left(\frac{4\sqrt{S}}{N_0 K}\right) \tag{4-86}$$

The function $h(x)$ when multiplied by $(\sqrt{\pi}/2)\exp(-x^2)$ becomes Dawson's integral, which converges to zero as $x \longrightarrow \infty$. Hence, the last term in (4-84) converges for large ρ, and it is possible to evaluate it using the infinite series given in (4-85). Having the p.d.f. of τ_e, one can now by straightforward algebraic manipulation compute its variance:

$$\sigma_{\tau_e}^2 = \int_{-pt_0/2}^{pt_0/2} \tau_e^2 p(\tau_e)\,d\tau_e$$
$$= \left[\frac{\exp(\rho t_0^2/2)}{\rho}\sqrt{\frac{2\pi}{\rho}}\left\{\operatorname{erf}\left(\sqrt{\frac{\rho t_0^2}{2}}\right) + (4\rho t_0^2 - 1)\exp(-\rho t_0^2) h\left(\sqrt{\frac{\rho t_0^2}{2}}\right)\right\}\right.$$
$$\left. - \frac{8}{\rho}\left[1 - \exp\left(-\frac{\rho t_0^2}{2}\right)\right] t_0 + \frac{2t_0^3}{3}\exp\left(-\frac{\rho t_0^2}{2}\right)\left(\frac{p^3 - 64}{8}\right)\right] \tag{4-87}$$

For large ρ, (4–87) simplifies to

$$\sigma_{\tau_e}^2 = \frac{\begin{array}{c}\operatorname{erf}(\sqrt{\rho t_0^2/2}) + (4\rho t_0^2 - 1)\exp(-\rho t_0^2) h(\sqrt{\rho t_0^2/2}) \\ - 4\sqrt{2\rho t_0^2/\pi}\exp(-\rho t_0^2/2)[1 - \exp(-\rho t_0^2/2)]\end{array}}{\rho[\operatorname{erf}(\sqrt{\rho t_0^2/2}) + \exp(-\rho t_0^2) h(\sqrt{\rho t_0^2/2})]} \tag{4-88}$$

which behaves like $1/\rho$ in the limit as ρ approaches infinity.

The implementation of the DLL cross-correlator can be greatly simplified if the input signal is first converted to binary form by passing it through a hard limiter. This allows the analog multipliers to be replaced by modulo 2 adders. The penalty associated with making this change is negligible for high input signal-to-noise ratios, since the signal component of the input is already in binary form. At low input signal-to-noise ratios, assuming a limiter characteristic of the form A sgn x, the signal component of the cross-correlation between the limiter output and the PN reference signal is given by

$$E\{A \operatorname{sgn}[\sqrt{S}\, s_{\mathrm{PN}}(t) + n(t)] s_{\mathrm{PN}}(t+\tau)\} \cong A\sqrt{\frac{2}{\pi}\left(\frac{S}{N_i}\right)} R_{\mathrm{PN}}(\tau) \tag{4-89}$$

where $N_i = N_0 B_i$ is the input noise power in the one-sided low-pass input bandwidth B_i. If no limiter were used at the input, but instead an ideal amplifier maintained the total power entering the phase detector constant at A^2, then the signal component of the cross- correlation would in this situation be

$$A\sqrt{\frac{S}{S+N_i}}R_{\text{PN}}(\tau) \cong A\sqrt{\frac{S}{N_i}}R_{\text{PN}}(\tau) \tag{4-90}$$

Comparing (4–89) and (4–90) we observe that the signal component of the cross-correlation with amplitude-limited input is smaller by a factor $\sqrt{2/\pi}$. Thus, as far as the output of the cross-correlation network is concerned, the phase detector characteristic $R_{\text{DLL}}(\tau)$ is modified by the factor $\sqrt{2/\pi(A^2/N_i)}$ when the input is first passed through the hard limiter. Concerning the effect of the limiter on the noise at the correlator network output, we first observe that the autocorrelation function $R_{no}(\tau)$ of the noise at the limiter output is given in terms of the autocorrelation function of $n(t)$ by

$$R_{no}(\tau) = \frac{2A^2}{\pi}\sin^{-1}\left(\frac{R_n(\tau)}{N_i}\right) \tag{4-91}$$

Since the DLL tracking bandwidth is quite narrow relative to the noise bandwidth, the effect of the above noise can be accounted for principally in terms of the value of its spectral density at the origin:

$$G_{no}(0) = \int_{-\infty}^{\infty} R_{no}(\tau)\,d\tau = \left(\frac{N_0}{2}\right)\left[\frac{2.2}{\pi}\left(\frac{A^2}{N_i}\right)\right] \tag{4-92}$$

Hence, to a first approximation, one may replace $N_0/2$ in (4-78) by $G_{no}(0)$ of above, and for a fixed tracking-loop bandwidth, the net reduction in loop signal-to-noise ratio ρ is merely a factor of 1/1.1 or 0.4 dB. Thus, we conclude that the use of a limiter in front of a DLL allows a tradeoff between a small degradation in theoretical noise performance and a large simplification in cross-correlation circuitry.

The Double-Loop Range Tracking Receiver

Another method for tracking a digital ranging signal is to structure the transmitted code so that it contains a clock component that is rapidly acquirable by a tracking loop in the ranging receiver. The term *clock* as used here implies a sequence of alternating ones and zeros or, in waveform terminology, a ± 1 deterministic square wave denoted by Ck(t). The clock becomes imbedded in the input signal format by transmitting the product of the clock and code symbol waveforms. An important parameter, as we shall discuss shortly, is the product of the frequency f_{Ck} of the clock signal by the code digit period t_0.

In the ranging receiver, we incorporate a double-loop tracking device (Fig. 4-12) whose operation is as follows. The inner or "clock" loop is a standard PLL, which rapidly locks up to the incoming clock component. The clock estimate is then used to time a code generator capable of producing local

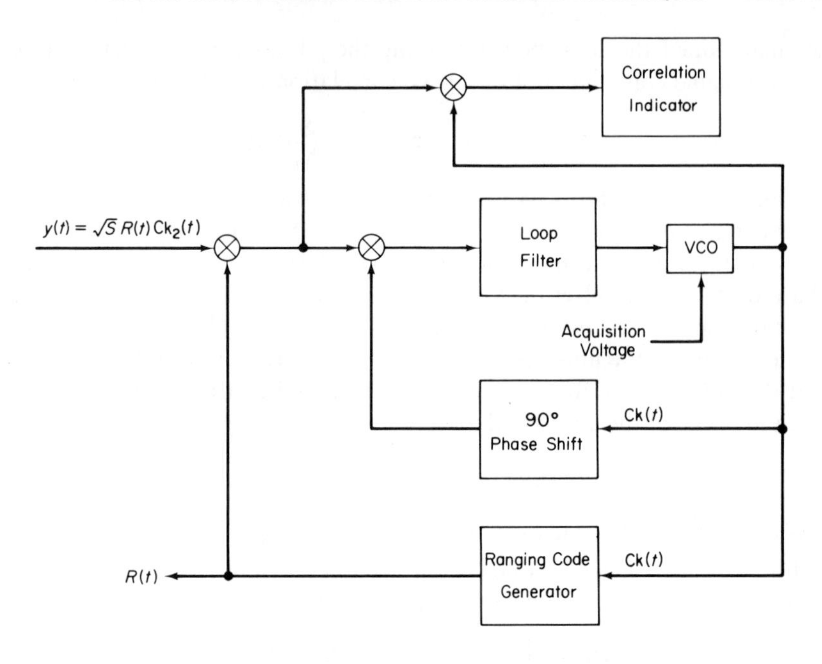

Fig. 4-12 A double-loop range tracking receiver

replicas of the transmitted ranging code components $R(t)$. This code generator also possesses the ability to shift these component code sequences one digit at a time. Starting then with the first component sequence, the outer or "code" loop forms the cross-correlation of the code generator output and the received code times clock. This process continues, each time advancing the phase of the locally generated code symbol waveform by one digit period for each cross-correlation calculation. When the local code replica is in phase with the corresponding code component of the received sequence, the output of the cross-correlator will contain a *maximum of clock component*; that is, if the cross-correlator output is correlated with the clock estimate produced by the inner loop, the result of this correlation is a maximum. If the inner loop were perfectly in lock with no phase error, the value of this maximum correlation (relative to unity) would be essentially equal to the normalized cross-correlation function between the total ranging code and the first component sequence evaluated at $\tau = 0$.

The above procedure is carried out for each component code in sequence until finally the entire ranging code is acquired. As each component code is acquired, the amount of clock component as measured by the correlation indicator in Fig. 4-12 increases until a value of unity is reached, when the entire received code signal and the total code generator output are in phase. At this point, the clock is completely demodulated. As the acquisition progresses—the clock is becoming more nearly completely demodulated—the effective signal to noise ratio in the clock-loop bandwidth increases. When acquisition is complete, the tracking function of the inner loop causes the outer loop to follow a change in phase of the incoming code, which is reflected as a change in phase of the clock.

The equivalent phase detector characteristic used for tracking in the clock loop is a function of the ratio of the clock frequency to the switching frequency of the code. Since this characteristic is approximately determined by the product of the individual clock and code correlation functions,[7] then for a high frequency clock (relative to a code digit period), the PN correlation function of the code acts as a low frequency amplitude modulation on the clock correlation function. The resulting phase detector characteristic will thus have a multitude of stable null points (coinciding with those of the clock correlation function) and hence is unsuitable as a tracking function.

The best choice of clock frequency is one in which one period of the clock is equal to two code-digit periods—$f_{Ck} = 1/2t_0$. This particular clock signal shall be denoted by $Ck_1(t)$. Assuming then that the input received signal is composed of the product of $Ck_1(t)$ and a PN code of long period [that is, $R(t) = s_{PN}(t)$, the equivalent phase detector characteristic for this case (Fig. 4–13) becomes

$$R_{DBL}(\tau) \triangleq \frac{1}{2pt_0}\int_0^{2pt_0} s_{PN}(t)Ck_1(t)s_{PN}(t+\tau)Ck_1\left(t - \frac{1}{4f_{Ck}} + \tau\right)d\tau$$

$$= \begin{cases} \left(\frac{p-1}{p}\right)\frac{\tau}{t_0} & |\tau| \le \frac{t_0}{2} \\ -\left(\frac{p-1}{p}\right)\frac{\tau}{t_0}\left[1 - \frac{t_0}{|\tau|}\right] & \frac{t_0}{2} \le |\tau| \le t_0 \\ (-1)^{m+1}\left(\frac{2}{p}\right)\frac{\tau}{t_0}\left[1 - \frac{mt_0}{|\tau|}\right] & mt_0 \le |\tau| \le (m+\frac{1}{2})t_0 \\ (-1)^{m+2}\left(\frac{2}{p}\right)\frac{\tau}{t_0}\left[1 - \frac{(m+1)t_0}{|\tau|}\right] & (m+\frac{1}{2})t_0 \le |\tau| \le (m+1)t_0 \end{cases} \quad m = 1, 2, \ldots, (p-2)$$

$$R_{DBL}(\tau) = -R_{DBL}(|\tau| - pt_0) \quad (p-1)t_0 \le |\tau| \le (2p-1)t_0 \tag{4-93}$$

We notice that the correlation function for the double-loop tracking device as specified by (4-93) is similar to that found for the DLL [see Eq. (4-70)]. In particular, the linear slopes of both tracking curves in the vicinity of the origin are essentially equal for large p. The primary difference between the two tracking curves is that $R_{DBL}(\tau)$ is inverted at odd multiples of pt_0 and hence its period is twice that of $R_{DLL}(\tau)$. The reason for this is that even though the PN code components are in phase at odd multiples of pt_0, the two clock components are half a period out of phase relative to their positions at even multiples of pt_0. These quasi-stable nulls at odd half periods of the

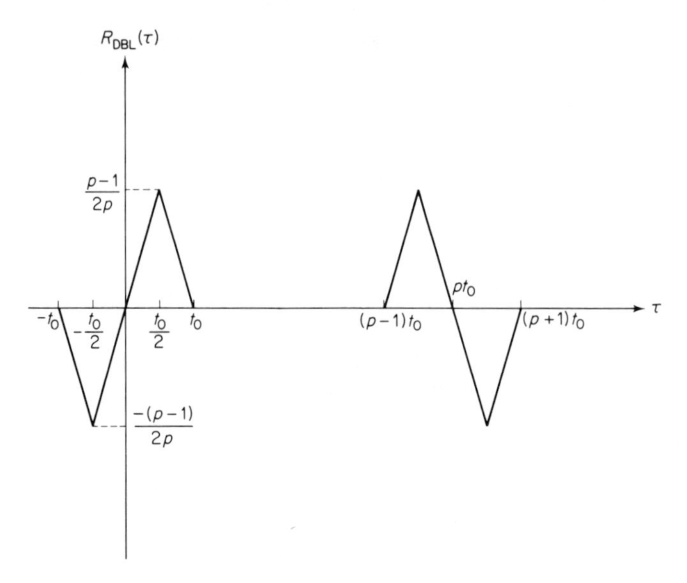

Fig. 4-13 Equivalent phase detector characteristic, $R_{DBL}(\tau)$, for the double-loop range tracking receiver

correlation function are often undesirable in a practical system. One way of solving this problem is to instrument the double-loop tracking device as in Fig. 4-14. Let $Ck_2(t)$ denote a clock at twice the frequency of $Ck_1(t)$. We transmit the code on this clock and hence the received signal at the input to the double-loop tracking device is now $y(t) = \sqrt{S} s_{PN}(t) Ck_2(t)$. The ideal reference signal at the first demodulator is the product of the code $s_{PN}(t)$ and the locally derived clock $Ck_1(t)$, which when multiplied by the input signal produces $Ck_1(t)$ shifted by one-quarter of a cycle. If this signal is now multiplied by $Ck_1(t)$ at the second multiplier, an equivalent tracking error curve is produced at the output of this multiplier whose functional form is identical to (4-93) except that the peak $(p - 1)/2p$ is now replaced by $(p + 1)/2p$ and the function is identically zero for $t_0 \leq |\tau| \leq (p - 1)t_0$. The first point is inconsequential for large p (which is where the equations apply anyway), since in either case the peak of the tracking curve is approximately 1/2. The important difference is that the tracking curve at odd multiples of pt_0 is no longer inverted and the small fluctuations between periods are totally eliminated.

From the above discussion, the similarity in loop error curves between the DLL and the double-loop receiver allows us to write down by inspection the p.d.f. of the timing error for the latter. In fact, for large p, (4-81) through (4-83) are valid for characterizing the double-loop receiver if the following

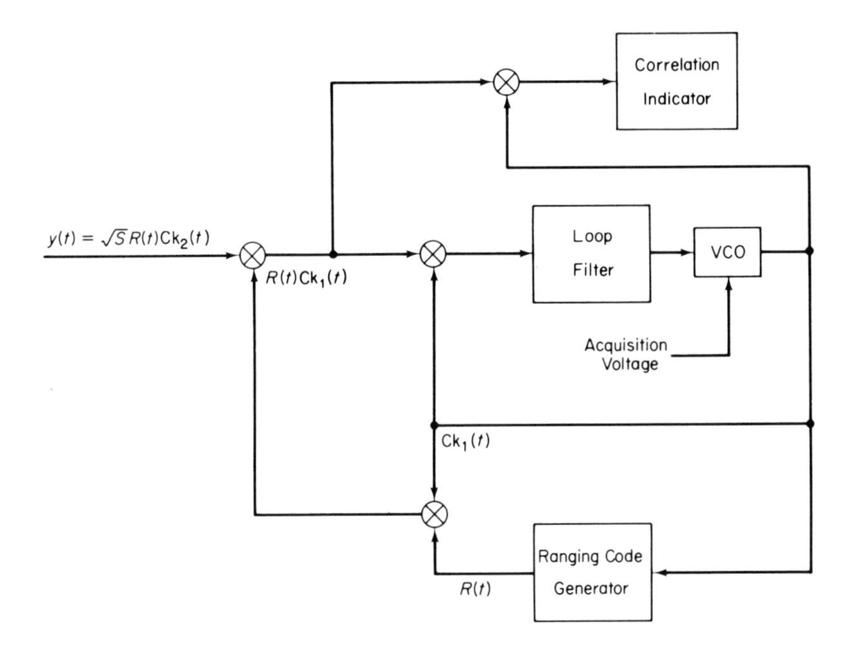

Fig. 4-14 An alternate mechanization for the double-loop range tracking receiver

substitutions are made:

$$
\begin{aligned}
\sqrt{S} &\longrightarrow 2\sqrt{S} \\
t_0 &\longrightarrow 2t_0 \\
N_0 &\longrightarrow \frac{N_0}{2} \\
R_{\mathrm{DLL}}(\tau_e) &\longrightarrow R_{\mathrm{DBL}}(\tau_e)
\end{aligned}
\tag{4-94}
$$

We also ignore the $(p + 1)/p$ factors in (4-83), since once again we are considering the case of large p. Making the substitutions of (4-94) in the loop signal-to-noise ratio ρ as defined by (4-86) does not change its value. Hence, the timing error variance as predicted by the linear theory for the double-loop receiver is identical to that of the DLL. It is important to note, however, that their performances in the nonlinear region of operation are quite different.

Finally, Table 4-2 summarizes various system applications which use the theory presented in this chapter.

TABLE 4-2 Various Ranging System Design Approaches

Ranging Signal	*Ranging System*	*Use or Application*
Fixed tone	Goddard range and range rate system[6,25]	Relay I satellite Synchronous Communications Satellite (SYNCOM) Eccentric Orbiting Geophysical Observatory (EGO) Navy's Omega navigation system[26]
Swept tone	Missile Trajectory Measurement System (MISTRAM)	Atlantic missile range to determine precise missile trajectories
PN codes	Mark I system[24,27] Modified Mark I system	Apollo Unified S-band system Air Force Space-Ground Link Subsystem (SGLS).
BINOR Codes	Proposed for an L-band navigation experiment on NASA's Application Technology Satellite E	

PROBLEMS

4-1. Consider the maximum *a posteriori* approach to epoch estimation discussed in Sec. 4-2.

a) Using Eqs. (4-3) and (4-4), show that the epoch estimate $\hat{\tau}_d$ satisfies the equation

$$R_s'(\tau_s - \hat{\tau}_d) = R_{ns}'(\hat{\tau}_d)$$

where the prime denotes differentiation with respect to the argument of the function.

b) Since for large signal-to-noise the estimate $\hat{\tau}_d$ is close to the true epoch τ_s, then expanding $R_s(\tau_s - \hat{\tau}_d)$ in a series in powers of $(\tau_s - \hat{\tau}_d)$, and keeping only terms of first order, show that the estimation error is given approximately by

$$\tau_e \triangleq \tau_s - \hat{\tau}_d = \frac{R_{ns}'(\hat{\tau}_d)}{R_s''(0)}$$

where $R_s''(0)$ is the second derivative of $R_s(\tau)$ evaluated at the origin.

c) Show that the mean and variance of the estimation error as approximated in part (b) are given by (4-6).

Hint: The Fourier transform of a periodic function is invariant to a shift of the function by an integral number of periods.

4-2. Assuming that a periodic ranging signal has the Fourier series expansion

$$s(t) = \sum_{n=-\infty}^{\infty} c_n \exp\left(jn\frac{2\pi}{T}t\right)$$

with a corresponding Fourier transform

$$S(\omega) = 2\pi \sum_{n=-\infty}^{\infty} c_n \delta\left(\omega - n\frac{2\pi}{T}\right)$$

show that (4-7) can be expressed in the form

$$\beta_s^2 = \frac{\sum_{n=-\infty}^{\infty} |c_n|^2 \left(n\frac{2\pi}{T}\right)^2}{\sum_{n=-\infty}^{\infty} |c_n|^2} = \frac{\sum_{n=-\infty}^{\infty} |c_n|^2 \left(n\frac{2\pi}{T}\right)^2}{S}$$

4-3. Consider a fixed-tone ranging system wherein the tone frequencies are geometrically related with $M = 5$. If the ranging signal is composed of a sum of eight sinusoidal tones with the lowest tone having a frequency of 8 Hz, then,

a) compute the maximum range (in meters) of the measurement.
b) If $S/N_0 = 10$ dB and the individual tone powers are made equal, compute the mean-squared error in range.

4-4. Suppose that in the ranging system of Prob. 4-3 the range accuracy is determined by a phase measurement on the highest frequency tone and the ambiguity is resolved by successive phase measurements on the lower frequency tones. Then, if the loop bandwidth of the PLL which tracks the highest frequency tone is $W_{LN} = 10$ Hz, and individual tone powers are again made equal,

a) compute the mean-squared error in range and compare your answer with part b) of Prob. 4-3.
b) If all PLLs are assumed to have the same loop bandwidth, compute the probability of incorrectly resolving the phase ambiguity on each of the seven lower frequency tones.

4-5. Consider a sequential, square-wave tone ranging system such as that employed by the Jet Propulsion Laboratory on the Mariner–Mars 1969 and 1971 missions. As in the sinusoidal case, the highest frequency tone is used for range measurement accuracy while the lower frequency tones resolve ambiguity. For range accuracy measurement, the input ranging signal, consisting of the propagation-delayed highest frequency tone plus noise, is cross-correlated with a replica of the highest frequency square wave and a 90 degree shifted version of it. These two cross-correlations provide the estimate of epoch, according to the relation:

$$\hat{\tau}_d = \frac{T_N}{4}\left[1 - \frac{R_{yi}(\tau_s)}{|R_{yi}(\tau_s)| + |R_{yq}(\tau_s)|}\right] \text{sgn}\,[R_{yq}(\tau_s)] \qquad |\tau_s| \leq \frac{T_N}{2}$$

where

$$R_{yi}(\tau_s) \triangleq \frac{1}{T_0}\int_0^{T_0} y(t)s_N(t)\,dt$$

$$R_{yq}(\tau_s) \triangleq \frac{1}{T_0}\int_0^{T_0} y(t)s_N(t + T_N/4)\,dt$$

and the received signal is characterized by

$$y(t) = s_N(t - \tau_s) + n(t)$$

In the above, T_0 is the correlation time and $s_N(t)$ is the highest frequency square-wave tone of period T_N and power S.

a) Show that the mean and variance of the in-phase and quadrature cross-correlations are given by

$$E\{R_{yi}(\tau_s)\} = \begin{cases} S\left(1 - \dfrac{4\tau_s}{T_N}\right) & 0 \leq \tau_s \leq \dfrac{T_N}{2} \\ S\left(1 + \dfrac{4\tau_s}{T_N}\right) & -\dfrac{T_N}{2} \leq \tau_s \leq 0 \end{cases}$$

$$E\{R_{yq}(\tau_s)\} = \begin{cases} \dfrac{4S}{T_N}\tau_s & |\tau_s| \leq \dfrac{T_N}{4} \\ S\left(2 - \dfrac{4\tau_s}{T_N}\right) & \dfrac{T_N}{4} \leq |\tau_s| \leq \dfrac{T_N}{2} \end{cases}$$

$$\sigma^2_{R_{yi}} = \sigma^2_{R_{yq}} = \frac{N_0 S}{2T_0}$$

b) Using the results of part a), show that for large signal-to-noise ratio and $\tau_s = 0$, the variance of the epoch estimate is given by

$$\sigma^2_{\hat{\tau}_d} = \frac{N_0 T_N^2}{32 S T_0}$$

(Hint: Use the simple approximation)

$$\sigma^2_{\hat{\tau}_d} = \left(\frac{\partial \hat{\tau}_d}{\partial R_{yi}}\right)^2 \sigma^2_{R_{yi}} + \left(\frac{\partial \hat{\tau}_d}{\partial R_{yq}}\right)^2 \sigma^2_{R_{yq}}$$

c) Find the equivalent mean-squared error in range $\sigma^2_{\hat{d}}$.

4-6. Consider measuring the range of a vehicle that is moving at constant velocity v. Assume that a separate channel is available for transmitting a carrier on which the corresponding Doppler shift $f_d = 2v/c$ can be measured.

a) If ranging information is transmitted as a *fixed-tone* signal, discuss how the equations of Sec. 4-3, which describe the fixed-tone ranging technique, must be modified to include Doppler effect.

b) Repeat part a) if the ranging information is transmitted as a *swept-tone* signal.

4-7. Consider the problem of cross-correlating a received signal, consisting of the product of a PN signal and a clock, with a locally generated replica of the PN signal and a phase-shifted version of the clock. Mathematically speaking, let $s_{PN}(t)$ be a PN signal of period pt_0 (p odd) and $C(t)$ be a unit power clock of period $2t_0$. The receiver forms the cross-correlation function

$$R(\alpha_t, \alpha_r, \delta) = \frac{1}{2pt_0}\int_0^{2pt_0} s_{pn}(t)C(t - \alpha_t t_0)s_{pn}(t + \delta t)C[t + (\delta - \alpha_r)t_0]\,dt$$

where α_t and α_r are respectively the normalized angles of the transmitted and locally-generated clocks, and δ is the normalized synchronization error between the received and locally-generated signals.

If δ is separated into its integer and fractional parts, i.e., $\delta = m + \Delta$, where m is an integer and $0 \leq \Delta \leq 1$, then using the periodic property of the clock and the two-level autocorrelation property of the PN code,

a) show that $R(\alpha_t, \alpha_r, \delta)$ can be expressed in the form

$$R(\alpha_t, \alpha_r, \delta) = \frac{(-1)^m}{p}\left[\frac{1}{t_0}\int_0^{t_0} C(t - \alpha_t t_0)C[t + (\Delta - \alpha_r)t_0]\,dt\right]$$
$$+ \left(\frac{p+1}{p}\right)(-1)^m \times \begin{cases} \dfrac{1}{t_0}\displaystyle\int_0^{(1-\Delta)t_0} C(t - \alpha_t t_0)C[t + (\Delta - \alpha_r)t_0]\,dt & m = lp \\ \dfrac{1}{t_0}\displaystyle\int_{(1-\Delta)t_0}^{t_0} C(t - \alpha_t t_0)C[t + (\Delta - \alpha_r)t_0]\,dt & m = lp - 1 \\ 0 & m \neq lp, m \neq lp - 1 \end{cases}$$

where l is an integer.

b) If $C(t)$ is a *square-wave* clock, i.e., $Ck_1(t)$, $\alpha_t = 0$, and $\alpha_r = -1/2t_0$, evaluate $R(0, -1/2t_0, \delta)$ and show that for large p the answer reduces to the correlation function for the double-loop tracking device as given by (4-93).

c) If $C(t)$ is a *sinusoidal* clock, i.e., $C(t) = \sqrt{2}\sin \pi t/t_0$, evaluate $R(0, -1/2t_0, \delta)$.

4-8. Suppose now that the transmitted PN signal $s_{\mathrm{PN}}(t)$ multiplies a unit power clock $C(t)$ of period t_0 and the receiver forms the cross-correlation function

$$R(\delta t_0) = \frac{1}{pt_0}\int_0^{pt_0} s_{\mathrm{PN}}(t)C(t)s_{\mathrm{PN}}(t + \delta t_0)\,dt$$

where δ is defined as in the previous problem. Again separating δ into its integer and fractional parts,

a) show that $R(\delta t_0)$ can be expressed in the form

$$R(\delta t_0) = -\frac{1}{p}\left[\frac{1}{t_0}\int_0^{t_0} C(t)\,dt\right]$$

$$+\left(\frac{p+1}{p}\right) \times \begin{cases} \dfrac{1}{t_0}\displaystyle\int_0^{(1-\Delta)t_0} C(t)\,dt & m = lp \\ \dfrac{1}{t_0}\displaystyle\int_{(1-\Delta)t_0}^{t_0} C(t)\,dt & m = lp - 1 \\ 0 & m \neq lp, m \neq lp - 1 \end{cases}$$

where l is an integer.

b) If $C(t)$ is a *square-wave* clock, i.e., $Ck_2(t)$, evaluate $R(\delta t_0)$. For large p, compare this result with that obtained in part b) of Prob. 4-7 and discuss any differences.

c) Evaluate $R(\delta t_0)$ when $C(t) = \sqrt{2} \sin 2\pi t/t_0$.

4-9. As an extension of the results of Prob. 4-8, consider the switched-carrier system discussed in Prob. 7-11 and assume that one of the channels carries ranging information. The signal input to the ranging receiver in one of the spacecrafts is now of the form

$$s_{\text{PN}}(t)C(t)\,\mathfrak{S}(t)$$

where $\mathfrak{S}(t) \triangleq \frac{1}{2} + \frac{1}{2}\mathfrak{S}w(t)$ is a 0 or 1 waveform defined in terms of the ± 1 switching waveform $\mathfrak{S}w(t)$. If the receiver now forms the cross-correlation function

$$R(\delta t_0) = \frac{1}{2pt_0}\int_0^{2pt_0} s_{\text{PN}}(t)C(t)\,\mathfrak{S}(t)s_{\text{PN}}(t + \delta t_0)\,dt$$

then assuming that $C(t)$ is selected as the digital clock $Ck_2(t)$ and $\mathfrak{S}w(t)$ as the phase-shifted digital clock $Ck_1(t - t_0/2)$,

a) Evaluate $R(\delta t_0)$.

b) Compare the result of part a) with the answer found in part b) of Prob. 4-8 and assess the effect of $\mathfrak{S}(t)$ on the correlation function.

4-10. Consider a ranging system wherein the demodulated ranging signal and additive noise are passed through a low-pass limiter before being acquired and tracked by the ranging receiver. If the ranging signal is modeled as a range code on a clock (i.e., a $\pm A$ waveform) and the low-pass limiter by a signum (sgn) function,

a) Show that the signal amplitude suppression factor is given by

$$\tilde{\alpha} \triangleq \frac{E\{s_0(t)\}}{E\{s_0(t)\}\,|_{\rho_i=\infty}} = \text{erf}\sqrt{\rho_i/2}$$

where $\rho_i = 2A^2/N_0W_i$ is the limiter input signal-to-noise ratio in the two-sided input bandwidth W_i and $s_0(t)$ denotes the limiter output.

b) Assuming that the ranging receiver makes use of the total low-pass limiter output, what is the signal-to-noise ratio at the limiter output? These results will be used in Chap. 7 where the performance of a turnaround transponder ranging system is considered as an application of the single-channel theory.

4-11. Calculate the effect of a single-pole, low-pass filter on the cross-correlation function $R_{yz}(\tau)$ where $x(t) = s_{PN}(t)Ck_1(t)$, $z(t) = s_{PN}(t)Ck_1(t + t_0/2)$, and $y(t)$ is the output of the low-pass filter when $x(t)$ is the input to this filter.

REFERENCES

1. EASTERLING, M., "A Long-Range Precision Ranging System," Jet Propulsion Laboratory, Pasadena, Calif., Technical Report No. 32–80 (July 10, 1961).
2. ———, "Methods for Obtaining Velocity and Range Information from CW Radars, "Jet Propulsion Laboratory, Pasadena, Calif., Technical Report No. 32–657 (September 1, 1964).
3. OTTEN, D. D., "A Satellite System for Radio Navigation," *AIAA First Communications Satellite Systems Conference* (October, 1968).
4. VAN TREES, H. L., *Detection, Estimation, and Modulation Theory, Part I*, John Wiley and Sons, Inc., New York, N.Y., 1968.
5. BAGHDADY, E. J., and KRONMILLER, G. C., JR., "The Goddard Range and Range Rate Tracking System: Concept, Design and Performance," *1965 International Space and Electronics Symposium Record* (November 2–4, 1965).
6. GOLDSTEIN, R., "Ranging with Sequential Components," Jet Propulsion Laboratory, Pasadena, Calif., SPS 37–52, Vol. II (1968) 46–49.
7. GOLOMB, S. W., et. al., *Digital Communications with Space Applications*, Prentice-Hall, Inc., Englewood Cliffs, N. J., 1964.
8. TITSWORTH, R. C., "Correlation Properties of Cyclic Sequences," Jet Propulsion Laboratory, Pasadena, Calif., Technical Report No. 32–388 (July 1, 1963).
9. ———, "Optimal Ranging Codes," *IEEE Transactions on Space Electronics and Telemetry*, Vol. SET-10, No. 1 (1964) 19–30.
10. STIFFLER, J. J., "Block Coding and Synchronization Studies: Rapid Acquisition Sequences," Jet Propulsion Laboratory, Pasadena, Calif., SPS 37–42, Vol. IV (1966) 191–197.
11. COOK, C. E., "Pulse Compression: Key to More Efficient Radar Transmission," *Proceedings of the IRE*, Vol. 48, No. 3 (1960) 310–316.
12. ———, and CHIN, J. E., "Linear FM Pulse Compression," *Space/Aeronautics*, Vol. 34 (1960) 124–129.
13. OHMAN, G. P., "Getting High Range Resolution with Pulse Compression Radar," Electronics, Vol. 33, No. 41 (1960) 53–57.

14. KLAUDER, J. R., PRICE, A. C., DARLINGTON, S., and ALBERSHEIM, W. J., "The Theory and Design of Chirp Radars," *Bell System Technical Journal*, Vol. 39, No. 4 (1960) 745–808.

15. DICKE, R. H., "Object Detection System," U. S. Patent 2, 624, 876, January 6, 1953.

16. DARLINGTON, S., "Pulse Transmission," U. S. Patent 2, 678, 997, May 18, 1954.

17. SKOLNIK, M. I., *Introduction to Radar Systems*, McGraw-Hill Book Co., New York, N. Y., 1962.

18. BERKOWITZ, R. S., *Modern Radar: Analysis, Evaluation, and System Design*, John Wiley & Sons Inc., New York, N. Y., 1965.

19. COOK, C. E., and BERNFELD, M., *Radar Signals*, Academic Press, New York, N. Y., 1967.

20. SPILKER, J. J., JR., and MAGILL, D. T., "The Delay-Lock Discriminator—An Optimum Tracking Device," *Proceedings of the IRE*, Vol. 49, No. 9 (1961) 1403–1416.

21. SPILKER, J. J., JR., "Delay-Lock Tracking of Binary Signals," *IEEE Transactions on Space Electronics and Telemetry*, Vol. SET-9, No. 1 (1963) 1–8.

22. LINDSEY, W. C., *Synchronization Systems in Communication and Control*, Prentice-Hall, Inc., Englewood Cliffs, N.J., 1972.

23. NISHIMURA, T., "The Mean-Squared Deviation of a Phase-Locked Loop Having a Triangular S-Curve," Jet Propulsion Laboratory, Pasadena, Calif., SPS 37–31, Vol. IV (1965) 311–315.

24. TAUSWORTHE, R. C., "Optimum Design of Turnaround Ranging Systems," *Proceedings of the National Telemetering Conference*, Houston, Texas (1965) 36–38.

25. HABIB, E. J., KRONMILLER, G. C., JR., ENGELS, P. D., and FRANKS, H. J., JR., "Development of a Range and Range Rate Spacecraft Tracking System," National Aeronautics and Space Administration, NASA TND-2093, Goddard Space Flight Center, Greenbelt, Md. (June, 1964).

26. PIERCE, J. A., "Omega", *IEEE Transactions on Aerospace and Electronic Systems*, Vol. AES-1, No. 3 (1965) 206–215.

27. LINDLEY, P. L., "The PN Technique of Ranging as Applied in the Ranging Subsystem Mark I," Jet Propulsion Laboratory, Pasadena, Calif., Technical Report No. 32–811 (November 15, 1965).

Chapter 5

PHASE-COHERENT DETECTION WITH PERFECT REFERENCE SIGNALS

5-1. INTRODUCTION

In this chapter we shall be concerned with the problem of *digital data transmission* through the communication system model shown in Fig. 5-1. In the form illustrated, the transmitter-receiver configuration represents a *coded digital communication* system. If the channel encoder and decoder blocks are omitted, the system in Fig. 5-1 represents an *uncoded digital communication* system. The word *system* will be used for either the coded or uncoded systems, with qualification only where there is danger of confusion.

The transmitter of Fig. 5-1 is characterized as follows. The *source* represents the *information* or *data* to be transmitted. The *source sampler* samples the source and applies these samples to an N-level *quantizer*. The quantizer output consists of one of N *source symbols*. Each source symbol is applied to the *source encoder*, which converts the source symbols into *data symbols*.* The set of data symbols comprises the *symbol alphabet*. Each data symbol is composed of $\log_r N$ *data digits* of numerical base, r. In general, the source encoder may be designed so as to remove source redundancy from the source samples. The *channel encoder* provides a one-to-one mapping of the data symbols into *code words* or *channel symbols*. The complete set of code words or channel symbols forms the *code dictionary*. In general, each code word is composed of a sequence of *channel digits* of base r. When the code words in the dictionary are all constrained to have the same *block length*, that is, the

* For example, the combination of source encoder and N-level quantizer could be an *analog-to-digital converter*.

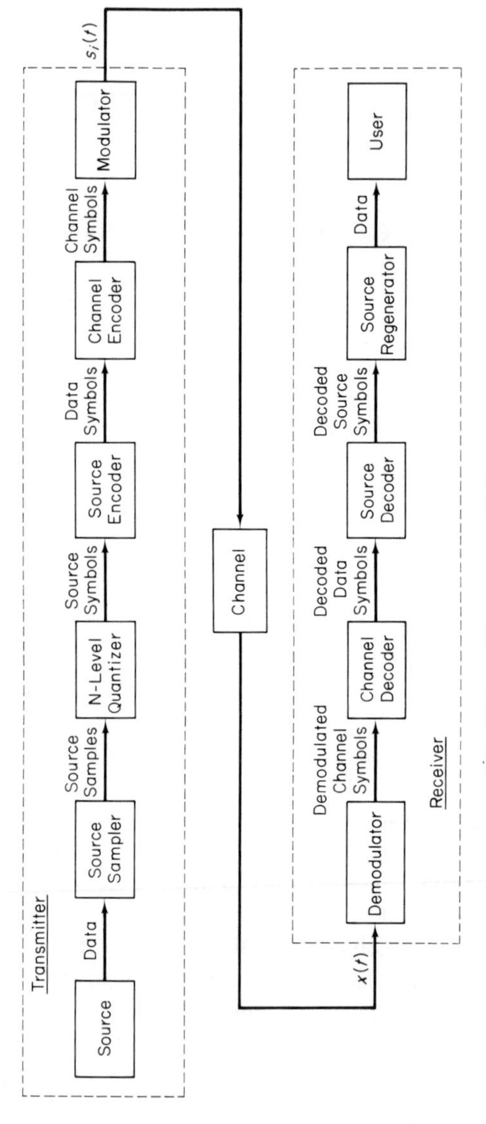

Fig. 5-1 Data transmission system

same number of channel digits, the code is called a *block code*. Usually the channel encoder inserts redundancy for purposes of *error control*—to reduce the number of errors that may occur during data transmission.

In an uncoded digital communication system, the function of the *modulator* is to provide a one-to-one mapping of the data (or channel) symbols into the set of *signals* $\{s_i(t), i = 1, 2, \ldots, N\}$. On the other hand, the modulator in a coded digital communication system provides a one-to-one mapping of the channel symbols into the set of signals $\{s_i(t), i = 1, 2, \ldots, N\}$. For example, if the source encoder converts each source symbol into n binary digits (*bits*), i.e. $r = 2$, then there are $N = 2^n$ possible data symbols available for application to the channel encoder. Thus, one needs $N = 2^n$ signals in the signaling set $\{s_i(t)\}$.

In either system, one signal from the set $\{s_i(t)\}$ is transmitted every T seconds so that the *transmission rate*, $\mathcal{R}_b$, is defined by

$$\mathcal{R}_b = \frac{\log_r N}{T} \quad \text{digits/sec} \tag{5-1}$$

or equivalently, the time required to transmit one digit is

$$T_b = \frac{T}{\log_r N} \quad \text{sec/digit} \tag{5-2}$$

Since this transmission rate actually occurs when the source symbols are equally probable and statistically independent, maximum information rate from the source is ensured.

In general, the signal $s_i(t)$ may be represented by

$$\begin{aligned} s_i(t) &= A_i(t) \sin(\omega_0 t + \theta_i(t) + \theta_0) \\ &= B_i(t) \cos(\omega_0 t + \theta_0) + C_i(t) \sin(\omega_0 t + \theta_0) \end{aligned} \tag{5-3}$$

where

$$A_i(t) = \sqrt{B_i^2(t) + C_i^2(t)} \tag{5-4}$$

$$\theta_i(t) = \tan^{-1}\left[\frac{B_i(t)}{C_i(t)}\right] \tag{5-5}$$

The waveforms $A_i(t)$, $B_i(t)$, $C_i(t)$, and $\theta_i(t)$ are all known functions of time, $f_0 = \omega_0/2\pi$ (in Hz) is the *carrier frequency*, and θ_0 is the *carrier phase*.

If either the amplitude, $A_i(t)$, the frequency, f_0, or the phase, $\theta_i(t)$, of $s_i(t)$ is varied discretely with known time functions, one has the various forms of *digital modulation*. When the amplitudes $A_i(t)$, $i = 1, 2, \ldots, N$, are pulse waveforms and ω_0 and θ_0 are constant, this discrete form of amplitude modulation is commonly referred to as *pulse-amplitude modulation* (PAM). Other pulse modulation techniques include *pulse-position modulation* (PPM) and *pulse-duration modulation* (PDM). In the former, data are conveyed by varying the position in time of a pulse waveform, while in the latter, data are conveyed by varying the duration of the pulse. If the frequency of signal $s_i(t)$ is varied discretely with both $A_i(t)$ and $\theta_i(t)$ held constant—that is, $\omega_0 =$

$2\pi(k + i)/T$ for $i = 1, 2, \ldots, N$, and k is any integer—the modulation technique is called *multiple frequency-shift-keying* (MFSK). If the phase $\theta_i(t) = 2(i - 1)\pi/N$ for all $i = 1, 2, \ldots, N$ and both $A_i(t)$ and ω_0 are held constant, we have *multiple phase-shift-keying* (MPSK). If $N = 2$, MFSK and MPSK are usually referred to as *frequency-shift-keying* (FSK) and *phase-shift-keying* (PSK), respectively. Finally, if the amplitude and phase of a constant frequency carrier are varied discretely, one has the hybrid of PAM and MPSK modulation called *combined amplitude-phase-shift-keying* (APSK).

At the receiver an inverse operation is performed. The demodulator interprets the received signal as a sequence of digits of base r. In an uncoded digital communication system, these digits are converted by the *source decoder* into source symbols which are regenerated in the source *regenerator* and made available to the user.

The receiver in a coded digital communication system differs from the uncoded case in that the *channel decoder* converts the sequence of digits of base r at the demodulator output into data symbols which, if no errors are made, correspond to the data symbols transmitted. The function of the source decoder here is the same as in an uncoded system. When the modulator and demodulator are included as part of the communication channel, then the diagram in Fig. 5-1 represents the communication system model usually defined from an information-theoretic viewpoint. The figure of merit for such a transmitting and receiving system is taken to be the probability of making a channel symbol error (*word error probability*) or the probability of a specific bit being in error (*bit error probability*).

It is significant to note that each block shown at the receiver requires some level of *synchronization* (timing information) with respect to the corresponding block located at the transmitter. In a *phase-coherent* system, the demodulator requires *carrier synchronization* (sync), that is, a precise knowledge of the transmitted carrier phase and frequency. [Systems that do not require carrier phase information are treated under the topic of *noncoherent detection* (see Chap. 10) and are sometimes called *noncoherent* or *phase-incoherent systems*.] It is also required that the channel decoder know the time instants when the signals that represent the channel symbols end and begin. This is called *channel symbol* or *word synchronization*. In order that the source decoder function properly, it must be able to separate the decoded digits from the channel decoder output into correct source symbols. Thus, *source symbol synchronization* is required. In an uncoded system, source symbol sync corresponds to word sync. Finally, the data regeneration requires *frame* or *block sync*, that is, the time instants that separate data samples from various sources.

We shall assume in this chapter that the receiver has exact knowledge of all levels of synchronization required for system operation—in other words, that transmitter and receiver are perfectly synchronized. In later chapters, we shall relax this constraint and discuss the effects of inexact synchronization on system performance.

5-2. THE BINARY AND N-ARY DECISION PROBLEM

In the simplest terms, decision making in statistical decision theory is viewed as a game in which there are two participants—the observer against nature, say. The states of nature are represented conveniently by the set of possible hypotheses or messages $\{m_i, i = 1, \ldots, N\}$, and the observer is allowed to perform an experiment, which in this case consists of transmitting one of a discrete set of specified signals $\{s_i(t), i = 1, 2, \ldots, N; 0 \leq t \leq T\}$, where t denotes time and T represents the duration of the signals $s_i(t)$. Which waveform is actually transmitted depends on the random message input. We assume that there is a one-to-one correspondence between the set of possible messages $\{m_i\}$ and the set of possible signals $\{s_i(t)\}$. For example, when the message $m = m_i$, the transmitter emits $s_i(t)$. The purpose of the observer is to make a *decision*, based on an observation of the received signal over some specified interval of time, as to which of the N possible signals was transmitted.

In what follows, we consider a received signal $y(t)$ observed during the interval $0 \leq t \leq T$. In the *binary decision* case, $y(t) = s_1(t) + n(t)$ or $s_2(t) + n(t)$ with either of two known signals, $s_1(t)$ and $s_2(t)$, transmitted and further corrupted by additive Gaussian noise, $n(t)$. For the N-ary decision case, one of N known signals is transmitted and is further corrupted by additive Gaussian noise, $n(t)$. If, for example, $s_i(t)$ is transmitted, we observe

$$y(t) = s_i(t) + n(t) \qquad i = 1, \ldots, N \tag{5-6}$$

In what follows, we assume that the signaling set $\{s_i(t)\}$ is represented by the set of signal vectors $\{\mathbf{s}_i, i = 1, 2, \ldots, N\}$ with the properties discussed in Appendix A.

It is well known[1-5] that whatever criterion of optimality we choose (minimum error probability, least cost, Neyman-Pearson, minimax, etc.), comparison of a likelihood ratio or its logarithm with a specified constant suggests the optimum decision rule.

The Binary Decision Problem

For the binary decision case, the *likelihood ratio* l is then the ratio of two Gaussian p.d.f.'s, $p_1(\mathbf{n}) = p(\mathbf{y} \mid \mathbf{s}_1)$ and $p_2(\mathbf{n}) = p(\mathbf{y} \mid \mathbf{s}_2)$, which are the respective conditional p.d.f.'s of receiving $\mathbf{y}$ under the hypothesis that $\mathbf{s}_1$ and $\mathbf{s}_2$ are transmitted. Using the results given in Appendix A, and assuming that $\mathbf{n}$, $\mathbf{s}_1$, and $\mathbf{s}_2$ are statistically independent, we obtain*

$$\begin{aligned} \ln l \triangleq \ln \frac{p(\mathbf{y} \mid \mathbf{s}_1)}{p(\mathbf{y} \mid \mathbf{s}_2)} &= \tfrac{1}{2}[(\mathbf{y} - \mathbf{s}_2), R_n^{-1}(\mathbf{y} - \mathbf{s}_2)] \\ &\quad - \tfrac{1}{2}[(\mathbf{y} - \mathbf{s}_1), R_n^{-1}(\mathbf{y} - \mathbf{s}_1)] \end{aligned} \tag{5-7}$$

---

* The notation $(\mathbf{x}, \mathbf{y})$ implies an inner product in function space, while $(\mathbf{x}, R_n^{-1}\mathbf{y})$ implies the inner product with the inverse operator R_n^{-1} acting on $\mathbf{y}$.

This ratio may be simplified to

$$\begin{aligned}\ln l &= [\mathbf{y}, R_n^{-1}(\mathbf{s}_1 - \mathbf{s}_2)] + \tfrac{1}{2}(\mathbf{s}_2, \mathbf{s}_2) - \tfrac{1}{2}(\mathbf{s}_1, \mathbf{s}_1) \\ &= [\mathbf{y}, R_n^{-1}(\mathbf{s}_1 - \mathbf{s}_2)] + \tfrac{1}{2}(E_2 - E_1)\end{aligned} \tag{5-8}$$

where $(\mathbf{s}_k, \mathbf{s}_k) = E_k$ is the energy in the kth signal. In (5-8), we have made use of the fact

$$[(\mathbf{s}_1 - \mathbf{s}_2), R_n^{-1}\mathbf{y}] = [\mathbf{y}, R_n^{-1}(\mathbf{s}_1 - \mathbf{s}_2)] \tag{5-9}$$

which is true because of the symmetric property of the covariance function $R_n(t_1, t_2) = R_n(t_2, t_1)$.

Since the energy terms are known constants, it is obvious that the optimum decision rule depends on the value of the inner product of (5-9). Interpreting both the inner product and the operator acting on the incoming vector **y**, in terms of the integrals in equations (A-3) and (A-4) of Appendix A, it is apparent that the inner product corresponds to a *matched filter detection*. To make our terminology more explicit, let **v** denote the vector $R_n^{-1}[\mathbf{s}_1 - \mathbf{s}_2]$. Then the *desired operation* of the receiver on **y** is

$$q \triangleq (\mathbf{y}, \mathbf{v}) \triangleq \int_0^T y(t)\, v(t)\, dt = \int_0^T y(t)\, h(T - t)\, dt \tag{5-10}$$

with the signal $v(t)$ defined by

$$v(t) \triangleq \int_0^T R_n^{-1}(t, \tau)[s_1(\tau) - s_2(\tau)]\, d\tau \tag{5-11}$$

or

$$s_1(t) - s_2(t) = \int_0^T R_n(t, \tau)\, v(\tau)\, d\tau \tag{5-12}$$

The kernel $R_n^{-1}(t, \tau)$ is inverse to the covariance function $R_n(t, \tau)$.

To this point our result is completely general in that the noise must only be Gaussian. For the case of interest here, the covariance function of the noise corresponds to a white Gaussian process; that is,

$$R_n(t, \tau) = \frac{N_0}{2}\delta(t - \tau) \tag{5-13}$$

Therefore

$$v(t) = \frac{N_0}{2}[s_1(t) - s_2(t)] \tag{5-14}$$

and

$$h(t) = \frac{N_0}{2}[s_1(T - t) - s_2(T - t)] \tag{5-15}$$

Thus, the desired operation on $y(t)$ corresponds to passing it through a filter whose impulse response $h(t) = v(T - t)$—that is, a filter matched to the known signal $v(t)$, hence the name *matched filter detection*. This mechanization is illustrated in Fig. 5-2a, where we have assumed that $E_1 = E_2$. From an alternate point of view, the optimum receiver may be implemented in terms of the cross-correlation operation indicated in Fig. 5-2b; hence the name *correlation detection*. If $s_1(t) = -s_2(t)$, we have the special case of

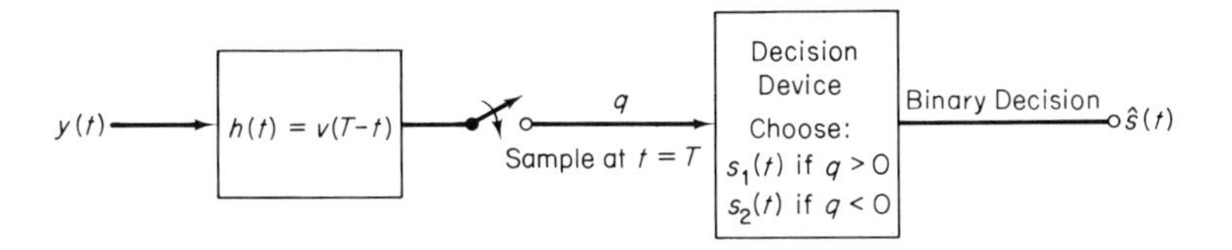

Fig. 5-2a Matched filter receiver for binary decision problem

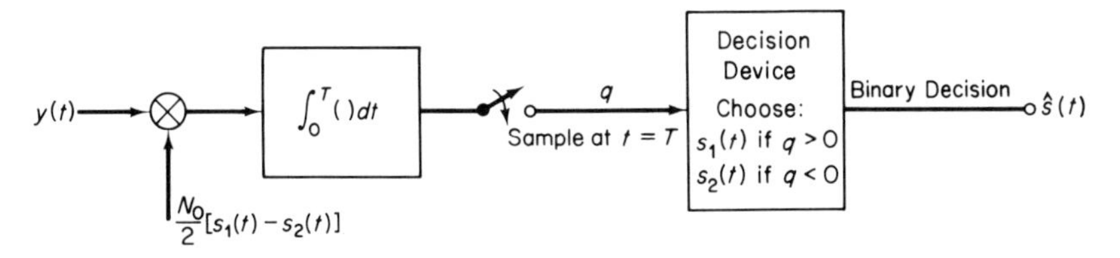

Fig. 5-2b Correlation receiver for binary decision problem

PSK. For either receiver mechanization, the output is sampled at $t = T$; if it is greater than zero (equal energy, equiprobable waveforms are assumed), $s_1(t)$ is announced while if the output is less than zero, $s_2(t)$ is announced.

The N-ary Decision Problem

The binary decision problem is readily extended to the N-ary case by considering one of N known signals to be transmitted. For minimum error probability we then compare the N *a posteriori* p.d.f.'s $p(\mathbf{s}_i \mid \mathbf{y})$, $i = 1, \ldots, N$, and select as the signal most likely to have been transmitted the one corresponding to the largest *a posteriori* probability. This is then the *maximum a posteriori probability* decision rule. Using Bayes' rule we have

$$p(\mathbf{s}_i \mid \mathbf{y}) = \frac{P(\mathbf{s}_i)\, p(\mathbf{y} \mid \mathbf{s}_i)}{p(\mathbf{y})} \tag{5-16}$$

The rule for selecting $s_i(t)$ as the transmitted signal then takes the two alternate forms

$$p(\mathbf{s}_i \mid \mathbf{y}) > p(\mathbf{s}_j \mid \mathbf{y}) \qquad \text{or} \qquad \frac{p(\mathbf{y} \mid \mathbf{s}_i)}{p(\mathbf{y} \mid \mathbf{s}_j)} > \frac{P(\mathbf{s}_j)}{P(\mathbf{s}_i)} \qquad \text{all } j \neq i \tag{5-17}$$

where $P(\mathbf{s}_i)$, $i = 1, 2, \ldots, N$, are the *a priori* probabilities of occurrence of the signals $s_i(t)$, $i = 1, 2, \ldots, N$. The second form of (5-17) involves the likelihood ratio and so represents the extension of the binary decision problem to that of N-ary decision. In the case of additive Gaussian noise, the ratio of the p.d.f.'s of (5-17) is given in (5-7) with $\mathbf{s}_i$ and $\mathbf{s}_j$ replacing $\mathbf{s}_1$ and $\mathbf{s}_2$, respectively. Taking logarithms we again obtain the inner products and find that the optimum N-ary decision rule (in the maximum *a posteriori*, minimum error probability sense) corresponds to comparing the simultaneous output at time $t = T$ of N filters. In the special case of white Gaussian noise, the optimum receiver for a set of equal energy, equiprobable signals $\{s_i(t)\}$ consists of N multipliers and integrators that compute the N quantities

$$q_i \triangleq (\mathbf{y}, \mathbf{s}_i) = \int_0^T y(t)\, s_i(t)\, dt \qquad i = 1, \dots, N \tag{5-18}$$

and selects the greatest quantity as corresponding to the signal most likely to occur. The earliest derivation of the above equation appears in the works of Kotelnikov[6] and Woodward.[7]

The mechanization of the optimum receiver can be realized with a bank of N multipliers and finite time integrators of the type shown in Fig. 5-3 with the

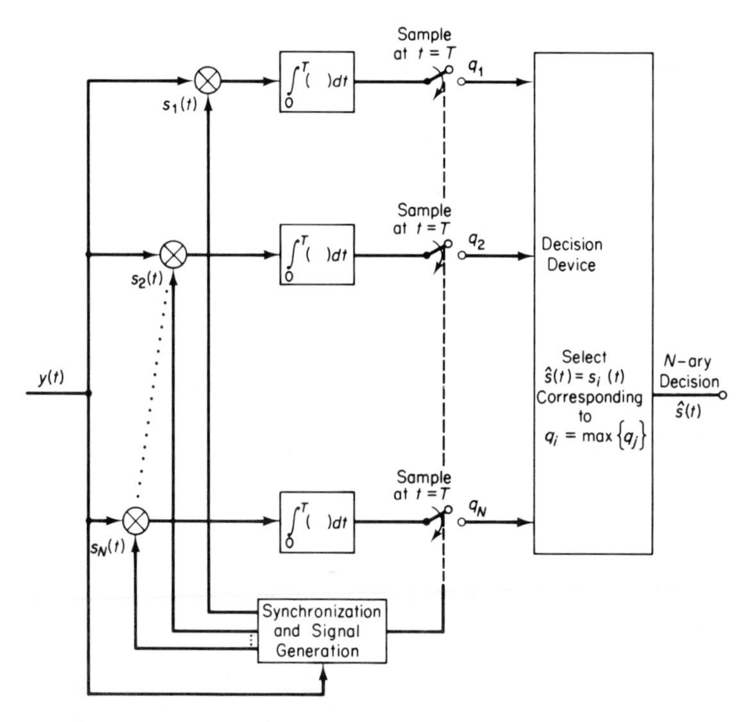

Fig. 5-3 Correlation receiver for N-ary decision problem

input to the multiplier now being $s_i(t)$. These devices are again referred to as *correlators*, and the process is *correlation detection*. The quantities q_i are referred to as the *correlator outputs*. At the conclusion of the signaling interval, a decision mechanism examines the outputs, determines the greatest, and announces that signal for which the output was greatest.

5-3. SIGNAL SET REPRESENTATION AND JOINT P.D.F. OF CORRELATOR OUTPUTS

Before we begin any analysis pertaining to system performance, it is necessary to develop a method for specifying the properties of the various signaling sets of greatest practical interest. It will be convenient to denote the set of normalized *signal inner products* λ_{ij} of the signal set $\{s_i(t)\}$ by the symmetric $N \times N$ matrix

$$\mathbf{\Lambda} \triangleq [\lambda_{ij}] \tag{5-19}$$

where $\mathbf{\Lambda}$ is nonnegative definite. The elements λ_{ij} of this matrix are defined for all i and j as*

$$\lambda_{ij} \triangleq \frac{1}{ST}\int_0^T s_i(t)\, s_j(t)\, dt = \frac{(\mathbf{s}_i, \mathbf{s}_j)}{ST} \tag{5-20}$$

where we have assumed that all N signals have equal energy:

$$E \triangleq ST = (\mathbf{s}_i, \mathbf{s}_i) = \int_0^T s_i^2(t)\, dt \qquad i = 1, 2, \ldots, N \tag{5-21}$$

Thus, the statistics of the correlator outputs $q_i = (\mathbf{y} \cdot \mathbf{s}_i)$ are now easily determined. Since the input to the receiver is Gaussian and the correlation operation is linear, the output samples q_i are Gaussian random variables. If we assume that $s_1(t)$ was transmitted, the mean of q_i becomes

$$\bar{q}_i = ST\lambda_{1i} \tag{5-22}$$

and the covariances become

$$q_{ij} \triangleq \overline{[(q_i - \bar{q}_i)(q_j - \bar{q}_j)]} = \frac{N_0 ST\lambda_{ij}}{2} \tag{5-23}$$

where the overbar denotes mathematical expectation. The covariance matrix $\mathbf{Q}$ of the correlator outputs is, therefore,

$$\mathbf{Q} = \left(\frac{N_0 ST}{2}\right)\mathbf{\Lambda} \tag{5-24}$$

The joint p.d.f. of the vector $\mathbf{q} \triangleq (q_1, q_2, \ldots, q_N)$, conditioned on the fact that $s_1(t)$ was transmitted, is represented by the multivariate Gaussian p.d.f.

$$p(\mathbf{q}\,|\,\mathbf{s}_1) = \frac{1}{(2\pi)^{N/2}|\mathbf{Q}|^{1/2}} \exp\left[-\tfrac{1}{2}(\mathbf{q} - \bar{\mathbf{q}})\mathbf{Q}^{-1}(\mathbf{q} - \bar{\mathbf{q}})^t\right] \tag{5-25}$$

---

* If the signals are represented by code vectors which consist of sequences of ones and minus ones or ones and zeros, then as in Chap. 4, the correlation coefficient between any two code vectors $\mathbf{x}$ and $\mathbf{y}$ is defined by

$$\lambda_{\mathbf{xy}} = \frac{A - D}{N}$$

where A is the number of term-by-term agreements and D is the number of term-by-term disagreements between $\mathbf{x}$ and $\mathbf{y}$.

where $|\mathbf{Q}|$ is the determinant of the matrix $\mathbf{Q}$, $\mathbf{Q}^{-1}$ is its inverse, and the t superscript denotes the transpose operation. This multivariate p.d.f. will be needed in evaluating system error probabilities.

Some examples of $\mathbf{\Lambda}$ for various signal sets are summarized in what follows.

Orthogonal Signal Sets

The set of signals $\{s_i(t), i = 1, 2, \ldots, N\}$ is said to be orthogonal if and only if

$$\lambda_{ij} = \begin{cases} 0 & i \neq j \\ 1 & i = j \end{cases} \tag{5-26}$$

The matrix $\mathbf{\Lambda}$ becomes the $N \times N$ identity matrix:

$$\mathbf{\Lambda}_O = I \tag{5-27}$$

Bi-orthogonal Signal Sets

This signal set can be obtained from an orthogonal set of $N/2$ signals by augmenting it with the negative of each signal. Here,

$$\lambda_{ij} = \begin{cases} 0 & i \neq j, |i - j| \neq N/2 \\ -1 & i \neq j, |i - j| = N/2 \\ 1 & i = j \end{cases} \tag{5-28}$$

Hence, the covariance matrix for bi-orthogonal signals can be written in the form

$$\mathbf{\Lambda}_{BO} = \begin{bmatrix} I & -I \\ -I & I \end{bmatrix} \tag{5-29}$$

where I is now the $N/2 \times N/2$ identity matrix. We note that the determinant of the matrix $\mathbf{\Lambda}_{BO}$ defined in (5-29) is zero; hence, the multivariate p.d.f. of (5-25) is singular—that is, $\mathbf{Q}^{-1}$ does not exist.

Transorthogonal or Regular Simplex Signal Sets

This set of signals is defined by the inner products*

$$\lambda_{ij} = \begin{cases} \dfrac{-1}{N-1} & i \neq j \\ 1 & i = j \end{cases} \tag{5-30}$$

---

* We note that a transorthogonal signal set is a special case of a *uniform-distance signal set* for which $\lambda_{ij} = \lambda$ for all $i \neq j$ and λ is bounded by $-1/(N-1) \leq \lambda \leq 1$.

The matrix $\mathbf{\Lambda}_R$ becomes

$$\mathbf{\Lambda}_R = \begin{bmatrix} 1 & \cdots & \frac{-1}{N-1} \\ \cdots & \cdots & \cdots \\ \frac{-1}{N-1} & \cdots & 1 \end{bmatrix} \tag{5-31}$$

Transorthogonal signals produce the minimum error probability that can be attained when one of N equiprobable, equal energy signals is transmitted over an infinite bandwidth channel disturbed by additive white Gaussian noise. Note that when N is large, (5-31) approximates (5-27); hence an orthogonal set of signals is essentially optimum when other constraints are absent. Notice that $\mathbf{\Lambda}_R$ is singular.

Polyphase Signal Sets

This signal set consists of N equally spaced vectors around the unit circle. The elements of the covariance matrix that defines this set of signals are given by

$$\lambda_{ij} = \begin{cases} \cos\left[\frac{2\pi}{N}(i-j)\right] & i \neq j \\ 1 & i = j \end{cases} \tag{5-32}$$

so that

$$\mathbf{\Lambda}_P = \begin{bmatrix} 1 & \cos\left(\frac{2\pi}{N}\right) & \cdots & \cos\left[\frac{2\pi(N-1)}{N}\right] \\ \cos\left(\frac{2\pi}{N}\right) & 1 & & \vdots \\ \vdots & & \ddots & \vdots \\ \cos\left[\frac{2\pi(N-1)}{N}\right] & \cdots & \cdots & 1 \end{bmatrix} \tag{5-33}$$

For $N = 2$ the signal set represents so-called *antipodal signals*, while for $N = 4$ we get a bi-orthogonal signal set. Notice also that this matrix is singular, since the sum of the elements in every column is zero; hence the matrix has zero determinant.

L-orthogonal Signal Sets

L-orthogonal signal sets[8,9] are generalizations of bi-orthogonal signal sets and consist of $N = ML$ equal energy, equal time-duration signals. A given L-orthogonal signal set has the following two properties:

1) the signal set may be divided into M disjoint subsets, each subset containing L signals and

2) signals from different subsets are orthogonal.

One instance of an L-orthogonal signal set is realized when each signal subset is chosen as a set of polyphase signals. A construction of such an L-orthogonal set, when M is an integer power of L, is given in Reed and Scholtz.[8] The signal correlation coefficients are defined by

$$\lambda_{ij} = \begin{cases} 0 & \left\lceil \frac{i-1}{L} \right\rceil \neq \left\lceil \frac{j-1}{L} \right\rceil \\ \cos\left[\frac{2\pi}{L}(i-j)\right] & \left\lceil \frac{i-1}{L} \right\rceil = \left\lceil \frac{j-1}{L} \right\rceil \\ 1 & i = j \end{cases} \tag{5-34}$$

The corresponding correlation matrix $\mathbf{\Lambda}_{LO}$ can be partitioned in the following form:

$$\mathbf{\Lambda}_{LO} = \begin{bmatrix} \mathbf{\Lambda}_P & \mathbf{0} & \cdots & \mathbf{0} \\ \mathbf{0} & \mathbf{\Lambda}_P & & \vdots \\ \vdots & & \ddots & \vdots \\ \mathbf{0} & \cdots & & \mathbf{\Lambda}_P \end{bmatrix} \tag{5-35}$$

where $\mathbf{\Lambda}_P$ is an $L \times L$ matrix as defined in (5-33) with N replaced by L, and $\mathbf{0}$ is the $L \times L$ null matrix whose elements are all zero. From an inspection of $\mathbf{\Lambda}_{LO}$, it is obvious that the signal set we have generated is L-orthogonal. Again, the special cases $L = 2$ and $L = 4$ yield bi-orthogonal signal sets. At this time, other special cases appear to be of little practical importance in the implementation of phase-coherent systems.

5-4. GENERATION OF BINARY SIGNAL SETS, $r = 2$

In this section we shall briefly discuss methods of generating orthogonal, bi-orthogonal, and transorthogonal signal sets for the binary case: $r = 2$ and $N = 2^n$. In general, the signal $s_i(t)$ has the representation

$$s_i(t) = \sum_{k=1}^{D} s_k^{(i)} \phi_k(t)$$

where $\{\phi_k(t)\}$ forms an orthonormal set of functions on a T-second interval, and D is the *dimensionality of the signal space*. Thus the signal $s_i(t)$ has the vector representation $\mathbf{s}_i \triangleq (s_1^{(i)}, s_2^{(i)}, \ldots, s_D^{(i)})$. The situation we are concerned with is depicted in Fig. 5-4 wherein the *data vector* $\mathbf{d}^{(i)} \triangleq (d_1^{(i)}, d_2^{(i)}, \ldots, d_n^{(i)})$, $i = 1, 2, \ldots, N$ arriving from the source encoder output of Fig. 5-1, is applied to the channel encoder. As a result of the application of this n-digit (bit) vector to the channel encoder, a D-digit code word $\mathbf{b}^{(i)} \triangleq (b_1^{(i)}, b_2^{(i)}, \ldots,$

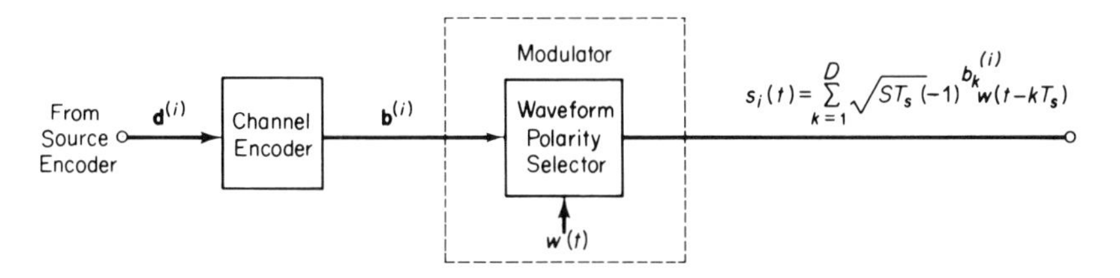

Fig. 5-4 Generation of the signal set $\{s_i(t): i = 1, 2, \ldots, N\}$ for the binary case, $r = 2$

$b_D^{(i)}$) is applied to the modulator of Fig. 5-4. When $\mathbf{b}^{(i)}$ is applied to the modulator, then $s_i(t)$ is transmitted into the channel.

Without loss in generality we assume that the components of the data and code word vectors are sequences of zeros and ones so that the waveform that corresponds to $\mathbf{d}^{(i)}$ or $\mathbf{b}^{(i)}$ can be envisioned as the pulse train illustrated in Fig. 5-5. When the sequence of zeros and ones represents a data vector, then the

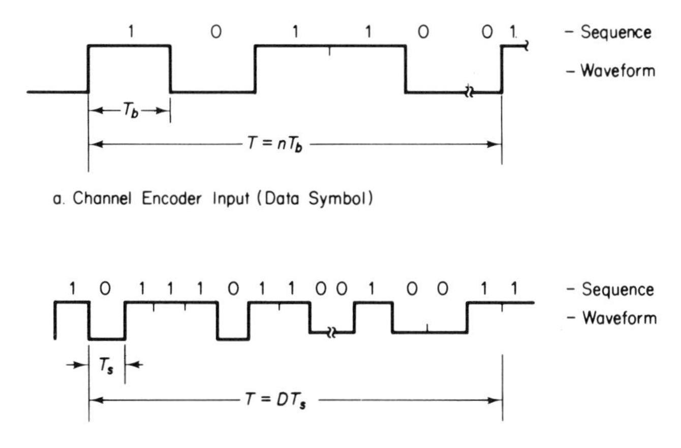

Fig. 5-5 Input-output sequences of channel encoder and the corresponding waveforms. a. Channel encoder input (data symbol). b. Channel encoder output (code word).

corresponding binary waveform in Fig. 5-5a has duration $T = nT_b$ seconds. Here T_b is the *time per data bit*. On the other hand, if the sequence of zeros and ones represents a code word, then the corresponding binary waveform in Fig. 5-5b has duration $T = DT_s$ where T_s is the *time per code word digit*. Therefore, with a single unit energy waveforms $w(t)$ of duration T_s seconds, one can generate the signal set $\{s_i(t)\}$, in which $s_i(t)$ is represented by

$$s_i(t) = \sum_{k=1}^{D} \sqrt{ST_s}\,(-1)^{b_k^{(i)}} w(t - kT_s)$$

From (5-20) it follows that the correlation properties of the signals are identical to those of the code words.

We shall now discuss a method of generating *code word dictionaries* ($\mathbf{b}^{(i)}, i = 1, 2, \ldots, N$) for representing orthogonal, bi-orthogonal, and trans-orthogonal signal sets.

Orthogonal Codes

Let $\mathbf{H}_n$ be a $2^n \times 2^n$ matrix defined inductively by

$$\mathbf{H}_n \triangleq \left[\begin{array}{c|c} \mathbf{H}_{n-1} & \mathbf{H}_{n-1} \\ \hline \mathbf{H}_{n-1} & \bar{\mathbf{H}}_{n-1} \end{array}\right] = \begin{bmatrix} \mathbf{b}^{(1)} \\ \mathbf{b}^{(2)} \\ \cdot \\ \cdot \\ \cdot \\ \mathbf{b}^{(N)} \end{bmatrix} \tag{5-36}$$

where

$$\mathbf{H}_1 \triangleq \begin{bmatrix} 0 & 0 \\ 0 & 1 \end{bmatrix} \tag{5-37}$$

and the dimensionality of the signal set $D = 2^n$. Here, the overbar denotes the complement of the "H" matrix, wherein ones are replaced by zeros and vice versa. Also, the rows of $\mathbf{H}_n$ represent the code vectors. Such a matrix is called a *Hadamard matrix*,[10] and it is easy to show that its rows and columns are orthogonal:

$$\lambda_{\mathbf{b}^{(i)}\mathbf{b}^{(j)}} \triangleq \frac{A(n) - D(n)}{N} = \lambda_{ij} = \begin{cases} 0 & i \neq j \\ 1 & i = j \end{cases} \tag{5-38}$$

Here $A(n)$ represents the number of term-by-term agreements and $D(n)$ represents the number of term-by-term disagreements between the two code vectors $\mathbf{b}^{(i)}$ and $\mathbf{b}^{(j)}$. The code word dictionary generated by (5-36) is known as a first-order $(2^n, n)$ *Reed-Muller code.*

Now that a code dictionary has been formed, all that remains to completely specify the code is to define the mapping between the data vectors $\{\mathbf{d}^{(i)}\}$ and the code word vectors $\{\mathbf{b}^{(i)}\}$.

Define a matrix $\mathbf{G}_n$ to be a $2^n \times n$ matrix whose ith row is the binary representation of the integer $i - 1$. The columns of $\mathbf{G}_n$, that is, $\mathbf{g}^{(1)t}, \mathbf{g}^{(2)t}, \ldots, \mathbf{g}^{(n)t}$, represent *code generators* of the code word dictionary $\mathbf{H}_n$. Also define the $2^n \times n$ *data symbol matrix*, $\mathbf{I}_n$, whose ith row $i = 1, 2, \ldots, N$ is the ith data vector $\mathbf{d}^{(i)}$. Since the elements of $\mathbf{d}^{(i)}$ correspond to the binary representation of the integer $i - 1$, we notice that $\mathbf{I}_n$ is identical to $\mathbf{G}_n$. In terms of the above matrix definitions, the code dictionary $\mathbf{H}_n$ is easily generated from

$$\mathbf{H}_n = \mathbf{G}_n \mathbf{I}_n^t \tag{5-39}$$

It is understood that the multiplication and addition operations involved in the matrix multiplication are reduced modulo 2.

For example, when $n = 3$, we have

$$\mathbf{G}_3 = \begin{bmatrix} 0 & 0 & 0 \\ 0 & 0 & 1 \\ 0 & 1 & 0 \\ 0 & 1 & 1 \\ 1 & 0 & 0 \\ 1 & 0 & 1 \\ 1 & 1 & 0 \\ 1 & 1 & 1 \end{bmatrix}; \quad \mathbf{I}_3^t = \begin{bmatrix} 0 & 0 & 0 & 0 & 1 & 1 & 1 & 1 \\ 0 & 0 & 1 & 1 & 0 & 0 & 1 & 1 \\ 0 & 1 & 0 & 1 & 0 & 1 & 0 & 1 \end{bmatrix} = [\mathbf{d}^{(1)t}\mathbf{d}^{(2)t}\cdots\mathbf{d}^{(8)t}]$$

so that

$$\mathbf{H}_3 = \left[\begin{array}{cccc|cccc} 0 & 0 & 0 & 0 & 0 & 0 & 0 & 0 \\ 0 & 1 & 0 & 1 & 0 & 1 & 0 & 1 \\ 0 & 0 & 1 & 1 & 0 & 0 & 1 & 1 \\ 0 & 1 & 1 & 0 & 0 & 1 & 1 & 0 \\ \hline 0 & 0 & 0 & 0 & 1 & 1 & 1 & 1 \\ 0 & 1 & 0 & 1 & 1 & 0 & 1 & 0 \\ 0 & 0 & 1 & 1 & 1 & 1 & 0 & 0 \\ 0 & 1 & 1 & 0 & 1 & 0 & 0 & 1 \end{array}\right] = \begin{bmatrix} \mathbf{b}^{(1)} \\ \mathbf{b}^{(2)} \\ \cdot \\ \cdot \\ \cdot \\ \mathbf{b}^{(8)} \end{bmatrix}$$

One mechanization of the channel encoder for a $(2^n, n)$ orthogonal code is illustrated in Fig. 5-6. A clock that generates a square wave of frequency $1/2T_s$ digits/sec drives a cascade of $n - 1$ *binary frequency dividers* whose

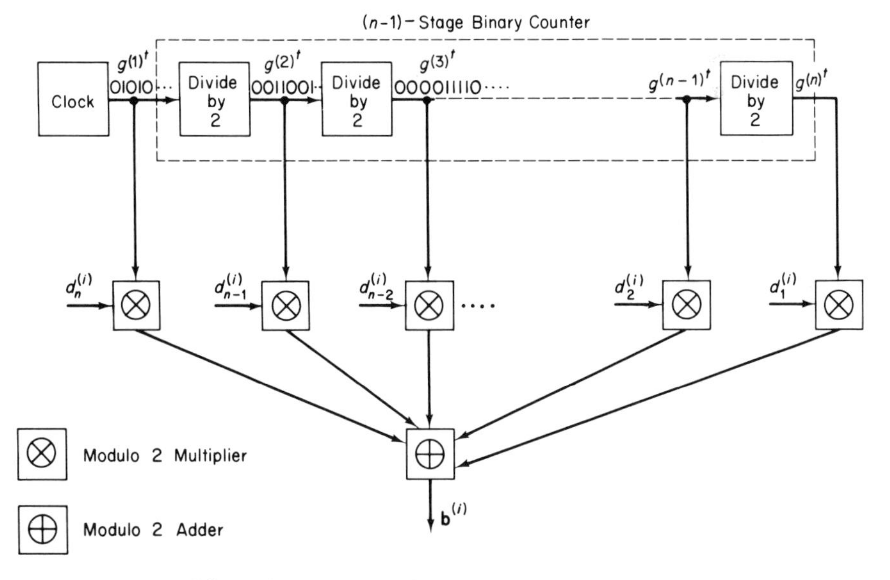

Fig. 5-6 Encoder for a $(2^n, n)$ orthogonal code

output waveform, together with the clock output waveform, corresponds to the code generator vectors. These output waveforms are inputted to n modulo-2 multipliers. Every T seconds, n data digits are fed into these same multipliers, and the corresponding code vector $\mathbf{b}^{(i)}$ is produced at the output of the modulo 2 adder.

Bi-orthogonal Codes

The code dictionary of a $(2^{n-1}, n)$ bi-orthogonal code is easily defined in terms of the dictionary of an orthogonal code through the partitioned matrix

$$\mathbf{B}_n = \begin{bmatrix} \mathbf{H}_{n-1} \\ \hline \overline{\mathbf{H}}_{n-1} \end{bmatrix} \tag{5-40}$$

Here the dimensionality of the signal set $D = 2^{n-1}$. The correlation coefficients for the rows are defined in (5-28). An example of a (32, 6) code is illustrated in Fig. 5-7, where the Hadamard design is clearly depicted.*

A mechanization of the channel encoder for a $(2^{n-1}, n)$ bi-orthogonal code similar to that given in Fig. 5-6 for the $(2^n, n)$ orthogonal code is illustrated in Fig. 5-8. We note that without the data digit $d_1^{(i)}$ the remainder of the diagram mechanizes a $(2^{n-1}, n - 1)$ orthogonal code. The presence of $d_1^{(i)}$ serves to output from the modulo-2 adder either the orthogonal code vector produced by the remaining data digits or its complement.

Transorthogonal Codes

A transorthogonal code dictionary can be generated from an orthogonal code dictionary by deleting the first row or column in $\mathbf{H}_n$. In practice, this can be accomplished by deleting the first digit of each code word generated by the encoder of Fig. 5-6.

5-5. PERFORMANCE CHARACTERIZATION OF PHASE-COHERENT RECEIVERS

In this section we characterize the performance of a wide class of perfectly synchronized phase-coherent communication systems which employ the signaling sets just discussed. The criterion of performance is the probability that the receiver will err in making its decision. This probability is just one minus the probabliity of correct detection. If we denote the probability of correct detection by $P_c(N)$, then the word error probability $P_E(N) = 1 - P_c(N)$ is determined in terms of P_{c_i}—the probability of correct de-

* This code was used in the Mariner 6 and 9 spacecraft telemetry systems to transmit television pictures of Mars back to Earth.

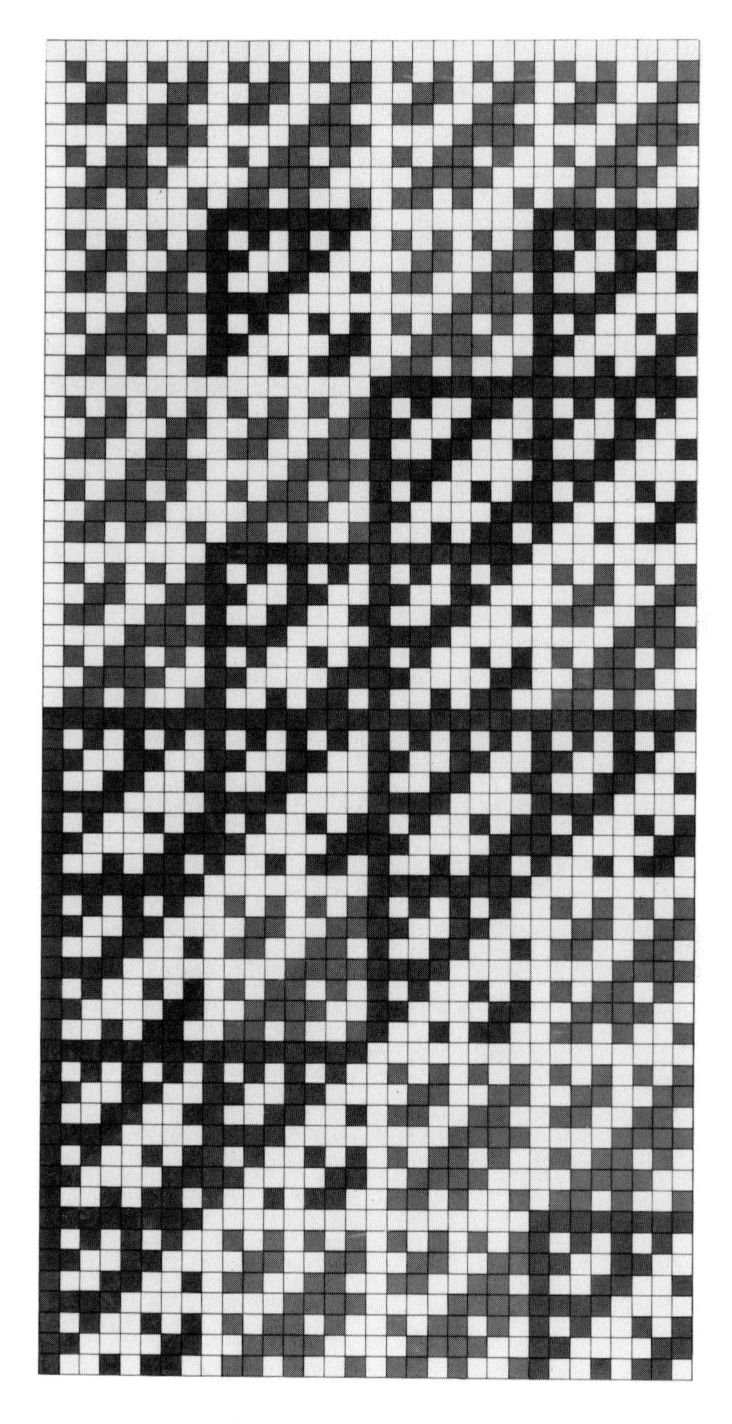

Fig. 5-7 A (32, 6) bi-orthogonal code (This is equivalent to a first-order Reed-Muller code and its complement)

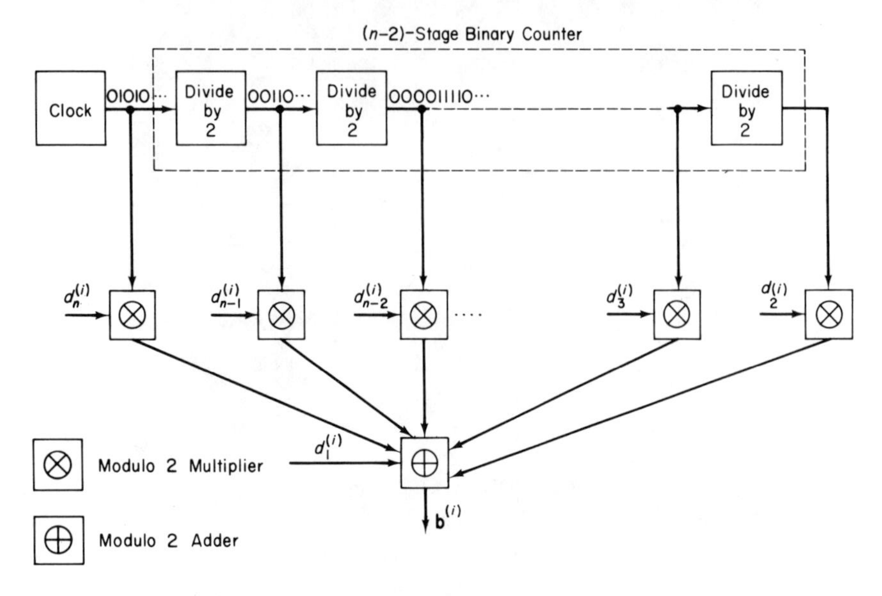

Fig. 5-8 Encoder for a $(2^n, n)$ bi-orthogonal code

tection when signal $s_i(t)$ was transmitted. Letting $P(\mathbf{s}_i)$ denote the *a priori* probability of $s_i(t)$, then

$$P_c(N) = \sum_{i=1}^{N} P(\mathbf{s}_i) P_{c_i} \tag{5-41}$$

where
$$P_{c_i} = \text{Prob}\,[(\mathbf{y}, \mathbf{s}_i) = \max_j\, (\mathbf{y}, \mathbf{s}_j) \,|\, \mathbf{s}_i] \tag{5-42}$$

Assuming without loss in generality that $s_1(t)$ was transmitted, then (5-42) is equivalent to evaluating

$$\begin{aligned} P_{c_1} &= \text{Prob}\,[q_2 \le q_1, q_3 \le q_1, \ldots, q_N \le q_1 \text{ for all possible values of } q_1 \,|\, \mathbf{s}_1] \\ &= \underbrace{\int_{-\infty}^{\infty} \int_{-\infty}^{q_1} \cdots \int_{-\infty}^{q_1}}_{N-1 \text{ fold}} p(\mathbf{q} \,|\, \mathbf{s}_1)\, d\mathbf{q} \end{aligned} \tag{5-43}$$

where $p(\mathbf{q} \,|\, \mathbf{s}_1)$ is the multivariate Gaussian p.d.f. given by (5-25). If we normalize the q_i's by introducing the change of variables

$$z_i = (q_i - \bar{q}_i)/\sqrt{q_{ii}} \tag{5-44}$$

then (5-43), may be rewritten as[11,12]

$$P_{c_1} = \int_{-\infty}^{\infty} \int_{-\infty}^{z_1 + \sqrt{2R_d}(1-\lambda_{12})} \cdots \int_{-\infty}^{z_1 + \sqrt{2R_d}(1-\lambda_{1N})} p(\mathbf{z} \,|\, \mathbf{s}_1)\, d\mathbf{z} \tag{5-45}$$

where $R_d \triangleq ST/N_0$, $\mathbf{z} \triangleq z_1, \ldots, z_N$, and

$$p(\mathbf{z} \,|\, \mathbf{s}_1) = \frac{\exp\,[-\frac{1}{2}\mathbf{z}\boldsymbol{\Lambda}^{-1}\mathbf{z}^t]}{(2\pi)^{N/2}\,|\boldsymbol{\Lambda}|^{1/2}} \tag{5-46}$$

with $|\boldsymbol{\Lambda}|$ the determinant of the matrix $\boldsymbol{\Lambda}$. Corresponding expressions may be obtained, assuming that $s_j(t)$ was transmitted, by changing the subscripts 1 to j throughout the expressions.

An example of considerable practical importance occurs when $N = 2$ and $P(\mathbf{s}_1) = P(\mathbf{s}_2) = 1/2$. In this case, the error probability as found from (5-41) and (5-45) becomes

$$P_E(2) = 1 - P_c(2) = \frac{1}{2}\operatorname{erfc}\left[\sqrt{R_d\left(\frac{1-\lambda}{2}\right)}\right] \tag{5-47}$$

where

$$\operatorname{erfc}(x) \triangleq \frac{2}{\sqrt{\pi}}\int_x^\infty \exp(-y^2)\,dy$$

and we have set $\lambda_{12} = \lambda$. The case where $\lambda = 0$ corresponds to coherent reception of FSK signals, while the case where $\lambda = -1$ corresponds to coherent reception of PSK signals. Thus, for $N = 2$ we have established the fact that the performance of a binary PSK system is 3 dB better than the orthogonal FSK signaling system. If the frequency spacing between the two FSK signals is chosen so as to minimize the error probability, then the performance of a coherent FSK system is only 2.2 dB inferor to that of PSK (see Prob. 5-5).

The Set of Equiprobable, Equal Energy, Orthogonal Signals

A typical set of signals that may be generated in practice for N-ary transmission consists of N non-overlapping sinusoidal pulses

$$s_i(t) = \begin{cases} \sqrt{2SN}\sin\left(\dfrac{\pi k N t}{T}\right); & \dfrac{(i-1)T}{N} \leq t \leq \dfrac{iT}{N} \\ 0 \quad \text{otherwise} & \end{cases} \tag{5-48}$$

where k is any integer and $i = 1, 2, \ldots, N$. This set of signals is orthogonal and useful in PPM systems. The correlation matrix for this set of signals is defined in (5-27).

Another orthogonal set of signals useful in MFSK signaling systems is the set of sinusoidal pulses

$$s_i(t) = \sqrt{2S}\sin\left(\frac{\pi(k+i)t}{T}\right); \qquad 0 \leq t \leq T \tag{5-49}$$

where $i = 1, 2, \ldots, N$. In (5-48) and (5-49) the carrier radian frequency is $\omega_0 = \pi k/T$. To ensure orthogonality of the signals of (5-49), the signal frequencies must be separated by $1/2T$ Hz so that the *effective single-sided bandwidth* of this set is $B_e = N/2T$ (see Prob. 5-3). [For noncoherent reception of MFSK, $B_e = N/T$; that is, the effective bandwidth is two times that required in a coherent system.] There are many other practical methods for generating orthogonal signal sets; some examples can be found in Ref. 10.

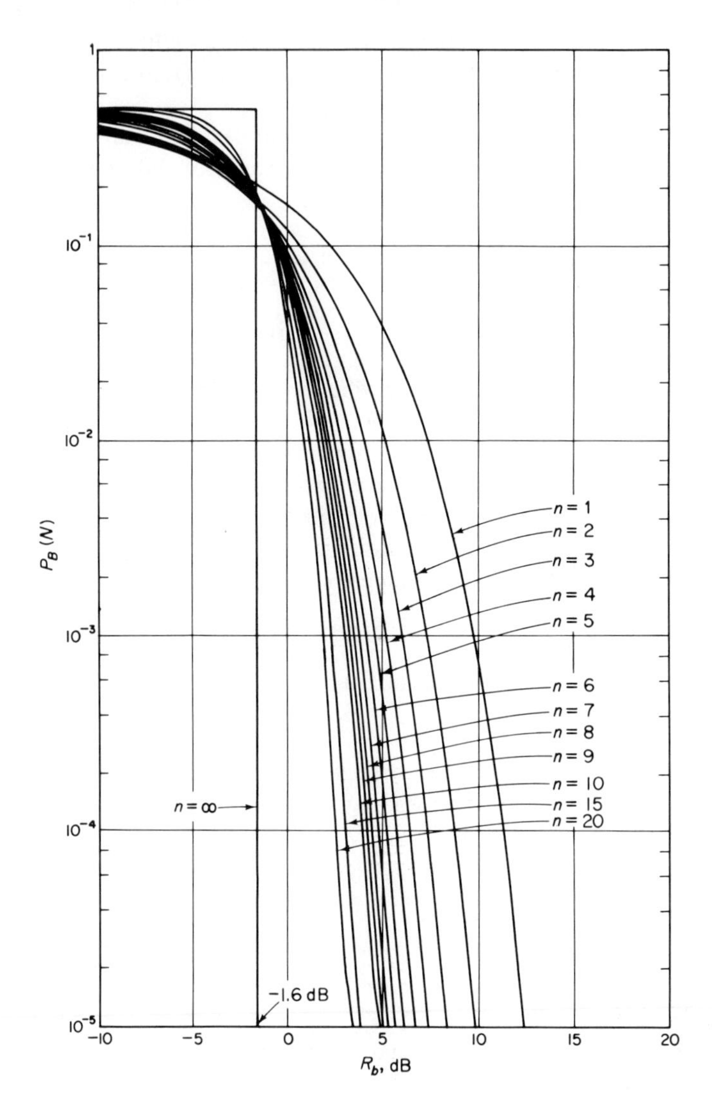

Fig. 5-9 Bit error probability performance of a system transmitting an orthogonal signal set

We now proceed to evaluate word and bit error probabilities for the orthogonal signal set. Making the change of variables $x = z_1/\sqrt{2}$ and $y_i = z_i/\sqrt{2}$ for $i = 2, 3, \ldots, N$ in (5-45) and using (5-27) and (5-41), the probability of correct detection of a channel symbol (code word) becomes

$$P_c(N) = \int_{-\infty}^{\infty} \frac{\exp(-x^2)}{\sqrt{\pi}} \left[\frac{1}{2}\operatorname{erfc}(-x - \sqrt{R_d})\right]^{N-1} dx \qquad (5\text{-}50)$$

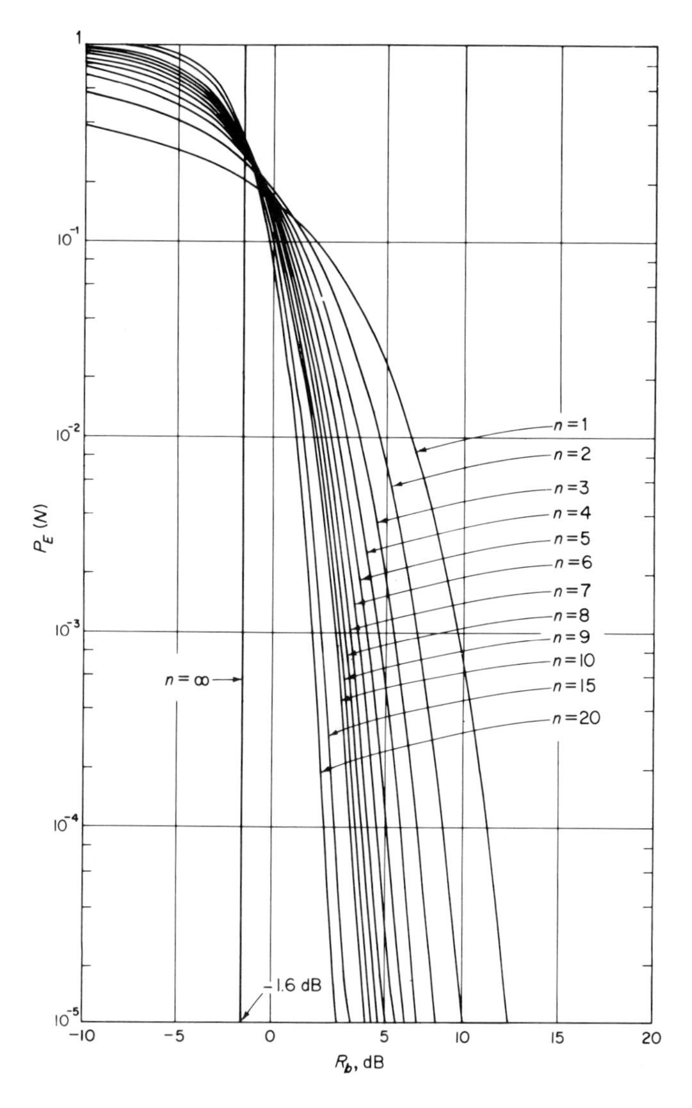

Fig. 5-10 Word error probability performance of a system transmitting an orthogonal signal set

Note that (5-50) specifies the performance of MFSK systems when $N > 2$, and the performance of a coherent FSK system when $N = 2$.

An upper union bound[3] for the word error probability $P_E(N) = 1 - P_c(N)$ is given by

$$P_E(N) \leq \left(\frac{N-1}{2}\right) \operatorname{erfc}\left(\sqrt{\frac{R_d}{2}}\right) \tag{5-51}$$

which becomes increasingly tight for fixed N as R_d is increased. For values of $P_E(N) \leq 10^{-3}$, the error probability is well approximated by the above bound. For $N = 2$, the bound of (5-51) becomes an equality.

The bit error probability $P_B(N)$ is of interest in the binary case, i.e., the signal base $r = 2$. In this case, the data symbols contain $n = \log_2 N$ data bits; thus, the probability of error for a specific data bit should be determined. If a word is incorrectly detected, then since $\lambda_{ij} = 0$ for $i \neq j$, the decision is equally likely to be made in favor of any one of the $N - 1$ incorrect signals. In terms of the n decoded data bits corresponding to the incorrectly detected word, the probability that exactly k of them are in error is

$$\frac{{}_nC_k}{\sum_{k=1}^{n} {}_nC_k}$$

where

$$ {}_nC_k \triangleq \frac{n!}{k!(n-k)!} \tag{5-52}$$

Hence, the average number of decoded data bits in error (given that an n-bit word of $N = 2^n$ digits has been detected incorrectly) is

$$\frac{\sum_{k=1}^{n} k({}_nC_k)}{\sum_{k=1}^{n} {}_nC_k} = \frac{n2^{n-1}}{2^n - 1}$$

so that the probability of a given data bit being in error (given that a word is in error) is*

$$P(B|W) = \frac{2^{n-1}}{2^n - 1} \tag{5-53}$$

Since the probability, $P(W|B)$, that a word is in error, given that a bit is in error, is one, then from Bayes' rule we can write

$$P_B(N) = \frac{P(B|W)P_E(N)}{P(W|B)} = \frac{2^{n-1}}{2^n - 1} P_E(N) \tag{5-54}$$

Here $P_E(N)$ is one minus the probability of correct word detection as given in (5-50). Table 5-1 gives the quantities $P_B(N)$ and $P_E(N)$ versus $R_b = R_d/\log_2 N$ for various values of $n = \log_2 N$. Figures 5-9 and 5-10 illustrate the results of these tabulations.

The Set of Equiprobable, Equal Energy, Bi-orthogonal Signals

An appropriate set of N signals which possess the correlation matrix of (5-29) consists of a set of $N/2$ orthogonal signals as specified by either

---

* An alternate argument which can be used to arrive at (5-53) is as follows. In any column of the data symbol matrix, $\mathbf{I}_n$, there are an equal number, that is, 2^{n-1}, of ones and zeros. Thus, if a code word is incorrectly detected, then for any given bit position (column of $\mathbf{I}_n$), there are 2^{n-1} out of a possible $2^n - 1$ ways in which that bit can be in error.

TABLE 5-1 Performance of Orthogonal Codes in White Gaussian Noise with Optimal (maximum-likelihood) Decoding (number of code words, $N = 2^n$)

n	R_b	$P_E(N)$	$P_B(N)$
1	0.080	0.38865	0.38865
1	0.125	0.36184	0.36184
1	0.180	0.33569	0.33569
1	0.245	0.31031	0.31031
1	0.320	0.28580	0.28580
1	0.405	0.26226	0.26226
1	0.500	0.23975	0.23975
1	0.605	0.21834	0.21834
1	0.720	0.19807	0.19807
1	0.845	0.17899	0.17899
1	0.980	0.16110	0.16110
1	1.125	0.14442	0.14442
1	1.280	0.12895	0.12895
1	1.445	0.11467	0.11467
1	1.620	0.10155	0.10155
1	1.805	0.089555	0.089555
1	2.000	0.078650	0.078650
1	2.205	0.068782	0.068782
1	2.420	0.059897	0.059897
1	2.645	0.051938	0.051938
1	2.880	0.044843	0.044843
1	3.125	0.038550	0.038550
1	3.380	0.032996	0.032996
1	3.645	0.028119	0.028119
1	3.920	0.023857	0.023857
1	4.205	0.020152	0.020152
1	4.500	0.016947	0.016947
1	4.805	0.014189	0.014189
1	5.120	0.011826	0.011826
1	5.445	0.0098122	0.0098122
1	5.780	0.0081048	0.0081048
1	6.125	0.0066642	0.0066642
1	6.480	0.0054547	0.0054547
1	6.845	0.0044445	0.0044445
1	7.220	0.0036048	0.0036048
1	7.605	0.0029103	0.0029103
1	8.000	0.0023389	0.0023389
1	8.405	0.0018710	0.0018710
1	8.820	0.0014897	0.0014897
1	9.245	0.0011807	0.0011807
1	9.680	0.00093142	0.00093142
1	10.125	0.00073136	0.00073136
1	10.580	0.00057159	0.00057159
1	11.045	0.00044463	0.00044463
1	11.520	0.00034426	0.00034426
1	12.005	0.00026529	0.00026529
1	12.500	0.00020348	0.00020348

TABLE 5-1 (cont.)

n	R_b	$P_E(N)$	$P_B(N)$
2	0.080	0.58621	0.39081
2	0.125	0.54145	0.36097
2	0.180	0.49624	0.33083
2	0.245	0.45115	0.30077
2	0.320	0.40677	0.27118
2	0.405	0.36362	0.24242
2	0.500	0.32222	0.21481
2	0.605	0.28299	0.18866
2	0.720	0.24627	0.16418
2	0.845	0.21234	0.14156
2	0.980	0.18136	0.12091
2	1.125	0.15344	0.10229
2	1.280	0.12857	0.08571
2	1.445	0.10669	0.07112
2	1.620	0.087666	0.058444
2	1.805	0.071331	0.047554
2	2.000	0.057467	0.038311
2	2.205	0.045839	0.030559
2	2.420	0.036201	0.024134
2	2.645	0.028305	0.018870
2	2.880	0.021911	0.014607
2	3.125	0.016792	0.011194
2	3.380	0.012740	0.008493
2	3.645	0.0095697	0.0063798
2	3.920	0.0071165	0.0047444
2	4.205	0.0052395	0.0034930
2	4.500	0.0038192	0.0025461
2	4.805	0.0027562	0.0018374
2	5.120	0.0019693	0.0013129
2	5.445	0.0013931	0.0009288
2	5.780	0.00097578	0.00065052
2	6.125	0.00067671	0.00045114
2	6.480	0.00046468	0.00030978
2	6.845	0.00031594	0.00021063
2	7.220	0.00021270	0.00014180
2	7.605	0.00014179	0.00009453
2	8.000	0.000093595	0.000062397
2	8.405	0.000061177	0.000040785
2	8.820	0.000039597	0.000026398
2	9.245	0.000025379	0.000016919
2	9.680	0.000016108	0.000010739
2	10.125	0.000010124	0.000006749
2	10.580	0.0000063010	0.0000042007
2	11.045	0.0000038835	0.0000025890
2	11.520	0.0000023703	0.0000015802
2	12.005	0.0000014327	0.0000009551
2	12.500	0.00000085751	0.00000057167

TABLE 5-1 (cont.)

n	R_b	$P_E(N)$	$P_B(N)$
3	0.080	0.71018	0.40581
3	0.125	0.65748	0.37570
3	0.180	0.60166	0.34380
3	0.245	0.54381	0.31075
3	0.320	0.48516	0.27724
3	0.405	0.42699	0.24399
3	0.500	0.37050	0.21172
3	0.605	0.31683	0.18105
3	0.720	0.26690	0.15251
3	0.845	0.22141	0.12652
3	0.980	0.18082	0.10333
3	1.125	0.14535	0.08306
3	1.280	0.11498	0.06570
3	1.445	0.089493	0.051139
3	1.620	0.068531	0.039161
3	1.805	0.051628	0.029502
3	2.000	0.038262	0.021864
3	2.205	0.027895	0.015940
3	2.420	0.020006	0.011432
3	2.645	0.014116	0.008066
3	2.880	0.0097989	0.0055994
3	3.125	0.0066927	0.0038244
3	3.380	0.0044980	0.0025703
3	3.645	0.0029748	0.0016999
3	3.920	0.0019363	0.0011064
3	4.205	0.0012405	0.0007088
3	4.500	0.00078228	0.00044702
3	4.805	0.00048566	0.00027752
3	5.120	0.00029686	0.00016963
3	5.445	0.00017867	0.00010210
3	5.780	0.00010590	0.00006051
3	6.125	0.000061812	0.000035321
3	6.480	0.000035535	0.000020306
3	6.845	0.000020122	0.000011498
3	7.220	0.000011224	0.000006414
3	7.605	0.0000061676	0.0000035244
3	8.000	0.0000033388	0.0000019079
3	8.405	0.0000017807	0.0000010176
3	8.820	0.00000093575	0.00000053471
3	9.245	0.00000048448	0.00000027685
3	9.680	0.00000024716	0.00000014123
3	10.125	0.00000012424	0.00000007099
3	10.580	0.000000061537	0.000000035164
3	11.045	0.000000030034	0.000000017162
3	11.520	0.000000014444	0.000000008254
3	12.005	0.000000006845	0.000000003912
3	12.500	0.000000003197	0.000000001827

TABLE 5-1 (cont.)

n	R_b	$P_E(N)$	$P_B(N)$
4	0.080	0.79366	0.42329
4	0.125	0.73939	0.39434
4	0.180	0.67875	0.36200
4	0.245	0.61310	0.32699
4	0.320	0.54422	0.29025
4	0.405	0.47419	0.25290
4	0.500	0.40514	0.21608
4	0.605	0.33914	0.18087
4	0.720	0.27793	0.14823
4	0.845	0.22284	0.11885
4	0.980	0.17473	0.09319
4	1.125	0.13393	0.07143
4	1.280	0.10032	0.05350
4	1.445	0.073422	0.039159
4	1.620	0.052497	0.027998
4	1.805	0.036667	0.019556
4	2.000	0.025019	0.013344
4	2.205	0.016678	0.008895
4	2.420	0.010863	0.005793
4	2.645	0.0069137	0.0036873
4	2.880	0.0043008	0.0022938
4	3.125	0.0026154	0.0013949
4	3.380	0.0015552	0.0008294
4	3.645	0.00090440	0.00048234
4	3.920	0.00051451	0.00027440
4	4.205	0.00028640	0.00015275
4	4.500	0.00015604	0.00008322
4	4.805	0.000083220	0.000044384
4	5.120	0.000043459	0.000023178
4	5.445	0.000022227	0.000011854
4	5.780	0.000011135	0.000005939
4	6.125	0.0000054655	0.0000029149
4	6.480	0.0000026285	0.0000014019
4	6.845	0.0000012389	0.0000006607
4	7.220	0.00000057228	0.00000030522
4	7.605	0.00000025913	0.00000013820
4	8.000	0.00000011502	0.00000006134
4	8.405	0.000000050050	0.000000026693
4	8.820	0.000000021352	0.000000011388
4	9.245	0.000000008931	0.000000004763
4	9.680	0.000000003663	0.000000001953
4	10.125	0.000000001473	0.000000000786
4	10.580	0.000000000581	0.000000000310
4	11.045	0.000000000225	0.000000000120
4	11.520	0.000000000085	0.000000000045
4	12.005	0.000000000032	0.000000000017
4	12.500	0.000000000012	0.000000000006

TABLE 5-1 (cont.)

n	R_b	$P_E(N)$	$P_B(N)$
5	0.080	0.85179	0.43963
5	0.125	0.79972	0.41276
5	0.180	0.73813	0.38097
5	0.245	0.66826	0.34491
5	0.320	0.59218	0.30564
5	0.405	0.51268	0.26461
5	0.500	0.43291	0.22344
5	0.605	0.35603	0.18376
5	0.720	0.28484	0.14701
5	0.845	0.22146	0.11430
5	0.980	0.16720	0.08630
5	1.125	0.12251	0.06323
5	1.280	0.087085	0.044947
5	1.445	0.060033	0.030985
5	1.620	0.040130	0.020712
5	1.805	0.026011	0.013425
5	2.000	0.016349	0.008438
5	2.205	0.0099670	0.0051442
5	2.420	0.0058946	0.0030424
5	2.645	0.0033829	0.0017460
5	2.880	0.0018847	0.0009728
5	3.125	0.0010197	0.0005263
5	3.380	0.00053598	0.00027663
5	3.645	0.00027382	0.00014133
5	3.920	0.00013602	0.00007020
5	4.205	0.000065729	0.000033925
5	4.500	0.000030910	0.000015954
5	4.805	0.000014151	0.000007304
5	5.120	0.0000063098	0.0000032567
5	5.445	0.0000027408	0.0000014146
5	5.780	0.0000011602	0.0000005988
5	6.125	0.00000047871	0.00000024708
5	6.480	0.00000019257	0.00000009939
5	6.845	0.000000075542	0.000000038989
5	7.220	0.000000028901	0.000000014916
5	7.605	0.000000010785	0.000000005566
5	8.000	0.000000003926	0.000000002026
5	8.405	0.000000001394	0.000000000720
5	8.820	0.000000000483	0.000000000249
5	9.245	0.000000000163	0.000000000084
5	9.680	0.000000000054	0.000000000028
5	10.125	0.000000000017	0.000000000009

TABLE 5-1 (cont.)

n	R_b	$P_E(N)$	$P_B(N)$
6	0.080	0.89297	0.45357
6	0.125	0.84515	0.42928
6	0.180	0.78515	0.39881
6	0.245	0.71366	0.36250
6	0.320	0.63275	0.32139
6	0.405	0.54569	0.27717
6	0.500	0.45663	0.23194
6	0.605	0.36997	0.18792
6	0.720	0.28973	0.14716
6	0.845	0.21898	0.11123
6	0.980	0.15956	0.08105
6	1.125	0.11199	0.05688
6	1.280	0.075667	0.038434
6	1.445	0.049200	0.024990
6	1.620	0.030780	0.015634
6	1.805	0.018528	0.009411
6	2.000	0.010733	0.005452
6	2.205	0.0059850	0.0030400
6	2.420	0.0032138	0.0016324
6	2.645	0.0016627	0.0008445
6	2.880	0.00082926	0.00042121
6	3.125	0.00039895	0.00020264
6	3.380	0.00018526	0.00009410
6	3.645	0.000083090	0.000042204
6	3.920	0.000036019	0.000018295
6	4.205	0.000015101	0.000007670
6	4.500	0.0000061268	0.0000031120
6	4.805	0.0000024069	0.0000012225
6	5.120	0.00000091602	0.00000046528
6	5.445	0.00000033788	0.00000017162
6	5.780	0.00000012084	0.0000006138
6	6.125	0.000000041916	0.000000021291
6	6.480	0.000000014105	0.000000007165
6	6.845	0.000000004606	0.000000002340
6	7.220	0.000000001460	0.000000000741
6	7.605	0.000000000449	0.000000000228
6	8.000	0.000000000134	0.000000000068
6	8.405	0.000000000039	0.000000000020
6	8.820	0.000000000011	0.000000000006

TABLE 5-1 (cont.)

n	R_b	$P_E(N)$	$P_B(N)$
7	0.080	0.92245	0.46485
7	0.125	0.87979	0.44336
7	0.180	0.82297	0.41472
7	0.245	0.75177	0.37885
7	0.320	0.66787	0.33656
7	0.405	0.57481	0.28967
7	0.500	0.47765	0.24070
7	0.605	0.38208	0.19255
7	0.720	0.29350	0.14791
7	0.845	0.21608	0.10889
7	0.980	0.15223	0.07672
7	1.125	0.10251	0.05166
7	1.280	0.065934	0.033227
7	1.445	0.040485	0.020402
7	1.620	0.023726	0.011957
7	1.805	0.013272	0.006688
7	2.000	0.0070887	0.0035722
7	2.205	0.0036162	0.0018224
7	2.420	0.0017631	0.0008885
7	2.645	0.00082217	0.00041432
7	2.880	0.00036697	0.00018493
7	3.125	0.00015692	0.00007908
7	3.380	0.000064348	0.000032427
7	3.645	0.000025327	0.000012763
7	3.920	0.0000095769	0.0000048262
7	4.205	0.0000034823	0.0000017548
7	4.500	0.0000012186	0.0000006141
7	4.805	0.00000041070	0.00000020697
7	5.120	0.00000013340	0.00000006723
7	5.445	0.000000041784	0.000000021056
7	5.780	0.000000012626	0.000000006363
7	6.125	0.000000003682	0.000000001856
7	6.480	0.000000001037	0.000000000522
7	6.845	0.000000000282	0.000000000142
7	7.220	0.000000000074	0.000000000037
7	7.605	0.000000000019	0.000000000009

TABLE 5-1 (cont.)

n	R_b	$P_E(N)$	$P_B(N)$
8	0.080	0.94367	0.47368
8	0.125	0.90643	0.45499
8	0.180	0.85368	0.42851
8	0.245	0.78413	0.39360
8	0.320	0.69871	0.35073
8	0.405	0.60097	0.30166
8	0.500	0.49669	0.24932
8	0.605	0.39294	0.19724
8	0.720	0.29660	0.14888
8	0.845	0.21305	0.10694
8	0.980	0.14535	0.07296
8	1.125	0.094040	0.047204
8	1.280	0.057644	0.028935
8	1.445	0.033456	0.016794
8	1.620	0.018382	0.009227
8	1.805	0.0095613	0.0047994
8	2.000	0.0047100	0.0023642
8	2.205	0.0021986	0.0011036
8	2.420	0.00097334	0.00048858
8	2.645	0.00040904	0.00020532
8	2.880	0.00016336	0.00008200
8	3.125	0.000062074	0.000031159
8	3.380	0.000022470	0.000011279
8	3.645	0.0000077584	0.0000038944
8	3.920	0.0000025583	0.0000012842
8	4.205	0.00000080660	0.00000040488
8	4.500	0.00000024341	0.00000012218
8	4.805	0.000000070375	0.000000035325
8	5.120	0.000000019509	0.000000009793
8	5.445	0.000000005189	0.000000002605
8	5.780	0.000000001325	0.000000000665
8	6.125	0.000000000325	0.000000000163
8	6.480	0.000000000077	0.000000000038
8	6.845	0.000000000017	0.000000000009

TABLE 5-1 (cont.)

n	R_b	$P_E(N)$	$P_B(N)$
9	0.080	0.95901	0.48044
9	0.125	0.92700	0.46441
9	0.180	0.87879	0.44025
9	0.245	0.81182	0.40671
9	0.320	0.72606	0.36374
9	0.405	0.62472	0.31297
9	0.500	0.51420	0.25760
9	0.605	0.40287	0.20183
9	0.720	0.29923	0.14991
9	0.845	0.21001	0.10521
9	0.980	0.13892	0.06959
9	1.125	0.086454	0.043311
9	1.280	0.050556	0.025328
9	1.445	0.027760	0.013907
9	1.620	0.014308	0.007168
9	1.805	0.0069239	0.0034687
9	2.000	0.0031469	0.0015765
9	2.205	0.0013444	0.0006735
9	2.420	0.00054044	0.00027075
9	2.645	0.00020467	0.00010253
9	2.880	0.000073122	0.000036633
9	3.125	0.000024684	0.000012366
9	3.380	0.0000078859	0.0000039507
9	3.645	0.0000023880	0.0000011964
9	3.920	0.00000068654	0.00000034394
9	4.205	0.00000018766	0.00000009401
9	4.500	0.000000048831	0.000000024463
9	4.805	0.000000012111	0.000000006067
9	5.120	0.000000002866	0.000000001436
9	5.445	0.000000000647	0.000000000324
9	5.780	0.000000000140	0.000000000070
9	6.125	0.000000000029	0.000000000014

TABLE 5-1 (cont.)

n	R_b	$P_E(N)$	$P_B(N)$
10	0.080	0.97013	0.48554
10	0.125	0.94296	0.47194
10	0.180	0.89940	0.45014
10	0.245	0.83565	0.41823
10	0.320	0.75046	0.37560
10	0.405	0.64647	0.32355
10	0.500	0.53044	0.26548
10	0.605	0.41206	0.20623
10	0.720	0.30152	0.15091
10	0.845	0.20700	0.10360
10	0.980	0.13291	0.06652
10	1.125	0.079642	0.039860
10	1.280	0.044469	0.022256
10	1.445	0.023117	0.011570
10	1.620	0.011185	0.005598
10	1.805	0.0050373	0.0025211
10	2.000	0.0021130	0.0010575
10	2.205	0.00082629	0.00041355
10	2.420	0.00030161	0.00015095
10	2.645	0.00010292	0.00005151
10	2.880	0.000032891	0.000016462
10	3.125	0.0000098622	0.0000049359
10	3.380	0.0000027800	0.0000013914
10	3.645	0.00000073820	0.00000036946
10	3.920	0.00000018500	0.00000009259
10	4.205	0.000000043834	0.000000021938
10	4.500	0.000000009835	0.000000004922
10	4.805	0.000000002093	0.000000001047
10	5.120	0.000000000423	0.000000000212
10	5.445	0.000000000081	0.000000000041
10	5.780	0.000000000015	0.000000000007

TABLE 5-1 (cont.)

n	R_b	$P_E(N)$	$P_B(N)$
15	0.080	0.99377	0.49690
15	0.125	0.98310	0.49157
15	0.180	0.95964	0.47984
15	0.245	0.91483	0.45743
15	0.320	0.84044	0.42023
15	0.405	0.73303	0.36653
15	0.500	0.59810	0.29906
15	0.605	0.45050	0.22526
15	0.720	0.30981	0.15491
15	0.845	0.19285	0.09643
15	0.980	0.10796	0.05398
15	1.125	0.054105	0.027053
15	1.280	0.024200	0.012100
15	1.445	0.0096437	0.0048220
15	1.620	0.0034218	0.0017109
15	1.805	0.0010813	0.0005407
15	2.000	0.00030468	0.00015235
15	2.205	0.000076689	0.000038345
15	2.420	0.000017287	0.000008644
15	2.645	0.0000035013	0.0000017507
15	2.880	0.00000063949	0.00000031976
15	3.125	0.00000010576	0.00000005288
15	3.380	0.000000015907	0.000000007954
15	3.645	0.000000002185	0.000000001093
15	3.920	0.000000000275	0.000000000138
15	4.205	0.000000000032	0.000000000016
20	0.080	0.99868	0.49934
20	0.125	0.99491	0.49745
20	0.180	0.98349	0.48175
20	0.245	0.95494	0.47747
20	0.320	0.89584	0.44792
20	0.405	0.79452	0.39726
20	0.500	0.65056	0.32528
20	0.605	0.48093	0.24046
20	0.720	0.31501	0.15750
20	0.845	0.18018	0.09009
20	0.980	0.089046	0.044523
20	1.125	0.037745	0.018872
20	1.280	0.013657	0.006829
20	1.445	0.0042068	0.0021034
20	1.620	0.0011020	0.0005510
20	1.805	0.00024556	0.00012278
20	2.000	0.000046629	0.000023314
20	2.205	0.0000075658	0.0000037829
20	2.420	0.0000010532	0.0000005266
20	2.645	0.00000012642	0.00000006321
20	2.880	0.000000013165	0.000000006583
20	3.125	0.000000001197	0.000000000599
20	3.380	0.000000000096	0.000000000048

(5-48) or (5-49) and their negatives. Thus, if $\{s_i(t), i = 1, 2, \ldots, N/2\}$ comprises a set of $N/2$ orthogonal signals, then the set of signal pairs $\{s_i(t), s_{i+N/2}(t) = -s_i(t); i = 1, 2, \ldots, N/2\}$ represents a *bi-orthogonal signal set*.

Since for the above set of signals $p(\mathbf{z}|\mathbf{s}_1)$ is a singular p.d.f., the dimensionality of the integration in (5-45) must be reduced in order to evaluate the probability of correct detection. An alternate procedure, which is also optimum for bi-orthogonal signals, computes the correlations q_i, $i = 1, 2, \ldots, N/2$, of (5-18), and the decision is then made on the basis of which q_i has the largest absolute value—the sign of q_i indicates whether $s_i(t)$ or $-s_i(t)$ was most likely transmitted (see Fig. 5-11). In contrast with Fig. 5-3, we note from Fig. 5-11 that only $N/2$ cross-correlators (matched filters) are required

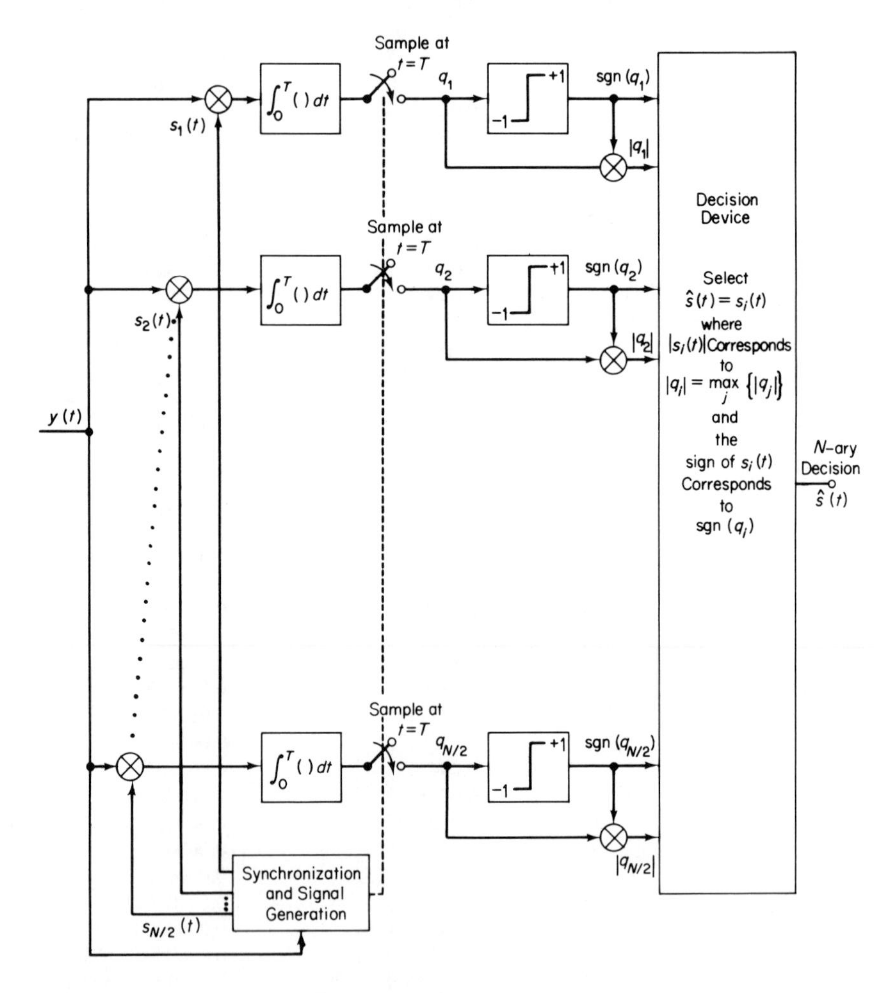

Fig. 5-11 Correlation receiver for reception of bi-orthogonal signals

in a system that transmits N signals. This fact represents one of the more important advantages of a bi-orthogonal signal set over an orthogonal signal set, which requires N cross-correlators at the receiver. Furthermore, for the same number of signals, the effective bandwidth for a bi-orthogonal set is $B_e = N/4T$, which is one-half that of the orthogonal set (see Prob. 5-8).

Based upon the above decision rule, the probability of correct detection when signal $s_1(t)$ was transmitted is

$$P_{c1} = \text{Prob}[q_1 > 0, |q_1| > (|q_2|, \ldots, |q_{N/2}|) | \mathbf{s}_1]$$
$$= \int_0^\infty \underbrace{\int_{-q_1}^{q_1} \cdots \int_{-q_1}^{q_1}}_{N/2-1 \text{ fold}} p(\mathbf{q} | \mathbf{s}_1) \, d\mathbf{q} \tag{5-55}$$

where $p(\mathbf{q} | \mathbf{s}_1)$ is the multivariate Gaussian p.d.f. of Eq. (5-25) corresponding to a set of $N/2$ orthogonal signals. Normalizing q_i in (5-55) by setting $z_i = (q_i - \bar{q}_i)/\sqrt{q_{ii}}$ and then letting $x = z_1/\sqrt{2}$ and $y_i = z_i/\sqrt{2}$ for all $i = 2, 3, \ldots, N/2$ gives the average probability of correct word detection; that is,

$$P_c(N) = \int_{-\sqrt{R_d}}^\infty \frac{\exp(-x^2)}{\sqrt{\pi}} \left[\int_{-x-\sqrt{R_d}}^{x+\sqrt{R_d}} \frac{\exp(-y^2)}{\sqrt{\pi}} \, dy \right]^{N/2-1} dx$$
$$= \int_{-\sqrt{R_d}}^\infty \frac{\exp(-x^2)}{\sqrt{\pi}} [1 - \text{erfc}(x + \sqrt{R_d})]^{N/2-1} \, dx \tag{5-56}$$

The word error probability is union bounded by

$$P_E(N) \leq \left(\frac{N-2}{2}\right) \text{erfc}\left(\sqrt{\frac{R_d}{2}}\right) + \frac{1}{2} \text{erfc}(\sqrt{R_d}) \tag{5-57}$$

which becomes increasingly tight for fixed N as R_d is increased.

In determining the bit error probability for the binary case, that is, $r = 2$ and $N = 2^n$, there are two kinds of word errors that can occur. An *error of the first kind* occurs when the negative $-s_1(t)$ of the transmitted signal $s_1(t)$ is selected. The probability of this event is denoted by $P_1(N)$ and is given by

$$P_1(N) = \text{Prob}[q_1 < 0, |q_1| > (|q_2|, \ldots, |q_{N/2}|) | \mathbf{s}_1]$$
$$= \int_{-\infty}^0 \int_{q_1}^{-q_1} \cdots \int_{q_1}^{-q_1} p(\mathbf{q} | \mathbf{s}_1) \, d\mathbf{q} \tag{5-58}$$

where $p(\mathbf{q} | \mathbf{s}_1)$ is the multivariate Gaussian p.d.f. used in (5-55). Normalizing q_i as before and introducing the same change of variables as that preceding (5–56) gives

$$P_1(N) = \int_{-\infty}^{-\sqrt{R_d}} \frac{\exp(-x^2)}{\sqrt{\pi}} [-1 + \text{erfc}(x + \sqrt{R_d})]^{N/2-1} \, dx \tag{5-59}$$

The probability, $P_2(N)$, of selecting any one of the $2^n - 2$ signals that is orthogonal to the transmitted signal corresponds to an *error of the second kind.* This probability is much greater than $P_1(N)$ and is given by the word error probability determined from (5-56) minus $P_1(N)$; that is,

$$P_2(N) = P_E(N) - P_1(N) \tag{5-60}$$

Assume that complementary data words (symbols) are encoded into complementary channel symbols (words) so as to minimize the probability that a word error will cause all data bits to be in error.* Then, if an error of the first kind is made, there are exactly n bits in error. Therefore, the conditional bit error probability given that an error of the first kind occurred is exactly one. When an error of the second kind is made, the average number of data bits in error is

$$\frac{\sum_{k=1}^{n-1} k({}_nC_k)}{\sum_{k=1}^{n-1} {}_nC_k} = \frac{n}{2}$$

so that the probability of a bit being in error given that an error of the second kind was committed is exactly equal to one-half.** The summations here extend only to $n - 1$ because errors of the first kind are not allowed. Based on the foregoing, the total bit error probability for bi-orthogonal codes is given by

$$P_B(N) = P_1(N) + P_2(N)/2 \tag{5-61}$$

The quantities $P_B(N)$ and $P_E(N)$ versus R_b for various values of $n = \log_2 N$ are given in Table 5-2 and are illustrated in Figs. 5-12 and 5-13.

The Set of Equiprobable, Equal Energy, Transorthogonal Signals

For the *transorthogonal* or *regular simplex* signals the word error probability may be obtained from the word error probability for orthogonal signals by noting that an orthogonal signal set can be produced from a transorthogonal signal set as follows. Let $\{s_i(t),\ i = 1, 2, \ldots, N\}$ be a set of transorthogonal signals with signal inner product defined by Eq. (5–30). Then,

---

* This statement implies that the mapping between the data symbol matrix, $\mathbf{I}_n$, and the code word dictionary should be as follows. The first 2^{n-1} rows of $\mathbf{I}_n$ (representing the binary numbers 0 through $2^{n-1} - 1$) are mapped one-to-one into the 2^{n-1} rows of $\mathbf{H}_{n-1}$, and the latter 2^{n-1} rows of $\mathbf{I}_n$ (representing the binary numbers 2^{n-1} through 2^n) are mapped one-to-one into the complement of the *mirror image* of $\mathbf{H}_{n-1}$. By mirror image of $\mathbf{H}_{n-1}$, we mean a matrix whose first row is the last row of $\mathbf{H}_{n-1}$, whose second row is the next to last row of $\mathbf{H}_{n-1}$, etc.

** An alternate argument to arrive at this conclusion is as follows. If an error of the second kind is committed (which by definition eliminates the possibility of an error of the first kind), then in any bit position of the remaining $2^n - 2$ possible incorrect data symbols, there are an equal number, that is, $2^{n-1} - 1$, of ones and zeros. Thus, if an error of the second kind occurs, the probability of a bit being in error is one-half.

TABLE 5-2 Performance of Bi-orthogonal Codes in White Gaussian Noise with Optimal (maximum-likelihood) Decoding (number of code words, $N = 2^n$)

n	R_b	$P_E(N)$	$P_B(N)$
1	0.080	0.34458	0.34458
1	0.125	0.30854	0.30854
1	0.180	0.27425	0.27425
1	0.245	0.24196	0.24196
1	0.320	0.21186	0.21186
1	0.405	0.18406	0.18406
1	0.500	0.15866	0.15866
1	0.605	0.13567	0.13567
1	0.720	0.11507	0.11507
1	0.845	0.096800	0.096800
1	0.980	0.080757	0.080757
1	1.125	0.066807	0.066807
1	1.280	0.054799	0.054799
1	1.445	0.044565	0.044565
1	1.620	0.035930	0.035930
1	1.805	0.028717	0.028717
1	2.000	0.022750	0.022750
1	2.205	0.017864	0.017864
1	2.420	0.013903	0.013903
1	2.645	0.010724	0.010724
1	2.880	0.0081975	0.0081975
1	3.125	0.0062097	0.0062097
1	3.380	0.0046612	0.0046612
1	3.645	0.0034670	0.0034670
1	3.920	0.0025551	0.0025551
1	4.205	0.0018658	0.0018658
1	4.500	0.0013499	0.0013499
1	4.805	0.00096760	0.00096760
1	5.120	0.00068714	0.00068714
1	5.445	0.00048342	0.00048342
1	5.780	0.00033693	0.00033693
1	6.125	0.00023263	0.00023263
1	6.480	0.00015911	0.00015911
1	6.845	0.00010780	0.00010780
1	7.220	0.000072348	0.000072348
1	7.605	0.000048096	0.000048096
1	8.000	0.000031671	0.000031671
1	8.405	0.000020658	0.000020658
1	8.820	0.000013346	0.000013346
1	9.245	0.0000085399	0.0000085399
1	9.680	0.0000054125	0.0000054125
1	10.125	0.0000033977	0.0000033977
1	10.580	0.0000021125	0.0000021125
1	11.045	0.0000013008	0.0000013008
1	11.520	0.00000079333	0.00000079333
1	12.005	0.00000047918	0.00000047918
1	12.500	0.00000028665	0.00000028665

TABLE 5-2 (cont.)

n	R_b	$P_E(N)$	$P_B(N)$
2	0.080	0.57042	0.34458
2	0.125	0.52188	0.30854
2	0.180	0.47329	0.27425
2	0.245	0.42538	0.24196
2	0.320	0.37883	0.21186
2	0.405	0.33424	0.18406
2	0.500	0.29214	0.15866
2	0.605	0.25293	0.13567
2	0.720	0.21690	0.11507
2	0.845	0.18423	0.09680
2	0.980	0.15499	0.08076
2	1.125	0.12915	0.06681
2	1.280	0.10660	0.05480
2	1.445	0.087145	0.044565
2	1.620	0.070570	0.035930
2	1.805	0.056608	0.028717
2	2.000	0.044983	0.022750
2	2.205	0.035410	0.017864
2	2.420	0.027614	0.013903
2	2.645	0.021333	0.010724
2	2.880	0.016328	0.008198
2	3.125	0.012381	0.006210
2	3.380	0.0093006	0.0046612
2	3.645	0.0069219	0.0034670
2	3.920	0.0051037	0.0025551
2	4.205	0.0037281	0.0018658
2	4.500	0.0026980	0.0013499
2	4.805	0.0019343	0.0009676
2	5.120	0.0013738	0.0006871
2	5.445	0.00096661	0.00048342
2	5.780	0.00067374	0.00033693
2	6.125	0.00046520	0.00023263
2	6.480	0.00031819	0.00015911
2	6.845	0.00021559	0.00010780
2	7.220	0.00014469	0.00007235
2	7.605	0.000096190	0.000048096
2	8.000	0.000063341	0.000031671
2	8.405	0.000041315	0.000020658
2	8.820	0.000026691	0.000013346
2	9.245	0.000017080	0.000008540
2	9.680	0.000010825	0.000005413
2	10.125	0.0000067953	0.0000033977
2	10.580	0.0000042249	0.0000021125
2	11.045	0.0000026016	0.0000013008
2	11.520	0.0000015867	0.0000007933
2	12.005	0.00000095837	0.00000047918
2	12.500	0.00000057330	0.00000028665

TABLE 5-2 (cont.)

n	R_b	$P_E(N)$	$P_B(N)$
3	0.080	0.70485	0.37158
3	0.125	0.65055	0.33877
3	0.180	0.59320	0.30590
3	0.245	0.53400	0.27326
3	0.320	0.47428	0.24126
3	0.405	0.41537	0.21033
3	0.500	0.35853	0.18092
3	0.605	0.30488	0.15346
3	0.720	0.25534	0.12827
3	0.845	0.21054	0.10562
3	0.980	0.17089	0.08564
3	1.125	0.13651	0.06837
3	1.280	0.10731	0.05371
3	1.445	0.083005	0.041534
3	1.620	0.063173	0.031603
3	1.805	0.047306	0.023661
3	2.000	0.034855	0.017431
3	2.205	0.025269	0.012636
3	2.420	0.018026	0.009014
3	2.645	0.012654	0.006328
3	2.880	0.0087425	0.0043714
3	3.125	0.0059446	0.0029723
3	3.380	0.0039786	0.0019893
3	3.645	0.0026213	0.0013107
3	3.920	0.0017002	0.0008501
3	4.205	0.0010858	0.0005429
3	4.500	0.00068280	0.00034140
3	4.805	0.00042282	0.00021141
3	5.120	0.00025786	0.00012893
3	5.445	0.00015489	0.00007744
3	5.780	0.000091638	0.000045819
3	6.125	0.000053407	0.000026704
3	6.480	0.000030663	0.000015331
3	6.845	0.000017343	0.000008672
3	7.220	0.0000096648	0.0000048324
3	7.605	0.0000053064	0.0000026532
3	8.000	0.0000028706	0.0000014353
3	8.405	0.0000015301	0.0000007651
3	8.820	0.00000080367	0.00000040184
3	9.245	0.00000041594	0.00000020797
3	9.680	0.00000021212	0.00000010606
3	10.125	0.00000010660	0.00000005330
3	10.580	0.000000052788	0.000000026394
3	11.045	0.000000025759	0.000000012880
3	11.520	0.000000012387	0.000000006193
3	12.005	0.000000005870	0.000000002935
3	12.500	0.000000002741	0.000000001370

TABLE 5-2 (cont.)

n	R_b	$P_E(N)$	$P_B(N)$
4	0.080	0.79185	0.40184
4	0.125	0.73688	0.37206
4	0.180	0.67553	0.33990
4	0.245	0.60920	0.30583
4	0.320	0.53976	0.27056
4	0.405	0.46931	0.23502
4	0.500	0.40005	0.20021
4	0.605	0.33403	0.16711
4	0.720	0.27300	0.13655
4	0.845	0.21828	0.10916
4	0.980	0.17065	0.08534
4	1.125	0.13041	0.06521
4	1.280	0.097387	0.048696
4	1.445	0.071058	0.035530
4	1.620	0.050652	0.025326
4	1.805	0.035274	0.017637
4	2.000	0.023999	0.011999
4	2.205	0.015953	0.007977
4	2.420	0.010363	0.005181
4	2.645	0.0065792	0.0032896
4	2.880	0.0040831	0.0020415
4	3.125	0.0024776	0.0012388
4	3.380	0.0014703	0.0007351
4	3.645	0.00085346	0.00042673
4	3.920	0.00048473	0.00024237
4	4.205	0.00026944	0.00013472
4	4.500	0.00014660	0.00007330
4	4.805	0.000078101	0.000039050
4	5.120	0.000040746	0.000020373
4	5.445	0.000020822	0.000010411
4	5.780	0.000010424	0.000005212
4	6.125	0.0000051133	0.0000025567
4	6.480	0.0000024580	0.0000012290
4	6.845	0.0000011580	0.0000005790
4	7.220	0.00000053476	0.00000026738
4	7.605	0.00000024207	0.00000012104
4	8.000	0.00000010743	0.00000005371
4	8.405	0.000000046738	0.000000023369
4	8.820	0.000000019937	0.000000009968
4	9.245	0.000000008338	0.000000004169
4	9.680	0.000000003419	0.000000001710
4	10.125	0.000000001375	0.000000000687
4	10.580	0.000000000542	0.000000000271
4	11.045	0.000000000210	0.000000000105
4	11.520	0.000000000079	0.000000000040
4	12.005	0.000000000030	0.000000000015
4	12.500	0.000000000011	0.000000000005

TABLE 5-2 (cont.)

n	R_b	$P_E(N)$	$P_B(N)$
5	0.080	0.85116	0.42736
5	0.125	0.79879	0.40033
5	0.180	0.73687	0.36891
5	0.245	0.66666	0.33356
5	0.320	0.59029	0.29525
5	0.405	0.51056	0.25533
5	0.500	0.43065	0.21535
5	0.605	0.35375	0.17688
5	0.720	0.28265	0.14133
5	0.845	0.21945	0.10973
5	0.980	0.16545	0.08272
5	1.125	0.12104	0.06052
5	1.280	0.085907	0.042954
5	1.445	0.059129	0.029564
5	1.620	0.039464	0.019732
5	1.805	0.025540	0.012770
5	2.000	0.016029	0.008015
5	2.205	0.0097579	0.0048789
5	2.420	0.0057630	0.0028815
5	2.645	0.0033031	0.0016516
5	2.880	0.0018380	0.0009190
5	3.125	0.00099333	0.00049667
5	3.380	0.00052160	0.00026080
5	3.645	0.00026623	0.00013312
5	3.920	0.00013215	0.00006607
5	4.205	0.000063812	0.000031906
5	4.500	0.000029990	0.000014995
5	4.805	0.000013723	0.000006862
5	5.120	0.0000061163	0.0000030581
5	5.445	0.0000026558	0.0000013279
5	5.780	0.0000011239	0.0000005619
5	6.125	0.00000046362	0.00000023181
5	6.480	0.00000018647	0.00000009323
5	6.845	0.000000073136	0.000000036568
5	7.220	0.000000027977	0.000000013989
5	7.605	0.000000010439	0.000000005220
5	8.000	0.000000003800	0.000000001900
5	8.405	0.000000001349	0.000000000675
5	8.820	0.000000000467	0.000000000234
5	9.245	0.000000000158	0.000000000079
5	9.680	0.000000000052	0.000000000026
5	10.125	0.000000000017	0.000000000008

TABLE 5-2 (cont.)

n	R_b	$P_E(N)$	$P_B(N)$
6	0.080	0.89275	0.44691
6	0.125	0.84480	0.42264
6	0.180	0.78465	0.39243
6	0.245	0.71300	0.35654
6	0.320	0.63192	0.31598
6	0.405	0.54474	0.27238
6	0.500	0.45560	0.22780
6	0.605	0.36893	0.18447
6	0.720	0.28873	0.14436
6	0.845	0.21807	0.10904
6	0.980	0.15878	0.07939
6	1.125	0.11136	0.05568
6	1.280	0.075181	0.037591
6	1.445	0.048845	0.024422
6	1.620	0.030534	0.015267
6	1.805	0.018365	0.009183
6	2.000	0.010630	0.005315
6	2.205	0.0059231	0.0029616
6	2.420	0.0031783	0.0015892
6	2.645	0.0016432	0.0008216
6	2.880	0.00081904	0.00040952
6	3.125	0.00039381	0.00019690
6	3.380	0.00018277	0.00009139
6	3.645	0.000081939	0.000040970
6	3.920	0.000035506	0.000017753
6	4.205	0.000014881	0.000007440
6	4.500	0.0000060358	0.0000030179
6	4.805	0.0000023706	0.0000011853
6	5.120	0.00000090203	0.00000045102
6	5.445	0.00000033267	0.00000016634
6	5.780	0.00000011896	0.00000005948
6	6.125	0.000000041261	0.000000020630
6	6.480	0.000000013884	0.000000006942
6	6.845	0.000000004533	0.000000002267
6	7.220	0.000000001437	0.000000000718
6	7.605	0.000000000442	0.000000000221
6	8.000	0.000000000132	0.000000000066
6	8.405	0.000000000038	0.000000000019
6	8.820	0.000000000011	0.000000000005

TABLE 5-2 (cont.)

n	R_b	$P_E(N)$	$P_B(N)$
7	0.080	0.92237	0.46134
7	0.125	0.87966	0.43989
7	0.180	0.82276	0.41140
7	0.245	0.75149	0.37575
7	0.320	0.66751	0.33376
7	0.405	0.57438	0.28719
7	0.500	0.47718	0.23859
7	0.605	0.38160	0.19080
7	0.720	0.29304	0.14652
7	0.845	0.21567	0.10783
7	0.980	0.15188	0.07594
7	1.125	0.10224	0.05112
7	1.280	0.065731	0.032865
7	1.445	0.040343	0.020171
7	1.620	0.023633	0.011817
7	1.805	0.013215	0.006607
7	2.000	0.0070550	0.0035275
7	2.205	0.0035977	0.0017988
7	2.420	0.0017534	0.0008767
7	2.645	0.00081735	0.00040868
7	2.880	0.00036470	0.00018235
7	3.125	0.00015591	0.00007795
7	3.380	0.000063915	0.000031957
7	3.645	0.000025151	0.000012575
7	3.920	0.0000095084	0.0000047542
7	4.205	0.0000034568	0.0000017284
7	4.500	0.0000012095	0.0000006047
7	4.805	0.00000040759	0.00000020380
7	5.120	0.00000013238	0.00000006619
7	5.445	0.000000041462	0.000000020731
7	5.780	0.000000012528	0.000000006264
7	6.125	0.000000003654	0.000000001827
7	6.480	0.000000001029	0.000000000514
7	6.845	0.000000000280	0.000000000140
7	7.220	0.000000000073	0.000000000037
7	7.605	0.000000000019	0.000000000009

TABLE 5-2 (cont.)

n	R_b	$P_E(N)$	$P_B(N)$
8	0.080	0.94364	0.47186
8	0.125	0.90638	0.45320
8	0.180	0.85360	0.42680
8	0.245	0.78401	0.39201
8	0.320	0.69855	0.34928
8	0.405	0.60077	0.30039
8	0.500	0.49647	0.24824
8	0.605	0.39271	0.19636
8	0.720	0.29638	0.14819
8	0.845	0.21286	0.10643
8	0.980	0.14519	0.07259
8	1 125	0.093918	0.046959
8	1.280	0.057557	0.028779
8	1.445	0.033399	0.016699
8	1.620	0.018346	0.009173
8	1.805	0.0095408	0.0047704
8	2.000	0.0046989	0.0023494
8	2.205	0.0021930	0.0010965
8	2.420	0.00097066	0.00048533
8	2.645	0.00040784	0.00020392
8	2.880	0.00016285	0.00008143
8	3.125	0.000061872	0.000030936
8	3.380	0.000022394	0.000011197
8	3.645	0.0000077312	0.0000038656
8	3.920	0.0000025491	0.0000012746
8	4.205	0.00000080363	0.00000040181
8	4.500	0.00000024250	0.00000012125
8	4.805	0.000000070108	0.000000035054
8	5.120	0.000000019435	0.000000009717
8	5.445	0.000000005169	0.000000002585
8	5.780	0.000000001320	0.000000000660
8	6.125	0.000000000324	0.000000000162
8	6.480	0.000000000076	0.000000000038
8	6.845	0.000000000017	0.000000000009

TABLE 5-2 (cont.)

n	R_b	$P_E(N)$	$P_B(N)$
9	0.080	0.95900	0.47951
9	0.125	0.92698	0.46350
9	0.180	0.87875	0.43938
9	0.245	0.81177	0.40589
9	0.320	0.72599	0.36299
9	0.405	0.62463	0.31232
9	0.500	0.51410	0.25705
9	0.605	0.40276	0.20138
9	0.720	0.29913	0.14956
9	0.845	0.20991	0.10496
9	0.980	0.13884	0.06942
9	1.125	0.086400	0.043200
9	1.280	0.050519	0.025260
9	1.445	0.027737	0.013868
9	1.620	0.014295	0.007147
9	1.805	0.0069165	0.0034582
9	2.000	0.0031432	0.0015716
9	2.205	0.0013427	0.0006714
9	2.420	0.00053969	0.00026985
9	2.645	0.00020436	0.00010218
9	2.880	0.000073008	0.000036504
9	3.125	0.000024644	0.000012322
9	3.380	0.0000078724	0.0000039362
9	3.645	0.0000023838	0.0000011919
9	3.920	0.00000068529	0.00000034265
9	4.205	0.00000018731	0.00000009365
9	4.500	0.000000048739	0.000000024370
9	4.805	0.000000012088	0.000000006044
9	5.120	0.000000002860	0.000000001430
9	5.445	0.000000000646	0.000000000323
9	5.780	0.000000000139	0.000000000070
9	6.125	0.000000000029	0.000000000014

TABLE 5-2 (cont.)

n	R_b	$P_E(N)$	$P_B(N)$
10	0.080	0.97012	0.48507
10	0.125	0.94295	0.47148
10	0.180	0.89939	0.44969
10	0.245	0.83563	0.41781
10	0.320	0.75043	0.37521
10	0.405	0.64643	0.32321
10	0.500	0.53040	0.26520
10	0.605	0.41201	0.20600
10	0.720	0.30147	0.15074
10	0.845	0.20695	0.10348
10	0.980	0.13288	0.06644
10	1.125	0.079618	0.039809
10	1.280	0.044453	0.022227
10	1.445	0.023107	0.011554
10	1.620	0.011179	0.005590
10	1.805	0.0050346	0.0025173
10	2.000	0.0021117	0.0010559
10	2.205	0.00082576	0.00041288
10	2.420	0.00030140	0.00015070
10	2.645	0.00010285	0.00005142
10	2.880	0.000032866	0.000016433
10	3.125	0.0000098541	0.0000049270
10	3.380	0.0000027776	0.0000013888
10	3.645	0.00000073754	0.00000036877
10	3.920	0.00000018483	0.00000009241
10	4.205	0.000000043793	0.000000021897
10	4.500	0.000000009826	0.000000004913
10	4.805	0.000000002091	0.000000001045
10	5.120	0.000000000422	0.000000000211
10	5.445	0.000000000081	0.000000000041
10	5.780	0.000000000015	0.000000000007

TABLE 5-2 (cont.)

n	R_b	$P_E(N)$	$P_B(N)$
15	0.080	0.99377	0.49689
15	0.125	0.98310	0.49155
15	0.180	0.95964	0.47982
15	0.245	0.91483	0.45742
15	0.320	0.84044	0.42022
15	0.405	0.73303	0.36651
15	0.500	0.59810	0.29905
15	0.605	0.45050	0.22525
15	0.720	0.30981	0.15491
15	0.845	0.19285	0.09643
15	0.980	0.10796	0.05398
15	1.125	0.054104	0.027052
15	1.280	0.024199	0.012100
15	1.445	0.0096436	0.0048218
15	1.620	0.0034217	0.0017109
15	1.805	0.0010813	0.0005407
15	2.000	0.00030468	0.00015234
15	2.205	0.000076687	0.000038344
15	2.420	0.000017287	0.000008643
15	2.645	0.0000035012	0.0000017506
15	2.880	0.00000063948	0.00000031974
15	3.125	0.00000010576	0.00000005288
15	3.380	0.000000015907	0.000000007953
15	3.645	0.000000002185	0.000000001092
15	3.920	0.000000000275	0.000000000138
15	4.205	0.000000000032	0.000000000016
20	0.080	0.99868	0.49934
20	0.125	0.99491	0.49745
20	0.180	0.98349	0.49175
20	0.245	0.95494	0.47747
20	0.320	0.89584	0.44792
20	0.405	0.79452	0.39726
20	0.500	0.65056	0.32528
20	0.605	0.48093	0.24046
20	0.720	0.31501	0.15750
20	0.845	0.18018	0.09009
20	0.980	0.089046	0.044523
20	1.125	0.037745	0.018872
20	1.280	0.013657	0.006829
20	1.445	0.0042068	0.0021034
20	1.620	0.0011020	0.0005510
20	1.805	0.00024556	0.00012278
20	2.000	0.000046628	0.000023314
20	2.205	0.0000075658	0.0000037829
20	2.420	0.0000010532	0.0000005266
20	2.645	0.00000012642	0.00000006321
20	2.880	0.000000013165	0.000000006583
20	3.125	0.000000001197	0.000000000599
20	3.380	0.000000000096	0.000000000048

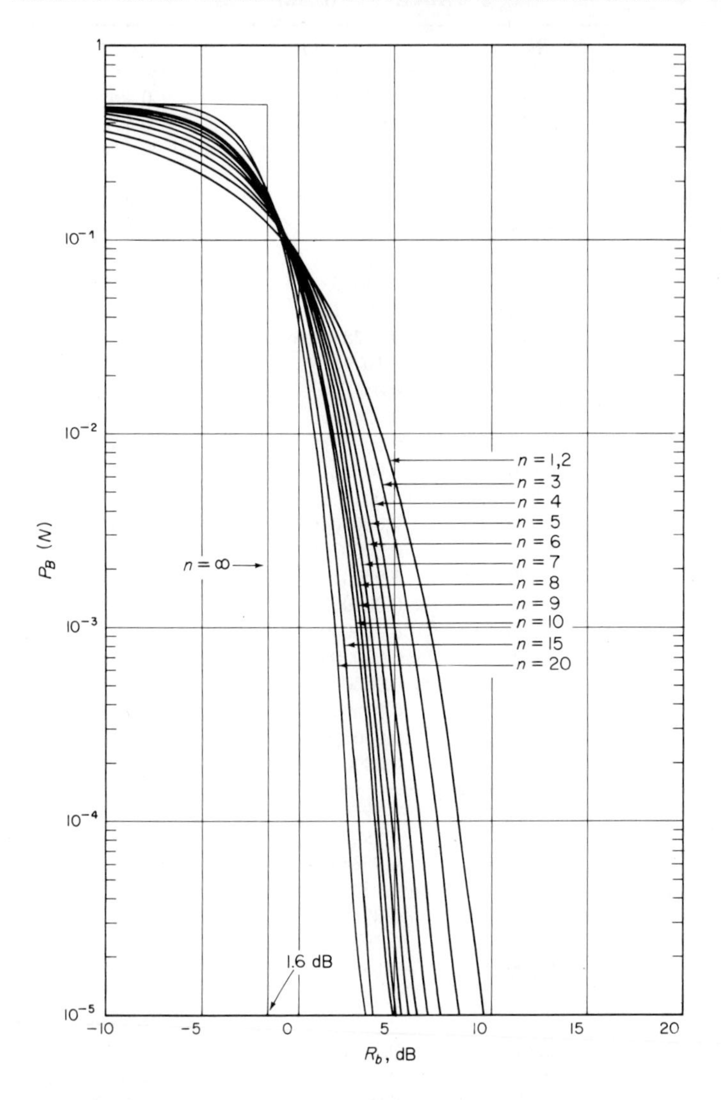

Fig. 5-12 Bit error probability performance of a system transmitting a bi-orthogonal signal set

the signal set

$$s_i'(t) = \begin{cases} s_i(t) & 0 \leq t < T \\ \sqrt{S} & T \leq t \leq TN/(N-1) \qquad i = 1, 2, \ldots, N \end{cases} \tag{5-62}$$

is orthogonal by virtue of the fact that each signal has energy $STN/(N-1)$ and

$$\lambda_{ij} = \frac{N-1}{STN}\left[\int_0^T s_i(t)s_j(t)\,dt + \int_T^{TN/(N-1)} S\,dt\right] = 0 \tag{5-63}$$

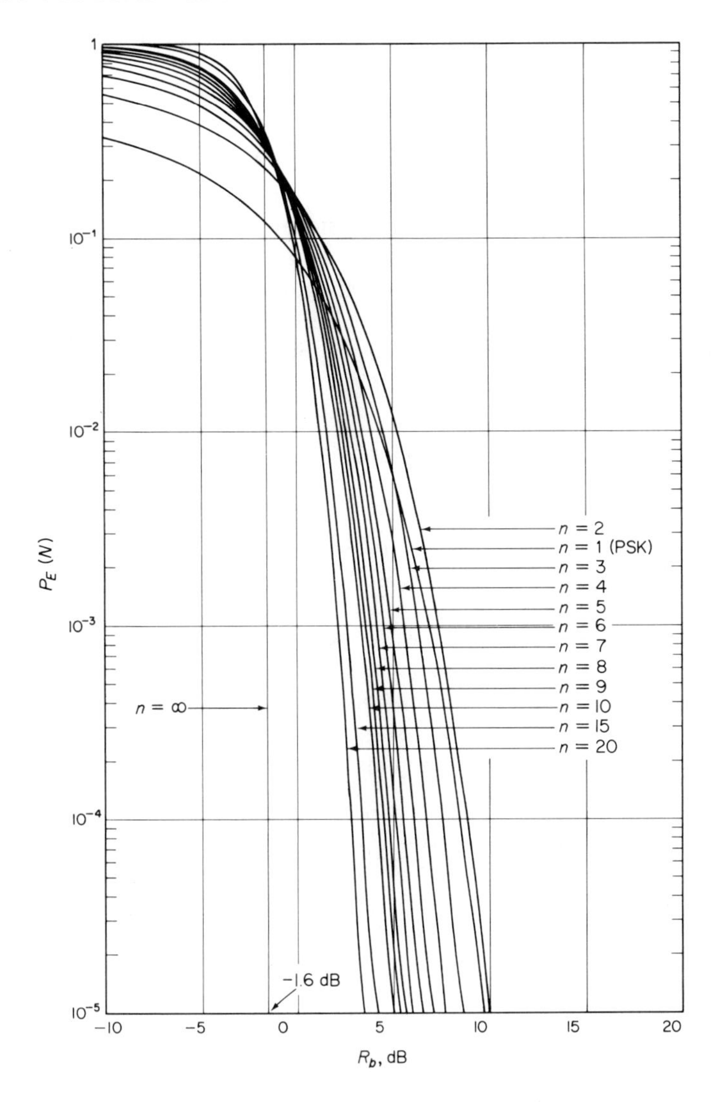

Fig. 5-13 Word error probability performance of a system transmitting a bi-orthogonal signal set.

Since the last $T/(N-1)$ seconds of all signals are the same, nothing is achieved by transmitting this portion of the signals. Hence, the word error probability of the transorthogonal signal set is equal to that resulting from the transmission of an orthogonal signal set with energy $STN/(N-1)$. Thus, the probability of correct detection for the same signal energy is given by

$$P_c(N) = \int_{-\infty}^{\infty} \frac{\exp(-x^2)}{\sqrt{\pi}} \left\{\frac{1}{2} \operatorname{erfc}\left[-x - \sqrt{R_d\left(\frac{N}{N-1}\right)}\right]\right\}^{N-1} dx \quad (5\text{-}64)$$

The error probability $P_E(N) = 1 - P_c(N)$ is union bounded by

$$P_E(N) \leq \left(\frac{N-1}{2}\right) \operatorname{erfc}\left[\sqrt{\frac{R_d}{2}\left(\frac{N}{N-1}\right)}\right] \tag{5-65}$$

which again becomes increasingly tight for fixed N as R_d is increased. Note that for $N = 2$, we have the performance of a PSK system given by (5-47) with $\lambda = -1$, which is a saving of 3 dB in signal-to-noise ratio performance over the orthogonal signaling system; for large N, the saving is negligible. Since if a word error is made for transorthogonal codes, it is equally likely to be in favor of any incorrect word, the bit error probability is related to the word error probability in the same way it was for orthogonal signals; see Eq. (5-54).

Limiting Error Probability Performance of Block Codes as $N \to \infty$

We are now in a position to relate the limiting error probability performance of block codes as the block length N approaches infinity to the notions of channel capacity, communication efficiency, and the Shannon limit introduced in Chap. 1. We shall deal with specific examples—orthogonal, biorthogonal, and regular simplex signal sets—although the conclusions drawn apply to a more general class of block codes.

To begin, let us consider the limiting ($N \to \infty$) error probability performance of a system transmitting orthogonal signals, where $P_E(N)$ is one minus $P_c(N)$ as given by (5-50). Hence,

$$\lim_{N\to\infty} P_E(N) = 1 - \int_{-\infty}^{\infty} \frac{\exp(-x^2)}{\sqrt{\pi}} \times \lim_{N\to\infty}\left[\frac{1}{2}\operatorname{erfc}\left(-x - \sqrt{R_b \log_r N}\right)\right]^{N-1} dx \tag{5-66}$$

One method for evaluating the above limit is to approximate the function in brackets by a unit step occurring at the point of inflection x_0:

$$G_O(x) \triangleq \left[\frac{1}{2}\operatorname{erfc}\left(-x - \sqrt{R_b \log_r N}\right)\right]^{N-1} \cong \begin{cases} 0 & x < x_0 \\ 1 & x > x_0 \end{cases} \tag{5-67}$$

where x_0 is the solution of

$$\frac{d^2 G_O(x)}{dx^2} = 0 \tag{5-68}$$

This approximation improves as N increases at fixed R_b. Solving (5-68) and assuming large N gives (see Prob. 5-8)

$$\operatorname{erfc}\left(x_0 + \sqrt{R_b \log_r N}\right) \cong \frac{2}{N-1} \tag{5-69}$$

Since $\operatorname{erfc}(-x) = 2 - \operatorname{erfc}(x)$, substituting (5-69) into (5-67) gives the value of $G_O(x)$ at its inflection point; that is,

$$G_O(x_0) = \left[1 - \frac{1}{N-1}\right]^{N-1} \cong \frac{1}{e} \tag{5-70}$$

To solve for x_0 more explicitly, we make use of the asymptotic expansion for erfc (x) as x becomes large. Hence from (5-69)

$$\frac{\exp\left[-(x_0 + \sqrt{R_b \log_r N})^2\right]}{\sqrt{\pi}\,[x_0 + \sqrt{R_b \log_r N}]} \cong \frac{2}{N-1} \tag{5-71}$$

Taking the natural logarithm of both sides gives

$$\begin{aligned}(x_0 + \sqrt{R_b \log_r N})^2 + \ln(x_0 + \sqrt{R_b \log_r N})\\ \cong \ln(N-1) - \ln(2\sqrt{\pi})\end{aligned} \tag{5-72}$$

which for large values of the argument $(x_0 + \sqrt{R_b \log_r N})$ and large N becomes

$$x_0 \cong \sqrt{\ln N} - \sqrt{R_b \log_r N} = \sqrt{\ln N} - \sqrt{\frac{R_b}{\ln r} \ln N} \tag{5-73}$$

Hence,

$$\lim_{N\to\infty} x_0 = \begin{cases} \infty & R_b < \ln r \\ 0 & R_b = \ln r \\ -\infty & R_b > \ln r \end{cases} \tag{5-74}$$

Substituting the unit step approximation for $G_O(x)$ into (5-66) and taking the limit as $N \longrightarrow \infty$ finally gives

$$\lim_{N\to\infty} P_E(N) = 1 - \tfrac{1}{2}\,\text{erfc}\,[\lim_{N\to\infty} x_0] = \begin{cases} 1 & R_b < \ln r \\ \frac{1}{2} & R_b = \ln r \\ 0 & R_b > \ln r \end{cases} \tag{5-75}$$

Allowing N (or n) to approach infinity is equivalent to considering the case of unconstrained bandwidth. Hence, *an orthogonal coding scheme produces error-free transmission in the unconstrained bandwidth, additive white Gaussian noise channel provided that the communication efficiency $R_b > \ln r$.* Comparing this statement with the discussion of Eq. (1-11) we observe that for $r = 2$, the orthogonally coded system approaches the Shannon limit under the constraints mentioned above.

For transorthogonal signals, the expression for the probability of correct detection as given by (5-64) approaches that of (5-50) as N becomes large. Hence, by inspection, the limiting error probability performance of a system transmitting a transorthogonal signal set is also specified by (5-75).

By a procedure similar to that used to arrive at the limiting error probability performance of orthogonally coded systems, one can evaluate the limit of (5–56) as $N \longrightarrow \infty$ and thus arrive at the corresponding limiting performance of a system transmitting a bi-orthogonal signal set. Approximating the function

$$G_{BO}(x) \triangleq [1 - \text{erfc}\,(x + \sqrt{R_b \log_r N})]^{N/2-1} \tag{5-76}$$

by a unit step at its point of inflection x_0 we find that for large N, x_0 must satisfy

$$\text{erfc}\,[x_0 + \sqrt{R_b \log_r N}] \cong \frac{1}{N/2 - 1} \tag{5-77}$$

and hence

$$G_{BO}(x_0) = \left[1 - \frac{1}{N/2 - 1}\right]^{N/2-1} \cong \frac{1}{e} \tag{5-78}$$

Following the same steps used to arrive at (5-71) through (5-73), we find that for large N, x_0 is once again approximately given by (5-73). Since x_0 is always greater than $-\sqrt{R_b \log_r N}$, the limit of one minus $P_c(N)$ as given by (5-56) as $N \longrightarrow \infty$ becomes identical to (5-75). Hence a bi-orthogonally coded system has the same limiting error probability performance as that demonstrated for systems using the orthogonal and transorthogonal signal sets.

The Set of Equiprobable, Equal Energy, Polyphase Signals (MPSK)

For multiple phase-shift-keying (MPSK), a convenient set of signals that may be generated is the *polyphase signal set*

$$s_i(t) = \sqrt{2S} \sin\left[\frac{\pi k t}{T} + \theta_i\right] \qquad 0 \leq t \leq T \tag{5-79}$$

where $\theta_i = 2(i - 1)\pi/N$ for $i = 1, 2, \ldots, N$ and k is any integer. The correlation matrix for this type of signaling has elements as given by (5-32). Notice that for $N = 2$, we get antipodal signals (PSK); the case $N = 4$ produces *quadriphase-shift-keying* (QPSK); and $N = 8$ yields *octaphase-shift-keying*. These three signaling sets are depicted in Fig. 5-14, along with the decision thresholds that are implemented in the optimum receiver.

Since the matrix $\mathbf{\Lambda}_p$ is singular, the dimensionality of the integration in (5-45) must be reduced in order to evaluate $P_c(N)$. A simpler method is to transform the problem into polar coordinates by representing the ith cross-correlator output in terms of its amplitude and phase.[13] Since $\theta_i = 2(i - 1)\pi/N$, then the optimum receiver of Fig. 5-3 computes, for all $i = 1, 2, \ldots, N$, the quantities

$$\begin{aligned}
q_i &= \sqrt{2S} \int_0^T y(t) \sin\left[\frac{\pi k t}{T} + \theta_i\right] dt \\
&= \sqrt{2S}\left[\cos\theta_i \int_0^T y(t) \sin\left(\frac{\pi k t}{T}\right) dt + \sin\theta_i \int_0^T y(t) \cos\left(\frac{\pi k t}{T}\right) dt\right] \\
&= V \cos\,[\theta_i - \eta]
\end{aligned} \tag{5-80}$$

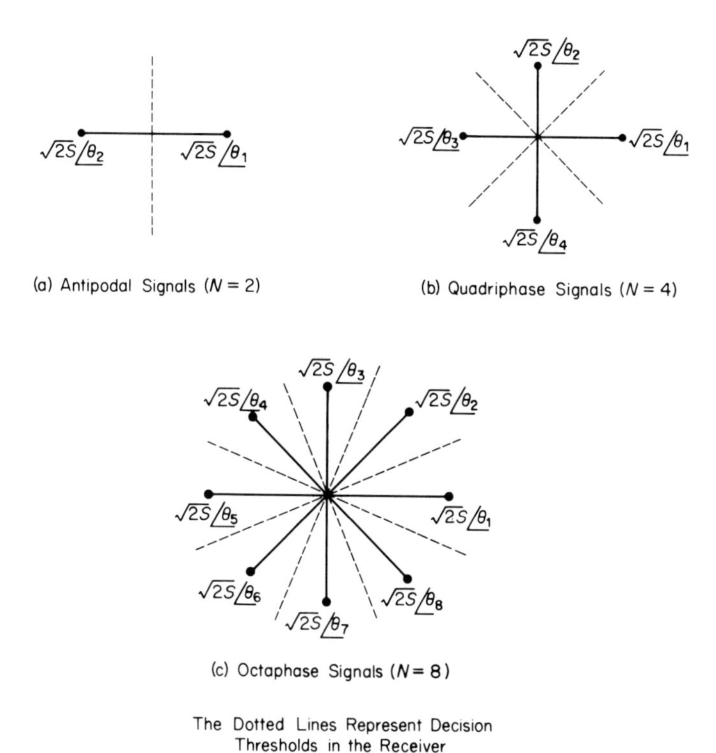

Fig. 5-14 The signal vectors for various polyphase signal sets

where the amplitude V and phase η are defined by

$$V \triangleq \left\{ \Bigg[\underbrace{\int_0^T y(t)\left[\sqrt{2S}\sin\frac{\pi k t}{T}\right]dt}_{V_s} \Bigg]^2 + \Bigg[\underbrace{\int_0^T y(t)\left[\sqrt{2S}\cos\frac{\pi k t}{T}\right]dt}_{V_c} \Bigg]^2 \right\}^{1/2} \tag{5-81}$$

$$\eta \triangleq \tan^{-1}\frac{V_c}{V_s} = \tan^{-1}\left\{ \frac{\displaystyle\int_0^T y(t)\left[\sqrt{2S}\cos\frac{\pi k t}{T}\right]dt}{\displaystyle\int_0^T y(t)\left[\sqrt{2S}\sin\frac{\pi k t}{T}\right]dt} \right\} \tag{5-82}$$

On the basis of this representation of q_i, it is important to note that the receiver of Fig. 5-3 requires N correlators, whereas an equivalent receiver (see Fig. 5-15) computes V_c and V_s and requires only two correlators. Since for any i, q_i represents a sample of a narrowband Gaussian process, the joint p.d.f. of

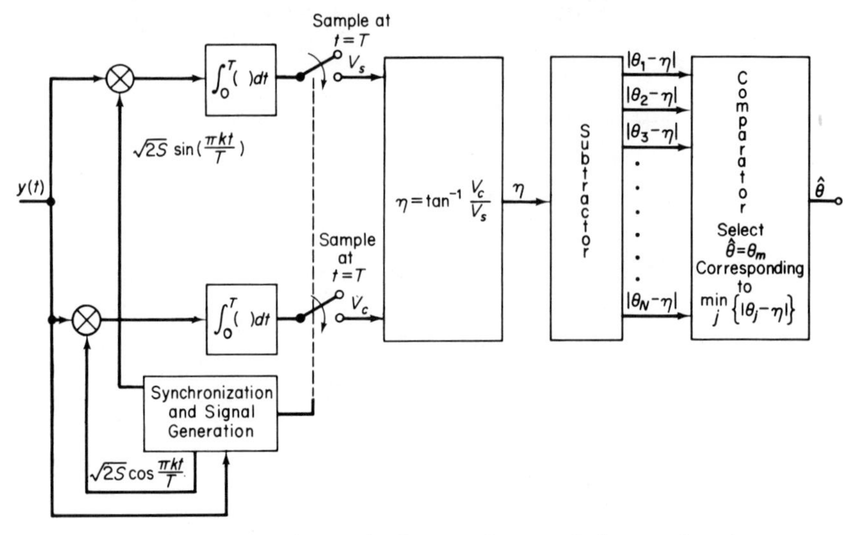

Fig. 5-15 Phase measuring embodiment of a correlation receiver for detecting polyphase signals

V and η, given that $s_i(t)$ was transmitted, is (see Prob. 5-9)

$$p(V,\eta)=\begin{cases}\dfrac{V}{\pi STN_0}\exp\left[-\dfrac{V^2-2STV\cos(\eta-\theta_i)+(ST)^2}{STN_0}\right] & 0\le V\le\infty \\ & 0\le\eta\le 2\pi \\ 0 & \text{elsewhere}\end{cases} \tag{5-83}$$

Letting $r=V/\sqrt{STN_0}$ and averaging over the random variable r gives

$$p(\eta)=\begin{cases}\displaystyle\int_0^\infty \frac{r}{\pi}\exp\{-[r^2-2r\sqrt{R_d}\cos(\eta-\theta_i)+R_d]\}\,dr & 0\le\eta\le 2\pi \\ 0 & \text{elsewhere}\end{cases} \tag{5-84}$$

Since q_i is a maximum when the distance $|\theta_i-\eta|$ is a minimum [see (5-80)], then for η in the region $(\theta_i-\pi/N,\ \theta_i+\pi/N)$, the inequality $|\theta_j-\eta|>|\theta_i-\eta|$ is valid for all $j\neq i$ and hence $q_i>q_j$ for all $j\neq i$. Based upon this result, the probability of correctly detecting the ith signal is the probability that η is in the region $[(2i-3)\pi/N,(2i-1)\pi/N]$; i.e.

$$P_{c_i}=\int_{(2i-3)\pi/N}^{(2i-1)\pi/N} p(\eta)\,d\eta \tag{5-85}$$

Letting $\psi\triangleq\eta-2(i-1)\pi/N$, we see that P_{c_i} is independent of i; hence the average probability of correct reception becomes (see Prob. 5-10)

$$P_c(N) = \frac{2}{\pi} \exp(-R_d) \int_0^\infty r \exp(-r^2) \times \left[\int_0^{\pi/N} \exp(2r\sqrt{R_d} \cos \psi)\, d\psi \right] dr \tag{5-86}$$

Introducing the changes of variables $u = r \cos \psi$, $v = r \sin \psi$ and noting that $v = u \tan \psi$, (5-86) reduces to

$$P_c(N) = \frac{2}{\pi} \int_0^\infty \exp\{-(u - \sqrt{R_d})^2\} \left[\int_0^{u \tan \pi/N} \exp(-v^2)\, dv \right] du \tag{5-87}$$

The corresponding word error probability is derived in Appendix B and can be written as

$$P_E(N) = \frac{N-1}{N} - \frac{1}{2} \operatorname{erf}\left[\sqrt{R_d} \sin \frac{\pi}{N}\right] - \frac{1}{\sqrt{\pi}} \int_0^{\sqrt{R_d} \sin \pi/N} \exp(-y^2) \operatorname{erf}\left(y \cot \frac{\pi}{N}\right) dy \tag{5-88}$$

For the case of PSK, $N = 2$ and the performance given by (5-88) reduces to (5-47) with $\lambda = -1$. It is also possible to evaluate the integral in (5-88) in closed form for $N = 4$, i.e. QPSK. The result is given by

$$P_E(4) = \operatorname{erfc}\left[\sqrt{\frac{R_d}{2}}\right] - \frac{1}{4} \operatorname{erfc}^2\left[\sqrt{\frac{R_d}{2}}\right] \tag{5-89}$$

In the region where $R_d \gg 1$, (5–89) is well approximated by

$$P_E(4) \cong \operatorname{erfc}\left[\sqrt{\frac{R_d}{2}}\right] \tag{5-90}$$

This says that system word error probability performance using four signals is approximately 3 dB worse than that of the FSK system discussed earlier; however the QPSK system requires only one-half the bandwidth.

For values of $P_E(N)$ less than 10^{-3}, the region of greatest interest in practice, the error probability for $N > 2$ is well approximated by [14,15] (also see Prob. 5-11)

$$P_E(N) \cong \operatorname{erfc}\left[\sqrt{R_d} \sin \frac{\pi}{N}\right] \tag{5-91}$$

This approximation becomes extremely tight for fixed N as R_d increases. Table 5-3 gives $P_E(N)$ as determined from (5-88) vs. R_b in dB for values of $N = 2, 4, 8, 16, 32$, and 64; these data are illustrated in Fig. 5-16. Such a set of error probability curves originally appeared in Ref. 16.

If one encodes the source symbols using a *Gray code*[17], then when $P_E(N)$ is acceptably small, the bit error probability is approximated by

$$P_B(N) \cong \frac{P_E(N)}{\log_2 N} \tag{5-92}$$

TABLE 5-3 Coherent Detection of MPSK*

R_b(dB)	$P_E(2)$	$P_E(4)$	$P_E(8)$
−5.00	.2132280	.3809898	.5856787
−4.00	.1861138	.3375893	.5456900
−3.00	.1583683	.2916561	.5014898
−2.00	.1306445	.2442210	.4533385
−1.00	.1037591	.1967522	.4018029
.00	.7864960 (−1)	.1511134	.3478009
1.00	.5628195 (−1)	.1093962	.2926161
2.00	.3750613 (−1)	.7360555 (−1)	.2378716
3.00	.2287841 (−1)	.4523339 (−1)	.1854530
4.00	.1250082 (−1)	.2484537 (−1)	.1373689
5.00	.5953867 (−2)	.1187229 (−1)	.9552945 (−1)
6.00	.2388290 (−2)	.4770877 (−2)	.6143974 (−1)
7.00	.7726742 (−3)	.1544752 (−2)	.3585831 (−1)
8.00	.1909072 (−3)	.3817786 (−3)	.1854315 (−1)
9.00	.3362664 (−4)	.6725278 (−4)	.8244400 (−2)
10.00	.3871521 (−5)	.7743434 (−5)	.3034185 (−2)
11.00			.8811869 (−3)
12.00			.1901351 (−3)
13.00			.2825011 (−4)
14.00			.2624343 (−5)
15.00			
16.00			
17.00			
18.00			
19.00			
20.00			
21.00			
22.00			
23.00			
24.00			
25.00			
26.00			
27.00			
28.00			

*The negative integer in parentheses following each entry in the table represents the power of ten by which the entry should be multiplied.

TABLE 5-3 (cont.)

$P_E(16)$	$P_E(32)$	$P_E(64)$
.7525811	.8604450	.9234688
.7253750	.8442718	.9143934
.6947273	.8259693	.9041281
.6604861	.8053696	.8925640
.6225698	.7823000	.8795776
.5809768	.7565766	.8650268
.5357994	.7280032	.8487480
.4872527	.6963782	.8305576
.4357215	.6615120	.8102553
.3818230	.6232501	.7876274
.3264683	.5815028	.7624510
.2709039	.5362812	.7345001
.2167101	.4877398	.7035539
.1657299	.4362284	.6694092
.1199032	.3823477	.6318989
.8099516 (−1)	.3270015	.5909158
.5024693 (−1)	.2714322	.5464452
.2803827 (−1)	.2172168	.4986063
.1370918 (−1)	.1661966	.4477017
.5682776 (−2)	.1203116	.3942723
.1915742 (−2)	.8133036 (−1)	.3391521
.4983953 (−3)	.5050061 (−1)	.2835079
.9365990 (−4)	.2821166 (−1)	.2288472
.1168884 (−4)	.1381341 (−1)	.1769638
	.5736071 (−2)	.1297958
	.1937969 (−2)	.8917717 (−1)
	.5055512 (−3)	.5650043 (−1)
	.9529383 (−4)	.3236575 (−1)
	.1187544 (−4)	.1635177 (−1)
		.7061428 (−2)
		.2505663 (−2)
		.6949747 (−3)
		.1412478 (−3)
		.1894742 (−4)

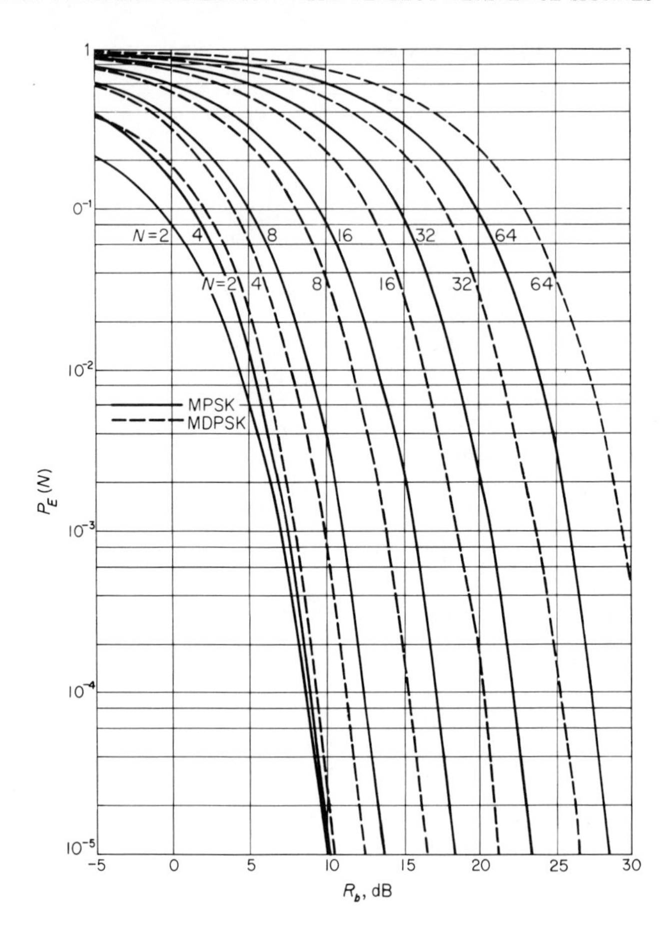

Fig. 5-16 Word error probability performance for MPSK and MDPSK systems

The validity of this approximation is justified as follows. For large signal-to-noise ratios, we can assume that the only significant word errors are those whose corresponding signal phases are adjacent to that of the transmitted signal. Since a Gray code has the property that in going from word-to-word you change only a single coordinate, then for a given transmitted signal (code word), an error in an adjacent code word is accompanied by one and only one bit error. Thus, assuming equiprobable transmitted signals, the average bit error probability is related to the average word error probability as in (5-92). For $N = 4$, Eq. (5-61) is an exact relationship; hence, for equal bit signal-to-noise ratios, QPSK and PSK have identical bit error probability performance. This result is significant when one realizes that QPSK offers an effective saving in bandwidth of 3 dB over PSK for the same $\mathcal{R}_b$.

The Set of Equiprobable, Equal Energy, L-orthogonal Signals

A set of N signals which have the correlation properties of Eq. (5-34) can be realized by combining the orthogonal signal representation of (5-49) with the polyphase signal representation of (5-79). Specifically, the set of $N = ML$ signals

$$s_i(t) = \begin{cases} \sqrt{2S}\sin\left[\dfrac{\pi(k+m)t}{T} + \theta_l\right]; & L = 1, 2 \\ \sqrt{2S}\sin\left[\dfrac{\pi(k+2m)t}{T} + \theta_l\right]; & L > 2 \end{cases} \tag{5-93}$$

is L-orthogonal (see Prob. 5-13). As in (5-48), k is again any integer, $\theta_l = 2(l-1)\pi/L$ and $i \triangleq (m-1)L + l$ with $m = 1, 2, \ldots, M$ and $l = 1, 2, \ldots, L$. Several other examples of L-orthogonal signal sets are given in Prob. 5-14.

Since for the above set of signals, (5-35) is singular and $p(\mathbf{z}\,|\,\mathbf{s}_1)$ of (5-46) is a singular p.d.f., the dimensionality of the integration in (5-45) must be reduced in order to evaluate the probability of correct detection. A procedure for doing so is described in the appendix of Reed and Scholtz.[8] The details of the derivation are quite involved, and hence we merely present the result. For $L > 2$, the probability of correct detection is given by*

$$P_c(N) = \iint_B \frac{\exp\{-(x_1^2 + x_2^2)\}}{\pi} \times \left[\iint_{A(x_1)} \frac{\exp\{-(y_1^2 + y_2^2)\}}{\pi}\, dy_1\, dy_2\right]^{M-1} dx_1\, dx_2 \tag{5-94}$$

where $A(x_1)$ and B are regions of integration (see Fig. 5-17). In particular, B is the set of points (x_1, x_2) in the wedge bounded by the straight lines

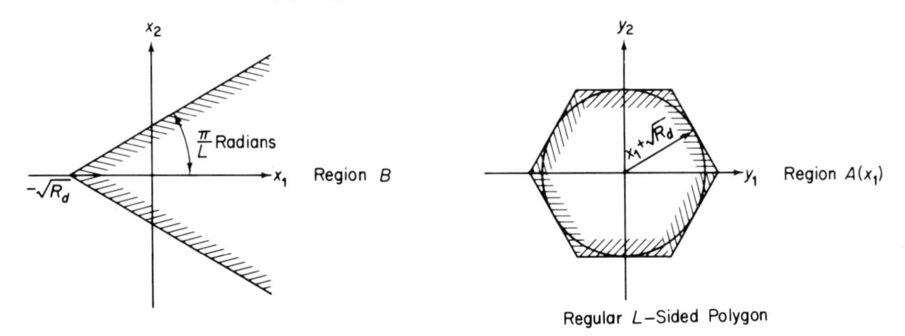

Fig. 5-17 Areas of integration for Eq. (5-94) (Example drawn for $L = 6$)

* For $L = 1$ and $L = 2$, we have the well known results for orthogonal and bi-orthogonal signaling respectively.

$$x_2 = (x_1 + \sqrt{R_d}) \tan \frac{\pi}{L}$$
$$x_2 = -(x_1 + \sqrt{R_d}) \tan \frac{\pi}{L} \tag{5-95}$$

and $A(x_1)$ is the set of points within an L-sided regular polygon circumscribed on a circle of radius $x_1 + \sqrt{R_d}$. Substituting (5-95) in (5-94) and making the change of variables $u_1 = x_1 + \sqrt{R_d}$ gives (for $L > 2$)

$$P_c(N) = \frac{2}{\pi} \int_0^\infty \exp[-(u_1 - \sqrt{R_d})^2] \left[\int_0^{u_1 \tan \pi/L} \exp[-x_2^2]\, dx_2 \right] \times \left[\iint_{A(u_1)} \frac{\exp\{-(y_1^2 + y_2^2)\}}{\pi} dy_1\, dy_2 \right]^{M-1} du_1 \tag{5-96}$$

Since any L-sided regular polygon can be subdivided into L equal area triangles, then from the symmetry of the double integral on $A(u_1)$, we get

$$\begin{aligned} P_c(N) &= \frac{2}{\pi} \int_0^\infty \exp[-(u_1 - \sqrt{R_d})^2] \left[\int_0^{u_1 \tan \pi/L} \exp[-x_2^2]\, dx_2 \right] \\ &\quad \times \left[L \int_0^{u_1} \int_{-y_1 \tan \pi/L}^{y_1 \tan \pi/L} \frac{\exp\{-(y_1^2 + y_2^2)\}}{\pi} dy_1\, dy_2 \right]^{M-1} du_1 \\ &= \frac{1}{\sqrt{\pi}} \int_0^\infty \exp[-(u_1 - \sqrt{R_d})^2] \operatorname{erf}\left(u_1 \tan \frac{\pi}{L}\right) \\ &\quad \times \left[\frac{L}{\sqrt{\pi}} \int_0^{u_1} \exp(-y_1^2) \operatorname{erf}\left(y_1 \tan \frac{\pi}{L}\right) dy_1 \right]^{M-1} du_1 \end{aligned} \tag{5-97}$$

The special cases of $L = 2$ and $L = 4$ both correspond to bi-orthogonal signal sets and hence (5-97) reduces to (5-56) [see Prob. 5-15]. When $M = 1$, we merely have a single set of polyphase signals, and by inspection (5-97) reduces to (5-87). For other combinations of M and L, the integrals in (5-97) must be evaluated by numerical integration on a general purpose digital computer.

Upper and lower bounds on the probability of word error $P_E(N) = 1 - P_c(N)$ may be readily calculated by circular approximation to the area of integration $A(u_1)$. In particular, an upper bound on $P_E(N)$ [lower bound on $P_c(N)$] is obtained by approximating $A(u_1)$ by its inscribed circle of radius u_1. Similarly, a lower bound on $P_E(N)$ may be calculated by approximating $A(u_1)$ by its circumscribed circle of radius, $u_1 \sec(\pi/L)$. Considering first the upper bound on $P_E(N)$, and letting $y_1 = r \cos \psi$, $y_2 = r \sin \psi$ gives

$$\begin{aligned} P_c(N) &\geq \frac{2}{\pi} \int_0^\infty \exp[-(u_1 - \sqrt{R_d})^2] \left[\int_0^{u_1 \tan \pi/L} \exp[-x_2^2]\, dx_2 \right] \\ &\quad \times \left[\int_0^{2\pi} \int_0^{u_1} \frac{r \exp\{-r^2\}}{\pi} dr\, d\psi \right]^{M-1} du_1 \\ &= \frac{2}{\pi} \int_0^\infty \exp[-(u_1 - \sqrt{R_d})^2] \left[\int_0^{u_1 \tan \pi/L} \exp[-x_2^2]\, dx_2 \right] \\ &\quad \times \{1 - \exp[-u_1^2]\}^{M-1} du_1 \end{aligned} \tag{5-98}$$

Expanding via the binomial expansion

$$\{1 - \exp[-u_1^2]\}^{M-1} = \sum_{k=0}^{M-1} {}_{M-1}C_k(-1)^k \exp[-ku_1^2] \tag{5-99}$$

Substituting (5-99) into (5-98) and making the changes of variables $u = u_1\sqrt{1+k}$, $v = x_2\sqrt{1+k}$ gives

$$P_c(N) \geq \sum_{k=0}^{M-1} \frac{(-1)^k}{(1+k)} {}_{M-1}C_k \exp\left\{-\frac{kR_d}{1+k}\right\} \times \left\{\frac{2}{\pi}\int_0^\infty \exp\left[-\left(u - \sqrt{\frac{R_d}{1+k}}\right)^2\right]\left[\int_0^{u \tan \pi/L} \exp\left\{-\frac{v^2}{1+k}\right\} dv\right] du\right\} \tag{5-100}$$

Recognizing that the $k = 0$ term in (5-100) corresponds to $P_c(L)|_{MPSK}$ as given by (5-87), we write the upper bound on $P_E(N)$ as

$$P_E(N) \leq P_E(L)\Big|_{\text{MPSK}} - \sum_{k=1}^{M-1} \frac{(-1)^k}{(1+k)} {}_{M-1}C_k \exp\left\{-\frac{kR_d}{1+k}\right\} \times \left\{\frac{2}{\pi}\int_0^\infty \exp\left[-\left(u - \sqrt{\frac{R_d}{1+k}}\right)^2\right]\left[\int_0^{u \tan \pi/L} \exp\left\{-\frac{v^2}{1+k}\right\} dv\right] du\right\} \tag{5-101}$$

where $P_E(L)|_{\text{MPSK}} \triangleq 1 - P_c(L)|_{\text{MPSK}}$. Similarly, one can derive a lower bound on $P_E(N)$ with the result

$$P_E(N) \geq P_E(L)\Big|_{\text{MPSK}} - \sum_{k=1}^{M-1} \frac{(-1)^k}{(1+k')} {}_{M-1}C_k \exp\left\{\frac{-k'R_d}{1+k'}\right\} \times \left\{\frac{2}{\pi}\int_0^\infty \exp\left[-\left(u - \sqrt{\frac{R_d}{1+k'}}\right)^2\right]\left[\int_0^{u \tan \pi/L} \exp\left\{-\frac{v^2}{1+k'}\right\} dv\right] du\right\} \tag{5-102}$$

where

$$k' \triangleq k \sec^2 \frac{\pi}{L} \tag{5-103}$$

The bounds of (5-101) and (5-102) can be simplified by reducing the double integration involving an infinite limit to a single integration having a finite limit. The procedure for doing this reduction parallels that given in Appendix B. The results are

$$P_E(N) \leq P_{E_u}(N) \triangleq P_E(L)\Big|_{\text{MPSK}} - \sum_{k=1}^{M-1} \frac{(-1)^k}{(1+k)} {}_{M-1}C_k \exp\left\{-\frac{kR_d}{1+k}\right\} \times \left[\frac{\theta}{\pi} + \frac{1}{2}\text{erf}\left(\sqrt{\frac{R_d}{1+k}} \sin\theta\right) + \frac{1}{\sqrt{\pi}}\int_0^{\sqrt{R_d/(1+k)}\sin\theta} \exp\{-y^2)\, \text{erf}\{y \cot\theta\}\, dy\right] \tag{5-104}$$

$$\theta \triangleq \tan^{-1}\left[\frac{\tan \pi/L}{\sqrt{1+k}}\right]$$

$$P_E(N) \geq P_{E_l}(N) \triangleq P_E(L)\Big|_{\text{MPSK}} - \sum_{k=1}^{M-1} \frac{(-1)^k}{(1+k')} {}_{M-1}C_k \exp\left\{-\frac{k'R_d}{1+k'}\right\}$$
$$\times \left[\frac{\theta'}{\pi} + \frac{1}{2}\operatorname{erf}\left(\sqrt{\frac{R_d}{1+k}}\sin\theta'\right)\right.$$
$$\left. + \frac{1}{\sqrt{\pi}} \int_0^{\sqrt{R_d/(1+k')}\sin\theta'} \exp\{-y^2\}\operatorname{erf}\{y\cot\theta'\}\,dy\right] \quad (5\text{-}105)$$
$$\theta' \triangleq \tan^{-1}\left[\frac{\tan \pi/L}{\sqrt{1+k'}}\right]$$

We note from (5-104) and (5-105) that for $M = 1$, the two bounds become equal and give the exact performance; that is, $P_E(L)|_{\text{MPSK}}$. Also, for large L (small M), the bounds are very tight since both the inscribed and circumscribed circles are good approximations to the L-sided polygon. Figure 5-18

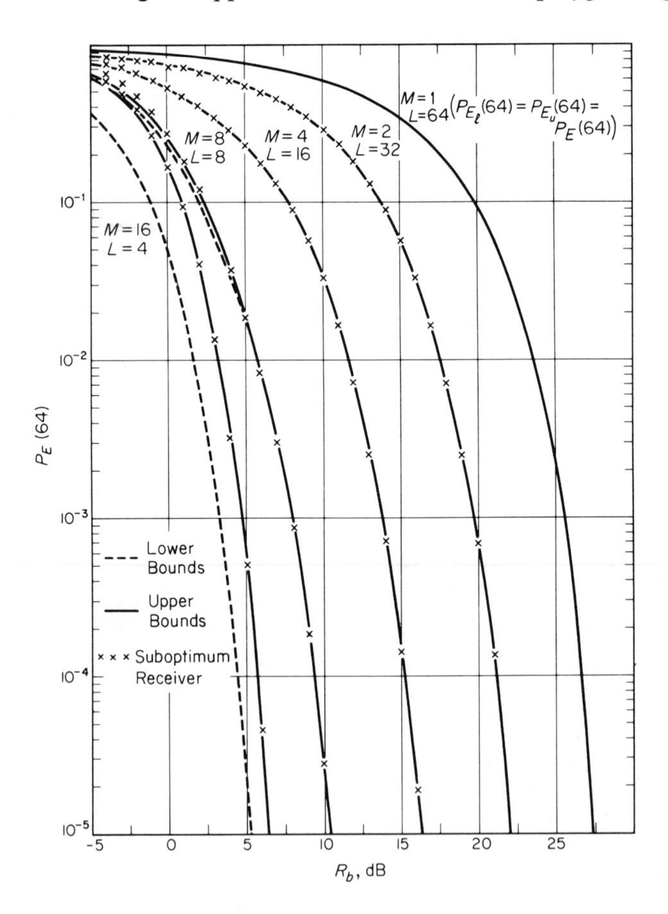

Fig. 5-18 Error probability performance for coherent detection of L-orthogonal signals ($N = 64$)

illustrates the upper and lower bounds of Eq. (5-104) and (5-105) respectively as functions of $R_b = R_d/\log_2 N$ in dB for $N = 64$ and various combinations of M and L.

Optimum detection of the set of signals in (5-93) requires the correlation receiver of Fig. 5-3 and yields the performance as given by Eq. (5-97). When N becomes large, the number of correlators required to implement the receiver of Fig. 5-3 becomes prohibitive. Hence, we investigate a suboptimum receiver that requires only $2M$ correlators and whose error probability performance is quite close to that of the optimum receiver.

The suboptimum receiver illustrated in Fig. 5-19 uses a noncoherent detector to determine the most probable index m characterizing the received signal [see Eq. (5-93)] and simultaneously makes a maximum-likelihood decision on the phase θ_l, or equivalently, the index l.[9] We note from (5-93) that the index i is correct if and only if m and l are correctly chosen. The

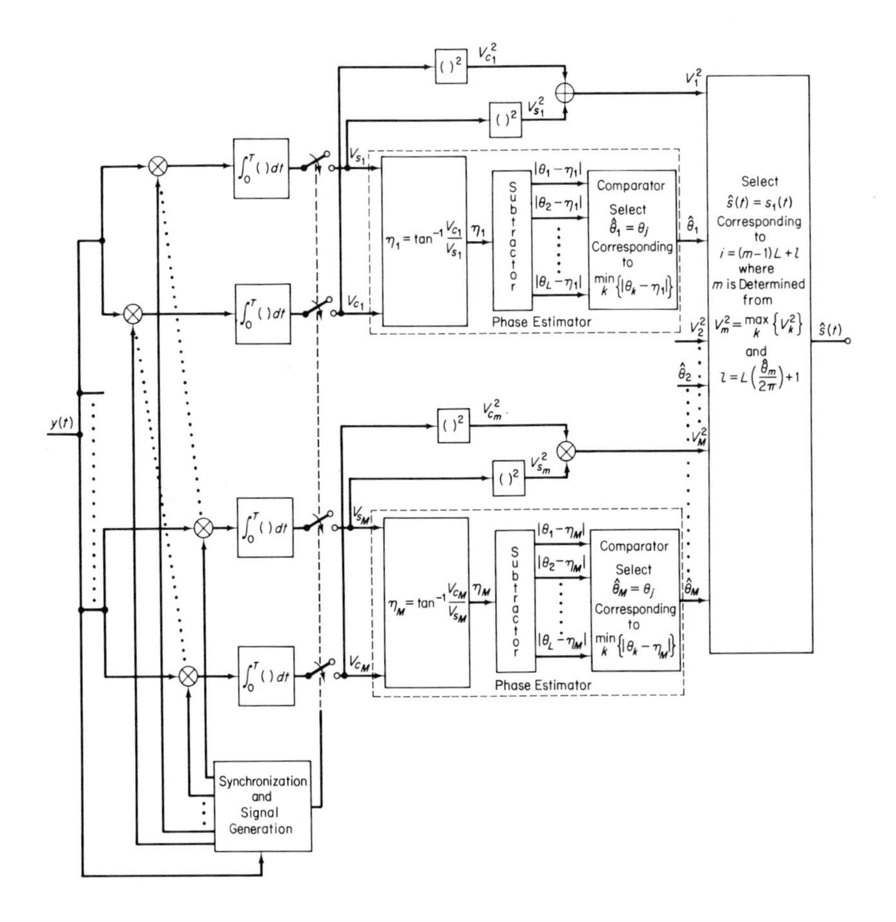

Fig. 5-19 A suboptimal receiver for coherent detection of L-orthogonal signals

probability that m is correct is the probability of the event $\{V_m^2 = \max_k V_k^2\}$ or, equivalently, the probability of choosing the correct frequency from a total of M choices. This probability $P_c(M)|_{\text{NON}}$ is evaluated in Chap. 10 and can be obtained by taking one minus the error probability performance as given by (10-16) with N replaced by M. Having selected the correct frequency, the conditional probability $P_c(L)|_{\text{MPSK}}$ of correctly choosing l (or, equivalently, the phase θ_l from a total of L choices) is given by (5-87) with N replaced by L. Thus the probability of correct reception is

$$P_c(N) = [P_c(M)|_{\text{NON}}][P_c(L)|_{\text{MPSK}}]$$

or, equivalently, the word error probability for the suboptimum receiver of Fig. 5-19 is given by

$$P_E(N) = P_E(M)|_{\text{NON}} + P_E(L)|_{\text{MPSK}} - [P_E(M)|_{\text{NON}}][P_E(L)|_{\text{MPSK}}] \tag{5-106}$$

The error probability performance of the suboptimum receiver as given by Eq. (5-106) is superimposed on the performance bounds for the optimum receiver in Fig. 5-18. For large R_b, the difference in performance between the optimum and suboptimum receivers is virtually negligible. Finally, note that the receiver implementation of Fig. 5-19 requires M noncoherent frequency detectors and L phase estimators. Since the phase estimate is obtained after the correct frequency has been chosen, an alternate implementation could employ only a single phase estimator whose input terminals are switched to those of the noncoherent detector corresponding to the chosen frequency.

5-6. COHERENT AND DIFFERENTIALLY COHERENT DETECTION OF DIFFERENTIALLY ENCODED MPSK SIGNALS

In this section, we shall present the theory of *differential encoding* by first discussing the practical problem which it overcomes. In the MPSK receiver of Fig. 5-15, it was assumed that the receiver was perfectly synchronized in frequency and had exact knowledge of the carrier reference phase. However, in practice, it is a fact that the receiver frequently does not have this exact carrier reference phase knowledge, although it may be capable of establishing a phase reference to within ϕ_a radians of the exact phase. Thus, something more must be done to resolve this phase ambiguity. One solution is to allocate a portion of the total transmitted power to a residual carrier. A second solution is to employ differential encoding.

Let $s^{(i)}(t) = \sqrt{2S}\sin(\omega_0 t + \theta^{(i)})$ represent the signal to be transmitted during the ith transmission interval $(i-1)T \leq t \leq iT$. Here $\theta^{(i)}$ denotes the transmitted signal phase which ranges over the set of allowable phases

$\{\theta_j = 2(j-1)\pi/N,\ j = 1, 2, \ldots, N\}$. Then at the conclusion of the ith transmission interval, the receiver makes an N-ary decision, say $\hat{\theta}^{(i)}$. This decision variable consists of the unambiguous part $\hat{\theta}_u^{(i)}$, plus arbitrary phase ϕ_a, that is, $\hat{\theta}^{(i)} = \hat{\theta}_u^{(i)} + \phi_a$. To resolve the phase ambiguity ϕ_a, assume that during the ith transmission interval it is desired to transmit θ_j. Then, if the information, θ_j, is encoded as $\theta^{(i)} - \theta^{(i-1)}$, and the receiver makes its decision at the conclusion of the ith transmission interval on the basis of the *difference decision variable* $\hat{\theta}^{(i)} - \hat{\theta}^{(i-1)} = (\hat{\theta}_u^{(i)} + \phi_a) - (\hat{\theta}_u^{(i-1)} + \phi_a) = \hat{\theta}_u^{(i)} - \hat{\theta}_u^{(i-1)}$, then clearly the phase ambiguity disappears. Inherent in the above is the assumption that the value of ϕ_a remains constant over this $2T$-second time interval. Notice that $N + 1$ differentially encoded symbols are needed to convey N symbols of information. In effect, one symbol is required to resolve the ambiguity.

As a first example, let $N = 2$ and suppose that we desire to transmit the information sequence $\theta_2, \theta_2, \theta_1, \theta_2$ with θ_1 arbitrarily chosen as a reference phase. Then, the differentially encoded sequence of transmitted phases is given by $\theta^{(1)} = \theta_1$, $\theta^{(2)} = \theta^{(1)} + \theta_2 = \theta_2$, $\theta^{(3)} = \theta^{(2)} + \theta_2 = \theta_1$, $\theta^{(4)} = \theta^{(3)} + \theta_1 = \theta_1$, $\theta^{(5)} = \theta^{(4)} + \theta_2 = \theta_2$, corresponding to the signal sequence $s^{(1)}(t) = s_1(t)$, $s^{(2)}(t) = s_2(t)$, $s^{(3)}(t) = s_1(t)$, $s^{(4)}(t) = s_1(t)$, $s^{(5)}(t) = s_2(t)$. In the absence of any errors in the received signal, the receiver would decode the modulo 2π difference of the received sequence $\theta^{(1)} = \theta_1, \theta^{(2)} = \theta_2, \theta^{(3)} = \theta_1, \theta^{(4)} = \theta_1, \theta^{(5)} = \theta_2$; that is, it would produce $\hat{\theta}^{(2)} - \hat{\theta}^{(1)} = \theta_2, \hat{\theta}^{(3)} - \hat{\theta}^{(2)} = \theta_2$, $\hat{\theta}^{(4)} - \hat{\theta}^{(3)} = \theta_1$, $\hat{\theta}^{(5)} - \hat{\theta}^{(4)} = \theta_2$ as the output. If the received symbols are phase-inverted to read $\theta^{(1)} = \theta_2, \theta^{(2)} = \theta_1, \theta^{(3)} = \theta_2, \theta^{(4)} = \theta_2$, $\theta^{(5)} = \theta_1$, then the output remains $\hat{\theta}^{(2)} - \hat{\theta}^{(1)} = \theta_2, \hat{\theta}^{(3)} - \hat{\theta}^{(2)} = \theta_2, \hat{\theta}^{(4)} - \hat{\theta}^{(3)} = \theta_1, \hat{\theta}^{(5)} - \hat{\theta}^{(4)} = \theta_2$. Thus, the 180° phase ambiguity has been automatically eliminated by the encoding and decoding procedure.

As a second example, let $N = 4$ and suppose that one wishes to send the message sequence $\theta_1, \theta_2, \theta_4, \theta_3, \theta_2, \theta_2$. Then, if θ_1 is chosen again as the reference phase, the corresponding sequence of transmitted phases is $\theta^{(1)} = \theta_1, \theta^{(2)} = \theta^{(1)} + \theta_1 = \theta_1, \theta^{(3)} = \theta^{(2)} + \theta_2 = \theta_2, \theta^{(4)} = \theta^{(3)} + \theta_4 = \theta_1, \theta^{(5)} = \theta^{(4)} + \theta_3 = \theta_3$, $\theta^{(6)} = \theta^{(5)} + \theta_2 = \theta_4$, $\theta^{(7)} = \theta^{(6)} + \theta_2 = \theta_1$. In the absence of errors, again the receiver would take the received sequence $\theta^{(1)} = \theta_1$, $\theta^{(2)} = \theta_1$, $\theta^{(3)} = \theta_2$, $\theta^{(4)} = \theta_1$, $\theta^{(5)} = \theta_3$, $\theta^{(6)} = \theta_4$, $\theta^{(7)} = \theta_1$ and differentially decode it into $\hat{\theta}^{(2)} - \hat{\theta}^{(1)} = \theta_1, \hat{\theta}^{(3)} - \hat{\theta}^{(2)} = \theta_2, \hat{\theta}^{(4)} - \hat{\theta}^{(3)} = \theta_4, \hat{\theta}^{(5)} - \hat{\theta}^{(4)} = \theta_3, \hat{\theta}^{(6)} - \hat{\theta}^{(5)} = \theta_2, \hat{\theta}^{(7)} - \hat{\theta}^{(6)} = \theta_2$.

When one employs differential encoding at the transmitter, then two different system mechanizations are possible. A digital communication system that employs differential encoding of the channel symbols and coherent detection of successive differentially encoded MPSK signals is called a *differentially encoded coherent MPSK system*; a digital communication system that employs differential encoding of MPSK signals and, which makes its decision on the basis of phase-difference detection, is called a *multiple-*

state, differentially coherent, phase-shift-keyed (MDPSK) *system.* In either case, the decision made during the ith transmission interval is correlated with the decision made during the $(i-1)$st interval; thus, errors made over two symbol times are correlated. Obviously, some degradation in system performance over coherent detection of MPSK signals will exist.

Error Probability for Coherent Detection of Differentially Encoded MPSK

Here, we shall assume that the carrier reference is established at the receiver by means of a suppressed carrier-tracking loop such as those discussed in Chap. 2. There we observed (in the absence of noise) that when an MPSK signal is demodulated, using coherent references derived directly from a suppressed carrier signal, an N-fold ambiguity concerning the true carrier phase existed in the reconstructed reference. This is due to the fact that in the interval $(0, 2\pi)$, there exist N stable lock points, say $\{\phi_a = \varphi_{lk};\ k = 1, 2, \ldots, N\}$.

When we differentially encode the data symbols and establish an ambiguous coherent reference signal by means of a suppressed carrier-tracking loop, then a pair of N-ary decisions $\hat{\theta}^{(i)}$ and $\hat{\theta}^{(i-1)}$ are required to decode one data symbol; that is, the phase difference $\hat{\theta}^{(i)} - \hat{\theta}^{(i-1)}$ modulo 2π represents the decoded signal phase corresponding to the ith transmission interval. The receiver that incorporates this detection procedure is illustrated in Fig. 5-20.

To evaluate the error probability performance of such a system, we begin by considering the possible ways of making a correct decision on the ith data symbol. Clearly, if both unambiguous phase estimates $\hat{\theta}_u^{(i-1)}$ and $\hat{\theta}_u^{(i)}$ represent correct decisions on the corresponding transmitted differentially encoded symbols, then the ith data symbol will be decoded correctly. The probability of this event is $P_{c_1}(N) \triangleq [P_c(N)|_{\text{MPSK}}]^2$ where $P_c(N)|_{\text{MPSK}}$ is given by (5-87). Further, if $\hat{\theta}_u^{(i-1)}$ and $\hat{\theta}_u^{(i)}$ are both shifted from their respective correct values $\theta^{(i-1)}$ and $\theta^{(i)}$ by $2k\pi/N$, $k = 1, 2, \ldots, N-1$, then the ith data symbol is still decoded correctly, since $\hat{\theta}_u^{(i)} - \hat{\theta}_u^{(i-1)}$ is independent of this shift. Denoting the probability of this event by $P_{E_k}^2(N)$, $k = 1, 2, \ldots, N-1$, then

$$P_{E_k}^2(N) = \text{Prob}\,[(2k-1)\pi/N \leq \eta^{(i-1)} - \varphi_{lk} - \theta^{(i-1)} \leq (2k+1)\pi/N,$$
$$(2k-1)\pi/N \leq \eta^{(i)} - \varphi_{lk} - \theta^{(i)} \leq (2k+1)\pi/N] \qquad (5\text{-}107)$$

where $\eta^{(i-1)}$ and $\eta^{(i)}$ are the measured phases in the $i-1$st and ith transmission intervals respectively. In order to evaluate $P_{E_k}^2(N)$ one must first establish the p.d.f.'s of $\eta^{(i)}$ and $\eta^{(i-1)}$. Following a procedure similar to that leading to (5-84), the joint p.d.f. of the measured amplitude $V^{(i)}$ and phase $\eta^{(i)}$ in the

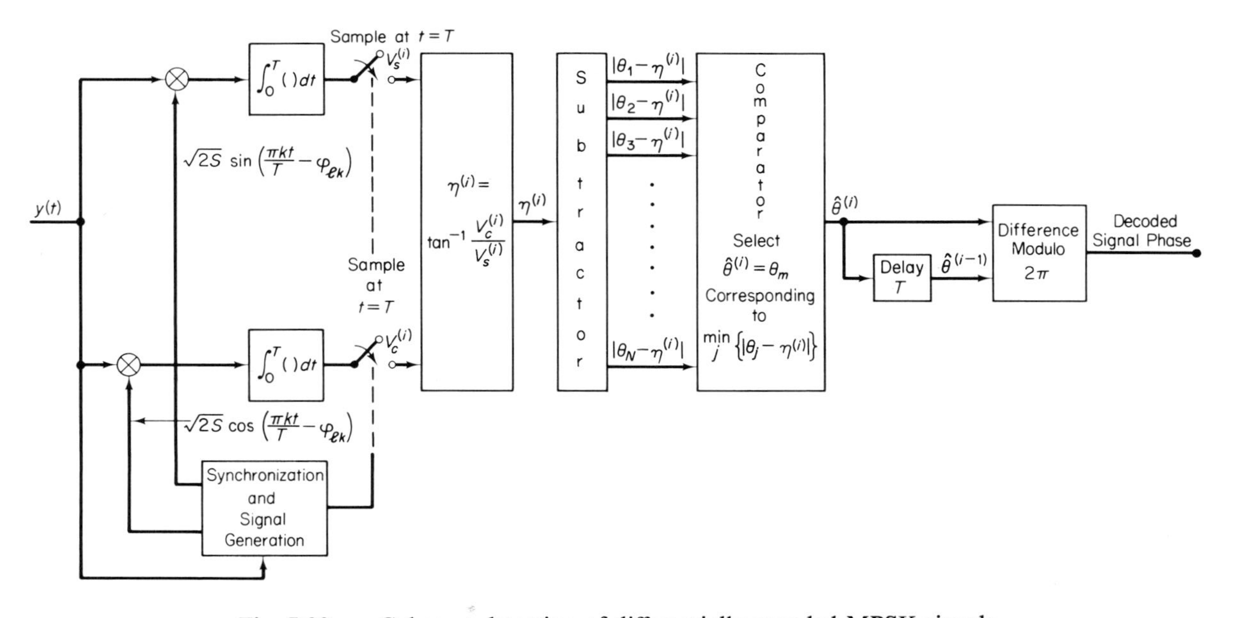

Fig. 5-20 Coherent detection of differentially encoded MPSK signals

ith transmission interval, given that $s^{(i)}(t)$ was transmitted, is

$$p(V^{(i)}, \eta^{(i)}) = \begin{cases} \dfrac{V^{(i)}}{\pi STN_0} \exp\left[-\dfrac{(V^{(i)})^2 - 2STV^{(i)}\cos(\eta^{(i)} - \theta^{(i)} - \varphi_{lk}) + (ST)^2}{STN_0}\right] & 0 \leq V^{(i)} \leq \infty \\ & 0 \leq \eta^{(i)} \leq 2\pi \\ 0 & \text{elsewhere} \end{cases} \tag{5-108}$$

and hence

$$\begin{aligned} p(\eta^{(i)}) &= \int_0^\infty \frac{r}{\pi} \exp\{-[r^2 - 2r\sqrt{R_d}\cos(\eta^{(i)} - \theta^{(i)} - \varphi_{lk}) + R_d]\}\,dr \\ p(\eta^{(i-1)}) &= \int_0^\infty \frac{r}{\pi} \exp\{-[r^2 - 2r\sqrt{R_d}\cos(\eta^{(i-1)} - \theta^{(i-1)} - \varphi_{lk}) + R_d]\}\,dr \end{aligned} \tag{5-109}$$

Since $\eta^{(i-1)}$ and $\eta^{(i)}$ are statistically independent, then

$$\begin{aligned} P_{E_k}^2(N) &= \int_{(2k-1)\pi/N}^{(2k+1)\pi/N} p_a(\eta_a^{(i-1)})\,d\eta_a^{(i-1)} \int_{(2k-1)\pi/N}^{(2k+1)\pi/N} p_a(\eta_a^{(i)})\,d\eta_a^{(i)} \\ &= \left[\int_{(2k-1)\pi/N}^{(2k+1)\pi/N} p_a(x)\,dx\right]^2 \end{aligned} \tag{5-110}$$

where $\eta_a^{(i-1)} \triangleq \eta^{(i-1)} - \varphi_{lk} - \theta^{(i-1)}$, $\eta_a^{(i)} \triangleq \eta^{(i)} - \varphi_{lk} - \theta^{(i)}$ and $p_a(\eta_a^{(i-1)}) \triangleq p(\eta^{(i-1)})$, $p_a(\eta_a^{(i)}) \triangleq p(\eta^{(i)})$.

Introducing the change of variables $u = r\cos x$, $v = r\sin x$ into (5-110) gives the result

$$\begin{aligned} P_{E_k}(N) &= \frac{1}{\pi}\int_0^\infty \left[\exp\{-(u - \sqrt{R_d})^2\} \int_{u\tan[(2k-1)\pi/N]}^{u\tan[(2k+1)\pi/N]} \exp(-v^2)\,dv\right] du \\ &= \frac{1}{N} + \frac{1}{4}\operatorname{erf}\left[\sqrt{R_d}\sin\frac{(2k+1)}{N}\pi\right] - \frac{1}{4}\operatorname{erf}\left[\sqrt{R_d}\sin\frac{(2k-1)}{N}\pi\right] \\ &\quad + \frac{1}{2\sqrt{\pi}}\int_0^{\sqrt{R_d}\sin[(2k+1)\pi/N]} \exp(-y^2)\operatorname{erf}\left[y\cot\frac{(2k+1)}{N}\pi\right] dy \\ &\quad - \frac{1}{2\sqrt{\pi}}\int_0^{\sqrt{R_d}\sin[(2k-1)\pi/N]} \exp(-y^2)\operatorname{erf}\left[y\cot\frac{(2k-1)}{N}\pi\right] dy \end{aligned} \tag{5-111}$$

Since the preceding events are mutually exclusive and exhaustive of the ways to make a correct decision, the probability of correct detection of the ith symbol is

$$P_c(N) = P_{c_1}(N) + \sum_{k=1}^{N-1} P_{E_k}^2(N) \tag{5-112}$$

TABLE 5-4 Coherent Detection of Differentially Encoded MPSK*

R_b(dB)	$P_E(2)$	$P_E(4)$	$P_E(8)$	$P_E(16)$	$P_E(32)$	$P_E(64)$
−5.00	.3358525	.5582829	.7288710	.8385675	.9082459	.9492678
−4.00	.3032477	.5139068	.6985254	.8187634	.8964317	.9426181
−3.00	.2668313	.4618597	.6631630	.7961605	.8829792	.9350824
−2.00	.2273609	.4024843	.6221921	.7707165	.8678204	.9266102
−1.00	.1861437	.3371837	.5749506	.7424362	.8508859	.9171369
.00	.1450368	.2686943	.5208287	.7113057	.8320886	.9065782
1.00	.1062962	.2010637	.4595580	.6771694	.8113153	.8948246
2.00	.7223538 (−1)	.1391205	.3917878	.6395509	.7884302	.8817463
3.00	.4472659 (−1)	.8738950 (−1)	.3195483	.5975031	.7632900	.8671986
4.00	.2469518 (−1)	.4875649 (−1)	.2464782	.5496689	.7357572	.8510263
5.00	.1183854 (−1)	.2353004 (−1)	.1773728	.4947303	.7056812	.8330679
6.00	.4765511 (−2)	.9506917 (−2)	.1172161	.4321816	.6727996	.8131595
7.00	.1544199 (−2)	.3085828 (−2)	.6978711 (−1)	.3630647	.6365323	.7911399
8.00	.3817452 (−3)	.7633274 (−3)	.3657022 (−1)	.2902692	.5957816	.7668585
9.00	.6725160 (−4)	.1344922 (−3)	.1638674 (−1)	.2182392	.5490130	.7401822
10.00		.1547676 (−4)	.6054531 (−2)	.1521479	.4947987	.7109839
11.00			.1761190 (−2)	.9670551 (−1)	.4326505	.6790607
12.00			.3801892 (−3)	.5489675 (−1)	.3637101	.6439297
13.00			.5645642 (−4)	.2713621 (−1)	.2909640	.6045891
14.00				.1131700 (−1)	.2189083	.5595249
15.00				.3825883 (−2)	.1527366	.5071894
16.00				.9962809 (−3)	.9717444 (−1)	.4468240
17.00				.1870888 (−3)	.5522873 (−1)	.3792061
18.00				.2303150 (−4)	.2734013 (−1)	.3069585
19.00					.1142232 (−1)	.2343184
20.00					.3869632 (−2)	.1664227
21.00					.1009664 (−2)	.1082103
22.00					.1888940 (−3)	.6315829 (−1)
23.00					.2106868 (−4)	.3230054 (−1)
24.00						.1404489 (−1)
25.00						.4996923 (−2)

*The negative integer in parentheses following each entry in the table represents the power of ten by which the entry should be multiplied.

Since we have assumed equiprobable data symbols, the above expression is true for all $i = 1, 2, \ldots, N$. The corresponding error probability is

$$P_E(N) = 2P_E(N)|_{\text{MPSK}}\left[1 - \frac{1}{2}P_E(N)|_{\text{MPSK}} - \frac{1}{2}\frac{\sum_{k=1}^{N-1} P_{E_k}^2(N)}{P_E(N)|_{\text{MPSK}}}\right] \tag{5-113}$$

For $N = 2$, we have the special case of coherent detection of differentially encoded PSK for which (5-113) reduces to

$$P_E(2) = 2P_E(2)|_{\text{PSK}}[1 - P_E(2)|_{\text{PSK}}] = \text{erfc}\sqrt{R_d}[1 - \tfrac{1}{2}\text{erfc}\sqrt{R_d}] \tag{5-114}$$

Another case for which (5-113) can be simplified is the quadriphase case, i.e. $N = 4$, for which

$$P_E(4) = 2\,\text{erfc}\sqrt{\frac{R_d}{2}} - 2\,\text{erfc}^2\sqrt{\frac{R_d}{2}} + \text{erfc}^3\sqrt{\frac{R_d}{2}} - \frac{1}{4}\text{erfc}^4\sqrt{\frac{R_d}{2}} \tag{5-115}$$

The error probability expression derived above is tabulated in Table 5-4 vs. $R_b = R_d/\log_2 N$ in dB for $N = 2, 4, 8, 16, 32$, and 64. The data in this table are illustrated in Fig. 5-22.

Error Probability for Differentially Coherent Detection of Differentially Encoded MPSK

In certain applications, it may be undesirable to implement a suppressed carrier-tracking loop so as to provide a carrier reference for coherent detection. One approach which cleverly avoids this issue is to implement the receiver of Fig. 5-21. This receiver employs differentially coherent detection of differentially encoded signals. We note that when the channel phase characteristic is constant for $2T$ seconds, the receiver of Fig. 5-21 can be shown to be the optimum *a posteriori* probability computing receiver of differentially encoded, equiprobable, equal energy signals. On the other hand, when the channel phase characteristic is constant for kT seconds, $k > 2$, an alternate differentially coherent detection scheme in which the present phase is compared not only with the preceding phase but with previous $(k - 1)$ phases can be implemented. The error probability performance of such a system would lie between that of the ideal MPSK system of Sec. 5-5 and the system analyzed in what follows.

The probability of error for the receiving system of Fig. 5-21 is the probability that the measured phase difference $\eta^{(i)} - \eta^{(i-1)}$ differs by more than π/N in absolute value from the phase difference $\theta^{(i)} - \theta^{(i-1)}$ of the transmitted signals; that is,

$$\begin{aligned} P_E(N) &= \text{Prob}\left[|\psi| = |(\eta^{(i)} - \eta^{(i-1)}) - (\theta^{(i)} - \theta^{(i-1)})| > \frac{\pi}{N}\right] \\ &= 1 - \int_{-\pi/N}^{\pi/N} p_\Psi(\psi)\,d\psi \end{aligned} \tag{5-116}$$

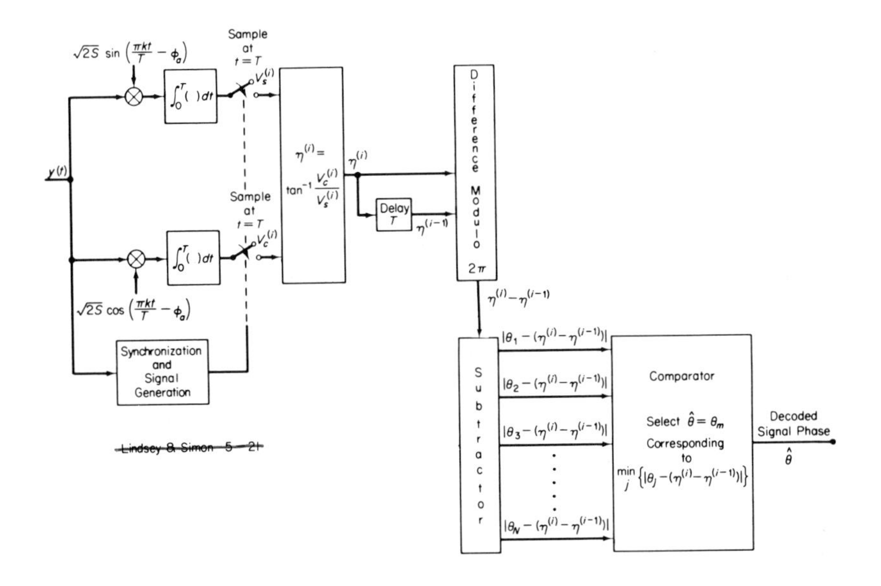

Fig. 5-21 Differentially coherent detection of differentially encoded MPSK signals

where $p_\Psi(\psi)$ is the p.d.f. of Ψ. In order to evaluate $p_\Psi(\psi)$, one must first establish the p.d.f.'s of $\eta^{(i)}$ and $\eta^{(i-1)}$. Comparing Figs. 5-20 and 5-21, we observe that these p.d.f.'s are characterized by Eq. (5-109) with φ_{lk} replaced by ϕ_a. Here ϕ_a is an arbitrary phase associated with the demodulation reference signals that are applied to the input correlators in Fig. 5-21. Letting $\eta_a^{(i)} = \eta^{(i)} - \theta^{(i)} - \phi_a$ and $\eta_a^{(i-1)} = \eta^{(i-1)} - \theta^{(i-1)} - \phi_a$ and noting that $\eta^{(i)}$ and $\eta^{(i-1)}$ are statistically independent, the p.d.f. of the difference phase $\psi = \eta_a^{(i)} - \eta_a^{(i-1)} = (\eta^{(i)} - \eta^{(i-1)}) - (\theta^{(i)} - \theta^{(i-1)})$ is

$$p_\Psi(\psi) = \int_{-\infty}^{\infty} p_a(\eta_a^{(i-1)})\, p_a(\psi + \eta_a^{(i-1)})\, d\eta_a^{(i-1)} \tag{5-117}$$

In (5-117) we have introduced the notation $p_a(\eta_a^{(i-1)}) \triangleq p(\eta^{(i-1)})$ and $p_a(\eta_a^{(i)}) \triangleq p(\eta^{(i)})$. Since $p_\Psi(\psi)$ is an even function of its argument, then $P_E(N)$ of (5-116) can be written as

$$P_E(N) = \int_{-\pi}^{\pi} p_\Psi(\psi)\, d\psi - \int_{-\pi/N}^{\pi/N} p_\Psi(\psi)\, d\psi = 2\int_{\pi/N}^{\pi} p_\Psi(\psi)\, d\psi \tag{5-118}$$

Using the method of characteristic functions, one may show that the p.d.f. of the phase difference ψ is given by[18]

$$p_\Psi(\psi) = \frac{1}{2\pi}\int_0^{\pi/2} (\sin\alpha)[1 + R_d(1 + \cos\psi\sin\alpha)] \times \exp[-R_d(1 - \cos\psi\sin\alpha)]\, d\alpha \tag{5-119}$$

The integration may be carried out using the theory of Bessel functions. Thus,

$$p_\Psi(\psi) = p_\Psi\left(\frac{\pi}{2}\right) + \frac{1}{4}R_d \exp(-R_d)\{I_1(R_d \cos\psi) + L_1(R_d \cos\psi) + \cos\psi[I_0(R_d\cos\psi) + L_0(R_d\sin\psi)]\} \tag{5-120}$$

where $I_n(x)$ and $L_n(x)$ are the modified Bessel function and Struve function, respectively. The exact word error probability may be found from (5-118) by numerical integration of $p_\Psi(\psi)$ on a digital computer. For large R_d the word error probability is approximately given by[18]

$$P_E(N) \cong \operatorname{erfc}[U] + \frac{U\exp(-U^2)}{4\sqrt{\pi}(1/8 + R_d)} \tag{5-121}$$

where

$$U = \sqrt{2R_d}\sin\left(\frac{\pi}{2N}\right) \tag{5-122}$$

This result is easy to evaluate on a digital computer.

Several other approximations for the probability of error have been developed and are worthy of mention. Arthurs and Dym[19] developed the approximation

$$P_E(N) \cong \operatorname{erfc}\left(\sqrt{R_d}\sin\frac{\pi}{\sqrt{2N}}\right) \tag{5-123}$$

This expression is not accurate for low signal-to-noise ratios, and the method of derivation does not provide a criterion of the quality of the approximation. Bussgang and Leiter[20] derived an upper bound to $P_E(N)$ which for large signal-to-noise ratios (greater than 5 dB) is itself a good approximation. The result is

$$P_E(N) \cong \operatorname{erfc}\left[\frac{R_d}{\sqrt{1+2R_d}}\sin\frac{\pi}{N}\right] \tag{5-124}$$

When R_d and N are both large, (5-124) resembles (5-123); however, for small signal-to-noise ratios (5-124) is superior.

If we let $N = 2$, the error probability can be exactly determined from (5-118) and (5-119) and is given by

$$P_E(2) = 1 - P_c(2) = \frac{1}{2}\exp(-R_d) \tag{5-125}$$

This result was obtained by Lawton[21] and Cahn[16] and represents the ideal performance of what is usually referred to as DPSK.

A practical system that uses 4–phase MDPSK is the Kineplex system developed by Collins Radio Company and described in Ref. 22. The evaluation of error probabilities in such a 4–phase MDPSK system was independently made by Lawton[23] and Cahn[24].

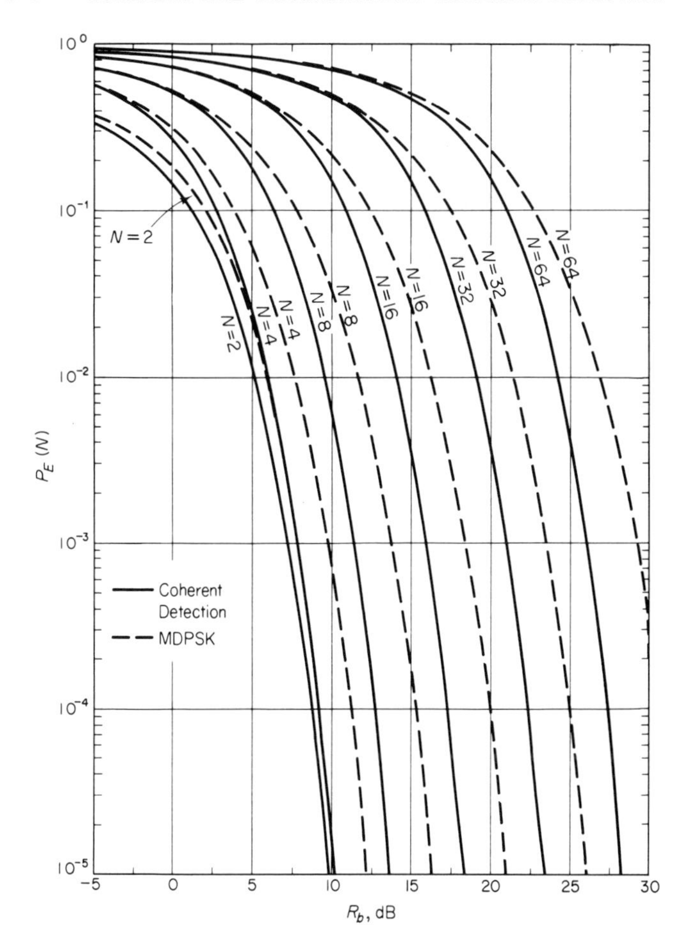

Fig. 5-22 Word error probability performance for coherent and differentially coherent detection of differentially encoded MPSK

The exact expression for error probability as given by (5-118) is given in Table 5-5 versus R_b in dB for $N = 2, 4, 8, 16, 32$, and 64. The data in this table are superimposed on the error probability curves for coherent detection of MPSK in Fig. 5-16 and also on the error probability curves for coherent detection of differentally encoded MPSK in Fig. 5-22.

A comparison of the error probabilities for coherent and differentially coherent reception of several of the signaling formats described thus far is given in Fig, 5-23. For completeness, we include the corresponding results for noncoherent FSK, which will be derived later in Chap. 10.

Finally, the principle of differential encoding can also be applied to an

TABLE 5-5 Differentially Coherent Detection of Differentially Encoded MPSK*

R_b(dB)	$P_E(2)$	$P_E(4)$	$P_E(8)$
−5.00	.3644467	.5605952	.7231345
−4.00	.3357950	.5210281	.6915373
−3.00	.3029055	.4757209	.6551711
−2.00	.2660411	.4249930	.6140029
−1.00	.2259423	.3697198	.5682252
.00	.1839397	.3114289	.5182698
1.00	.1419795	.2522997	.4647956
2.00	.1024848	.1950121	.4086669
3.00	.6798899 (−1)	.1424245	.3509490
4.00	.4055754 (−1)	.9711721 (−1)	.2929335
5.00	.2116461 (−1)	.6090552 (−1)	.2361733
6.00	.9332812 (−2)	.3445888 (−1)	.1824704
7.00	.3329212 (−2)	.1715878 (−1)	.1337626
8.00	.9094045 (−3)	.7285818 (−2)	.9188380 (−1)
9.00	.1775196 (−3)	.2534215 (−2)	.5822380 (−1)
10.00	.2269997 (−4)	.6863767 (−3)	.3337061 (−1)
11.00	.1704224 (−5)	.1357942 (−3)	.1687569 (−1)
12.00		.1810654 (−4)	.7298485 (−2)
13.00		.1470238 (−5)	.2595356 (−2)
14.00			.7221801 (−3)
15.00			.1477453 (−3)
16.00			.2054934 (−4)
17.00			.1761502 (−5)
18.00			
19.00			
20.00			
21.00			
22.00			
23.00			
24.00			
25.00			
26.00			
27.00			
28.00			
29.00			
30.00			

* The negative integer in parentheses following each entry in the table represents the power of ten by which the entry should be multiplied.

TABLE 5-5 (cont.)

$P_E(16)$	$P_E(32)$	$P_E(64)$
.8372487	.9080373	.9492429
.8168953	.8961134	.9425715
.7935355	.8825124	.9350082
.7670593	.8671511	.9265000
.7373997	.8499374	.9169791
.7045084	.8307522	.9063541
.6683332	.8094378	.8945075
.6288127	.7857989	.8812985
.5859018	.7596145	.8665668
.5396224	.7306538	.8501362
.4901276	.6986922	.8318159
.4377657	.6635268	.8114017
.3831400	.6249964	.7886774
.3271625	.5830070	.7634185
.2710868	.5375660	.7353972
.2165025	.4888255	.7043910
.1652596	.4371331	.6701952
.1192988	.3830878	.6326406
.8037485 (−1)	.3275926	.5916175
.4970457 (−1)	.2718897	.5471083
.2763042 (−1)	.2175573	.4992283
.1344911 (−1)	.1664388	.4482751
.5546586 (−2)	.1204786	.3947841
.1860180 (−2)	.8144989 (−1)	.3395846
.4821662 (−3)	.5060358 (−1)	.2838440
.9075201 (−4)	.2832486 (−1)	.2290835
.1149423 (−4)	.1394713 (−1)	.1771377
	.5874747 (−2)	.1300261
	.2049377 (−2)	.8968899 (−1)
	.5691962 (−3)	.5755463 (−1)
	.1192418 (−3)	.3404384 (−1)
	.1739514 (−4)	.1836026 (−1)
	.1577954 (−5)	.8848493 (−2)
		.3666603 (−2)
		.1230086 (−2)
		.3092488 (−3)

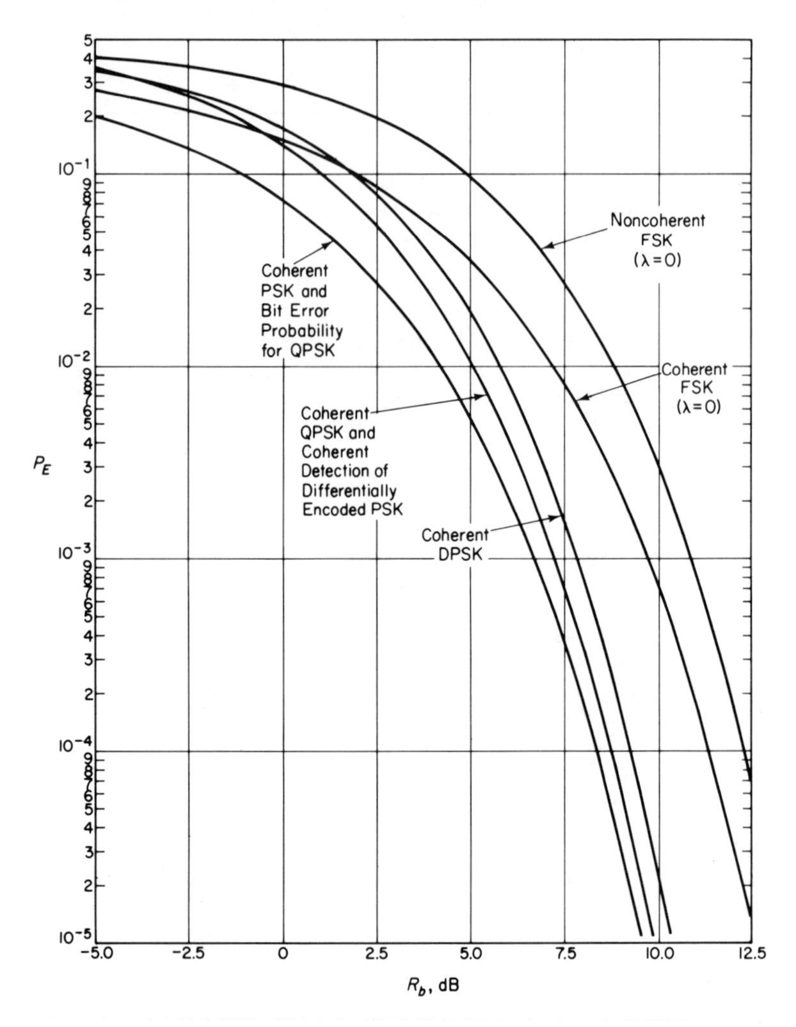

Fig. 5-23 A comparison of error probabilities for coherent, differentially coherent, and noncoherent reception of several signaling formats

L-orthogonal signal set. Using the representation of (5-93), the L phases corresponding to each of the M frequencies are differentially encoded. A suboptimum receiver for the differentially encoded L-orthogonal signal set would still use a noncoherent detector to determine the most probable frequency; however, the phase estimate would be made by coherent or differentially coherent detectors such as those in Figs. 5-20 and 5-21 respectively. The configuration of such a receiver and its associated error probability performance are given in Prob. 5-22.

5-7. BIT ERROR PROBABILITY FOR DIFFERENTIALLY ENCODED DATA AND COHERENT DETECTION OF ORTHOGONAL AND BI-ORTHOGONAL SIGNALS

In a block-coded system that uses all of the transmitter power for data transmission and reconstructs reference signals from the modulated signal, a two-fold ambiguity exists if the modulation is binary PSK. To resolve this ambiguity, one can differentially encode the data symbols (now in binary digits) prior to applying them to the channel encoder and remove the ambiguity at the receiver by taking the modulo 2 difference between successive bits of the decoder output. Since errors are correlated for at most two bits, the bit error probability for an orthogonal system is given by

$$P_{BD}(N) = 2P_B(N)[1 - P_B(N)] \tag{5-126}$$

where $P_B(N)$ is defined in Eq. (5-54). For bi-orthogonal signals, the same expression holds except that $P_B(N)$ of (5-126) is defined by (5-61). Using (5-126) and Tables 5-1 and 5-2, one can evaluate the bit error probability for a block-coded system with differentially encoded data symbols.

5-8. CONVOLUTIONAL CODES

With technology advancing as rapidly as it is, the engineer is continually faced with the challenge of designing a communication system having a better performance than those already in practice. To a majority of engineers today, the phrase "those already in practice" implies systems using a block-coded signaling format (such as those discussed thus far in this chapter) because chronologically these were the first to be investigated. Since the early years of coding, however, a great deal of theory has been developed that now permits the engineer to design a more efficient communication system in the Shannon sense. In particular, over the last two decades, a large body of literature has been developed on the subject of *convolutional codes*, since in many applications these codes have proved to be equal or superior to block codes both in performance and ease of implementation.[25-30]

For our purposes, the best way to explain the physics of a convolutional code is to compare its mechanization to that of a particular subclass of block codes known as *parity-check codes*. A parity-check code transforms an input message of K bits into an M-digit output code word, where each of the M output code word digits consists of a parity check (modulo 2 linear sum) on some subset of the K input bits. By varying the M individual subsets of K input bits that result in each of the M output digits, many different parity-check codes can be generated, each of which maps a K-component input vector (a K-tuple) uniquely into an M-component output vector. A typical encoder for such a system might consist of a K-stage input shift register, an M-stage output shift register, and M modulo 2 adders, each with a single output and K input connections. In each word interval T, a K-component input vector is serially fed into the input register, while at the same time an M-component code word corresponding to the previous input K-tuple is serially shifted out of the output register. Then, since in each T-sec interval the *total* contents of both registers are shifted out, each input K-bit message affects only one output code word.

In a convolutional code, only an input register is used, and some $v \ll M$ modulo 2 adders are connected to its stages. The outputs of these summers are sampled sequentially (typically by a commutator) after every shift of $d \leq K$ input bits into the input register. Hence, in general, several (K/d) output sequences of v digits each will be affected by the same d input information bits. This overlap effect of the input on the output is precisely what distinguishes convolutional codes from the less constrained, parity-check block codes.

The specific connections between the v modulo 2 adders and the K shift register stages of a convolutional encoder may be specified by a $v \times K$ *matrix of parity-check equations*, where the jth row of this matrix (the jth parity-check equation) represents the connections of the K shift register stages to the jth modulo 2 summer. The elements of the matrix are zeros and ones, with zero representing the absence of a connection and one the presence of a connection. As an example, if the jth row of the matrix is 10110, then the jth modulo 2 summer is connected to the first, third, and fourth shift register stages.

Encoding Procedure (Terminated-Tree Structure)

In the literature, one finds that convolutional codes, much like block codes, are described as being *systematic* or *nonsystematic*. In particular, a *systematic convolutional code* is one for which the first d out of every v output code digits are identical to the d input information bits that cause that particular v-digit output sequence. A typical channel encoder for a binary systematic convolutional code is illustrated in Fig. 5-24. The operation of this device is as follows. Information bits are fed into a K-stage shift register, d at a time; we assume for simplicity that K is a multiple of d. The first d output digits are obtained from direct taps on the first d states of the register; hence they are identical to the d information bits. The remaining v–d code digits are formed from linear combinations of the contents of the register stages. The particular connections between the v–d modulo 2 adders and the K shift register stages determine the structure of the code word remainder (the last v–d digits of each code word). After a total of L groups of d information bits have been transmitted through the encoder, K additional zeros, d at a time, are shifted into the re-

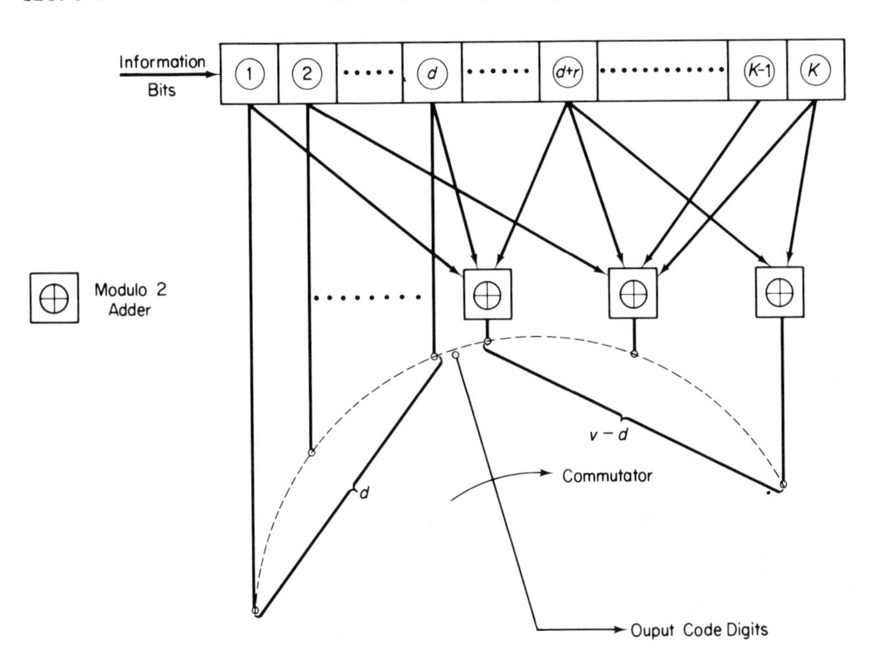

Fig. 5-24 Systematic convolutional encoder

gister to clear its contents.* The rate (in input information bits/output digits) of such a code is given by $\mathcal{R}_C = Ld/[v(Ld + K)/d]$. Ordinarily $L \gg K/d$ and hence $\mathcal{R}_C \cong d/v$. Herein lies one of the main advantages of convolutional codes over block codes: long streams of information bits can be continuously encoded without the necessity of grouping them into blocks of a certain length. If, however, the data coming out of the source appear naturally in block form, the convolutional code can be converted into a block code by terminating each message with the proper number of zeros necessary to clear the register, as mentioned above.

Binary codes generated as above possess a tree structure terminating in 2^{Ld} branches. The terminus of each of these branches corresponds to a unique path through the tree and gives the output code word generated by the input bits found along that path. A typical example of such a code tree is given in Fig. 5-25a for $K = 4$, $L = 2$, $d = 2$, $v = 4$, and the modulo 2 adder connections of the encoder in Fig. 5-25b.

---

* Actually, it is only necessary to insert $K–d$ additional zeros into the register following the last d bits of the message, since these last d bits will be shifted out of the register by the first d bits of the next message sequence. Furthermore, in continuous operation, it is not necessary to clear the register contents at all. However, it is convenient at this point to think of transmitting only a single Ld bit message and then returning the shift register to its initial state in preparation for the next message.

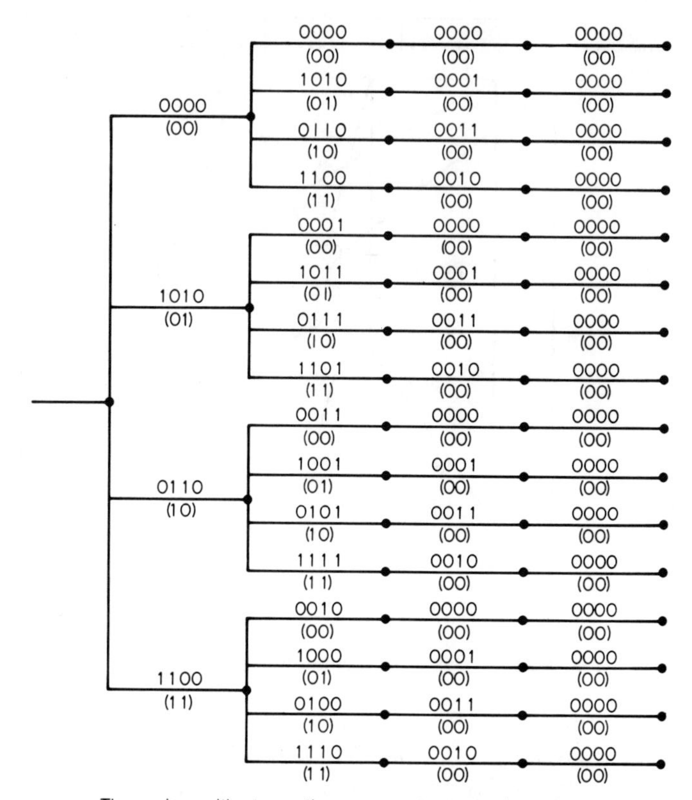

The numbers without parentheses represent output code words.
The numbers in parentheses indicate input information bits

(a)

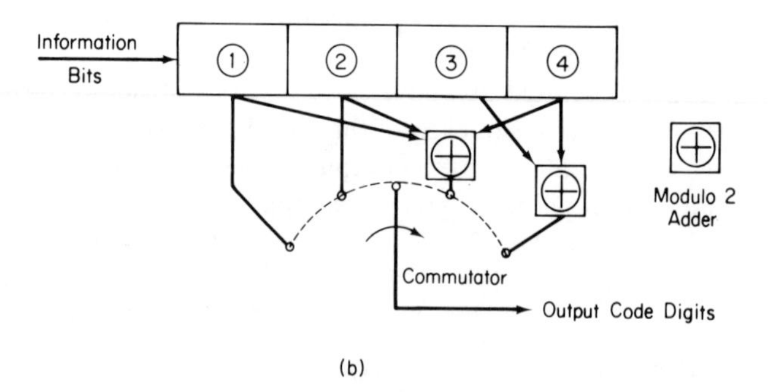

(b)

Fig. 5-25 An example of the tree representation and corresponding 4-stage convolutional encoder

A *nonsystematic convolutional code* represents a mapping of an input K-tuple into an output M-tuple in which none of the M output digits are constrained to correspond to input information bits. Hence the encoder for a binary nonsystematic convolutional code consists of a K-stage shift register and v modulo 2 adders, each having specific connections to the register stages (Fig. 5–26). Typically, for a nonsystematic code, $d = 1$ (one

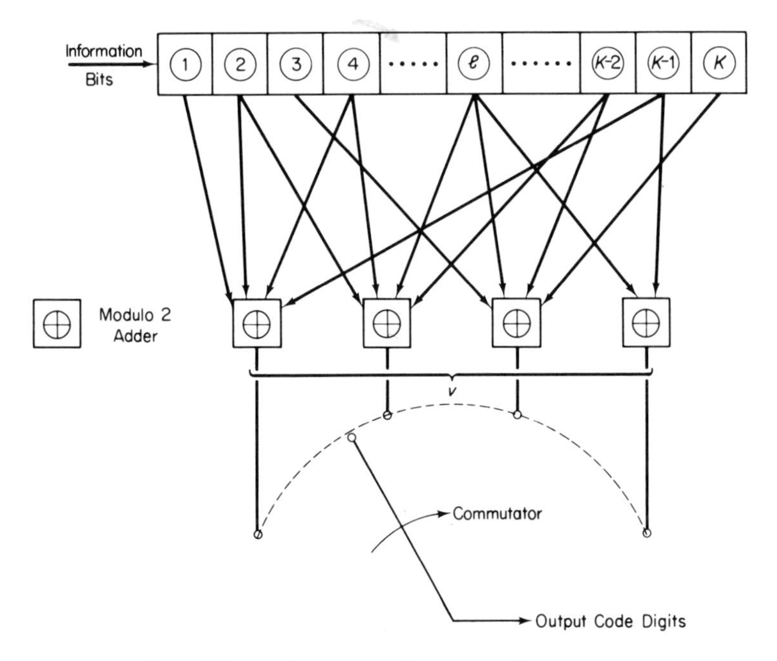

Fig. 5-26 Nonsystematic convolutional encoder

information bit at a time is shifted into the register), hence the code rate in this case is given approximately by $\mathcal{R}_C \cong 1/v$. Again, a tree structure with 2^{Ld} terminating branches is appropriate for characterizing this type of convolutional code. An example of such a convolutional code and its accompanying tree is illustrated in Fig. 5-27 for $K = 3$, $d = 1$, $v = 2$, and $L = 2$. We do not clear out the register with zeros this time so that we may use this example in what follows to illustrate a fundamental property of the basic tree structure.

In studying convolutional codes, an important parameter is the *constraint span of the code*, which is equivalent to the number of *states* of the shift register and denoted by K_0. Assuming that in each input time interval we shift d binary bits into the encoder, then the number of *shift register states* is defined here to be equal to the number of input time intervals required before

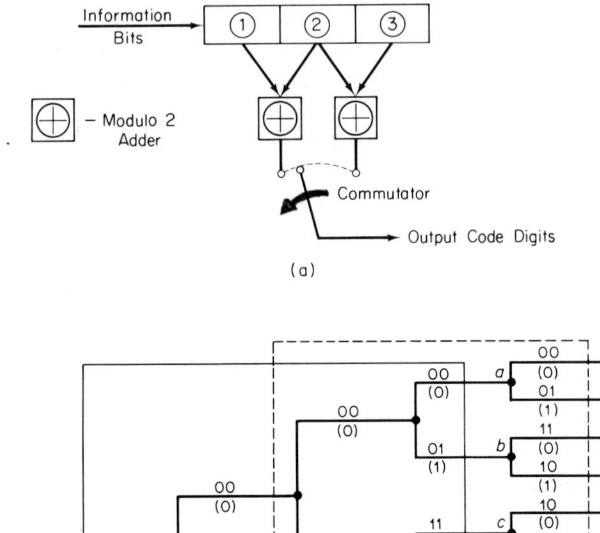

Fig. 5-27 An example of the repetitive nature of the tree diagram associated with a convolutional encoder

the first input bit transmitted leaves the register. Thus, since d bits are inputted in each time interval, K_0 is equal to K/d. Many authors call the quantity K_0 the *constraint length* of the code. Others define constraint length as $K_0 - 1$ or $K_0 v$. In this sense the literature is not consistent. Here we define constraint length L_c in bits by $L_c = K_0 v$.

The Trellis and State Diagram Representations

The concept of the equivalent *trellis* representation (introduced by Forney[28]) of a convolutional code tree is best illustrated with a simple example such as the nonsystematic encoder and its associated terminated-tree structure in Fig. 5-27. Since for this example $d = 1$, transmission of the $(K + 1)$st input bit causes the first input bit to drop out of the register; thus the $(K + 1)$st output sequence is independent of the first transmitted bit. Hence, the tree structure is repetitive after K input bits; that is, each part of the tree enclosed in dotted lines is identical to that part contained in the solid line rectangle. This property of the tree allows it to be redrawn as a *trellis* (Fig. 5-28), wherein the corresponding nodes in each of the dot-enclosed tree sec-

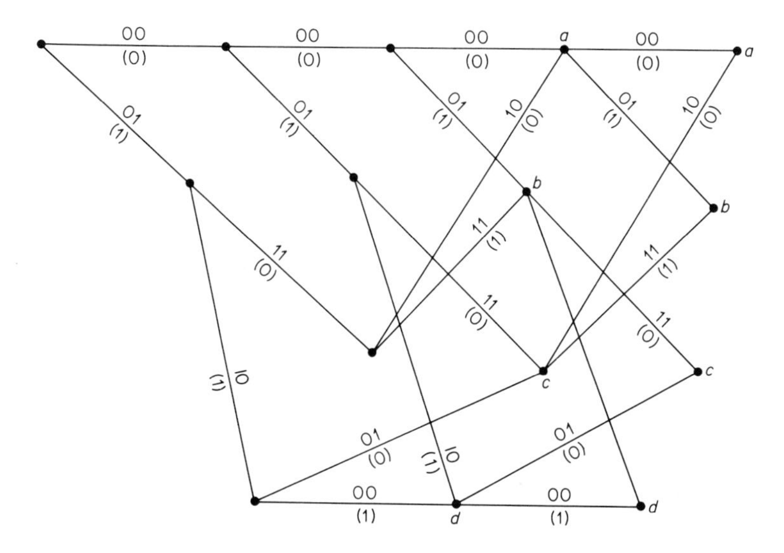

Fig. 5-28 The trellis representation of a convolutional encoder.

tions leading to the identical pair of output terminating branches have been merged together—for example, a with a, b with b, and so forth. For longer input sequences, the trellis would continue in the same repetitive pattern, since now each time a new information bit is added, we drop out the Kth previous bit from the register. More specifically, for Fig. 5-27b, the four nodes labeled a at the output of the fourth branch would be merged together with the same happening for nodes b, c, and d. In general then, after K input bits, the convolutional encoder takes on one of 2^{K-1} distinct "states" (merged nodes). From that point on, each transmission of an input bit produces a v-digit output sequence that uniquely causes a system transition to one of two possible states out of the total of 2^{K-1} states. Furthermore, to determine the states of the system at any point in the input sequence (after K bits), one need examine only the previous $K - 1$ bits. Hence, we can represent the trellis of Fig. 5-28 by a *state diagram*, where in general we associate each state with the previous $K - 1$ input bits leading to it and the transitions between states by the output sequence produced by the input bit causing the transition.

Figure 5-29 illustrates the state diagram for the trellis of Fig. 5-28. Note

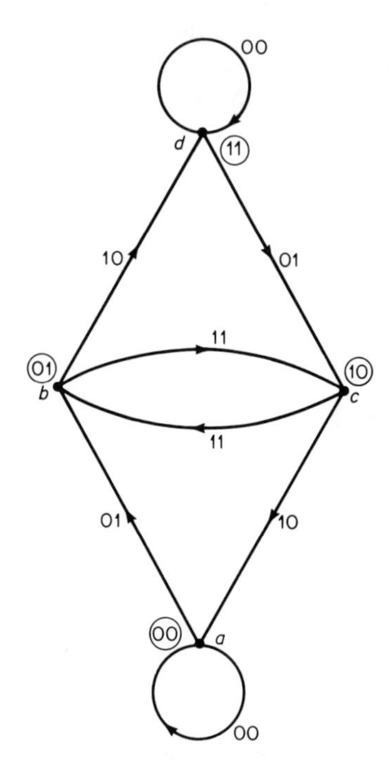

Fig. 5-29 The state diagram representation of a convolutional encoder

that the structure of the trellis and hence the state diagram are independent of v, the number of modulo 2 adders, and their connections with the K-stage shift register. On the other hand, the forementioned structure is quite dependent on d—the number of input digits shifted into the register during each transmission time. Since each node in the tree now has 2^d $(d \geq 1)$ output branches and K/d stages in the tree are required before repetition sets in, the total number of states in the trellis or state diagram becomes 2^{K-d}, with each state characterized by an input sequence of length $K - d$ bits. Figure 5-30 illustrates an example of a binary convolutional encoder and its associated state diagram for $K = 4$, $d = 2$, and $v = 3$.

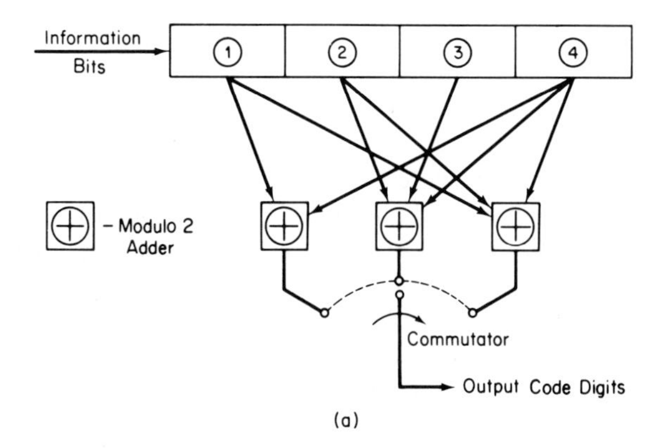

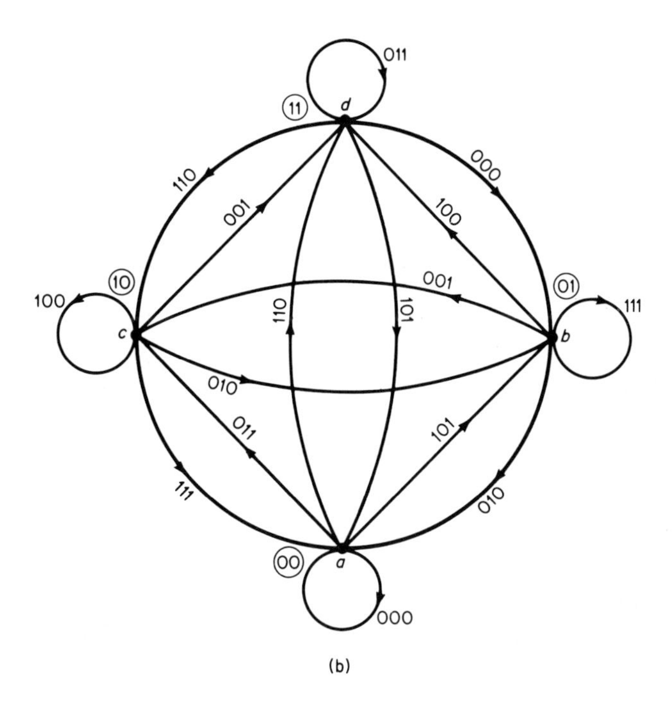

Fig. 5-30 An example of a 4-stage convolutional encoder and its equivalent state diagram

Maximum-Likelihood Decoding of Convolutional Codes (The Viterbi Algorithm)

Several techniques for decoding convolutional codes are presently available, each having its own particular advantages and disadvantages. The one of most interest here (as was the case for the block codes discussed earlier in this chapter) is the so-called *maximum-likelihood decoder*, which uses the inequality of (5-17) in arriving at its decision rule. The specific form of this decision rule depends on the channel through which the convolutional code is to be transmitted and the *a priori* probabilities of the input messages. For example, for the additive white Gaussian noise channel with PSK modulation of the ± 1 input code digits onto a carrier and equally likely input messages, (5-18) applies if T is interpreted as the total time for transmission of a convolutional code word.* Letting $d_{jk}^{(i)}$ denote the kth transmitted code digit of the convolutional encoder output sequence corresponding to the jth input of d bits, and $p_s(t)$ denote the fundamental pulse shape of the PSK modulation, then $s_i(t)$ of (5-6) is defined as

$$s_i(t) = \sum_{j=1}^{L} \sum_{k=1}^{v} d_{jk}^{(i)} p_s\{t - [(k-1) + (j-1)v]T_s\}$$

where $T_s = (d/v)T_b$ is the transmitted code digit time. Substituting $s_i(t)$ of above in (5-18) and simplifying gives

$$\begin{aligned} q_i &= \sum_{j=1}^{L} \left[\sum_{k=1}^{v} \int_{[k-1+(j-1)v]T_s}^{[k+(j-1)v]T_s} y(t)\, d_{jk}^{(i)} p_s\{t - [(k-1) + (j-1)v]T_s\}\, dt \right] \\ &= \sum_{j=1}^{L} \left[\sum_{k=1}^{v} y_{jk} d_{jk}^{(i)} \right] \end{aligned} \tag{5-127}$$

where we recognize that $T = vLT_s$ and

$$\begin{aligned} y_{jk} &= \int_{[k-1+(j-1)v]T_s}^{[k+(j-1)v]T_s} y(t) p_s\{t - [(k-1) + (j-1)v]T_s\}\, dt \\ &= d_{jk}^{(i)} E_s + \int_{[k-1+(j-1)v]T_s}^{[k+(j-1)v]T_s} n(t) p_s\{t - [(k-1) + (j-1)v]T_s\}\, dt \end{aligned} \tag{5-128}$$

with E_s the energy per code digit. As before, the decoder chooses the input message sequence that yields the largest value of q_i. The brackets in (5-127) indicate that the computation of q_i will be done one tree branch at a time as we sequentially step through the tree.

The brute force approach to maximum-likelihood decoding is to compute the test statistic (often called *metric*) q_i for each complete path in the tree and then choose that path for which q_i is the largest. Ordinarily, the transmitted (and hence received) sequence might be quite long, resulting in a large number of calculations and comparisons in applying the above procedure. A decoding algorithm proposed by Viterbi,[27] subsequently shown to be maximum-likelihood by Forney,[28] makes use of the repetition property of the convolutional code tree to reduce this number of comparisons.

---

* Note that it is not necessary to restrict ourselves to a particular channel with a specific form of modulation as we have done here. The results that follow apply to a general class of memoryless channels, which include additive white Gaussian noise channels using polyphase and MFSK modulations. Nevertheless, for simplicity of explanation we shall continue to use the additive white Gaussian model with PSK modulation as described above.

More specifically, from the equivalent trellis representation of the code tree, we observe that after the first K_0 branches one can immediately determine which of the 2^d paths merging at a particular node, say a, has the greatest likelihood of being the transmitted sequence, regardless of the continuation of the path beyond that point. For each such group of paths, the contribution to q_i due to any given set of tree branches beyond the K_0th branch is the same. Hence, the path whose leading K_0 branches yield the largest contribution to q_i "survives" over the other paths entering the same node. The algorithm procedure then goes as follows. Starting at the end of the K_0th tree branch, one computes for each of the 2^{K-d} nodes a surviving path from the 2^d possible paths entering that particular node based on the maximum contribution to q_i up to this point. In going from the K_0th to the $K_0 + 1$st step in the code tree, each surviving path splits into 2^d branches. Thus, the number of possible paths to consider at the end of the $K_0 + 1$st step is 2^K. However, we note that this is the same number that existed at the K_0th step; in fact, the node structure is also the same. Hence, we once again determine a surviving path for each node—based now, however, on the metric computed for the first $K_0 + 1$ tree branches. Actually, it is not necessary to recompute the entire metric function each time we step through the code tree. By storing the metrics calculated for the nth step, we can merely increment these by the additional contributions accompanying the transition to the $n + 1$st step. Alternately, in terms of the state diagram, we associate with each state a storage device that remembers both the path that has survived thus far and the value of the metric accumulated along it. Continuing as above, we sequentially step through the code tree, each time saving 2^{K-d} surviving paths. The algorithm is truncated by forcing into the convolutional encoder a specific sequence of $K - d$ binary input digits that correspond uniquely to one of the possible system states. Thus, the very last step in the decision process is to choose the surviving path that terminates at the above forced state.

Two other contributions relating to the optimality of the Viterbi algorithm are worthy of mention. Viterbi and Odenwalder[29] give a simple proof of the equivalence of the algorithm to a maximum-likelihood decoding procedure that compares the likelihood functions of all paths at the end of the L-branch tree. Omura,[30] on the other hand, takes a state-space approach to the problem based upon the method of dynamic programming. The advantage of the latter contribution is the motivation that it provides for examining convolutional encoders in terms of their algebraic structure.[31]

Other Methods of Decoding Convolutional Codes

We have been concerned with the optimal procedure for decoding a convolutional code—namely, maximum-likelihood decoding. Although implementation of the Viterbi algorithm results theoretically in the minimum error probability, several other more practical considerations, discussed below, are important in choosing and designing a good decoding scheme.

One problem immediately facing the design engineer is the complexity of implementation, in terms of storage and computation, associated with the Viterbi algorithm when the constraint span K_0 becomes large (greater than 10). These long constraint spans become necessary when one desires very low error probabilities at high data rates. Hence, a good deal of attention has been paid to designing short constraint span codes so that the Viterbi decoding algorithm can be applied in the implementation of a communication system.[32–34] A second difficulty with maximum-likelihood decoding is that theoretically the decoded path is not chosen until the very last step in the tree, and hence we cannot decode any of the bits until we have examined all of the received code digits. The implication of this statement is that the length of path history that must be stored for a long input data sequence becomes formidable. In practice, this problem is partially circumvented by truncating the path memory after say M_t bits (tree branches) have been accumulated and by deciding on a bit whose corresponding surviving path M_t branches forward yields the highest metric out of the 2^{K_0} possibilities. A bit error will thus occur only if the bit selected is from an incorrect path that ultimately would have been eliminated if memory truncation had not been applied, and if this incorrect path differed from the correct one M_t branches back. If M_t is chosen to be several times as large as K_0, then the probability of this joint event is very small.

Sequential Decoding. It is for these and other reasons that one is interested in good suboptimal decoding schemes. One such possibility is that of *sequential decoding*[35–37] for which several algorithms have been suggested.[35–40] The basic approach common to all of these algorithms is described as follows. Starting with the first step in the tree, we tentatively hypothesize that the correct path begins with the tree branch (out of in general 2^d possibilities) whose corresponding v code digits yield the largest metric with respect to the received signal in the first time interval. Having then tentatively decoded the first d input information bits, we use this hypothesis to reduce the number of choices available on the next d bits—the 2^d tree branches at the second step in the tree that emanate from the branch tentatively chosen at the first step. By comparing the code digits of these 2^d tree branches to the received signal in the second transmission interval and choosing that one that again yields the largest metric, we tentatively select the second branch of the path in the tree, which we shall ultimately choose as correct. This digit-by-digit (sequential) decoding procedure continues as long as the number of errors between the received sequence and the hypothesized transmitted sequence remains small. Once the decoder makes an incorrect hypothesis at some step in the tree, however, it will inevitably be forced into making subsequent choices between v-digit sequences bearing no relation to the received sequence in the corresponding time interval. Presumably, the decoder will be able to recognize this at some time after the incorrect hypothesis was made and will retrace its steps by going back and making alternate hypotheses, this time hopefully decoding correctly.

Although in principle the concept appears to be quite simple, the choice of an appropriate strategy to decide when the decoder should discontinue its present tentative path and return to change old hypotheses, and how far back it should go, is indeed a complex problem. Suffice it to say that a good search algorithm for sequential decoding is one that requires a minimum amount of computation and ultimately selects the correct path in the tree with high probability. Several algorithms have been suggested that were introduced, chronologically, by Wozencraft (1957),[35] Fano (1963),[39] and Jelinek (1969).[40] It is not our intention here to pursue these decoding schemes in detail but merely to mention their existence and refer the reader to the appropriate references.

For large values of constraint span K_0, sequential decoders offer a significant cost advantage over the maximum-likelihood decoder. Whereas the cost of a sequential decoder is essentially independent of K_0, that of a Viterbi decoder increases exponentially with K_0. Unfortunately, the picture is not as bright when one considers the problems associated with storing the received data during high noise periods, where the decoder is making many hypotheses changes and hence large numbers of computations.[41] When the relation between the data rate, decoding speed, and ambient noise conditions is such that the *buffer* storage provided for the received data *overflows*, the decoding process cannot continue with small error, since data lost during the overflow condition cannot be recovered. Hence, the buffer overflow problem is fundamental in the design of a sequential decoder. Furthermore, the overflow event is the primary concern in the design of a sequential decoder, since for large K_0 undetected errors in the absence of overflow are much less frequent than are overflows themselves.

Several models for determining the probability of buffer overflow have been suggested in the literature.[42–45] The amount of searching required to get back on the right track in terms of the number of incorrect branches searched per decoded bit is a random variable whose average value is bounded for data rates less than $\mathcal{R}_0$ and is unbounded for rates greater than $\mathcal{R}_0$. This quantity $\mathcal{R}_0$ is the so-called *computational cutoff rate* of a sequential decoder. Herein lies a fundamental disadvantage of a sequential decoder: the decoding speed in branches per second must be 10 to 20 times faster than the incoming data rate to combat buffer overflow, and hence the maximum bit rate is limited[32]. In the maximum-likelihood decoder, on the other hand, decoding progresses at the received data rate. Another disadvantage is that once the decoding operation has been interrupted by buffer overflow, resynchronization (usually by inserting a known, single constraint-length sequence into the data stream periodically) is necessary before decoding can continue. Despite these significant disadvantages, the feasibility of sequential decoding has clearly been demonstrated in a wide number of applications.[46–48] Also, practical methods for combating the buffer overflow problem have been suggested for systematic convolutional codes.

Feedback Decoding. Another type of suboptimal decoding is *feedback decoding*, a special case of which is *threshold decoding*.[49] Although the performance of Viterbi and sequential decoding is generally superior to that of feedback decoding, the latter has the distinct advantage of being easy to implement. The operation of a feedback decoder is described as follows. The first L_t branches (L_t, a parameter of the decoder) are examined, and the path yielding the largest metric with respect to the first $L_t v$ code digits of the received sequence is selected. At this point, we decode the first information bit corresponding to the first tree branch on the maximum metric path chosen. This decision uniquely moves us to a specific node at the second step in the tree. Starting there, we consider all paths of length L_t branches emanating from that node and again make a maximum metric selection—now with respect to the sequence of $L_t v$ received code digits whose first is the $v + 1$st. We then proceed to decode the second information bit. This process continues in the manner just described. The technique is called *feedback decoding* because each decoding decision on an information bit is fed back to affect future decisions. Hence, decoding schemes of this type possess the undesirable property of error propagation caused by incorrect decisions. For a large class of convolutional codes, however, this phenomenon does not present a serious problem.

It should be clear that the decoding accuracy of a feedback decoding scheme improves as the parameter L_t increases. Unfortunately, however, the complexity of the decision device also grows as L_t gets large. Threshold decoding provides a partial solution to this paradox, but it is only useful for a specially constructed class of convolutional codes and only for moderate values of L_t. Further details of this specific type of feedback decoder are well documented.[49] One final note relative to feedback decoders is that they are particularly applicable to burst error channels.

We have discussed several different schemes for optimally or suboptimally decoding convolutional codes. Many other hybrid type decoding schemes have been developed that make use of the basic decoding techniques discussed here; a few examples can be found in Refs. 50 through 52.

Error Probability Performance of Convolutional Codes

Success in finding explicit expressions for the error probability performance of certain classes of block codes is unfortunately not readily achieved when studying convolutional codes. In analyzing convolutional codes, we must presently be content with finding bounds on the error probability performance achievable with a given decoding scheme. Using these bounds as guidelines, one attempts next to find the "best" code—in the sense of minimum error probability—for a given K_0 or, since this is usually difficult if not impossible, to find "good" codes that give low error probability performance.

The error probability performance of any code (block or convolutional) is intimately related to its *distance structure*—the relative distances among its code word members. The distance between two code words is proportional to the number of code digits in which they differ. One should not conclude that maximizing the minimum code word distance guarantees that error probability is simultaneously minimized, although low error probability is definitely correlated with large minimum distance. An outline of the theory

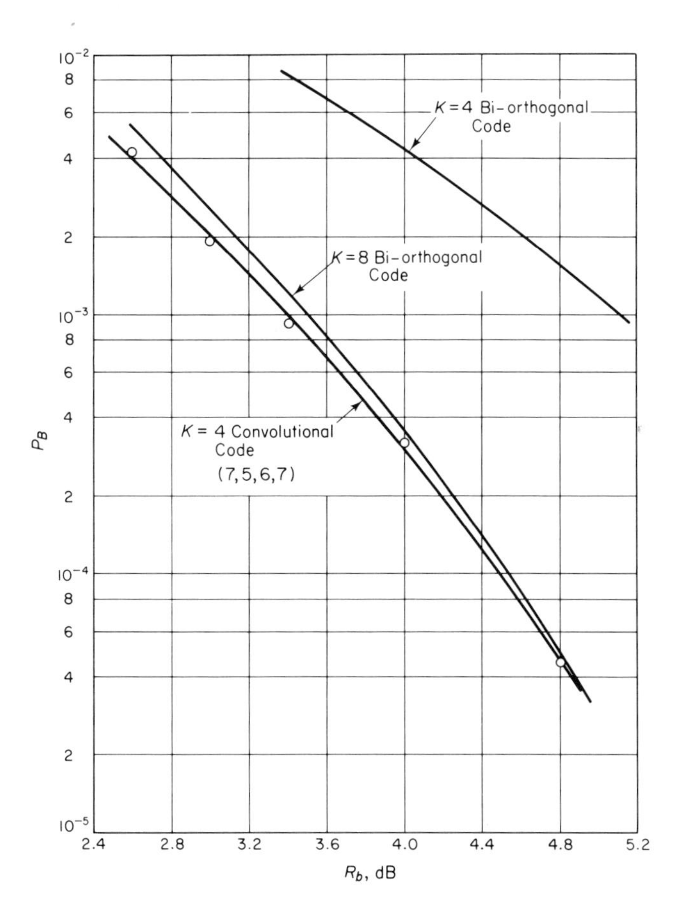

Fig. 5-31 Bit error probabilities for $K = 4$ convolutional and bi-orthogonal codes and a comparison with a $K = 8$ bi-orthogonal code (Courtesy of J. A. Heller).

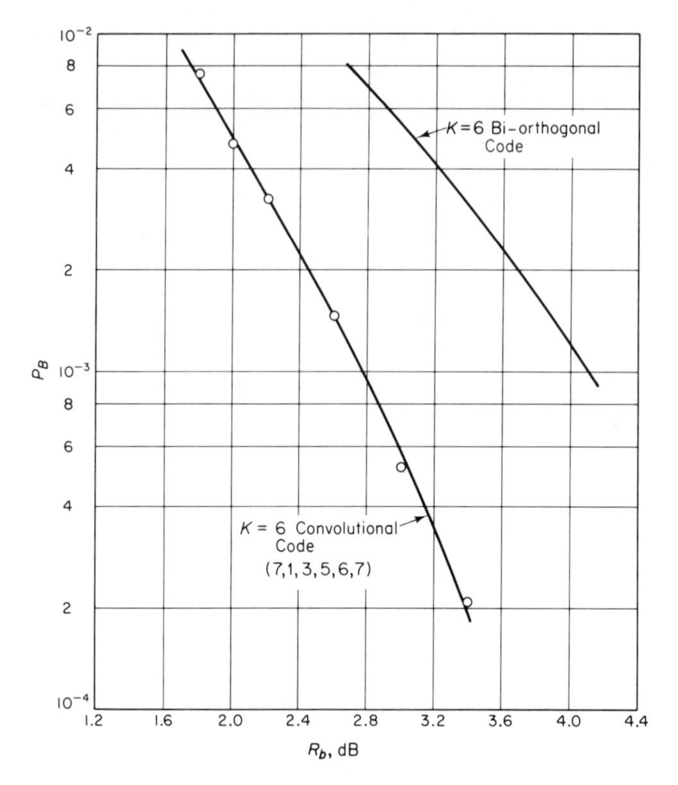

Fig. 5-32 Bit error probabilities for $K = 6$ convolutional and bi-orthogonal codes (Courtesy of J. A. Heller).

relative to the distance structure of convolutional codes is given in Appendix C along with other criteria for the selection of good codes.

Ad hoc schemes have been proposed that are useful in generating codes with good distance structure. In particular, Heller[32] suggests a code-generating scheme that for small constraint spans produces codes with a minimum distance $d_{\min}$ (defined in Appendix C) equal to or near his upper bound on this quantity. Using a Viterbi decoding algorithm simulated on an XDS 930 digital computer, experimental error probability performance results were obtained for $d = 1$, $\mathcal{R}_c = 1/3$, and values of $K = 4$, 6, and 8.[34] These results are presented in Figs. 5-31 through 5-33, where for comparison purposes the performance of a K-bit bi-orthogonal block code is also shown. Note in Fig. 5-31 that a $K = 8$ bi-orthogonal code is needed to give approximately the same error probability performance as Heller's $K = 4$ Viterbi-decoded convolutional code. The K-digit number associated with each convolutional code in the figures is the octal representation of the columns of the code's matrix of parity-check equations. For example,

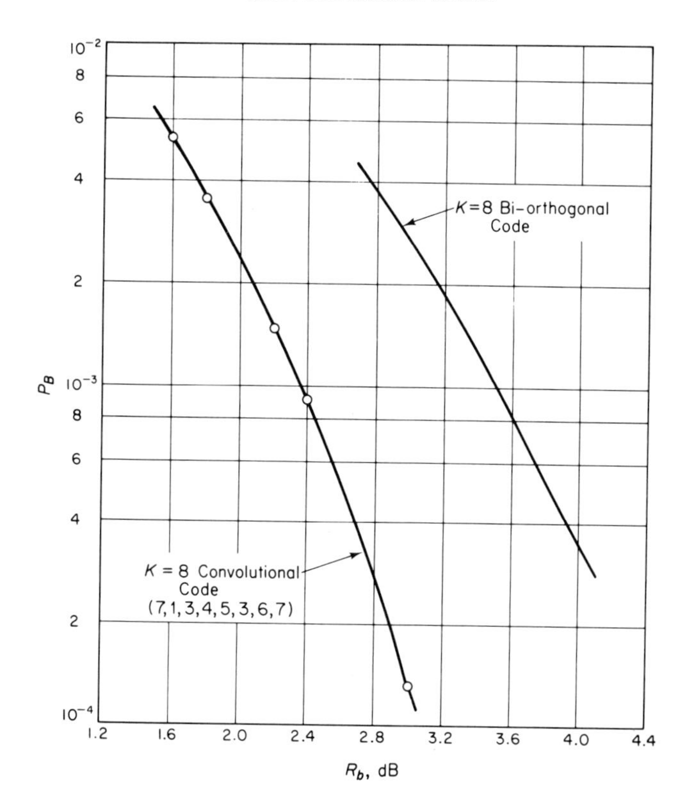

Fig. 5-33 Bit error probabilities for $K = 8$ convolutional and bi-orthogonal codes (Courtesy of J. A. Heller).

the 4-digit number '7567' represents a $K = 4$, $v = 3$ convolutional code having as its parity-check equation matrix:

$$\begin{pmatrix} 1111 \\ 1011 \\ 1101 \end{pmatrix}$$

The more direct approach of using error probability performance to influence the design of a code construction algorithm has recently been adopted by Layland.[61] The proposed algorithm makes changes in the code's parity-check equations to effect the steepest descent in bit error probability performance, as computed from a union bound such as that given in Appendix D. Using the same Viterbi decoding algorithm and random noise generator Heller did in arriving at the results of Figs. 5-31 through 5-33, the simulated performance of Layland's so called "hill-climbing" algorithm has been computed for $d = 1$ and values of $K = 3$ to $K = 10$ (Fig. 5-34). The K-digit octal numbers accompanying the legend of this figure have the same significance as discussed for the preceding three figures. Note that the $K = 3$ curve corresponds to a $v = 2$ code. Furthermore, Laylands' "optimum" codes for $K = 3$ and $K = 4$ are identical to those obtained by Heller's algorithm, which is based upon minimum distance.

Also plotted in Fig. 5-34 is the union bound of Appendix D as applied to

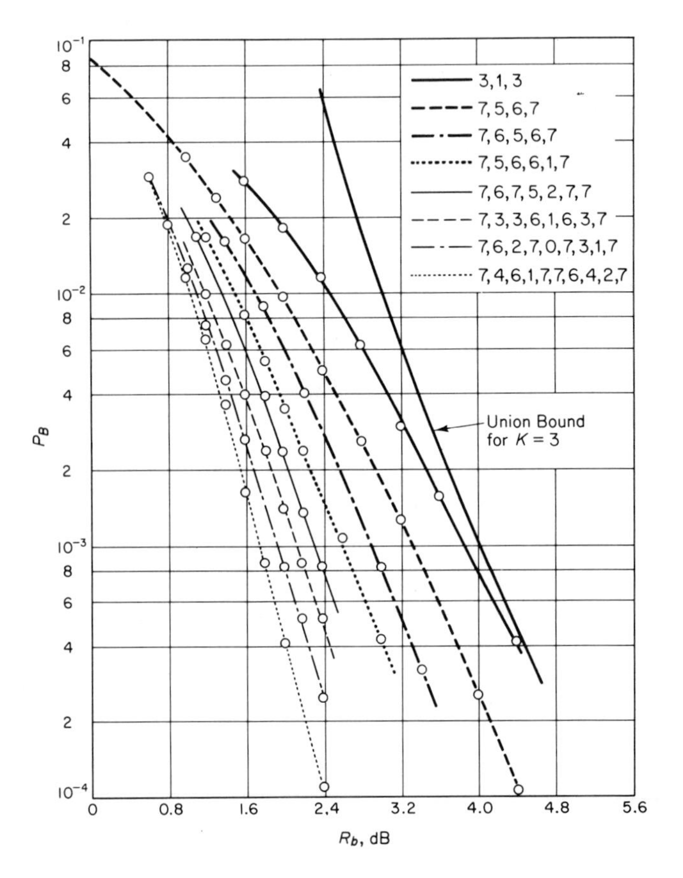

Fig. 5-34 Bit error probabilities of eight convolutional codes, $K = 3$ to $K = 10$ (Courtesy of J. W. Layland).

the distance structure of the $K = 3$ code. For this special case, Eq. (D-6) reduces to[53]

$$P_B < \frac{\frac{1}{2}\,\mathrm{erfc}\sqrt{5R_b/2}}{[1 - 2\exp(-R_b/2)]^2} \tag{5-129}$$

We observe that for large R_b the union bound is quite tight but deviates considerably from the simulated results at low energy-to-noise ratios. As alluded to in Appendix D, this behavior is typical of union bounds of this type. Furthermore, the closeness of the bound to the simulated results at high energy-to-noise ratios is not surprising in view of the fact that Layland's codes are designated at a point where $\exp[-R_b/v] = 0.1$, which for $v = 2$ gives $R_b = 6.62$ dB.

The variation of bit error probability with code rate at a fixed constraint span is indicated in Fig. 5-35, where simulated results based on Layland's

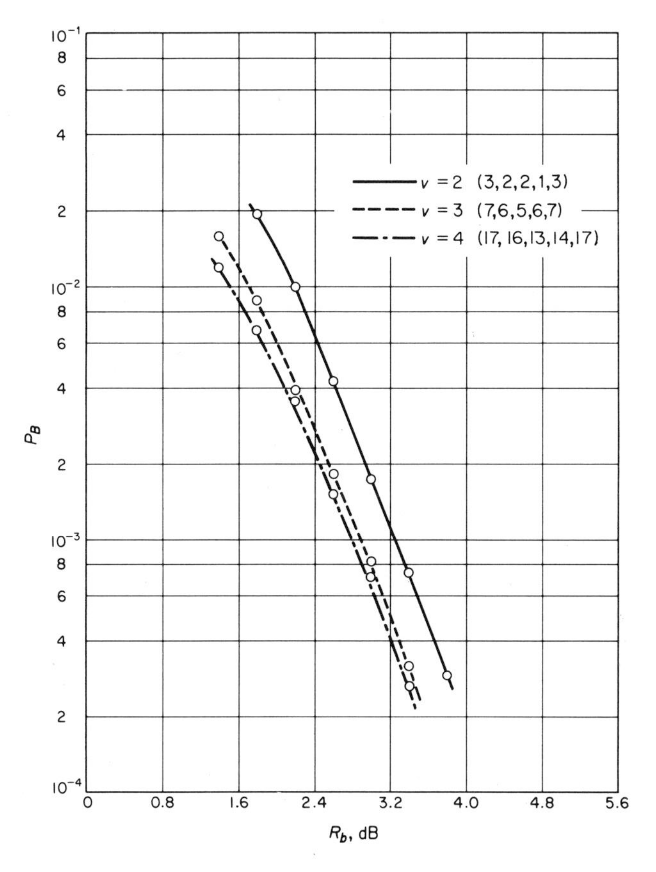

Fig. 5-35 Bit error probabilities of $K = 5$ codes of rate $\frac{1}{4}$, $\frac{1}{3}$, and $\frac{1}{2}$ (Courtesy of J. W. Layland).

algorithm are given for $K = 5$ codes transmitted at rates of 1/4, 1/3, and 1/2.

Another code-construction algorithm proposed by Odenwalder[52] in some sense represents a compromise between those of Heller and Layland—at least insofar as error probability performance is concerned. This can be observed by plotting the simulated performance results for all three algorithms on one grid, as in Fig. 5-36 for $K = 8$ and $v = 3$. Qualitatively, the three

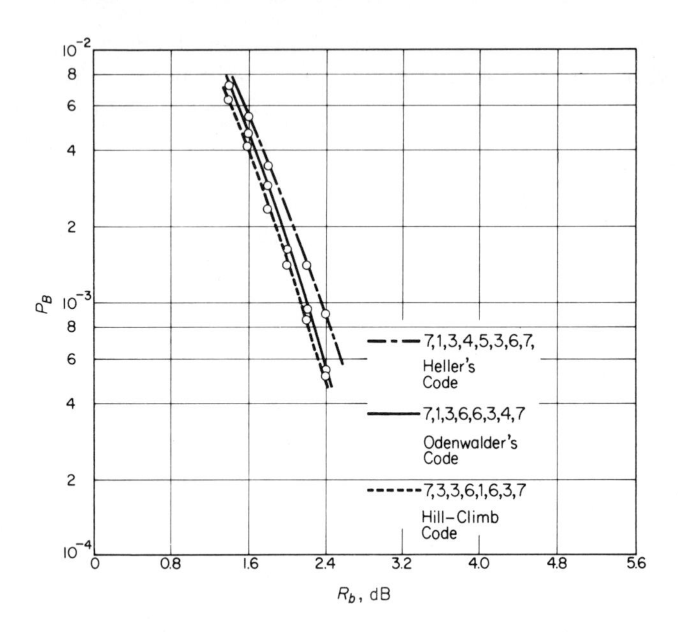

Fig. 5-36 Bit error probabilities for three $K = 8$, $v = 3$ codes (Courtesy of J. W. Layland).

algorithms rank in performance as indicated in the figure. Heller constructs his code solely on the basis of maximizing the minimum code word distance. Odenwalder goes one step beyond this by, in addition to maximizing $d_{\min}$, minimizing the number of code words at this distance. Finally, Layland, by designing according to the union bound on error probability, is in effect looking at the entire distance polynomial $T(D)$ [see Eq. (C–3)].

5-9. SELF-SYNCHRONIZABLE CODES

In the previous sections, we have considered various encoding and decoding procedures of practical interest with emphasis on obtaining good error probability performance. The task of generating block or convolutional codes that achieve this desired performance is relatively simple for a communications engineer. Frequently, the detection process is performed on a word-by-word basis, and since the system must be in word synchronization to operate properly, it is necessary in such applications to choose a set of code words that also have good word-synchronization properties. Consequently, considerable effort has been expended on the problem of developing *synchronizable* or *self-synchronizable codes.* A synchronizable code has the property that punctuation (comma or no comma, with a comma indicating that the next symbol is the beginning of a new code word) at a given position in a code symbol stream can always be determined by observing at most s code symbols in the neighborhood of the sync position in question.

A large body of literature[62-70] on synchronizable block codes for the noiseless channel is available to the coding theorist; however, the sophisticated codes included in this compendium appear to require table look-up in their mechanization. Several codes not dependent on commas for separation have been proposed for solution to the word-synchronization problem. Among these is the *comma-free code*, which has the property that no overlap of any two code words can itself be a code word.[62] Thus, if the code words $c_1, c_2, \ldots, c_M$ and $d_1, d_2, \ldots, d_M$ are members of a comma-free code dictionary, then $c_j c_{j+1} \ldots c_M d_1 d_2 \ldots d_{j-1}$ cannot be a code word for any $1 < j \leq M$. This is equivalent to the statement that if a code dictionary of M symbol words is comma free, then by definition, any succession of $2M - 1$ symbols occurring in an arbitrary length sequence of observed symbols must contain one and only one dictionary code word of length M.

The concept of comma freedom can be phrased in terms of the correlation properties of its code word members. As discussed in Chap. 4 and earlier in this chapter, the normalized correlation between two binary sequences, say $\{c_i\}$ and $\{d_i\}$, can be expressed as

$$\lambda_{cd}(k) = \frac{A(k) - D(k)}{A(k) + D(k)} = 1 - \frac{2D(k)}{M} \tag{5-130}$$

where $A(k)$ is the number of agreements and $D(k)$ is the number of disagreements between corresponding components of the sequences $\{c_i\}$ and $\{d_{i+k}\}$. In terms of the above nonoverlap condition then, the cross-correlation of a comma-free code word and any M-digit sequence formed from a combination of two comma-free code words must be less than $1 - 2/M$; that is, there must be at least one disagreement between the two. A generalization of this condition is used as follows to define a *comma-free code of index* P_{cf}. If we consider the set of correlations λ_k, $k = 1, 2, \ldots, N$, between the code word members and an arbitrary sequence of M digits, which is not a code word, then we require that

$$\max_k |\lambda_k| = \max_k \left| 1 - \frac{2D(k)}{M} \right|$$

be minimized. If the values of $D(k)$ yielding the maximum magnitude λ_k are denoted P_{cf} and $M - P_{cf}$, then clearly for all k

$$P_{cf} \leq D(k) \leq M - P_{cf}$$

$$|\lambda_k| \leq \left| 1 - \frac{2P_{cf}}{M} \right| \tag{5-131}$$

Hence, $\max_k |\lambda_k|$ is minimized by maximizing P_{cf}, where from (5-131) P_{cf} is upper bounded by $M/2$. If the set of N correlations between the N code word members and all possible M-digit sequences that are not code words satisfies the inequalities in (5-131), then the code is said to be comma free of index P_{cf}. Furthermore, if P_{cf} is maximized, then the code set is optimum in the comma-free sense. Briefly then, a comma-free code of index P_{cf} is one for which all possible M-tuples that could arise from erroneous synchronization of the data stream agree or disagree with every code word of the code in at least P_{cf} positions.

The search for codes having a large comma-free index represents a difficult problem; however, Stiffler[71] shows how one can construct comma-free codes from the orthogonal and bi-orthogonal codes discussed earlier in this chapter. The technique employed involves the addition of an appropriate vector to each code word in the dictionary. The new code is called a *coset* of the original code and has the same correlation properties as the original code. For example, the $N \times N$ matrix of orthogonal n-bit code words $\mathbf{H}_n$ [see (5-36)] can be made comma free by adding a *comma-free vector* $\mathbf{V}_n$ modulo 2 to each row (code word) of the matrix. For example, the vector $\mathbf{V}_n$ may be chosen to be a particular phase shift of a PN sequence of length $2^n - 1$ plus one adjoining binary digit, making the vector 2^n components long. Table 5-6 represents a summary of indices of comma freedom obtained by this method. Note that the number of bits n must be at least 4 to achieve comma freedom.

TABLE 5-6 Achievable Indices of Comma Freedom for n-bit Orthogonal Codes

n	P_{cf}
4	2
5	6
6	14
7	34

The corresponding comma-free vectors are

$\mathbf{V}_4 =$ 00100011 1101011**1**

$\mathbf{V}_5 =$ **1**0001101 11010100 00100101 10011111

$\mathbf{V}_6 =$ 00011100 10010110 11101100 110101**0**0
11111100 00010000 11000101 00111101

$\mathbf{V}_7 =$ **1**1111111 00000010 00001100 00101000
11110010 00100011 11101010 01111101
00001110 00100100 00111110 11011110
11000110 10010111 01110011 00101010

where the boldface zero or one represents the binary digit which supplements the PN vector. If $\mathbf{K}_n = \mathbf{V}_n \oplus \mathbf{H}_n$ denotes the n-bit orthogonal comma-free code matrix generated as above, then an $(n + 1)$-bit bi-orthogonal

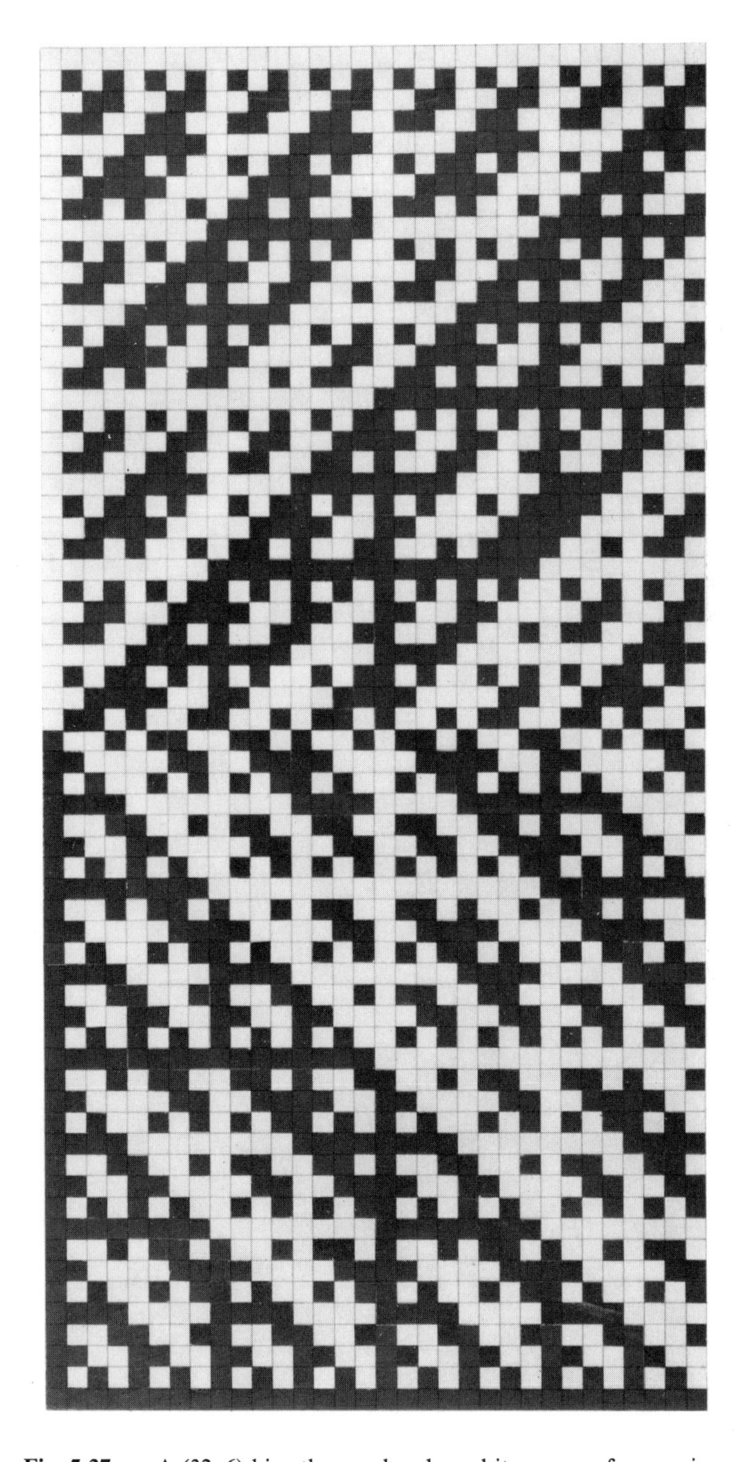

Fig. 5-37 A (32, 6) bi-orthogonal code and its comma-free version

Fig. 5-37 (Cont.)

comma-free code can be formed by augmenting $\mathbf{K}_n$ with its complement. Thus, Table 5-6 also applies to $(n + 1)$-bit bi-orthogonal codes generated as above. Illustrated in Fig. 5-37 is a (32, 6) bi-orthogonal code and its equivalent comma-free code obtained by adding (modulo 2) the vector $\mathbf{V}_5$ above.

Later efforts[72,73] produced comma-free vectors $\mathbf{V}_n$ whose comma-free indices were greater than the values given in Table 5-6. For example, using enumerative techniques, 32 distinct 6-bit bi-orthogonal comma-free codes of index 7 were found for use in the Mariner Mars 1969 high data rate telemetry system. One such comma-free vector is given by

$$\mathbf{V}_5 = 10101100100111000101000000000000,$$

and the corresponding code generated is optimum in the comma-free sense. Using Stiffler's comma-free vector of index 14 for 7-bit bi-orthogonal codes, a gradient-type computer search (which of course is not necessarily exhaustive) was performed and resulted in the determination of several hundred comma-free vectors with index 16 but none of index 17 or higher.[73] One such comma-free vector of index 16 is

$$\begin{aligned}\mathbf{V}_6 = {} & 00001100\ 00000110\ 11111100\ 10010100 \\ & 11011100\ 00010000\ 11000011\ 00111001\end{aligned}$$

The self-synchronizability of convolutional codes has only recently been studied by Layland.[74] The first observation made is that as with block codes, no convolutional code in its standard form is branch synchronizable. Hence, the output sequence of a convolutional encoder must be modified before transmission in order for branch synchronization to be known to the decoder. In a manner analogous to that of adding the comma-free property to an orthogonal or bi-orthogonal block code, some convolutional codes can be made synchronizable by the modulo 2 addition of a periodic sequence to the encoder output. Layland gives necessary conditions for a convolutional code to be made branch synchronizable by this technique. If the code cannot be made synchronizable—if it does not meet the necessary conditions, for example—it is still possible to ensure synchronization of the decoder by statistical means, providing the data source is sufficiently random. Simulation results given in Ref. 74 support this thesis.

APPENDIX A
ABSTRACT VECTOR SPACE CONCEPTS

In this appendix we summarize certain analogies between abstract vector spaces and the so-called "waveform or signal space." A vector space ϑ is defined as a set of vectors **a**, **b**, **c**, etc., satisfying the following properties:

1. $\mathbf{a} + \mathbf{b} = \mathbf{b} + \mathbf{a} = \mathbf{c}$, with **c** another vector in the space ϑ
2. $\mathbf{a} + (\mathbf{b} + \mathbf{c}) = \mathbf{a} + \mathbf{b} + \mathbf{c}$

3. $x(\mathbf{a} + \mathbf{b}) = x\mathbf{a} + x\mathbf{b}$, with x a real scalar
4. $x(y\mathbf{a}) = (xy)\mathbf{a}$, with x and y as real scalars
5. $\mathbf{0}$ is a zero (null) vector defined such that for every $\mathbf{a}$, $0\mathbf{a} = \mathbf{0}$. (A-1)

In particular, the vectors we shall be considering will be the space of all real-valued functions $a(t)$ that are Lebesgue square-integrable over the interval $(0, T)$; that is,

$$\int_0^T a^2(t)\, dt < \infty$$

We also define a *scalar* or *inner product* $(\mathbf{a}, \mathbf{b})$ with the following properties:

1. $(\mathbf{a}, \mathbf{b}) = (\mathbf{b}, \mathbf{a})$
2. $(x\mathbf{a} + y\mathbf{b}, \mathbf{c}) = x(\mathbf{a}, \mathbf{c}) + y(\mathbf{b}, \mathbf{c})$
3. $(\mathbf{a}, \mathbf{a}) = |\mathbf{a}|^2 > 0$ (if $\mathbf{a}$ is not the zero vector). (A-2)

The inner product is a scalar function of the vectors $\mathbf{a}$ and $\mathbf{b}$, and is the extension to abstract spaces of the projection of $\mathbf{a}$ on $\mathbf{b}$ in three-dimensional Euclidean space. In the case at hand, where $a(t)$ and $b(t)$ are defined over the interval $(0, T)$, the inner product is a real number:

$$(\mathbf{a}, \mathbf{b}) = \int_0^T a(t)\, b(t)\, dt \tag{A-3}$$

The final concept from the theory of abstract vector spaces necessary to this discussion relates to the operator R, which transforms one vector $\mathbf{a}$ into another vector $\mathbf{b} = R\mathbf{a}$ in the same space. A linear operator or transformation is one for which $R(x\mathbf{a} + y\mathbf{b}) = xR\mathbf{a} + yR\mathbf{b}$—for our purposes,

$$R\mathbf{a} = \int_0^T R(t, \tau)\, a(\tau)\, d\tau = b(t) \tag{A-4}$$

An inverse operator R^{-1} can be shown to exist if and only if $R\mathbf{a} = \mathbf{0}$ implies $\mathbf{a} = 0$. Then $R(R^{-1}\,\mathbf{a}) = \mathbf{a}$ and $R^{-1}(R\mathbf{a}) = \mathbf{a}$. With both R and R^{-1} defined, we have $RR^{-1} = I = R^{-1}R$, with I the identity operator defined by $I\mathbf{a} = \mathbf{a}$ for every vector $\mathbf{a}$. We assume the existence of an inverse operator throughout.

Consider now the random function $n(t)$, $0 \leq t \leq T$, derived from a (possibly nonstationary) zero mean Gaussian process. Its covariance function $R_n(t, \tau)$ is assumed known. Then the p.d.f. of the vector $\mathbf{n}$ is given by

$$p(\mathbf{n}) = K'_0 \exp\left\{-\frac{1}{2}[(\mathbf{n} - \mathbf{a}), R_n^{-1}(\mathbf{n} - \mathbf{a})]\right\} \tag{A-5}$$

where K'_0 is a constant, $\mathbf{a}$ is the expected value of $\mathbf{n}$, and R_n^{-1} the inverse operator corresponding to $R_n(t, \tau)$. Thus, defining a vector $\mathbf{h}$ by $\mathbf{h} = R_n^{-1}(\mathbf{n} - \mathbf{a})$ and assuming the existence of the inverse operator, we have

$$n(t) - a(t) = \int_0^T R_n(t, \tau)\, h(\tau)\, d\tau \tag{A-6}$$

Alternately, the time function $h(t)$ is given by

$$h(t) = \int_0^T R_n^{-1}(t, \tau)[n(\tau) - a(\tau)]\, d\tau \tag{A-7}$$

APPENDIX B
DERIVATION OF THE WORD ERROR PROBABILITY FOR POLYPHASE SIGNALS

The probability of error for polyphase signals is obtained by taking one minus the probability of correct detection as given by (5-87); i.e.,

$$P_E(N) = 1 - \frac{2}{\pi} \int_0^\infty \exp\{-(u - \sqrt{R_d})^2\} \left[\int_0^{u \tan \pi/N} \exp(-v^2)\, dv \right] du \tag{B-1}$$

Letting $z = u - \sqrt{R_d}$,

$$P_E(N) = 1 - \frac{2}{\pi} \int_{-\sqrt{R_d}}^\infty f(\sqrt{R_d}, z)\, dz \tag{B-2}$$

where

$$f(\sqrt{R_d}, z) = \exp(-z^2) \int_0^{(z+\sqrt{R_d}) \tan \pi/N} \exp(-v^2)\, dv \tag{B-3}$$

Taking the partial derivative of $P_E(N)$ with respect to $\sqrt{R_d}$, we get

$$\begin{aligned} \frac{\partial P_E(N)}{\partial \sqrt{R_d}} &= -\frac{2}{\pi} \left[\int_{-\sqrt{R_d}}^\infty \frac{\partial f(\sqrt{R_d}, z)}{\partial \sqrt{R_d}}\, dz + f(\sqrt{R_d}, -\sqrt{R_d}) \right] \\ &= -\frac{2}{\pi} \int_{-\sqrt{R_d}}^\infty \frac{\partial f(\sqrt{R_d}, z)}{\partial \sqrt{R_d}}\, dz \end{aligned} \tag{B-4}$$

From (B-3),

$$\begin{aligned} \frac{\partial f(\sqrt{R_d}, z)}{\partial \sqrt{R_d}} &= \exp(-z^2) \tan \frac{\pi}{N} \exp\left[-\left\{ (z + \sqrt{R_d}) \tan \frac{\pi}{N} \right\}^2 \right] \\ &= \tan \frac{\pi}{N} \exp\left[-\left\{ z^2 \sec^2 \frac{\pi}{N} + 2z\sqrt{R_d} \tan^2 \frac{\pi}{N} + R_d \tan^2 \frac{\pi}{N} \right\} \right] \end{aligned} \tag{B-5}$$

Completing the square in (B-5) gives

$$\begin{aligned} \frac{\partial f(\sqrt{R_d}, z)}{\partial \sqrt{R_d}} &= \tan \frac{\pi}{N} \exp\left\{ -R_d \sin^2 \frac{\pi}{N} \right\} \\ &\quad \times \exp\left\{ -\sec^2 \frac{\pi}{N} \left(z + \sqrt{R_d} \sin^2 \frac{\pi}{N} \right)^2 \right\} \end{aligned} \tag{B-6}$$

Substituting (B-6) into (B-4), we get

$$\begin{aligned} \frac{\partial P_E(N)}{\partial \sqrt{R_d}} &= -\frac{2}{\pi} \tan \frac{\pi}{N} \exp\left\{ -R_d \sin^2 \frac{\pi}{N} \right\} \\ &\quad \times \int_{-\sqrt{R_d}}^\infty \exp\left\{ -\sec^2 \frac{\pi}{N} \left(z + \sqrt{R_d} \sin^2 \frac{\pi}{N} \right)^2 \right\} dz \end{aligned} \tag{B-7}$$

Letting $t = (z + \sqrt{R_d} \sin^2 \pi/N) \sec \pi/N$,

$$\frac{\partial P_E(N)}{\partial \sqrt{R_d}} = -\frac{2}{\pi} \sin\frac{\pi}{N} \exp\left\{-R_d \sin^2\frac{\pi}{N}\right\} \int_{-\sqrt{R_d}\cos\pi/N}^{\infty} \exp\{-t^2\}\, dt$$

$$= -\frac{1}{\sqrt{\pi}} \sin\frac{\pi}{N} \exp\left\{-R_d \sin^2\frac{\pi}{N}\right\}\left[1 + \operatorname{erf}\left(\sqrt{R_d}\cos\frac{\pi}{N}\right)\right] \quad \text{(B-8)}$$

Integrating (B-8) with respect to $\sqrt{R_d}$ between 0 and $\sqrt{R_d}$, we get

$$P_E(N) - P_E(N)|_{\sqrt{R_d}=0} = -\frac{1}{2}\operatorname{erf}\left(\sqrt{R_d}\sin\frac{\pi}{N}\right)$$

$$-\frac{\sin\pi/N}{\sqrt{\pi}} \int_0^{\sqrt{R_d}} \exp\left\{-R_d \sin^2\frac{\pi}{N}\right\} \operatorname{erf}\left(\sqrt{R_d}\cos\frac{\pi}{N}\right) d\sqrt{R_d} \quad \text{(B-9)}$$

Letting $y = \sqrt{R_d} \sin \pi/N$ and noting that $P_E(N)|_{\sqrt{R_d}=0} = (N-1)/N$, (B-9) simplifies to the desired result

$$P_E(N) = \frac{N-1}{N} - \frac{1}{2}\operatorname{erf}\left[\sqrt{R_d}\sin\frac{\pi}{N}\right]$$

$$-\frac{1}{\sqrt{\pi}} \int_0^{\sqrt{R_d}\sin\pi/N} \exp(-y^2)\operatorname{erf}\left(y\cot\frac{\pi}{N}\right) dy \quad \text{(B-10)}$$

APPENDIX C
THE DISTANCE STRUCTURE OF CONVOLUTIONAL CODES AND OTHER CRITERIA FOR THE SELECTION OF GOOD CODES

A study of the error probability performance associated with a particular decoding scheme requires a knowledge of the distance properties of convolutional codes. By this we mean the number of code digits in which one code word differs from all other code words in the transmitted set. Since convolutional codes are group codes, we may without any loss in generality choose any code word in the set as reference and determine the distance structure of the code set in terms of the distance of all other code words from the reference. For simplicity, we shall choose the all zeros code word (corresponding to transmission of all zero code digits) as our reference. Our discussion will center on the case where an output sequence is generated from the encoder for each single input bit, that is, $d = 1$. Extension of the results to higher values of d is left as an exercise for the reader.

The two questions of interest in a study of the distance structure of a convolutional code are

(a) what is the set of distances, $\{d_i\}$—from the all zeros code word—of all the paths that merge with the all zeros path for the first time at step j in the trellis diagram (j arbitrary), and
(b) how many of these paths correspond to each distance value, d_i.

By "merge for the first time," we mean all paths that have deviated from the all zeros path only one time previously (possibly at the first node in the tree) and have returned to the all zeros path for the first time at step j. The answers to these questions are derived from a study of the closed loop transfer function, $T(D, W, N)$, of a modified version of the state diagram regarded as a signal flow graph[53]. It suffices for our purposes here to present the result in the form

$$T(D, W, N) = \sum_{k=d_{\min}}^{\infty} D^k \mathcal{P}_k(W, N) \tag{C-1}$$

where $d_{\min}$ is the minimum of the set $\{d_i\}$ often called "minimum free distance," k ranges over the set $\{d_i\}$, and $\mathcal{P}_k(W, N)$ is a polynomial in W and N with coefficients c_{rk}, that is,

$$\mathcal{P}_k(W, N) = \sum_r c_{rk} W^{a_{rk}} N^{b_{rk}} \tag{C-2}$$

The meaning of (C-1) and (C-2) is as follows: for each group (possibly one) of paths at distance k from the all zeros path, there are c_{rk} of length a_{rk}, and each of these differs in b_{rk} *input* bits from the all zeros. The summation over r runs over as many different length paths as there are at distance k. The length of a path is given by the number of branches along the path between the time it first deviated from the all zeros path and the first time it returns at step j. If, for example, we are only interested in the total number of paths at distance k irrespective of their composition in terms of length or input bits, then we may evaluate $\mathcal{P}_k(W, N)$ at $W = N = 1$, and (C-1) simplifies to

$$T(D) \triangleq T(D, W, N)\big|_{W=N=1} = \sum_{k=d_{\min}}^{\infty} n_k D^k \tag{C-3}$$

where

$$n_k = \sum_r c_{rk}$$

is the total number of paths at distance k. Strictly speaking, the above expressions are valid only for $j \longrightarrow \infty$; hence for finite j we must truncate the series to exclude terms with $a_{rk} > j$.

Before presenting bounds for convolutional codes, it is desirable to understand what is meant by a good code in terms of the properties it must possess. The simplest definition of a good code is one that results in small error probability. The performance of a specific convolutional code on an additive Gaussian noise channel (or for that matter any memoryless channel) is intimately related to the relative distances between its member code words, in particular the minimum free distance, $d_{\min}$. Since a convolutional code which is terminated after L information bits is a group code, this complete set of relative distances is equivalent to the set of Hamming weights of all nonzero code words. We shall refer to this set of weights as the *weight spectrum* of the code. Then, clearly, the larger the elements of the weight spectrum, the smaller the error probability. Furthermore, two codes having the identical weight spectrum will have the same error probability performance when transmitted over a constant memoryless channel and maximum-likelihood decoded. Hence, we define two codes as *equivalent* if for every finite L they have the same weight spectrum. Thus, in looking for the best convolutional code, it is only necessary to consider one code from each equivalence class. This reduction of the total number of possible codes to one per equivalence class provides a first step in simplifying the search for good convolutional codes.

The next step in this search is to eliminate all codes that produce *catastrophic error propagation* upon being decoded. Catastrophic error propagation, as defined by Massey,[54] is the event whereby a finite number of channel errors results in an infinite number of decoding mistakes when an infinitely long message is transmitted over a binary-input, binary-output channel. Necessary and sufficient conditions have been developed for detecting when a convolutional code will exhibit catastrophic error propagation. Perhaps the simplest of these in terms of understanding is obtained by examining the state diagram and noticing if any closed path exists in the diagram whose output sequence is all zeros. We are assuming here, as before, that the transmitted sequence is the all zeros sequence. We know, of course, that the all zeros state will always have a self-loop whose corresponding output sequence is all zeros. This, however, does not cause catastrophic error propagation; hence we exclude this case automatically.

To illustrate a necessary and sufficient condition, consider the binary convolutional encoder of Fig. 5-27 and its corresponding state diagram of Fig. 5-29. Consider the possibility that at some step in the decoding process, we begin to deviate from the correct path (corresponding to looping around state a) returning again n steps later, where n might possibly go to infinity. An example of such a deviant path in terms of states traversed might be $a\ b\ d\ d\ d \ldots d\ c\ a$. The corresponding received and decoded sequences are 01100000 . . . 0110 and 1111 . . . 00, respectively. Assuming this received sequence corresponds to the only deviation from the correct all zeros path, then a finite number of channel errors could conceivably cause the decoder to choose this overall incorrect path as its decoded output sequence. The corresponding number of input bit errors could be arbitrarily large and is, in fact, equal to the number of times we traverse the self-loop around state d plus two.

The key to such catastrophic error propagation is that there occurs in the state diagram a state (in this case d) around which a self-loop exists whose corresponding output digits are all zero. This is caused by the fact that the number of connections of the shift register stages to each modulo 2 adder is *even* (in this case two connections to each stage). Hence, when the shift register stages are continually filled with ones, an even parity check on each adder forces the output sequence to be all zeros. Thus, we conclude that for binary tree codes, catastrophic error propagation will occur if the encoder is such that each adder has an even number of connections to the stages of the shift register.

Other necessary and sufficient conditions have been found by Massey and Sain[55] and Odenwalder[52]. In particular, we define the subgenerator polynomial of a convolutional code by

$$g_i(D) = g_{i1}D^{K-1} + g_{i2}D^{K-2} + \cdots g_{iK} \qquad 1 \leq i \leq v \tag{C-4}$$

where $g_{ij} = 1$ if the ith modulo 2 adder is connected to the jth shift register stage and $g_{ij} = 0$ if no such connection exists. Then, Massey and Sain have shown that catastrophic errors will result if and only if any two of these subgenerator polynomials have a common polynomial factor. More recently, Odenwalder states that catastrophic error propagation can occur if and only if the code is initially in any nonzero state and a shift of $K - 1$ input bits into the register causes all $(K - 1)v$ output digits to be zero.

Now we can begin to understand the difference between systematic and nonsystematic codes, whose definitions were introduced earlier. The fact is that systematic codes can never exhibit catastrophic error propagation. The reason for this is that any closed loop path in the state diagram (other than the self-loop around the all zero state) must contain at least one branch generated by a nonzero input data bit; hence the leading digit of the corresponding output sequence is also nonzero. This advantage of systematic codes over nonsystematic codes should not be carried too far, since only a small fraction of nonsystematic codes is catastrophic. For binary tree codes this fraction is given by $1/(2^v - 1)$.

The final criterion that we shall discuss in choosing good convolutional codes is that of maximizing the minimum free distance. One point to note is that nonsystematic codes which do not exhibit catastrophic error propagation have an equal or larger maximum value of d_{min} than systematic codes. To illustrate this point, Table C-1[53] gives the maximum value of d_{min} for systematic and nonsystematic binary codes (with no catastrophic error propagation) and $K = 2, 3, 4, 5$.

TABLE C-1 Maximum Values of d_{min}

K	*Systematic*	*Nonsystematic*
2	3	3
3	4	5
4	4	6
5	5	7

As the constraint span, K, grows larger, so does the separation between the maximum values of d_{min} for systematic and nonsystematic codes. In fact, for asymptotically large K, a systematic code of constraint span, K performs approximately the same as a nonsystematic, noncatastrophic, code of constraint span, $K(1 - \mathcal{R}_C)$[56]. A similar reduction in effective constraint span (or length) was noted earlier by Bussgang,[57] who proved his result for a specific type of suboptimum decoder.

Upper bounds on the minimum free distance have been derived by several people. McEliece and Rumsey[58] obtained an upper bound on minimum code word distance for systematic codes which was later stated for nonsystematic codes by Heller[32] and is given by

$$d_{\min} \leq \frac{2^L}{2^L - 1}\left(\frac{(L + K - 1)v}{2}\right) \tag{C-5}$$

This bound is valid for any L but becomes totally useless in the form given above for arbitrarily large L. However, since the bound as given by (C-5) is also valid for nonzero sequences of length $h < L$, then the tightest bound can be obtained by minimizing on h the functional in (C-5) with L replaced by h. Hence, for arbitrarily large L,

$$d_{\min} \leq \min_h \frac{2^h}{2^h - 1}\left(\frac{(h + K - 1)v}{2}\right) \tag{C-6}$$

This minimum occurs at a value of h which varies as log K; hence in the limit for large K,

$$d_{\min} \leq \frac{Kv}{2} \tag{C-7}$$

Letting d_H denote the largest integer less than or equal to the right-hand side of (C-6), then when d_H is odd, Odenwalder[52] has found under certain conditions an improved (tighter) bound on $d_{\min}$. Specifically, defining

$$d_E = \left\{\frac{2^{h_0}}{2^{h_0} - 1}\left(\frac{(h_0 + K - 1)v + 1}{2}\right)\right\} \tag{C-8}$$

where h_0 is the value of h which minimizes the right-hand side of (C-6)—and the braces in (C-8) denote the largest integer less than or equal to the quantity within them—then if d_H is odd, and $d_E < d_H + 1$, it follows that

$$d_{\min} \leq d_H - 1 \tag{C-9}$$

The quantity d_E as defined in (C-8) has the significance of an upper bound on the minimum distance of a code whose input sequence is of length h_0 and whose corresponding code words are increased in length by a single parity-check digit.[59]

An alternate procedure for improving (C-6) has been found by Odenwalder and is given as follows. If d_H is odd and

$$d_H > \frac{2^{h_0}}{2^{h_0} - 1}\left(\frac{(h_0 + K - 1)v - 1 + 2^{1-h_0}}{2}\right) \tag{C-10}$$

then the tighter bound of (C-9) applies.

APPENDIX D
ERROR PROBABILITY BOUNDS FOR MAXIMUM-LIKELIHOOD DECODING OF AN ARBITRARY CONVOLUTIONAL CODE

When considering the general question of error probability for convolutional codes, the two basic quantities of interest are

(a) *the first-event error probability*, P_{E1}, the probability that the first incorrect path chosen as a survivor occurs at step j, and

(b) *the bit error probability*, P_B, the ratio of the number of decoded bit errors to the total number of transmitted bits.

At present, we consider only upper bounds on error probability for an arbitrary convolutional code transmitted by PSK over the additive white Gaussian noise channel without regard for the goodness of the code.

Using a union bounding argument applied to the total number of surviving erroneous paths at step j (irrespective of whether they survived at previous steps in the tree), upper bounds for first event and bit error probability may be derived. These are given by[52,53]

$$P_{E1} < \sum_{k=d_{\min}}^{\infty} \frac{1}{2} n_k \operatorname{erfc} \sqrt{kR_s} \tag{D-1}$$

and

$$P_B < \sum_{k=d_{\min}}^{\infty} \frac{1}{2} m_k \operatorname{erfc} \sqrt{kR_s} \tag{D-2}$$

where R_s is the ratio of the energy per code word digit to the noise spectral density.

In these equations, $1/2 \operatorname{erfc} \sqrt{kR_s}$ represents the probability of error in comparing the correct path to an incorrect one which differs in k digits from it, $\{n_k\}$ are the coefficients of $T(D)$ as in (C-3), and $\{m_k\}$ are the coefficients of the derivative of $T(D, 1, N)$ with respect to N, that is,

$$\frac{\partial}{\partial N}[T(D, W, N)|_{W=1}]\bigg|_{N=1} \triangleq \frac{\partial T(D, N)}{\partial N}\bigg|_{N=1} = \sum_{k=d_{\min}}^{\infty} m^k D^k \tag{D-3}$$

Letting $k = d_{\min} + l$ and recognizing that for $x \geq 0$, $y \geq 0$,

$$\operatorname{erfc} \sqrt{\frac{x+y}{2}} \leq \exp(-y/2) \operatorname{erfc} \sqrt{\frac{x}{2}} \tag{D-4}$$

(D-1) and (D-2) may be simplified (yielding looser bounds) to[52,53]

$$P_{E1} < \frac{1}{2} \operatorname{erfc} \sqrt{d_{\min} R_s} \exp(d_{\min} R_s) T(D)|_{D=\exp(-R_s)} \tag{D-5}$$

and

$$P_B < \frac{1}{2} \operatorname{erfc} \sqrt{d_{\min} R_s} \exp(d_{\min} R_s) \frac{\partial T(D, N)}{\partial N}\bigg|_{N=1, D=\exp(-R_s)} \tag{D-6}$$

The code word digit signal-to-noise ratio R_s can be related to the bit signal-to-noise ratio R_b, through the rate, $\mathcal{R}_C$, by

$$R_b = \frac{R_s}{\mathcal{R}_C} \tag{D-7}$$

As previously mentioned, all of these bounds can be extended to general memoryless channels but with somewhat weaker results. In particular, the first two factors of (D-5) and (D-6)—whose product is always less than one—are replaced by unity, and the value of D at which evaluation of $T(D)$ and $\partial T(D, N)/\partial N$ are to be performed takes on a more general definition in terms of the transition probability densities of the channel.

One problem with union bounds such as those given is that they are reasonably tight only for high energy-to-noise ratios. In order to obtain useful bounds that have physical significance at lower R_b, one must resort to Gallager-type bounds.[60]

A second problem associated with applying these particular union bounds to codes with large constraint spans is manifested in the complexity of the calculation required to determine $T(D, N)$. Although we have not explicitly discussed the technique of this calculation, we can envision this problem when we recall the exponential growth with K_0 of the number of states in the state diagram. Hence, except for very specific encoder configurations, direct evaluation of $T(D, N)$ from the state diagram is hopelessly unwieldy.

Two further points are worthy of note. First, from (D-1) and (D-5), we note that the bounds for first-event error probability at the jth step are independent of j because we have not attempted to truncate $T(D)$ as is necessary when j is finite. Thus, strictly speaking, the bounds for P_{E1} as given by (D-1) and (D-5) apply for infinitely long codes. Since these expressions are only bounds, however, they are still valid for finite j, and their tightness obviously improves as j increases. The second point concerns the way in which first-event error probability relates to the code word error probability, denoted by P_E. For an L-bit message sequence, the probability of not choosing the correct path through the tree is given by the sum of the first-event error probabilities at step j, $j = 1, 2, \ldots, L$, since the corresponding events are mutually exclusive. Hence, a union bound on P_E can be obtained simply by multiplying the union bounds on P_{E1} by L.

PROBLEMS

5-1. For the N-ary decision problem, show that the mean and covariance of the correlator output, under the assumption that $s_1(t)$ was transmitted, are given by (5-22) and (5-23) respectively.

5-2. For phase-coherent reception of equiprobable, equal energy signals, the probability of correct word detection is found by substituting

(5-45) in (5-41) and letting $P(\mathbf{s}_i) = 1/N$ for $i = 1, 2, \ldots, N$. Show that when $N = 2$, the corresponding word error probability, $P_E(2)$, is given by (5-47).

5-3. Consider an MFSK system that employs the signal set

$$s_i(t) = \sqrt{2S}\sin\left[\left(\frac{\pi k}{T} + 2\pi i\Delta f\right)t\right] \qquad i = 1, 2, \ldots, N$$

a) Show that the minimum frequency spacing that produces orthogonality is $\Delta f = 1/2T$; hence, the effective bandwidth occupancy of the signal set is $B_e = N/2T$.

b) Using this orthogonal signal set, define a bi-orthogonal signal set and show that it's effective bandwidth occupancy is $B_e = N/4T$.

5-4. If a binary signaling scheme is characterized by the transmitted signal set

$$s_i(t) = A(t)\cos\{[\omega_0 + \pi(-1)^i\Delta f]t\} \qquad 0 \le t \le T, \quad i = 1, 2$$

then show that the signal cross-correlation coefficient is given by

$$\lambda_{12} = \frac{\int_0^T A^2(t)\cos 2\pi\Delta f t\, dt}{\int_0^T A^2(t)\, dt}$$

Neglect double frequency terms.

5-5. In binary FSK systems, the amplitude function of Prob. 5-4 can be characterized in various ways. When $A(t) = \sqrt{2S}$ for $0 \le t \le T$,

a) show that

$$\lambda_{12} = \frac{\sin 2\pi\Delta f T}{2\pi\Delta f T}$$

b) What is the minimum frequency shift, Δf, that yields orthogonal signals?

c) What value of Δf minimizes the error probability, $P_E(2)$ as given by (5-47)?

d) For the value of λ_{12} found in part c), determine how many dB in signal-to-noise ratio a PSK system outperforms a coherent FSK system.

5-6. Evaluate λ_{12} of Prob. 5-4 if

a) $$A(t) = \begin{cases} \exp(-Wt) & 0 \le t \le T \\ 0 & \text{elsewhere} \end{cases}$$

b) $$A(t) = \begin{cases} \exp[-a^2(t - T/2)^2] & 0 \le t \le T \\ 0 & \text{elsewhere} \end{cases}$$

Assume $aT \gg 1$ so that most of the tail of $A(t)$ lies in the interval $0 \le t \le T$.

5-7. Starting with the general expression for the probability of correct word detection as given by (5-41) together with (5-43) or (5-55), derive (5-50) and (5-56) corresponding to the special cases of equal energy, equiprobable, orthogonal and bi-orthogonal signal sets, respectively.

5-8. Verify Eq. (5-69) for large N. [Hint: Use the asymptotic expansion of the complementary error function for large argument; that is, for $x \gg 1$, $\text{erfc}(x) \cong \exp(-x^2)/\sqrt{\pi}\,x$. Also make use of the relation $\text{erfc}(-x) = 2 - \text{erfc}(x)$].

5-9. By establishing the joint p.d.f. $p(V_c, V_s)$ and using the relations $V = \sqrt{V_c^2 + V_s^2}$ and $\eta = \tan^{-1} V_c/V_s$, show that $p(V, \eta)$ is given by (5-83). Averaging out random variable V, reduce your result to (5-84).

5-10. Establish (5-86) by substituting (5-84) into (5-85) and averaging over the equal *a priori* probabilities of the transmitted signals. Show that (5-86) reduces to (5-87), and the average word error probability is given by (5-88).

5-11. For large signal-to-noise ratio conditions, Gilbert[15] has found an approximate formula for the average word error probability of a maximum-likelihood receiver in terms of the geometrical configuration of the transmitted signals in signal space. Applying his formula to an MPSK system, the N signals are represented as uniformly spaced points on the circumference of a circle of radius R and

$$P_E(N) \cong \frac{N_s}{2}\,\text{erfc}\left[\sqrt{R_d}\left(\frac{r_{\min}}{2R}\right)\right]$$

where $r_{\min}$ is the smallest of the set of $N(N-1)/2$ Euclidean distances between pairs of signal points, and N_s is determined as follows.

For each signal $s_i(t)$, let N_s denote the number of signals that are at the minimum distance $r_{\min}$. Then,

$$N_s = \left(\sum_{i=1}^{N} N_{s_i}\right)\Big/N$$

a) Show that $r_{\min} = 2R \sin \pi/N$

b) For $N > 2$, show that $N_s = 2$

c) From the results of parts a) and b), verify Eq. (5-91).

5-12. If in Prob. 5-11, N is also assumed to be large, show that for a given average word error probability, an increase in N by a factor of 2 requires a 6 dB increase in R_d.

5-13. A general representation for the L-orthogonal signal set is given by[8,9]

$$\begin{aligned} s_i(t) &= \sqrt{S}\,[x_m(t)\cos\theta_l + y_m(t)\sin\theta_l] \qquad 0 \le t \le T \\ m &= 1, 2, \ldots, M \\ l &= 1, 2, \ldots, L \\ i &= (m-1)L + l \end{aligned}$$

with $\theta_l = 2(l-1)\pi/L$ and

$$\frac{1}{T}\int_0^T x_j(t)x_n(t)\,dt = \frac{1}{T}\int_0^T y_j(t)y_n(t)\,dt = \begin{cases}1 & j=n\\ 0 & j\neq n\end{cases}$$

$$\frac{1}{T}\int_0^T x_j(t)y_n(t)\,dt = 0 \text{ for all } j \text{ and } n.$$

If

$$x_m(t) = \sqrt{2}\sin\left[\frac{\pi(k+m)t}{T}\right] \qquad L = 1, 2$$

$$y_m(t) = \sqrt{2}\cos\left[\frac{\pi(k+m)t}{T}\right] \qquad m = 1, 2, \ldots, M$$

and

$$x_m(t) = \sqrt{2}\sin\left[\frac{\pi(k+2m)t}{T}\right] \qquad L > 2$$

$$y_m(t) = \sqrt{2}\cos\left[\frac{\pi(k+2m)t}{T}\right] \qquad m = 1, 2, \ldots, M$$

show that the signal set $\{s_i(t)\}$ is L-orthogonal. Assume that k is any integer.

5-14. a) In terms of the L-orthogonal signal representation of Prob. 5-13, let $\{x_m(t)\}$ be a set of M non-overlapping sinusoidal pulses defined over the interval $0 \le t \le T/2$ and $\{y_m(t)\}$ be a set of M non-overlapping sinusoidal pulses defined over the interval $T/2 \le t \le T$; i.e.,

$$x_m(t) = 2\sqrt{M}\sin\left(\frac{2\pi kMt}{T}\right);\ \frac{(m-1)T}{2M} \le t \le \frac{mT}{2M}$$

$$y_m(t) = 2\sqrt{M}\sin\left(\frac{2\pi kMt}{T}\right);\ \frac{(M+m-1)T}{2M} \le t \le \frac{(M+m)T}{2M}$$

$$m = 1, 2, \ldots, M$$

Show that the resulting signal set $\{s_i(t)\}$ is L-orthogonal.

b) If M is a power of 2, then another pair of orthogonal signal sets for $\{x_m(t)\}$ and $\{y_m(t)\}$ can be modelled after the Hadamard design for an orthogonal code. In particular, we define

$$x_m(t) = \sum_{i=1}^{2M}(-1)^{b_i^{(m)}}w\left(t - \frac{(i-1)T}{2M}\right)$$

$$y_m(t) = \sum_{i=1}^{2M}(-1)^{b_i^{(M+m)}}w\left(t - \frac{(i-1)T}{2M}\right) \qquad m = 1, 2, \ldots, M \quad 0 \le t \le T$$

where $w(t)$ is an arbitrary waveform of energy $T/2M$ and non-zero only over the interval $0 \le t \le T/2M$, and $b_i^{(m)}$ is the ith element in the mth row of a $2M \times 2M$ Hadamard matrix. Show that the resulting signal set $\{s_i(t)\}$ is L-orthogonal.

5-15. Show that for the special cases of $L = 2$ and $L = 4$, the error probability performance of an L-orthogonal signal set of $N = ML$ signals defined by (5-93) is equivalent to that of a bi-orthogonal signal set of N signals.

5-16. Consider a set of N equiprobable (but not necessarily equal energy) signals, where each signal in the set has the two-dimensional representation

$$s_i(t) = \sqrt{\frac{2}{T}}\, a_i \cos \omega_0 t - \sqrt{\frac{2}{T}}\, b_i \sin \omega_0 t \qquad \begin{array}{l} 0 \leq t \leq T \\ i = 1, 2, \ldots, N \end{array}$$

or equivalently is described by the vector

$$\mathbf{s}_i = a_i + jb_i$$

Such an N-ary signal set can be used to characterize a combined amplitude and phase-shift-keyed modulation technique[75-78] (see Prob. 5-17). If a signal from this set is transmitted over an additive, white Gaussian noise channel and detected by a maximum-likelihood receiver, then an upper bound on the word error probability performance of this receiver has been derived by Gilbert[15] and has the form

$$P_E(N) \leq \frac{1}{N} \sum_{i=1}^{N} \sum_{\substack{j=1 \\ i \neq j}}^{N} \frac{1}{2} \operatorname{erfc}\left[\frac{|\mathbf{s}_i - \mathbf{s}_j|}{2\sqrt{N_0}}\right]$$

a) Show that the average power of the signal set is

$$S_{\text{ave}} = \frac{1}{N} \sum_{i=1}^{N} \frac{a_i^2 + b_i^2}{T}$$

b) Denoting the average signal power to average noise power ratio by $R_{d_{\text{ave}}} = S_{\text{ave}}T/N_0$, show that the Gilbert upper bound can be written in the form

$$P_E(N) \leq \frac{1}{N} \sum_{i=1}^{N} \sum_{\substack{j=1 \\ i \neq j}}^{N} \frac{1}{2} \operatorname{erfc}\left[\sqrt{R_{d_{\text{ave}}}}\left(\frac{|\mathbf{s}_i - \mathbf{s}_j|}{2\sqrt{\frac{1}{N}\sum_{i=1}^{N} |\mathbf{s}_i|^2}}\right)\right]$$

5-17. a) Evaluate the Gilbert bound of part b) of Prob. 5-16 for these signal sets.

b) For an average signal-to-noise ratio $R_{d_{\text{ave}}} = 10$ dB, compare the word error probability performance of these four signal sets [using the bounds evaluated in part a)] to that of a polyphase signal set with eight signals. For the polyphase signal set, use the approximate expression for average word error probability given by (5-91).

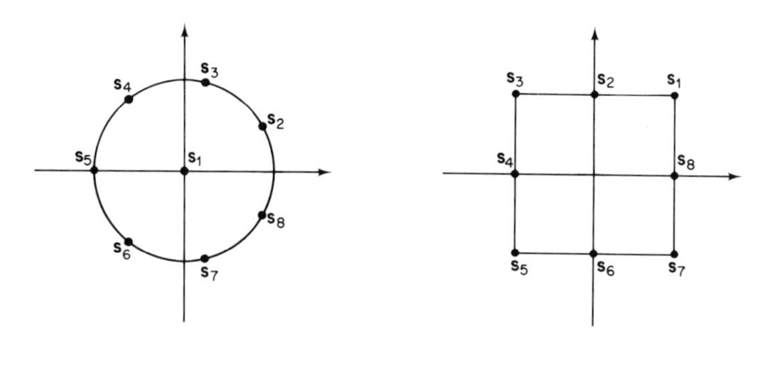

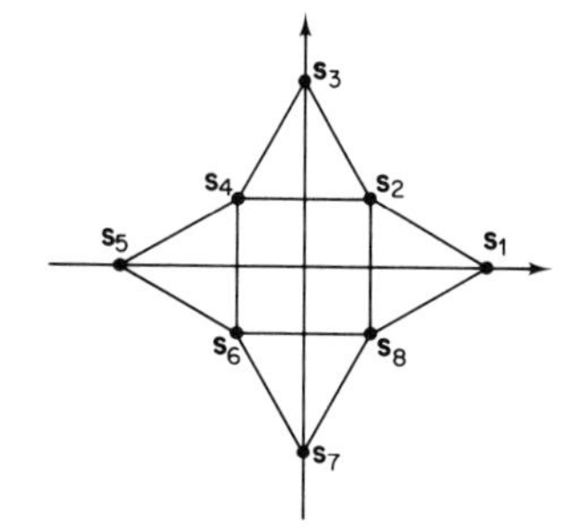

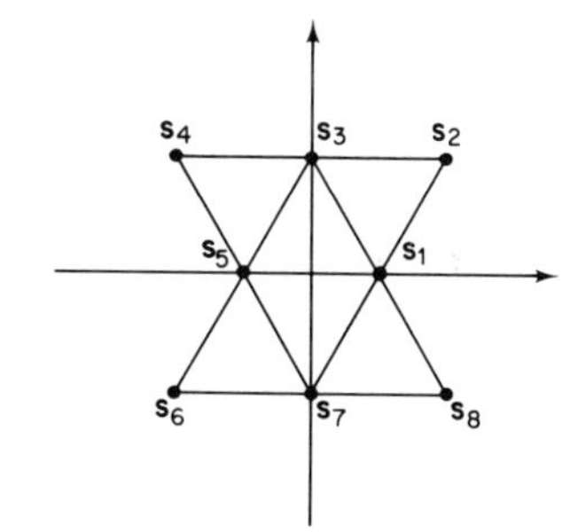

5-18. a) Following an approach similar to that given in Appendix B, and starting with (5-109) and (5-110) verify (5-111).

b) Starting with (5-112), verify (5-113) and its special cases as given by (5-114) and (5-115).

5-19. If the approximation technique of Prob. 5-11 is applied to a system employing coherent detection of differentially encoded MPSK signals, show that for large signal-to-noise ratio, the average word error probability can be approximated by

$$P_E(N) = 2\,\mathrm{erfc}\left(\sqrt{R_d}\sin\frac{\pi}{N}\right)\left[1 - \mathrm{erfc}\left(\sqrt{R_d}\sin\frac{\pi}{N}\right)\right]$$

Assume that the only significant word errors are those which are adjacent (in signal space) to the transmitted signal.

5-20. If in Prob. 5-19, N is also assumed large, show that for a given average word error probability, an increase in N by a factor of 2 requires a 6 dB increase in R_d. Assume that the given error probability is chcsen such that $\mathrm{erfc}(\sqrt{R_d}) \sin \pi/N) \ll 1$.

5-21. Consider a *continuous phase binary FSK* (CPFSK) system wherein the phase is made continuous from bit to bit. In such a system the transmitted signal during the ith bit time is characterized by

$$s_k(t) = \sqrt{2S}\cos\{[\omega_0 + \pi(-1)^k \Delta f]t + \phi_i\} \qquad iT \leq t \leq (i+1)T \quad k = 1, 2$$

where ϕ_i is the carrier phase at the beginning of the ith interval and is assumed known. Because of the phase continuity constraint, the initial phase of the sinusoid corresponding to a particular bit depends on previous data bits; hence, the optimum decision rule requires observing the received signal over more than a single bit interval. For an observation over L bit times, beginning with the ith bit, a maximum-likelihood decision on the ith bit is obtained by correlating the received waveform with the 2^L possible CPFSK waveforms. Bit i is then selected as the corresponding bit in the highest correlated waveform. This procedure also determines ϕ_{i+1} which, by identical processing, is used to determine bit $i + 1$. Letting $L = 2$ and $s_{jk}(t)$, denote the transmitted signal corresponding to the data sequence jk where $j, k = 0, 1$.

a) Show that the probability of incorrectly detecting bit i, given that 01 is the transmitted sequence in the ith and $(i + 1)$st intervals, is upper bounded by

$$P(01, 10) + P(01, 11)$$

where $P(jk, mn)$ denotes the probability that the received signal correlates better with $s_{mn}(t)$ than with $s_{jk}(t)$ given that $s_{jk}(t)$ was transmitted.

b) Generalize the result of part a) to show that a union bound on the average bit error probability is given by

$$P_B = \frac{1}{4} \sum_{j=0}^{1} \sum_{\substack{k=0 \\ j \neq m}}^{1} \sum_{n=0}^{1} P(jk, mn)$$

c) If, for simplicity, we let $i = 0$ and consider the observation interval $0 \leq t \leq 2T$, then show that $P(jk, mn)$ is given by

$$P(jk, mn) = \tfrac{1}{2}\operatorname{erfc}\sqrt{R_d(1 - \lambda_{jkmn})}$$

where $R_d \triangleq ST/N_0$ and

$$\lambda_{jkmn} \triangleq \frac{\int_0^{2T} s_{jk}(t)s_{mn}(t)\,dt}{2ST}$$

d) Evaluate λ_{jkmn} for the eight combinations of j, k, m, and n of interest in computing P_B.

e) If for any Δf, $\lambda_{\max}$ is the maximum of all λ_{jkmn} found in part d), show that

$$P_B < \operatorname{erfc}\sqrt{R_d(1 - \lambda_{\max})}$$

Further details pertaining to the performance of continuous phase binary FSK systems can be found in Ref. 79.

5-22. The suboptimum receiver of Fig. 5-19 can be generalized to handle the L-orthogonal signal set representation of Prob. 5-13 and coherent or differentially coherent detection of the phase θ_l. In particular, we consider three possible configurations for the phase estimator in the mth

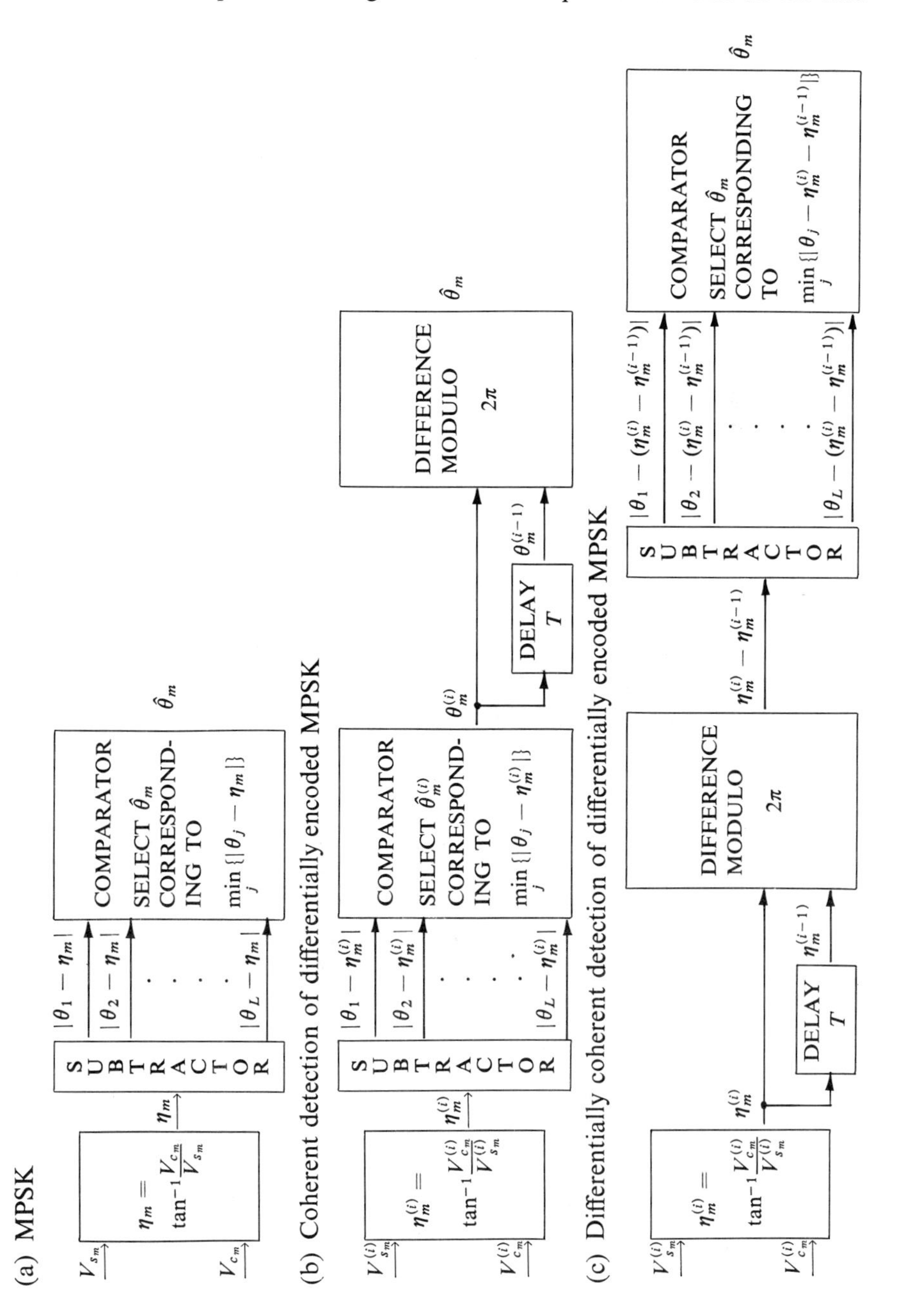

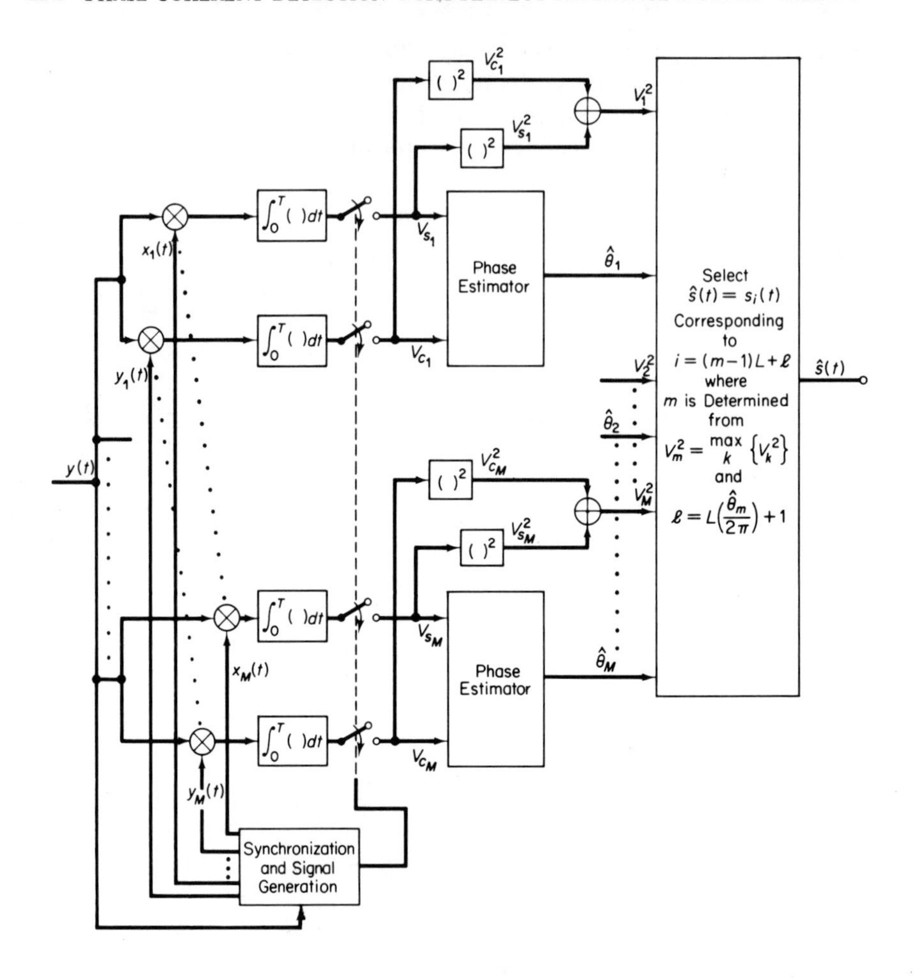

branch corresponding respectively to the cases of a) MPSK (maximum-likelihood detection of θ_l), b) coherent MPSK with differential encoding of θ_l, and c) differentially coherent MPSK (MDPSK) with differential encoding of θ_l. The error probability performance of case a) was considered in the text and is given by (5-106). Show that for cases b) and c), the average word error probability is given by

$$P_E(N) = P_E(M)|_{NON} + P_E(L)|_\theta - [P_E(M)|_{NON}][P_E(L)|_\theta]$$

where for case b), $P_E(L)|_\theta$ is given by (5-113) with N replaced by L and for case c), $P_E(L)|_\theta$ is given by (5-118) with N replaced by L.

5-23. Let P_l and P_u represent lower and upper bounds respectively on the average word error probability performance of an N-ary modulation system. Show that $P_l/\log_2 N$ and P_u are respectively lower and upper bounds on the system's bit error probability performance.[80]

5-24. For an N-ary modulation system with equal energy signals, the rank of the correlation matrix of the signal set (often called the *dimensionality of the signal space*) can be used to define system bandwidth. In particular, if $D \leq N$ is the rank of the $N \times N$ correlation matrix, then for large D, Pollak and Landau[81] have shown that $D \cong 2BT$ where T is the signal duration and B is our measure of system bandwidth. Based upon this definition of bandwidth,

a) show that $B = N/2T$ for an orthogonal signal set and $B = N/4T$ for a bi-orthogonal set. Note that these results are equivalent to the effective bandwidth occupancy of these signal sets as in Prob. 5-3.
b) Show that for the transorthogonal signal set, $B = (N - 1)/2T$.
c) For a polyphase signal set, show that

$$B = \begin{cases} 1/2T & N = 2 \\ 1/T & N > 2 \end{cases}$$

d) For an L-orthogonal signal set made up of M frequencies with L phases per frequency, show that for $L \geq 4$, $B = M/T$.

5-25. Consider an N-ary modulation system with symmetry such that for a given row of the correlation matrix, any other row can be obtained by a permutation of the given row. If the transmitted signals are equiprobable, then lower and upper bounds on the average word error probability performance are given by[15,80]

$$\frac{1}{2} \operatorname{erfc}\sqrt{K_N R_b/2} \leq P_E(N) \leq \frac{(N-1)}{2} \operatorname{erfc}\sqrt{K_N R_b/2}$$

where

$$K_N \triangleq (1 - \lambda_{\max}) \log_2 N$$

with $\lambda_{\max}$ the maximum of the off-diagonal correlation coefficients.

a) Evaluate K_N for orthogonal, bi-orthogonal, transorthogonal, polyphase, and L-orthogonal modulations.
b) Using the bandwidths found in Prob. 5-24, compute the ratio $\mathcal{R}_b/B$ for the signal sets in part a).
c) For $N = 64$, make a system comparison among the signal sets being considered by tabulating K_{64} vs. $\mathcal{R}_b/B$. For the L-orthogonal signal set, choose $L = 2^l$ where $l = 1, 2, \ldots, 6$.

5-26. Consider a convolutional code whose matrix of parity-check equations is given by

$$\begin{pmatrix} 1 & 0 \\ 1 & 1 \end{pmatrix}$$

a) Sketch the convolutional encoder.
b) Assuming that one information bit at a time is transmitted through the encoder, what is the code rate? Assume that a large number of information bits have been transmitted through the encoder.
c) Draw the code tree, trellis, and state diagrams for the convolutional code generated by the parity-check equation matrix given.
d) For the input information bit sequence 1101101110, what is the output digit sequence? Assume that the contents of the shift register are initially cleared, that is, all zeros.

5-27. Repeat Prob. 5-26 if the convolutional code is generated by the parity-check equation matrix

$$\begin{pmatrix} 1 & 1 & 1 & 1 \\ 1 & 1 & 0 & 1 \end{pmatrix}$$

REFERENCES

1. Van Trees, H. L., *Detection, Estimation and Modulation Theory, Part I*, John Wiley & Sons, Inc., New York, N. Y., 1968.

2. Weber, C. L., *Elements of Detection and Signal Design*, McGraw-Hill Book Company, New York, N. Y., 1968.

3. Wozencraft, J. M., and Jacobs, I. M., *Principles of Communication Engineering*, John Wiley & Sons, Inc., New York, N. Y., 1965.

4. Helstrom, C. W., *Statistical Theory of Signal Detection*, Pergamon Press, New York, N. Y., 1960.

5. Davenport, W. B., and Root, W. L., *An Introduction to the Theory of Random Signals and Noise*, McGraw-Hill Book Company, New York, N. Y., 1958.

6. Kotelnikov, V. A., *The Theory of Optimum Noise Immunity*, McGraw-Hill Book Company, New York, N. Y., 1959.

7. Woodward, P. M., *Probability and Information Theory With Applications to Radar*, McGraw-Hill Book Company, New York, N. Y., 1954.

8. Reed, I. S., and Scholtz, R. A., "*N*-Orthogonal Phase-Modulated Codes," *IEEE Transactions on Information Theory*, Vol. IT-12, No. 3 (July, 1966) 388–395.

9. Stiffler, J. J., and Viterbi, A. J., "Performance of *N*-Orthogonal Codes," *IEEE Transactions on Information Theory*, Vol. IT-13, No. 3 (July, 1967) 521–522.

10. Golomb, S. W., et. al., *Digital Communications With Space Applications*, Prentice-Hall, Inc., Englewood Cliffs, N. J., 1964.

11. Viterbi, A. J., "On Coded Phase-Coherent Communications," *IRE Transactions on Space Electronics and Telemetry*, Vol. SET-7, No. 1 (March, 1961) 3–14.

12. ———, *Principles of Coherent Communications*, McGraw-Hill Book Company, New York, N. Y., 1966.

13. Fleck, J. T., and Trabka, E. A., "Embodiments of the Maximum Likelihood Receiver for Detection of Coherent Pulsed Phase Shift Keyed Signals in the Presence of Additive White Gaussian Noise," Detect Memo No. 5A in *Investigation of Digital Data Communication Systems*, Report No. UA-1420-S-1, J. G. Lawton, ed., Cornell Aeronautical Laboratory, Inc., Buffalo, N. Y., January, 1961. Available as ASTIA Document No. AD 256 584.

14. ———, "Comments on *M*-State Coherent Digital Communications Systems," Detect Memo No. 15 in *Investigation of Digital Data Communication Systems*, Report No. UA-1420-S-1, J. G. Lawton, ed., Cornell Aeronautical Laboratory, Inc., Buffalo, N. Y., January, 1961. Available as ASTIA Document No. AD 256 584.

15. Gilbert, E. N., "A Comparison of Signaling Alphabets," *Bell System Technical Journal*, Vol. 31 (May, 1952) 504–522.

16. Cahn, C. R., "Performance of Digital Phase-Modulation Communication Systems," *IRE Transactions on Communications Systems*, Vol. CS-7, No. 1 (May, 1959) 3–6.

17. Lucky, R. W., Salz, J., and Weldon, E. J., Jr., *Principles of Data Communication*, McGraw-Hill Book Company, New York, N. Y., 1968.

18. Fleck, J. T., and Trabka, E. A., "Error Probabilities of Multiple-State Differentially Coherent Phase Shift Keyed Systems in the Presence of White, Gaussian Noise," Detect Memo No. 2A in *Investigation of Digital Data Communication Systems*, Report No. UA-1420-S-1, J. G. Lawton, ed., Cornell Aeronautical Laboratory, Inc., Buffalo, N. Y., January, 1961. Available as ASTIA Document No. AD 256 584.

19. Arthurs, E., and Dym, H., "On the Optimum Detection of Digital Signals in the Presence of White Gaussian Noise—A Geometric Interpretation and a Study of Three Basic Data Transmission Systems," *IRE Transactions on Communication Systems*, Vol. CS-10, No. 4 (December, 1962) 336–372, also TM 3250, Mitre Corporation, Bedford, Mass.

20. Bussgang, J. J., and Leiter, M., "Error Rate Approximations for Differential Phase-Shift Keying," *IEEE Transactions on Communication Systems*, Vol. CS-12, No. 1 (March, 1964) 18–27, also SR-103, Mitre Corporation, Bedford, Mass.

21. Lawton, J. G., "Comparison of Binary Data Transmission Systems," *Proceedings of the National Convention on Military Electronics*, (1958) 54–61,

22. Doelz, M., Heald, E., and Martin, D., "Binary Data Transmission Techniques for Linear Systems," *Proceedings of the IRE*, Vol. 45 (May, 1957) 656–661.

23. Lawton, J. G., "Theoretical Error Rates of Differentially Coherent Binary and Kineplex Data Transmission Systems," *Proceedings of the IRE* (Correspondence), Vol. 47 (February, 1959) 333–334.

24. Cahn, C. R., "Comparison of Coherent and Phase-Comparison Detection of a Four-Phase Signal," *Proceedings of the IRE* (Correspondence), Vol. 47 (September, 1959) 1662.

25. ELIAS, P., "Coding for Noisy Channels," *IRE Convention Record*, Part 4 (1955) 37–46.

26. YUDKIN, H. L., "Channel State Testing in Information Decoding," Ph. D. Dissertation, Massachusetts Institute of Technology, Cambridge, Mass. 1964.

27. VITERBI, A. J., "Error Bounds for Convolutional Codes and an Asymptotically Optimum Decoding Algorithm," *IEEE Transactions on Information Theory*, Vol. IT-13, No. 2 (April, 1967) 260–269.

28. FORNEY, G. D., JR., "Coding System Design for Advanced Solar Missions," Final Report on Contract NAS 2–3637 (submitted to NASA Ames Research Center), Watertown, Mass., Codex Corp. (December, 1967).

29. VITERBI, A. J., and ODENWALDER, J. P., "Further Results on Optimal Decoding of Convolutional Codes," *IEEE Transactions on Information Theory*, Vol. IT-15, No. 6 (November, 1969) 732–734.

30. OMURA, J. K., "On the Viterbi Decoding Algorithm," *IEEE Transactions on Information Theory*, Vol. IT-15, No. 1 (January, 1969) 177–179.

31. FORNEY, G. D., JR., "Convolutional Codes I: Algebraic Structure," *IEEE Transactions on Information Theory*, Vol. IT-16, No. 6 (November, 1970) 720–738.

32. HELLER, J. A., "Short Constraint Length Convolutional Codes," Jet Propulsion Laboratory, Pasadena, Calif. SPS 37–54, Vol. III (December, 1968) 171–177.

33. JELINEK, F., and BAHL, L. R., "Maximum Likelihood and Sequential Decoding of Short Constraint Length Convolutional Codes," *Proceedings of the Seventh Annual Allerton Conference on Circuit and System Theory*, Monticello, Ill., (October, 1969) 130–139.

34. HELLER, J. A., "Improved Performance of Short Constraint Length Convolutional Codes," Jet Propulsion Laboratory, Pasadena, Calif., SPS 37–56, Vol. III (April, 1969) 83.

35. WOZENCRAFT, J. M., "Sequential Decoding for Reliable Communication," *IRE Convention Record*, Part 2 (1957) 11–25, also Technical Report 325, MIT Research Laboratory of Electronics, Cambridge, Mass.

36. WOZENCRAFT, J. M. and REIFFEN, B., *Sequential Decoding*, Technology Press of the Massachusetts Institute of Technology and John Wiley & Sons, Inc., New York, N. Y., 1961.

37. REIFFEN, B., "Sequential Decoding for Discrete Input Memoryless Channels," *IRE Transactions on Information Theory*, Vol. IT-8, No. 2 (April, 1962) 208–220.

38. JACOBS, I. M., "Sequential Decoding for Efficient Communication from Deep Space," *IEEE Transactions on Communication Technology*, Vol. COM-15, No. 4 (August, 1967) 492–501.

39. FANO, R. M., "A Heuristic Discussion of Probabilistic Decoding," *IEEE Transactions on Information Theory*, Vol. IT-9, No. 2 (April, 1963) 64–74.

40. JELINEK, F., "A Fast Sequential Decoding Algorithm Utilizing a Stack," *IBM Journal of Research and Development*, Vol. 13, No. 6 (November, 1969) 675–685.

41. SAVAGE, J. E., "Sequential Decoding—The Computational Problem," *Bell System Technical Journal*, Vol. 45 (January, 1966) 149–176.

42. ———, "The Distribution of the Sequential Decoding Computation Time," *IEEE Transactions on Information Theory*, Vol. IT-11, No. 2 (April, 1966) 143–147.

43. JORDAN, K. L., "The Performance of Sequential Decoding in Conjunction with Efficient Modulation," *IEEE Transactions on Communication Technology*, Vol. COM-14, No. 3 (June, 1966) 283–297.

44. JACOBS, I. M., and BERLEKAMP, E. R., "A Lower Bound to the Distribution of Computation for Sequential Decoding," *IEEE Transactions on Information Theory*, Vol. IT-13, No. 2 (April, 1967) 167–174.

45. HELLER, J. A., "A Model for Buffer Overflow in Sequential Decoding," Jet Propulsion Laboratory, Pasadena, Calif., SPS 37–56, Vol. III (April, 1969) 78–83.

46. FALCONER, D. D., and NIESSEN, C. W., *Simulation of Sequential Decoding for a Telemetry Channel*, Quarterly Progress Report, MIT Research Laboratory of Electronics, Cambridge, Mass. (January, 1966) 183–193.

47. GRAY, R. M., "Simulation of Sequential Decoding with Decision-Directed Channel Measurement," S. M. Thesis, Massachusetts Institute of Technology, Cambridge, Mass., 1966.

48. HELLER, J. A., "Description and Operation of a Sequential Decoder Simulation Program," Jet Propulsion Laboratory, Pasadena, Calif., SPS 37–58, Vol. III (August, 1969) 36–42.

49. MASSEY, J. L., *Threshold Decoding*, MIT Press, Cambridge, Mass., 1963.

50. FALCONER, D. D., "A Hybrid Sequential and Algebraic Decoding Scheme," Ph.D. Dissertation, Massachusetts Institute of Technology, Cambridge, Mass., 1966.

51. HUBBARD, F. L., and JELINEK, F., "Practical Sequential Decoding and a Simple Hybrid Scheme," *Proceedings of the First International Conference on System Sciences*, Honolulu, Hawaii (January, 1968) 478–481.

52. ODENWALDER, J. P., "Optimal Decoding of Convolutional Codes," Ph.D. Dissertation, University of California at Los Angeles, Los Angeles, Calif., 1970.

53. VITERBI, A. J., "Convolutional Codes and Their Performance in Communication Systems," *IEEE Transactions on Communication Technology* (*Special Issue on Error Correcting Codes*), Vol. COM-19, No. 5 (October, 1971) 751–772.

54. MASSEY, J. L., "Catastrophic Error-Propagation in Convolutional Codes," *Proceedings of the Eleventh Midwest Symposium on Circuit Theory*, Notre Dame, Ind. (May, 1968).

55. MASSEY, J. L., and SAIN, M. K., "Inverses of Linear Sequential Circuits," *IEEE Transactions on Computers*, Vol. C-17, No. 2 (April, 1968) 330–337.

56. HELLER, J. A., and BUCHER, E. A., "Error Probability Bounds for Systematic Convolutional Codes," *IEEE Transactions on Information Theory*, Vol. IT-16, No. 2 (March, 1970) 219–224.

57. BUSSGANG, J. J., "Some Properties of Binary Convolutional Code Generators," *IEEE Transactions on Information Theory*, Vol. IT-11, No. 1 (January, 1965) 90–100.

58. McEliece, R., and Rumsey, H. C., "Capabilities of Convolutional Codes," Jet Propulsion Laboratory, Pasadena, Calif., SPS 37–50, Vol. III (April, 1968) 248–251.

59. Layland, J., and McEliece, R., "An Upper Bound on the Free Distance of a Tree Code," Jet Propulsion Laboratory, Pasadena, Calif., SPS 37–62, Vol. III (April, 1970) 63–64.

60. Gallager, R. G., "A Simple Derivation of the Coding Theorem and Some Applications," *IEEE Transactions on Information Theory*, Vol. IT-11, No. 1 (January, 1965) 3–18.

61. Layland, J., "Performance of Short Constraint Length Convolutional Codes and a Heuristic Code-Construction Algorithm," Jet Propulsion Laboratory, Pasadena, Calif., SPS 37–64, Vol. II (August, 1970) 40–44.

62. Golomb, S. W., Gordon, B., and Welch, L. R., "Comma-Free Codes," *Canadian Journal of Mathematics*, Vol. 10 (1958), 202–209.

63. Kendall, W. B., and Reed, I. S., "Path-Invariant Comma-Free Codes," *IRE Transactions on Information Theory*, Vol. IT-8, No. 4 (October, 1962) 350–355.

64. Jiggs, B. H., "Recent Results in Comma-Free Codes," *Canadian Journal of Mathematics*, Vol. 15 (1963), 178–187.

65. Eastman, W. L., and Even, S., "On Synchronizable and PSK-Synchronizable Block Codes," *IEEE Transactions on Information Theory*, Vol. IT-10, No. 4 (October, 1964) 351–356.

66. Eastman, W. L., "On the Construction of Comma-Free Codes," *IEEE Transactions on Information Theory*, Vol. IT-11, No. 2 (April, 1965) 263–266.

67. Golomb, S. W., and Gordon, B., "Codes with Bounded Synchronization Delay," *Information and Control*, Vol. 8 (August, 1965) 355–376.

68. Scholtz, R. A., "Codes with Synchronization Capability," *IEEE Transactions on Information Theory*, Vol. IT-12, No. 2 (April, 1966) 135–142.

69. ———, "A Noiseless Coding Theorem for Synchronizable Codes," *Proceedings of the First International Conference on System Sciences*, Honolulu, Hawaii (January, 1968) 786–788.

70. ———, "Maximal and Variable Word-Length Comma-Free Codes," *IEEE Transactions on Information Theory*, Vol. IT-15, No. 2 (March, 1969) 300–306.

71. Stiffler, J. J., "Self-Synchronizing Binary Telemetry Codes," Ph.D. Dissertation, California Institute of Technology, Pasadena, Calif., 1962.

72. Baumert, L. D., and Rumsey, H. C., Jr., "The Index of Comma Freedom for the Mariner Mars 1969 High Data Rate Telemetry Code," Jet Propulsion Laboratory, Pasadena, Calif., SPS 37–46, Vol. IV (August, 1967) 221–226.

73. ———, "The Maximum Indices of Comma Freedom for the High Data Rate Telemetry Codes," Jet Propulsion Laboratory, Pasadena, Calif., SPS 37–51, Vol. III (June, 1968) 215–217.

74. Layland, J. W., "Synchronizability of Convolutional Codes," Jet Propulsion Laboratory, Pasadena, Calif., SPS 37–64, Vol. II (August, 1970) 44–50.

75. Cahn, C. R., "Combined Digital Phase and Amplitude Modulation Communication Systems", *IRE Transactions on Communications Systems*, Vol. CS-8, No. 3 (September, 1960) 150–154.

76. CAMPOPIANO, C. N., and GLAZER, B. G., "A Coherent Digital Amplitude and Phase Modulation Scheme", *IRE Transactions on Communications Systems*, Vol. CS-10, No. 1 (March, 1962) 90–95.

77. LUCKY, R. W., and HANCOCK, J. C., "On the Optimum Performance of N-ary Systems having Two Degrees of Freedom", *IRE Transactions on Communications Systems*, Vol. CS-10, No. 2 (June, 1962) 185–192.

78. THOMAS, C. M., "Amplitude-Phase-Keying with M-ary Alphabets: A Technique for Bandwidth Reduction", *Proceedings of the International Telemetering Conference*, Los Angeles, Calif. (October, 1972) 289–300.

79. PELCHAT, M. G., DAVIS, R. C., and LUNTZ, M. B., "Coherent Demodulation of Continuous Phase Binary FSK Signals", *Proceedings of the International Telemetering Conference*, Washington, D.C. (September, 1971) 181–190.

80. JACOBS, I., "Comparison of M-ary Modulation Systems", *Bell System Technical Journal*, Vol. 46, No. 5 (May, 1967) 843–864.

81. LANDAU, H. J., and POLLAK, H. O., "Prolate Spheroidal Wave Functions, Fourier Analysis and Uncertainty-III: The Dimension of the Space of Essentially Time- and Band-Limited Signals", *Bell System Technical Journal*, Vol. 41, No. 4 (July, 1962) 1295–1336.

Chapter 6

PHASE-COHERENT DETECTION WITH NOISY REFERENCE SIGNALS

6-1. INTRODUCTION

In studying the problem of ideal coherent detection in the previous chapter, we have tacitly assumed that the communication system is *perfectly synchronized.* By this we mean the receiver possessed an exact knowledge of the instants in time at which modulation might or might not change its state, as well as an exact replica of the transmitted carrier and/or subcarrier reference. The problem of deriving at the receiver a knowledge of the instants in time at which modulation may change state is referred to as *symbol synchronization*; *carrier* (subcarrier) *synchronization* deals with the receiver's attempt to obtain estimates of the phase and frequency of the transmitted carrier (subcarrier).

Unfortunately, in any practical communication system the sychronization signals needed to perform a coherent detection are not known exactly since they are derived at the receiver in the presence of noise. For example, the carrier reference might be derived at the receiver by means of a PLL; consequently, the reference phase and frequency are random. Hence, the sychronization signals required in the operation of a coherent detector are said to be *noisy* in that they possess random time and phase jitter.

The immediate effect of a *noisy reference signal* is degradation in system performance from the ideal. More specifically, a noisy carrier or subcarrier reference used for demodulating the data will cause an increase in error probability relative to the results derived in the previous chapter. Thus, our intent is to develop a theory that will enable us to modify these ideal detection results so as to apply to a practical communication system.

Frequently, a communications engineer translates the increase in error probability due to noisy synchronization references to an equivalent increase in signal power required to produce the same error probability as obtainable in a perfectly synchronized system. This increase in required signal power is referred to as the *noisy reference loss*. The uncertainty of the instants at which modulation changes state is referred to as *symbol sync jitter* and causes additional degradation in system performance above that resulting from noisy carrier reference signals. Evaluation of the performance degradation due to symbol sync jitter is treated in Chap. 9.

This chapter begins with a study of the p.d.f.'s of the decision variables that appear at the data detector output when a noisy synchronization reference is present in the receiver. After this, we assess the probability that a binary decision is in error for two carrier-tracking methods—specifically the PLL and the squaring or Costas loop mechanizations. System performance is evaluated for three cases of great practical interest: (1) loop phase error varies slowly over the duration of a symbol, (2) phase error varies rapidly during a symbol interval, and (3) phase error varies at a rate somewhere in between cases (1) and (2). The system parameter used as an indicator of transition between cases (1) and (2) is the ratio of system data rate to the bandwidth of the carrier-tracking loop. Finally, the error probability performance of N-ary signal sets to be coherently detected with noisy reference signals is treated. The results given again follow directly from their ideal counterparts in the previous chapter.

6-2. SYSTEM MODEL

As discussed in Chap. 5, optimum detection (in the sense of minimizing error probability) of deterministic signals in white Gaussian noise requires the receiver to perform a cross-correlation of the received signal, $x(t)$, with each possible transmitted signal $s_j(t)$; $j = 1, 2, \ldots N$, and make a decision as to what was transmitted by choosing the largest result. Basically, this cross-correlation operation can be performed by a multiplier-integrator combination with readout at time $t = T$ where T is the duration of the kth transmitted signal, or by a matched filter with readout at $t = T$. For a perfectly matched filter, the carrier reference signal, $r(t)$, is assumed to be exactly in phase with the signal component of the received waveform—that is, $r(t) = \sqrt{2}\cos(\omega_0 t + \theta)$—and the readout instances are known *a priori* at the receiver. Fig. 6-1 depicts this situation for the case $N = 2$.

A functional diagram of a correlation detector for binary signals that uses a noisy carrier reference for detection is illustrated in Fig. 6-2. The detector shown is of the integrate-and-dump type and can be mechanized with an operational amplifier of sufficient stability. The difference between the correlation detector of Fig. 6-1 and that of Fig. 6-2 is exhibited by the relation-

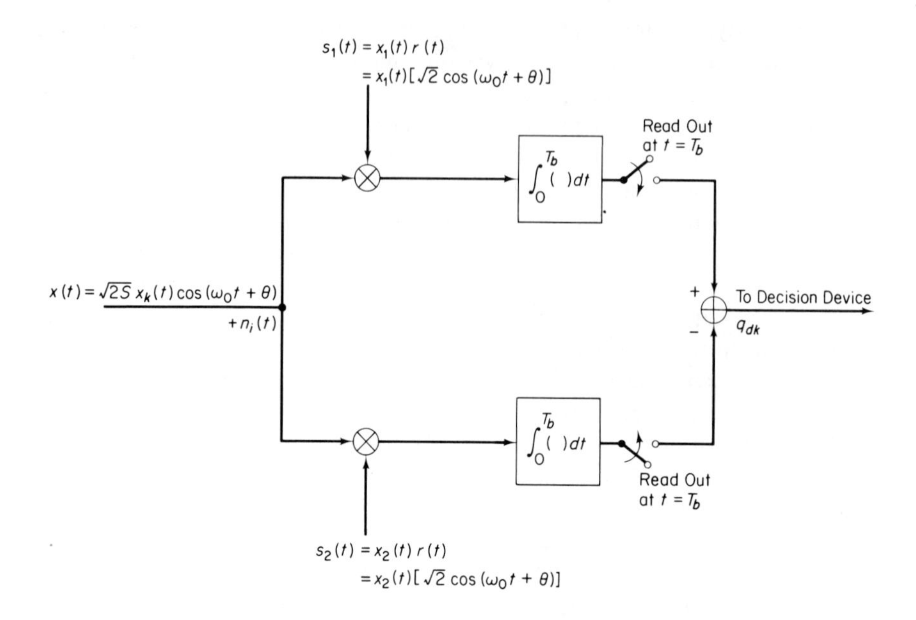

Fig. 6-1 Coherent detector; no mismatching, i.e., $\varphi = 0$

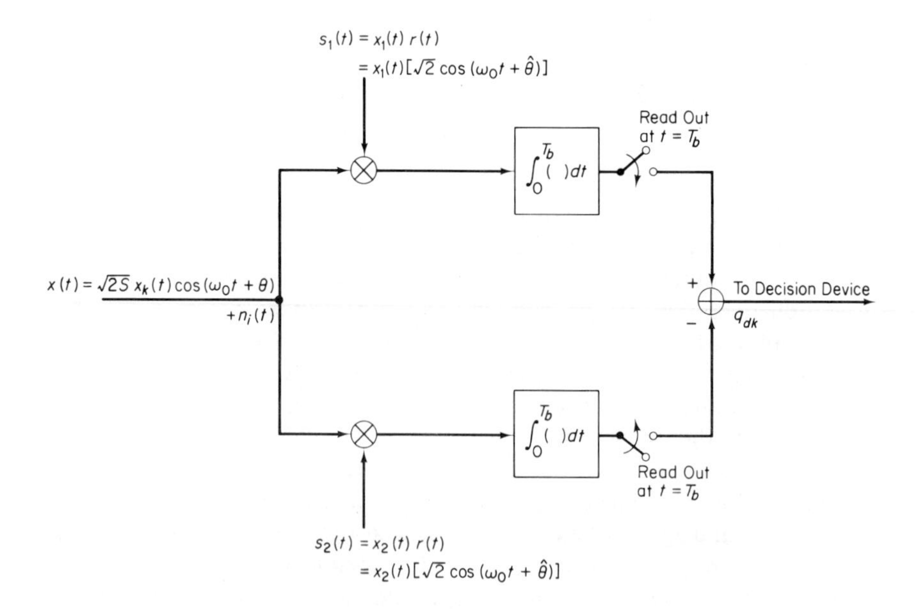

Fig. 6-2 Coherent detector; mismatching of phases by $\varphi = \theta - \hat{\theta}$

ship between the received and stored carrier phases. To perform ideal coherent detection, we must have $\theta = \hat{\theta}$ for all time. In practice, however, this is rarely the case since the channel continuously changes the phase of the transmitted signal relative to that of the received signal. Thus, one usually derives the reference signal, $r(t)$, at the receiver and uses the result in the detection process as though it were noise-free.

Proceeding now to a mathematical statement of the problem, we are given a situation where it is desired to transmit digital data over the additive Gaussian noise channel. The one-sided spectral density of the additive noise, $n_i(t)$, is again denoted by N_0 w/Hz. The received signal is represented by

$$x(t) = \sqrt{2S}x_k(t) \cos [\omega_0 t + \theta(t)] + n_i(t) \tag{6-1}$$

where $\theta(t)$ is the random phase shift (a slowly varying quantity) introduced by the channel and $x_k(t)$ represents the ± 1 modulation corresponding to the kth signal in the transmitted signaling alphabet. The signal, $x(t)$, is demodulated at the receiver by means of a reference carrier generated by the second-order PLL of Chap. 2. This reference carrier is taken to be

$$r(t) = \sqrt{2} \cos [\omega_0 t + \hat{\theta}(t)] \tag{6-2}$$

where $\hat{\theta}(t)$ is an estimate of phase shift introduced by the channel, and its accuracy is dependent on the power allocated to the auxiliary carrier. The product of $r(t)$ and $x(t)$ (with double frequency terms neglected) is

$$a(t) = \sqrt{S}\, x_k(t) \cos \varphi(t) + N[t, \varphi(t)] \tag{6-3}$$

where $\varphi(t)$ is the phase error process, i.e. $\varphi(t) = \theta(t) - \hat{\theta}(t)$ and $N[t, \varphi(t)]$ is defined by Eq. (2-7). Cross-correlating $a(t)$ with each $x_j(t), j = 1, 2, \ldots, N$ produces a set of random variables $\{q_j, j = 1, 2, \ldots, N\}$ from which an N-ary decision must be made. The decision strategy selected should be identical with that used for ideal coherent detection, namely, choose that random variable having the largest magnitude as corresponding to the most likely transmitted signal.

For binary signals—that is, $N = 2$—one can base his decision on the sample statistic of the *differenced cross-correlator output*—that is, the value obtained by sampling (at the conclusion of the signaling interval) the output of the correlator which is matched to $x_1(t)$ and the correlator which is matched to $x_2(t)$, then differencing the result. If $x_1(t) = -x_2(t)$, only one correlator is required.

6-3. DIFFERENCED CROSS-CORRELATOR OUTPUT STATISTICS ($N = 2$)

The approach we shall take is to use the method of conditional probabilities—that is, we shall develop expressions for the particular parameter, distribution, or moment of interest, conditioned on the system phase error,

$\varphi(t)$, being fixed over a T_b-second interval.[1-2] By averaging over this condition, which is random, we determine the behavior of interest.

We begin our analysis by considering the p.d.f. of the decision variable

$$q_{dk} \triangleq \int_0^{T_b} a(t)[x_1(t) - x_2(t)]\, dt \tag{6-4}$$

conditioned on a specific T_b-second segment of the random process $\varphi(t)$ and under the hypothesis that $x_k(t)$ was transmitted. Substituting (6-3) into (6-4), we observe that this conditional p.d.f., denoted by $p(q_{dk} \mid \varphi)$, is Gaussian with conditional mean and variance, respectively, given by

$$\mu_k = (-1)^{k+1}\sqrt{S}\left[\int_0^{T_b} x_k^2(t) \cos \varphi(t)\, dt - \int_0^{T_b} x_1(t)x_2(t) \cos \varphi(t)\, dt\right]$$

$$\sigma_k^2 = \frac{N_0}{2}\int_0^{T_b} [x_1(t) - x_2(t)]^2\, dt; \qquad k = 1, 2 \tag{6-5}$$

If $x_k(t)$, $k = 1, 2$ is a two-level (± 1) waveform, then

$$\mu_k = (-1)^{k+1}\sqrt{S}[1 - \lambda_{12}(\varphi)]\int_0^{T_b} \cos \varphi(t)\, dt$$

$$\sigma_k^2 = N_0 T_b[1 - \lambda_{12}(0)]; \qquad k = 1, 2 \tag{6-6}$$

The *random signal correlation coefficient*, $\lambda_{12}(\varphi)$, is defined analogously to (5-20) by

$$\lambda_{12}(\varphi) \triangleq \frac{\dfrac{1}{T_b}\displaystyle\int_0^{T_b} x_1(t)x_2(t) \cos \varphi(t)\, dt}{\dfrac{1}{T_b}\displaystyle\int_0^{T_b} \cos \varphi(t)\, dt} \tag{6-7}$$

Hence, when $\varphi(t)$ is identically equal to zero over the interval $0 \le t \le T_b$, (6-7) reduces to the ideal coherent result of (5-20)—that is, $\lambda_{12}(0) = \lambda_{12}$.

If the decision variable is normalized such that $q_{dk} = \sigma_k q'_{dk}$, then

$$p(q'_{dk} \mid \varphi) = \frac{1}{\sqrt{2\pi}} \exp\left[-\frac{(q'_{dk} - \mu_k/\sigma_k)^2}{2}\right] \tag{6-8}$$

It is easy to show that system performance can be studied by assuming that the decision variable has mean $\sqrt{2}\mu_1/\sigma_1$ and unit variance under the hypothesis that signal plus noise is present, and zero mean and unit variance in the presence of noise alone. Thus, the conditional p.d.f.'s under these two hypotheses become

$$p(q_{d1} \mid \varphi) = \frac{1}{\sqrt{2\pi}} \exp\left[-\frac{(q_{d1} - \sqrt{2}\mu_1/\sigma_1)^2}{2}\right]$$

and (6-9)

$$p(q_{d2} \mid \varphi)] = \frac{1}{\sqrt{2\pi}} \exp\left[-\frac{q_{d2}^2}{2}\right] = p(q_{d2})$$

where for simplicity of notation we have dropped the primes on the q_{dk}'s. Observe that in the presence of noise only, the conditional p.d.f. $p(q_{d2} | \varphi)$ is independent of system phase error.

It is of fundamental interest to study the properties of the output p.d.f. $p(q_{d1})$, when the phase error can be considered constant over a symbol interval. Also, since $\cos \varphi(t) = \cos \phi(t)$, then from (6-9) we have that

$$p(q_{d1}) = \int_{-\pi}^{\pi} p(\phi) p(q_{d1} | \phi) \, d\phi \tag{6-10}$$

where $p(\phi)$ is given by (2-16). For this special case (6-6) simplifies to

$$\begin{aligned} \mu_k &= (-1)^{k+1} \sqrt{S}\, T_b (1 - \lambda_{12}) \cos \phi \\ \sigma_k^2 &= N_0 T_b (1 - \lambda_{12}); \qquad k = 1, 2 \end{aligned} \tag{6-11}$$

hence,

$$\frac{\mu_1}{\sigma_1} = \sqrt{2R_{b\lambda}} \cos \phi; \quad R_{b\lambda} \triangleq \frac{ST_b}{N_0}\left(\frac{1 - \lambda_{12}}{2}\right) = R_b\left(\frac{1 - \lambda_{12}}{2}\right) \tag{6-12}$$

The integration in (6-10) is best accomplished, numerically, on a digital computer. Figures 6-3 and 6-4 illustrate the differenced cross-correlator output p.d.f. for various values of α, β, and two values of $R_{b\lambda}$. The specific values of $R_{b\lambda}$ were chosen such that error probability is 10^{-3} (typical for telemetry systems) and 10^{-5} (typical for command systems) when the system is perfectly synchronized.

The significance of the parameters α and β deserves some explanation. As discussed in Chap. 2, a second-order loop with imperfect integrating filter frequently must operate in the presence of frequency detuning—$\Omega_0 = \omega - \omega_0 \neq 0$. When this occurs the p.d.f. $p(\phi)$ becomes asymmetric, owing to the fact that the loop slips more cycles in one direction than the other. The net effect on loop performance is that the variance of the phase error is increased beyond its value when $\beta = 0$. In practice, this manifests itself in a lowering of the loop threshold. In phase-coherent detection wherein the carrier reference is provided by a PLL, a nonzero loop detuning, $\beta \neq 0$, produces an asymmetric p.d.f. for the decision variable. The consequence is that more errors in the decision process are made in the direction of the sign of Ω_0 than in the opposite direction. Hence, it behooves the communications engineer to establish procedures that can be used to reduce β to zero. When $\Omega_0 \neq 0$, the engineer frequently says the loop is *stressed*. Finally, the parameter α can be taken to be a measure of loop signal-to-noise ratio in the sense that the larger the value of α, the smaller are the deleterious effects due to noisy references and vice versa.

In practice, the moments of (6-10) are of interest. In particular, the mean of the differenced cross-correlator outputs is given by

$$\overline{q_{d1}} = \int_{-\infty}^{\infty} \int_{-\pi}^{\pi} q_{d1} p(q_{d1} | \phi) p(\phi) \, d\phi \, dq_{d1} \tag{6-13}$$

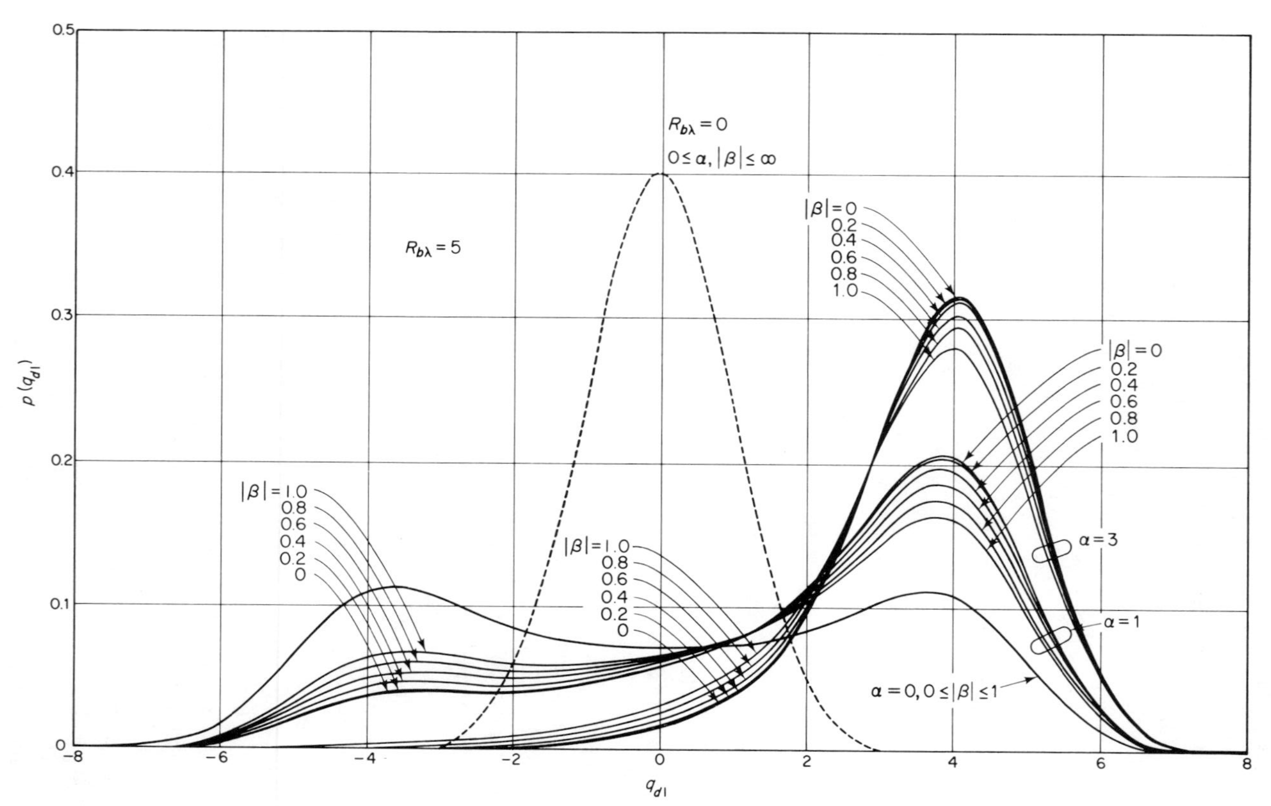

Fig. 6-3 Probability density function of cross-correlator output statistic; $R_{b\lambda} = 5$

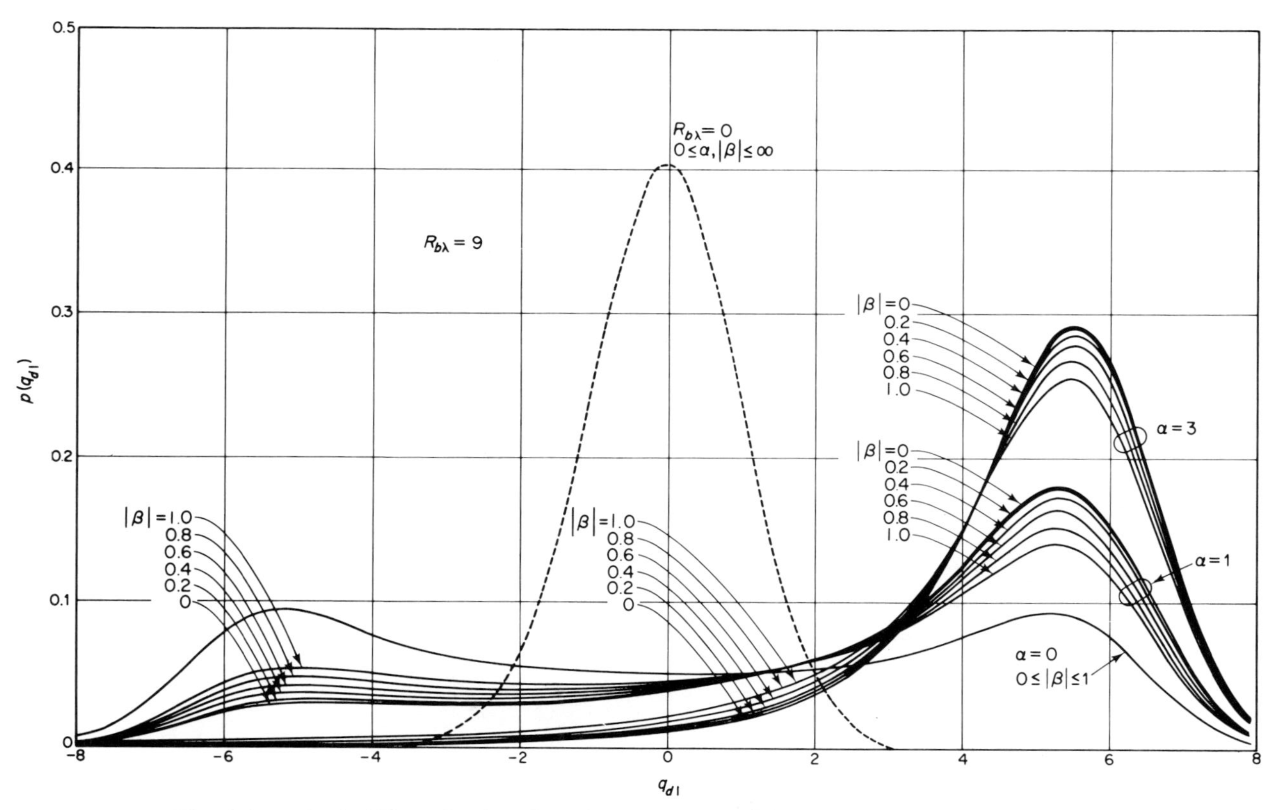

Fig. 6-4 Probability density function of cross-correlator output statistic; $R_{b\lambda} = 9$

Substituting (6-9) and (2-16) into (6-13), and carrying out the integration yields

$$\overline{q_{d1}} = \sqrt{4R_{b\lambda}}\,(\overline{\cos\phi}) = \sqrt{4R_{b\lambda}}\,Re\left\{\frac{I_{1-j\beta}(\alpha)}{I_{-j\beta}(\alpha)}\right\} \tag{6-14}$$

The mean-squared value of the p.d.f., $p(q_{d1})$, is determined from

$$\overline{q_{d1}^2} = \int_{-\infty}^{\infty}\int_{-\pi}^{\pi} q_{d1}^2\, p(q_{d1}\,|\,\phi)p(\phi)\,d\phi \tag{6-15}$$

Direct substitution of (6-9), and (2-16) into (6-15) gives

$$\overline{q_{d1}^2} = 1 + 4R_{b\lambda}\,\overline{\cos^2\phi} = 1 + 2R_{b\lambda} + 2R_{b\lambda}\,Re\left\{\frac{I_{2-j\beta}(\alpha)}{I_{-j\beta}(\alpha)}\right\} \tag{6-16}$$

Thus, the variance of $p(q_{d1})$ may be obtained from (6-14) and (6-16):

$$\sigma_{q_{d1}}^2 = \overline{q_{d1}^2} - (\overline{q_{d1}})^2 = 1 + 2R_{b\lambda} + 2R_{b\lambda}\,Re\left\{\frac{I_{2-j\beta}(\alpha)}{I_{-j\beta}(\alpha)}\right\} - 4R_{b\lambda}\left[Re\left\{\frac{I_{1-j\beta}(\alpha)}{I_{-j\beta}(\alpha)}\right\}\right]^2 \tag{6-17}$$

Figures 6-5 and 6-6 illustrate, as a function of α, the effects produced by $|\beta|$

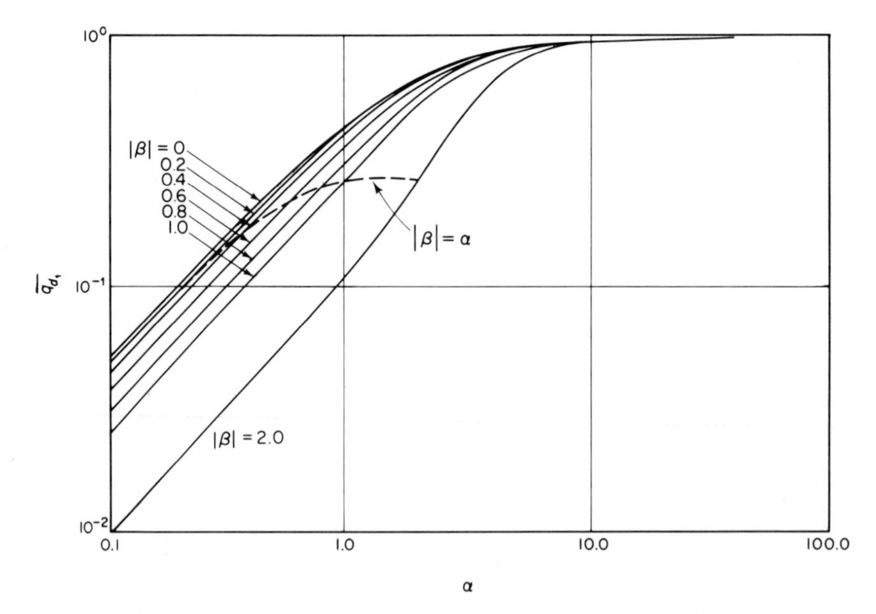

Fig. 6-5 Normalized mean of cross-correlator output statistic versus α

on the normalized mean and variance of the p.d.f. $p(q_{d1})$. Normalization is taken with respect to the values of the mean and variance of $p(q_{d1})$, corresponding to perfect synchronization. The dashed lines on the curves correspond to the situation $|\beta| = \alpha$.

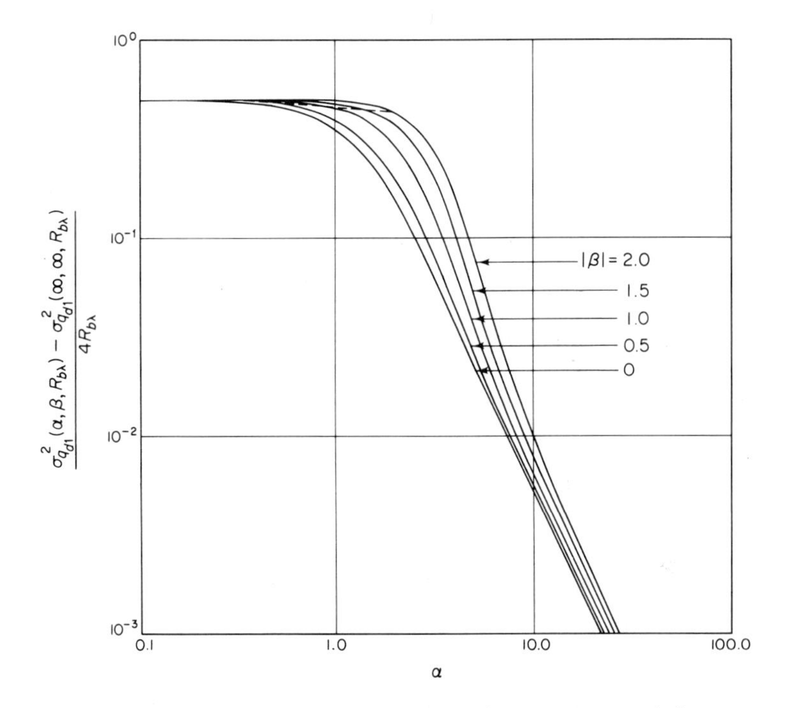

Fig. 6-6 Normalized variance of cross-correlator output statistic versus α

6-4. PERFORMANCE OF THE DATA DETECTOR, $N = 2$ (CARRIER TRACKING WITH A PLL)

We now turn to the problem of specifying the probability that, given the outputs of the cross-correlators matched to $x_1(t)$ and $x_2(t)$, the decision based on these outputs is in error. If we assume that $x_1(t)$ and $x_2(t)$ have equal *a priori* probabilities, then the probability of choosing $x_1(t)$ when in fact $x_2(t)$ was transmitted is identical to the probability of choosing $x_2(t)$ when the transmitter output was $x_1(t)$. Thus, without loss of generality, we assume that $x_1(t)$ was transmitted and compute the probability that, based on the output of the data detector, $x_2(t)$ is decided upon. This probability of error is one minus the probability that the decision is correct—that is, the output of the filter matched to signal $x_2(t)$ is less than the output of the filter matched to $x_1(t)$. This conditional probability of error is given by

$$P_E(\phi) = 1 - \int_{-\infty}^{\infty} p(q_{d1} \mid \phi)\, dq_{d1} \int_{-\infty}^{q_{d1}} p(q_{d2} \mid \phi)\, dq_{d2}. \tag{6-18}$$

Substituting (6-9) into (6-18) and simplifying the result yields

$$P_E(\phi) = \tfrac{1}{2} \operatorname{erfc}\left[\sqrt{R_{b\lambda}}\, Y(\phi)\right] \tag{6-19}$$

where

$$Y(\phi) \triangleq \left(\frac{1 - \lambda_{12}(\phi)}{1 - \lambda_{12}(0)}\right) \frac{1}{T_b} \int_0^{T_b} \cos \phi(t)\, dt \tag{6-20}$$

Equation (6-19) indicates the conditional error probability can be calculated, at least numerically, if the p.d.f. of the random variable, $Y(\phi)$, is known. Unless some simplifying assumptions are made, however, computation of the p.d.f. of $Y(\phi)$ appears to be formidable. In what follows, we consider three special cases of practical interest.

Phase Error Constant During the Symbol Interval

Under the assumption of constant phase in the symbol interval, the conditional error probability as given in (6-19) becomes

$$P_E(\phi) = \tfrac{1}{2} \operatorname{erfc} [\sqrt{R_{b\lambda}} \cos \phi] \tag{6-21}$$

while the average error probability is written as

$$P_E = \frac{1}{2} \int_{-\pi}^{\pi} p(\phi) \operatorname{erfc} [\sqrt{2R_{b\lambda}} \cos \phi]\, d\phi \tag{6-22}$$

Substitution of (2-16) into (6-22) yields

$$P_E = \frac{1}{2} \int_{-\pi}^{\pi} \int_{\phi}^{\phi+2\pi} \operatorname{erfc} [\sqrt{R_{b\lambda}} \cos \phi] \frac{\exp [\beta(\phi - x) + \alpha(\cos \phi - \cos x)]}{4\pi^2 \exp [-\pi\beta] |I_{j\beta}(\alpha)|^2}\, dx\, d\phi \tag{6-23}$$

Carrying out the integration (see Appendix A) gives

$$P_E = \frac{1}{2}\left\{1 - \sqrt{\frac{R_{b\lambda}}{\pi}} \exp\left(-\frac{R_{b\lambda}}{2}\right) \sum_{k=0}^{\infty} \epsilon_k (-1)^k I_k\left(\frac{R_{b\lambda}}{2}\right) \times \left[\frac{Re\{b_{2k+1-j\beta}\}}{2k+1} - \frac{Re\{b_{2k-1-j\beta}\}}{2k-1}\right]\right\} \tag{6-24}$$

where

$$b_{2k+1-j\beta} = \frac{I_{2k+1-j\beta}(\alpha)}{I_{-j\beta}(\alpha)}$$
$$b_{2k-1-j\beta} = \frac{I_{2k-1-j\beta}(\alpha)}{I_{-j\beta}(\alpha)} \tag{6-25}$$

The preceding expression for error probability indicates several important results. For example, when $R_{b\lambda}$ approaches infinity, (6-24) reduces to

$$P_E = \frac{1}{2}\left\{1 - \frac{1}{\pi} \sum_{k=0}^{\infty} \epsilon_k (-1)^k \left[\frac{Re\{b_{2k+1-j\beta}\}}{2k+1} - \frac{Re\{b_{2k-1-j\beta}\}}{2k-1}\right]\right\} \tag{6-26}$$

This shows that the performance of the receiver exhibits an irreducible error probability depending upon both β and α. In the limit as α approaches infinity—perfect carrier synchronization for finite β—P_E approaches zero. Note that as $R_{b\lambda}$ approaches infinity and $\pi/2 \leq \phi \leq \pi$, $1/2 \operatorname{erfc} [\sqrt{R_{b\lambda}} \cos \phi]$ becomes equal to unity. Hence, from (6-22), the irreducible error is equivalent to the probability that the magnitude of the phase error is greater than $\pi/2$.

The average error probability as a function of $R_{b\lambda}$ for various values of α and β is plotted in Figs. 6-7, 6-8, and 6-9. In these figures the effects of the

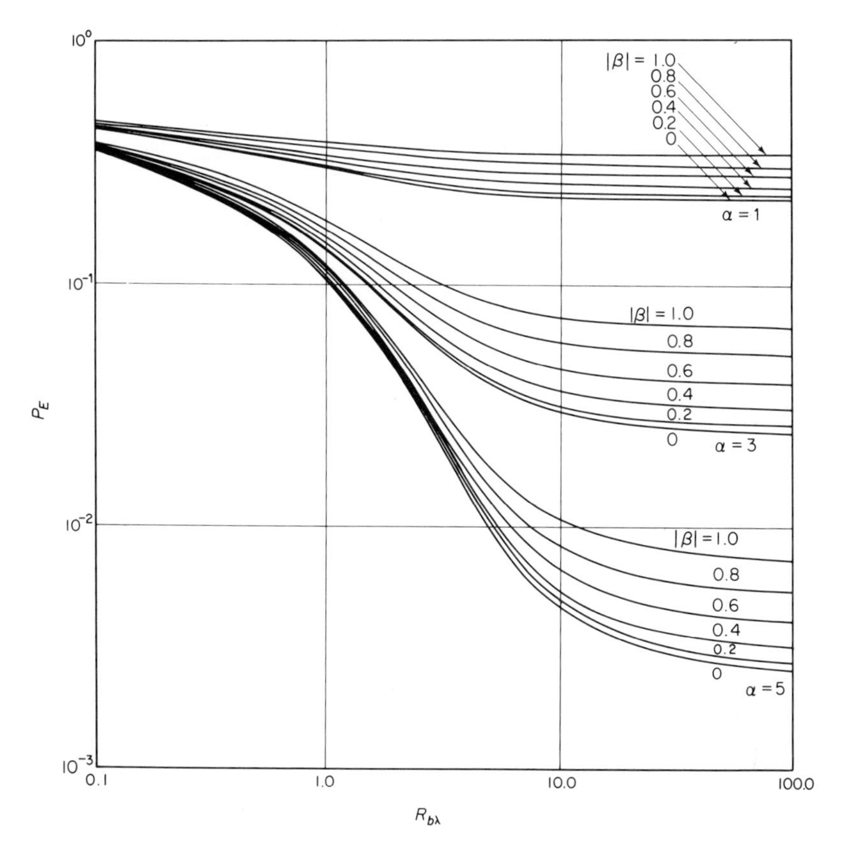

Fig. 6-7 Average error probability performance of a coherent detector with noisy carrier synchronization reference signals; slowly-varying phase error case; $\alpha = 1, 3, 5$

presence of loop detuning are clearly exhibited as well as the effects of the noisy reference $r(t)$. It is clear that in any operational system care must be taken to zero out as much as possible any static phase error by tuning the tracking loop's VCO.

Let us now examine the region of operation for which the previous results are applicable. A measure of the variation of the random process $\varphi(t)$ relative to T_b is given by its normalized correlation time[3]

$$\frac{\tau_c}{T_b} = \frac{1}{T_b \sigma_\varphi^2} \int_0^\infty |R_\varphi(\tau)|\, d\tau \tag{6-27}$$

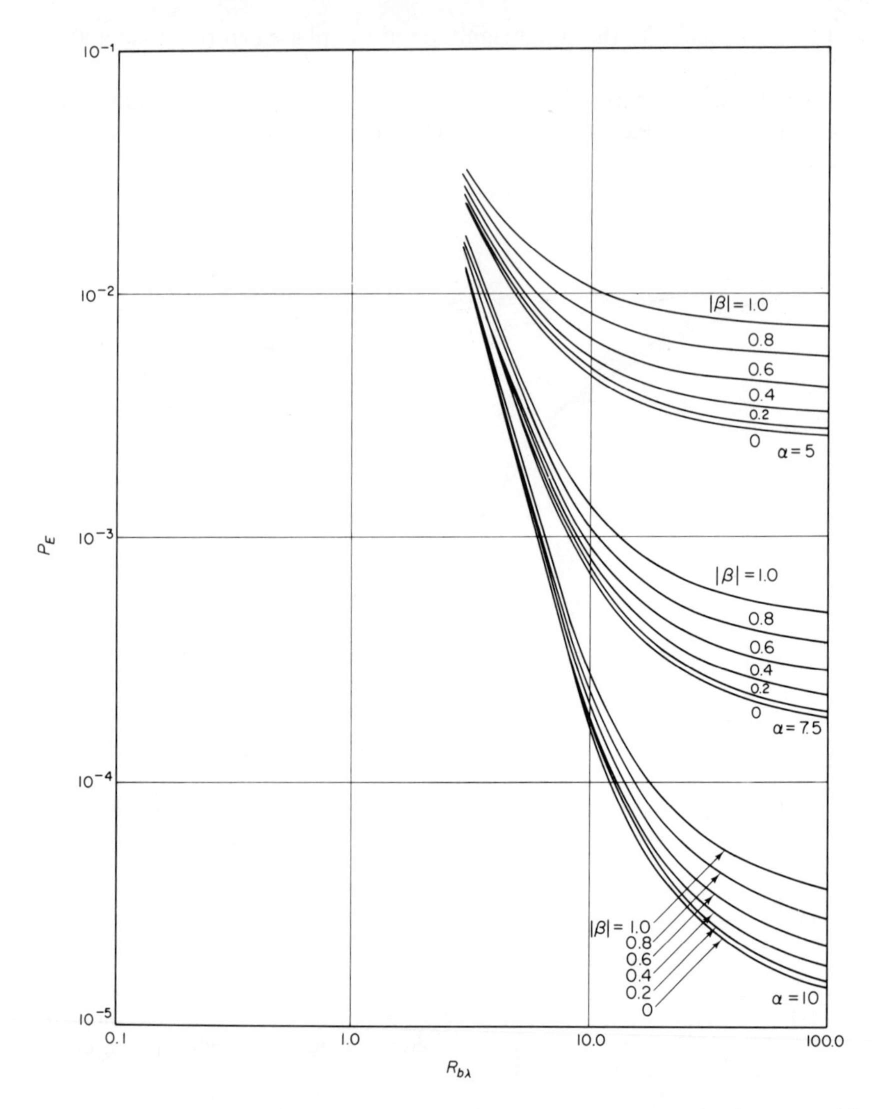

Fig. 6-8 Average error probability performance of a coherent detector with noisy carrier synchronization reference signals; slowly-varying phase error case; $\alpha = 5, 7.5, 10$

When this quantity is large relative to unity, we say the phase error varies slowly over the symbol interval T_b. In terms of an actual system design, it is more convenient to talk about the loop bandwidth, W_L, which is a measure of $1/\tau_c$. In terms of the reciprocal time-bandwidth parameter $\delta = 2/W_L T_b$, it has been shown[4] that whenever $\delta > 8$, conditions for a slowly varying phase error process are, for all practical purposes, satisfied.

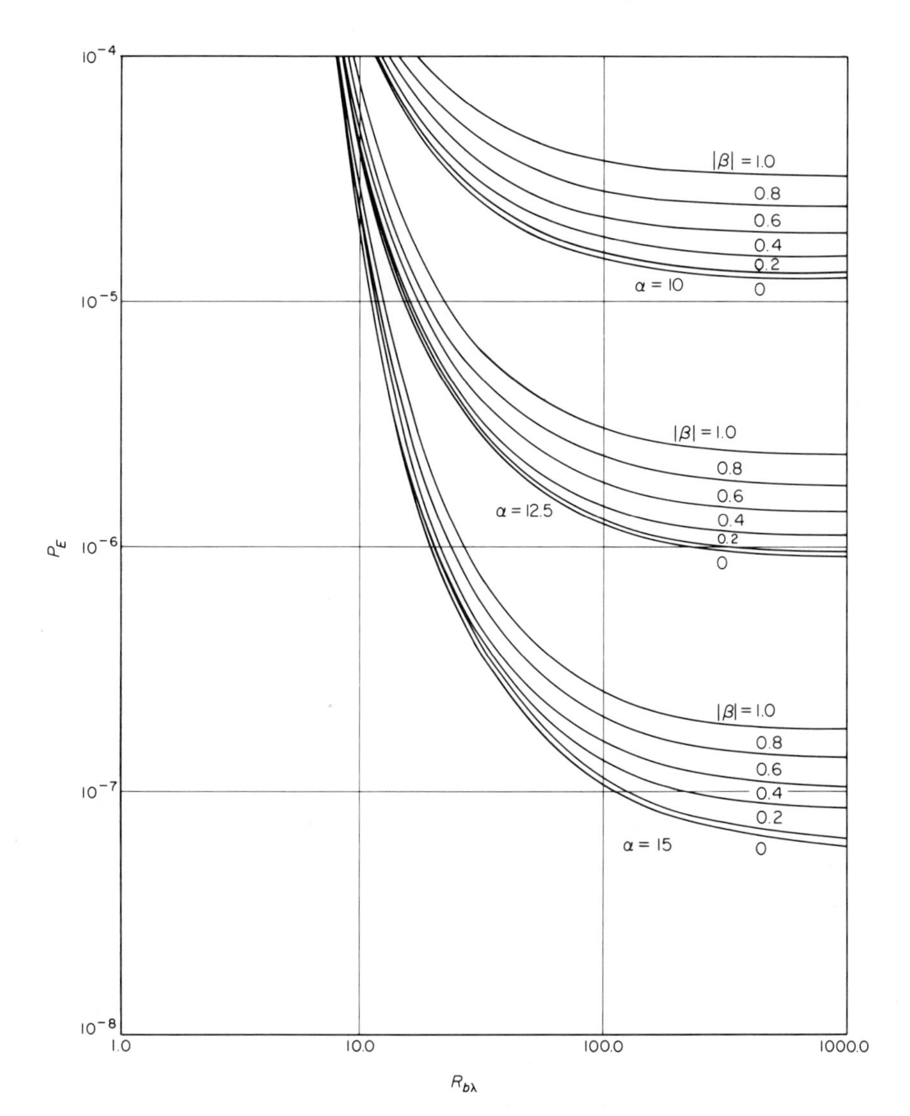

Fig. 6-9 Average error probability performance of a coherent detector with noisy carrier synchronization reference signals; slowly-varying phase error case; $\alpha = 10, 12.5, 15$

Finally, we summarize in Table 6-1 rigorous upper and lower bounds[5] for P_E as given in Eq. (6-23) for the special case, $\beta = 0$. The particular bound to be used in a given application depends upon the ratio of α to $R_{b\lambda}$, which is denoted by k. Also tabulated in Table 6-1 are approximate upper and lower bounds for the ratio of P_E as given in Eq. (6-23) for a particular value of α to the error probability for the perfectly synchronized system ($\alpha = \infty$). This

TABLE 6-1 Upper and Lower Bounds for P_E when $\beta = 0$

k Values	*Strict Bounds on P_E*	*Approximate Bounds on* $[P_E; \alpha]/[P_E; \infty]$	*Notation* $k = \dfrac{\alpha}{R_{b\lambda}}$
Lower Bounds Good for all $k \geq 2$	$\dfrac{\frac{c}{2}\{\exp[(k-1)R_{b\lambda}]\}\left[1 - \text{erfc}\left(\frac{\pi}{2\sqrt{2}}G\right)\right]}{\sqrt{2R_{b\lambda}}\pi I_0(kR_{b\lambda}G)}$	$[1 - (2/k)(1 - 4/\pi^2)]^{-1/2}$	$G^2 = [k - 2(1 - 4/\pi^2)]R_{b\lambda}$ $c = 2/(1 + \sqrt{1 + 2/R_{b\lambda}})$
Upper Bounds $k \geq 2$ Use for k near 2	$\dfrac{\exp[(k-1)R_{b\lambda}]}{4I_0(kR_{b\lambda})}R_{b\lambda}^{-1/4}(1 + 0.57/\sqrt{R_{b\lambda}})$	$(1 + 0.57/\sqrt{R_{b\lambda}})\pi R_{b\lambda}{}^{3/4}$	$C = 2/(1 + \sqrt{1 + 4/\pi R_{b\lambda}})$
Use for $2 \leq k \leq 4$	$\dfrac{\Theta(H)\left[1 - \text{erfc}\left(\frac{\pi}{2\sqrt{2}}H\right)\right]}{2\pi\sqrt{2R_{b\lambda}}HI_0(kR_{b\lambda})}C\{\exp[(k-1)R_{b\lambda}]\}$	$\Theta(H)\left[1 - \text{erfc}\left(\frac{\pi}{2\sqrt{2}}H\right)\right] \times (1 - 2/k)^{-1/2}$	$H^2 = (k - 2)R_{b\lambda}$
Use for large k	$\dfrac{C[\exp(R_{b\lambda})]I_0[(k-2)R_{b\lambda}]}{\sqrt{4\pi R_{b\lambda}}I_0(kR_{b\lambda})}$	$(1 - 2/k)^{-1/2}$	$\Theta(H) = 1 + \dfrac{\pi/4}{(k-2)C} \times \left\{1 - \sqrt{\frac{\pi}{2}}\left[\frac{H\exp(-\pi^2H/8)}{1 - \text{erfc}(\pi H/2\sqrt{2})}\right]\right\}$

ratio, which represents the increase in system error probability due to noisy references, is denoted in the table by $[P_E; \alpha]/[P_E; \infty]$.

Phase Error Varies Rapidly Over the Symbol Interval

In general, the conditional error probability described by Eq. (6-19) must be integrated over the p.d.f. of the random variable $Y(\phi)$. We saw that when $\phi(t)$ is essentially constant over a symbol interval, then the average error probability is obtained by averaging Eq. (6-19) over the p.d.f. of ϕ, $p(\phi)$. In other applications of phase-coherent communications, the phase error might be assumed to vary rapidly over the duration of T_b seconds;—that is, the correlation time of the phase error process is short in comparison to the symbol interval. In order to simplify matters, we further restrict ourselves to the case where signals $x_1(t)$ and $x_2(t)$ have constant envelope

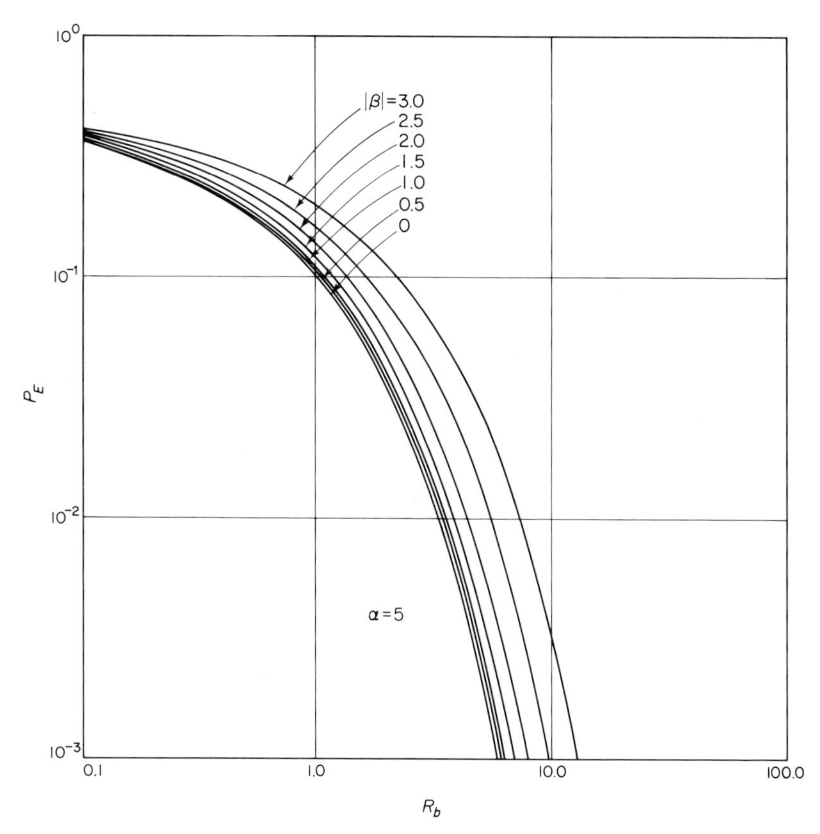

Fig. 6-10 Average error probability performance of a coherent detector with noisy carrier synchronization reference signals; rapidly-varying phase error case; $\alpha = 5$

over the symbol interval — that is, $x_1(t) = -x_2(t) = 1; 0 \leq t \leq T_b$. Then, from (6-7) and (6-20),

$$Y(\phi) = \frac{1}{T_b}\int_0^{T_b} \cos \phi(t)\, dt \tag{6-28}$$

Since in the rapidly varying phase error case the random variable $Y(\phi)$ as given above is a good measure of the true time average of the function $\cos \phi(t)$, and if $\phi(t)$ is an ergodic process, the time average may be replaced by the statistical mean. Thus,

$$Y(\phi) \cong \overline{\cos \phi} \tag{6-29}$$

and the system error probability is approximated by

$$P_E \cong \tfrac{1}{2} \operatorname{erfc}\,[\sqrt{R_b}\,(\overline{\cos \phi})] \tag{6-30}$$

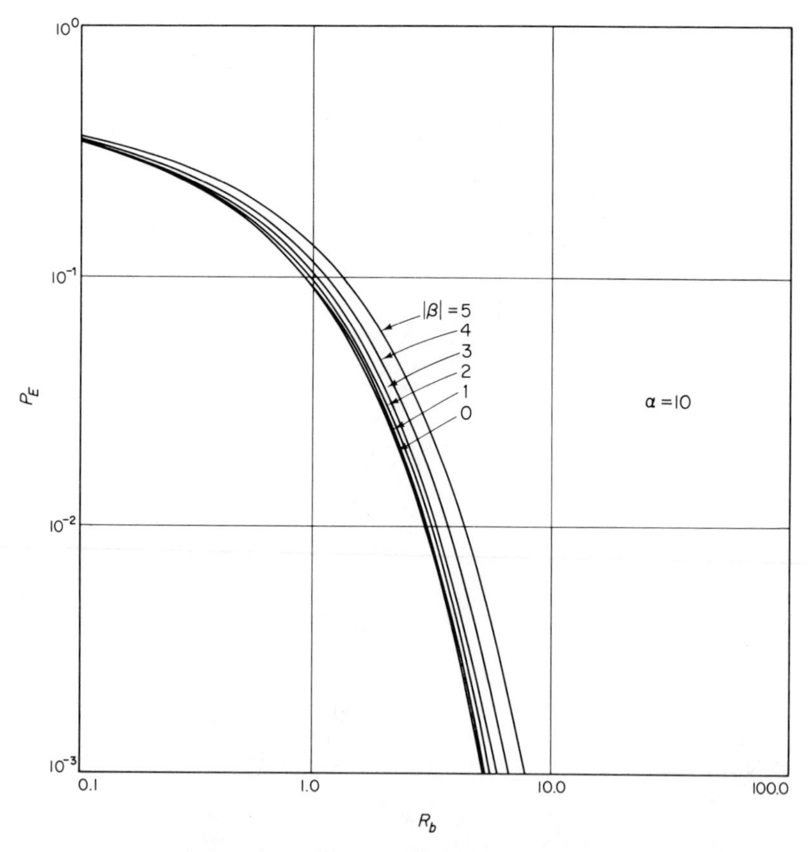

Fig. 6-11 Average error probability performance of a coherent detector with noisy carrier synchronization reference signals; rapidly-varying phase error case; $\alpha = 10$

where $\overline{\cos\phi}$ is given by (2-22) with $n = 1$ and $R_{b\lambda} = R_b$. This equation is plotted versus R_b for various values of α and β in Figs. 6-10 and 6-11. Notice here that an irreducible error does *not* exist as R_b is increased at fixed α and β.

Phase Error Varies Moderately Over the Symbol Interval

When the correlation time of the phase error process, relative to T_b, is such that neither of the extreme cases as expressed by (6-21) and (6-30) is valid, then one must seek an approximate solution for the p.d.f. of the random variable $Y(\phi)$ as given in (6-20). One technique which is valid in many practical situations is to employ the simplifying assumption that the phase error is small enough so that the linear PLL theory can be used. The consequences of this assumption are twofold: first, from the linear PLL theory we know that $\phi(t)$ is a Gaussian process. Second, assuming once again that $x_1(t) = -x_2(t)$, then (6-28) applies. Then, for high loop signal-to-noise ratios $Y(\phi)$ is approximated by

$$Y(\phi) \cong 1 - \frac{1}{2T_b}\int_0^{T_b} \phi^2(t)\,dt \tag{6-31}$$

and the computation of it's p.d.f. $p_Y(y)$, reduces to computing the p.d.f. of the integral of the square of a Gaussian process.

For the case of a first-order loop (and approximately for a second-order loop with zero detuning), the correlation function of $\phi(t)$ is

$$R_\phi(\tau) = \frac{1}{\rho}\exp\left[-2W_L|\tau|\right] \tag{6-32}$$

where ρ, the signal-to-noise ratio in the loop bandwidth, and W_L, the loop bandwidth, are defined in (2-17) and (2-14), respectively. Assuming, then, a Gaussian distribution for $\phi(t)$ with correlation function as in (6-32), the p.d.f. $p_Z(z)$ of the random variable

$$Z(\phi) \triangleq \frac{\rho}{T_b}\int_0^{T_b} \phi^2(t)\,dt = 2\rho[1 - Y(\phi)] \tag{6-33}$$

is well approximated by[4,6]

$$p_Z(z) = \frac{1}{\sqrt{\pi\delta}}\exp(2/\delta)z^{-3/2}\exp\left[-\frac{1}{\delta}\left(z + \frac{1}{z}\right)\right] \quad ; \quad 0 \le z \le \infty \tag{6-34}$$

Averaging (6-19) over the random variable $Z(\phi)$, we find

$$P_E = \frac{1}{2}\int_0^\infty p_Z(z)\,\text{erfc}\left[\sqrt{R_b}\left(1 - \frac{z}{2\rho}\right)\right]dz \tag{6-35}$$

The error probability as computed from (6-35) is plotted versus R_b in Fig. 6-12 for various values of $\alpha = \rho$, $\beta = 0$, and $\delta = 0.5$.

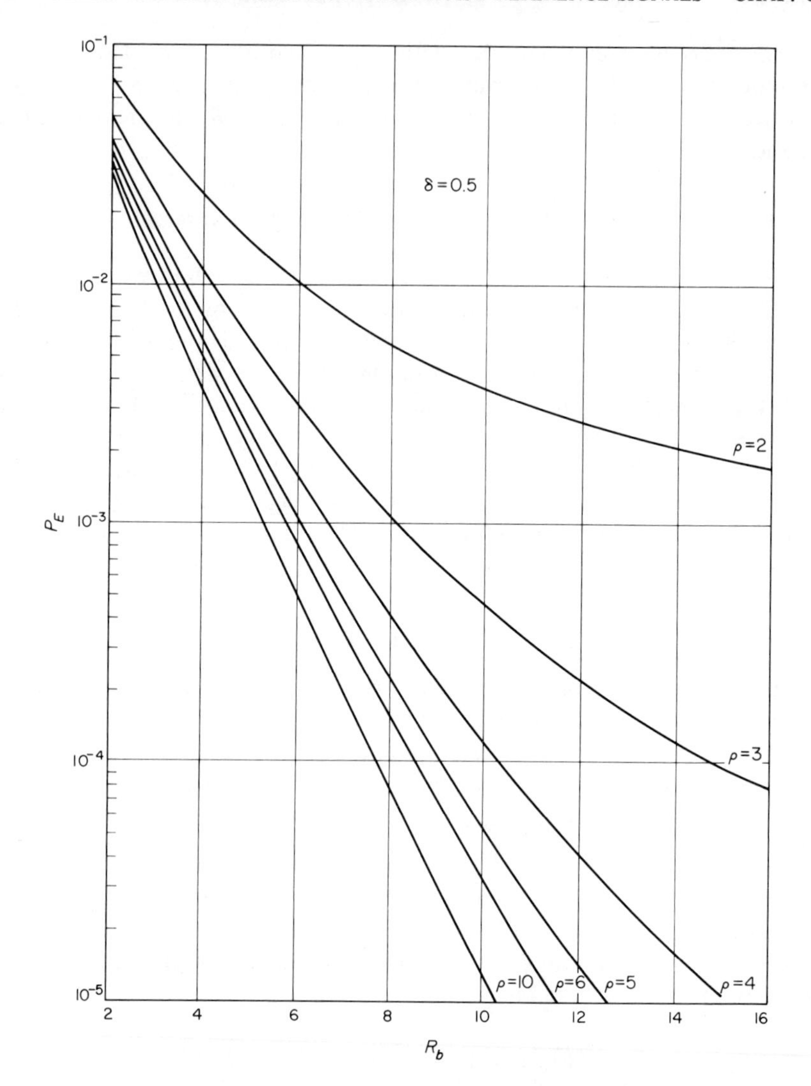

Fig. 6-12 Average error probability performance of a coherent detector with noisy carrier synchronization reference signals; moderately-varying phase error case; $\delta = .5$

6-5. PERFORMANCE OF THE DATA DETECTOR, $N = 2$ (SUPPRESSED CARRIER TRACKING WITH A SQUARING LOOP OR COSTAS LOOP)

A convenient and attractive way to provide carrier synchronization from a suppressed-carrier signal is to use the squaring loop or Costas loop discussed in Chap. 2. Since in either of these mechanizations the loop tracks a double-frequency term, then the VCO output must be frequency divided by two to provide a noisy reference for demodulation of the data off the carrier. Thus, the effective phase error, ϕ, for purposes of computing the error probability performance of the data detector, is one half of the actual phase error being tracked by the loop. Letting $p(2\phi)$ denote the p.d.f. of the loop phase error

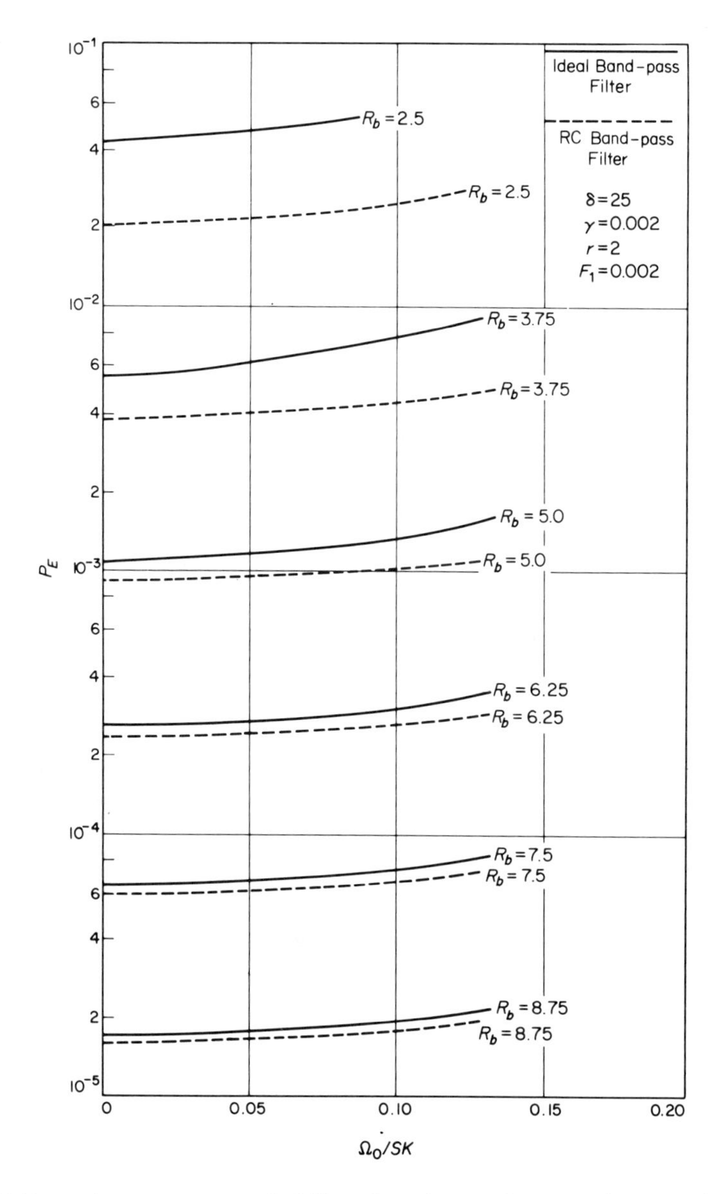

Fig. 6-13 Average error probability of a coherent detector whose synchronization reference signals are provided by a squaring or Costas loop

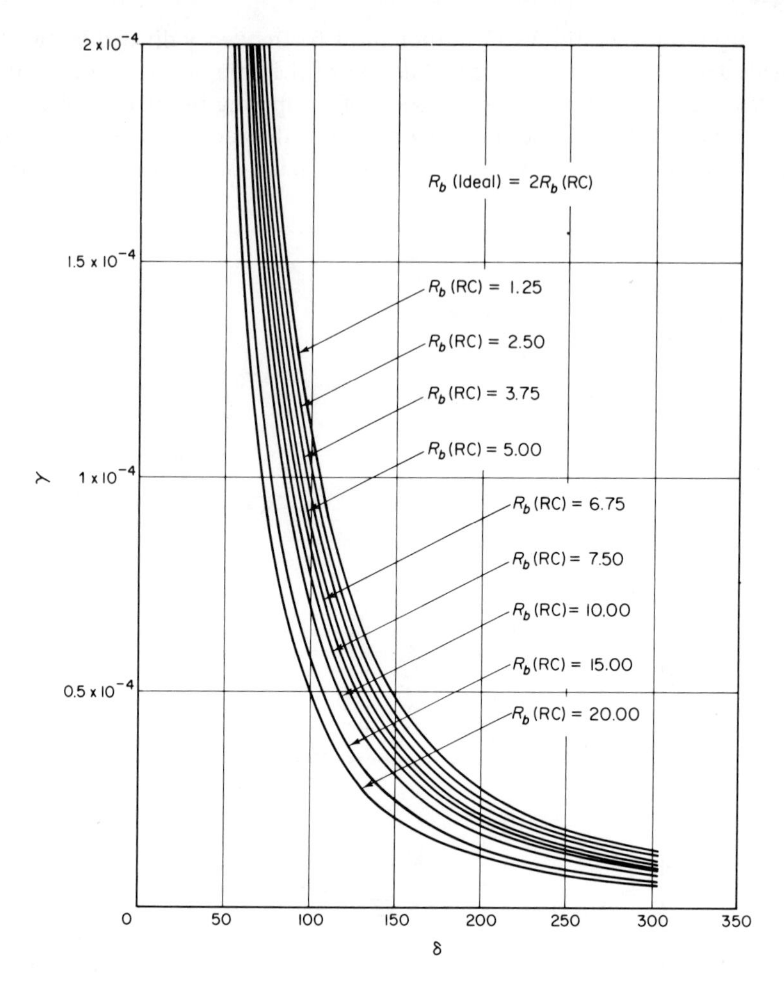

Fig. 6-14 The ratio of loop bandwidth to input bandwidth as a function of the ratio of symbol rate to loop bandwidth with bit signal-to-noise ratio as a parameter

with α and β as defined in (2-75), then by a simple transformation of variables, the p.d.f. of ϕ, $p_d(\phi)$, becomes

$$p_d(\phi) = 2p(2\phi) \qquad |\phi| \leq \frac{\pi}{2} \tag{6-36}$$

Assuming, for example, the slowly varying phase error case and that a "genie" is present at the receiver who is capable of perfectly resolving the 180 degree phase ambiguity, then the average symbol error probability is[7,8]

TABLE 6-2 Upper and Lower Bounds for P_E

l Values	*Exact Bounds on P_E*	*Approximate Bounds on P_E* $\left[\left(l - \frac{1}{2}\right)R_b > 4\right]$	*Notation* $l = \frac{\rho'}{R_b}$
Lower Bound $l > \frac{1}{2}$	$\frac{ce^{-R_b/2} I_0\left[\left(l - \frac{1}{2}\right)R_b\right]}{\sqrt{4\pi R_b} I_0(lR_b)}$	$\frac{ce^{-R_b}\left(1 - \frac{1}{2l}\right)^{-1/2}}{\sqrt{4\pi R_b}}$	$c = 2/(1 + \sqrt{1 + 2/R_b})$
Upper Bound $l > \frac{1}{2}$	$\frac{Ce^{-R_b/2} I_0\left[\left(l - \frac{1}{2}\right)R_b\right]}{\sqrt{4\pi R_b} I_0(lR_b)} + \frac{(\pi/4)^3 \exp[(l-1)R_b]}{16R_b\left(l - \frac{1}{2}\right)^{3/2} I_0(lR_b)}$	$\frac{Ce^{-R_b}\left(1 - \frac{1}{2l}\right)^{-1/2}}{\sqrt{4\pi R_b}} + \frac{(\pi/4)^4 (l/2)^{1/2} \exp(-R_b)}{\left(l - \frac{1}{2}\right)^{3/2} \sqrt{4\pi R_b}}$	$C = 2/(1 + \sqrt{1 + 4/\pi R_b})$

$$P_E = \int_{-\pi/2}^{\pi/2} P_E(\phi) p_d(\phi) \, d\phi$$
$$= \int_{-\pi}^{\pi} P_E\left(\frac{\Phi}{2}\right) p(\Phi) \, d\Phi \tag{6-37}$$

where $P_E(\phi)$ is given by (6-21) and $\Phi \triangleq 2\phi$. Although a "genie" is never available in a practical receiver, the above result is useful in that it represents an upper bound on performance.

Since the power in the data signal is used here for providing the carrier synchronization reference ($S = P_c$), the input signal-to-noise ratio in the input bandwidth $\rho_i = 2S/N_0W_i$ can be related to R_b through

$$\rho_i = \tfrac{1}{2} R_b \delta \gamma \tag{6-38}$$

where γ is defined in (2-81).

Figure 6-13 plots P_E, as defined by (6-37), versus Ω_0/SK with R_b as a parameter for $\lambda_{12} = -1$, $F_1 = 0.002$, $\delta = 25$, $\gamma = 0.002$, $r = 2$, and with RC and ideal band-pass filter characteristics. The effective signal-to-noise ratio in the loop bandwidth, ρ', is defined in terms of ρ_i (hence, R_b, δ, and γ) in (2-81). These curves can also be used for other combinations of values γ and δ. To determine these, Fig. 6-14 plots γ versus δ with R_b as a parameter for RC and ideal band-pass filters.

As in the first part of Sec. 6-4, the average error probability computed from (6-37) can be upper- and lower-bounded[5]. Exact and approximate results (for $\rho' > 4$) are given in Table 6-2.

6-6. DATA DETECTION PERFORMANCE OF BLOCK-CODED SYSTEMS

It is a simple matter at this point to extend the previous results to the class of block codes outlined in Chap. 5. As observed, a noisy coherent reference signal reduces the effective signal-to-noise ratio of the data detector. For the case where the phase error is essentially constant over the duration of a single code word, the error probability expressions for block-coded signal sets derived in the previous chapter are valid here as conditional word error probabilities if R_d is replaced by $R_d \cos^2 \phi$. For example, for a matched filter receiver with perfect word and symbol sync and a transmitted orthogonal signal set, the conditional word error probability is given by

$$P_E(\phi) = 1 - \int_{-\infty}^{\infty} \frac{\exp(-x^2)}{\sqrt{\pi}} \left[\frac{1}{2} \operatorname{erfc}(-x - \sqrt{R_d} \cos \phi)\right]^{2^n - 1} dx \tag{6-39}$$

where n is the number of bits per code word.

To illustrate the point graphically, Figs. 6-15 through 6-17 plot the word error probability for the suppressed carrier case as computed from (6-37) and (6-39) versus the ratio of energy per bit to noise spectral density, $R_b \triangleq R_d/n$, for $n = 6$, using various values of γ and δ as parameters. The case $\delta \doteq \infty$

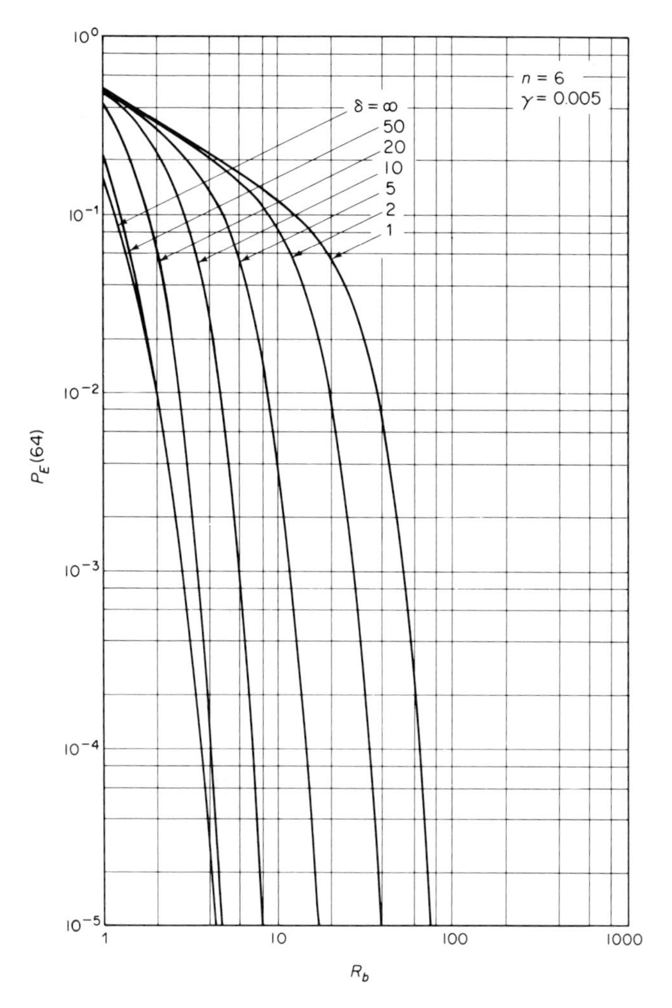

Fig. 6-15 Average word error probability performance of an orthogonally-coded system whose synchronization reference signals are provided by a squaring or Costas loop; $n = 6$, $\gamma = .005$

corresponds to performance in the presence of perfect carrier synchronization signals. A squaring loop with zero detuning and an ideal presquaring filter are assumed in the above illustrations; furthermore, for simplicity of computation we have set $\alpha = \rho'$.

Aside from the case where the carrier reference is derived from a Costas loop, our discussion thus far has not assumed any physical relation between the reference carrier-tracking loop and the biphase-modulated data signal. Hence, S and P_c were allowed to vary independently of each other. In certain physical systems, the data signal may be phase-modulated on the carrier from which the noisy synchronization reference is to be derived. In such cases

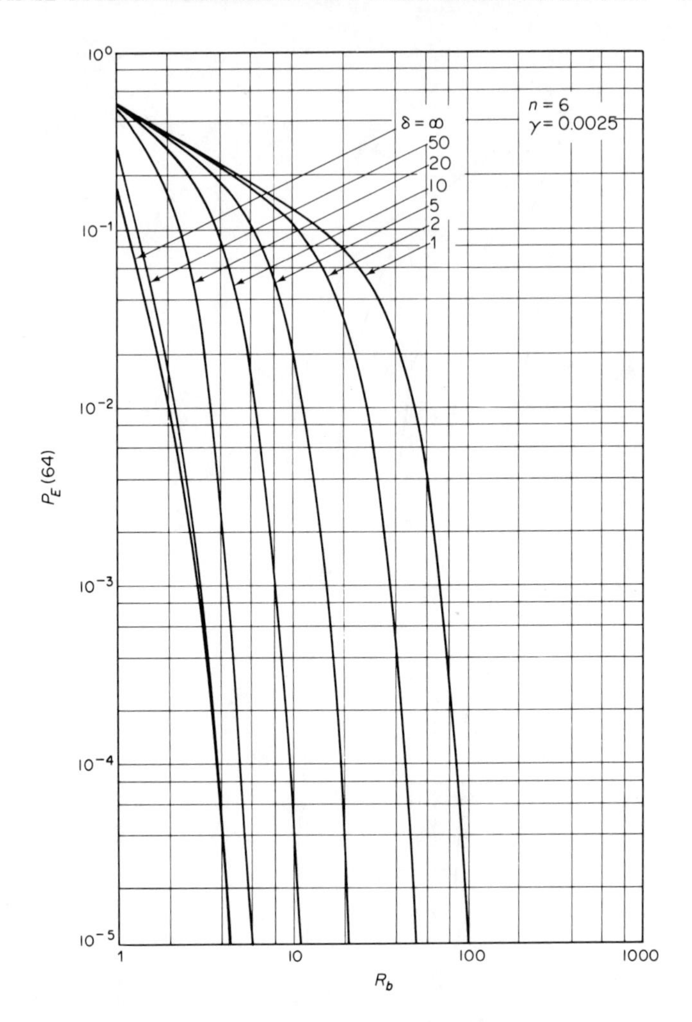

Fig. 6-16 Average word error probability performance of an orthogonally-coded system whose synchronization reference signals are provided by a squaring or Costas loop; $n = 6$, $\gamma_. = .0025$

S and P_c might be tied together through the modulation factor, m, of the channel ($S = [(1 - m^2)/m^2]P_c$ in a single-channel system). Details of this system and generalizations of it are considered in Chap. 7. The important point to note is that the results derived here are not tied to a particular system configuration; hence, to apply them, one simply uses the values of S and P_c [from which one can solve for α via Eq. (2-17)] as specified by the particular design requirements. If loop detuning, Ω_0, is present, one must determine the equivalent value of β from Eq. (2–17) before entering the curves presented in this chapter.

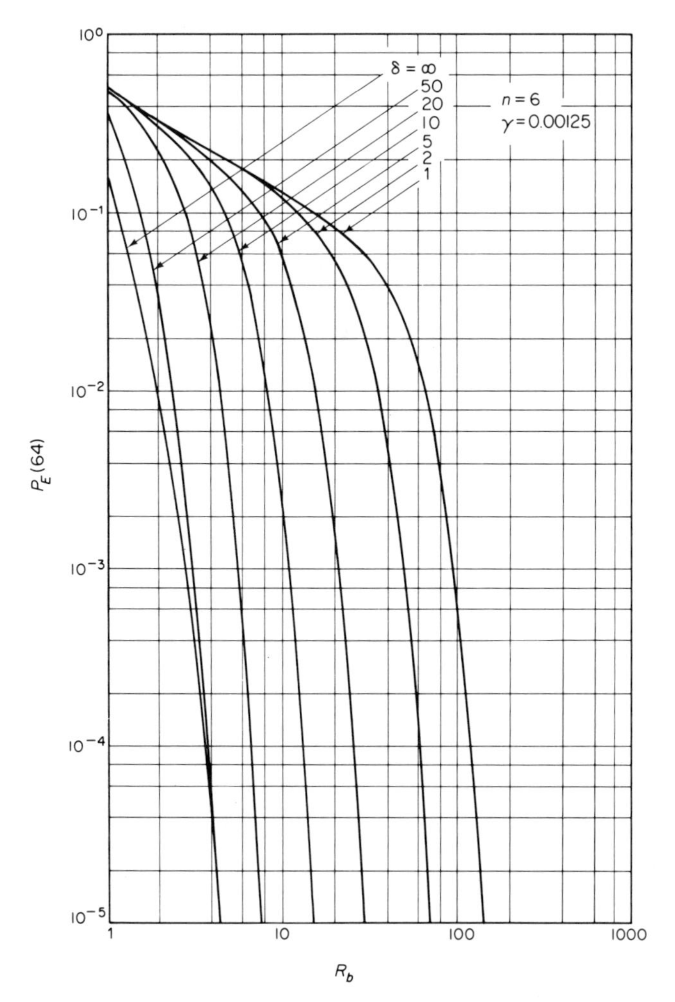

Fig. 6-17 Average word error probability performance of an orthogonally-coded system whose synchronization reference signals are provided by a squaring or Costas loop; $n = 6$, $\gamma = .00125$

6-7. THE NOISY REFERENCE PROBLEM FOR DETECTION OF POLYPHASE SIGNALS

The first part of the problem is, as before, to establish an expression for the probability of error conditioned on a given phase error. From this expression, and from the p.d.f. of the phase error associated with the carrier-tracking loop, one can assess the average word error probability performance of an MPSK communication system.

Following the method of analysis given in Sec. 5-5, the generalization of

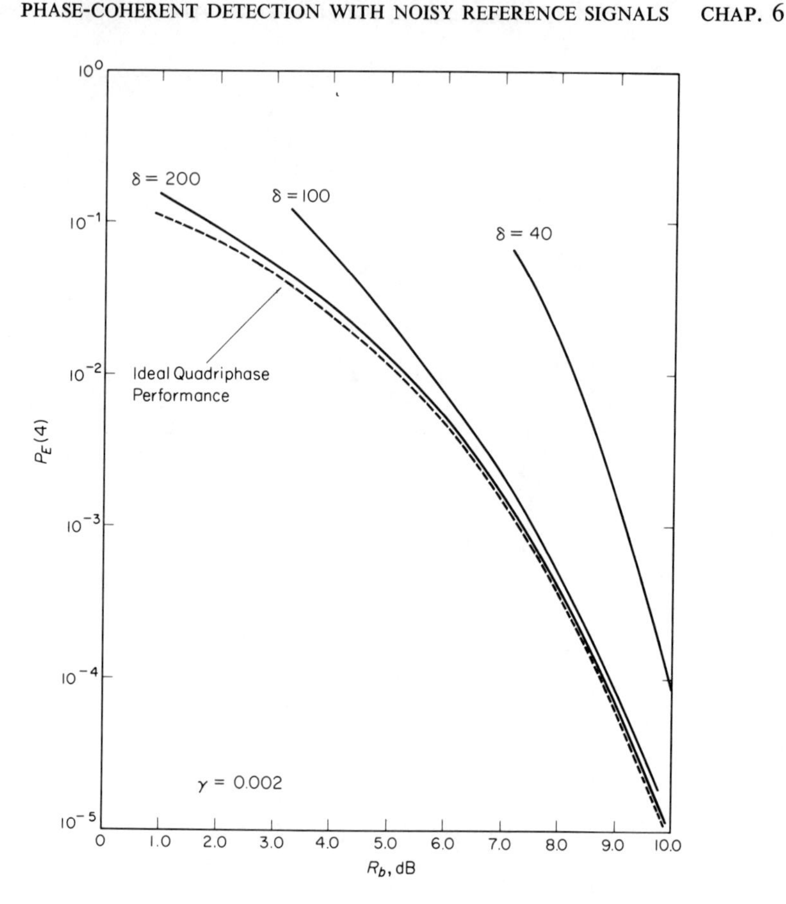

Fig. 6-18 Average word error probability performance of a quadriphase system whose synchronization reference signals are provided by a 4-phase Costas loop; $\gamma = .002$

(5-83) to the case where the synchronization references are noisy becomes

$$p(V, \eta \mid \phi) = \frac{V}{\pi STN_0} \exp\left[\frac{-V^2 + 2STV\cos\left(\eta - 2(i-1)\frac{\pi}{N} - \phi\right) - (ST)^2}{STN_0}\right]; \quad \begin{array}{l} 0 \leq V \leq \infty \\ 0 \leq \eta \leq 2\pi \end{array} \tag{6-40}$$

Hence, the conditional probability of correct detection is

$$P_c(\phi) = \frac{1}{\pi} \int_0^\infty \left[\exp\{-(u - \sqrt{R_d}\cos\phi)^2\} \times \int_{-u\tan\pi/N}^{u\tan\pi/N} \exp\{-(v - \sqrt{R_d}\sin\phi)^2\}\, dv\right] du \tag{6-41}$$

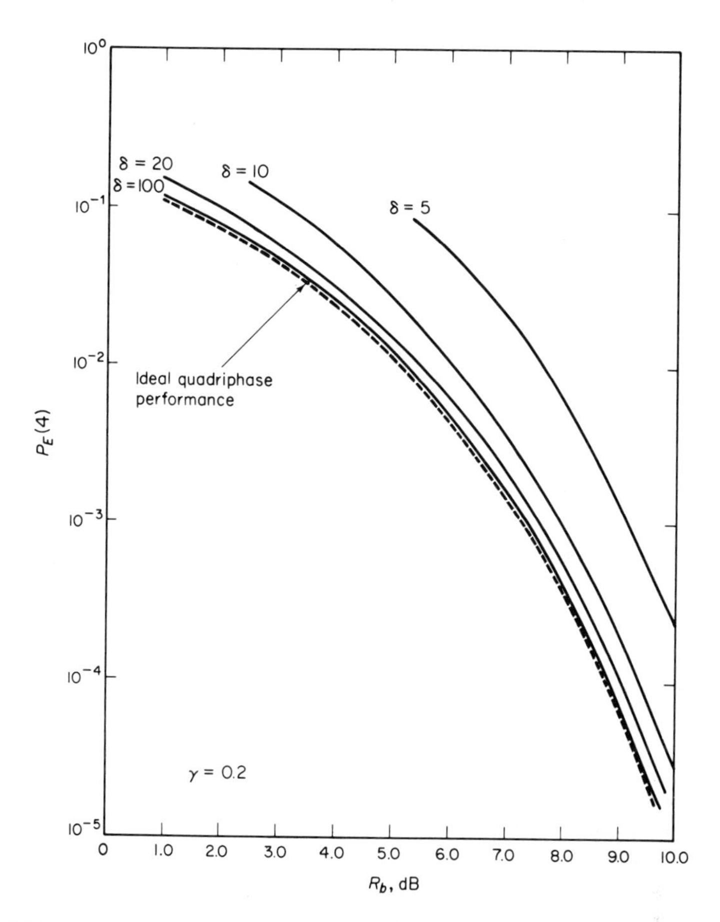

Fig. 6-19 Average word error probability performance of a quadriphase system whose synchronization reference signals are provided by a 4-phase Costas loop; $\gamma = .02$

or equivalently, the conditional word error probability is given by

$$\begin{aligned} P_E(\phi) = \frac{N-1}{N} &- \frac{1}{4}\operatorname{erf}\left[\sqrt{R_d}\sin\left(\frac{\pi}{N}+\phi\right)\right] - \frac{1}{4}\operatorname{erf}\left[\sqrt{R_d}\sin\left(\frac{\pi}{N}-\phi\right)\right] \\ &- \frac{1}{2\sqrt{\pi}}\int_0^{\sqrt{R_d}\sin(\pi/N+\phi)} \exp(-y^2)\operatorname{erf}\left[y\cot\left(\frac{\pi}{N}+\phi\right)\right]dy \\ &- \frac{1}{2\sqrt{\pi}}\int_0^{\sqrt{R_d}\sin(\pi/N-\phi)} \exp(-y^2)\operatorname{erf}\left[y\cot\left(\frac{\pi}{N}-\phi\right)\right]dy \end{aligned} \tag{6-42}$$

Unlike the block-coded case, the results given above are *not* obtained by simply replacing R_d by $R_d \cos^2 \phi$ in the equivalent ideal detection equations.

If now, for example, the noisy reference signals are established by the Nth power loop or N-phase Costas loop, then, analogous to (6-37), the average word error probability is

$$P_E(N) = \int_{-\pi/N}^{\pi/N} P_E(\phi) N p(N\phi)\, d\phi$$
$$= \int_{-\pi}^{\pi} P_E\left(\frac{\Phi}{N}\right) p(\Phi)\, d\Phi \qquad (6\text{-}43)$$

where $p(N\phi)$ denotes the loop phase error with α and β as defined in (2-113).

A graphic portrayal of the above is given in Figs. 6-18 and 6-19 where average word error probability is plotted versus the ratio of energy per bit to noise spectral density, $R_b = R_d/\log_2 N$, for $N = 4$, $\gamma = 0.002$ and 0.2, respectively, and various values of δ. Zero detuning and an RC filter are also assumed in these figures. Other parameters for the Nth power tracking loop are $F_1 = 0.002$ and $r = 2$. As for the binary case the curves assume that the N-fold ambiguity inherent in the suppressed carrier-tracking loop is perfectly resolved.

As discussed in Chap. 5, differential encoding of the transmitted signal phases can be used in practice to resolve this ambiguity. The noisy reference problem which arises in such a system is discussed next.

6-8. COHERENT DETECTION OF DIFFERENTIALLY ENCODED MPSK WITH SUPPRESSED CARRIER TRACKING

As discussed in Chaps. 2 and 5, differential encoding is often employed for resolving ambiguities associated with suppressed carrier-tracking loops. Thus, it is of interest to compute the average word error probability performance of a coherent receiver which employs such a loop.

For coherent detection of differentially encoded MPSK, a calculation similar to that given in Sec. 5-6 results in an expression for conditional word error probability; viz.,

$$P_E(\phi) = 1 - [P_c(\phi)|_{MPSK}]^2 - \sum_{k=1}^{N-1} P_{E_k}^2(\phi) \qquad (6\text{-}44)$$

where $P_c(\phi)|_{MPSK}$ is defined in (6-41) and

$$P_{E_k}(\phi) = \frac{1}{\pi}\int_0^{\infty}\left[\exp\{-(u - \sqrt{R_d}\cos\phi)^2\}\right.$$
$$\left.\times \int_{u\tan[(2k-1)\pi/N]}^{u\tan[(2k+1)\pi/N]} \exp\{-(v - \sqrt{R_d}\sin\phi)^2\}\, dv\right] du$$
$$= \frac{1}{N} + \frac{1}{4}\operatorname{erf}\left[\sqrt{R_d}\sin\left(\frac{(2k+1)}{N}\pi - \phi\right)\right]$$
$$- \frac{1}{4}\operatorname{erf}\left[\sqrt{R_d}\sin\left(\frac{(2k-1)}{N}\pi - \phi\right)\right]$$

$$+\frac{1}{2\sqrt{\pi}}\int_0^{\sqrt{R_d}\sin[(2k+1)\pi/N-\phi]}\exp(-y^2)\,\mathrm{erf}\left[y\cot\left(\frac{(2k+1)}{N}\pi-\phi\right)\right]dy$$
$$-\frac{1}{2\sqrt{\pi}}\int_0^{\sqrt{R_d}\sin[(2k-1)\pi/N-\phi]}\exp(-y^2)\,\mathrm{erf}\left[y\cot\left(\frac{(2k-1)}{N}\pi-\phi\right)\right]dy \tag{6-45}$$

Furthermore,

$$P_E(\phi)=P_E\left(\phi+\frac{2(k-1)\pi}{N}\right);\ k=1,2,\ldots,N \tag{6-46}$$

which is necessary for successful ambiguity resolution. If once again we assume the slowly varying phase error case, then substituting (6-44) into (6-43) gives the desired result.

For the special case of coherent detection of differentially encoded PSK,

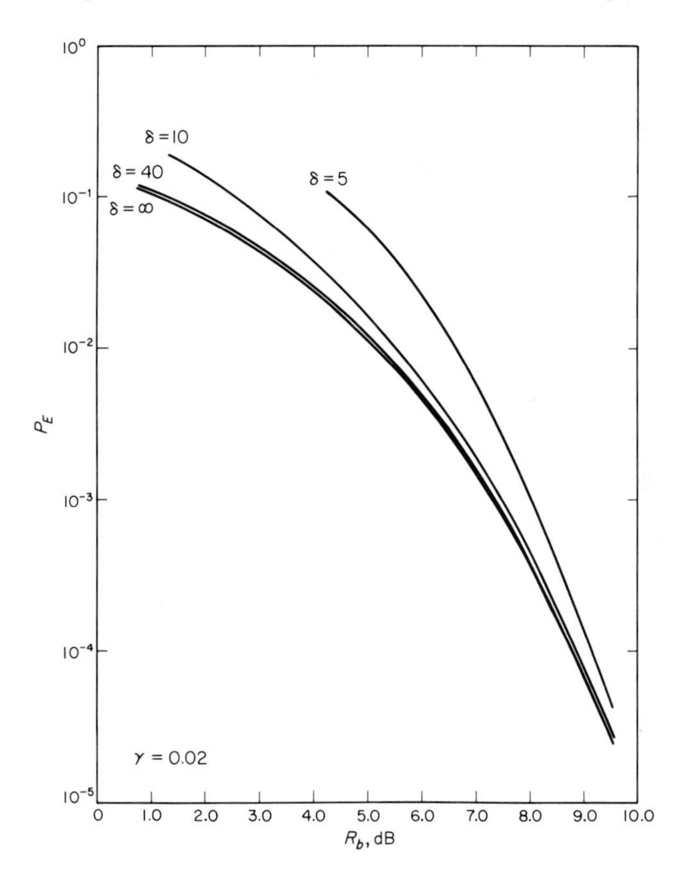

Fig. 6-20 Average error probability performance for coherent detection of differentially encoded PSK where the synchronization reference signals are provided by a Costas loop

Eq. (6-44) simplifies to

$$P_E(\phi) = \text{erfc}\,(\sqrt{R_b}\cos\phi)[1 - \tfrac{1}{2}\,\text{erfc}\,(\sqrt{R_b}\cos\phi)] \tag{6-47}$$

Figure 6-20 illustrates the performance of a coherent receiver which employs a Costas loop for tracking, and differentially encoded PSK signaling for ambiguity resolution. The data presented in this figure is obtained by substituting (6-47) into (6-37) with γ being held fixed and δ allowed to vary.

We conclude this section with a comparison of the word error probability performance for coherent detection of MPSK (with perfect ambiguity resolution) and differentially encoded MPSK (see Fig. 6-21). Included also are the corresponding results (taken from Fig. 5-22) for differentially coherent detec-

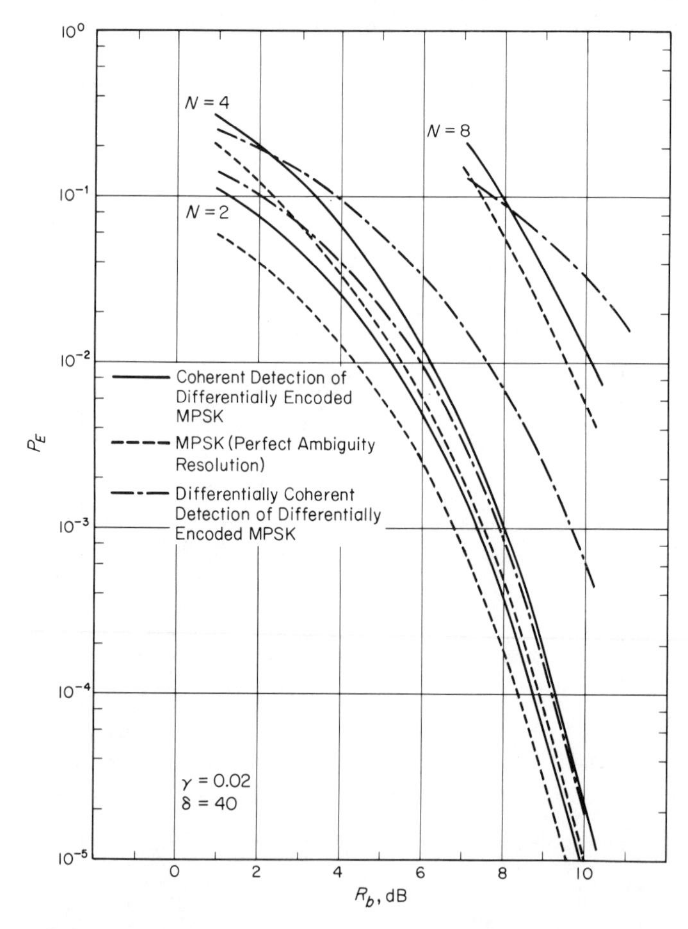

Fig. 6-21 A comparison of the average word error probability performance for coherent detection of MPSK and differentially encoded MPSK, and differentially coherent detection of differentially encoded MPSK

tion of differentially encoded MPSK. For $N = 2$, we observe that the differentially coherent receiver gives poorer performance for all values of R_b than does the coherent receiver. For $N = 4$, because of the large noisy reference loss at small R_b, the differentially coherent receiver actually gives better performance than the coherent receiver. In fact, for $N = 8$, we see that the performance of the differentially coherent receiver can even surpass the performance of a coherent receiver with perfect ambiguity resolution. However, for both $N = 4$ and $N = 8$, as R_b increases, the differentially coherent receiver performance deteriorates rapidly.

6-9. WORD ERROR PROBABILITY PERFORMANCE OF A SUBOPTIMUM L-ORTHOGONAL RECEIVER WITH NOISY REFERENCE SIGNALS

By now, we have established the necessary theory to characterize the word error probability performance of the suboptimum L-orthogonal receiver of Fig. 5-19 when the synchronization reference signals are noisy. Recalling that the portion of the receiver that detects the frequency of the transmitted signal is noncoherent, hence independent of the loop phase error, we need only be concerned with establishing the conditional probability of error associated with the phase estimation. This conditional error probability, $P_E(L; \phi)$, is however, identical to that of an MPSK receiver of L signals. Thus, the conditional error probability of the L-orthogonal receiver is[10]

$$P_E(\phi) = P_E(M)|_{\text{NON}} + P_E(L; \phi) - [P_E(M)|_{\text{NON}}][P_E(L; \phi)] \tag{6-48}$$

where $P_E(L; \phi)$ is given by (6-42) with N replaced by L. For the more general configuration as given in Prob. 5-22, Eq. (6-48) still applies if $P_E(L; \phi)$ is appropriately defined. For the phase estimator of case a), we again use (6-42) for $P_E(L; \phi)$, while for the phase estimator of case b), Eq. (6-44) applies with N replaced by L. Assuming that the M carrier-tracking loops are of the Lth power loop or L-phase Costas loop type as discussed in Chap. 2, and all have identical loop parameters, then the average word error probability of the suboptimum receiver is given by (6-43) with N replaced by L.

APPENDIX A
A SERIES SOLUTION FOR AVERAGE ERROR PROBABILITY, P_E

Using a result given by Luke[9] a simple change of variables gives

$$\operatorname{erf}(z\cos\phi) = \frac{z\exp(-z^2/2)}{\sqrt{\pi}}\sum_{k=0}^{\infty}(-1)^k\epsilon_k I_k\left(\frac{z^2}{2}\right) \times\left[\frac{\cos(2k+1)\phi}{2k+1} - \frac{\cos(2k-1)\phi}{2k-1}\right] \tag{A-1}$$

Recalling that

$$\operatorname{erfc}(y) = 1 - \operatorname{erf}(y) \tag{A-2}$$

then, substituting Eqs. (A-1) and (A-2) into (6-22) gives

$$\begin{aligned} P_E &= \int_{-\pi}^{\pi} p(\phi)\frac{1}{2}\left\{1 - \sqrt{\frac{R_{b\lambda}}{\pi}}\exp\left(-\frac{R_{b\lambda}}{2}\right)\sum_{k=0}^{\infty}(-1)^k\epsilon_k I_k\left(-\frac{R_{b\lambda}}{2}\right)\right. \\ &\quad \left.\times\left[\frac{\cos(2k+1)\phi}{2k+1} - \frac{\cos(2k-1)\phi}{2k-1}\right]\right\}d\phi \\ &= \frac{1}{2} - \frac{1}{2}\sqrt{\frac{R_{b\lambda}}{\pi}}\exp\left(-\frac{R_{b\lambda}}{2}\right)\sum_{k=0}^{\infty}(-1)^k\epsilon_k I_k\left(\frac{R_{b\lambda}}{2}\right) \\ &\quad \times\left[\frac{\overline{\cos(2k+1)\phi}}{2k+1} - \frac{\overline{\cos(2k-1)\phi}}{2k-1}\right] \end{aligned} \tag{A-3}$$

Using the results obtained for the circular moments of $p(\phi)$ in Chap. 2,

$$\begin{aligned} P_E &= \frac{1}{2}\left\{1 - \sqrt{\frac{R_{b\lambda}}{\pi}}\exp\left(-\frac{R_{b\lambda}}{2}\right)\sum_{k\infty 0}^{\infty}(-1)^k\epsilon_k I_k\left(\frac{R_{b\lambda}}{2}\right)\right. \\ &\quad \left.\times\left[\frac{Re\{b_{2k+1-j\beta}\}}{2k+1} - \frac{Re\{b_{2k-1-j\beta}\}}{2k-1}\right]\right\} \end{aligned} \tag{A-4}$$

where

$$\begin{aligned} b_{2k+1-j\beta} &= \frac{I_{2k+1-j\beta}(\alpha)}{I_{-j\beta}(\alpha)} \\ b_{2k-1-j\beta} &= \frac{I_{2k-1-j\beta}(\alpha)}{I_{-j\beta}(\alpha)} \end{aligned} \tag{A-5}$$

PROBLEMS

6-1. Evaluate the signal correlation coefficient $\lambda_{12}(\varphi)$ of (6-7) if the data sequence is
a) NRZ-L
b) RZ
c) Bi-phase Level (Manchester Code)

6-2. Starting with Eqs. (6-13) and (6-15), derive the mean $\overline{q_{d1}}$ and variance $\sigma^2_{q_{d1}}$ of the differenced cross-correlator output q_{d1} as given by (6-14) and (6-17) respectively.

6-3. Consider a PSK system that derives its carrier synchronization reference from a PLL. If the loop operates with zero detuning and its signal-to-noise ratio corresponds to $\alpha = 5$, compute the noisy reference loss at an average error probability of 10^{-2}. If the loop signal-to-noise ratio is increased so that $\alpha = 10$, by how much has the noisy reference loss been reduced? For both cases, assume that the phase error is essentially constant over the symbol interval.

6-4. Repeat Prob. 6-3 if the phase error may be assumed to vary rapidly over the symbol interval.

6-5. For the system of Prob. 6-3, assume that $\alpha = 5$ and $R_{b\lambda} = 1$. Using the results in Table 6-1, compute approximate lower and upper bounds on the ratio $[P_E; \alpha]/[P_E; \infty]$ and compare these bounds with the exact result as obtained from Fig. 6-7.

6-6. In Sec. 6-4, the statement was made that the irreducible error probability is equivalent to the probability that the magnitude of the phase error is greater than $\pi/2$, that is,

$$P_E = 1 - \int_{-\pi/2}^{\pi/2} p(\phi)d\phi$$

If the p.d.f. of ϕ is written in the Fourier series form

$$p(\phi) = \frac{1}{2\pi} \sum_{n=-\infty}^{\infty} C_n[\cos n\phi + j \sin n\phi]$$

with

$$C_n = \overline{\cos n\phi} - j\,\overline{\sin n\phi}.$$

a) show that the irreducible error probability can be expressed in the alternate form

$$P_E = \frac{1}{2} - \frac{2}{\pi} \sum_{n=1,3,5\ldots}^{\infty} \frac{(-1)^{(n-1)/2}}{n} Re\{b_{n-j\beta}\}$$

b) Show that the result of part a) agrees with that obtained in Appendix A

6-7. For the case of bi-orthogonal signals discussed in Chap. 5, develop an expression for the conditional word error probability $P_E(\phi)$, analogous to (6-39). Assume that the phase error varies slowly over the word time $T = nT_b$.

6-8. A telemetry system is to be designed so as to use a (64, 6) orthogonal code. Carrier synchronization is to be accomplished with a squaring loop in which a "genie" is used to perfectly resolve the 180-degree phase ambiguity. Assuming that the average word error probability is 10^{-2}, plot the noisy reference loss versus δ for $\gamma = .00125$, .0025, and .005. Repeat for $P_E(64) = 10^{-5}$.

6-9. Show that the conditional word error probability given by (6-44), together with (6-45), is independent of a phase shift of $2(k-1)\pi/N$, $k = 1, 2, \ldots, N$, and thus, the performance of the receiver is independent of the phase ambiguities present in the suppressed carrier-tracking loop.

REFERENCES

1. LINDSEY, W. C., "Phase-Shift-Keyed Signal Detection with Noisy Reference Signals," *IEEE Transactions on Aerospace and Electronic Systems*, Vol. AES-2, No. 4 (July, 1966) 393–401.

2. Lindsey, W. C., and Simon, M. K., "The Effect of Loop Stress on the Performance of Phase-Coherent Communication Systems," *IEEE Transactions on Communication Technology*, Vol. COM-18, No. 5 (October, 1970) 569–588.

3. Lindsey, W. C., *Synchronizations Systems in Communication and Control*, Prentice-Hall, Inc., Englewood Cliffs, N. J., 1972.

4. Blake, I. F., and Lindsey, W. C., "Effects of Phase-Locked Loop Dynamics on Phase-Coherent Communications," Jet Propulsion Laboratory, Pasadena, Calif., SPS 37–54, Vol. III (December, 1968) 192–195.

5. Aein, J. M., "Coherency for the Binary Symmetric Channel," *IEEE Transactions on Communication Technology*, Vol. COM-18, No. 4 (August, 1970) 344–352.

6. Hayes, J., and Lindsey, W. C., "Power Allocation—Rapidly Varying Phase Error," *IEEE Transactions on Communication Technology*, Vol. COM-17, No. 2 (April, 1969) 323–326.

7. Didday, R. L., and Lindsey, W. C., "Subcarrier Tracking Methods and Communication System Design," *IEEE Transactions on Communication Technology*, Vol. COM-16, No. 4 (August, 1968) 541–550.

8. Lindsey, W. C., and Simon, M. K., "The Performance of Suppressed Carrier Tracking Loops in the Presence of Frequency Detuning," *Proceedings of the IEEE*, Vol. 58, No. 9 (September, 1970) 1315–1321.

9. Luke, Y. L., *Integrals of Bessel Functions*, McGraw-Hill Book Company, New York, N. Y., 1962.

10. Lindsey, W. C., and Simon, M. K., "L-Orthogonal Signal Transmission and Detection," *IEEE Transactions on Communication Technology*, Vol. COM-20, No. 5 (October, 1972) 953–960.

Chapter 7

DESIGN OF ONE-WAY AND TWO-WAY PHASE-COHERENT COMMUNICATION SYSTEMS

7-1. INTRODUCTION

In the design of phase-coherent communication systems, which possess the tracking and data acquisition capabilities discussed in Chaps. 1 and 3, the total transmitted power frequently must be apportioned among several signals. Aside from providing for the power necessary for carrier tracking, the transmitter must allot a portion of its total power to signals that convey data to the receiver, and frequently signals that represent timing or synchronizing information.

The purpose of this chapter is to develop a theory that will enable the communications engineer to optimally determine this power apportionment for various practical phase modulation formats. The first sections concern the analysis and design of one-way and two-way *single-channel systems*[1]. By single channel we mean a system in which the data signal and the synchronization signal used to operate the data detector are embodied in a composite signal—that is, a data modulated subcarrier, which, in turn, phase modulates the RF carrier. For such a system all of the available transmitter sideband power is allocated to this composite signal. The ratio of power in the carrier to total power transmitted plays an important role in the design and specification of system performance. Following our discussion of single-channel systems, we consider the analysis and design of *two-channel systems*[2,3]—those in which two biphase modulated subcarriers are used to phase modulate the RF carrier. Systems in which two channels are required are common in practice. For example, one channel might be used to transmit data, while the other might be used to synchronize the data demodulator.

We refer to such a modulation format as data/sync. Alternately, each of the two channels might be used for both data transmission and data demodulator synchronization purposes. In this case, we refer to the modulation format as data 1/data 2. The final section of the chapter generalizes our results to include coded and uncoded *multichannel systems*[4].

7-2. OPTIMAL DESIGN OF SINGLE-CHANNEL SYSTEMS

Basic System Model

We begin the discussion by reviewing the general model of a two-way communication system introduced in Chap. 3 (see Fig. 3-1) with particular emphasis here on the addition of a signal which gives the data transmission capability. This augmented form of the two-way communication system combining data transmission and tracking capabilities is illustrated in Fig. 7-1.

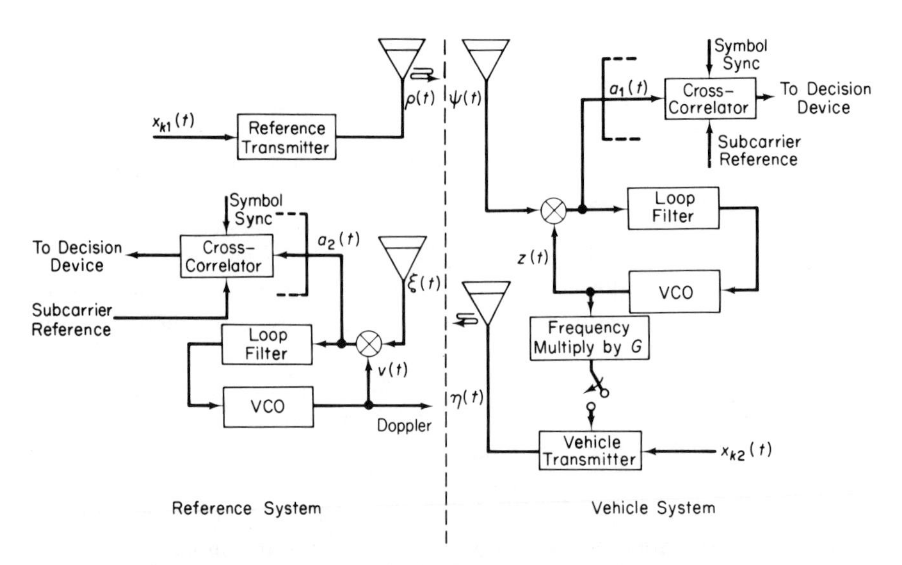

Fig. 7-1 A two-way single-channel communication system

The overall system is composed of two parts: a vehicle system and a ground station or reference system. These systems interact so as to acquire, transmit, and receive data of several types: tracking data, engineering data, scientific data, and command data.

Stored at the reference transmitter are two equiprobable, equal-energy, equal-duration baseband waveforms, $s_{k1}(t)$; $k = 1, 2$, which represent the states of the information to be transmitted to the vehicle. This information could be in the form of a command signal or a pseudo-noise signal which might be used for ranging purposes. The baseband signal is then modulated onto a subcarrier as indicated in Fig. 1-4 to form the data-modulated subcarrier signal $x_{k1}(t)$; $k = 1, 2$.

Analogous to our definition of correlation for baseband signals as given by Eq. (5–20), we define the normalized correlation coefficient λ_1 between the modulated subcarrier signals $x_{11}(t)$ and $x_{21}(t)$ as

$$\lambda_1 = \frac{1}{T_1} \int_0^{T_1} x_{11}(t) x_{21}(t)\, dt$$

and, for convenience, we assume that signals $x_{11}(t)$ and $x_{21}(t)$ are normalized such that

$$\int_0^{T_1} x_{k1}^2(t)\, dt = T_1; k = 1, 2$$

When the data signal is a sequence of ± 1's, and the subcarrier is a unit power square wave, then $x_{k1}(t)$ is a ± 1 digital waveform with $\lambda_1 = -1$. For this case, the data-modulated subcarrier is used to phase modulate the RF carrier; hence, we transmit

$$p(t) = \sqrt{2P_1} \sin [\omega_0 t + \theta_0 + (\cos^{-1} m_1) x_{k1}(t)] \tag{7-1}$$

where P_1 is the total power radiated, m_1 is the reference system *modulation factor and* $\theta_{m_1} \triangleq \cos^{-1} m_1$ *is the angle modulation index*. On the other hand, for $-1 < \lambda_1 \leq 0$, we transmit another signal

$$p(t) = \sqrt{2m_1^2 P_1} \sin [\omega_0 t + \theta_0] + \sqrt{2(1 - m_1^2)P_1}\, x_{k1}(t) \cos [\omega_0 t + \theta_0] \tag{7-2}$$

which is a bit more difficult to generate. However, since for $x_{k1}(t) = \pm 1$, (7-2) is mathematically equivalent to (7–1), we shall use (7–2) as our characterization of the transmitted signal for all $-1 \leq \lambda_1 \leq 0$, and (7–1) specifically for the case $\lambda_1 = -1$. Also, from (7–2) it is clear that the parameter m_1 represents the square root of the ratio of the power in the carrier to the total power radiated—that is, $m_1 \triangleq (P_{c1}/P_1)^{1/2}$.

The channel introduces an unknown phase shift to the transmitted waveform and further disturbs $p(t)$ with additive white Gaussian noise $n_1(t)$ of single-sided spectral density N_{01} w/Hz. Thus, we observe in the vehicle, for $\lambda_1 = -1$, the following:

$$\psi(t) = \sqrt{2P_1} \sin [\omega_1 t + (\cos^{-1} m_1) x_{k1}(t) + \theta_0 + \theta_1] + n_1(t)$$

The vehicle tracks the carrier components in $\psi(t)$ by means of a narrowband PLL. The output $z(t)$ of the VCO is used as a coherent reference in demodulating $\psi(t)$. The vehicle reference waveform $z(t)$ is conveniently taken to be

$$z(t) = \sqrt{2} \cos [\omega_1 t + \hat{\theta}_1(t)]$$

where $\hat{\theta}_1(t)$ is the PLL estimate of $\theta_0 + \theta_1$ in the presence of noise. The additive noise process $n_1(t)$ may be represented as in (3–3). After neglecting the double-frequency components, multiplication of $\psi(t)$ with $z(t)$ produces

$$a_1(t) = \sqrt{(1 - m_1^2)P_1}\, x_{k1}(t) \cos \varphi_{v1}(t) + N_1[t, \varphi_{v1}(t)] \tag{7-3}$$

where $N_1[t, \varphi_{v1}(t)]$ is approximately low-pass white Gaussian noise of single-sided spectral density N_{01} w/Hz, and $\varphi_{v1}(t) = \theta_0 + \theta_1 - \hat{\theta}_1(t)$ is the vehicle system phase error. We have also neglected the additional low-frequency component $(m_1^2 P_1)^{1/2} \sin \varphi_{v1}(t)$ since, in practice, the data symbols are modulated on a subcarrier such that the significant spectral components in the baseband modulation do not lie in the vicinity of the origin. Hence, we are implying the presence of a band-pass filter centered around the subcarrier frequency.

The decision in the vehicle is made in favor of that signal which gives rise to the largest cross-correlation; the vehicle demodulator computes

$$q_{dk} = \int_0^{T_1} a_1(t)[x_{11}(t) - x_{21}(t)]\, dt \tag{7-4}$$

and compares the result with zero. If $q_{dk} > 0$, x_{11} is announced, and if $q_{dk} < 0$, x_{21} is announced.

In the reverse direction—transmission of data back to the reference system—the output of the vehicle's VCO is used as a carrier for transmission of one of two equiprobable, equal-energy, equal-duration baseband waveforms, $s_{k2}(t)$; $k = 1, 2$. The unit power data-modulated subcarriers $x_{12}(t)$ and $x_{22}(t)$ corresponding to these baseband signals have a normalized correlation coefficient λ_2 over the time interval $(0, T_2)$. In this case, the output of the vehicle is conveniently represented by the following:

$$\eta(t) = \sqrt{2P_2} \sin [\omega_{10} t + (\cos^{-1} m_2) x_{k2}(t) + \hat{\theta}_{10}(t)] \tag{7-5}$$

for $\lambda_2 = -1$, and

$$\eta(t) = \sqrt{2m_2^2 P_2} \sin [\omega_{10} t + \hat{\theta}_{10}(t)] + \sqrt{2(1 - m_2^2)P_2}\, x_{k2}(t) \cos [\omega_{10} t + \hat{\theta}_{10}(t)]$$

for $-1 < \lambda_2 \leq 0$. Here m_2 is a modulation factor representing the square root of the ratio of power in the carrier to the total power radiated: $m_2 \triangleq (P_{c2}/P_2)^{1/2}$, $\theta_{m_2} \triangleq \cos^{-1} m_2$ is the corresponding angle modulation index, and $\omega_{10} \triangleq G\omega_1$, $\hat{\theta}_{10}(t) = G\hat{\theta}_1(t)$ with G the static phase gain of the vehicle transponder.

The downlink channel (assumed to be statistically independent from the uplink channel) further perturbs $\eta(t)$ by inserting an unknown phase shift θ_2 and additive white Gaussian noise $n_2(t)$ of single-sided spectral density N_{02} w/Hz. Thus, the reference receiver observes

$$\xi(t) = \sqrt{2P_2} \sin [\omega_2 t + (\cos^{-1} m_2) x_{k2}(t) + \hat{\theta}_{10}(t) + \theta_2] + n_2(t)$$

where $n_2(t)$ is modeled as in (3-7).

The ground receiver tracks the carrier component in $\xi(t)$ for purposes of Doppler measurement and for performing the demodulation of the data. We denote the output of the reference VCO by

$$v(t) = \sqrt{2} \cos [\omega_2 t + \hat{\theta}_2(t)]$$

where $\hat{\theta}_2(t)$ is the estimate of phase of the observed carrier component. Multiplying $\xi(t)$ by $v(t)$ and neglecting the double-frequency components gives

$$a_2(t) = \sqrt{(1 - m_2^2)P_2}x_{k2}(t) \cos \varphi_{r2}(t) + N_2[t, \varphi_{r2}(t)] \tag{7-6}$$

where $\varphi_{r2}(t) = \hat{\theta}_{10}(t) + \theta_2 - \hat{\theta}_2(t)$ is the reference system phase error. Again $N_2[t, \varphi_{r2}(t)]$ is approximately Gaussian and white with a low-pass single-sided spectral density N_{02} w/Hz.

To recapitulate, we see that the design engineer has at his disposal several communication parameters. For the uplink he has the total power radiated P_1, the single-sided noise spectral density N_{01}, uplink data rate $\mathcal{R}_1 = T_1^{-1}$, vehicle carrier-tracking loop bandwidth W_{L1}, and modulation factor $m_1 = (P_{c1}/P_1)^{1/2}$. The corresponding downlink parameters are P_2, N_{02}, $\mathcal{R}_2 = T_2^{-1}$, W_{L2}, and $m_2 = (P_{c2}/P_2)^{1/2}$. In the following section, we relate these parameters and determine the value of m_n that minimizes the probability of error P_{En} for a fixed data rate-to-carrier-tracking loop bandwidth ratio, say $\mathcal{R}_n/W_{Ln}$ $(n = 1, 2)$.

Probability Density Functions for the System Phase Errors

The p.d.f.'s of the system phase errors are of great importance in specifying the performance of the two-way system. In Chap. 3, we presented the reference system phase error p.d.f. [see (3-20)] for such a two-way system, but in the absence of any data modulation. When data signals are phase modulated on RF carriers, the phase error p.d.f. is unchanged in form and depends only on the carrier component of the received signal since the modulation being centered around the subcarrier frequency is not tracked by the narrowband PLL. Hence, (3-20) together with (3-17) and (3-18) still applies. We review the results here, rewriting them in terms of the system modulation factors and signal correlation coefficients. First,

$$p(\phi_{r2}) = \frac{I_0[|\alpha_1 + \rho_2 \exp(j\phi_{r2})|]}{2\pi I_0(\alpha_1)I_0(\rho_2)} \qquad |\phi_{r2}| \le \pi \tag{7-7}$$

where α_1 and ρ_2 are related to the uplink and downlink parameters through

$$\alpha_1 \triangleq \frac{\rho_1}{G^2 K_R(r_1, r_2, \xi)}$$

$$\rho_n \triangleq \frac{2m_n^2 P_n}{N_{0n} W_{Ln}} = m_n^2 \delta_n \frac{P_n T_n}{2N_{0n}}(1 - \lambda_n) = m_n^2 \delta_n R_n \qquad n = 1, 2 \tag{7-8}$$

with
$$R_n \triangleq \frac{P_n T_n}{2N_{0n}}(1 - \lambda_n)$$
$$\delta_n \triangleq \frac{4\mathcal{R}_n}{W_{Ln}(1-\lambda_n)} = \frac{4}{T_n W_{Ln}(1-\lambda_n)} \qquad n = 1, 2 \tag{7-9}$$

In certain cases of practical interest $G^2 K_R(r_1, r_2, \xi)$ is approximately one, so that α_1 is, to a good approximation, the signal-to-noise ratio existing in the vehicle tracking loop bandwidth. On the other hand, ρ_2 is the signal-to-noise ratio existing in the reference system tracking loop.

The p.d.f. of the phase error in the vehicle system is given by

$$p(\phi_{v1}) = \frac{\exp[\rho_1 \cos\phi_{v1}]}{2\pi I_0(\rho_1)} \qquad |\phi_{v1}| \leq \pi \tag{7-10}$$

By setting $\alpha_1 = \infty$ in (7-7), one obtains an expression for the p.d.f. of the one-way reference receiver phase error, ϕ_{r1}, which is identical to (7-10) if ρ_1 is replaced by ρ_2. Hence, studying the one-way performance of the downlink is equivalent to studying the performance of the vehicle system. In what follows, we shall implicitly assume this equivalence when we refer to one-way system performance, and for simplicity of notation ϕ_{r1} shall be used instead of ϕ_{v1}.

Demodulator Output Statistics and System Performance

We have established the way in which information is conveyed in one-way and two-way single-channel communication systems. Now it is necessary to develop analytical expressions that specify the performance of each system and to relate the various parameters that enter into a particular design. In order to shorten the subsequent material, we draw heavily upon previous notation and results. The approach is the same as that taken in Chap. 6 where the method of conditional probabilities is employed to develop expressions for the particular parameter, p.d.f., or moment of interest, conditioned on the fact that the system phase error is fixed at ϕ_{rn} rad. Thus, by averaging over the condition, which is random, we easily determine the behavior of interest as the system phase error takes on all possible values.

We begin our analysis by specifying the statistics at the output of the data demodulator. As already pointed out in (7-3) and (7-6), a system phase error of ϕ_{rn} radians effectively produces a reduction in signal strength by $\cos\phi_{rn}$. Further, when ϕ_{rn} is constant over T_n, the output of the demodulator is Gaussian since the input is Gaussian. Hence, as was done in Chap. 6, we consider equivalent normalized conditional output statistics of the demodulator under the hypotheses of signal plus noise and of noise alone. For the signal plus noise case,

$$p_n(q_{d1} | \phi_{rn}) = \frac{1}{\sqrt{2\pi}} \exp\left[-\frac{(q_{d1} - \sqrt{\Psi_n}\cos\phi_{rn})^2}{2}\right] \qquad n = 1, 2 \tag{7-11}$$

where the parameter Ψ_n is given by

$$\Psi_n = 4(1 - m_n^2)R_n = 2(1 - m_n^2)\left(\frac{P_n T_n}{N_{0n}}\right)(1 - \lambda_n) \qquad n = 1, 2 \tag{7-12}$$

In the presence of noise only,

$$p_n(q_{d2} \mid \phi_{rn}) = \frac{1}{\sqrt{2\pi}} \exp\left(-\frac{q_{d2}^2}{2}\right) = p_n(q_{d2}) \qquad n = 1, 2 \tag{7-13}$$

The output statistics $p_n(q_{d1})$ may be obtained by averaging (7-11) over the p.d.f. of the system phase error as given by (7-10) or (7-7) for $n = 1, 2$, respectively.

The moments of these p.d.f.'s are of practical interest. In particular. the mean of the demodulator output (differenced cross-correlator outputs) is given by

$$q_{d1} = \int_{-\infty}^{\infty} \int_{-\pi}^{\pi} q_{d1} p_n(q_{d1} \mid \phi_{rn}) p(\phi_{rn}) d\phi_{rn} dq_{d1} \qquad n = 1, 2 \tag{7-14}$$

Substituting (7-11) and (7-10) or (7-7) into (7-14) and carrying out the necessary integration gives

$$\bar{q}_{d1} = \sqrt{\Psi_n} \prod_{i=1}^{n} \frac{I_1(\alpha_i)}{I_0(\alpha_i)} \qquad n = 1, 2 \tag{7-15}$$

where, for brevity of notation, we have set ρ_2 equal to α_2. The mean-squared value of the demodulator output may be determined from

$$\overline{q_{d1}^2} = \int_{-\infty}^{\infty} \int_{-\pi}^{\pi} q_{d1}^2 \, p(\phi_{rn}) p_n(q_{d1} \mid \phi_{rn}) d\phi_{rn} dq_{d1} \tag{7-16}$$

which, upon carrying out the necessary integration, gives

$$\overline{q_{d1}^2} = 1 + \frac{\Psi_n}{2} + \frac{\Psi_n}{2} \prod_{i=1}^{n} \frac{I_2(\alpha_i)}{I_0(\alpha_i)} \qquad n = 1, 2 \tag{7-17}$$

Thus, the variance of the demodulator output (differenced cross-correlator outputs) may be obtained from (7-15) and (7-17) as

$$\sigma_{q_{d1}}^2 = 1 + \frac{\Psi_n}{2} + \frac{\Psi_n}{2} \prod_{i=1}^{n} \frac{I_2(\alpha_i)}{I_0(\alpha_i)} - \Psi_n \left[\prod_{i=1}^{n} \frac{I_1(\alpha_i)}{I_0(\alpha_i)}\right]^2 \qquad n = 1, 2 \tag{7-18}$$

Implicit in (7-15), (7-17), and (7-18) is the fact that when $n = 1$, α_1 must be replaced by ρ_1. In practice, these expressions for the moments may be used to check a particular design.

We now turn to the problem of calculating the probability that the demodulator will err in making its decision. The conditional error probability, the condition being that the system phase error ϕ_{rn} is constant over T_n, is given by an expression analogous to (6-18)—that is,

$$P_E(\phi_{rn}) = 1 - \int_{-\infty}^{\infty} p_n(q_{d1} \mid \phi_{rn}) dq_{d1} \int_{-\infty}^{q_{d1}} p_n(q_{d2} \mid \phi_{rn}) dq_{d2} \tag{7-19}$$

Substituting (7-11) and (7-13) into (7-19) and simplifying the results yields

$$P_E(\phi_{rn}) = \tfrac{1}{2} \operatorname{erfc}\left[\sqrt{(1 - m_n^2)R_n} \cos \phi_{rn}\right] \qquad n = 1, 2 \tag{7-20}$$

The average error rates for the reference and vehicle data demodulators, P_{En}, $n = 1, 2$, are easily obtained by averaging (7-20) over the phase error p.d.f.'s as defined in (7-10) and (7-7) for $n = 1, 2$, respectively. These integrals can be expressed as infinite sums via the method outlined in Appendix A of Chap. 6. The results are

$$P_{En} = \frac{1}{2}\left\{1 - \sqrt{\frac{R_n(1-m_n^2)}{\pi}} \exp\left[-\frac{R_n}{2}(1-m_n^2)\right] \times \sum_{k=0}^{\infty} \epsilon_k(-1)^k \left[\frac{b_{2k+1}^{(n)}}{2k+1} - \frac{b_{2k-1}^{(n)}}{2k-1}\right] I_k\left[\frac{R_n}{2}(1-m_n^2)\right]\right\} \tag{7-21}$$

where $$b_l^{(n)} = \prod_{i=1}^{n} \frac{I_l(\alpha_i)}{I_0(\alpha_i)} \qquad n = 1, 2$$

and $\epsilon_k = 1$ if $k = 0$, and $\epsilon_k = 2$ if $k \neq 0$. Again, for $n = 1$, α_1 must be replaced by ρ_1. It is of interest to consider the error probability performance of the reference system ($n = 2$) in the limit as R_2 approaches infinity. Equivalently, letting α_2 approach infinity in (7–21), we find that

$$P_{E2} = \frac{1}{2}\left\{1 - \frac{1}{\pi}\sum_{k=0}^{\infty} \epsilon_k(-1)^k \left[\frac{b_{2k+1}^{(n)}}{2k+1} - \frac{b_{2k-1}^{(n)}}{2k-1}\right]\right\} = \text{Prob}\left[|\phi_{rn}| > \frac{\pi}{2}\right] \tag{7-22}$$

showing that the performance of the reference system exhibits an irreducible error probability that depends on the signal-to-noise ratio in the vehicle carrier-tracking loop; the damping coefficients r_1 and r_2; the ratio, ξ, as defined following (3-18); and G. A plot of this irreducible error versus ξ for $G = 1$, $r_1 = r_2 = 2$ and ρ_1 as a parameter is given in Fig. 7-2. In the limit as ρ_1 approaches infinity—perfect phase measurement in the vehicle—we note that P_{E2} approaches zero. The existence of this irreducible error rate is unattractive and degrades performance of the overall system.

Design Characteristics

In this section, design trends are presented enabling the communications engineer to select such parameters as the system data rate, tracking loop bandwidth, modulation factor, and the like, to minimize probability of error.

Careful analysis of the parameters involved in Eq. (7-21) indicates that when α_1 and the data rate-to-loop bandwidth ratio δ_2 are fixed, a plot of the error rate P_{E2} versus the square of the modulation factor $m_2 = (P_{c2}/P_2)^{1/2}$ (using R_2 as a running parameter) has a minimum for some $0 \leq m_2 \leq 1$, and thus an optimum m_2 and error rate exist.

To illustrate this point graphically, we consider the one-way performance characteristic: P_{E2} evaluated at $\alpha_1 = \infty$, or equivalently just P_{E1}. Figure 7-3 represents the P_{E1} versus m_1^2 characteristic. Notice that this characteristic assumes its minimum value at $m_1 = m_{1\,\text{opt}}$ and increases to one-half on either

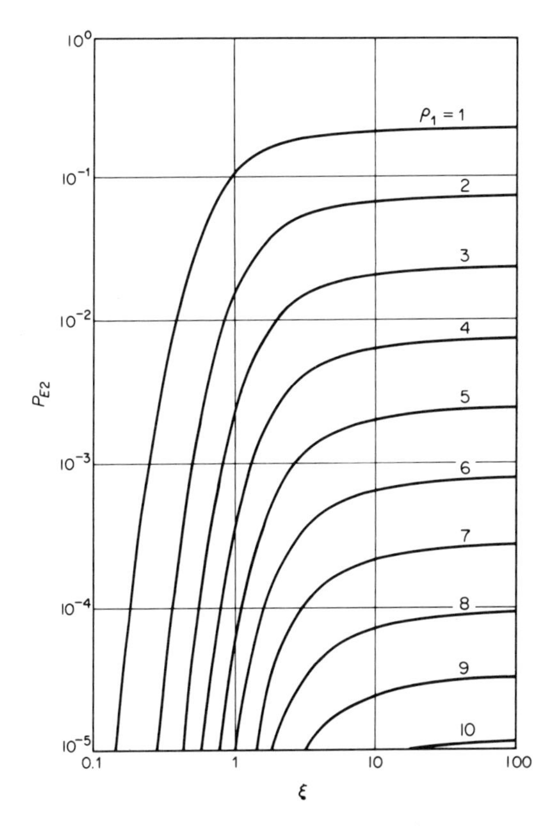

Fig. 7-2 Irreducible error probability P_{E2} versus ξ for various values of ρ_1; $r_1 = r_2 = 2$; $G = 1$

side of $m_{1\,\text{opt}}$. This interesting behavior may be explained as follows. For $m_1 < m_{1\,\text{opt}}$ the tradeoff between power in the carrier and power in the data is overexpended in the data, and the performance of the carrier-tracking loop suffers. Thus, although the data signal-to-noise ratio is large, the noisy reference loss is dominating. For $m_1 > m_{1\,\text{opt}}$ the tradeoff between power in the carrier and power in the data is overexpended in the carrier, such that the performance of the demodulator suffers at the expense of a smaller phase error in the carrier-tracking loop. Here the noisy reference loss is small, but so is the data signal-to-noise ratio.

Figure 7-4 represents two-way performance of the ground or reference system. In this case we have set α_1 equal to 9 dB. The behavior of P_{E2} versus m_2^2 is similar to that obtained for the one-way system. The major difference is that the P_{E2} versus m_2^2 characteristic exhibits a bottoming behavior for large values of R_2, which is due to the presence of additive noise on the uplink. This bottoming behavior can be eliminated by using a clean carrier

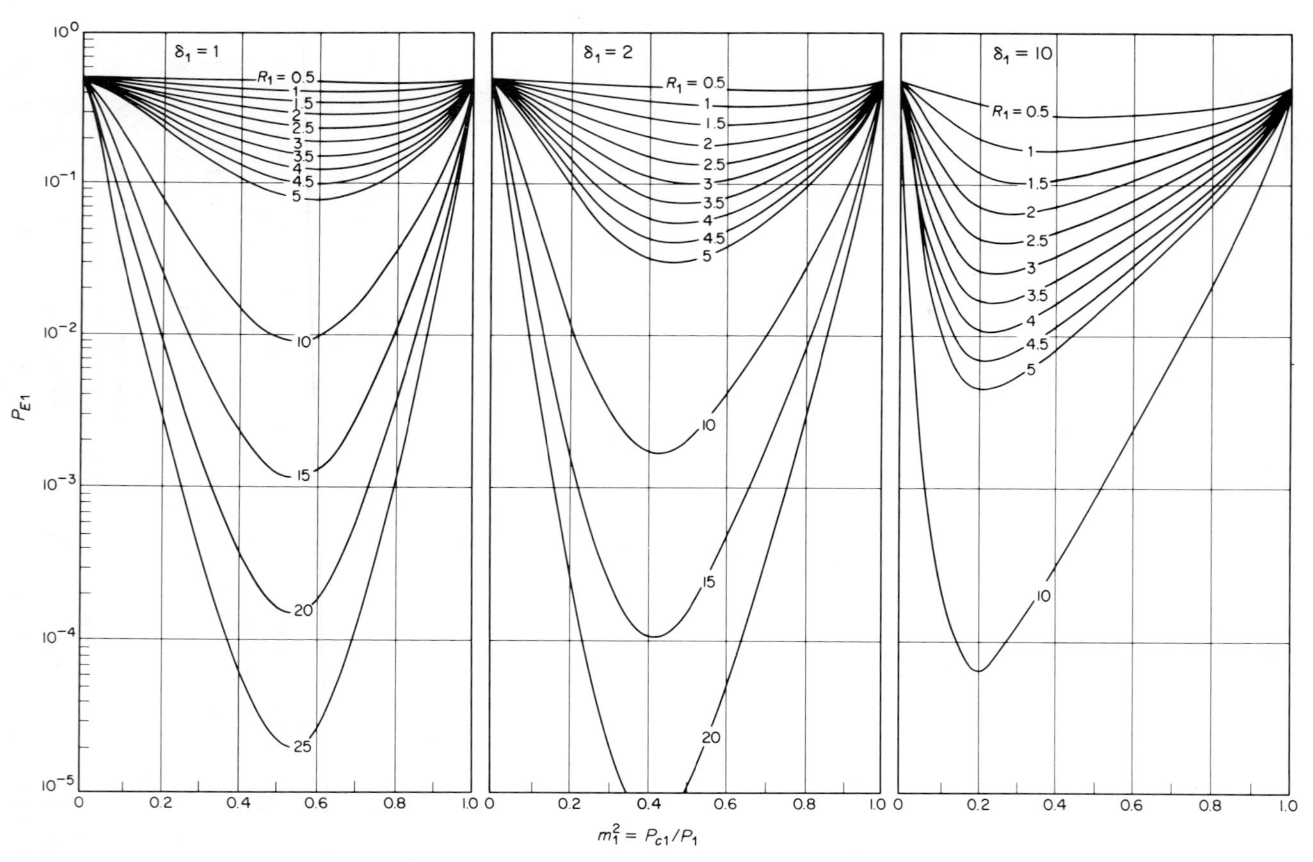

Fig. 7-3(a) Error probability versus modulation factor squared (one-way link); $\delta_1 = 1, 2, 10$

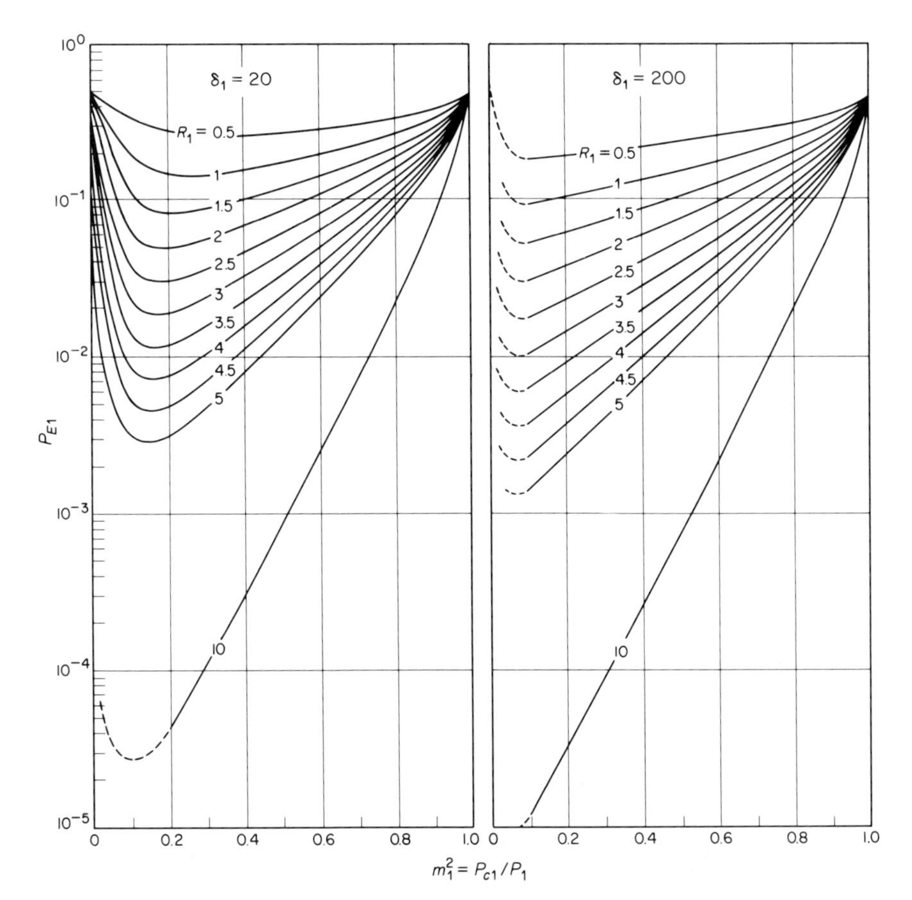

Fig. 7-3(b) Error probability versus modulation factor squared (one-way link); $\delta_1 = 20, 200$

reference in the vehicle or by increasing the uplink signal-to-noise ratio to a point where phase jitter in the vehicle's carrier-tracking loop becomes negligible.

The results given in Figs. 7-5 and 7-6 represent the optimized performance of the system. These curves were extracted from Figs. 7-3 and 7-4 by taking that value of P_{En} for which $m_n = m_{n_{\text{opt}}}$ and plotting the result versus R_n for fixed δ_n and α_n. Using these results, the design engineer may predict the performance of one-way and two-way systems, given the system parameters, or he may carry out a particular design, given the required performance constraints.

In practice, it is convenient to have an approximate formula which specifies the optimum modulation factor, $m_{n_{\text{opt}}}$, as a function of the various system parameters. Assuming that α_n is large enough so that linear PLL

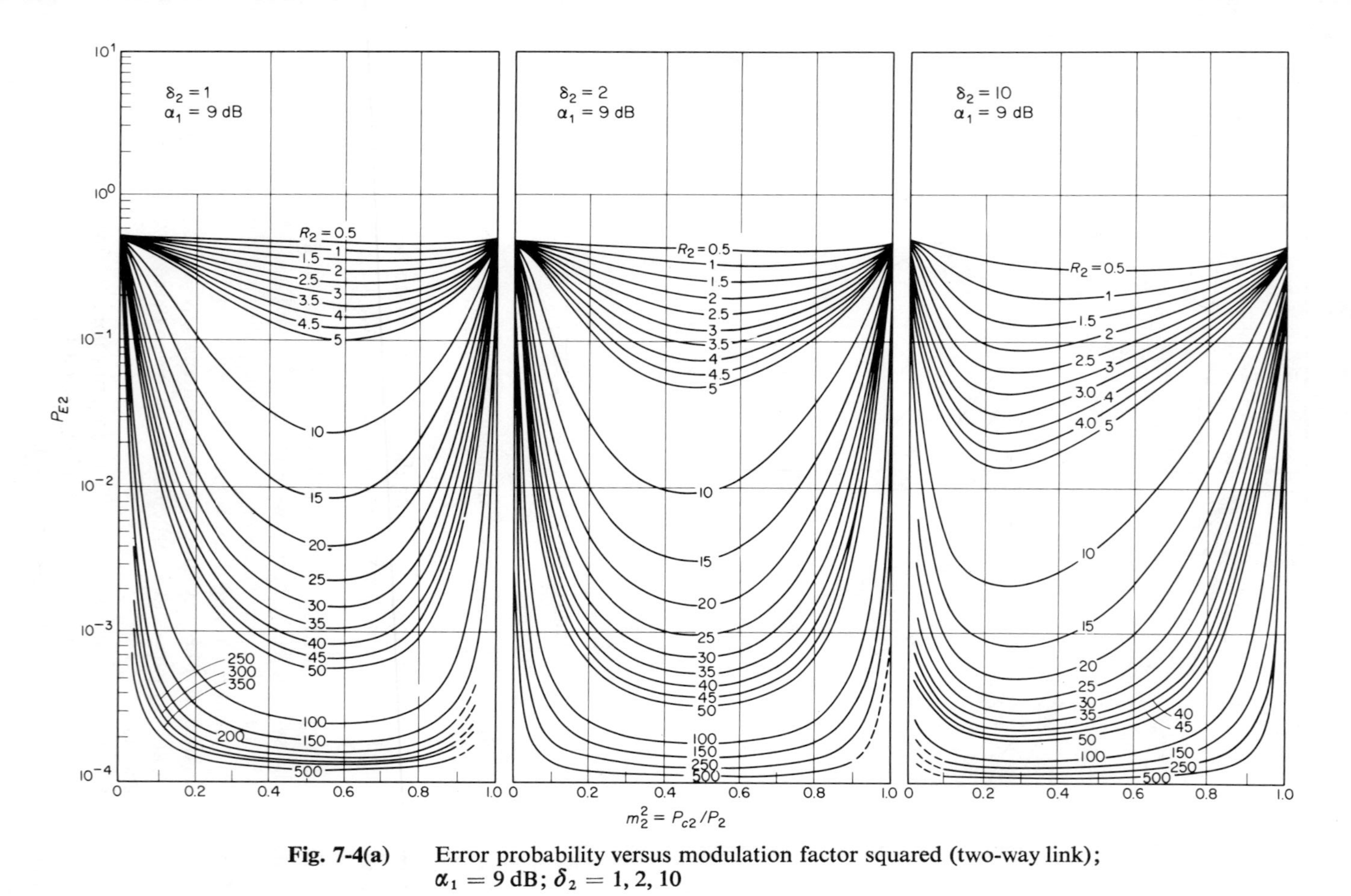

Fig. 7-4(a) Error probability versus modulation factor squared (two-way link); $\alpha_1 = 9$ dB; $\delta_2 = 1, 2, 10$

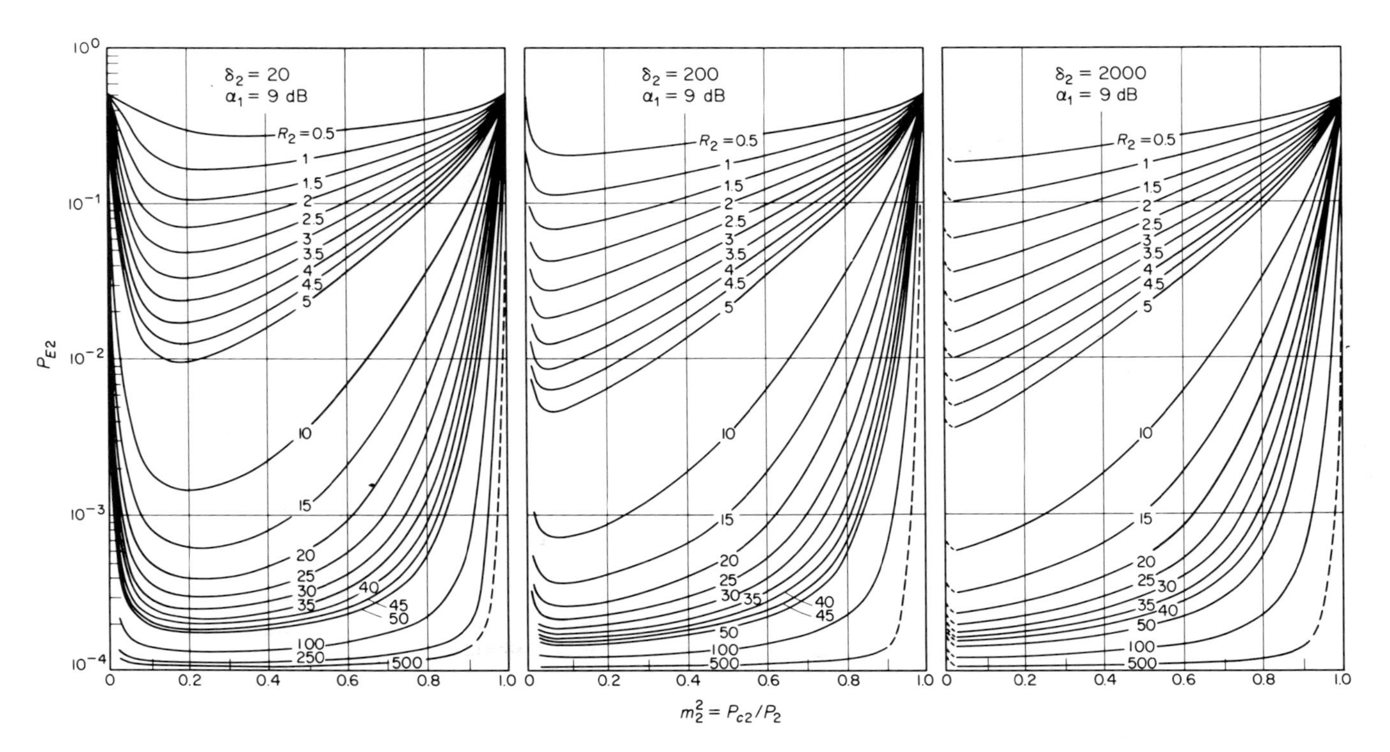

Fig. 7-4(b) Error probability versus modulation factor squared (two-way link); $\alpha_1 = 9$ dB; $\delta_2 = 20, 200, 2000$

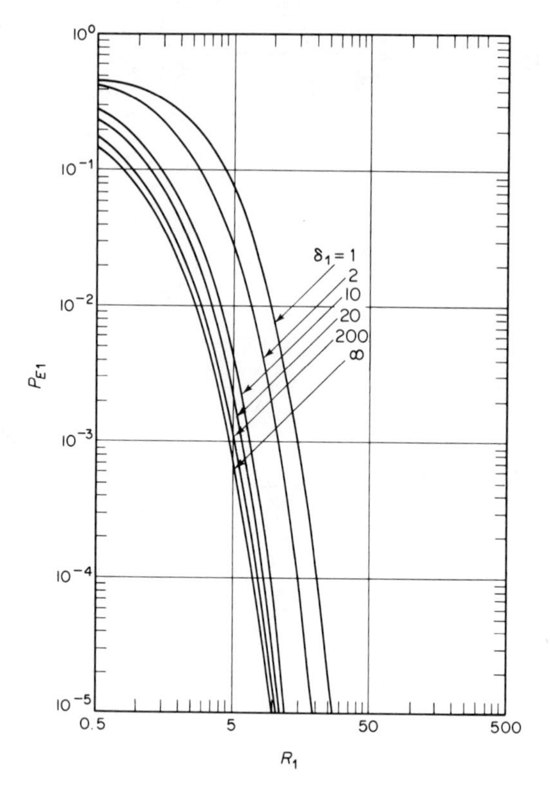

Fig. 7-5 Error probability versus signal-to-noise ratio for optimum modulation factor (one-way link)

theory applies, then P_{En} is given by

$$P_{En} \cong \frac{1}{\sqrt{2\pi \sum_{j=1}^{n} \alpha_j^{-1}}} \int_{-\infty}^{\infty} \exp\left(\frac{-\phi_{rn}^2}{2 \sum_{j=1}^{n} \alpha_j^{-1}}\right) \times \tfrac{1}{2} \operatorname{erfc}[\sqrt{(1-m_n^2)R_n} \cos \phi_{rn}] \, d\phi_{rn} \tag{7-23}$$

where for $n = 1$, α_1 is replaced by ρ_1.

For $n = 2$ (the error probability performance of the reference system), differentiating Eq. (7-23) with respect to m_2 and equating the result to zero yields

$$E\left\{\exp\left[-R_2(1-m_2^2)\cos^2\left(\frac{y}{\sqrt{\alpha_1^{-1} + m_2^2\delta_2 R_2}}\right)\right] \times \frac{m_2\delta_2 R_2\sqrt{1-m_2^2}}{(\alpha_1^{-1} + m_2^2\delta_2 R_2)^{3/2}}\, y \sin\left(\frac{y}{\sqrt{\alpha_1^{-1} + m_2^2\delta_2 R_2}}\right) - \sqrt{\frac{m_2^2}{1-m_2^2}} \cos\left(\frac{y}{\sqrt{\alpha_1^{-1} + m_2^2\delta_2 R_2}}\right)\right\} = 0 \tag{7-24}$$

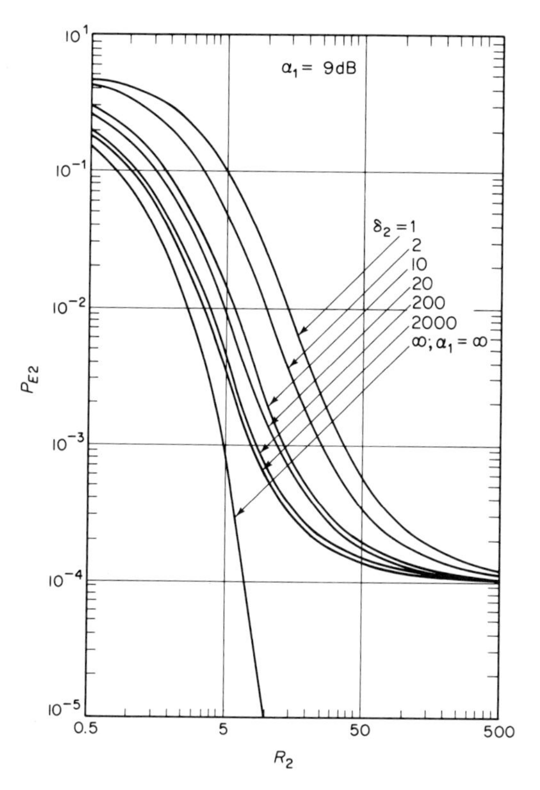

Fig. 7-6 Error probability versus signal-to-noise ratio for optimum modulation factor (two-way link)

where $E\{\cdot\}$ denotes the statistical expectation of the quantity in the parenthesis, and y is a zero mean, unit variance Gaussian random variable. Carrying out this expectation and solving for the value of m_2^2 that produces the minimum gives

$$m_{2_{\text{opt}}}^2 = \frac{(\delta_2/2)(2R_2 - 1) - \alpha_1^{-1}(1 + \delta_2)}{2\delta_2 R_2(1 + \delta_2/2)} + \frac{\sqrt{[(\delta_2/2)(2R_2 - 1) - \alpha_1^{-1}(1 + \delta_2)]^2 + 2\delta_2(1 + \delta_2/2)[\delta_2 R_2 + \alpha_1^{-1}(2R_2 - \alpha_1^{-1})]}}{2\delta_2 R_2(1 + \delta_2/2)} \tag{7-25}$$

For the one-way system, an approximate value of $m_{1_{\text{opt}}}^2$ is obtained by letting α_1 approach infinity in (7-25), or in terms of the vehicle performance parameters,

$$m_{1_{\text{opt}}}^2 = \frac{2R_1 - 1 + \sqrt{(2R_1 - 1)^2 + 8R_1(1 + \delta_1/2)}}{4R_1(1 + \delta_1/2)} \tag{7-26}$$

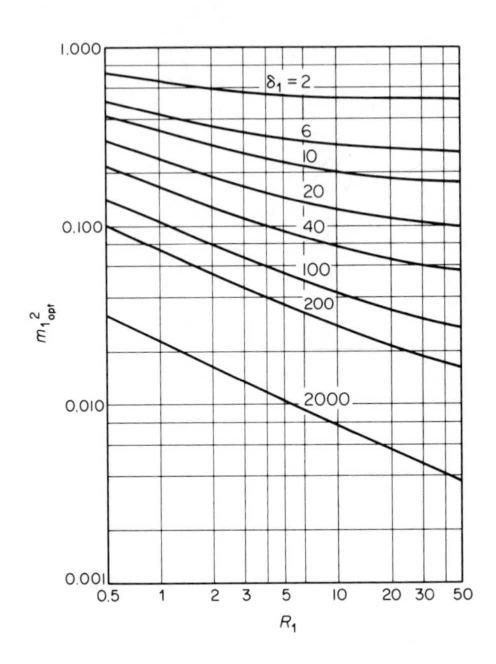

Fig. 7-7 Optimum modulation factor squared versus R_1 for various values of δ_1 (approximate formula used)

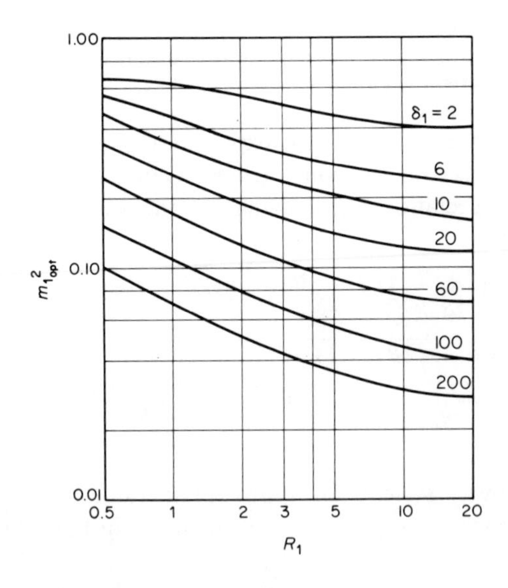

Fig. 7-8 Optimum modulation factor squared versus R_1 for various values of δ_1 (exact computations)

Figure 7-7 is a plot of $m_{1_{opt}}^2$ as determined from (7-26) versus R_1 for various values of the parameter δ_1. For comparison purposes, Fig. 7-8 gives the corresponding exact results as determined computationally from Eq. (7-21) and Fig. 7-3.

Thus far we have considered only the situation where the phase error varies slowly relative to the channel data rate. For the case of rapidly varying phase error, the average error probability P_{En} is given by (6-30), which in terms of system modulation factors becomes

$$P_{E_n} = \tfrac{1}{2} \operatorname{erfc} [\sqrt{(1 - m_n^2)R_n} (\overline{\cos \phi_{rn}})] \tag{7-27}$$

where

$$\overline{\cos \phi_{rn}} = \prod_{i=1}^{n} \frac{I_1(\alpha_i)}{I_0(\alpha_i)} \tag{7-28}$$

and for $n = 1$, α_1 is replaced by ρ_1. Substituting (7-28) in (7-27) and differentiating P_{E1} with respect to m_1 gives a transcendental equation whose solution is the value of m_1 which minimizes the vehicle's error probability:

$$1 + 2(1 - m_1^2)R_1 \left\{1 + \frac{I_2(\rho_1)}{I_0(\rho_1)} - 2\left[\frac{I_1(\rho_1)}{I_0(\rho_1)}\right]^2\right\} = 1 + \frac{2}{\delta_1}\left[\frac{I_1(\rho_1)}{I_0(\rho_1)}\right] \tag{7-29}$$

Assuming this value of m_1, the value of m_2 which minimizes P_{E2} is the solution of an equation identical to (7-29) with ρ_1 replaced by α_2 and all other one subscripts replaced by two's. The square of the optimum modulation factor, m_1, is plotted versus R_1 in Fig. 7-9 for various values of δ_1. The corresponding minimum values of P_{E1} are plotted versus R_1 in Fig. 7-10 with again δ_1 as a parameter.

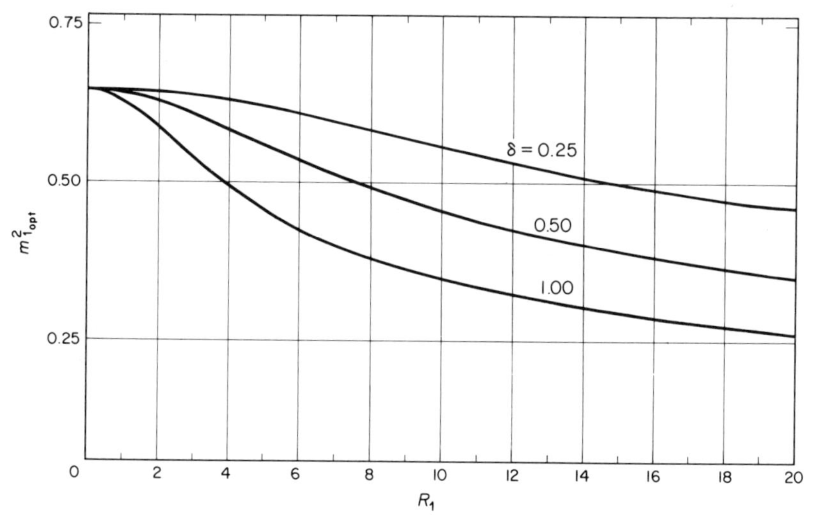

Fig. 7-9 Optimum modulation factor squared versus R_1 for various values of the parameter δ_1 (rapidly-varying phase error)

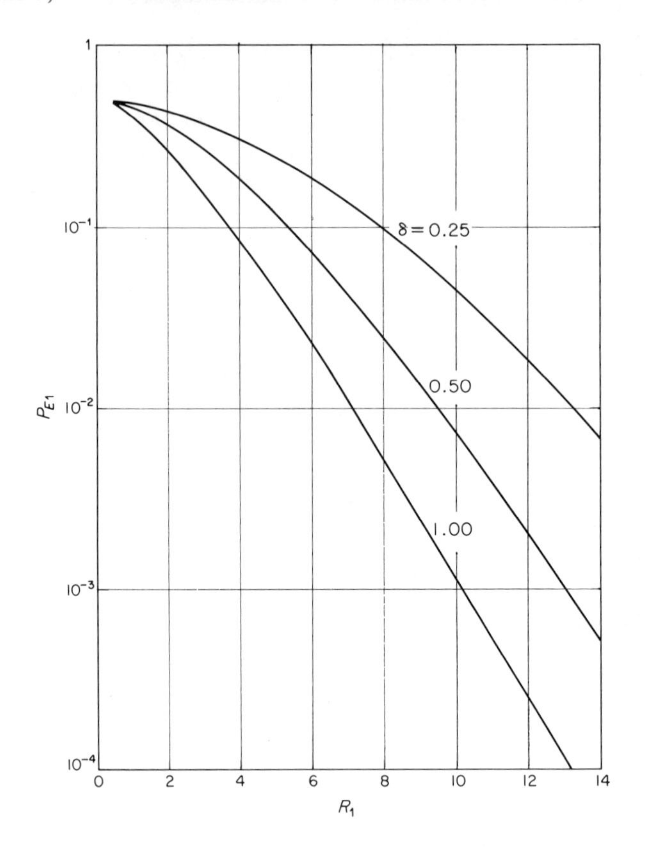

Fig. 7-10 Error probability versus signal-to-noise ratio for optimum modulation factor (rapidly-varying phase error)

A similar analysis can be carried out for the case when the phase error p.d.f. is characterized by Eq. (6-34). The details of the solution along with curves of optimum m_1^2 and minimum P_{E1} are given explicitly in Ref. 5.

Suboptimum Design

In the section on *design characteristics,* we based the system design on attaining minimum error probability by selecting the value of m that yielded the minimum error rate, given a fixed total transmitter power and data rate. Another design criterion, which is perhaps not as difficult to carry out, is to select the value of m that yields the maximum signal-to-noise ratio at the demodulator output. By output signal-to-noise ratio we mean

$$\text{SNR}_n \triangleq \frac{(\bar{q}_{d1})^2}{2\sigma^2_{q_{d1}}} \qquad n = 1, 2 \tag{7-30}$$

where $\bar{q}_{d1}$ and $\sigma^2_{q_{d1}}$ are defined by (7–15) and (7–18), respectively. Making these substitutions in (7–30) yields

$$\mathrm{SNR}_2 = \frac{2(1 - m_2^2)R_2[I_1(m_2^2\delta_2R_2)/I_0(m_2^2\delta_2R_2)]^2[I_1(\alpha_1)/I_0(\alpha_1)]^2}{1 + 2(1 - m_2^2)R_2\{1 + [I_2(m_2^2\delta_2R_2)I_2(\alpha_1)]/[I_0(m_2^2\delta_2R_2)I_0(\alpha_1)] - 2[I_1(m_2^2\delta_2R_2)/I_0(m_2^2\delta_2R_2)]^2[I_1(\alpha_1)/I_0(\alpha_1)]^2\}} \tag{7-31}$$

for the signal-to-noise ratio at the output of the reference receiver demodulator, while the signal-to-noise ratio at the output of the vehicle demodulator becomes

$$\mathrm{SNR}_1 = \frac{2(1 - m_1^2)R_1[I_1(\rho_1)/I_0(\rho_1)]^2}{1 + 2(1 - m_1^2)R_1\{1 + [I_2(\rho_1)/I_0(\rho_1)] - 2[I_1(\rho_1)/I_0(\rho_1)]^2\}} \tag{7-32}$$

with

$$\rho_1 = m_1^2\delta_1R_1 \tag{7-33}$$

It is clear that the maximum value of (7-32) is dependent on m_1 and R_1 for a fixed δ_1. This dependence is indicated in Fig. 7-11 for a value of $\delta_1 = 2$.

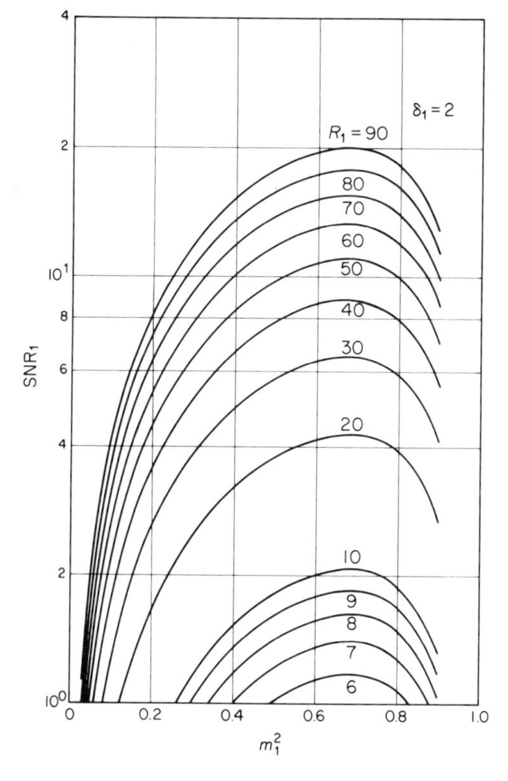

Fig. 7-11 Demodulator output signal-to-noise ratio versus modulation factor squared for various values of the parameter R_1

System Performance as a Function of the Carrier-Tracking Loop Signal-to-Noise Ratio

Of further importance in the design of two-way communication systems is understanding how additive noise on the uplink affects the performance of the reference system. We present here a result that illustrates the performance of the reference system demodulator as a function of vehicle carrier-tracking loop performance.

It is clear from (7-8) that ρ_2, which we have since denoted by α_2, can be rewritten as

$$\alpha_2 = \frac{m_2^2}{1 - m_2^2}\delta_2 R_{d2} \qquad R_{d2} = \frac{(1 - m_2^2)PT_2}{2N_{02}}(1 - \lambda_2) = (1-m_2^2)R_2 \tag{7-34}$$

where R_{d2} is the signal-to-noise ratio of the data in the reference system. Then the error probability performance of the reference system becomes

$$P_{E2} = \frac{1}{2}\left\{1 - \sqrt{\frac{R_{d2}}{\pi}}\exp\left(-\frac{R_{d2}}{2}\right)\sum_{k=0}^{\infty}\epsilon_k(-1)^k\left[\frac{b_{2k+1}^{(2)}}{2k+1} - \frac{b_{2k-1}^{(2)}}{2k-1}\right]I_k\left(\frac{R_{d2}}{2}\right)\right\} \tag{7-35}$$

with

$$b_l^{(2)} = \frac{I_l(\alpha_1)I_l(\alpha_2)}{I_0(\alpha_1)I_0(\alpha_2)}$$

Figure 7-12 is a plot of P_{E2} as determined from Eq. (7-35) versus α_1 with

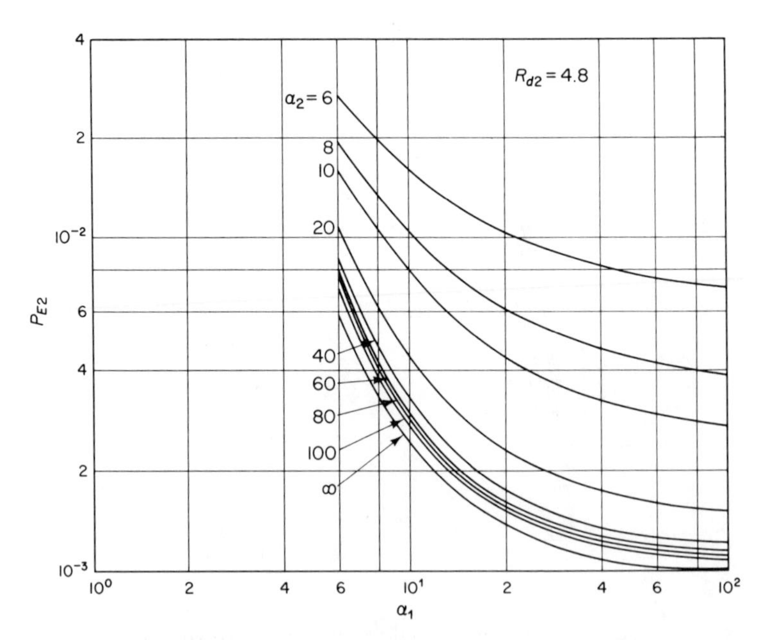

Fig. 7-12 Error probability P_{E2} versus α_1 with α_2 as a parameter (R_{d2} fixed)

R_{d2} fixed and α_2 as a parameter. Varying α_2 at fixed R_{d2} is equivalent to varying δ_2 (or m_2). The curves are drawn for a data signal-to-noise ratio R_{d2}, which produces an error probability of 10^{-3} when the system is perfectly synchronized. Thus, we see that for $\alpha_1 \geq 40$, the performance of the reference system is essentially unaffected by RF carrier phase jitter in the vehicle.

An Application of the Single-Channel Theory to a Turnaround Transponder Ranging System

As an example of the theory developed in the previous sections, we consider a *turnaround transponder ranging system* where the data modulation consists of ranging information. On the uplink, the composite ranging waveform $D(t) = R(t)\text{Ck}(t)$ biphase modulates the transmitted carrier. Here $R(t)$ denotes the ranging code and $\text{Ck}(t)$, the clock. Also, the equations describing the uplink in the first part of Sec. 7-2 remain valid if $x_{k1}(t)$ is replaced by $D(t)$. Thus, the signal $a_1(t)$ of Eq. (7-3) serves as input to the turnaround transponder in the vehicle system. This signal is amplified by a video amplifier and hard limited to maintain a constant signal level from the video amplifier even when only noise is present. We assume that the gain of the hard limiter is such that this constant signal level is always normalized to one watt.

Denote the output of this limiter by $\hat{D}(t)$. The signal $\hat{D}(t)$ is therefore composed of two components: one contains the code transmitted on the uplink possessing a power of $\tilde{\alpha}^2$ watts, and the other is a filtered version of the uplink noise and possesses a total power of $(1 - \tilde{\alpha}^2)$ watts. Here, $\tilde{\alpha} = \text{erf}\sqrt{(1 - m_1^2)P_1/N_{01}W_1}$ is the signal amplitude suppression factor corresponding to a limiter whose input is a digital signal (see Problem 10 of Chap. 4).* Thus, we have

$$\hat{D}(t) = \tilde{\alpha}R(t)Ck(t) + \sqrt{1 - \tilde{\alpha}^2}n_l(t) \tag{7-36}$$

where $n_l(t)$ is a unit variance noise process whose equivalent two-sided noise bandwidth W_1 is determined by the portion of the vehicle receiver preceding the limiter.

Replacing $x_{k2}(t)$ by $\hat{D}(t)$ in the equations of the downlink completes the analogy between the basic system model discussed earlier and the present example. The signal, $a_2(t)$, is applied to the ground system's ranging receiver

* For simplicity of what follows, we have ignored the phase jitter in the vehicle carrier-tracking loop. If this effect is to be accounted for, then the signal amplitude suppression factor must be redefined as

$$\tilde{\alpha} = \text{erf}\left[\sqrt{\frac{(1 - m_1^2)P_1}{N_{01}W_1}} \cos \phi_{v1}\right]$$

and the conditional error probabilities of (7-37) and (7-41) are now also conditioned on the vehicle system phase error ϕ_{v1}.

where the ranging code is acquired, whereupon the output of the ranging receiver is applied to the range detector for final range tally.

In designing a turnaround ranging system, one issue that the communications engineer must contend with is how to allocate the total transmitter power optimally between the downlink carrier and the ranging channel. If the modulation factor m_2 is chosen so that too much power is devoted to the ranging channel, then the carrier-tracking loop cannot track the carrier component and the system will not operate. On the other hand, too much power in the carrier leaves too little in the ranging channel and the ranging receiver cannot acquire the code. These are the opposing factors and, obviously, there must be a "best" way to apportion the total transmitter power. Consequently, the purpose of this section is to develop mathematically tractible expressions which enable the communications engineer to select the parameter $m_2^2 = P_{c2}/P_2$. The criterion that we use to select this "best" is that of minimizing the probability of incorrectly selecting the code phase.

To make the situation more definite, suppose that we must search a range uncertainty interval Ω to find the proper value of the code phase. Assuming that the code is a PN sequence of length p, the range uncertainty or ambiguity interval is divided into p parts (range cells or code phases) each of which is one code digit period t_0 in range. Thus, $\Omega = pt_0$ where $1/t_0$ also represents the clock rate of the modulating sequence. Assuming that an initial synchronization or clock lock is in effect, then each range cell must be examined for some time interval to ascertain if it is the proper cell or not. Mathematically, this situation is analogous to one in which one of p signal vectors is sent over a Gaussian channel, and we wish to determine which signal was sent with a minimum probability of error. Presuming that the ranging receiver performs essentially a cross-correlation operation, then the probability of selecting the incorrect code phase may be easily determined. The method we employ is to fix the reference system phase error and compute the conditional probability of selecting the incorrect code phase. Thus, by averaging over the phase error p.d.f., we determine the average probability of incorrectly selecting the code phase.

It has been shown[6] that the probability of incorrect alignment associated with PN sequences may be calculated as for the case of orthogonal sequences if the signal energy or power is increased by the factor $1 + 1/p$. Proceeding from these considerations, and the results in Chap. 4, we obtain the following expression for the conditional probability of incorrect alignment:

$$P_{E2}(\phi_{r2}) = 1 - \int_{-\infty}^{\infty} \frac{\exp(-x^2)}{\sqrt{\pi}} \left[\frac{1}{2}\operatorname{erfc}(-x - A_2 \cos\phi_{r2})\right]^p dx \qquad (7\text{-}37)$$

where

$$A_2 = \sqrt{\frac{\tilde{\alpha}^2[(p+1)/p](1-m_2^2)(P_2 t_0/N_{02})}{1 + 2(1-m_2^2)(1-\tilde{\alpha}^2)P_2(\cos\phi_{r2})^2/N_{02}W_1}} \qquad (7\text{-}38)$$

To arrive at Eq. (7-37), we must assume that the spectral density of the noise component, $n_l(t)$, which is transmitted to the ground, is essentially flat over a bandwidth that is wide in comparison with the bandwidth of the ranging receiver. This assumption is compatible with systems of practical interest. Further, the parameter A_2 is well approximated—over the range of signal-to-noise ratios where the system may be expected to operate satisfactorily—by

$$A_2 \cong \sqrt{\tilde{\alpha}^2\left(\frac{p+1}{p}\right)(1-m_2^2)\frac{P_2 t_0}{N_{02}}} \tag{7-39}$$

since practical considerations dictate that

$$\frac{2(1-m_2^2)(1-\tilde{\alpha}^2)P_2}{N_{02}W_1}(\overline{\cos\phi_{r2}})^2 \ll 1 \tag{7-40}$$

The significance of this assumption is that the noise at the output of the limiter is not harmful in itself, but, because the total output power of the limiter is constant, the signal power available in the downlink ranging channel is reduced. From a practical viewpoint, if (7-40) is not valid, then a turnaround system essentially loses its ability to perform a range measurement with great accuracy.

The average probability of incorrect code alignment is obtained by averaging the conditional error probability of (7-37) over the phase error p.d.f. of (7-7) with $\lambda_2 = -1$, T_2 replaced by t_0, and P_2 by $[\tilde{\alpha}^2(p+1)/p]P_2$ in (7-8) and (7-9). Furthermore, since the range of interest in P_{E2} is for values of signal-to-noise ratio that yield $P_{E2} \leq 10^{-3}$, one may use an upper union bound as an approximation:

$$P_{E2}(\phi_{r2}) \cong \frac{p}{2}\operatorname{erfc}\left(\frac{A_2}{\sqrt{2}}\cos\phi_{r2}\right) \tag{7-41}$$

Using Eq. (7–41) in place of (7–20) for $n = 2$ and assuming linear PLL theory for the p.d.f. of ϕ_{r2}, one can determine an optimum modulation factor $m_{2\,\text{opt}}$ from (7-25) by replacing R_2 with $\tilde{\alpha}^2[(p+1)/p]\,[P_2 t_0/N_{02}]$. This value of $m_{2\,\text{opt}}$ when substituted in (7-41) determines the minimum probability of incorrect alignment.

To select the "best" value of m_1 for the uplink design, one must specify the criterion upon which this "best" is to be evaluated. The criterion that immediately comes to mind is to select m_1 on the basis of the vehicle acquiring the code with minimum probability of incorrect alignment. Accepting this as the uplink design philosophy, one may proceed on the basis of the formulas given for the choice of $m_{2\,\text{opt}}$ with the appropriate change of subscripts and A_1 approximately defined by

$$A_1 \cong \sqrt{\left(\frac{p+1}{p}\right)(1-m_1^2)\frac{P_1 t_0}{N_{01}}} \tag{7-42}$$

7-3. DESIGN OF TWO-CHANNEL SYSTEMS

In this section, we shall be concerned with communication systems that phase modulate an RF carrier with two modulated data subcarriers. The frequency spectra of these two subcarriers must be selected appropriately, since the modulation technique requires that they be nonoverlapping (lie in separate portions of the RF channel) and situated outside the pass band of the RF carrier-tracking loop. This particular concept is extremely important in assuring successful demodulation of telemetry and/or command data while simultaneously performing the tracking function. Figure 7-13 illustrates a typical

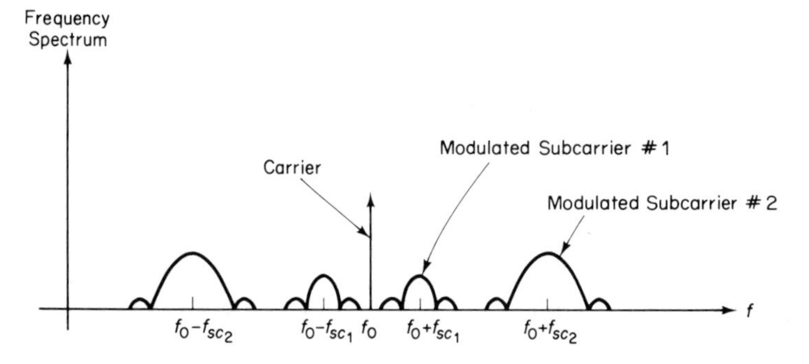

Fig. 7-13 Frequency spectrum of an RF carrier which is phase modulated by two modulated subcarriers

frequency spectrum of the transmitted signal in a two-channel system. Some examples of telecommunication systems that employed the two-channel approach are the Ranger, Mariner, Lunar Orbiter, Surveyor, Pioneer, and Apollo.

Basic System Model

The system block diagram for a two-channel, two-way communication link is illustrated in Fig. 7–1 with the understanding that two modulated data subcarriers—namely, $x_{kn}(t)$, $X_{kn}(t)$, $n = 1, 2$—are now incident on the vehicle and reference transmitters. Thus, the reference transmitter emits the phase-modulated waveform

$$p(t) = \sqrt{2P_1} \sin [\omega_0 t + (\cos^{-1} m_{11})x_{k1}(t) + (\cos^{-1} m_{21})X_{k1}(t)] \tag{7-43}$$

where $\cos^{-1} m_{11}$ and $\cos^{-1} m_{21}$ are angle modulation indices that allocate the proper amount of available sideband power to each signal. Assuming for simplicity that θ_0—the random phase of the reference transmitter—equals zero, we observe in the vehicle the Doppler-shifted, phase-shifted, noise-corrupted waveform

$$\psi(t) = \sqrt{2P_1} \sin [\omega_1 t + (\cos^{-1} m_{11})x_{k1}(t) + (\cos^{-1} m_{21})X_{k1}(t) + \theta_1] + n_1(t) \tag{7-44}$$

Using a simple trigonometric expansion, together with the facts that $x_{k1}(t) = \pm 1$ and $X_{k1}(t) = \pm 1$, Eq. (7-44) may be written as follows:

$$\begin{aligned}\psi(t) = {} & \underbrace{\sqrt{2P_{c1}} \sin (\omega_1 t + \theta_1)}_{\text{carrier component}} \\ & + \underbrace{\sqrt{2P_{11}} x_{k1}(t) \cos (\omega_1 t + \theta_1)}_{\substack{\text{modulated subcarrier} \\ \text{component 1}}} \\ & + \underbrace{\sqrt{2P_{21}} X_{k1}(t) \cos (\omega_1 t + \theta_1)}_{\substack{\text{modulated subcarrier} \\ \text{component 2}}} \\ & - \underbrace{\sqrt{2P_{l1}} X_{k1}(t) x_{k1}(t) \sin (\omega_1 t + \theta_1) + n_1(t)}_{\substack{\text{cross-modulation} \\ \text{loss component}}}\end{aligned} \tag{7-45}$$

In practice, power in the cross-modulation loss component is usually spread over a wide bandwidth and thus may be neglected when compared with the power in the noise process $n_1(t)$. The powers in the received carrier component, the modulated subcarrier components, and the cross-modulation loss component are, respectively,

$$\begin{aligned} P_{c1} &= m_{21}^2 m_{11}^2 P_1 \\ P_{11} &= m_{21}^2 (1 - m_{11}^2) P_1 \\ P_{21} &= m_{11}^2 (1 - m_{21}^2) P_1 \\ P_{l1} &= (1 - m_{11}^2)(1 - m_{21}^2) P_1 \end{aligned} \tag{7-46}$$

and the total power transmitted is

$$P_1 = P_{11} + P_{21} + P_{c1} + P_{l1} \tag{7-47}$$

It is obvious that m_{11} and m_{21} must be less than unity.

In the vehicle, the carrier component in $\psi(t)$ is tracked by means of a narrowband PLL and the coherent reference so produced is used to demodulate the two subcarriers—that is,

$$\begin{aligned} a_1(t) = {} & \underbrace{\sqrt{P_{11}}\, x_{k1}(t) \cos \varphi_{v1}(t)}_{\substack{\text{modulated} \\ \text{subcarrier 1}}} \\ & + \underbrace{\sqrt{P_{21}}\, X_{k1}(t) \cos \varphi_{v1}(t)}_{\substack{\text{modulated} \\ \text{subcarrier 2}}} + N_1[t, \varphi_{v1}(t)] \end{aligned} \tag{7-48}$$

In Eq. (7-48), we neglected all double-frequency components, the term $\sqrt{P_{c1}} \sin \varphi_{v1}(t)$, and the cross-modulation loss component. All these terms produce negligible effects on the ultimate data decisions. On the downlink we transmit

$$\eta(t) = \sqrt{2P_2} \sin [\omega_{10}t + (\cos^{-1} m_{12})x_{k2}(t) + (\cos^{-1} m_{22})X_{k2}(t) + \hat{\theta}_{10}(t)] \tag{7-49}$$

where $\cos^{-1} m_{12}$ and $\cos^{-1} m_{22}$ are angle modulation indices that allocate the proper amount of sideband power to the modulated subcarriers $x_{k2}(t)$ and $X_{k2}(t)$, respectively. On the ground we observe the Doppler-shifted, phase-shifted, noise-corrupted waveform

$$\xi(t) = \sqrt{2P_2} \sin [\omega_2 t + (\cos^{-1} m_{12})x_{k2}(t) + (\cos^{-1} m_{22})X_{k2}(t) + \hat{\theta}_{10}(t) + \theta_2] + n_2(t) \tag{7-50}$$

A simple trigonometric expansion in (7–50) leads to

$$\begin{aligned}\xi(t) = {} & \underbrace{\sqrt{2P_{c2}} \sin [\omega_2 t + \theta_2 + \hat{\theta}_{10}(t)]}_{\text{carrier component}} + \underbrace{\sqrt{2P_{12}}x_{k2}(t) \cos [\omega_2 t + \theta_2 + \hat{\theta}_{10}(t)]}_{\text{modulated subcarrier component 1}} \\ & + \underbrace{\sqrt{2P_{22}}X_{k2}(t) \cos [\omega_2 t + \theta_2 + \hat{\theta}_{10}(t)]}_{\text{modulated subcarrier component 2}} \\ & - \underbrace{\sqrt{2P_{l2}}X_{k2}(t)x_{k2}(t) \sin [\omega_2 t + \theta_2 + \hat{\theta}_{10}(t)]}_{\text{cross-modulation loss component}} + n_2(t)\end{aligned} \tag{7-51}$$

where

$$\begin{aligned}P_{c2} &= m_{22}^2 m_{12}^2 P_2 \\ P_{12} &= m_{22}^2(1 - m_{12}^2)P_2 \\ P_{22} &= m_{12}^2(1 - m_{22}^2)P_2 \\ P_{l2} &= (1 - m_{12}^2)(1 - m_{22}^2)P_2\end{aligned} \tag{7-52}$$

If we multiply the observed waveform $\xi(t)$ with the ground receiver's PLL estimate $v(t)$ and neglect similar components as before, we have

$$a_2(t) = \underbrace{\sqrt{P_{12}}x_{k2}(t) \cos \varphi_{r2}(t)}_{\text{modulated subcarrier 1}} + \underbrace{\sqrt{P_{22}}X_{k2}(t) \cos \varphi_{r2}(t)}_{\text{modulated subcarrier 2}} + N_2[t, \varphi_{r2}(t)] \tag{7-53}$$

Carrier-Tracking Loop Performance

With regard to the carrier-tracking performance of the vehicle and reference PLL's the expressions derived in Chap. 3 for the statistics of the phase errors ϕ_{v1} and ϕ_{r2} apply directly to the two-channel system. It is instructive to rewrite ρ_n of (3-17) in terms of the system modulation factors, namely,

$$\rho_n = \frac{2m_{2n}^2 m_{1n}^2 P_n}{N_{0n} W_{Ln}} \qquad n = 1, 2 \tag{7-54}$$

The variance of the ground receiver phase error, $\sigma_{\phi_{r2}}^2$, is then given by (3–25) and is plotted in Fig. 7-14 versus the parameter $m_2^2 \triangleq m_{22}^2\, m_{12}^2 = P_{c2}/P_2$.

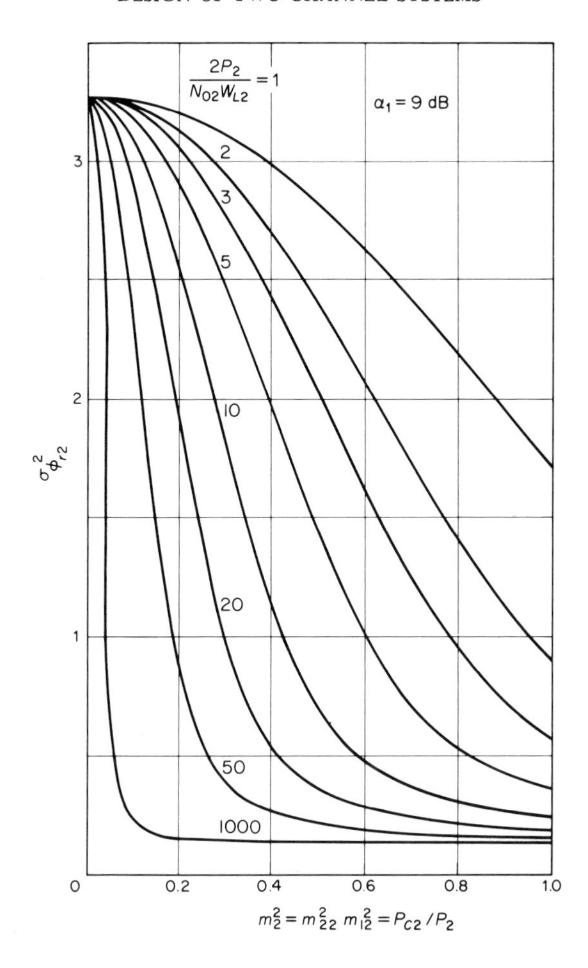

Fig. 7-14 Variance of the ground receiver phase error versus m_2^2 for various values of total downlink signal-to-noise ratio in the ground receiver's carrier-tracking loop

This function is illustrated for various values of $2P_2/N_{02}W_{L2} = \rho_2/m_2^2$. As for the single-channel case, the curves are drawn for $\alpha_1 = 9$ dB. Figure 7-15 illustrates the variance of the phase error $\sigma^2_{\phi_{v1}}$ in the vehicle as a function of the parameter $m_1^2 \triangleq m_{21}^2 m_{11}^2 \triangleq P_{c1}/P_1$ and for various values of $2P_1/N_{01}W_{L1}$. In carrying out a particular design, one may determine the value of m_n^2, $n = 1, 2$, which is required to yield a specified value of $\sigma^2_{\phi_{rn}}$, given the carrier power-to-noise density ratios and bandwidths of the carrier-tracking loops.*

* Again, ϕ_{r1} is used for ϕ_{v1} to make the notation more compact.

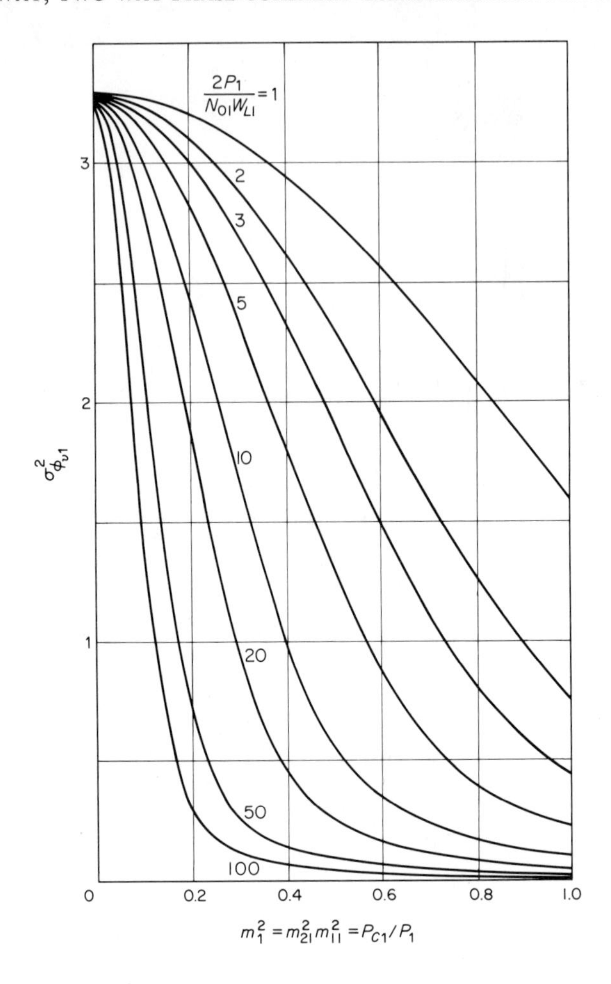

Fig. 7-15 Variance of the vehicle receiver phase error versus m_1^2 for various values of total uplink signal-to-noise ratio in the vehicle receiver's carrier-tracking loop

Power Allocation and Selection of Modulation Factors for Two-Channel Systems (Data/Sync)

Here, we make use of the results given in previous parts to present a method whereby the modulation factors m_{dn}, m_{sn}, $n = 1, 2$ may be specified.* The method is based upon the constraint that the maximum mean-squared RF and sync phase errors cannot exceed prespecified values.

---

* For convenience, we use m_{dn} and m_{sn} to denote data and sync modulation factors, respectively, rather than m_{1n} and m_{2n}.

The ratio of power in the carrier to total power transmitted

$$m_n^2 = \frac{P_{cn}}{P_n} = m_{sn}^2 m_{dn}^2 \triangleq k_{1n}^2 \qquad n = 1, 2 \tag{7-55}$$

may be determined from Fig. 7-14 (for two-way systems) and from Fig. 7-15 (for one-way systems) by specifying the maximum value that the mean-squared RF phase error is allowed to assume and the total signal-to-noise ratio in the carrier-tracking loop bandwidth.

Specifically, the abscissa of the point whose ordinate is the prespecified maximum value of $\sigma_{\phi_{rn}}^2$ and which is located on the curve corresponding to the given value of $2P_n/N_{0n}W_{Ln}$, is the required ratio. Once this ratio has been determined, the remaining equivalent power

$$P_{rn}\,(\overline{\cos\phi_{rn}})^2 = (1 - m_{sn}^2 m_{dn}^2) P_n\,(\overline{\cos\phi_{rn}})^2 \qquad n = 1, 2 \tag{7-56}$$

must be apportioned between the data and sync signals, and the cross-modulation loss. The reason that an equivalent power must be defined is due to the fact that the RF phase error affects the power remaining in the data and sync signals.

Since in many applications, the symbol sync error is much smaller than the subcarrier sync phase error, the former may be neglected in determining the performance of the vehicle's data detector. Hence, in any particular design the communications engineer may specify the maximum value that the mean-squared subcarrier sync phase error may assume and the power in the sync channel may be determined by using the curves in Fig. 7–15 at the subcarrier level. Use of this figure, however, requires not only a knowledge of the maximum value that the mean-squared subcarrier sync phase error, say σ_{sn}^2, may assume, but also the bandwidth of the subcarrier-tracking loop W_{sn} and the noise spectral density N_{0n}.

Given this information, one computes the quantity

$$\frac{2P_{rn}(\overline{\cos\phi_{rn}})^2}{N_{0n}W_{sr}} = \frac{2(1 - k_{1n}^2)P_n(\overline{\cos\phi_{rn}})^2}{N_{0n}W_{sn}} \qquad n = 1, 2 \tag{7-57}$$

and uses this in place of $2P_1/N_{01}W_{L1}$ in Fig. 7-15. The abscissa of the point generated by the intersection of the line $\sigma_{sn}^2 = \sigma_{sn\,\max}^2$ with the curve whose parameter is $2P_{rn}\,(\overline{\cos\phi_{rn}})^2/N_{0n}W_{sn}$ gives the ratio of equivalent sync power to equivalent remaining power:

$$\frac{P_{sn}\,(\overline{\cos\phi_{rn}})^2}{P_{rn}\,(\overline{\cos\phi_{rn}})^2} = \frac{m_{dn}^2(1 - m_{sn}^2)}{1 - m_{sn}^2 m_{dn}^2} \triangleq k_{2n}^2 \qquad n = 1, 2 \tag{7-58}$$

From k_{1n}^2 and k_{2n}^2 as specified by Eqs. (7-55) and (7-58), respectively, one can solve for the modulation factors, m_{sn} and m_{dn}:

$$m_{sn} = \frac{k_{1n}}{\sqrt{k_{1n}^2 + k_{2n}^2 - k_{1n}^2 k_{2n}^2}} \tag{7-59}$$

$$m_{dn} = \sqrt{k_{1n}^2 + k_{2n}^2 - k_{1n}^2 k_{2n}^2} = \frac{k_{1n}}{m_{sn}} \tag{7-60}$$

There remains the problem of relating $(\overline{\cos \phi_{rn}})^2$ to the carrier-tracking loop parameters. From the nonlinear theory, $\overline{\cos \phi_{rn}}$ is given by (7-28), where for $n = 2$,

$$\begin{aligned} \alpha_1 &= \frac{\rho_1}{G^2 K_R(r_1, r_2, \xi)} \\ \rho_n &= \frac{2k_{1n}^2 P_n}{N_{0n} W_{Ln}} \qquad n = 1, 2 \end{aligned} \tag{7-61}$$

and for $n = 1$, α_1 is replaced by ρ_1. For design purposes, $(\overline{\cos \phi_{rn}})^2$ is plotted versus $m_n^2 = k_{1n}^2$ in Figs. 7-16 and 7-17 for various values of $2P_n/N_{0n}W_{Ln}$ and $n = 1, 2$, respectively.

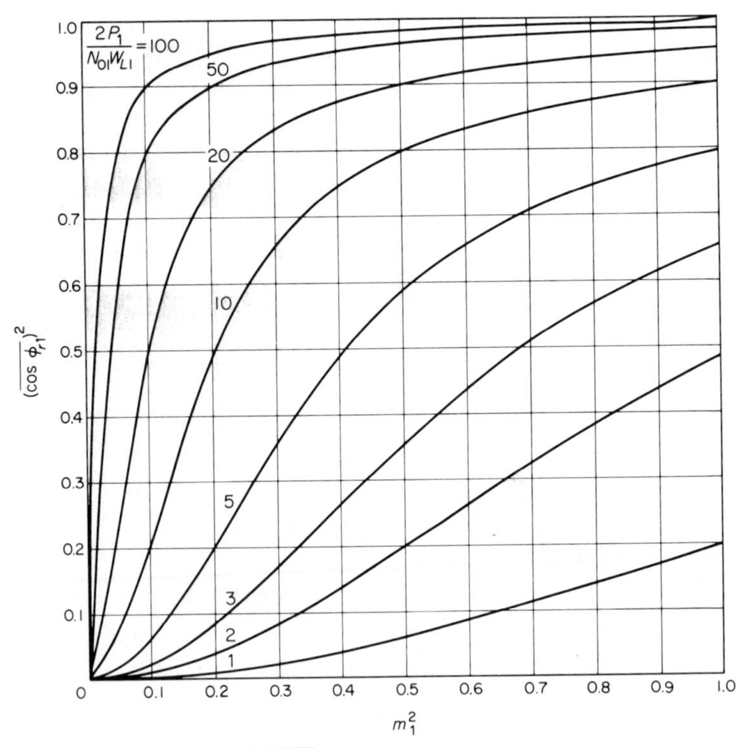

Fig. 7-16 The quantity $(\overline{\cos \phi_{r1}})^2$ versus m_1^2 for various values of $2P_1/N_{01}W_{L1}$

Determination of System Data Rate for a Given Bit Error Probability

The final parameter to be determined is the data rate $\mathcal{R}_n = T_n^{-1}$, $n = 1, 2$. This may be selected from Fig. 7-18, given a certain bit error proba-

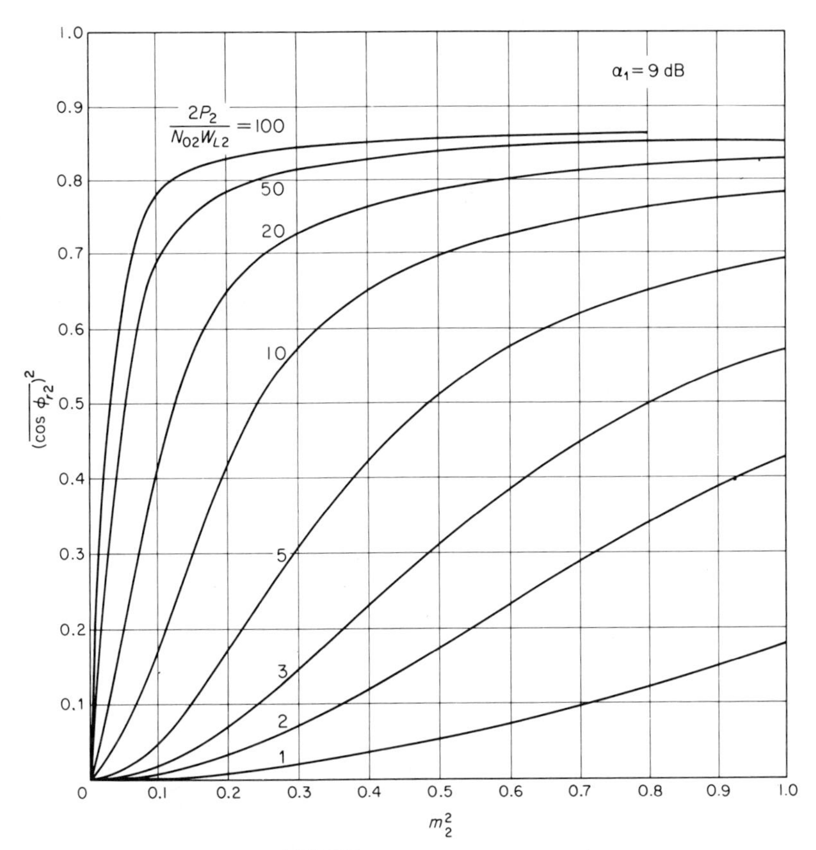

Fig. 7-17 The quantity $(\overline{\cos \phi_{r2}})^2$ versus m_2^2 for various values of $2P_2/N_{02}W_{L2}$; $\alpha_1 = 9$ dB.

bility, say, P_{En}. From Eqs. (7-57) and (7-58), the signal-to-noise ratio in the subcarrier-tracking loop is given by

$$\gamma_n \triangleq \frac{2P_{sn}(\overline{\cos \phi_{rn}})^2}{N_{0n}W_{sn}} = k_{2n}^2\left[\frac{2P_{rn}(\overline{\cos \phi_{rn}})^2}{N_{0n}W_{sn}}\right] \tag{7-62}$$

The abscissa of the point whose ordinate is the given value of P_{En} and which is located on the curve corresponding to the value of γ_n found in (7-62) yields a value of

$$R_{dn}(\overline{\cos \phi_{rn}})^2 = \frac{P_{dn}T_n}{N_{0n}}(\overline{\cos \phi_{rn}})^2 = m_{sn}^2(1 - m_{dn}^2)\frac{P_nT_n}{N_{0n}}(\overline{\cos \phi_{rn}})^2 \tag{7-63}$$

Since R_{dn}, $(\overline{\cos \phi_{rn}})^2$, P_{dn}, N_{0n}, m_{sn} and m_{dn} in Eq. (7-63) are known, the system data rate $\mathcal{R}_n = T_n^{-1}$ may be selected to provide for this signal-to-noise ratio. From (7-63) we have

$$\mathcal{R}_n = \frac{1}{T_n} = \frac{m_{sn}^2(1 - m_{dn}^2)P_n(\overline{\cos \phi_{rn}})^2}{N_{0n}[R_{dn}(\overline{\cos \phi_{rn}})^2]} \tag{7-64}$$

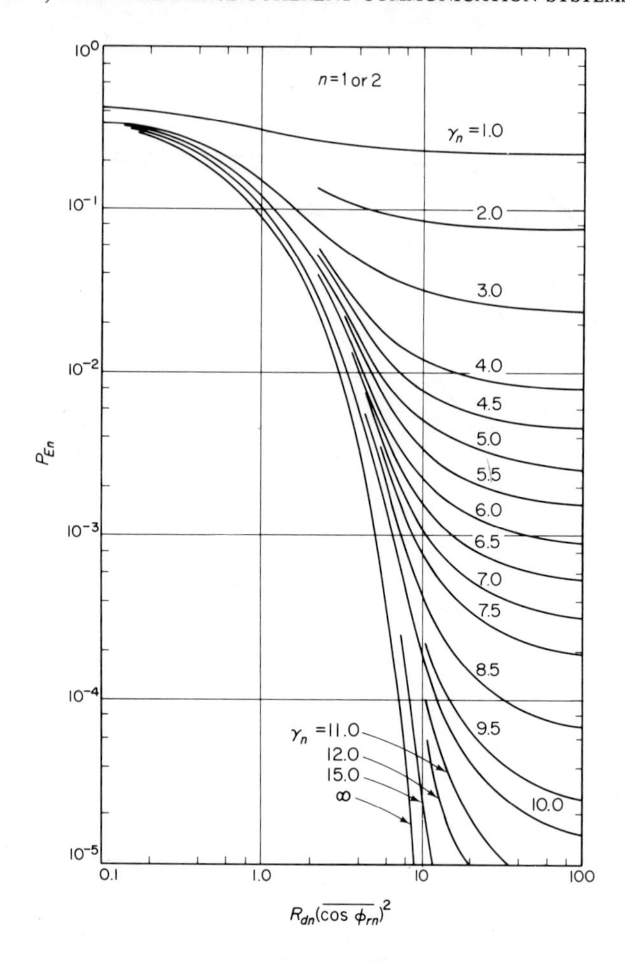

Fig. 7-18 Error probability P_{En} versus $R_{dn}(\overline{\cos \phi_{rn}})^2$ for various values of signal-to-noise ratio in the subcarrier-tracking loop

Note that the procedure used to arrive at (7-63) is simply an application of the noisy reference problem treated in the previous chapter where the noisy reference signal is derived from the subcarrier-tracking loop.

Power Allocation and Selection of Modulation Factors for Two-Channel Systems (Data 1/Data 2)

We present here a method for selecting modulation factors in a two-channel system where each channel represents data information and each data-bearing signal carries its own sync information. We return to our original notation by replacing subscripts d and s of the previous part by the numerals 1 and 2 to correspond to data channels 1 and 2, respectively. For example, the modulation factor m_{12} represents information transmitted along data channel 1 on the downlink.

The criterion of optimization is to minimize the error probabilities in both channels of the up- and downlinks, assuming perfect subcarrier demodulation and symbol sync timing. We begin by writing the four expressions for the error probabilities just mentioned—that is,

$$\begin{aligned}
P_{E11} &= \frac{1}{2}\int_{-\pi}^{\pi} p(\phi_{v1})\,\mathrm{erfc}\left[\sqrt{m_{21}^2(1-m_{11}^2)\frac{P_1T_{11}}{N_{01}}}\cos\phi_{v1}\right]d\phi_{v1} \\
P_{E21} &= \frac{1}{2}\int_{-\pi}^{\pi} p(\phi_{v1})\,\mathrm{erfc}\left[\sqrt{m_{11}^2(1-m_{21}^2)\frac{P_1T_{21}}{N_{01}}}\cos\phi_{v1}\right]d\phi_{v1} \\
P_{E12} &= \frac{1}{2}\int_{-\pi}^{\pi} p(\phi_{r2})\,\mathrm{erfc}\left[\sqrt{m_{22}^2(1-m_{12}^2)\frac{P_2T_{12}}{N_{02}}}\cos\phi_{r2}\right]d\phi_{r2} \\
P_{E22} &= \frac{1}{2}\int_{-\pi}^{\pi} p(\phi_{r2})\,\mathrm{erfc}\left[\sqrt{m_{12}^2(1-m_{22}^2)\frac{P_2T_{22}}{N_{02}}}\cos\phi_{r2}\right]d\phi_{r2}
\end{aligned} \tag{7-65}$$

where $p(\phi_{v1})$ and $p(\phi_{r2})$ are defined in (7-10) and (7-7), respectively. The first subscript on P_E and T refers to the particular data channel, whereas the second subscript corresponds, as before, to the vehicle or reference systems.

It is clear from the first two equations of (7–65) that the error probability performance of the two vehicle data channels does not depend upon m_{12} and m_{22}. Furthermore, to get an analytical solution, we will once again assume that ρ_1 is large enough so that the linear PLL theory applies. Following a procedure similar to that which arrives at (7-26) we find that the optimum values of m_{11} and m_{21} must satisfy

$$\begin{aligned}
m_{11}^2 &= \frac{2R_{21}-1+\sqrt{(2R_{21}-1)^2+8R_{21}(1+\delta_{11}/2)}}{4R_{21}(1+\delta_{11}/2)} \\
m_{21}^2 &= \frac{2R_{11}-1+\sqrt{(2R_{11}-1)^2+8R_{11}(1+\delta_{21}/2)}}{4R_{11}(1+\delta_{21}/2)}
\end{aligned} \tag{7-66}$$

where

$$\begin{aligned}
R_{21} &= \frac{m_{21}^2P_1T_{11}}{N_{01}} \qquad R_{11} = \frac{m_{11}^2P_1T_{21}}{N_{01}} \\
\delta_{21} &= \frac{2}{W_{L1}T_{21}} \qquad \delta_{11} = \frac{2}{W_{L1}T_{11}}
\end{aligned} \tag{7-67}$$

Using the above values of m_{11} and m_{21} in the last two equations of (7-65), one can find by differentiation the values of m_{12} and m_{22} that minimize

P_{E12} and P_{E22}. The solutions satisfy the equations

$$m_{22}^2 = \frac{(\delta_{22}/2)(2R_{12}-1)-\alpha_1^{-1}(1+\delta_{22})}{2\delta_{22}R_{12}(1+\delta_{22}/2)} + \frac{\sqrt{\begin{array}{l}[(\delta_{12}/2)(2R_{12}-1)-\alpha_1^{-1}(1+\delta_{22})]^2 \\ \quad + 2\delta_{22}(1+\delta_{22}/2)[\delta_{22}R_{12}+\alpha_1^{-1}(2R_{12}-\alpha_1^{-1})]\end{array}}}{2\delta_{22}R_{12}(1+\delta_{22}/2)}$$

$$m_{12}^2 = \frac{(\delta_{12}/2)(2R_{22}-1)-\alpha_1^{-1}(1+\delta_{12})}{2\delta_{12}R_{22}(1+\delta_{12}/2)} + \frac{\sqrt{\begin{array}{l}[(\delta_{12}/2)(2R_{22}-1)-\alpha_1^{-1}(1+\delta_{12})]^2 \\ \quad + 2\delta_{12}(1+\delta_{12}/2)[\delta_{12}R_{22}+\alpha_1^{-1}(2R_{22}-\alpha_1^{-1})]\end{array}}}{2\delta_{12}R_{22}(1+\delta_{12}/2)} \tag{7-68}$$

where

$$R_{12} = \frac{m_{12}^2 P_2 T_{22}}{N_{02}} \qquad R_{22} = \frac{m_{22}^2 P_2 T_{12}}{N_{02}}$$
$$\delta_{12} = \frac{2}{W_{L2}T_{12}} \qquad \delta_{22} = \frac{2}{W_{L2}T_{22}} \tag{7-69}$$

and

$$\alpha_1 = \left[\frac{2m_{11}^2 m_{12}^2 P_1}{N_{01}W_{L1}}\right]\left[\frac{1}{G^2K_R(r_1, r_2, \xi)}\right] \tag{7-70}$$

with m_{11} and m_{21} being the solutions of (7-66). Indeed the solutions to (7-66) and (7-68) are difficult to come by, but do in principle exist as can be evidenced by plotting the surfaces P_{Eij} versus m_{ij} and m_{jj} for $i, j = 1, 2$. We have not attempted to solve these equations simultaneously but nevertheless present them to illustrate a possible optimization procedure for a system with two data channels.

An Improved Modulation-Demodulation Technique for Certain Systems with Two Data Channels

Frequently in two-channel communication systems, the allocation of power at the transmitter between the carrier and data channels is such that the power lost in the intermodulation product is greater than that available for one of the data channels. As an example, consider a downlink telemetry two-channel system, such as that used in the Mariner Mars 1969 application, where science information is transmitted over data channel 1 at a rate of 16.2 kilobits/sec and engineering information over data channel 2 at $33\frac{1}{3}$ bits/sec. The angle modulation index on the science channel is $\theta_{m_{12}} = \cos^{-1} m_{12} = 65°$. Thus, from (7-52),

$$\frac{P_{I2}}{P_{22}} = \frac{1-m_{12}^2}{m_{12}^2} = (\tan\theta_{m_{12}})^2 = 4.6 \tag{7-71}$$

that is, the power lost in the intermodulation product is 6.63 dB greater than the power in the low-rate engineering channel.

One technique for reducing this loss is to transmit the intermodulation signal, the product of the two data-channel signals, on data channel 2. Modulating the carrier in such a way makes it possible to interchange the powers previously allocated to the intermodulation product and data channel 2.* To see this, we write the Doppler-shifted, phase-shifted, noise-corrupted downlink signal received on the ground in a form similar to (7-50):

$$\begin{aligned}\xi(t) = \sqrt{2P_2} \sin [\omega_2 t + (\cos^{-1} m_{12}) x_{k2}(t) \\ + (\cos^{-1} m_{22}) x_{k2}(t) X_{k2}(t) + \hat{\theta}_{10}(t) + \theta_2] + n_2(t)\end{aligned} \qquad (7\text{-}72)$$

Then, substituting the product $x_{k2}(t)\, X_{k2}(t)$ for $X_{k2}(t)$ in (7–51), and noting that $x_{k2}^2(t) = 1$, (7-72) decomposes into

$$\begin{aligned}\xi(t) = \underbrace{\sqrt{2\tilde{P}_{c2}} \sin [\omega_2 t + \hat{\theta}_{10}(t) + \theta_2]}_{\text{carrier component}} + \underbrace{\sqrt{2\tilde{P}_{12}} x_{k2}(t) \cos [\omega_2 t + \hat{\theta}_{10}(t) + \theta_2]}_{\text{data component 1}} \\ - \underbrace{\sqrt{2\tilde{P}_{22}} X_{k2}(t) \sin [\omega_2 t + \hat{\theta}_{10}(t) + \theta_2]}_{\text{data component 2}} \\ + \underbrace{\sqrt{2\tilde{P}_{l2}} x_{k2}(t) X_{k2}(t) \cos [\omega_2 t + \hat{\theta}_{10}(t) + \theta_2]}_{\text{cross-modulation loss component}} + n_2(t)\end{aligned} \qquad (7\text{-}73)$$

where

$$\begin{aligned}\tilde{P}_{c2} &= m_{12}^2 m_{22}^2 P_2 \\ \tilde{P}_{12} &= m_{22}^2 (1 - m_{12}^2) P_2 \\ \tilde{P}_{22} &= (1 - m_{12}^2)(1 - m_{22}^2) P_2 \\ \tilde{P}_{l2} &= m_{12}^2 (1 - m_{22}^2) P_2\end{aligned} \qquad (7\text{-}74)$$

Notice that if the carrier is suppressed by choosing one of the modulation factors equal to zero, then if the other modulation factor is set equal to .5, the data rates on both subcarriers are equal, and the subcarrier frequencies themselves are equal to the data rate, the interplex modulation technique reduces to the familiar case of quadriphase signaling discussed in Chap. 5.

From (7-52), we observe that

$$\begin{aligned}\tilde{P}_{c2} &= P_{c2} \qquad \tilde{P}_{12} = P_{12} \\ \tilde{P}_{22} &= P_{l2} \qquad \tilde{P}_{l2} = P_{22}\end{aligned} \qquad (7\text{-}75)$$

Hence,

$$\frac{\tilde{P}_{l2}}{\tilde{P}_{22}} = \frac{1}{(\tan \theta_{m_{12}})^2} = .217 \qquad (7\text{-}76)$$

The receiver that provides for demodulation of (7-73) has one major difference from that used for demodulating (7-50)—namely, the carrier needed for demodulating data component 2 must be in quadrature with the carrier that demodulates data component 1. This quadrature carrier would be derived in the receiver by phase shifting the VCO output by 90°. Thus, the only modifications required in the receiver would be the addition of a 90°

*This modulation technique has been referred to as *interplex* in Ref. 7.

phase-shift network and an analog multiplier (phase detector) for demodulating data component 2 with the quadrature carrier reference.

7-4. DESIGN OF MULTICHANNEL SYSTEMS

We conclude this chapter by generalizing the results of the previous sections to the case of a *multichannel*, frequency-multiplexed, phase-modulated communication system.[3] In such a system, many signals with non-overlapping frequency spectra are frequency multiplexed with the combined signal being used to phase modulate the RF carrier. The application of such a technique to a system where, for example, several signals are transmitted from a satellite to Earth by means of a single saturated power amplifier and antenna, is very attractive from the point of view of system simplicity.

The transmitter is assumed to be an ideal phase modulator which modulates an RF carrier with L phase-modulated, sinusoidal data subcarriers, and in general, M binary-valued data subcarriers. The receiver is a superheterodyne PLL whose output is applied to one of L subcarrier extractors. Each subcarrier extractor consists of a subcarrier-tracking loop, a timing loop, and a data detector, which operates as a cross-correlator. Results are presented that allow the design engineer to allocate the total power among L modulated data subcarriers, M binary-valued data subcarriers, and the carrier signal. The total power in the distortion component is computed. Finally, a method is given that allows the communications engineer to select the data rate and modulation factor of each data subcarrier that minimize the probability that the data detector will err in the decision process. The results are sufficiently general that they may be used in designing block-coded systems.

Basic System Model

A simplified block diagram of a frequency-multiplexed PM communication system is shown in Figs. 7-19 and 7-20. Figure 7-19 illustrates the transmitter, while Fig. 7-20 shows the receiver mechanization. Briefly, the transmitter phase modulates the RF carrier with the set of phase-modulated, sinusoidal data subcarriers $\{S_l(t),\ l = 1, 2, \ldots, L\}$ and the set $\{T_m(t),\ m = 1, \ldots, M\}$ of binary-valued data subcarriers to produce the output

$$\rho(t) = \sqrt{2P} \sin [\omega_0 t + \theta(t)] \tag{7-77}$$

where

$$\theta(t) = \sum_{l=1}^{L} S_l(t) + \sum_{m=1}^{M} T_m(t) \tag{7-78}$$

The subcarrier signals are phase modulated by the individual N-level digital data streams, $s_{kl}(t)$, where the index $k = 1, 2, \ldots, N$ denotes which signal

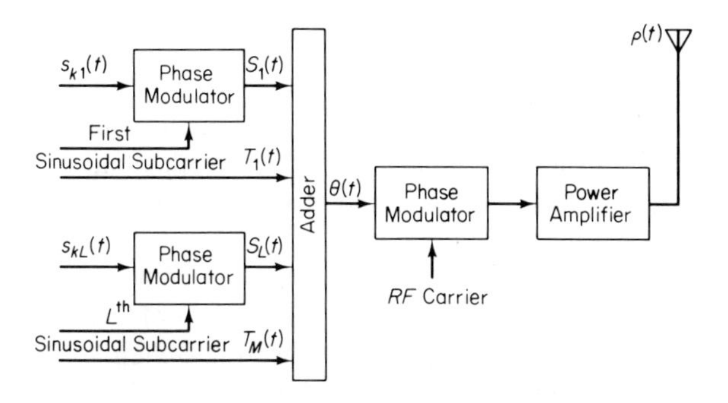

Fig. 7-19 Transmitter mechanization

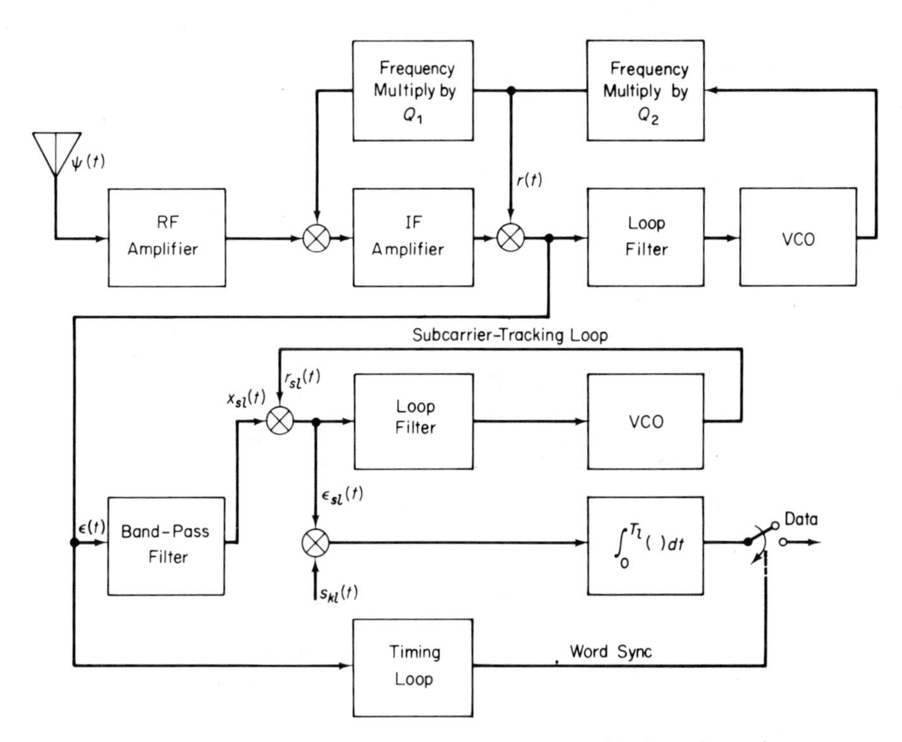

Fig. 7-20 Receiver mechanization for the lth data channel

$s_{kl}(t)$ is transmitted over the lth channel $l = 1, 2, \ldots, L$. Hence, the lth data-modulated subcarrier is conveniently represented by

$$S_l(t) = \sqrt{2}\, p_l \cos[\omega_l t + (\cos^{-1} m_l)s_{kl}(t)] \tag{7-79}$$

The modulation factor m_l apportions the total power of the lth data-modulated sinusoidal subcarrier between the subcarrier itself and the sideband modulation $s_{kl}(t)$. For convenience it is assumed, without loss in generality.

that

$$T_m(t) = \tau_m b_m(t) \qquad m = 1, \ldots, M \tag{7-80}$$

where $b_m(t)$ is a sequence of ± 1's, and τ_m is a multiplying constant. It is further assumed that the data signals $s_{kl}(t)$ occur with equal probability and exist for a time duration of T_l sec, and that the sequence $\{b_m(t)\}$ has zero average value.

In general, some of the $T_m(t)$ signals in Eq. (7-78) could be used to transmit binary data. In this case the $T_m(t)$ become the product of a data sequence and an unmodulated square-wave subcarrier. The single-channel mechanization discussed earlier is a good example of such a system. For this special case, we would have $L = 0$, $M = 1$.

Another special case of practical interest concerns the use of communication satellites to transmit L_1 data sources. Here, one would need to phase modulate the carrier with only one signal from the set $\{T_m(t)\}$—that is, $L = L_1$, $M = 1$. This signal could be used for purposes of synchronizing each receiver.

For the sake of generality, however, we shall treat the case of M binary-valued signals and let the design engineer select signaling configurations for his particular application. Also, the problem of specifying individual subcarrier frequencies will not be treated here as such a choice would be made after the particular system mission had been defined.

The lth receiver (Fig. 7–20) observes the phase- and frequency-shifted transmitted signal plus the additive noise $n_1(t)$,

$$\psi(t) = \sqrt{2P} \sin [\omega_1 t + \theta(t) + \theta_1] + n_1(t) \tag{7-81}$$

and must recover from it the lth data stream $s_{kl}(t)$. This is accomplished by means of a superheterodyne PLL receiver, which tracks the carrier component of the received signal and generates the receiver's reference signal $r(t)$. Subcarrier extractors, each consisting of a subcarrier tracking loop, the data demodulator (a cross-correlator), and a symbol synchronization loop, are used to obtain the desired information from the PLL demodulator output,

$$\epsilon(t) = \sqrt{P} \sin [\theta(t) + \varphi(t)] + N[t, \varphi(t)] \tag{7-82}$$

For the purpose of simplicity, all amplifier and multiplier gains in Fig. 7–20 are normalized to unity. Also, variations in $\varphi(t)$ are assumed small with respect to the minimum of the set $\{1/T_l\}$.

To select modulation factors for such a system, it is important to understand the distribution of total transmitter power among the various modulation terms, in particular the average power remaining in the carrier component P_c, the average power remaining in the lth sinusoidal subcarrier after product demodulation $P_s(l)$, the average power remaining in the mth binary-valued data subcarrier after product demodulation $P_t(m)$, and the total average power P_d lost due to the fact that the receiver's carrier-tracking loop is not a perfect phase detector.

Distribution of Transmitter Power Among the Various Modulation Terms

The transmitted signal of Eq. (7-77) may be rewritten as

$$\rho(t) = \sqrt{2P}\, Im\, \{\exp(j\omega_0 t) \times \exp[j\theta(t)]\} \tag{7-83}$$

where Im denotes the imaginary part of the quantity in the brackets. Now

$$\exp[j\theta(t)] = \prod_{l=1}^{L} \prod_{m=1}^{M} \exp\{j\sqrt{2}\, p_l \cos[\omega_l t + (\cos^{-1} m_l) s_{kl}(t)]\} \exp[j\tau_m b_m(t)] \tag{7-84}$$

and $\exp(jz \cos\theta)$ can be expressed in a Fourier-Bessel series

$$\exp(jz\cos\theta) = \sum_{l=-\infty}^{\infty} j^l J_l(z) \exp(jl\theta) \tag{7-85}$$

Applying Eq. (7-85) to (7-84) gives

$$\exp\{j\sqrt{2}\, p_l \cos[\omega_l t + (\cos^{-1} m_l) s_{kl}(t)]\} = \sum_{s=-\infty}^{\infty} j^s J_s(\sqrt{2}\, p_l) \exp\{js[\omega_l t + (\cos^{-1} m_l) s_{kl}(t)]\} \tag{7-86}$$

and, since $b_m(t) = \pm 1$,

$$\exp[j\tau_m b_m(t)] = \cos\tau_m + j b_m(t) \sin\tau_m. \tag{7-87}$$

Making use of (7-86) and (7-87) in (7-84) yields

$$\exp[j\theta(t)] = \prod_{l=1}^{L} \left\{ \sum_{s_l=-\infty}^{\infty} j^{s_l} J_{s_l}(\sqrt{2}\, p_l) \exp[js_l\{\omega_l t + (\cos^{-1} m_l) s_{kl}(t)\}] \right\} \times \prod_{m=1}^{M} \left\{ \sum_{l_m=0}^{1} j^{l_m} \cos\left[\tau_m b_m(t) - \frac{\pi l_m}{2}\right] \right\} \tag{7-88}$$

The power in the carrier component may be determined by selecting those terms in the previous expression for which $s_l = l_m = 0$. Thus,

$$\frac{P_c}{P} = \prod_{l=1}^{L} J_0^2(\sqrt{2}\, p_l) \prod_{m=1}^{M} \cos^2\tau_m. \tag{7-89}$$

If for any l, $\sqrt{2}\, p_l = 2.4048$, then $J_0(\sqrt{2}\, p_l) = 0$ and there is no power in the carrier component. Also, if $\tau_m = \pi/2$ for any m, the carrier is completely suppressed.

Proceeding as before, the demodulated signal can be expressed as

$$\epsilon(t) = \sqrt{P}\, Im\, \{\exp[j\varphi(t)] \times \exp[j\theta(t)]\} \tag{7-90}$$

where $\exp[j\theta(t)]$ is given by (7–88). To obtain an expression for the power in the ith data subcarrier, we set $s_l = 0$ for $l \neq i$, $s_i = 1$, and $l_m = 0$ for all m. Thus, the ratio of average normalized power in the ith modulated data subcarrier channel (subcarrier and subcarrier sidebands) to total transmitted power is given by

$$\frac{P_s(i)}{P} = 2\left[J_1^2(\sqrt{2}\,p_i)\prod_{\substack{l=1\\l\neq i}}^{L} J_0^2(\sqrt{2}\,p_l)\prod_{m=1}^{M}\cos^2\tau_m\right][\overline{\cos\phi}]^2 \tag{7-91}$$

where ϕ is the modulo 2π reduced RF tracking error.

Similarly, to find the power in the ith binary-valued data subcarrier, we set $s_l = 0$ for all l, $l_m = 0$ for all $i \neq m$, and $l_i = 1$. Thus, the ratio of the average normalized power in the ith binary-valued data subcarrier to the total transmitted power becomes

$$\frac{P_t(i)}{P} = \left[\sin^2\tau_i\prod_{l=1}^{L} J_0^2(\sqrt{2}\,p_l)\prod_{\substack{m=1\\m\neq i}}^{M}\cos^2\tau_m\right][\overline{\cos\phi}]^2 \tag{7-92}$$

For illustration purposes the power components $\sin^2 x$, $\cos^2 x$, $J_0^2(\sqrt{2}\,x)$, and $2J_1^2(\sqrt{2}\,x)$ are plotted in dB versus x in Fig. 7-21.

Finally, the ratio of average normalized power in the distortion component—that is, power lost due to intermodulation terms—to the total transmitted power is given by

$$\frac{P_d}{P} = 1 - \frac{P_c}{P} - \sum_{l=1}^{L}\frac{P_s(l)}{P} - \sum_{m=1}^{M}\frac{P_t(m)}{P} \tag{7-93}$$

These rather general results will be useful in what follows.

As a first example of the use of these results, consider the problem of combining many signals for transmission from a satellite to Earth. Here $M = 1$, L is large, and it is assumed that $p_l = p/\sqrt{L}$ for all l. Then, for large L

$$\begin{aligned}\frac{P_c}{P} &= \cos^2\tau_1\exp(-p^2)\\ \frac{P_s(i)}{P} &= \frac{p^2}{L}\cos^2\tau_1\exp(-p^2)[\overline{\cos\phi}]^2\\ \frac{P_t(1)}{P} &= \sin^2\tau_1\exp(-p^2)[\overline{\cos\phi}]^2\end{aligned} \tag{7-94}$$

for all $i = 1, \ldots, L$ since

$$J_0(x) \cong 1 - \left(\frac{x}{2}\right)^2 \qquad \text{and} \qquad J_1(x) \cong \frac{x}{2}$$

for small x. From Eq. (7-94) the maximum power that can be obtained in the data subcarrier channel is obtained when $p = 1$; hence,

$$P_{s_{\max}} = \frac{P}{eL}(\cos^2\tau_1)[\overline{\cos\phi}]^2 \tag{7-95}$$

As a second example of these results, we assume that L and M are both large with $p_l = p/\sqrt{L}$ for all l and $\tau_m = \tau/\sqrt{M}$ for all $m = 1\ldots, M$. The normalized powers in the carrier, each sinusoidal data subcarrier chan-

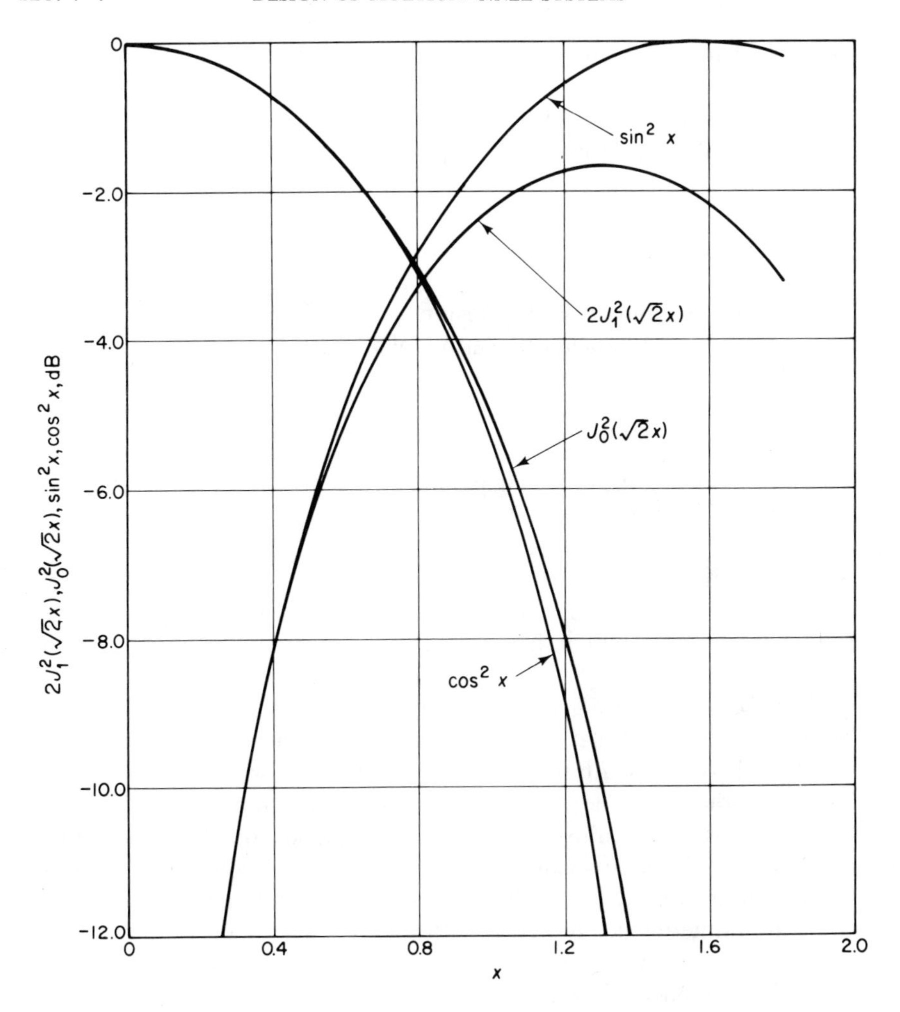

Fig. 7-21 The functions $2J_1^2(\sqrt{2}\,x)$, $J_0^2(\sqrt{2}\,x)$, $\sin^2 x$, and $\cos^2 (x)$ in dB

nel, and each binary-valued data subcarrier then become

$$\frac{P_c}{P} = \exp\left[-(p^2 + \tau^2)\right]$$

$$\frac{P_s}{P} = \frac{p^2}{L} \exp\left[-(p^2 + \tau^2)\right]\left[\overline{\cos \phi}\right]^2 \tag{7-96}$$

$$\frac{P_t}{P} = \frac{\tau^2}{M} \exp\left[-(p^2 + \tau^2)\right]\left[\overline{\cos \phi}\right]^2$$

Again the values of τ and p, which maximize power in the data subcarriers and signal set $\{T_m(t)\}$, are obtained by setting $p = \tau = 1$. This means that the maximum signal power detectable by means of the product demodulator is approximately $1/e$ times less than the power obtainable by means of a perfect phase detector.

Choice of Parameters in the Design of Multichannel Satellite-to-Earth Links (L Large, $M = 1$)

To recapitulate, the following parameters are to be selected in the design: the set of data rates $\{\mathcal{R}_l = T_l^{-1}\}$, τ, and the set of modulation factors $\{m_l\}$. The value of $[\overline{\cos \phi}]^2$ depends on the total signal-to-noise ratio in the RF carrier-tracking loop. This function has been studied in Sec. 7–3 where the plot of Fig. 7–16 is now used in designing a multichannel system. For simplicity we begin by considering the special case of binary signals—that is, $N = 2$.

The Case of Binary Signals. In the extraction of data from the lth receiver output, it is assumed that this output is filtered by a band-pass filter, which selects and passes the lth modulated data subcarrier without distortion and suppresses or rejects the other L-1 modulated data subcarriers. Of course, this filter passes the additive noise $N[t, \varphi(t)]$ and a portion of the power in the signal distortion component. In the analysis that follows, this distortion component is neglected since, in practice, it represents only a small fraction of the total interference power. It is further assumed that the bandwidth of this band-pass filter—the filter that precedes the subcarrier-tracking loop in Fig. 7-20—is sufficiently wide that the noise $N[t, \varphi(t)]$ looks essentially white (at the input) to the subcarrier-tracking loop. For the present, it is assumed that the symbol synchronization loop extracts the necessary symbol sync from the receiver output with negligible jitter, and hence provides the multiplier of the cross-correlator with a perfect phase reference. This assumption is justified from a practical point of view since a bandwidth of one Hz or less is certainly practical in this type of tracking loop. Also, there is essentially no Doppler on the signal component of $x_{sl}(t)$ in Fig. 7-20. A signal-to-noise ratio of 20 dB, which gives essentially perfect symbol sync, requires that the power in the timing signal be 100 times the noise spectral density. Thus, the modulation index, τ, may be chosen to establish this signal-to-noise ratio in the symbol synchronization loop.

Under these assumptions the problem of demodulation of data may be restated as follows. Observed at the input to the lth subcarrier extractor is the equivalent noise-corrupted signal

$$x_{sl}(t) = \sqrt{2P_s(l)} \sin [\omega_l t + (\cos^{-1} m_l) s_{kl}(t)] + N[t, \varphi(t)] \tag{7-97}$$

where $k = 1,2$, and $l = 1, \ldots, L$. This may be written, using an appropriate trigonometric identity, as

$$x_{sl}(t) = \sqrt{2P_{cs}(l)} \sin \omega_l t + \sqrt{2(1 - m_l^2)P_s(l)} s_{kl}(t) \cos \omega_l t + N[t, \varphi(t)] \tag{7-98}$$

where $m_l^2 = P_{cs}(l)/P_s(l)$ is a constant which allocates the total power $P_s(l)$ in the lth modulated data subcarrier between the unmodulated subcarrier component and the sidebands. The subcarrier-tracking loop tracks the subcarrier component by means of a PLL and produces the reference signal $r_{sl}(t) = \sqrt{2} \cos (\omega_l t + \varphi_{sl})$ where φ_{sl} is the estimate of the subcarrier phase. Thus, the output of the subcarrier-tracking loop, $\epsilon_{sl}(t) \triangleq x_{sl}(t)\, r_{sl}(t)$ is

$$\epsilon_{sl}(t) = \sqrt{2(1 - m_l^2)P_s(l)}\, s_{kl}(t) \cos \varphi_{sl} + N[t, \varphi(t), \varphi_{sl}(t)] \tag{7-99}$$

where all double-frequency components have been neglected, and where $N[t, \varphi(t), \varphi_{sl}(t)]$ is approximately a bandlimited white Gaussian noise process with single-sided spectral density, N_0 w/Hz. The conditional error probability, given ϕ_{sl}, is therefore expressed by

$$P_{El}(\phi_{sl}) = \tfrac{1}{2} \operatorname{erfc} [\sqrt{(1 - m_l^2)R_s(l)} \cos \phi_{sl}] \tag{7-100}$$

where

$$R_s(l) \triangleq \frac{P_s(l)T_l}{N_0} \tag{7-101}$$

and for the special case under consideration

$$P_s(l) = 2[J_1(\sqrt{2}\, p_l) \prod_{\substack{i=1 \\ i \neq l}}^{L} J_0^2(\sqrt{2}\, p_i) \cos \tau_1][\overline{\cos \phi}]^2 \tag{7-102}$$

If L is large and the data channels are equally weighted—that is, $p_l = p/\sqrt{L}$—then $P_s(l)$ is given by Eq. (7-94) and has a maximum value as specified in (7-95).

Recognizing the resemblance between the above equations for the lth data subcarrier and those of the one-way single-channel system discussed in Sec. 7-2, one can immediately deduce the optimum design of the multichannel system. If the p.d.f. of the lth channel subcarrier phase error, $p(\phi_{sl})$ is characterized by an equation similar to Eq. (7-10), then averaging $P_{El}(\phi_{sl})$ over this p.d.f. gives the average error probability in the lth data channel, which is plotted in Fig. 7-3 versus m_l^2 with the proper changes in notation. If the linear PLL theory is used to characterize $p(\phi_{sl})$, then the square of the optimum modulation factor for the lth data subcarrier channel is given by (7-26) and is plotted in Fig. 7-7.

At this point, other comparisons should be obvious, hence we curtail the discussion and go on to the design of block-coded systems. We shall pursue in detail the case of orthogonal codes, although the theory to be developed could apply equally to other block-coded systems described in Chap. 5.

The Design of Block-Coded Systems, $N = 2^n$. The problem may be described as follows: the waveforms $s_{kl}(t)$ are assumed to exist for $T_l = nT_{bl}$ seconds, contain equal energies, and are orthogonal; that is,

$$\int_0^{nT_{bl}} s_{kl}(t)s_{jl}(t) = nT_{bl}\delta_{kj} \tag{7-103}$$

where $\delta_{kj} = 0$ if $k \neq j$, and $\delta_{kj} = 1$ if $k = j$. The output, $\epsilon_{sl}(t)$ of the lth data channel subcarrier-tracking loop is cross-correlated with the N possible transmitted signals $s_{kl}(t)$, $k = 1, 2, \ldots, N$, to form the set of statistics

$$q_{kl} = \int_0^{nT_{bl}} \epsilon_{sl}(t)s_{kl}(t)\,dt; \qquad k = 1, 2, \ldots, N \tag{7-104}$$

The output of the decoder is those n bits which, if encoded, would produce that $s_{kl}(t)$ which yields the largest q_{kl}.

As 2^n cross-correlators are required to decode an n-bit orthogonal code, the complexity of the decoder becomes impractical for n of about eight or greater. Also, the complexity of the decoder and the maximum bit rate at which the decoder will operate are major factors in the design of the decoder. In this text, we do not investigate techniques for reducing the decoder complexity or for increasing the maximum bit rate at which the decoder will operate. The interested reader is referred to material contained in Refs. 8, 9, and 10.

The conditional probability of correct word detection, is given by Eq. (5-50), which in the present notation becomes

$$P_{cl}(N; \phi_{sl}) = \int_{-\infty}^{\infty} \frac{\exp(-x^2)}{\sqrt{\pi}}\left[\frac{1}{2}\operatorname{erfc}(-x - A_l\cos\phi_{sl})\right]^{N-1} dx \tag{7-105}$$

where $\qquad A_l = \sqrt{n(1 - m_l^2)R_{sb}(l)} \qquad R_{sb}(l) = \dfrac{R_s(l)}{n} = \dfrac{P_s(l)T_{bl}}{N_0}$

By averaging (7-105) over $p(\phi_{sl})$ one finds the average word error rate of the lth data channel. For code words containing $n = 5$ and 8 bits of information, Figs. 7-22 and 7-23, respectively, depict the average word error rate versus the square of the modulation factor m_l, with $R_{sb}(l)$ and $\delta_{sb}(l)$ as parameters. The corresponding plots of minimum error rate versus $R_{sb}(l)$ are illustrated in Figs. 7-24 and 7-25 with $\delta_{sb}(l)$ as a parameter; $\delta_{sb}(l)$ represents the ratio of the lth channel bit rate $\mathcal{R}_b = 1/T_{bl}$ to the one-sided bandwidth of the lth channel subcarrier-tracking loop.

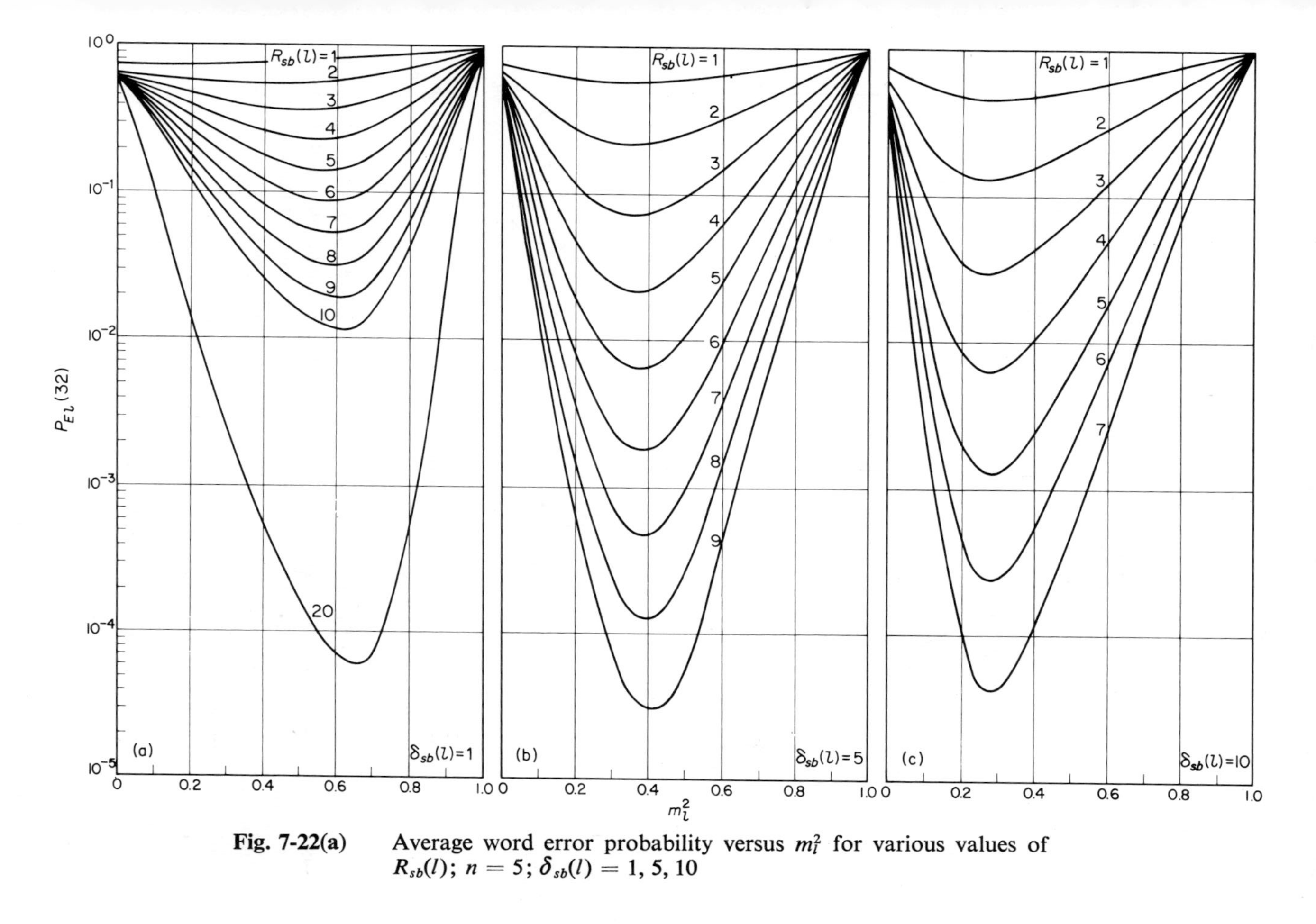

Fig. 7-22(a) Average word error probability versus m_l^2 for various values of $R_{sb}(l)$; $n = 5$; $\delta_{sb}(l) = 1, 5, 10$

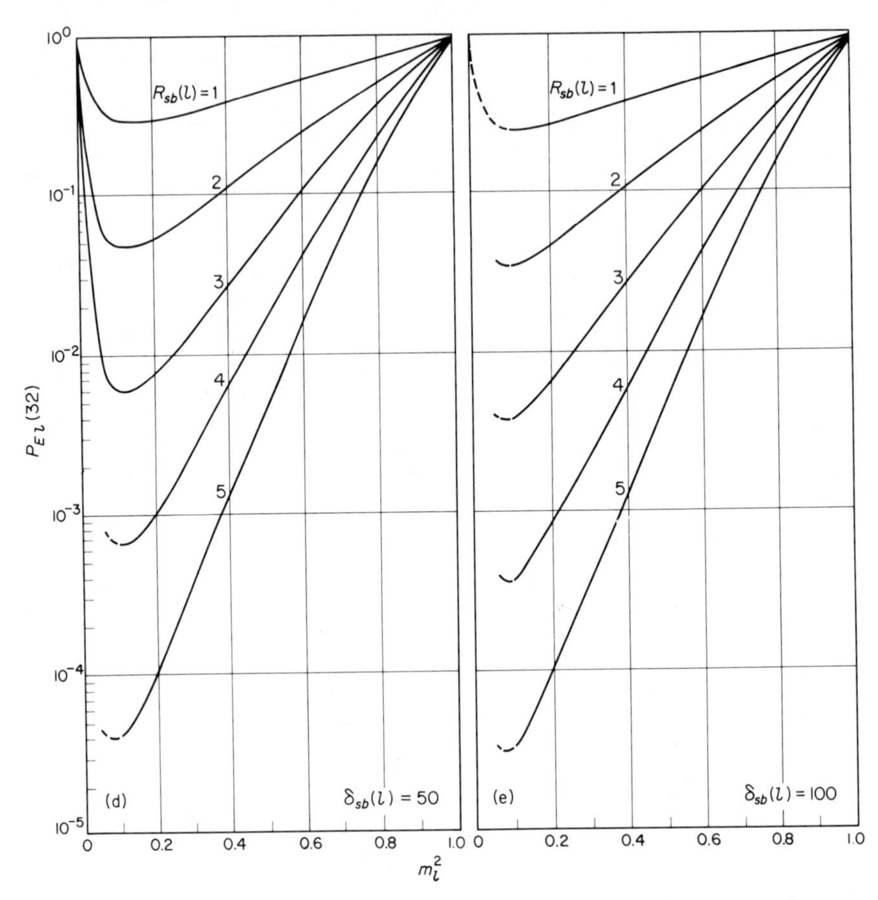

Fig. 7-22(b) Average word error probability versus m_l^2 for various values of $R_{sb}(l)$; $n = 5$; $\delta_{sb}(l) = 50, 100$

For $n \geq 5$, the performance of a block-coded, digital communication system using bi-orthogonal codes is essentially the same as one that uses orthogonal codes. Hence, for $n \geq 5$, the results that follow may also be applied to the design of systems whose code dictionaries are made up of bi-orthogonal codes.

In practice, an approximate formula, which allows the communications engineer to select the optimum value of the modulation factor m_l for various degrees of coding, is desirable. One proceeds to develop such an expression by noting that for large N

$$P_{cl}(N; \phi_{sl}) \cong 1 - \left(\frac{N-1}{2}\right) \operatorname{erfc}\left[\frac{A_l}{\sqrt{2}} \cos \phi_{sl}\right] \tag{7-106}$$

Averaging $P_{cl}(N; \phi_{sl})$ of (7-106) over $p(\phi_{sl})$ assuming linear PLL theory,

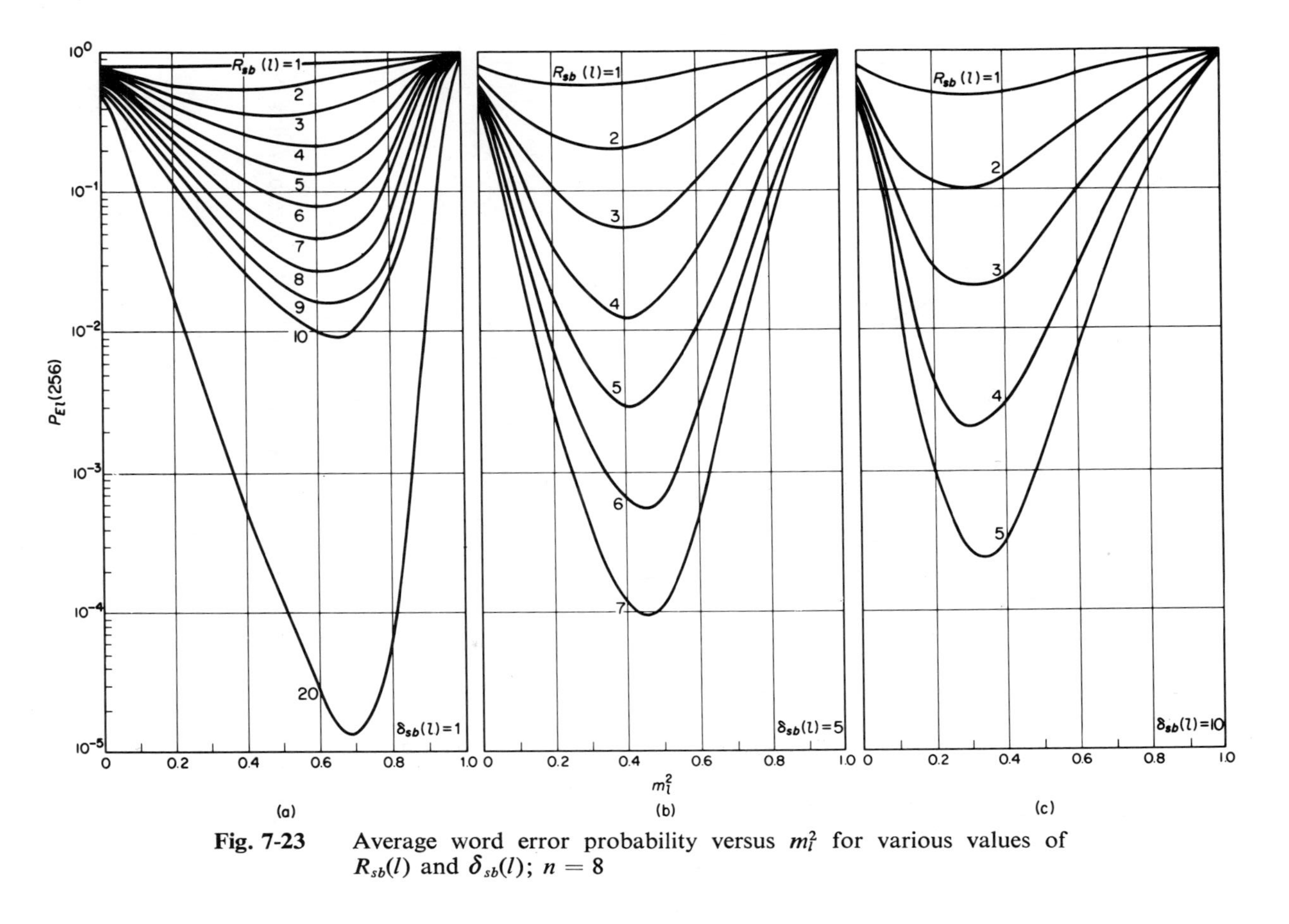

Fig. 7-23 Average word error probability versus m_l^2 for various values of $R_{sb}(l)$ and $\delta_{sb}(l)$; $n = 8$

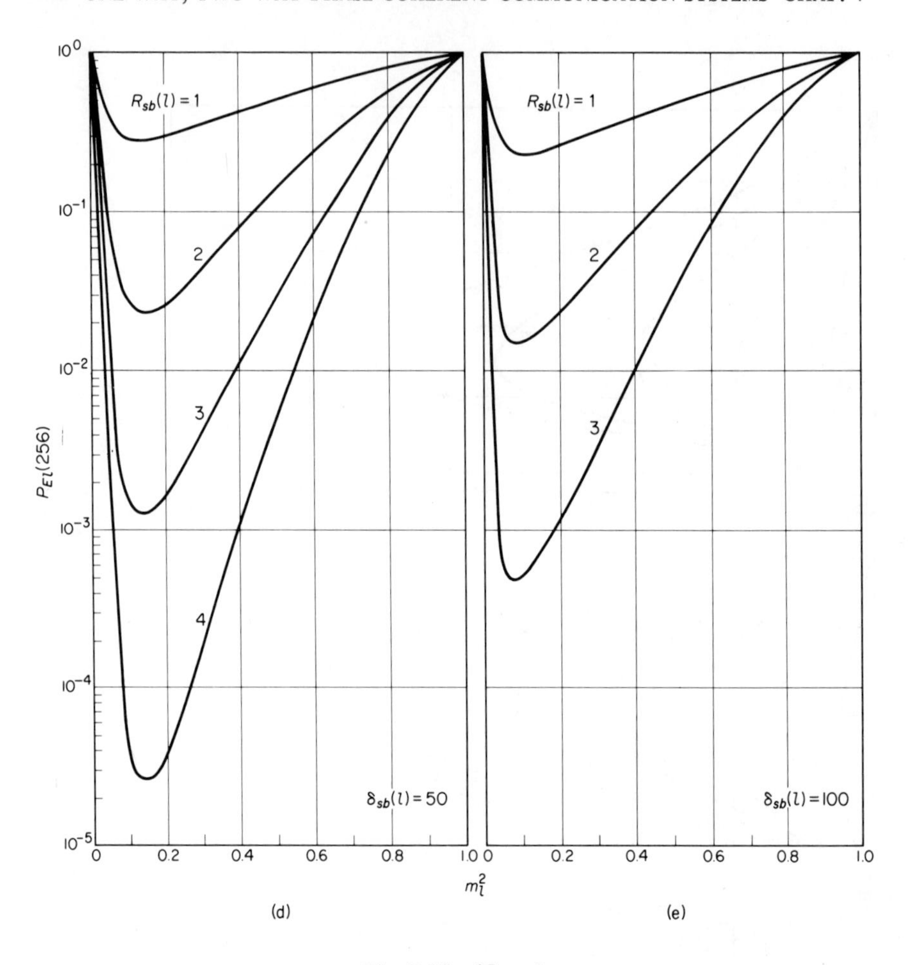

Fig. 7-23 (Cont.)

differentiating with respect to m_l, and equating the result to zero, yields (for block codes)*

$$m_{l_{\text{opt}}}^2 = \frac{nR_l - 1 + \sqrt{(nR_l - 1)^2 + 4nR_l(1 + \delta_l/n)}}{2nR_l(1 + \delta_l/n)} \tag{7-107}$$

The corresponding minimum error probability is obtained by substituting (7-107) in (7-106) and averaging over the p.d.f. of the subcarrier phase error. The accuracy of (7-107) has been checked and shown to agree quite closely with exact results obtained from Figs. 7-22 and 7-23.

---

* For simplicity of notation we have set $R_{sb}(l) = R_l$ and $\delta_{sb}(l) = \delta_l$.

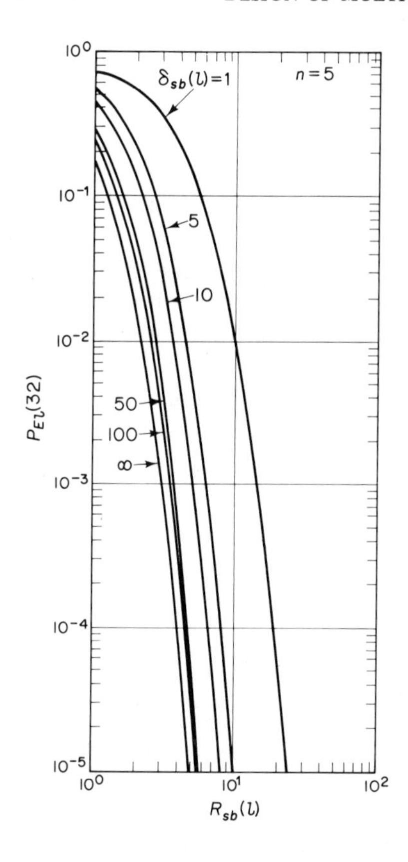

Fig. 7-24 Average word error probability versus $R_{sb}(l)$ for various values of $\delta_{sb}(l)$ when the power is divided optimally; $n = 5$

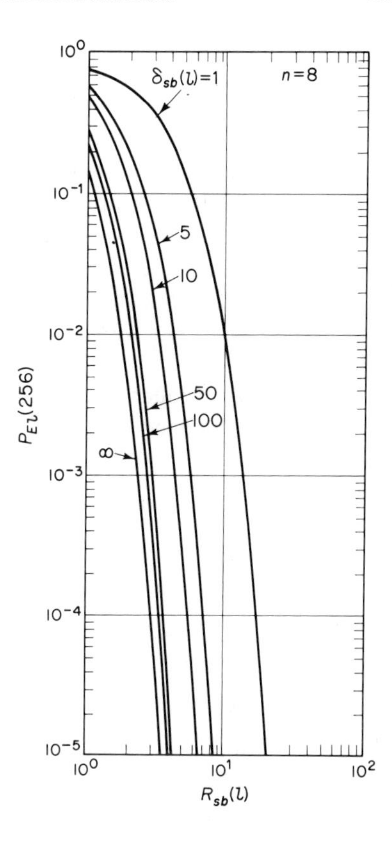

Fig. 7-25 Average word error probability versus $R_{sb}(l)$ for various values of $\delta_{sb}(l)$ when the power is divided optimally; $n = 8$

Choice of Parameters in the Design of a Deep Space-to-Earth, Block-Coded Communication System ($L = 0$, $M = 1$)

For the binary case, the design and performance of such a system has been developed in considerable detail in Sec. 7-2. In the case of block codes, the results may be extracted from those given previously by redefining certain parameters. In order to make this change in parameters obvious, the transmitted signal may be reduced to

$$\rho(t) = \sqrt{2P} \sin [\omega_0 t + T_1(t)] \tag{7-108}$$

Here the signal $T_1(t)$ is given by

$$T_1(t) = \tau_1 b_1(t) = \tau_1 s_k(t)\mathfrak{S}(t)$$
$$\tau_1 = (\cos^{-1} m_1) \tag{7-109}$$

where $s_k(t)$ is the kth code word, $k = 1, \ldots, N$, of duration $T = nT_b$, which amplitude modulates the unit power square-wave subcarrier $\mathfrak{S}(t)$. It is clear that in this development the transmission of a synchronization signal has been neglected. However, as noted in Chap. 5, block codes offer an added advantage in that they can be made comma-free or self-synchronizing. Consequently, symbol and word sync may be obtained from the time structure of $T_1(t)$.

Since $\mathfrak{S}(t)$ is a sequence of ± 1s and the signals $s_k(t)$ are ± 1s, the signal $b_1(t)$ is also a sequence of ± 1s. Thus, the receiver observes the noise-corrupted signal

$$\begin{aligned}\psi(t) &= \sqrt{2m_1^2 P}\sin(\omega_1 t + \theta_0) \\ &\quad + \sqrt{2(1 - m_1^2)P}b_1(t)\cos(\omega_1 t + \theta_0) + n_1(t) \\ &= \sqrt{2P_c}\sin(\omega_1 t + \theta_0) + \sqrt{2P_1}b_1(t)\cos(\omega_1 t + \theta_0) + n_1(t)\end{aligned} \tag{7-110}$$

where $m_1^2 = P_c/P$ and $P_1 = (1 - m_1^2)P$. The input to the data detector and symbol synchronization loop becomes

$$\psi(t)r(t) = \sqrt{(1 - m_1^2)P}b_1(t)\cos\varphi + N[t, \varphi(t)] \tag{7-111}$$

and the conditional probability of correct reception may, therefore, be written as in (7-105) with the appropriate changes of notation:

$$\begin{aligned}\phi_{sl} &\longrightarrow \phi \\ P_{cl}(N; \phi_{sl}) &\longrightarrow P_c(N; \phi) \\ A_l &\longrightarrow A_1 \\ m_l &\longrightarrow m_1 \\ R_{sb}(l) &\longrightarrow R_b = \frac{PT_b}{N_0}\end{aligned}$$

Hence, it is clear that the graphical results given in the previous part also apply to the design of this system if $\delta_{sb}(l)$ is now replaced by δ_b—that is, the ratio of bit rate $\mathcal{R}_b = 1/T_b$ to the one-sided bandwidth of the RF carrier-tracking loop.

It is clear from our discussions thus far, that a general analytical approach to determine the modulation factors that optimize system design for arbitrary L and M necessitates lengthy solutions of simultaneous equations using a digital computer. Other approaches to the problem of optimum power division, wherein the theoretical expressions for the power ratios in the various channels are subjected to a graphical, rather than an analytical approach, are given in Refs. 11 and 12.

PROBLEMS

7-1. A single-channel two-way communication system operates with the parameters of Prob. 3-1, with $G = 1$, and $\lambda_1 = \lambda_2 = -1$.

a) Find the value of ρ_1 that gives an irreducible error probability P_{E2} equal to 6×10^{-3}.

b) What is the corresponding value of α_1?

c) If $\alpha_1 = 9$ dB, and the data rate on the downlink is $\mathcal{R}_2 = 60$ bits/sec, find the optimum modulation factor m_2 for a downlink signal-to-noise ratio $R_2 = 10$.

d) What is the corresponding minimum value of error probability P_{E2}?

Repeat parts c) and d) with $\mathcal{R}_2 = 1200$ bits/sec.

7-2. Repeat part c) of Problem 7-1 using the approximate formula for $m_{2\,\text{opt}}$ as given in Eq. (7-25). Compare your results.

7-3. Starting with Eq. (7-24),

a) derive (7-25), which gives an approximate expression for the value of m_2^2 which minimizes P_{E2}.

b) show that by letting $\alpha_1 = \infty$, and changing all *two* subscripts to *ones*, Eq. (7-25) reduces to (7-26).

7-4. For the communication system of Prob. 7-1,

a) find the optimum modulation factor m_1 corresponding to an uplink data rate $\mathcal{R}_1 = 90$ bits/sec and an uplink signal-to-noise ratio $R_1 = 10$.

b) What is the corresponding minimum value of error probability P_{E1}?

7-5. Repeat part a) of Prob. 7-4 using the approximate formula for $m_{1\,\text{opt}}$ as given in Eq. (7-26). Compare your results.

7-6. For the two-way communication system of Prob. 7-1,

a) find the optimum modulation factor m_1 for an uplink signal-to-noise ratio $R_1 = 10$ and an uplink data rate $\mathcal{R}_1 = 2.25$ bits/sec. (Assume the rapidly-varying phase error case).

b) What is the corresponding minimum value of P_{E1}?

c) Assuming the value of m_1 found in part a), find the value of m_2 that minimizes P_{E2} if $\mathcal{R}_2 = 4.5$ bits/sec and $R_2 = 6$.

d) What is the corresponding minimum value of P_{E2}?

7-7. Assume the communication system of Prob. 3-1 is suboptimally designed such that the signal-to-noise ratio at the demodulator output is to be maximized. If the uplink data rate is $\mathcal{R}_1 = 18$ bits/sec,

a) find the optimum modulation factor m_1 for an uplink signal-to-noise ratio $R_1 = 20$.
b) What is the corresponding maximum value of SNR_1?

7-8. It is of interest to understand how additive noise on the uplink of a two-way communication system affects the reference system's error probability performance. If, in the communication system of Prob. 7-1, the downlink data signal-to-noise ratio is $R_{d2} = 4.8$, the total downlink signal-to-noise ratio $R_2 = 6.4$, and the downlink data rate $\mathcal{R}_2 = 75$ bits/sec, find the values of P_{E2} for uplink signal-to-noise ratios $\alpha_1 = 10$ and $\alpha_1 = 100$.

7-9. Consider a two-channel system whose total transmitted power P_1 is allocated as in Eq. (7-47).
a) If the powers in the two data channels P_{11} and P_{21} are fixed by the error rate requirements on these channels, show that the value of carrier power P_{c1} which minimizes the total transmitted power is given by $P_{c1} = \sqrt{P_{11}P_{21}}$.
b) Show that the corresponding minimum value of P_1 as found by the procedure of part a) is given by $P_1 = (\sqrt{P_{11}} + \sqrt{P_{21}})^2$.
c) Express the result obtained in part b) as a relation involving only the modulation factors m_{11} and m_{21}.
d) Rewrite the relation found in part c) as a condition on the modulation angles $\theta_{m_{11}} = \cos^{-1} m_{11}$ and $\theta_{m_{21}} = \cos^{-1} m_{21}$.

7-10. As an example of a two-channel system, consider the downlink on the Mariner Mars 1969 space mission. A high-rate channel allows transmission of encounter science data from the spacecraft to Earth at a rate of 16.2 kilobits/sec. The uncoded science information is encoded into a (32, 6) bi-orthogonal code, and the resulting symbols are phase-modulated onto a 259.2 kHz square-wave subcarrier. Uncoded engineering data at a rate of 33-1/3 bits/sec is phase-modulated onto a 24 kHz square-wave subcarrier and transmitted over the engineering data channel. The angle modulation indices for the two channels are: $\theta_{m_{12}} = 65°$ (science) and $\theta_{m_{22}} = 12°$ (engineering).
a) Find the ratios of carrier power to total transmitted power P_{c2}/P_2 and the cross-modulation loss to total transmitted power P_{I2}/P_2.
b) If $P_2/N_{02} = 46.14$ dB, find the bit signal-to-noise ratios in the science and engineering channels; that is, $P_{12}T_{12}/N_{02}$ and $P_{22}T_{22}/N_{02}$.
c) Repeat parts a) and b) with the interplex modulation technique applied to the engineering channel.

7-11. It is desired to use a single transmitter (for example, Klystron amplifier, waveguide equipment, and antenna) to transmit two-channel information on separate carriers to a pair of spacecraft. One possibility, which eliminates the problem of undesirable intermodulation products caused by transmitter nonlinearity, is to switch the excitation alternately between the two carrier frequencies so that at any given time the Klystron and antenna see only one phase-modulated carrier. If this is done, then the ground transmitter's output signal can be written in the form

$$\rho(t) \triangleq \rho_1(t) + \rho_2(t)$$

with

$$\rho_1(t) = [\tfrac{1}{2} + \tfrac{1}{2}\mathfrak{S}w(t)]\sqrt{P_1}\sin[\omega_{01}^{(1)}t + (\cos^{-1} m_{11}^{(1)})x_{k_1}(t) + (\cos^{-1} m_{21}^{(1)})X_{k_1}(t)]$$

and

$$\rho_2(t) = [\tfrac{1}{2} + \tfrac{1}{2}\mathfrak{S}w(t)]\sqrt{P_1}\sin[\omega_{02}^{(2)}t + (\cos^{-1} m_{11}^{(2)})x_{k_1}(t) + (\cos^{-1} m_{21}^{(2)})X_{k_1}(t)]$$

where the superscript in parentheses denotes the particular spacecraft and $\mathfrak{S}w(t)$ is the carrier switching signal—that is, a ± 1 digital waveform.

a) Using trigonometric expansions, write the signals $\rho_1(t)$ and $\rho_2(t)$ in terms of their in-phase and quadrature components.
b) Repeat part a) when the switching signal $\mathfrak{S}w(t)$ is specifically chosen equal to one of the modulated subcarriers, for example, $x_{k_1}(t)$, and identify the ratio of power in each of the signal components to the total power.
c) Repeat part a) when the switching signal $\mathfrak{S}w(t)$ is chosen as the intermodulation product $x_{k_1}(t)X_{k_1}(t)$.
d) Repeat parts b) and (c) when the intermodulation product $x_{k_1}(t) \times X_{k_1}(t)$ is transmitted in place of $X_{k_1}(t)$.

7-12. Discuss several possibilities that would indicate how the interplex modulation scheme might be applied in a system with *three* data-modulated, square-wave subcarriers. In each case, write the ground transmitter's output signal in terms of its in-phase and quadrature components and identify the ratio of the power associated with each component to the total power transmitted.

REFERENCES

1. Lindsey, W. C., "Optimal Design of One-Way and Two-Way Coherent Communication Links," *IEEE Transactions on Communication Technology*. Vol. COM-14, No. 4 (August, 1966) 418–431.
2. ———, "Determination of Modulation Indexes and Design of Two-Channel Coherent Communication Systems," *IEEE Transactions on Communication Technology*, Vol. COM-15, No. 2 (April, 1967) 229–237.

3. SERGO, J. R., and HAYES, J. F. "Optimum Power Allocation for Phase and Synchronization Error in a Two Channel System," *Proceedings of the National Telemetering Conference*, (1969) 306–310.

4. ———, "Design of Block-Coded Communication Systems," *IEEE Transactions on Communication Technology*, Vol. COM-15, No. 4 (August, 1967) 525–534.

5. HAYES, J. F., and LINDSEY, W. C., "Power Allocation—Rapidly Varying Phase Error," *IEEE Transactions on Communication Technology*, Vol. COM-17, No. 2 (April, 1969) 323–326.

6. GOLOMB, S. W., et. al., *Digital Communications with Space Applications*, Prentice-Hall, Inc., Englewood Cliffs, N. J. (1964) 100.

7. BUTMAN, S. and TIMOR, U., "Interplex—An Efficient Multichannel PSK/PM Telemetry System," *IEEE Transactions on Communication Technology*, Vol. COM-20, No.3 (June, 1972) 415–419.

8. KOERNER, M.A., "Decoding Techniques for Block-Coded Digital Communication Systems," Jet Propulsion Laboratory, Pasadena, Calif., SPS 37–17, Vol. IV (1962) 71–73.

9. ———, "A Decoding Algorithm for Reed-Muller codes," Jet Propulsion Laboratory, Pasadena, Calif., SPS 37–38, Vol. IV (1966) 210–217.

10. GREEN, R. R., "A Serial Orthogonal Decoder," Jet Propulsion Laboratory, Pasadena, Calif., SPS 37–39, Vol. IV (1966) 257–262.

11. FOLEY, Y. K., GAUMOND, B. J., and WITHERSPOON, J. T., "Optimum Power Division for Phase-Modulated Deep-Space Communications Links," *IEEE Transactions on Aerospace and Electronic Systems*, Vol. AES-3, No.3 (May, 1967) 400–409.

12. KADAR, I., "A Simplified Optimum Selection of Modulation Indices for Multitone Phase Modulation," *Proceedings of the National Electronics Conference*, (1968) 444–449.

Chapter 8

DESIGN AND PERFORMANCE OF PHASE-COHERENT SYSTEMS PRECEDED BY BAND-PASS LIMITERS

8-1. INTRODUCTION

In previous chapters, we developed a theory for assessing the performance of tracking and data acquisition systems whose carrier demodulation reference is derived from a PLL. As discussed in Chap. 2, such phase-coherent communication systems quite often incorporate a band-pass limiter in the IF section of the carrier-tracking loop to limit the dynamic range of the input signal plus noise, thereby reducing fluctuations of loop parameters with input level. Thus, we are motivated to study the effect on system error probability performance of preceding the PLL with a BPL.[1,2] In what follows, we shall draw heavily upon the material and notation in Sec. 2-5 of Chap. 2 and the appropriate sections of Chaps. 6 and 7.

We shall begin with an analysis of the noisy reference performance of one-way systems which employ coherent PSK, DPSK and differentially encoded PSK. Demodulators of such signaling formats are typical of data detection systems used in the Mariner and Pioneer projects. Following this, we shall generalize our results to the case of block-coded communication systems. Finally we consider the optimum design of systems which employ BPL's with the single-channel case serving as a detailed example.

8-2. THE NOISY REFERENCE PROBLEM IN COHERENT SYSTEMS PRECEDED BY A BAND-PASS LIMITER

To review, the noisy reference problem in phase-coherent communications is that of demodulating a data-bearing signal of average power S with a carrier reference signal of average power P_c derived from a source that

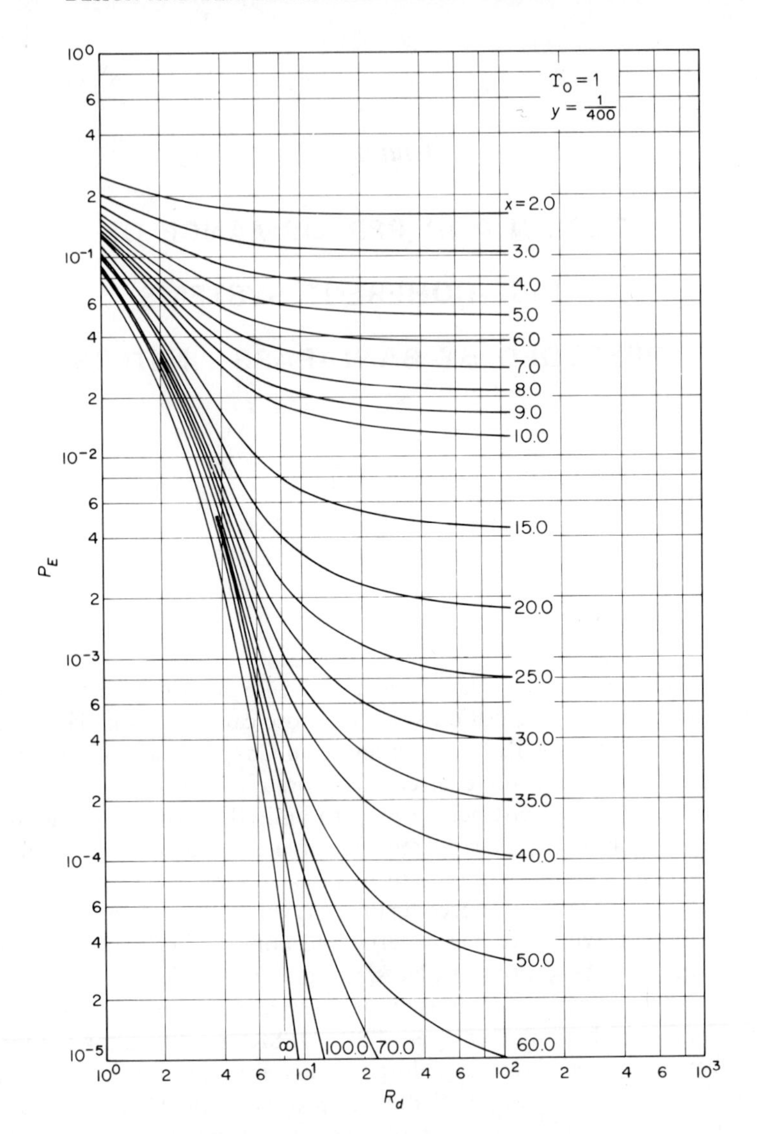

Fig. 8-1 Error probability versus data signal-to-noise ratio; $r_0 = 2$; $\Upsilon_0 = 1$; PSK signaling (slowly-varying phase error)

possesses a random phase uncertainty. This was discussed in full in Chap. 6, and the only change to be introduced here is that of preceding the PLL (the source of the noisy reference) with a BPL. Also, since our intention is merely to illustrate the technique involved, we deal only with extreme cases of phase error variation relative to symbol time—slowly and rapidly varying

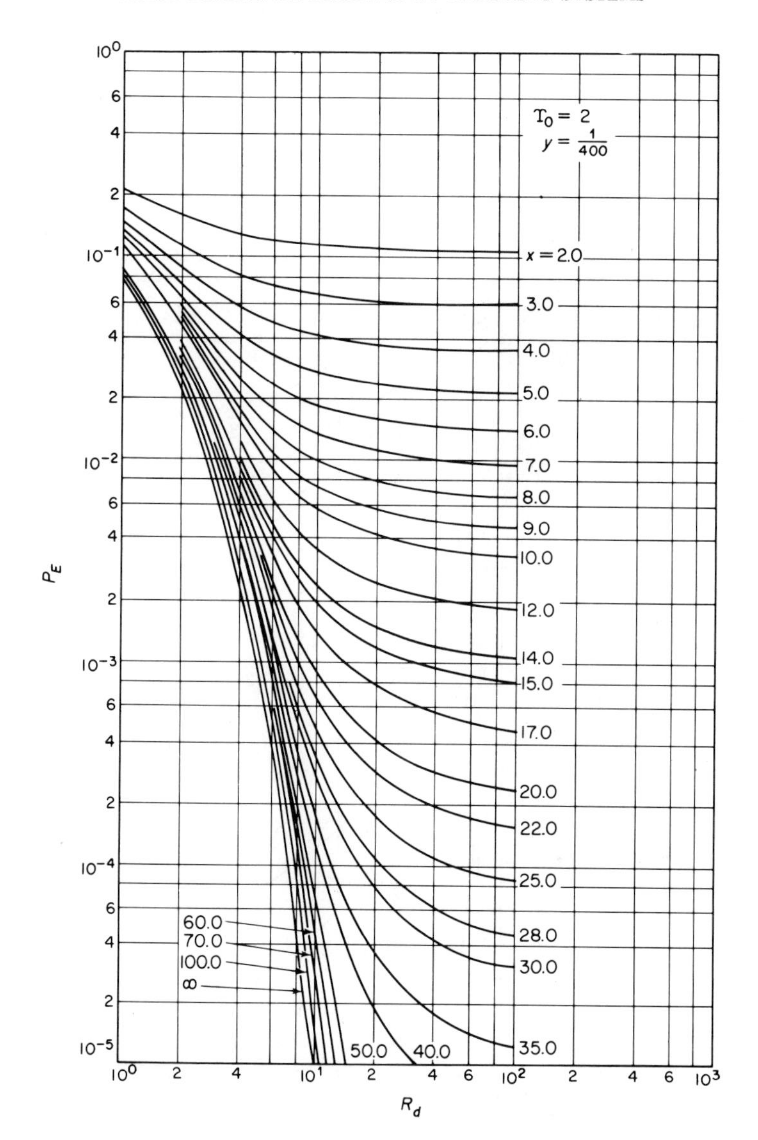

Fig. 8-2 Error probability versus data signal-to-noise ratio; $r_0 = 2$; $\Upsilon_0 = 2$; PSK signaling (slowly-varying phase error)

phase error. The case of moderately varying phase error follows directly from the theory to be presented here and that given in Chap. 6.[3]

As we have seen on several occasions, the average error probability of the data detector is computed by averaging the conditional error probability—conditioned on the noisy reference's phase error—over the p.d.f. of the phase

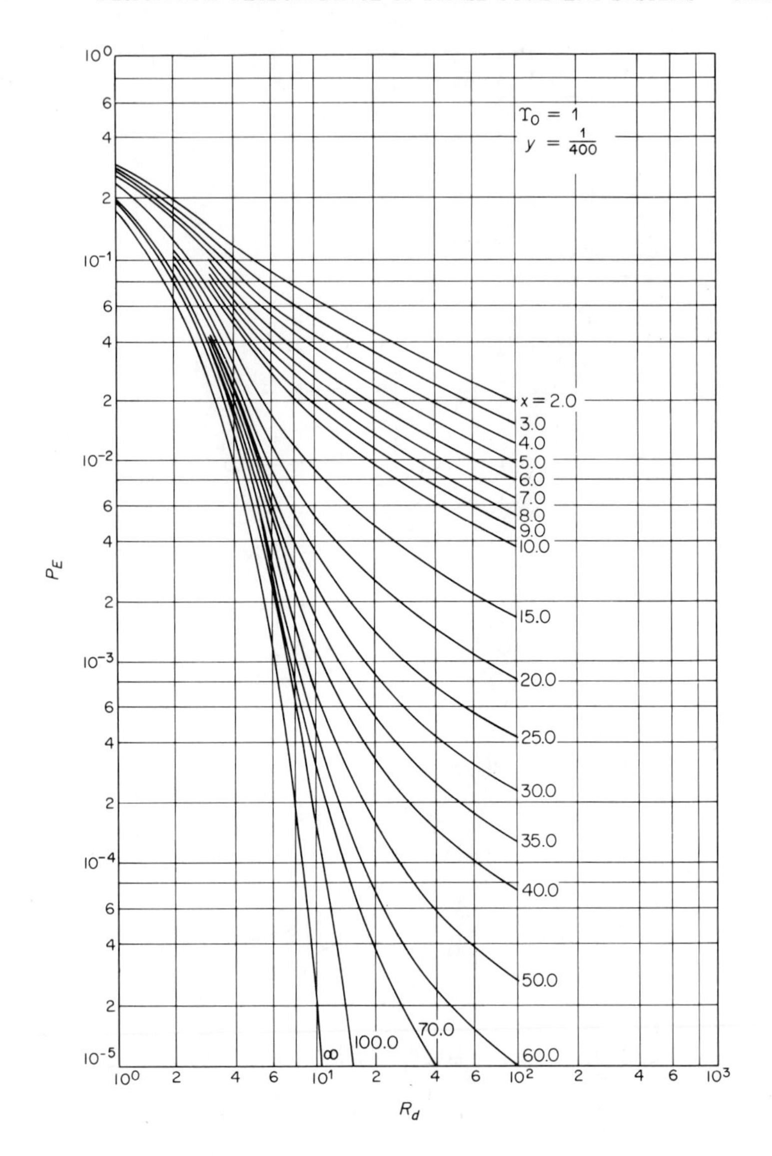

Fig. 8-3 Error probability versus data signal-to-noise ratio; $r_0 = 2$; $\Upsilon_0 = 1$; DPSK signaling (slowly-varying phase error)

error itself. Since the conditional error probability of the detector is solely dependent on the signal transmission format along with the type of detector used, and not on the source of the noisy reference signal, the expressions previously given for conditional error probability in Chap. 6 are directly applicable here, namely,

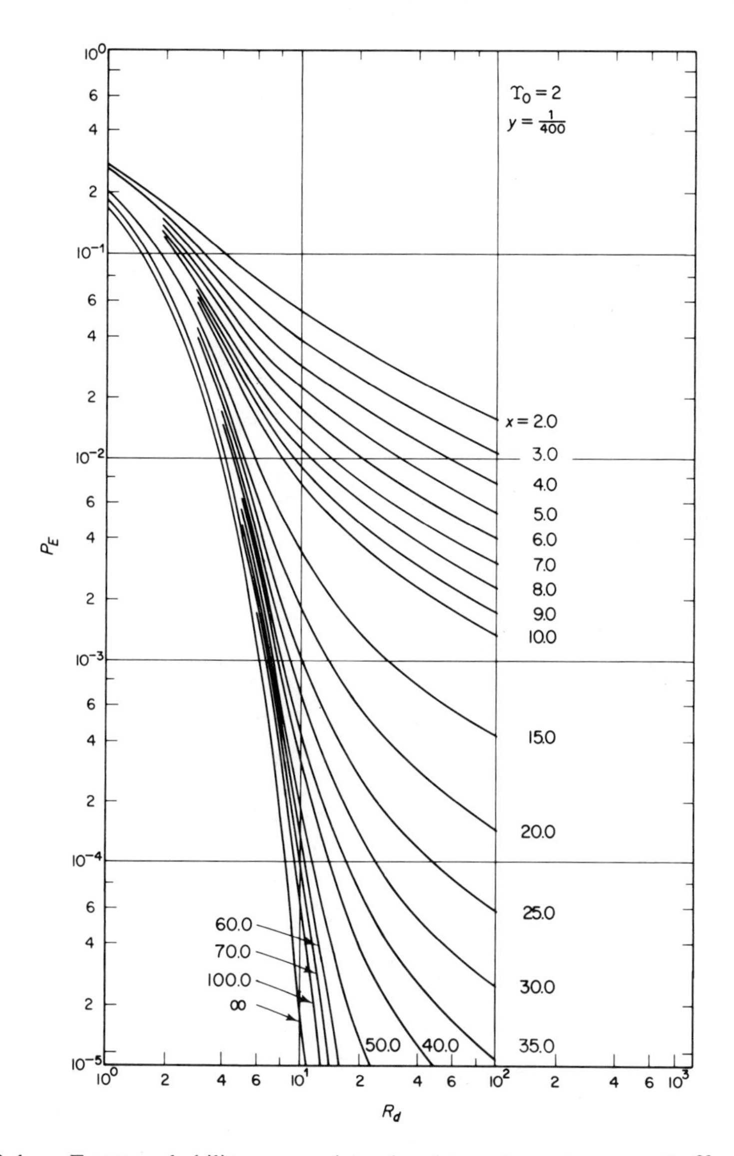

Fig. 8-4 Error probability versus data signal-to-noise ratio; $r_0 = 2$; $\Upsilon_0 = 2$; DPSK signaling (slowly-varying phase error)

$$\begin{aligned} P_E(\phi) &= \tfrac{1}{2}\operatorname{erfc}(\sqrt{R_d}\cos\phi) && \text{PSK} \\ P_E(\phi) &= \tfrac{1}{2}\exp(-R_d\cos^2\phi) && \text{DPSK} \\ P_E(\phi) &= \operatorname{erfc}(\sqrt{R_d}\cos\phi)[1 - \tfrac{1}{2}\operatorname{erfc}(\sqrt{R_d}\cos\phi)] && \text{Differentially Encoded PSK} \end{aligned} \tag{8-1}$$

The phase error p.d.f., on the other hand, directly depends on the presence

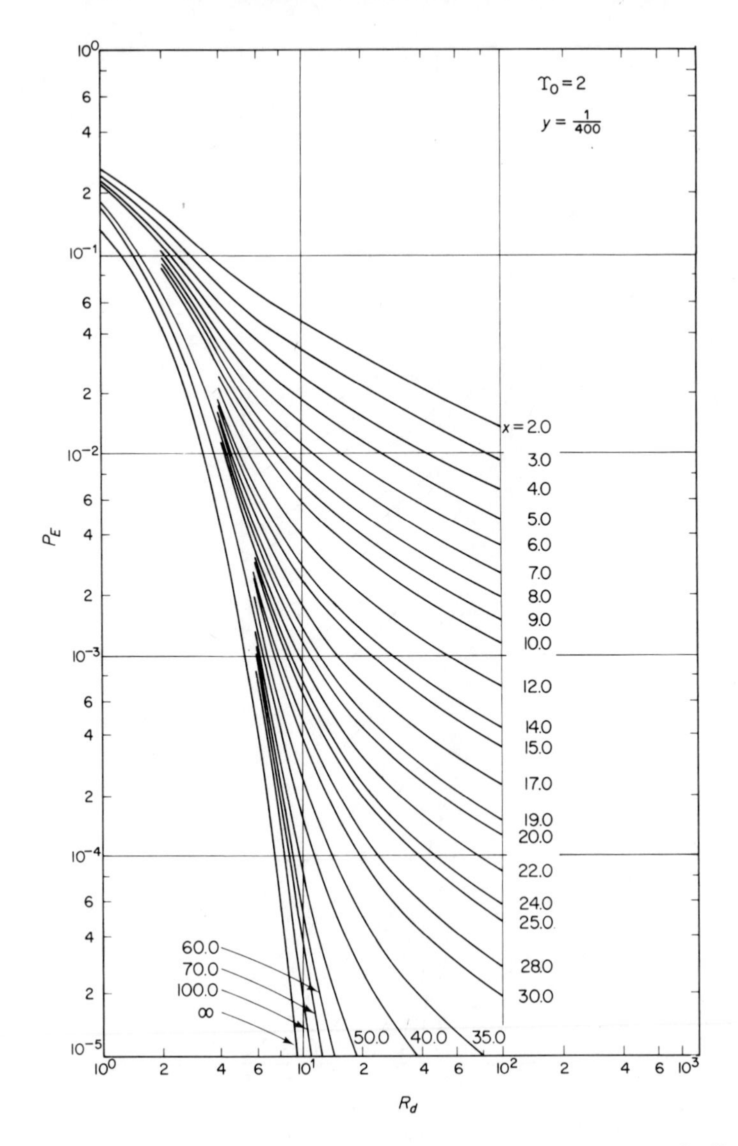

Fig. 8-5 Error probability versus data signal-to-noise ratio; $r_0 = 2$; $\Upsilon_0 = 2$; differentially encoded PSK signaling (slowly-varying phase error)

of a BPL. To a first-order approximation, this p.d.f. is given by (See Chap. 2)

$$p(\phi) = \frac{\exp(\rho \cos \phi)}{2\pi I_0(\rho)} \qquad |\phi| \leq \pi \tag{8-2}$$

where

$$\rho = \frac{x}{\Gamma_p}\left(\frac{1 + r_0}{1 + r_0/\mu}\right) \tag{8-3}$$

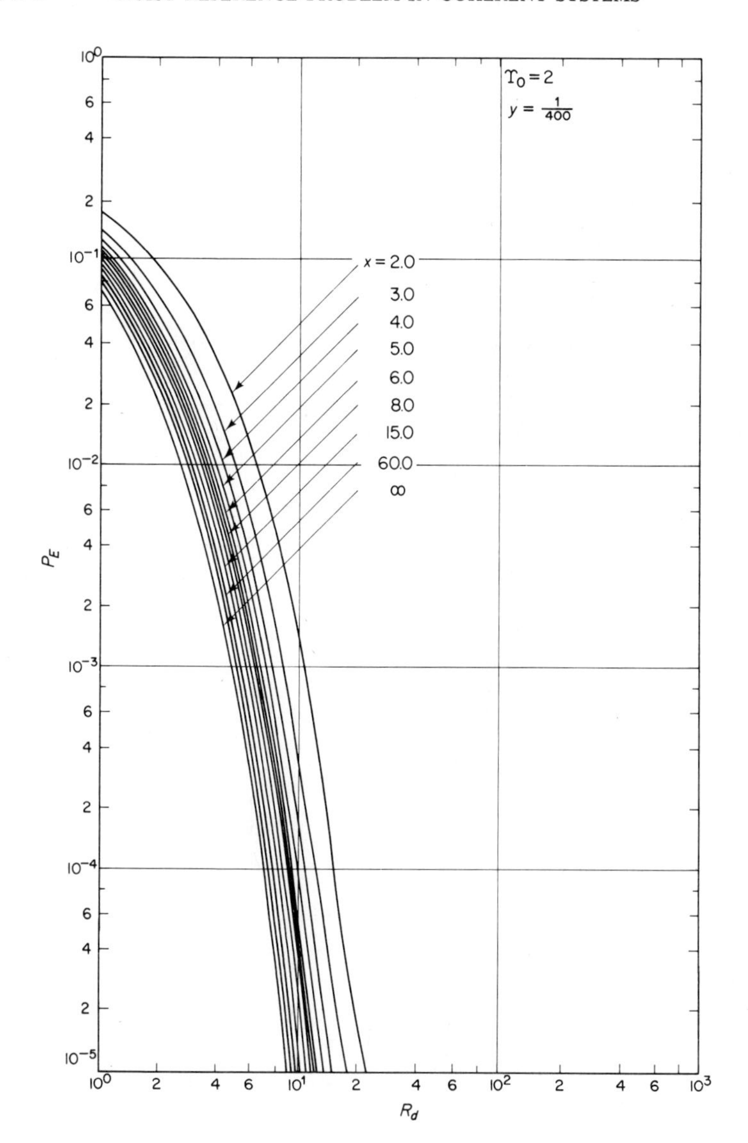

Fig. 8-6 Error probability versus data signal-to-noise ratio; $r_0 = 2$; $\Upsilon_0 = 2$; PSK signaling (rapidly-varying phase error)

with $x = 2P_c/N_0 W_{L0}$ and Γ_p approximated by

$$\Gamma_p \cong \frac{1 + xy}{\Gamma_0^{-1} + xy} \tag{8-4}$$

where Γ_0 is the value of Γ_p at $x = 0$.

In this chapter we shall be slightly more general than in Chap. 2 in that

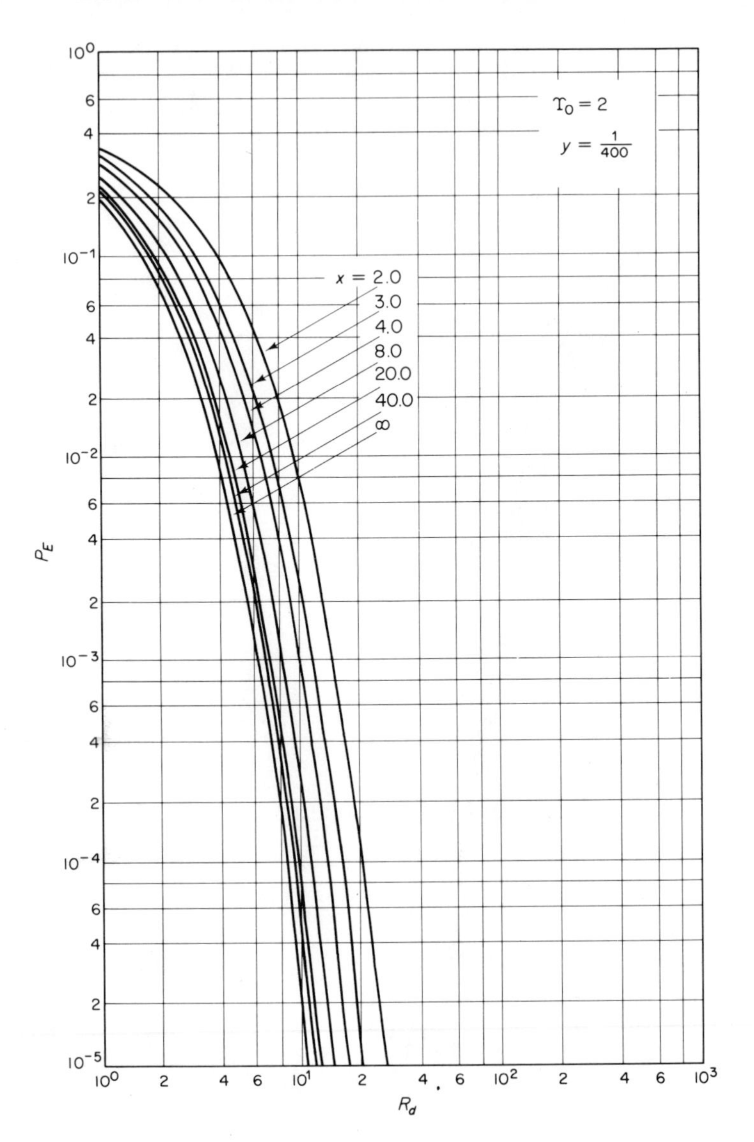

Fig. 8-7 Error probability versus data signal-to-noise ratio; $r_0 = 2$; $\Upsilon_0 = 2$; DPSK signaling (rapidly-varying phase error)

the signal-to-noise ratio in the design point loop bandwidth shall be allowed to take on several values—that is,

$$\Upsilon_0 = \frac{2P_{c0}}{N_0 W_{L0}} = \text{constant}$$

Hence, the parameter μ defined in Eq. (2-62) can be written in terms of

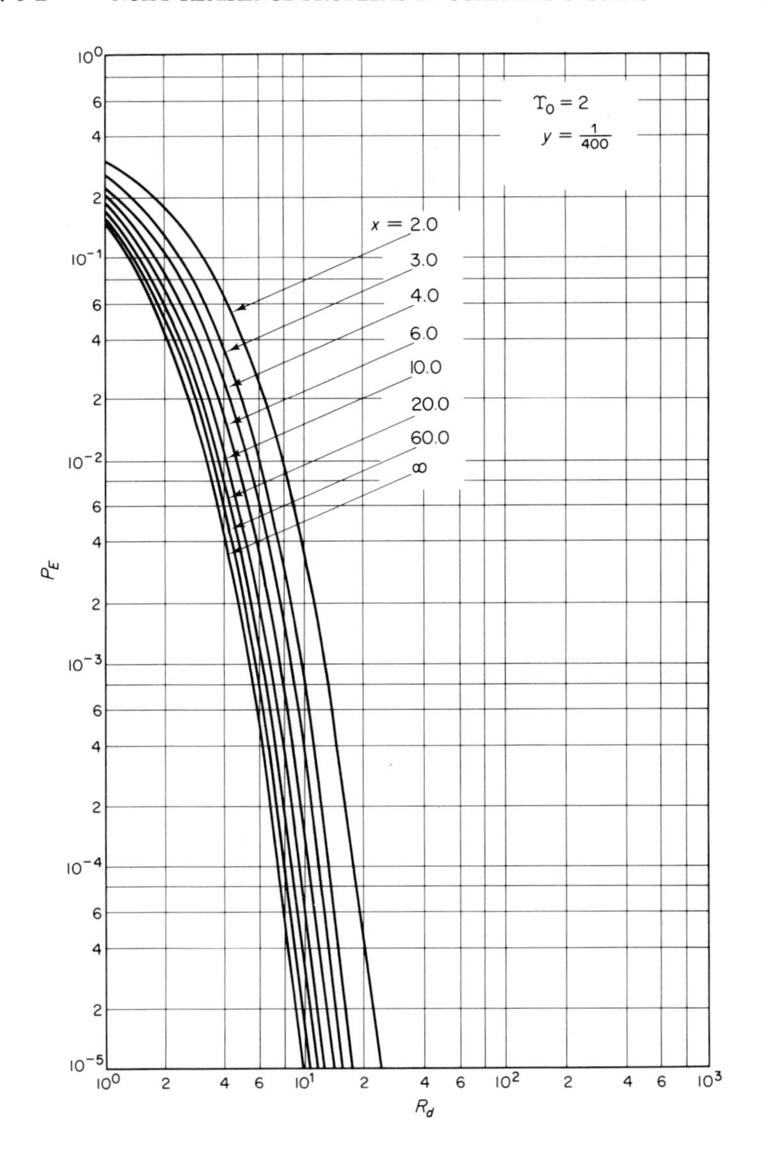

Fig. 8-8 Error probability versus data signal-to-noise ratio; $r_0 = 2$; $\Upsilon_0 = 2$; differentially encoded PSK signaling (rapidly-varying phase error)

Υ_0: namely,

$$\mu = \frac{\sqrt{\Upsilon_0}\exp(-\Upsilon_0 y/2)[I_0(\Upsilon_0 y/2) + I_1(\Upsilon_0 y/2)]}{\sqrt{x}\ \exp(-xy/2)[I_0(xy/2) + I_1(xy/2)]} \tag{8-5}$$

$$\cong \sqrt{\frac{(.7854\Upsilon_0 y + 0.4768\Upsilon_0^2 y^2)(1 + 1.024xy + 0.4768x^2y^2)}{(.7854xy + 0.4768x^2y^2)(1 + 1.024\Upsilon_0 y + 0.4768\Upsilon_0^2 y^2)}}$$

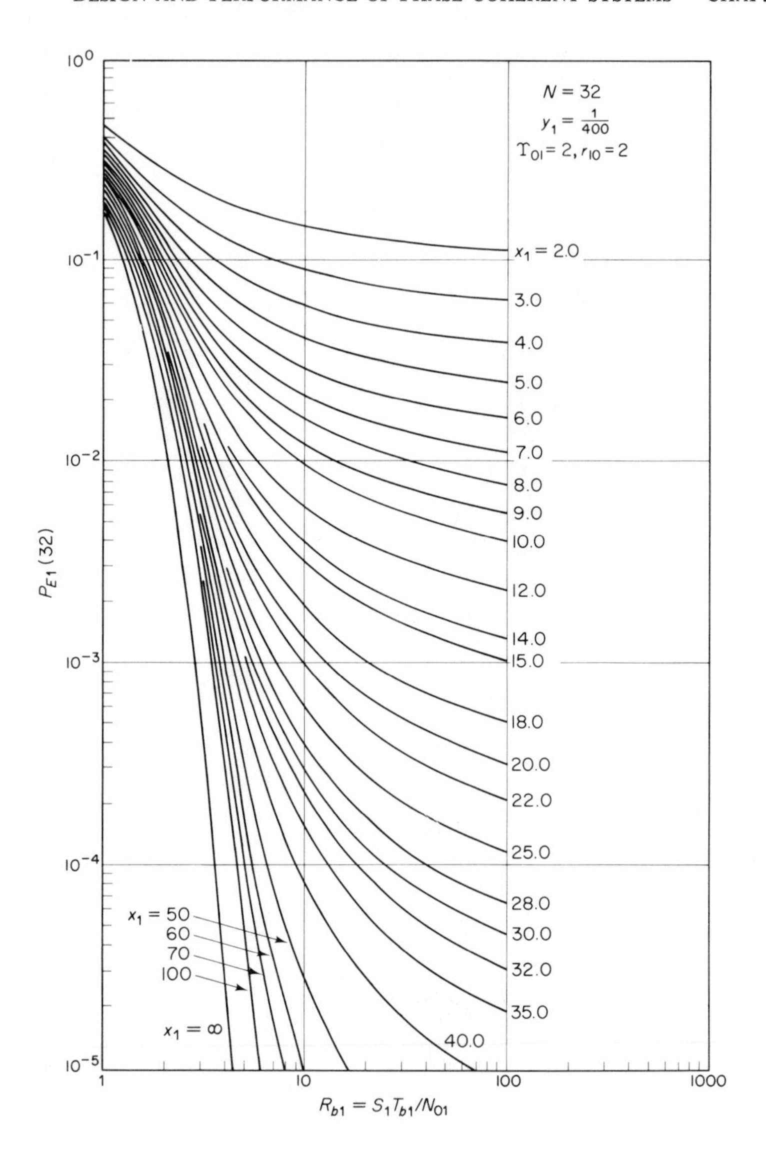

Fig. 8-9 Word error probability versus signal-to-noise ratio per bit; one-way system; 5-bit orthogonal code

with $y = W_{L0}/W_i$. The approximate formulas in (8-4) and (8-5) are useful for purposes of system design. Averaging (8-1) over the p.d.f. of (8-2) is best performed numerically on a digital computer. The results of these computations are illustrated in Figs. 8-1 through 8-5 where P_E is plotted versus $R_d = ST/N_0$ for fixed values of Υ_0 and y, with x as a parameter. We note from these figures that the error probability performance of a given

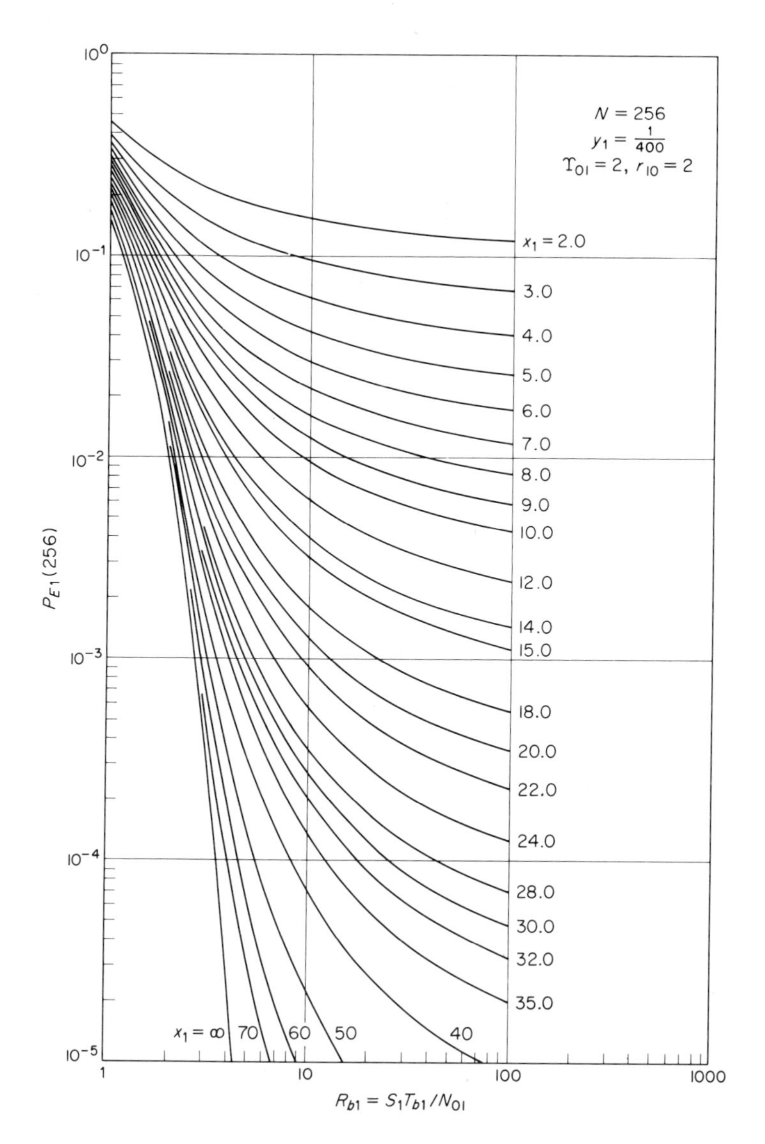

Fig. 8-10 Word error probability versus signal-to-noise ratio per bit; one-way system; 8-bit orthogonal code

signaling format improves as Υ_0, the design point signal-to-noise ratio, is increased. This is based upon the following reasoning: the parameter μ as defined by Eq. (8-5) is a monotonically increasing function of Υ_0; hence, from (8-3), ρ also increases with increasing Υ_0, thus causing attendant reduction in noisy reference loss or, correspondingly, an improvement in average error probability performance.

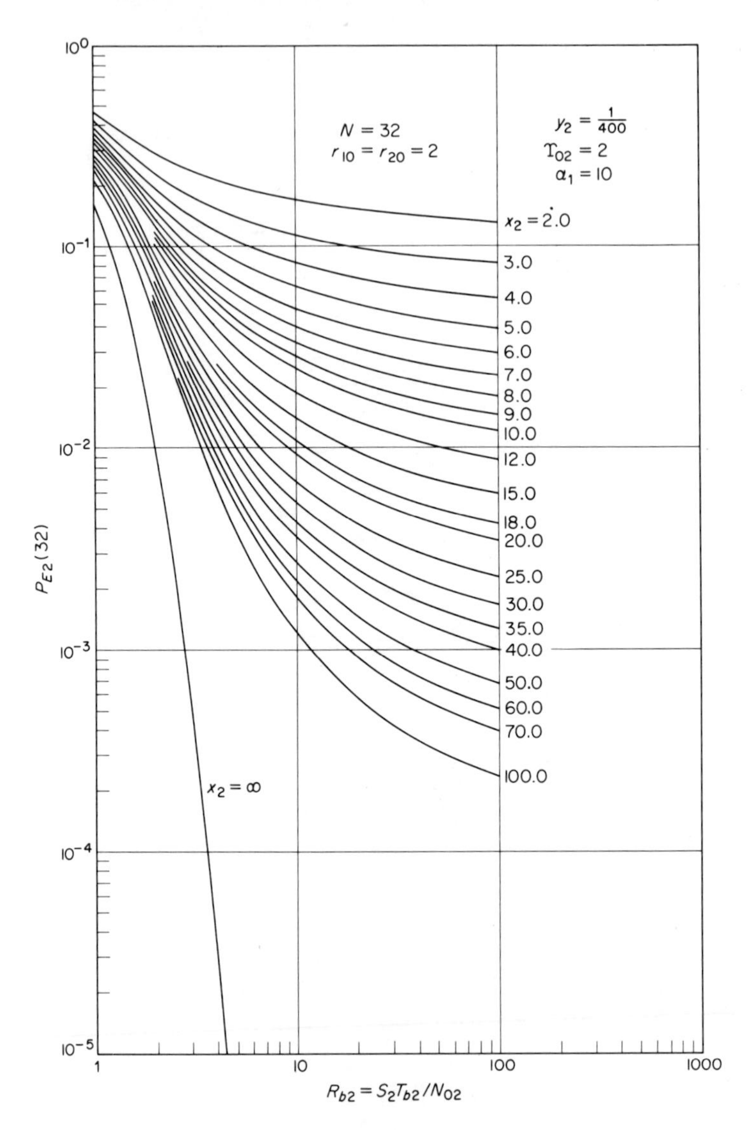

Fig. 8-11 Word error probability versus signal-to-noise ratio per bit; two-way system; 5-bit orthogonal code; $\alpha_1 = 10$

When the reference signal phase error varies rapidly over a symbol interval (the ratio of the data rate to the bandwidth of the carrier-tracking loop is much less than unity), expressions for average error probability in the three types of data detectors under consideration are obtained by replacing $\cos\phi$ by $\overline{\cos\phi} \cong I_1(\rho)/I_0(\rho)$ in Eq. (8-1). These results are illustrated in Figs. 8-6 through 8-8 where P_E is plotted versus R_d for fixed values of Υ_0 and

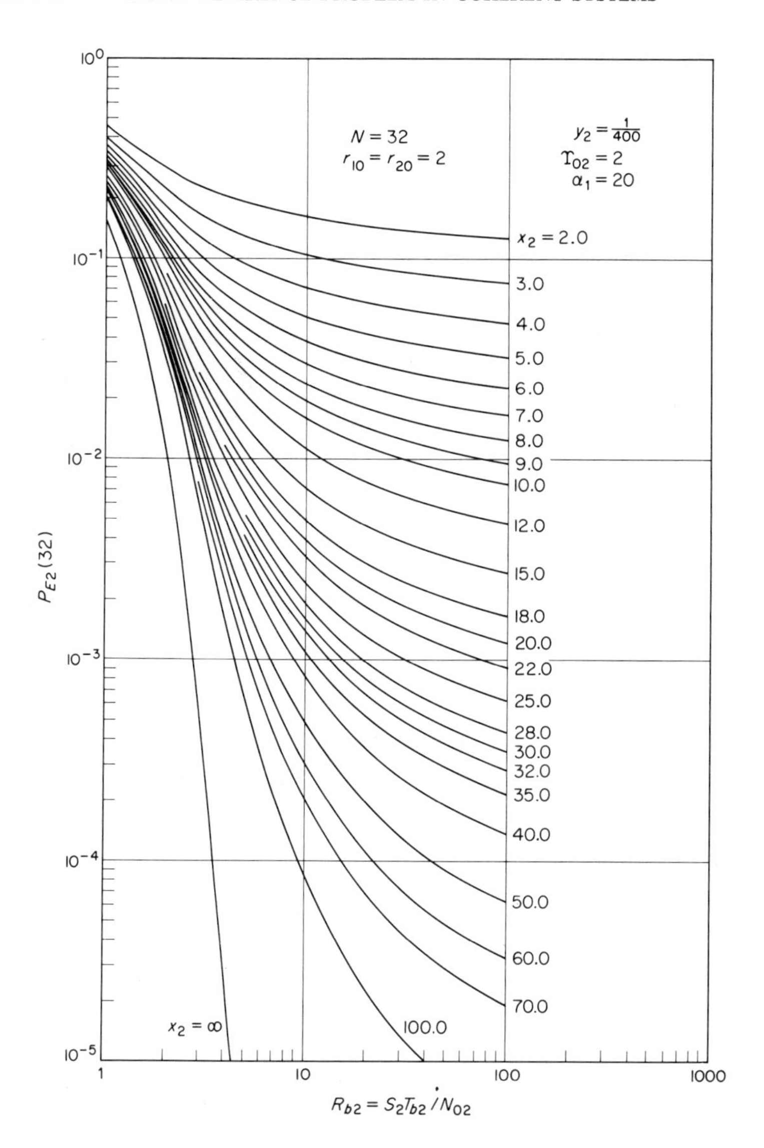

Fig. 8-12 Word error probability versus signal-to-noise ratio per bit; two-way system; 5-bit orthogonal code; $\alpha_1 = 20$

y, with x as a parameter. The curves of Figs. 8-1 through 8-8 can also be used as upper and lower bounds on the error probability performance of PSK, DPSK, and differentially encoded PSK systems having an arbitrary data rate and loop bandwidth. An approximate procedure for the bandwidth-time product region between the two extremes presented here has been discussed in Chap. 6 where the BPL was assumed absent. Equivalent results for the

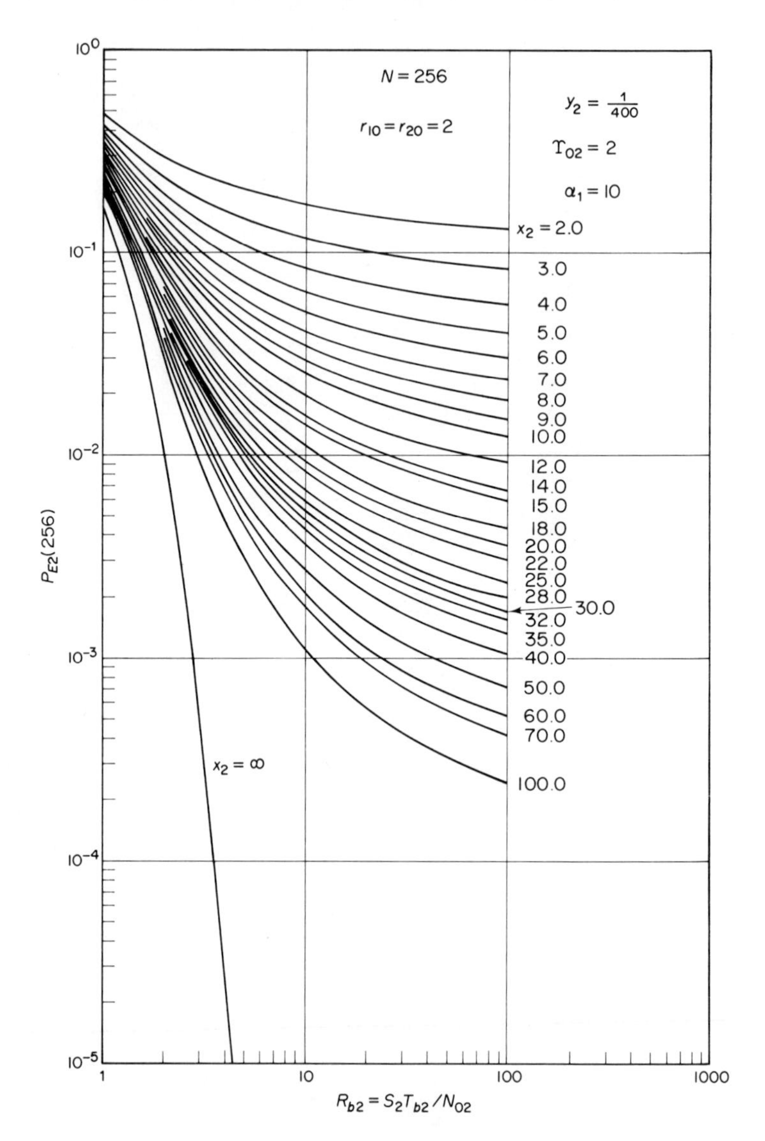

Fig. 8-13 Word error probability versus signal-to-noise ratio per bit; two-way system; 8-bit orthogonal code; $\alpha_1 = 10$

case where a BPL precedes the carrier-tracking loop are discussed in detail in Ref. 3.

At this point, it is relatively straightforward to generalize the block-coded results given in Sec. 7-4 to the case where a BPL precedes the PLL. Again, the conditional error probability is unaffected by the presence of the BPL; hence, Eq. (7-105) applies directly at the carrier level with $(1 - m_n^2)R_n$

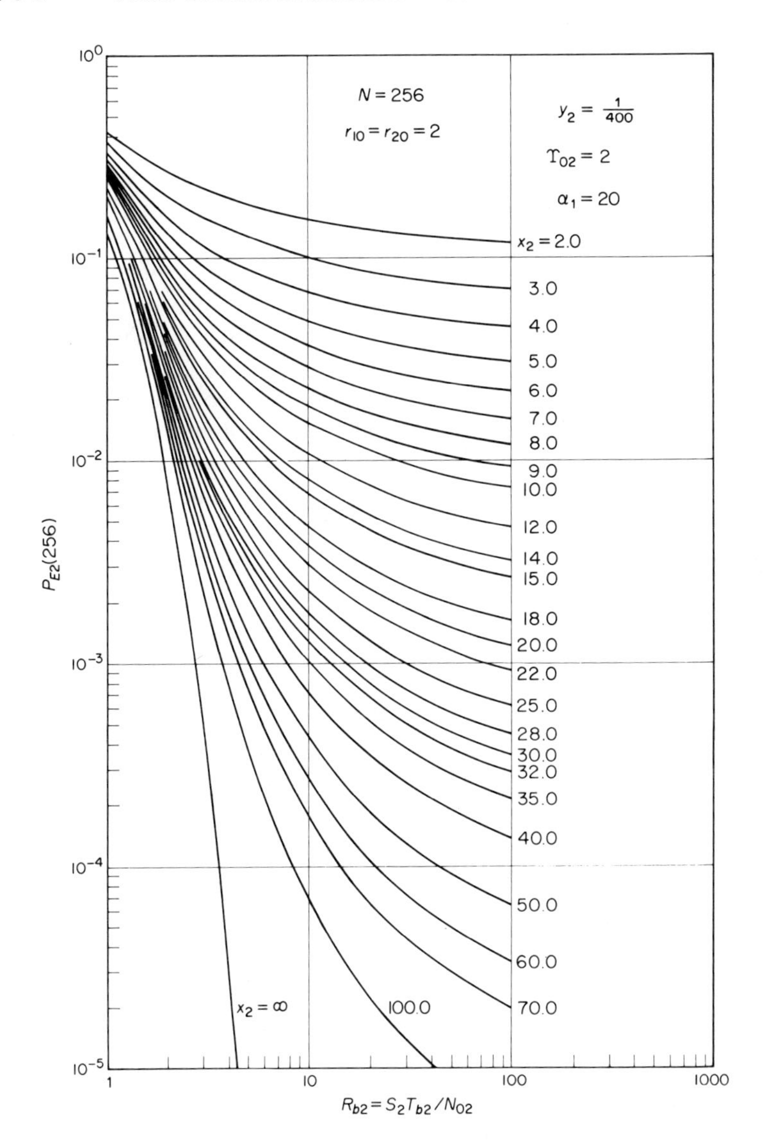

Fig. 8-14 Word error probability versus signal-to-noise ratio per bit; two-way system; 8-bit orthogonal code; $\alpha_1 = 20$

replaced by $R_{d_n} = S_n T_n/N_{0n}$ and the subscript n is once again used to distinguish one- and two-way systems rather than a particular data channel. The one-way performance obtained by averaging (7-105) over the vehicle phase error p.d.f.'s [(7-7) and (7-10)] is summarized in Figs. 8-9 and 8-10 for code words containing 5 and 8 bits of information, respectively. These figures depict word error probabilities versus signal-to-noise ratio in the design point bandwidth of the carrier-tracking loop. Clearly, system performance depends on the choice of a design point Υ_0 in the carrier-tracking loop. For purposes of presentation, the choice taken is that which corresponds to the design point in the Deep Space Network—$r_0 = 2$, $\Upsilon_0 = 2$, and $y = 1/400$. As $x \to \infty$, corresponding to the case of perfect RF sync, the deleterious effects of a noisy phase reference disappear and perfect coherent detection is possible.

Two-way system performance for 5- and 8-bit orthogonal codes is illustrated in Figs. 8-11 through 8-14 for various values of α_{1l} and x_2. The same carrier-tracking loop design point is used for this case as was used for the one-way case. We notice in this sequence of figures that as the signal-to-noise ratio in the ground receiver's design point loop bandwidth x_2 increases without limit, the deleterious effects of the uplink noise introduce an irreducible error probability. This irreducible error, which depends on the amount of carrier phase jitter introduced by the vehicle's carrier-tracking loop, is identically analogous to that produced in systems that do not employ band-pass limiters. The formula for computing this irreducible error $P_{ir}(N)$ is given by

$$\begin{aligned} P_{ir}(N) &= \lim_{x \to \infty} P_E(N) \\ &= 2 \int_{\pi/2}^{\pi} p(\phi)\, d\phi \end{aligned} \tag{8-6}$$

—the probability that the magnitude of the phase error exceeds $\pi/2$. Hence, $P_{ir}(N)$ is independent of the code and depends only on the design of the carrier-tracking loops (including the BPLs), the available power in the carrier components, and the channel noise. Thus, for given channel conditions and fixed loop parameters, large transmitter output power capabilities are once again desirable.

8-3. OPTIMUM DESIGN OF SINGLE-CHANNEL SYSTEMS EMPLOYING A BAND-PASS LIMITER

In Chap. 7, we discussed the optimal division of power in a single-channel system between a single carrier and a sideband data channel. The technique employed minimized error probability for a fixed total transmitted energy-to-noise ratio and ratio of data rate to carrier-tracking loop bandwidth. When a BPL precedes the PLL this approach becomes a good deal more tedious because the limiter suppression factor depends on the modulation factor that one is trying to optimize. Nevertheless, we shall develop the theory necessary to treat this problem and leave detailed numerical calculations to the reader. We then present a simpler approach to the power allocation problem, based on minimizing the total energy-to-noise ratio at the transmitter output.

System Design Philosophies

There are essentially two philosophies with regard to recovering the modulation from a single-channel system that employs a BPL in its RF tracking receiver: the first, which is typical of present-day design of vehicle transponders for recovering command information, is to extract the modulation directly from the RF phase detector output. We know from the results of Chap. 2 that the signal at this point suffers a loss in signal-to-noise ratio due to the suppression effect of the BPL preceding the loop. Thus, the second philosophy suggests that the information should be removed from the RF carrier prior to transmission through the limiter. This technique is typically employed for recovering ranging information in the vehicle. A block diagram representation of these two approaches is given in Fig. 8-15.

Another problem that often arises is that the bandwidth of the narrow-

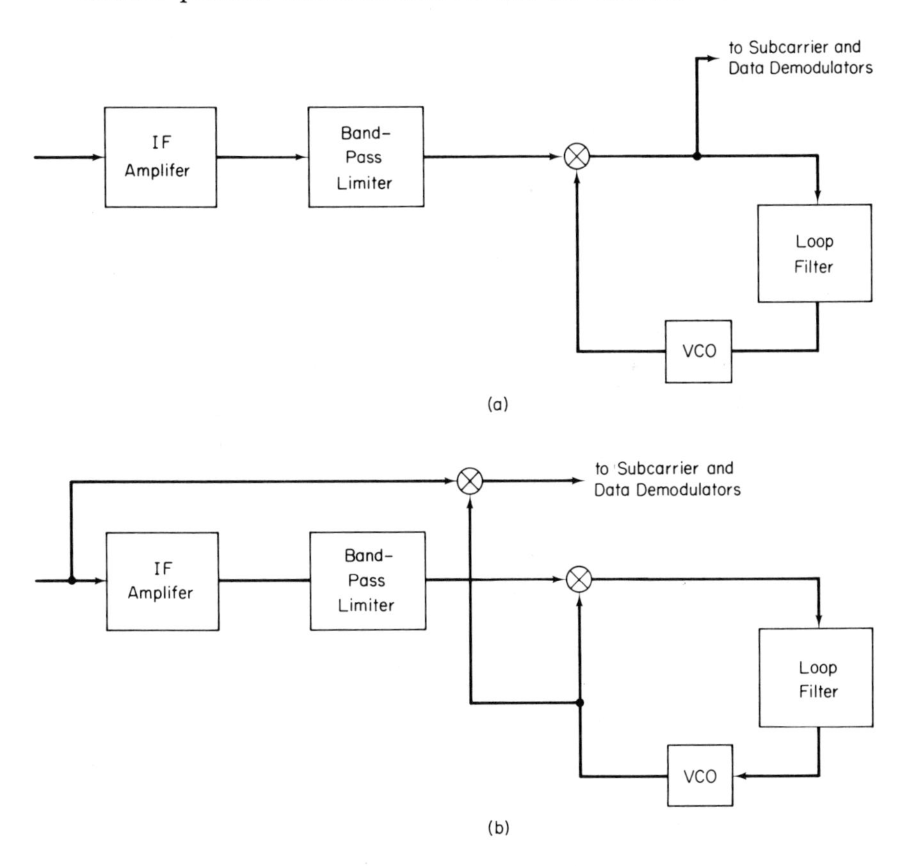

Fig. 8-15 Two methods for recovering the modulation from a single-channel system which employs a BPL in its RF tracking receiver

band amplifier at the input to the BPL (typically less than 10 kHz) might be sufficiently narrow to filter out all or part of the modulation, thus causing further degradation of performance.[4] This particularly occurs if the information is modulated on a subcarrier. In summary, the three cases we shall consider are

1. The total modulation passes through the BPL and is recovered from the output of the phase detector in the RF tracking loop.
2. The total modulation passes through the BPL; however, the modulation is recovered from the output of a separate phase detector whose inputs are the modulated carrier at the BPL input and the PLL reference signal.
3. The input filter to the BPL completely filters out the modulation, with the modulation being recovered as in case 2.

In cases 1 and 2, the BPL suppression factor, and thus the effective loop signal-to-noise ratio, are functions of the *total* power-to-noise ratio in the input bandwidth. In case 3, the BPL loss is increased (relative to case 2) since it is now only a function of the carrier component of the total power-to-noise ratio in the input bandwidth. Hence, with all other parameters unchanged, the effective loop signal-to-noise ratio is reduced, and the noisy reference loss increased. As we shall see, this effect produces a minor degradation in case 3 relative to case 2.

System Model

The development of the theory for the single-channel system with a BPL preceding its RF tracking loop parallels that described in Sec. 2-5 of Chap. 2. In particular, the first zone limiter output can now be expressed in the form

$$\begin{aligned} z_1(t) &= (4\tilde{\alpha}_1/\pi)\sin[\omega_0 t + (\cos^{-1} m_1)x_k(t) + \theta(t)] \\ &\quad + \sqrt{2}\{N_A(t)\cos[\omega_0 t + \theta(t)] - N_B(t)\sin[\omega_0 t + \theta(t)]\} \end{aligned} \tag{8-7}$$

where $x_k(t)$ is the digital modulation and

$$\begin{aligned} N_A(t) &= \frac{2\sqrt{2}}{\pi}\Bigg\{-\sqrt{(1-m_1^2)}\tilde{\alpha}_1\, x_k(t) \\ &\quad + \frac{\sqrt{(1-m_1^2)P}x_k(t) + N_c(t)}{\sqrt{[\sqrt{m_1^2 P} - N_s(t)]^2 + [\sqrt{(1-m_1^2)P}x_k(t) + N_c(t)]^2}}\Bigg\} \\ N_B(t) &= \frac{2\sqrt{2}}{\pi}\Bigg\{m_1\tilde{\alpha}_1 \\ &\quad - \frac{\sqrt{m_1^2 P} - N_s(t)}{\sqrt{[\sqrt{m_1^2 P} - N_s(t)]^2 + [\sqrt{(1-m_1^2)P}x_k(t) + N_c(t)]^2}}\Bigg\} \end{aligned} \tag{8-8}$$

with $P \triangleq S + P_c$ denoting the total signal power input to the BPL. The p.d.f.'s of the normalized, zero mean processes $(\pi/2\sqrt{2})N_A(t)$ and

$(\pi/2\sqrt{2})N_B(t)$ are given, respectively, by

$$p_A(N) = \frac{1}{2\pi\sqrt{1-[N+\sqrt{\tilde{\alpha}_1^2(1-m_1^2)}x_k(t)]^2}}\exp(-\rho_i)$$
$$\times\left\{2+\sum_{j=1}^{2}\sqrt{\pi\rho_i}\,Y_j\exp(\rho_i Y_j^2)[1+\operatorname{erf}(\sqrt{\rho_i}\,Y_j)]\right\}$$
$$\text{for } -1-\sqrt{\tilde{\alpha}_1^2(1-m_1^2)}x_k(t)\le N\le 1-\sqrt{\tilde{\alpha}_1^2(1-m_1^2)}x_k(t)$$

and

$$p_B(N) = \frac{1}{2\pi\sqrt{1-(-N+\sqrt{\tilde{\alpha}_1^2 m_1^2})^2}}\exp(-\rho_i)$$
$$\times\left[2+\sum_{j=1}^{2}\sqrt{\pi\rho_i}\,Z_j\exp(\rho_i Z_j^2)(1+\operatorname{erf}\sqrt{\rho_i}\,Z_j)\right]$$
$$\text{for } \tilde{\alpha}_1 m_1 - 1\le N\le \tilde{\alpha}_1 m_1 + 1 \tag{8-9}$$

where

$$\begin{aligned}
Y_1 &= \sqrt{m_1^2}\sqrt{1-[N+\sqrt{\tilde{\alpha}_1^2(1-m_1^2)}x_k(t)]^2}\\
&\quad+\sqrt{(1-m_1^2)}x_k(t)\sqrt{N+\sqrt{\tilde{\alpha}_1^2(1-m_1^2)}x_k(t)}\\
Y_2 &= -\sqrt{m_1^2}\sqrt{1-[N+\sqrt{\tilde{\alpha}_1^2(1-m_1^2)}x_k(t)]^2}\\
&\quad+\sqrt{(1-m_1^2)}x_k(t)\sqrt{N+\sqrt{\tilde{\alpha}_1^2(1-m_1^2)}x_k(t)}\\
Z_1 &= \sqrt{m_1^2}(-N+\sqrt{\tilde{\alpha}_1^2 m_1^2})\\
&\quad+\sqrt{1-m_1^2}x_k(t)\sqrt{1-(-N+\sqrt{\tilde{\alpha}_1^2 m_1^2})^2}\\
Z_2 &= \sqrt{m_1^2}(-N+\sqrt{\tilde{\alpha}_1^2 m_1^2})\\
&\quad-\sqrt{1-m_1^2}x_k(t)\sqrt{1-(-N+\sqrt{\tilde{\alpha}_1^2 m_1^2})^2}
\end{aligned} \tag{8-10}$$

The output of the phase detector with reference signal phase as given in Eq. (2-5) is (neglecting double-frequency terms):

$$\epsilon(t) = \sqrt{\frac{8}{\pi^2}\tilde{\alpha}_1^2 m_1^2}\sin\varphi(t)+\sqrt{\frac{8}{\pi^2}\tilde{\alpha}_1^2(1-m_1^2)}x_k(t)\cos\varphi(t)$$
$$+N_A(t)\cos\varphi(t)-N_B(t)\sin\varphi(t) \tag{8-11}$$

Using arguments similar to those presented in Sec. 2-5, the total noise power at the phase detector output is approximately given by Eq. (2-57), where σ_A^2 and σ_B^2 are now defined by

$$\sigma_A^2 = \frac{8}{\pi^2}\left\{(1-m_1^2)\left[1-\frac{1-\exp(-\rho_i)}{2\rho_i}-\tilde{\alpha}_1^2\right]+m_1^2\left[\frac{1-\exp(-\rho_i)}{2\rho_i}\right]\right\}$$
$$\sigma_B^2 = \frac{8}{\pi^2}\left\{m_1^2\left[1-\frac{1-\exp(-\rho_i)}{2\rho_i}-\tilde{\alpha}_1^2\right]+(1-m_1^2)\left[\frac{1-\exp(-\rho_i)}{2\rho_i}\right]\right\} \tag{8-12}$$

If, in addition, the input signal-to-noise ratio ρ_i is small, then Eqs. (2-59) through (2-64) also apply to the single-channel case under consideration.

Computation of Error Probability Performance

In computing the error probability performance of the data detector, care must be exercised in applying the Gaussian theory developed in Chaps. 5 and 6. In cases 2 and 3, the noise affecting the data detection process is clearly Gaussian, since it has not been transmitted through the BPL. In case 1, the noise entering the matched filter (the phase detector output) is not Gaussian, and in general its p.d.f. must be determined from those of $N_A(t)$ and $N_B(t)$ as given in (8-9). However, since the effective bandwidth of this noise is much wider than $1/T$, the integrate-and-dump action of the matched filter acts as a narrowband filter, and the output statistic is approximately Gaussian. Hence, based on these assumptions, one can apply Gaussian theory to all three cases if the data signal power and noise spectral density S_d and N_{0d}, respectively, which affect the calculation of the data signal-to-noise ratio $R_d \triangleq S_dT/N_{0d}$, are clearly defined.

Case 1. If the data signal applied to the matched filter is taken directly from the phase detector output, then [see (2-59) through (2-64)]

$$S_d = \frac{8}{\pi^2}(1 - m_1^2)\tilde{\alpha}_1^2 \qquad N_{0d} = N_{0e} = \frac{2\sigma_e^2}{W_e} \tag{8-13}$$

An expression for N_{0e} in terms of system design parameters can be obtained using the definition of Γ_p as in Eq. (2-48) with the result

$$N_{0e} = \frac{2m_1^2}{W_{L0}}\left(\frac{8\tilde{\alpha}_1^2\Gamma_p}{\pi^2 x}\right) \tag{8-14}$$

Combining (8-13) and (8-14),

$$R_d = \left(\frac{1 - m_1^2}{m_1^2}\right)\frac{x}{2\Gamma_p\delta_0} \tag{8-15}$$

with

$$\delta_0 \triangleq \frac{2}{W_{L0}T} \tag{8-16}$$

In this case also, the input signal-to-noise ratio ρ_i, which affects the computation of the phase error p.d.f. $p(\phi)$, is given by

$$\rho_i = \frac{xy}{m_1^2} \tag{8-17}$$

Case 2. When the data signal is obtained by demodulating the BPL input with the PLL reference, then

$$S_d = (1 - m_1^2)P \qquad N_{0d} = N_0 \tag{8-18}$$

Hence,

$$R_d = \left(\frac{1 - m_1^2}{m_1^2}\right)\frac{x}{2\delta_0} \tag{8-19}$$

with the parameter ρ_i as given by (8-17).

Case 3. For the case where the data signal is obtained as in case 2, but the input filter to the BPL completely filters out the modulation, then R_d is still given as in (8-19), but ρ_i now becomes

$$\rho_i = xy \tag{8-20}$$

As an example of the theory just developed, Table 8-1 compares the error

TABLE 8-1

Case	ρ, *dB*	R_d, *dB*	P_E
1	4.914	8.196	1.6314×10^{-3}
2	4.914	8.842	7.6516×10^{-4}
3	4.912	8.842	7.6634×10^{-4}

probability obtained from these three cases for the following set of parameters, which is typical of a command link for deep space application: $\Upsilon_0 = 1$, $r_0 = 2$, $W_{L0} = 18$ Hz, $W_i = 20 \times 10^3$ Hz, $T = .25$, $\tau_2/\tau_1 = .002$, $m_1^2 = 0.7$, and $x = 9$ dB. Since $\delta_0 = 0.444$, the low-rate formula for P_E as given by Eq. (6-30) has been invoked. The difference between the error probabilities in cases 1 and 2 is strictly due to the effect of the BPL loss on R_d, since the noisy reference loss (approximately 0.2 dB) is the same for both. This can be seen from the fact that the signal-to-noise ratio ρ in the PLL is the same in both cases. One concludes from this comparison that for the parameters fixed as chosen, one pays an error probability penalty of a factor of two or greater by extracting the data signal from the PLL phase detector output. The difference between cases 2 and 3 in error probability is negligibly small due to the reduction in the effective loop signal-to-noise ratio caused by filtering out the modulation sideband power. Here, the penalty is strictly due to a change in the noisy reference loss, since R_d is the same for both cases, and is so small that it will be indistinguishable in practical receiver mechanizations.

The Selection of an Optimum Modulation Factor

Using the definitions for ρ_i and R_d given for each of the three cases and for ρ as defined by Eq. (2-64), one may apply the techniques of Sec. 7-2 to the situation here, once again determining the modulation factor that minimizes error probability for a fixed total transmitted signal-to-noise ratio. Unfortunately, in cases 1 and 2 here, the calculations become quite a bit more complex than those in Chap. 7 because of the dependence of $\tilde{\alpha}_1$ (and hence ρ) on the modulation factor to be optimized. In fact, an approximate expression for m_{opt}, such as Eq. (7-26), is difficult if not impossible to achieve.

A simpler approach is to reverse the procedure essentially and to divide the power between carrier and sideband to minimize the total energy-to-noise ratio PT/N_0 at the transmitter for a fixed error probability in the data channel and ratio of data rate-to-design point loop bandwidth.[5] This procedure allows the design engineer to constrain *a priori* the minimum signal-to-noise ratio in the design point bandwidth x_{min} that he is willing to accept, based on tracking performance considerations, and to seek an optimum power split subject to that constraint. If the minimum total transmitted energy-to-noise ratio turns out to occur for a value of x less than x_{min}, the system designer may at his discretion decide to select the value of PT/N_0 corresponding to x_{min}.

The curves in Figs. 8-1 through 8-10 are helpful in setting up this design philosophy for a one-way system operating under the conditions of case 3. More specifically, one chooses *a priori* the parameters Υ_0, δ_0, and the type of signaling. This directs him to one specific figure. Then, for a given error probability P_E, one finds an infinite set of paired coordinate values (x, R_d). For each pair of values, the ratio of the total energy-to-noise spectral density may be found from*

$$\frac{PT}{N_0} = R'_d + \frac{x}{\delta_0} \tag{8-21}$$

To illustrate the procedure, consider a command application wherein the command data is obtained by demodulating the BPL input with the PLL reference signal (case 2). Plotted in Fig. 8-16 is PT/N_0 in dB versus x in dB for $P_E = 10^{-5}$, and parameter values $\tau_2/\tau_1 = 0.002$, $r_0 = 2$, $\Upsilon_0 = 2$, $W_i = 9$ kHz, $W_{L0} = 18$ Hz, $1/T = 4$ bits/sec. The minimum of this curve represents the best power split condition.

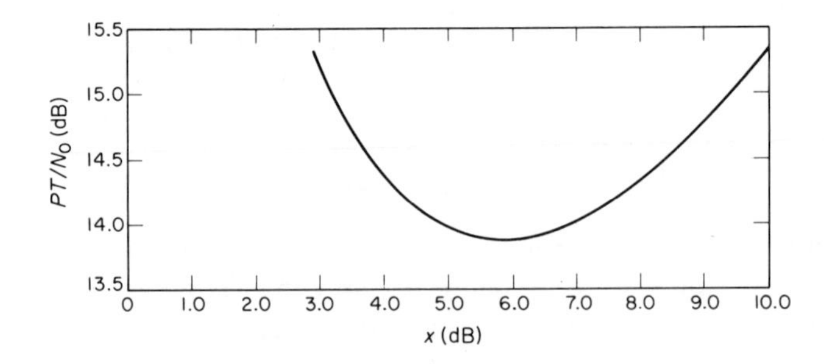

Fig. 8-16 Total power-to-noise ratio versus carrier signal-to-noise ratio in the design point loop bandwidth for a fixed design point—parameters are typical of command application

* For case 1, $R'_d = R_d\Gamma_p$; whereas for cases 2 and 3, $R'_d = R_d$.

This procedure can also be applied to a two-way system, using Figs. 8-11 and 8-12. Here one minimizes the total energy-to-noise ratio in the reference receiver for a fixed error probability in its data channel and a fixed value of signal-to-noise ratio in the vehicle's carrier-tracking loop (see Prob. 8-4).

Optimization of Performance as a Function of Design Point

We have just seen that for a given design point, an optimum trade-off exists between the power allocated to the carrier and sideband signals in the sense of minimum total transmitter energy-to-noise ratio. Another aspect of the optimum power allocation problem considers the locus of these minima as the design point is varied, and this information is used to select an optimum design point.[6] As an example, consider the plot of PT/N_0 in dB versus x in dB illustrated in Fig. 8-16. We observe that the minimum value of PT/N_0 occurs at $x = 5.8$ dB, which is greater than Υ_0. If Υ_0 is now increased in value, all other parameters held fixed, one finds that the minimum PT/N_0 together with the value of x at which it occurs both continue to decrease. Thus, it appears, at least at first glance, that continued improvement in performance (in the sense of minimum PT/N_0) can be had simply by raising the design point indefinitely. Of course, this cannot be true in practice; hence, some additional constraint must be placed on the problem to counteract this anomaly.

One practical consideration is that the value of x chosen for the final system design must be greater than some minimum value $x_{\min}$ where $x_{\min}$ is related to the design point by $x_{\min} = K_0\Upsilon_0$. When Υ_0 increases, a point

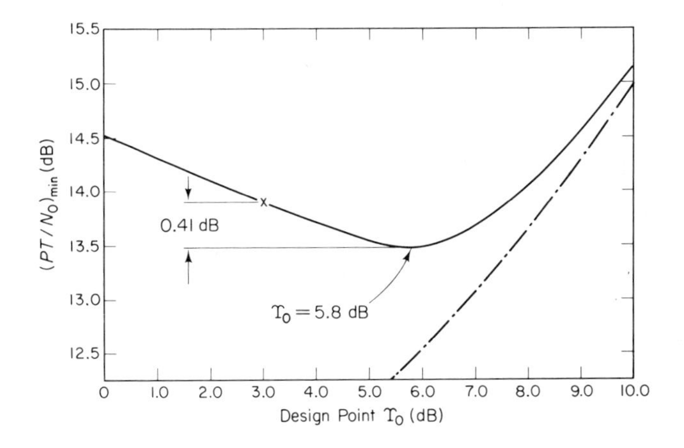

Fig. 8-17 Minimum total power-to-noise ratio as a function of design point—parameters are the same as those in Fig. 8-16

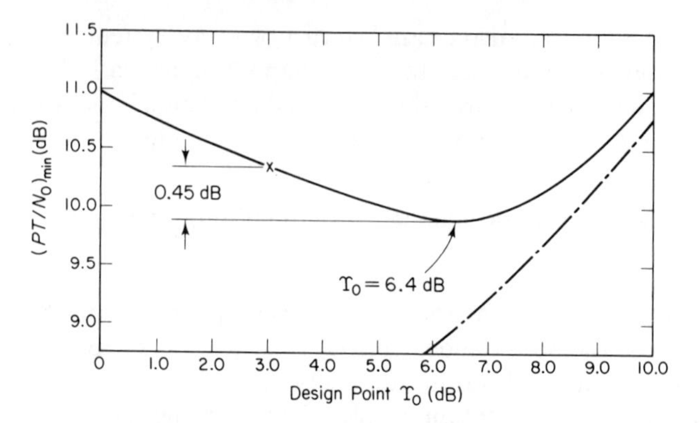

Fig. 8-18 Minimum total power-to-noise ratio as a function of design point—parameters are typical of telemetry applications; Case No. 1

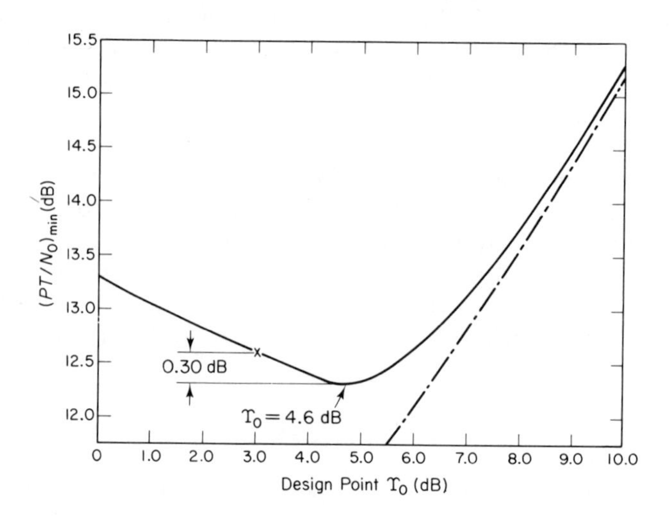

Fig. 8-19 Minimum total power-to-noise ratio as a function of design point—parameters are the same as those in Fig. 8-18; Case No. 2

is eventually reached at which the value of x at the minimum of PT/N_0 becomes equal to x_{min}. From that point on, the value of PT/N_0 at $x = x_{min}$ is selected as the best operating condition. Figure 8-17 illustrates a plot of $(PT/N_0)_{min}$ versus Υ_0 for case 2 and the same parameters as those used in arriving at Fig. 8-16.* Also assumed in Fig. 8-17 is that x_{min} is chosen at the design point itself (that is, $K_0 = 1$). The physical significance of such a choice

---

* The results for case 1 where the data signal is derived directly from the phase detector output are negligibly different from those of case 2. The difference is strictly due to the effect of the BPL loss on R_d because the noisy reference loss (approximately 0.2 dB) is the same for both. Hence, Fig. 8-17 can also be thought to apply to case 1.

is that the operating point loop bandwidth can never fall below its value at the design point; that is, W_{L0}. One notes from Fig. 8-17 that an improvement of 0.41 dB can be obtained by allowing the design point Υ_0 to be increased from its previously chosen value of 3 dB to 5.8 dB, the point at which the minimum value of $(PT/N_0)_{\min}$ is achieved.

Assuming case 3—that is, the single data channel is on a subcarrier that is outside the bandwidth of the input IF amplifier—Figs. 8-18 and 8-19 illustrate the choice of an optimum design point for two sets of parameters typical of telemetry systems and $P_E = 10^{-3}$. In both Figs. 8-18 and 8-19, $\tau_2/\tau_1 = 0.002$, $r_0 = 2$, $W_i = 4$ kHz, and $1/T = 8\frac{1}{3}$ bits/sec; in Fig. 8-18, $W_{L0} = 12$ Hz; in Fig. 8-19, $W_{L0} = 48$ Hz.

Indicated by dashed lines on Figs. 8-17, 8-18, and 8-19 is the asymptotic behavior of Eq. (8-21) as x becomes large, corresponding to the case where the noisy reference due to the RF carrier becomes negligible. This asymptote satisfies the equation

$$\frac{PT}{N_0} = [\text{erfc}^{-1}(2P_E)]^2 + \frac{K_0\Upsilon_0}{2\delta_0} \tag{8-22}$$

PROBLEMS

8-1. Consider the Mariner Mars 1969 two-channel telemetry downlink whose parameters are given in Prob. 7-10. If the ground receiver's design point loop bandwidth is $W_{L0} = 12$ Hz, and the design point is further characterized by $\Upsilon_0 = 2$, and $y = 1/400$,

a) Find the signal-to-noise ratio in the ground receiver's design point bandwidth $x = 2P_c/N_0W_{L0}$.

b) Evaluate the parameter μ and loop bandwidth W_L.

c) Using the bit signal-to-noise ratios found in part b) of Problem 7-10, find the corresponding error probabilities in the science and engineering channels.

8-2. A one-way coherent communication system is to be designed using the parameters given below:

Total Received Power P_2	-153.3 dBm
Received Noise Spectral Density N_{02}	-180.4 dBm/Hz
Carrier Channel	
Carrier Power/Total Received Power P_{c2}/P_2	-7.1 dB
Received Carrier Power P_{c2}	______
Design Point Signal-to-Noise Ratio Υ_{02}	3 dB
Design Point Loop Damping Parameter r_{20}	2
Design Point Loop Bandwidth W_{02}	10 dB-Hz (10 Hz)
IF Filter Bandwidth W_{i2}	4 kHz

Signal-to-Noise Ratio in the Design Point Loop Bandwidth x_2	______
Data Channel	
Data Power/Total Received Power S_2/P_2	-1 dB
Received Data Power S_2	______
System Data Rate $\mathcal{R}_{b2} = 1/T_{b2}$	______
(Assume a (32, 5) orthogonal code with a word error probability $P_{E2} = 10^{-2}$)	______

Fill in the missing entries in the above mission specification table.

8-3. Consider a one-way system operating under the conditions of case 3 and whose error probability performance is illustrated in Fig. 8-2. Plot a curve of PT/N_0 in dB versus x in dB and determine the system parameters for the best power split condition.

8-4. It is desired to extend the technique of optimizing performance as a function of design point to a two-channel system. Show that the optimization equation analogous to Eq. (8-21) is given by

$$\frac{PT_1}{N_0} = R_{d1} + \left(\frac{T_1}{T_2}\right)R_{d2} + \frac{x}{\delta_{01}} + \frac{R_{d1}R_{d2}(T_1/T_2)}{(x/\delta_{01})}$$

If, as before Υ_0 alone is allowed to vary subject to the constraint $x \geq x_{\min}$, then show that the asymptotic behavior of the above optimization equation for large x satisfies an equation similar to (8-22), namely,

$$\frac{PT_1}{N_0} = [\operatorname{erfc}^{-1}(2P_{E1})]^2 + \left(\frac{T_1}{T_2}\right)[\operatorname{erfc}^{-1}(2P_{E2})]^2 + \frac{K_0\Upsilon_0}{2\delta_{01}}$$
$$+ \frac{[\operatorname{erfc}^{-1}(2P_{E1})]^2[\operatorname{erfc}^{-1}(2P_{E2})]^2(T_1/T_2)}{(K_0\Upsilon_0)/2\delta_{01}}$$

REFERENCES

1. Lindsey, W. C., "Performance of Phase-Coherent Receivers Preceded by Band-pass Limiters" *IEEE Transactions on Coumunication Technology*, Vol. COM-16, No. 2 (April, 1968) 245–251.
2. ———, "Block Coding for Space Communications," *IEEE Transactions on Communication Technology*, Vol. COM-17, No. 2 (April, 1969) 217–225.
3. Blake, I. F., and Lindsey, W. C., "Effects of Phase-Locked Loop Dynamics on Phase-Coherent Communications," Jet Propulsion Laboratory, Pasadena, Calif., SPS 37–54, Vol. III (December, 1968) 192–195.
4. Simon, M. K., "The Effect of Limiter Suppression on Command Detection Performance," Jet Propulsion Laboratory, Pasadena, Calif., SPS 37–63, Vol. III (June, 1970) 66–70.
5. Jean, F., Private communication.

6. Simon, M. K., "On the Selection of an Optimum Design Point for Phase-Coherent Receivers Employing Band-Pass Limiter," *IEEE Transactions on Communication Technology*, Vol. COM-20, No. 3 (April, 1972) 210–214.

Chapter 9

SYMBOL SYNCHRONIZATION AND ITS EFFECTS ON DATA DETECTION

9-1. INTRODUCTION

In previous chapters, a framework has been developed for use in designing a coherent communication system. Thus far, we have discussed two aspects of the all-important synchronization (sync) problem—*carrier* and *subcarrier* sync—since in most applications they are the principle sources of performance degradation. The further deterioration of performance due to lack of perfect sync information such as symbol, word, and frame sync has tacitly been assumed negligible relative to the above contributions.

This chapter ties together our discussion of coherent communication over the additive Gaussian noise channel by presenting various philosophies for achieving *symbol synchronization*.* For our purposes in this chapter, the term *symbol synchronization* will be used to describe the problem of estimating the instants in time at which the modulation—that is, the waveform corresponding to the multiplication of each channel code digit by the fundamental pulse shape—can change its state. In engineering practice, symbol synchronization frequently refers to the establishment or recovery of the *clock* waveform present in the observed data. Our emphasis will be on ways of obtaining symbol sync and their ultimate effect on the performance of the data detector. The other two aspects of the synchronization problem—word and frame sync—usually involve coding techniques, transmission of a synchronization pattern or word, or data storage schemes such as comma-

---

* Symbol synchronization (symbol sync) is frequently referred to in the literature as bit or character sync. In practice, the term symbol sync is preferable over bit sync since it allows for the consideration of coded or uncoded communication systems.

free codes, Barker sequences, and pulse stuffing.[1,2] Methods of obtaining word sync that do not involve placing additional sync symbols into the data format will be briefly discussed. Such methods offer the advantage of maximum efficiency in transmitted data rate.

The problem of symbol synchronization deals with the estimation of the *time-of-arrival* or *epoch* of the received data symbols. Word synchronization, on the other hand, is concerned with separating the detected data symbols into proper groups called "words." Finally, organizing the words into so-called "frames" is the last step in the synchronization hierarchy. As an example, an element of a TV picture might correspond to a single data symbol, a line of TV information could be considered a word; the entire collection of lines corresponding to a single complete TV picture would be the corresponding frame. The simplest demonstration of a loss of sync would be the horizontal pulling or vertical rolling familiar to most who watch TV.

There are primarily two important factors that must be considered in implementing a coherent communication system requiring symbol synchronization. First, it is necessary to maintain a high efficiency in the data detection process; inaccurate symbol sync directly reduces the probability of making correct decisions. Second, for a given amount of power available for transmission of data and sync information, the synchronization scheme should require as little of this energy as possible. Clearly, a conflict exists between establishing a good symbol sync reference and one with minimum energy. This is one of the tradeoffs that faces the design engineer.

Basically, there are two schools of thought with regard to providing symbol synchronization in a coherent digital communication system. On the one hand, a separate channel is allotted for sending the synchronizing signal; in the other case, the sync information appears on the same channel as the data symbol sequence. Communication systems that provide sync information over a separate channel rely heavily on the fact that the phase distortion in the sync and data channels is the same. In practice, the coherence of these channels is at best an approximation. Nevertheless, for the additive Gaussian noise channel, this approximation is quite good.

The most common form of *data-derived symbol sync* is where the receiver includes a device that extracts the synchronization directly from the information-bearing signal. The presence of adequate transitions (zero crossings) in the data symbol sequence is helpful to the success of this type of operation, although schemes have been developed for operation at low signal-to-noise ratios where the difficulty in detecting transitions is of no consequence. The principal advantage of extracting sync from the data-bearing signal is that no additional power expenditure and frequency spectrum are required. Thus, all the transmitter sideband power is available to the data and sync signals simultaneously.

In this chapter, we first present methods by which the symbol synchronizing signal can be derived from the information-bearing signal. Included in this discussion is a derivation of the maximum *a posteriori* (MAP) estimator of symbol sync, several symbol synchronizer configurations motivated by the MAP estimator, and the design of symbol synchronizers based on the waveshape (pulse shape) of the synchronizing signal. The performance of this latter class of synchronizers is characterized in terms of the measures *mean-squared symbol sync jitter* and *mean time to first loss of synchronization.* We then present methods for providing symbol sync over a separate channel concluding with a discussion of the effect of symbol sync error on overall system error probability performance.

9-2. SYMBOL SYNCHRONIZATION FROM THE DATA-BEARING SIGNAL

The process of extracting symbol synchronization from the information-bearing signal is often referred to as *data-derived synchronization.* The most natural question that one might ask at first is: What is the optimum symbol synchronizer? The answer to this question, of course, depends highly upon the criterion of optimality selected. The MAP estimation criterion represents one approach to the problem of estimating symbol sync. Unfortunately, this approach avoids a direct attack on the important problems associated with symbol sync acquisition, that is, minimum acquisition time, maximum acquisition range, and maximum probability of acquisition.

The Maximum *a Posteriori* (MAP) Estimator of Symbol Sync

The problem of estimating epoch of a received signal plus noise can be approached using the theory of maximum *a posteriori* estimation of an unknown parameter in Gaussian noise.[2,3] Assuming complete knowledge of the past history of the received signal, the MAP estimate of the present symbol sync position can be derived.[4] A less restrictive application of the theory involves finding the estimate of epoch assuming memory only over a finite number of past symbol periods.[5] In other words, we observe the received signal over an interval of length K symbols during which the timing misalignment between the incoming symbols and the locally generated symbol timing clock is assumed constant but unknown. Based on this observation alone, we arrive at an estimate of the unknown symbol sync. The choice of the parameter K in practice is obviously determined by the rate of variation of the channel phase characteristic—that is, K cannot be chosen larger than that value for which the channel phase characteristic can be assumed constant. We furthermore assume that the symbol length T is known (and remains constant) and that a stored replica of the transmitted signal waveshape is available at the receiver.* Although we are primarily interested in the case of rectangular signal pulses—an NRZ data sequence (see Chap. 1)—for the sake of generality, we derive the MAP estimator of symbol sync assuming an arbitrary time-limited pulse waveshape $p_s(t)$ of energy ST. Thus, the theory presented could be applied to the synchronization of *Manchester coded* data (see Prob. 9-1). The choice of the signal waveshape which optimizes a specific performance criterion is treated as a separate problem and is discussed later.

The input signal to the symbol synchronizer can be characterized as

$$y(t) = \sum_{k=0}^{K} s(t; a_k, \epsilon) + n(t) = s(t, \epsilon) + n(t) \tag{9-1}$$

where a_k is the polarity (± 1) of the kth transmitted symbol $s(t; a_k, \epsilon)$ with random epoch ϵ assumed to be constant for KT sec. In terms of the basic signal waveshape $p_s(t)$, which is defined to be nonzero only over the interval $(0, T)$, the kth received symbol is written as

$$s(t; a_k, \epsilon) = a_k p_s[t - (k - 1)T - \epsilon] \tag{9-2}$$

Since $n(t)$ is assumed to be white Gaussian noise, then it can be expanded in any set of complete orthonormal basis functions with random coefficients that are Gaussian and uncorrelated. We choose the set that is convenient for representation of the signal pulse and denote this set by $\{\psi_i(t)\}$. Also, we define the kth subinterval $T_k(\epsilon)$ by $(k - 1)T + \epsilon \leq t \leq kT + \epsilon$ and t is further restricted to lie in the interval $(0, KT)$. From this we note that the 0th and Kth subintervals are truncated to lengths of ϵ and $T - \epsilon$, respectively (Fig. 9-1). Then, partitioning the received signal and noise into $K + 1$ segments corresponding respectively to their values in the $K + 1$ subintervals, we can express Eq. (9-1) together with (9-2) in the orthonormal series form

$$\begin{aligned} y(t) &= \sum_{k=0}^{K} y_k(t) = \sum_{k=0}^{K}\left[\sum_{i=1}^{M} y_{ik}\psi_i(t)\right] \\ &= \sum_{k=0}^{K}\left[a_k \sum_{i=1}^{M} p_{ik}(\epsilon)\psi_i(t) + \sum_{i=1}^{M} n_{ik}\psi_i(t)\right] \end{aligned}$$

where M is the dimensionality of the set $\{\psi_i(t)\}$ and

$$\left.\begin{aligned} y_{ik} &= \int_{T_k(\epsilon)} y(t)\psi_i(t)\,dt \\ p_{ik}(\epsilon) &= \int_{T_k(\epsilon)} p_s[t - (k - 1)T - \epsilon]\psi_i(t)\,dt \\ n_{ik} &= \int_{T_k(\epsilon)} n(t)\psi_i(t)dt \end{aligned}\right\} \tag{9-3}$$

* To simplify the notation here, we denote the symbol time by T, rather than T_s, without any loss in generality.

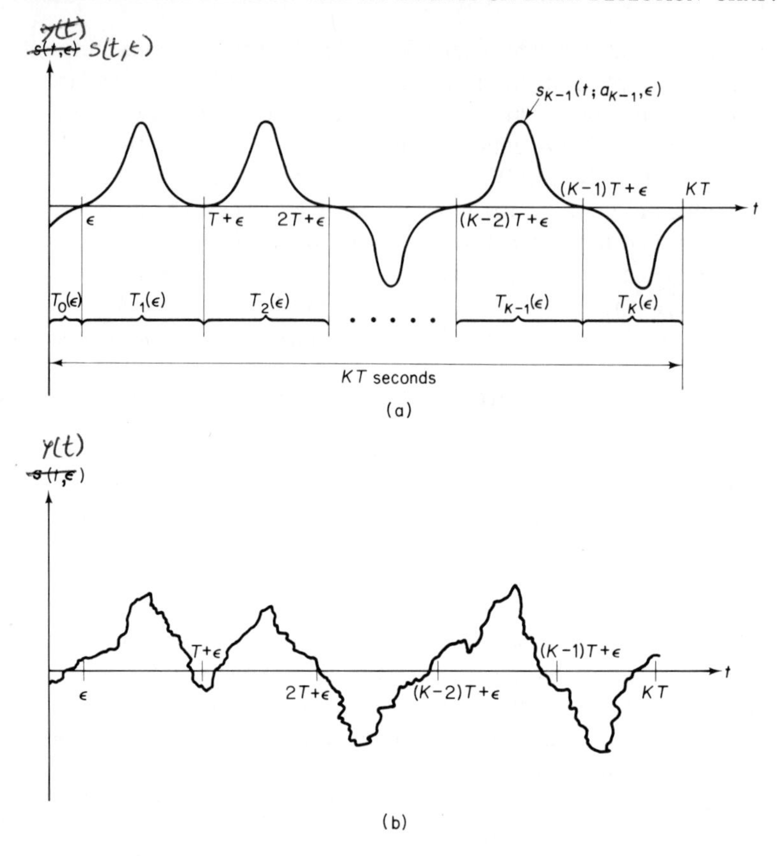

Fig. 9-1 Received signal without and with noise

It is convenient to arrange the orthonormal series weighting coefficients in row vectors denoted respectively be $\mathbf{y}_k$, $\mathbf{p}_k(\epsilon)$, and $\mathbf{n}_k$; $k = 0, 1, 2, \ldots, K$. Then, the ith component of the received signal vector $\mathbf{y}_k \triangleq a_k\mathbf{p}_k(\epsilon) + \mathbf{n}_k$, the signal vector $\mathbf{p}_k(\epsilon)$, and the noise $\mathbf{n}_k$ are defined by the orthogonality relationships in (9-3). From the above, $\mathbf{y}_k$ is conditionally Gaussian on a_k and ϵ with mean $a_k\mathbf{p}_k(\epsilon)$.

Letting $\mathbf{a}$ denote the data vector whose components are $a_0, a_1, a_2, \ldots, a_K$, $\mathbf{y} \triangleq (\mathbf{y}_0, \mathbf{y}_1, \ldots, \mathbf{y}_K)$, and noting that the random epoch and symbol sequence are independent, one obtains from Bayes' rule,

$$p(\epsilon \mid \mathbf{y}) = \frac{p(\epsilon)}{p(\mathbf{y})} \int_{\mathbf{a}} p(\mathbf{y} \mid \epsilon, \mathbf{a}) p(\mathbf{a}) \, d\mathbf{a} \tag{9-4}$$

where $p(\epsilon)$, $p(\mathbf{y})$, and $p(\mathbf{a})$ are all *a priori* p.d.f.'s of the corresponding variables in parentheses. Since the transmitted symbols are assumed to be independent, their joint p.d.f. $p(\mathbf{a})$ can be expressed as

$$p(\mathbf{a}) = \prod_{k=0}^{K} p(a_k) \tag{9-5}$$

where for equiprobable symbols

$$p(a_k) = \begin{cases} \frac{1}{2} & a_k = +1 \\ \frac{1}{2} & a_k = -1 \end{cases} \qquad k = 0, 1, 2, \ldots, K \tag{9-6}$$

so that the integral in (9-4) is in fact a $K + 1$-fold summation. Nevertheless, we shall maintain the integral notation for simplicity. Furthermore, from the independence properties of the noise vectors,

$$p(\mathbf{y} \mid \epsilon, \mathbf{a}) = \prod_{k=0}^{K} p(\mathbf{y}_k \mid \epsilon, a_k) \tag{9-7}$$

Hence, combining (9-4) through (9-7) gives

$$p(\epsilon \mid \mathbf{y}) = \frac{p(\epsilon)}{p(\mathbf{y})} \int_{\mathbf{a}} \prod_{k=0}^{K} p(\mathbf{y}_k \mid \epsilon, a_k) p(a_k)\, d\mathbf{a} \tag{9-8}$$

Assuming that the random epoch to be estimated is equally likely to occur anywhere in the interval $(-T/2, T/2)$—that is, its p.d.f. is uniform—and $p(\mathbf{y})$ is independent of ϵ, then maximizing $p(\epsilon \mid \mathbf{y})$ with respect to ϵ is equivalent to maximizing

$$\int_{\mathbf{a}} \prod_{k=0}^{K} p(\mathbf{y}_k \mid \epsilon, a_k) p(a_k) d\mathbf{a} = \prod_{k=0}^{K} \int_{a_k} p(\mathbf{y}_k \mid \epsilon, a_k) p(a_k) da_k \tag{9-9}$$

with respect to ϵ. The conditional p.d.f. $p(\mathbf{y}_k \mid \epsilon, a_k)$ is M-dimensional Gaussian, with a diagonal covariance matrix $(N_0/2)I$ where I is the identity matrix, that is,

$$p(\mathbf{y}_k \mid \epsilon, a_k) = \frac{1}{[2\pi(N_0/2)]^{M/2}} \exp \left\{ \frac{[\mathbf{y}_k - a_k \mathbf{p}_k(\epsilon)][\mathbf{y}_k - a_k \mathbf{p}_k(\epsilon)]^t}{2(N_0/2)} \right\} \tag{9-10}$$

Averaging over the p.d.f. of a_k as given in (9-6) and factoring out all terms that are independent of ϵ gives

$$\int_{a_k} p(\mathbf{y}_k \mid \epsilon, a_k) p(a_k) da_k = C \cosh \left[\frac{2}{N_0} \mathbf{y}_k \mathbf{p}_k^t(\epsilon) \right] \tag{9-11}$$

where C is a constant independent of ϵ. The maximum *a posteriori* estimate of ϵ—that is, $\hat{\epsilon}$—is then the value of ϵ that maximizes

$$f(y, \epsilon) = \prod_{k=0}^{K} \cosh \left[\frac{2}{N_0} \mathbf{y}_k \mathbf{p}_k^t(\epsilon) \right] \tag{9-12}$$

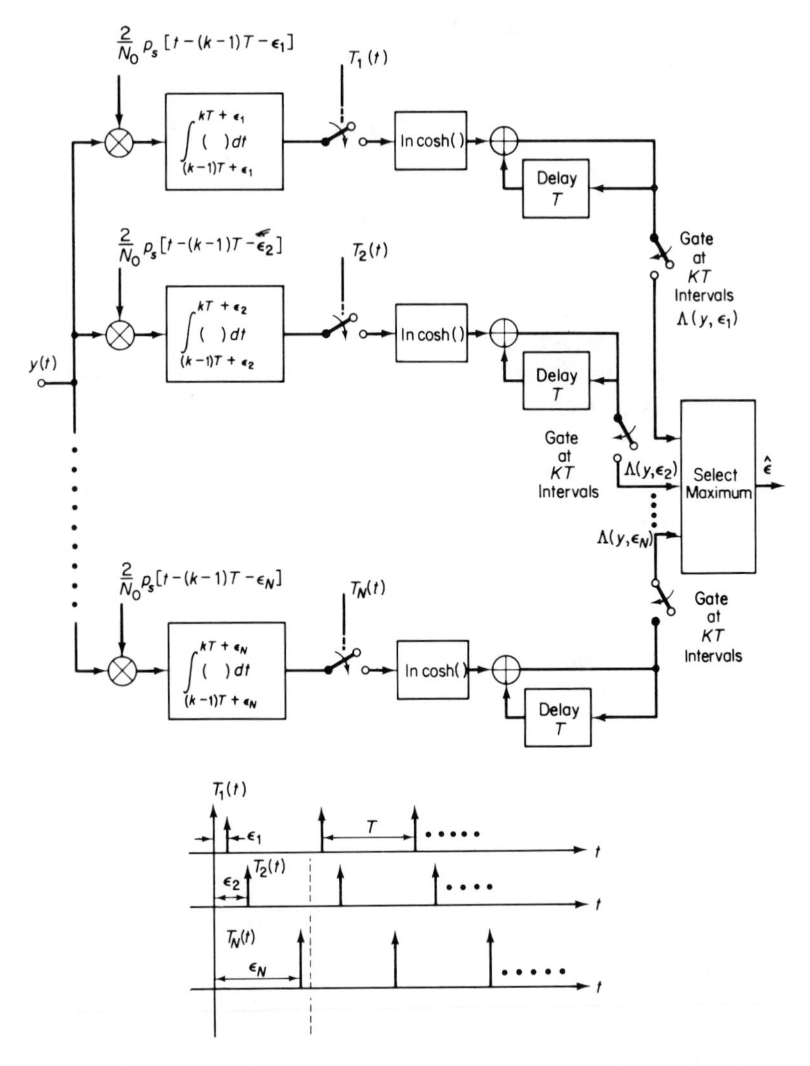

Fig. 9-2 MAP symbol synchronizer for arbitrary pulse waveshape

or equivalently

$$f(y, \epsilon) = \prod_{k=0}^{K} \cosh \left\{ \frac{2}{N_0} \int_{T_k(\epsilon)} y(t) p_s[t - (k - 1)T - \epsilon] \, dt \right\} \tag{9-13}$$

Since the logarithm is a monotonic function of its argument, we can equiva-

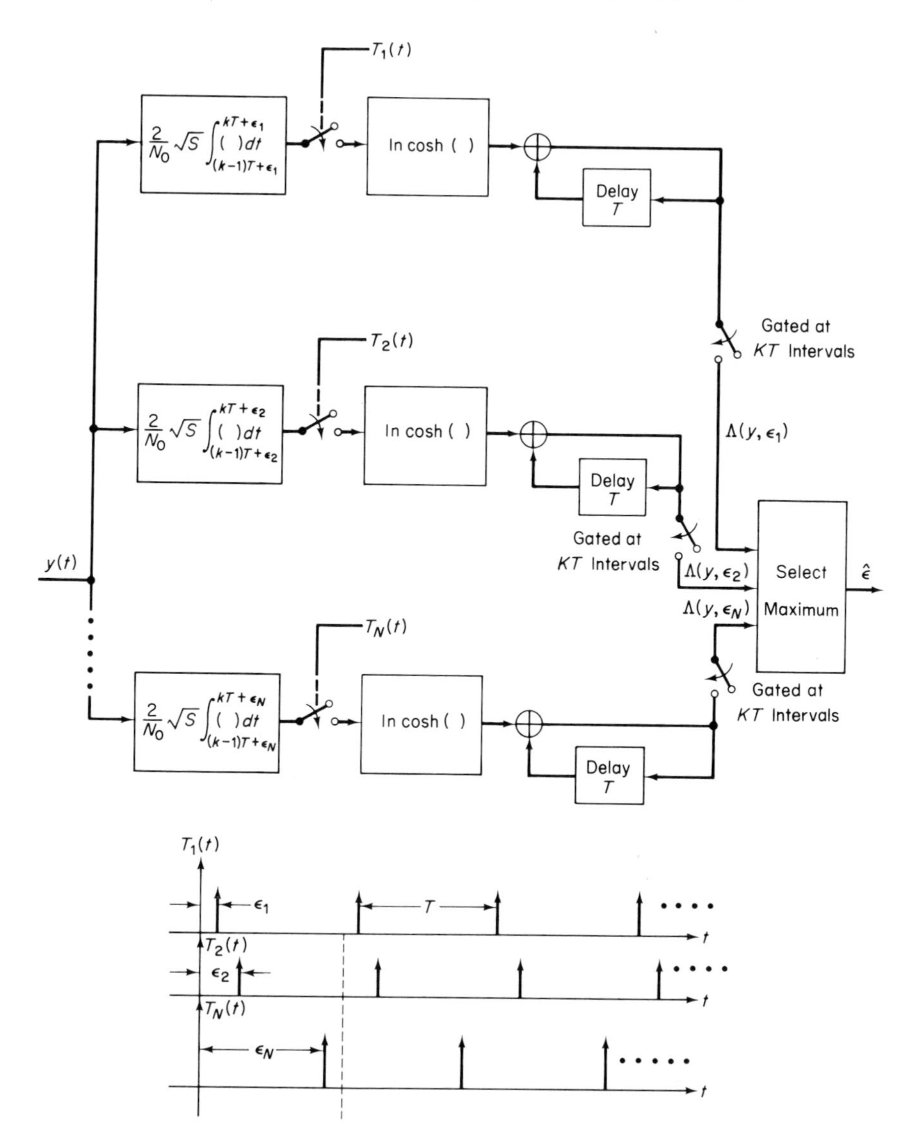

Fig. 9-3 MAP symbol synchronizer for rectangular pulse waveshape (NRZ data)

lently select that value of ϵ which maximizes

$$\Lambda(y, \epsilon) \triangleq \ln f(y, \epsilon) = \sum_{k=0}^{K} \ln \cosh\left[\frac{2}{N_0}\int_{T_k(\epsilon)} y(t)p_s[t-(k-1)T-\epsilon]\,dt\right] \tag{9-14}$$

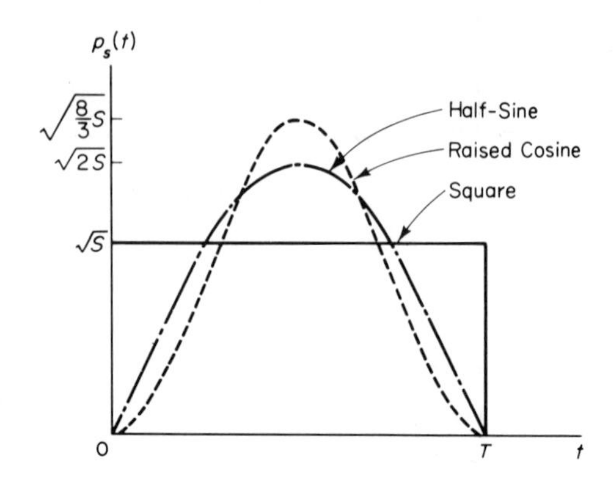

Half-Sine $$p_s(t) = \begin{cases} \sqrt{2S}\,\sin(\pi/T)t, & t \in (0,T) \\ 0, & t \notin (0,T) \end{cases}$$

Raised Cosine $$p_s(t) = \begin{cases} \sqrt{\frac{2}{3}S}[1 - \cos(2\pi/T)t], & t \in (0,T) \\ 0, & t \notin (0,T) \end{cases}$$

Square $$p_s(t) = \begin{cases} \sqrt{S} & t \in (0,T) \\ 0, & t \notin (0,T) \end{cases}$$

Fig. 9-4 Three possible symbol waveforms of equal energy

A practical interpretation of the above result in the time domain is to cross-correlate the received signal with a stored replica $p_s(t)$ in each subinterval, take the log hyperbolic cosine of this result and accumulate these values over all subintervals. The value of ϵ that yields the largest accumulated value is then declared the best estimate, $\hat{\epsilon}$. Actually, ϵ takes on a continuum of values between $-T/2$ and $T/2$, and hence to practically apply the above technique one must quantize ϵ into a number of levels N and perform a search procedure over these allowable epoch positions. Hence, the degree of accuracy desired in estimating ϵ dictates the number of quantization levels. We also note that in the 0th and Kth subintervals we use for correlation only those pieces of $p_s(t)$ corresponding to the last ϵ and the first $T - \epsilon$ sec, respectively.

The symbol synchronizer that is based on the above interpretation of (9-14) will be referred to as the *MAP symbol synchronizer* (Fig. 9-2). This type of physical realization is nothing more than an implementation of a maximum likelihood test among N hypotheses; thus, for a given value of K, the MAP symbol synchronizer represents an *open loop* device. If the sig-

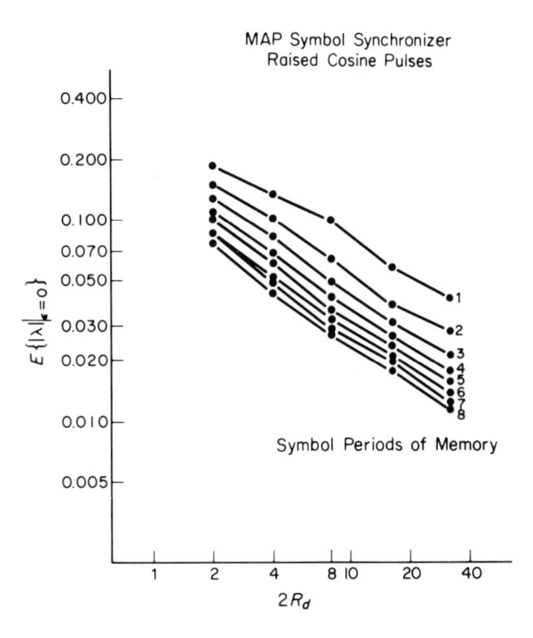

Fig. 9-5 Performance of the MAP symbol synchronizer in terms of the mean of the magnitude of the normalized sync error conditioned on zero input epoch versus signal-to-noise ratio with memory time as a parameter (Courtesy of P. A. Wintz and E. J. Luecke)

naling format is NRZ data, then the MAP synchronizer of Fig. 9-2 reduces to that illustrated in Fig. 9-3. These figures include the locally generated timing pulses used to start and terminate the N integrators all operating in a parallel fashion.

From a practical engineering standpoint, the MAP symbol synchronizer is an unsatisfactry solution to the symbol sync problem because of the large number of integrate and dump circuits required to achieve accurate bit timing. Despite its complexity, the performance of the MAP synchronizer is worth studying since it may be used as a standard against which the performance of easier to implement synchronizer configurations may be judged.

Clearly, the statistics of the random variable $\hat{\epsilon}$ or, alternately, the normalized symbol synchronization error $\lambda \triangleq (\epsilon - \hat{\epsilon})/T$ are difficult to come by. Wintz and Luecke[6] evaluate by Monte Carlo methods the performance of the MAP synchronizer in terms of the magnitude of its first conditional moment $E\{|\lambda|_{\epsilon=0}\}$ as a function of the signal-to-noise ratio $R_d = ST/N_0$ and the memory time in symbols K. The results were computed for three

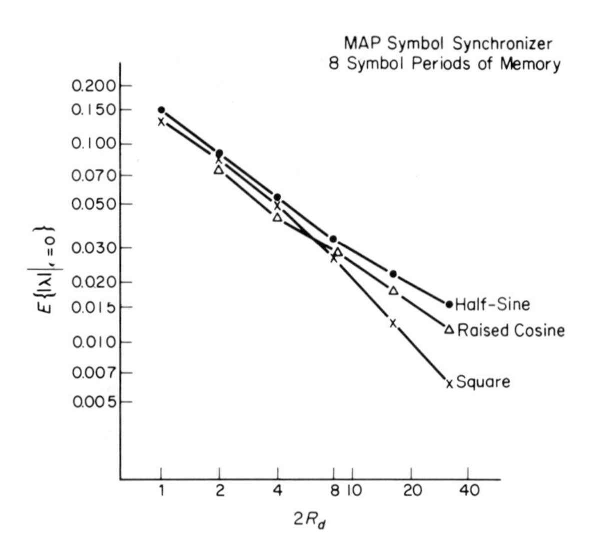

Fig. 9-6 Performance of the MAP symbol synchronizer in terms of the mean of the magnitude of the normalized sync error conditioned on zero input epoch versus signal-to-noise ratio for the three symbol waveforms of Fig. 9-4 (Courtesy of P. A. Wintz and E. J. Luecke)

different pulse waveforms $p_s(t)$: the half-sinusoid, raised cosine, and rectangular shapes (Fig. 9-4). Results for the raised cosine signal pulse shape are illustrated in Fig. 9-5 for values of $K = 1, 2, \ldots, 8$. Qualitatively, for this case $E\{|\lambda|_{\epsilon=0}\}$ varies inversely as the square root of K and the square root of $2R_d$. Assuming a memory time of eight symbol periods, Fig. 9-6 illustrates a comparison of the performance of the MAP symbol synchronizer in terms of $E\{|\lambda|_{\epsilon=0}\}$ for the three pulse shapes of Fig. 9-4.

Several Symbol Synchronizer Configurations Motivated by the MAP Estimation Approach

Open Loop Realizations. Here we describe several suboptimum realizations of the MAP symbol synchronizer, all of which achieve a reduction of implementation complexity. The simplest of these, at least in concept, is the serial realization illustrated in Fig. 9-7. Here, the N integrate-and-dump circuits of Fig. 9-2, which operate in parallel on a KT segment of the received signal, are replaced by a single integrate-and-dump, which serially processes the received signal over an interval of length NKT. During each successive interval of length K symbols, one of N different phases of the local symbol

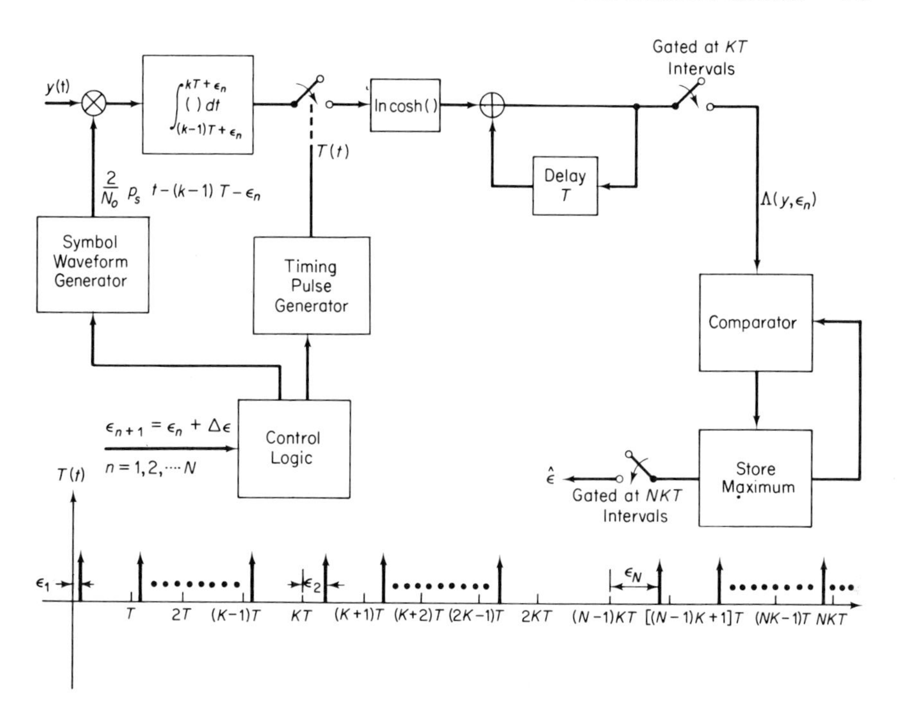

Fig. 9-7 Serial realization of a symbol synchronizer based on the MAP estimation technique

timing is selected for testing and hence after NK symbols of the received signal have been processed, all possible phases have been tested. At this point, the particular phase of the local symbol timing that resulted in the largest output from the single integrator estimator is chosen as the best estimate of symbol sync. Clearly, such a serial implementation increases the average time to acquire symbol sync by approximately a factor of N. Also, a device for storage of a magnitude value must be provided since after each K symbols the present integrator estimator output and associated symbol timing phases are compared with the largest past output and its timing phase.

One technique for getting around the requirement for a magnitude storage device is to use a sequential search procedure wherein each estimator output $\Lambda(y, \epsilon_i)$ is compared to a threshold Ξ, and an "in-sync" or "out-of-sync" decision is made depending on whether this output exceeds the threshold (Fig. 9-8). Starting with an assumed timing phase ϵ_i, arbitrarily chosen from the set of N allowable values, $\Lambda(y, \epsilon_i)$ is computed for the first K symbols in the received sequence. If $\Lambda(y, \epsilon_i) \geq \Xi$, then ϵ_i is chosen as the correct symbol sync condition and the test terminates. If $\Lambda(y, \epsilon_i) < \Xi$, then the next consecutive timing phase ϵ_{i+1} is switched into the test configuration and the search

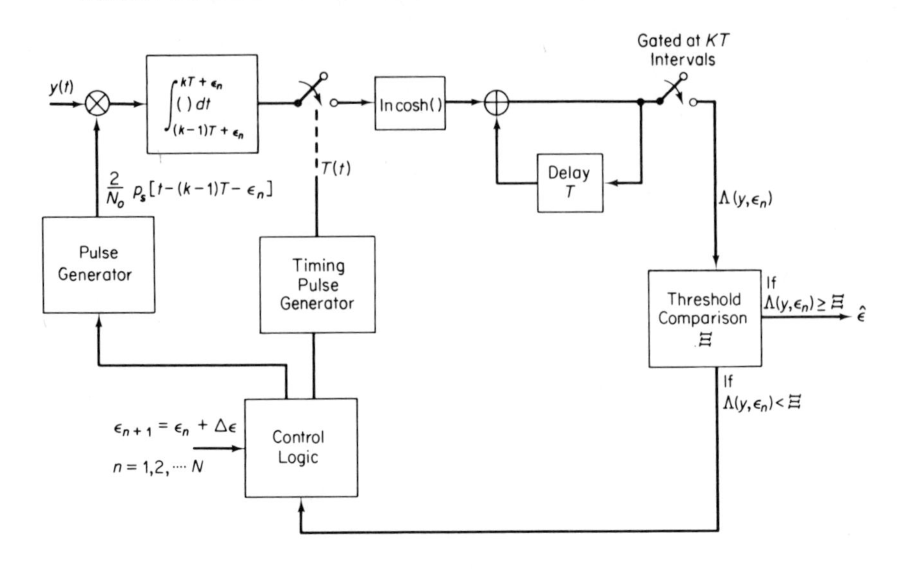

Fig. 9-8 A sequential search algorithm for estimating symbol sync

continues—that is, $\Lambda(y, \epsilon_{i+1})$ is computed over the next K symbol intervals. Since a *yes* or *no* decision is made after each K symbols and the test continues only if the answer is *no*, there is obviously no requirement for a magnitude storage device.

Two types of error are possible at each decision. A *false alarm error* occurs when ϵ_i is not the correct symbol sync but $\Lambda(y, \epsilon_i) \geq \Xi$. A *false dismissal error* corresponds to the situation where ϵ_i is the correct symbol sync but is not detected as such because $\Lambda(y, \epsilon_i) < \Xi$. The probabilities of occurrence of the false alarm and false dismissal events are useful in computing the average time to acquire symbol sync. The threshold Ξ can in theory then be chosen so as to minimize this average acquisition time.[5]

Closed Loop Realizations. Thus far we have discussed only *open loop* synchronizers, which approximate the performance of the MAP symbol synchronizer illustrated in Fig. 9-2. In most practical applications, an estimate of symbol sync is continuously required, due to the fact that ϵ varies slowly with time; therefore, it is desired to have an estimator that continuously (possibly in discrete intervals of time) updates itself—that is, a *closed loop symbol synchronizer*. We start by looking for a realization of such a closed loop synchronizer which is motivated by the MAP estimation approach.

First, we consider an alternate interpretation of Eq. (9-14). Differentiating (9-14) with respect to ϵ and equating to zero gives a transcendental equation whose solution is the MAP estimator, ϵ; that is,

$$\begin{aligned}\left.\frac{\partial\Lambda(y,\epsilon)}{\partial\epsilon}\right|_{\epsilon=\hat{\epsilon}} = \sum_{k=0}^{K}\Bigg[\frac{2}{N_0}\Bigg\{\int_{T_k(\hat{\epsilon})} y(t)\frac{\partial p_s[t-(k-1)T-\hat{\epsilon}]}{\partial\hat{\epsilon}}\,dt \\ + y(kT+\hat{\epsilon})p_s(T) - y[(k-1)T+\hat{\epsilon}]p_s(0)\Bigg\}\Bigg] \\ \times \tanh\left\{\frac{2}{N_0}\int_{T_k(\hat{\epsilon})} y(t)p_s[t-(k-1)T-\hat{\epsilon}]\,dt\right\} \\ = 0 \qquad (9\text{-}15)\end{aligned}$$

For any estimate of ϵ other than the MAP estimate, the function $\partial\Lambda(y, \epsilon)/\partial\epsilon$ will be either positive or negative depending on whether $\epsilon < \hat{\epsilon}$ or $\epsilon > \hat{\epsilon}$ and hence can be used to provide search direction. A simple example of a closed-loop symbol synchronizer that incorporates $\partial\Lambda(y, \epsilon)/\partial\epsilon$ as an error signal is illustrated in Fig. 9-9, where for simplicity we have assumed that $p_s(0)$

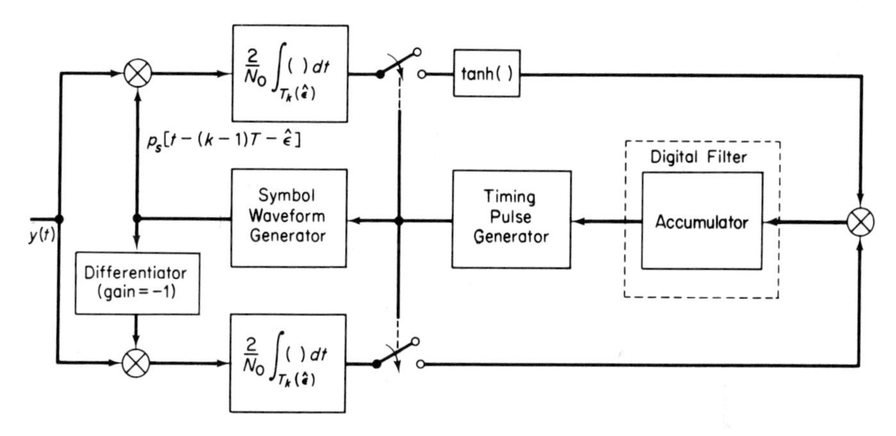

Fig. 9-9 A closed loop symbol synchronizer motivated by the MAP estimation approach

$= p_s(T) = 0$. The phase of the timing pulse generator, which controls the start and termination of the integrate-and-dump circuits, is bumped every T sec by an amount proportional to the magnitude of $\partial\Lambda(y, \epsilon)/\partial\epsilon$ and in a direction based on the sign of $\partial\Lambda(y, \epsilon)/\partial\epsilon$ as computed for the previous KT sec. Clearly, the closed loop dynamics of the symbol synchronizer shown produce a p.d.f. of the normalized synchronization error λ, which is quite different from that obtained for the MAP symbol synchronizer discussed previously.

As a further simplification of Fig. 9-9, one might replace the accumulator by a digital filter whose memory is chosen relative to the value of K.[7] Whereas before, the input to the timing pulse generator was a linear combination of the previous $K + 1$ outputs of the multiplier, we now have an accumulation of the infinite past of these quantities each weighted by the impulse response of the digital filter. By adjusting the memory of this filter, the input to the timing pulse generator can be made to essentially reflect only the previous $K + 1$ multiplier outputs. Since in practice the signal epoch ϵ will never remain constant in time, a weighted sum of the data that accentuates the recent ones and attenuates the others should be more appropriate than a pure linear combination; thus, the motivation for using a digital filter.

Other closed loop configurations motivated by the MAP approach can be obtained by approximating the ln cosh x nonlinearity for large and small values of its argument; i.e.,

$$\ln\cosh x \cong \begin{cases} \dfrac{|x|}{2} & |x| \gg 1 \\[2ex] \dfrac{x^2}{2} & |x| \ll 1 \end{cases} \tag{9-16}$$

Hence, from Eq. (9-14),

$$\Lambda(y, \epsilon) \cong \begin{cases} \displaystyle\sum_{k=0}^{K} \frac{1}{N_0} \left| \int_{T_k(\epsilon)} y(t) p_s[t - (k-1)T - \epsilon]\, dt \right| & \text{for large } R_d \\[3ex] \displaystyle\sum_{k=0}^{K} \frac{1}{2N_0^2} \left[\int_{T_k(\epsilon)} y(t) p_s[t - (k-1)T - \epsilon]\, dt \right]^2 & \text{for small } R_d \end{cases} \tag{9-17}$$

Furthermore, if we approximate the derivative $\partial\Lambda(y, \epsilon)/\partial\epsilon$ by the difference function

$$\frac{\Lambda(y, \epsilon + \frac{1}{2}\Delta\epsilon) - \Lambda(y, \epsilon - \frac{1}{2}\Delta\epsilon)}{\Delta\epsilon} \tag{9-18}$$

then,

$$\frac{\partial\Lambda(y, \epsilon)}{\partial\epsilon} \cong \frac{1}{N_0 \Delta\epsilon} \sum_{k=0}^{K} \left\{ \left| \int_{T_k(\epsilon + \Delta\epsilon/2)} y(t) p_s\left[t - (k-1)T - \epsilon - \frac{\Delta\epsilon}{2}\right] dt \right| - \left| \int_{T_k(\epsilon - \Delta\epsilon/2)} y(t) p_s\left[t - (k-1)T - \epsilon + \frac{\Delta\epsilon}{2}\right] dt \right| \right\} \quad \text{for large } R_d \tag{9-19}$$

and

$$\frac{\partial \Lambda(y, \epsilon)}{\partial \epsilon} \cong \frac{1}{2N_0^2 \Delta\epsilon} \sum_{k=0}^{K} \left\{ \left[\int_{T_k(\epsilon+\Delta\epsilon/2)} y(t) p_s\left[t-(k-1)T-\epsilon-\frac{\Delta\epsilon}{2}\right] dt \right]^2 - \left[\int_{T_k(\epsilon-\Delta\epsilon/2)} y(t) p_s\left[t-(k-1)T-\epsilon+\frac{\Delta\epsilon}{2}\right] dt \right]^2 \right\} \quad \text{for small } R_d \tag{9-20}$$

By using the above approximations to $\partial\Lambda(y, \epsilon)/\partial\epsilon$ as error signals in closed loop configurations, we are led to the symbol synchronizers illustrated in Fig. 9-10. Once again one may replace the accumulator by a digital filter with the same interpretation as given for Fig. 9-9. Symbol synchronizers of the type illustrated in Fig. 9-10 are referred to as *early-late gate symbol syn-*

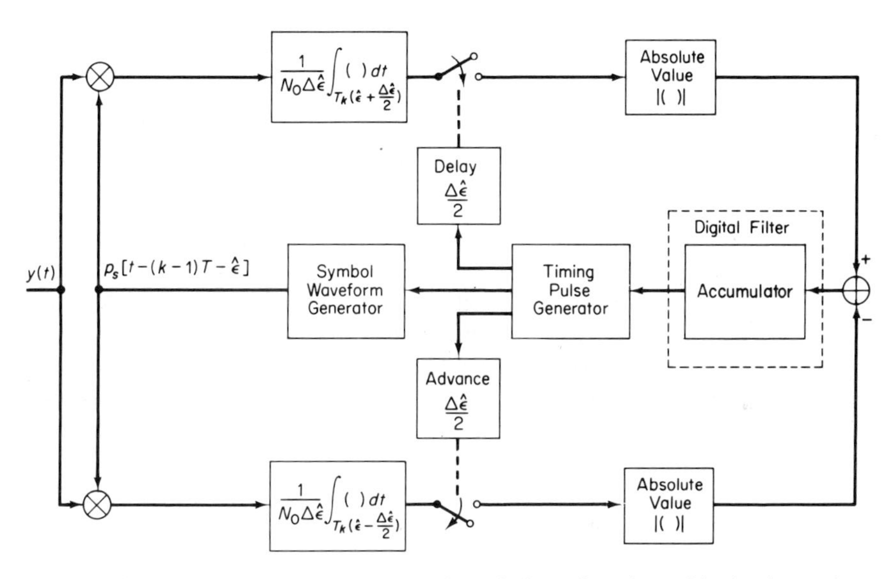

Fig. 9-10(a) An early-late gate type of symbol synchronizer with absolute value type of nonlinearity

chronizers and will be discussed in more detail later on in the chapter. The parameter $\Delta\epsilon$ can be varied to optimize system performance.

Before going on to the signal design aspect of the synchronization problem, we briefly make mention of two other contributions to the self-synchronization problem which make use of the correlation principle and which are thus of interest in low signal-to-noise ratio systems. Van Horn[8] suggests symbol-by-symbol correlation of the received signal with N delayed versions of a reference signal derived from a square-wave oscillator operating at the

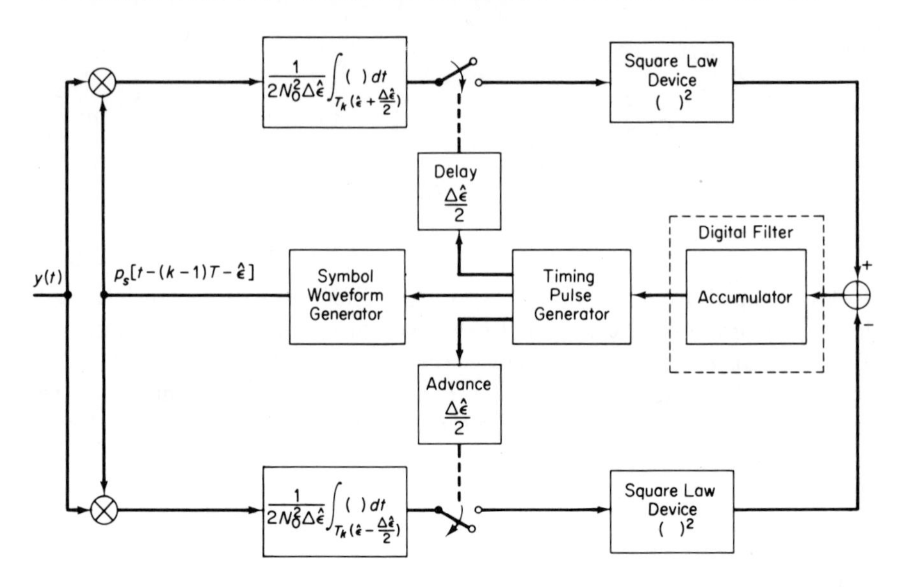

Fig. 9-10(b) An early-late gate type of symbol synchronizer with square-law type of nonlinearity

symbol rate of the input. Hence, a bank of cross-correlators is required as is a decision device which selects the maximum cross-correlation value. For $N = 2$, a difference of cross-correlation values may be used as a decision criterion. The output of the decision device is used as a control voltage in a feedback loop to vary the frequency of the reference oscillator. This configuration is quite similar to an early-late gate symbol synchronizer discussed later in this chapter. Van Horn makes no claims regarding the optimality of such a system but merely examines its performance when the input symbol stream is a *Manchester coded* signal (see Chap. 1). This code is used extensively in telemetry links because of its increased number of transitions per unit time (at the expense of increased bandwidth) relative to other codes operating at the same symbol rate. Wintz and Hancock[9] consider an adaptive approach to the symbol synchronization problem by analyzing the performance of an epoch estimator-correlator sync system with an N-ary FSK signaling format. In this system, a given waveshape is amplitude modulated onto a carrier with one of N possible frequencies and transmitted over a channel with unknown delay and known phase shift. If the phase shift is unknown, it is assumed that it can be adequately estimated by a PLL. Hence, the job at the receiver is to estimate from the data-bearing signal its epoch (time of arrival) and use this quantity, the channel delay, to decide via a correlation detection which of the N signals (carrier frequencies) was in fact transmitted. The detection part of the receiver consists then of a bank of matched filters, which are optimum, of course, only for perfect synchronization. The performance (error probability as a function of signal-to-noise ratio and rms sync error) of the estimator-correlator receiver as a unit is not dependent on the structure of the delay estimator. Several other receivers of the estimator-correlator type that have received significant attention are the Rake[10] and Kineplex[11,12] systems.

The Effect of Signal Waveshape on the Design of Symbol Synchronizers

Here we present some results that will motivate the study of several symbol synchronizers whose performance is highly dependent on the shape of the equivalent cross-correlation function between the input signal and a locally generated reference signal. We consider only those synchronizers whose performance can be determined from a PLL-type configuration with a phase detector characteristic specified by an arbitrary periodic nonlinearity such as $g(\lambda)$ (see Fig. 9-11). Hence, a problem of fundamental interest concerns the optimum design of signals for PLLs, where *optimum* depends on the performance measure chosen to represent the system behavior. For example, during the *signal acquisition mode*, the performance indices are *acquisition time* and *range*, and *probability of acquisition.* After symbol sync has been acquired, attention is focused on the tracking mode where measures such as

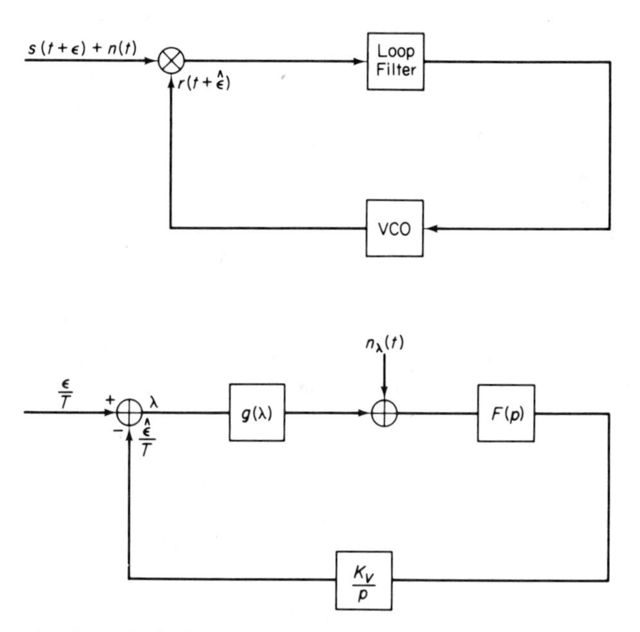

Fig. 9-11 A phase-locked loop with arbitrary periodic nonlinearity and its equivalent mathematical model

mean-squared sync error, mean time to first loss of synchronization and symbol slip rate[13] become significant. Regardless of the performance criterion chosen, optimum signal designs for an arbitrary loop filter and loop detuning are difficult to come by. Thus, one must consider special cases of the symbol synchronizer model illustrated in Fig. 9-11 that yield tractible results.

*Minimization of the Area Under the Tail of the Synchronization Error P.D.F.** The signal design problem for a first-order PLL and zero detuning has been considered by several authors.[14-16] The significance of the zero detuning assumption in a symbol synchronizer corresponds to having at the receiver a perfect match between the data symbol period and that of the reference generator. Hence, we will assume that both the synchronizer input signal $s(t + \epsilon)$ and the generated reference $r(t + \hat{\epsilon})$ are both of period T. The cross-correlation of $s(t + \epsilon)$ and $r(t + \hat{\epsilon})$ is defined by

$$g(\lambda) \triangleq \frac{1}{T} \int_{-T/2}^{T/2} s(t + \epsilon) r(t + \hat{\epsilon})\, dt \qquad |\lambda| \leq \frac{1}{2} \tag{9-21}$$

and has a period equal to unity. Furthermore, by constraining $s(t)$ and $r(t)$ to have unit power, a straightforward application of Schwarz' inequality reveals that

$$|g(\lambda)| \leq 1 \qquad -\tfrac{1}{2} \leq \lambda \leq \tfrac{1}{2} \tag{9-22}$$

Letting $p(\lambda)$ denote the p.d.f. of λ, then under the assumption of a first-order PLL and the constraint of (9-22), the cross-correlation function that minimizes the area

$$\int_{\lambda_0}^{1/2} p(\lambda)\, d\lambda \tag{9-23}$$

for all λ_0 between 0 and 1/2 is given by[14,15]

$$g(\lambda) = \operatorname{sgn}[\sin(2\pi\lambda)] \tag{9-24}$$

Since the canonical form of the p.d.f. $p(\lambda)$ for an Nth-order PLL with zero detuning has the same form as that of the first-order loop,[13] the nonlinearity that minimizes the area under the tail of $p(\lambda)$ is also given by Eq. (9-24) for all N. We note here that the loop nonlinearity of (9-24) is physically unrealizable, since it requires a reference signal with infinite average power.

The mean-squared error attainable with a second-order loop and the nonlinearity of (9-24) can be derived from $p(\lambda)$—namely,[13]

$$\begin{aligned} p(\lambda) &= \frac{\exp\left[-2\pi\alpha \int_{-1/2}^{\lambda} g(\eta)\, d\eta\right]}{\int_{-1/2}^{1/2} \exp\left[-2\pi\alpha \int_{-1/2}^{\lambda} g(\eta)\, d\eta\right] d\lambda} \\ &= 2\pi\alpha \frac{\exp[\pi\alpha(1 - 2|\lambda|)]}{2[\exp(\pi\alpha) - 1]} \end{aligned} \tag{9-25}$$

---

* This criteria is suggested by the fact that such an approach forces a large fraction of the total probability to lie in the vicinity of the true value of the signal epoch.

where α is defined in (2-17). Thus, the minimum mean-squared symbol sync error is

$$(\sigma_\lambda^2)_{\min} = \int_{-1/2}^{1/2} \lambda^2 p(\lambda)\, d\lambda$$
$$= \left(\frac{1}{2\pi}\right)^2 \frac{2}{\alpha^2}\left[\frac{1 - \exp(-\pi\alpha)[1 + \pi\alpha + (\pi\alpha)^2/2]}{1 - \exp(-\pi\alpha)}\right] \quad (9\text{-}26)$$

which for high signal-to-noise ratios ($\alpha \cong \rho$ and $\rho \gg 1$) becomes

$$(\sigma_\lambda^2)_{\min} = \left(\frac{1}{2\pi}\right)^2 \frac{2}{\rho^2} \quad (9\text{-}27)$$

For low signal-to-noise ratios ($\rho < 1$), Eq. (9-26) reduces to

$$(\sigma_\lambda^2)_{\min} = \left(\frac{\pi\alpha}{12}\right)\left[\frac{1}{\exp(\pi\alpha) - 1}\right] \quad (9\text{-}28)$$

Minimization of the k*th Absolute Central Moment of the Synchronization Error P.D.F.* The selection of a best periodic nonlinearity $g(\lambda)$ of Fig. 9-11 depends highly on the choice of the criterion used for optimization. In the previous discussion, we chose to minimize the area under the tails of $p(\lambda)$ giving rise to a nonrealizable loop nonlinearity. We now consider selecting the optimum cross-correlation function by minimizing the kth absolute central moment of the unnormalized synchronization error p.d.f. defined by

$$L_k \triangleq \int_{-1/2}^{1/2} |\lambda|^k \exp\left\{-2\pi\alpha \int_{-1/2}^{\lambda} g(\eta)\, d\eta\right\} d\lambda \qquad k = 0, 1, 2, \cdots \quad (9\text{-}29)$$

Such an approach, as we shall see, yields a physically realizable nonlinearity. As before, we consider first the case where both the received and reference signals are unconstrained in waveshape but have unit power and are periodic with period T. It is convenient to introduce the change of variables $t_n = (2\pi/T)t$ and think in terms of 2π periodicity. Hence, $s(t_n)$ and $r(t_n)$ can be represented by Fourier series expansions of the form,

$$s(t_n) = \sum_{k=1}^{\infty} 2s_k \cos(kt_n + \psi_k) \quad \text{and} \quad r(t_n) = \sum_{k=1}^{\infty} 2r_k \cos(kt_n + \phi_k) \quad (9\text{-}30)$$

where

$$2\sum_{k=1}^{\infty} r_k^2 = 2\sum_{k=1}^{\infty} s_k^2 = 1 \quad (9\text{-}31)$$

The problem is solved by finding the cross-correlation function $g(\lambda)$ of the received and reference signal expansions of (9-30), substituting this quantity into (9-29), and using the method of LaGrange multipliers to minimize L_k subject to the constraints of (9-31). Of the four parameters characterizing the kth harmonic of $s(t_n)$ and $r(t_n)$—that is, s_k, ψ_k, r_k, and ϕ_k—only $\delta_k \triangleq \psi_k - \phi_k$ and $s_k = r_k$ are needed to find an optimum solution. The results are summarized as follows:

1. $\delta_k = \pm\pi/2$ is a sufficient (but not necessary) condition for an extremum of L_k to exist.
2. the optimum Fourier coefficients $\{s_k\}$ are the solutions to the set of equations

$$\frac{\partial L_k}{\partial s_n} = s_n\left\{2\lambda_1 - 4\alpha \int_{-\pi}^{\pi} |\phi|^k\left(\frac{1-\cos n\phi}{n}\right) \times \exp\left[-\alpha \sum_{m=1}^{\infty} 2s_m^2\left(\frac{1-\cos m\phi}{m}\right)\right] d\phi\right\} = 0 \tag{9-32}$$

$$\frac{\partial L_k}{\partial (s_n^2)} = \frac{1}{2s_n}\frac{\partial L_k}{\partial s_n} \geq 0 \qquad n = 1, 2, \ldots \tag{9-33}$$

where λ_1 is a LaGrange multiplier to be determined from the unit power constraints. Equation (9-33) is necessary to exclude possible saddlepoints from being accepted as valid solutions to the minimization of L_k. Furthermore, it can be shown[16] that the number of harmonics in $s(t_n)$ and $r(t_n)$ is finite; that is, the functions are strictly bandlimited. The solution of (9-32) and (9-33) for arbitrary k and α requires numerical iteration on a digital computer. A typical cross-correlation function for $k = 2$, $\alpha = 32$, and 11 nonzero signal components is illustrated in Fig. 9-12. Note that this cross-correlation

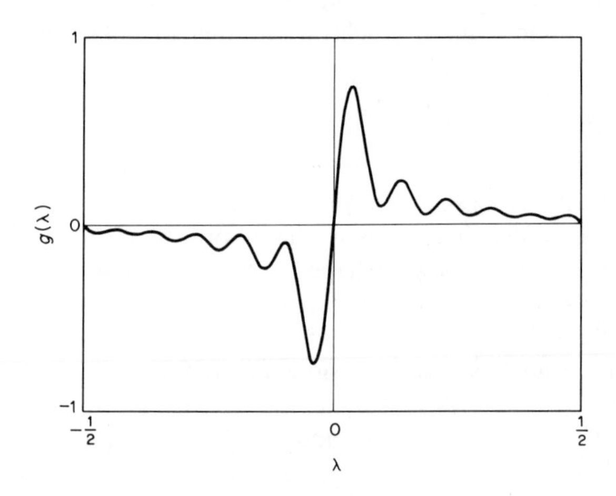

Fig. 9-12 A typical cross-correlation function for $k = 2$, $\alpha = 32$, and 11 nonzero signal components (Courtesy of J. W. Layland)

function is reminiscent of the phase detector characteristic obtained for the delay-locked loop of Chap. 4. The normalized mean-squared symbol sync error $(2\pi)^2\sigma_\lambda^2$ corresponding to the application of the nonlinearity of Fig. 9-12 in a PLL with zero detuning is plotted versus α in Fig. 9-13. For comparison purposes, we indicate by dashed lines the mean-squared symbol sync error

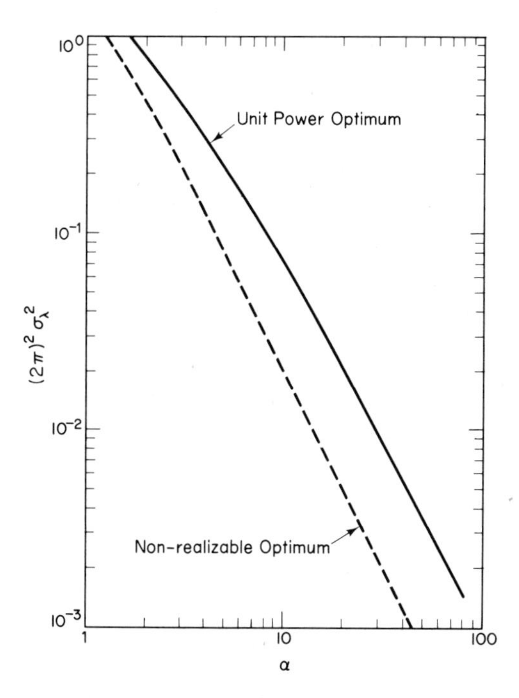

Fig. 9-13 Variance of the normalized symbol sync error vs α for various cross-correlation functions (Courtesy of J. W. Layland)

corresponding to the nonrealizable solution as described by (9-26). For large α, both curves of Fig. 9-13 appear to have the same asymptotic behavior with α; that is, σ_λ varies as $1/\alpha^2$.

A unique analytical solution to (9-32) and (9-33) can be found for the special case of small α. The result is $s_1 = r_1 = 1/\sqrt{2}$, $s_i = r_i = 0$ for $i = 2, 3, \ldots$, which is the PLL solution. Unfortunately, this solution does not guarantee a local or global minimum of the function L_k for larger values of α although it is a solution to (9-32) alone for any α.

Maximization of the Synchronization Error P.D.F. at the Origin for a Unit Power Square-Wave Input Signal. Another optimization problem that makes direct use of the moments L_k defined by (9-29) as a minimization criterion consists of finding the best unit-power reference waveform under the constraint of a unit-power square-wave synchronizer input signal. This problem has perhaps more practical significance than the former in that the modulation waveshape is often predetermined by considerations other than symbol synchronizer performance. Later on, we shall give an example of a particular symbol synchronizer implementation whose design is motivated by this approach.

The problem to be solved then consists of finding the reference signal

$r(t)$ that minimizes L_k when $s(t)$ is defined by $s(t) = 1$ for $0 \leq t \leq T/2$ and $s(t) = -1$ for $T/2 \leq t \leq T$. In particular, we find specific solutions for the case $k = 0$ and different values of α. Minimizing L_0 is equivalent to maximizing the value of the sync error p.d.f. at the origin. Making the transformation of variables from t to t_n and applying the calculus of variations gives an equation for $r(t_n)$ whose solution must again be found iteratively; that is,

$$\begin{aligned} r(t_n) = K_0 \Bigg\{ & \int_{t_0}^{\pi/2} (\phi - t_n) W(\phi) d\phi + \left(\frac{\pi}{2} - t_n\right) W\left(\frac{\pi}{2}\right) \\ & \times \int_0^{\pi/2} \exp\left[-\frac{2\alpha}{\pi} \int_0^{\pi/2} \left[\frac{\pi}{2} - \max\{\xi, \eta\}\right] r(\eta)\, d\eta\right] d\xi \\ & + W\left(\frac{\pi}{2}\right) \int_0^{\pi/2} \left[\frac{\pi}{2} - \max\{\xi, \eta\}\right] \\ & \times \exp\left[-\frac{2\alpha}{\pi} \int_0^{\pi/2} \left(\frac{\pi}{2} - \max\{\xi, \eta\}\right) r(\eta)\, d\eta\right] d\xi \Bigg\} \end{aligned} \tag{9-34}$$

where
$$W(\phi) = \exp\left[-\frac{2\alpha}{\pi} \int_0^{\phi} (\phi - \eta) r(\eta)\, d\eta\right] \tag{9-35}$$

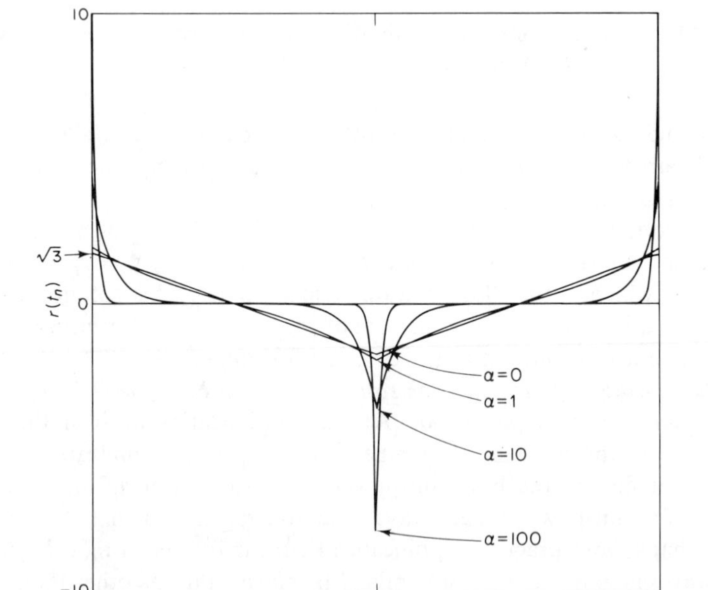

Fig. 9-14 Optimum unit power reference signal for various values of α (Courtesy of J. W. Layland)

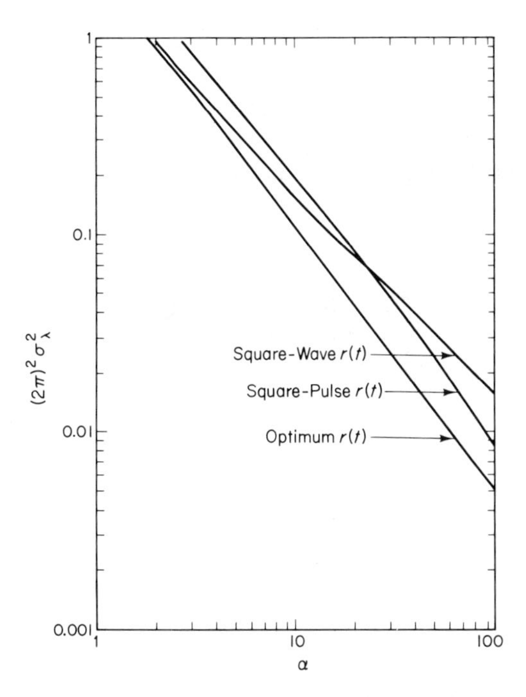

Fig. 9-15 Variance of the normalized symbol sync error vs α for various reference waveforms (Courtesy of J. W. Layland)

and K_0 is a normalization constant chosen to satisfy the unit power constraint. The optimum reference waveform as given by (9-34) is plotted in Fig. 9-14 for values of $\alpha = 0, 1, 10, 100$. We note for $\alpha = 0$ that the optimum reference waveform is triangular in shape whereas for α large, it resembles an alternating train of narrow pulses. It is important to note that these pulses do not in the limit of $\alpha \longrightarrow \infty$ become a train of alternating delta functions since the unit power constraint on $r(t_n)$ is incorporated directly in the calculus of variations approach. The corresponding normalized mean-squared symbol sync error is plotted versus α in Fig. 9-15, which also includes curves representing the behavior of the PLL when the reference signal is: (1) a unit-power square wave, and (2) a unit-power square-pulse approximation to the optimum $r(t_n)$ of (9-34). The width of this square pulse approximation is chosen equal to the half-amplitude width of the optimum reference signal as given by (9-34). The asymptotic performance for large α may be summarized as follows: for the constrained optimum of (9-34) and its square-pulse approximation, the mean-squared symbol sync error varies as $1/\alpha^{3/2}$, whereas for the square-wave reference waveform, σ_λ^2 is proportional to $1/\alpha$. We recall that when $s(t)$ and $r(t)$ are both allowed to vary in waveshape with α, then the asymptotic behavior of σ_λ^2 varies as $1/\alpha^2$.

The Digital Data Transition Tracking Loop (DTTL)*

As an example of the theory presented in this section, we consider the design of a *data transition tracking loop* (DTTL) such as that used as a symbol synchronizer in the high-rate telemetry receiver of the Mariner Mars 1969 mission.[17-21] A block diagram of such a loop is given in Fig. 9-16.

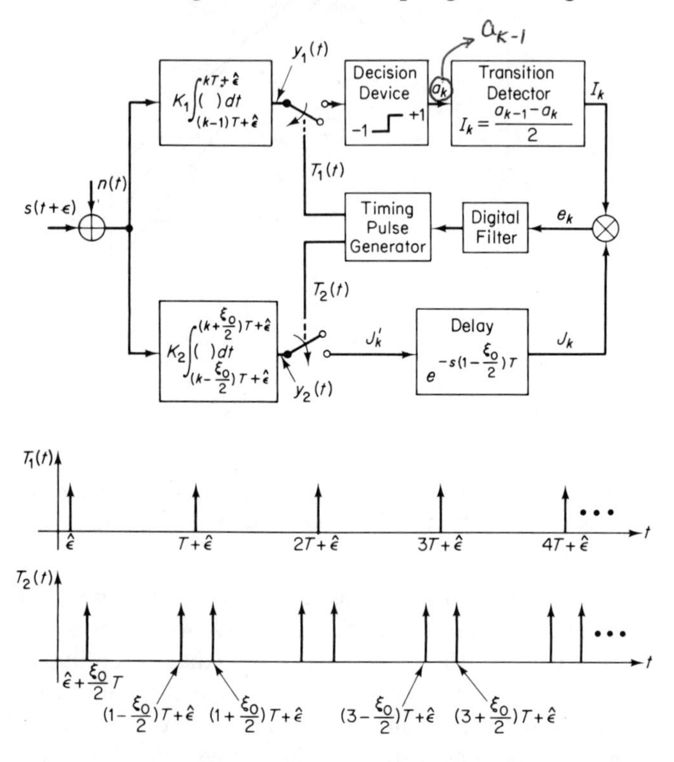

Fig. 9-16 The digital data transition tracking loop (DTTL)

The input noise-free signal $s(t, \epsilon)$ is a random pulse train as characterized by (9-1) together with (9-2), where $p_s(t)$ is now specifically the rectangular pulse defined by $p_s(t) = \sqrt{S}$ for $0 \leq t \leq T$, $p_s(t) = 0$ for all other t. The sum of signal plus noise $y(t)$ is passed through two parallel branches, which are triggered by a timing pulse generator according to a digitally filtered version of an error signal formed from the product of the branch outputs. Furthermore, the two branches are held at a fixed phase relationship with one another by the timing generator. Basically, the in-phase branch monitors the

---

* U.S. Patent No. 3,626,298; "Transition Tracking Bit Synchronization System", T. O. Anderson, W. J. Hurd, W. C. Lindsey, Dec. 7, 1971.

polarity of the actual transitions of the input data and the quadrature (mid-phase) branch obtains a measure of the lack of synchronization. The particular way in which these two pieces of information are derived and combined to synchronize the loop is described below.

The input signal is passed through in-phase and quadrature integrate and dump circuits. The output of the in-phase integrate and dump is sampled at intervals of T and a $\pm$ decision is made corresponding to each input symbol. The decision device is simply a hard limiter with a signum function input versus output characteristic. The transition detector then examines two adjacent decisions a_{k-1}, a_k and records an output I_k according to the following rules:

$$\begin{aligned} &\text{if } a_k = a_{k-1}, \quad \text{then } I_k = 0 \\ &\text{if } a_k = -1,\ a_{k-1} = +1, \quad \text{then } I_k = +1 \\ &\text{if } a_k = +1,\ a_{k-1} = -1, \quad \text{then } I_k = -1 \end{aligned} \tag{9-36}$$

The output J'_k of the quadrature integrate and dump is also sampled at intervals of T and must be delayed before multiplication with the appropriate I_k. As we shall see shortly, an improvement in steady-state mean-squared sync error and mean time to first slip performance can be obtained by integrating in the quadrature branch only over a portion of the symbol interval (e.g., $\xi_0 T$; $0 \leq \xi_0 \leq 1$). Then, for proper loop operation, the delay in the quadrature branch must be chosen equal to $(1 - \xi_0/2)T$. As an example, for full symbol integration in the quadrature branch corresponding to true mid-phase sampling in that branch, the required delay for correct formation of the error signal $e_k \triangleq I_k J_k$ is $T/2$. Actually, for many practical applications, the loop bandwidth-symbol time product is much less than unity; hence, the delay factor $\exp[-sT(1 - \xi_0/2)]$ has a negligible effect on the analysis of the loop operation and will be omitted in what follows. The error signal e_k is digitally filtered with the resulting output being used to control the instantaneous frequency of the timing pulse generator. Thus, an estimate $\hat{\epsilon}$ of the input random epoch is formed. An example of the appropriate waveforms in the loop is illustrated in Fig. 9-17 for a typical input sequence where perfect synchronization $\lambda = 0$, no additive noise, and mid-phase sampling have been assumed. Figures 9-18 and 9-19 illustrate the loop waveforms for $\xi_0 = 1/2$ and out-of-sync conditions $\lambda \leq \xi_0/2$ and $\lambda \geq \xi_0/2$, respectively.

Two other implementations of the symbol synchronizer of Fig. 9-16 have been considered. In one case each branch is split into two subbranches, each having an integrator.[18] The integrators need not be dumped now, since one can be resetting while the other is integrating. The outputs of the two subbranches are then multiplexed to give the resultant output waveform for that branch. In fact, when $\xi_0 \leq 1/2$, one subbranch from each branch can be eliminated, since the waveforms will no longer overlap in time.

In another implementation, the output of the integrator is processed before decision by removing the previously integrated value, a procedure

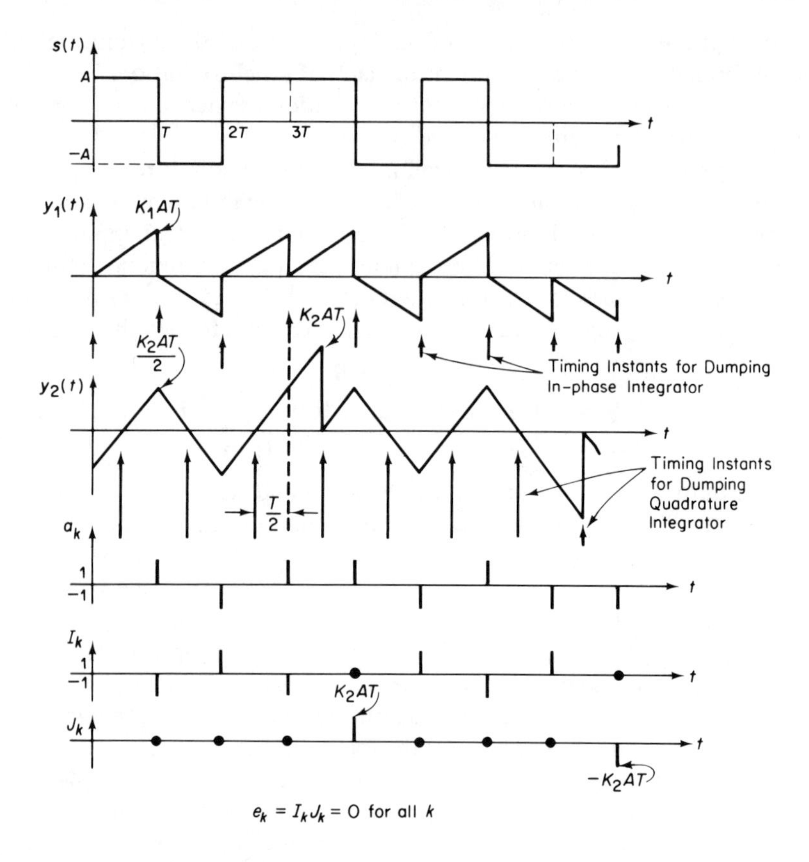

Fig. 9-17 Loop waveforms for a typical input sequence—perfect synchronization; mid-phase sampling ($\xi_0 = 1$) and no noise

equivalent to dumping the integrator after each sampling instant. Its application in the symbol sync loop being considered here is discussed in Ref. 19.

Under the assumptions that the input timing offset ϵ is essentially constant over a large number of symbol intervals and that the loop response is very slow with respect to a symbol interval ($W_L T \ll 1$), the symbol synchronizer of Fig. 9-16 can be modeled as a continuous PLL, see Fig. 9-11. Determining the steady-state performance of this loop relies on finding: (1) the loop cross-correlation function $g(\lambda)$, and (2) the two-sided spectral density $S(\omega, \lambda)$, of the equivalent additive noise $n_\lambda(t)$ at the output of the loop nonlinearity $g(\lambda)$. Based on the assumptions made above, the mean and variance of the error random variable e_k (Fig. 9-16) can be determined assuming λ to be fixed. In effect, then, we consider many records of the discrete random variable e_k at fixed λ and call the average value of this ensemble $g(\lambda)$ and its spectrum $S(\omega, \lambda)$. The actual spectrum $S(\omega)$ of the additive noise is then obtained by averaging $S(\omega, \lambda)$ over the p.d.f. $p(\lambda)$ of λ,

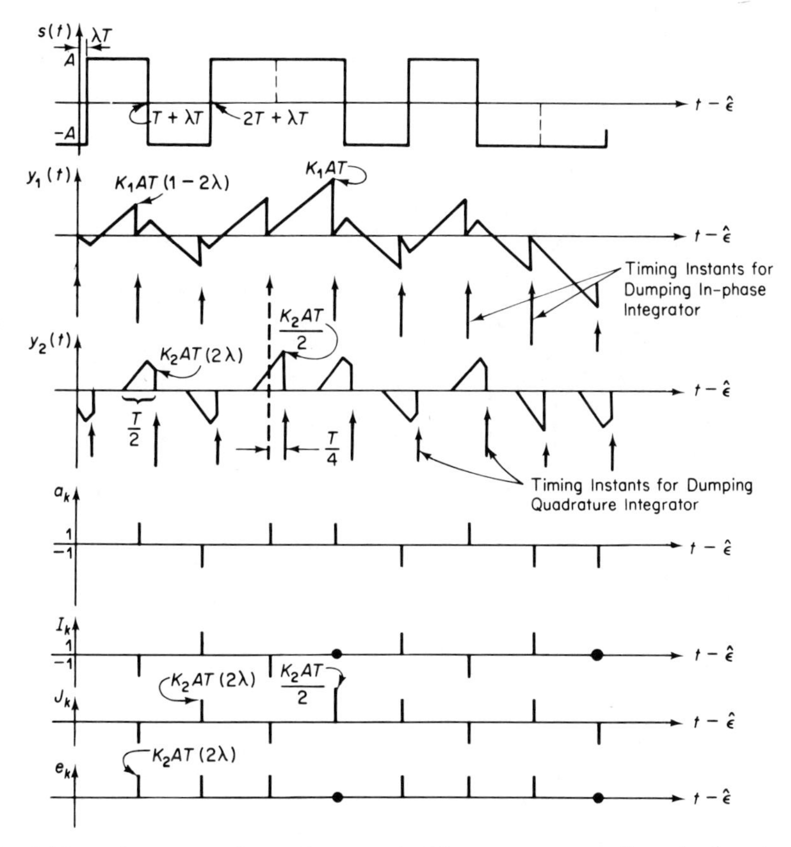

Fig. 9-18 Loop waveforms for a typical input sequence—$\xi_0 = \frac{1}{2}$; $\lambda \leq \frac{1}{4}$; no noise

$$S(\omega) = \int_{-1/2}^{1/2} S(\omega, \lambda) p(\lambda)\, d\lambda \tag{9-37}$$

With reference to Fig. 9-16, the preceding statements can be expressed mathematically:

$$\begin{aligned} g(\lambda) &= E_{n,s}\{e_k \mid \lambda\} \\ S(\omega, \lambda) &= \mathfrak{F}[E_{n,s}\{e_k e_{k+m} \mid \lambda\} - g^2(\lambda)] \\ &= \mathfrak{F}\{R(m, \lambda)\} \end{aligned} \tag{9-38}$$

where $E_{n,s}$ represents the conditional expectation on λ *both* with respect to the noise and the signal (symbol sequence) and the symbol $\mathfrak{F}$ denotes the discrete Fourier transform operation. The autocorrelation function $R(m, \lambda)$ has the following properties:

1. $R(m, \lambda)$ has nonzero value only at $m = 0, \pm 1$. Thus, the spectrum $S(\omega, \lambda)$ consists of the sum of a constant and a sinusoidal component with period $\omega_0 = 2\pi/T$. Since, as before, it is assumed that

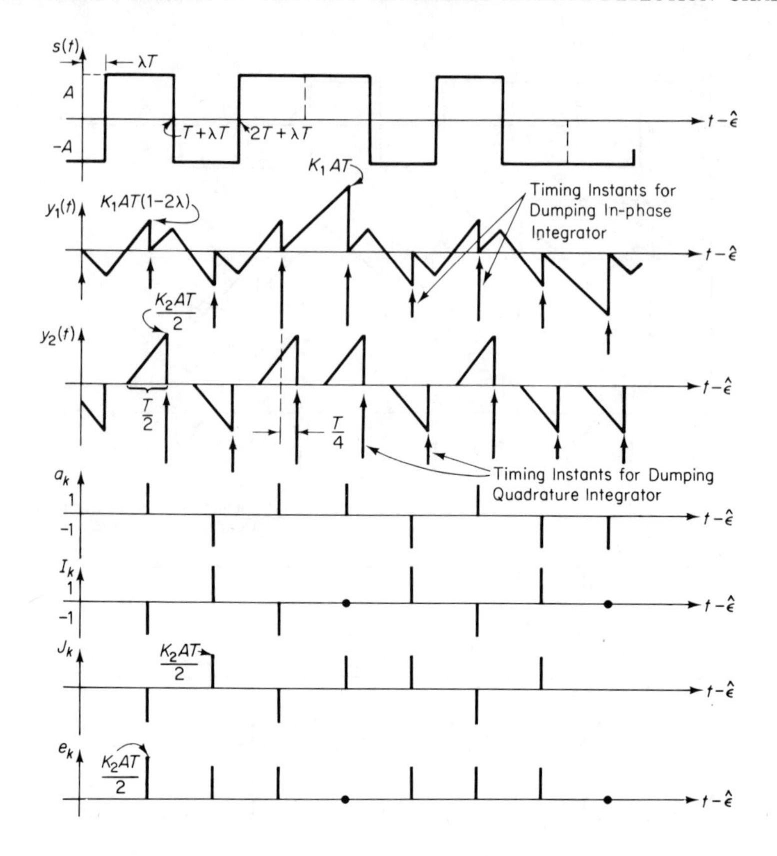

Fig. 9-19 Loop waveforms for a typical input sequence—$\xi_0 = \frac{1}{2}$; $\lambda \geq \frac{1}{4}$; no noise

$W_L T \ll 1$, it is sufficient then to consider only its value at zero frequency—that is, $S(0, \lambda)$—and assume a flat spectrum of this value for all ω of interest.

2. Since $R(-m, \lambda) = R(m, \lambda)$ for all m,

$$h(\lambda) \triangleq \frac{S(0, \lambda)}{(K_2)^2 \xi_0 (N_0 T/4)} = \frac{R(0, \lambda) + 2R(1, \lambda)}{(K_2)^2 \xi_0 (N_0 T/4)} \tag{9-39}$$

From evaluation of $g(\lambda)$ and $h(\lambda)$,[20]

$$\frac{g(\lambda)}{K_2 AT} \triangleq g_n(\lambda)$$

$$= \begin{cases} \lambda \operatorname{erf}[\sqrt{R_d}(1 - 2\lambda)] - \frac{1}{8}[\xi_0 - 2\lambda]\{\operatorname{erf}(\sqrt{R_d}) - \operatorname{erf}[\sqrt{R_d}(1 - 2\lambda)]\} & \text{for } \lambda \leq \xi_0/2 \\ \frac{\xi_0}{2} \operatorname{erf}[\sqrt{R_d}(1 - 2\lambda)] & \text{for } \xi_0/2 \leq \lambda \leq 1/2 \end{cases} \tag{9-40}$$

and

$$
\begin{aligned}
h(\lambda) = 1 &+ \frac{R_d}{2}\left[\xi_0 + \frac{12\lambda^2}{\xi_0}\right] - \frac{3}{8\pi}\xi_0 \exp^2(-R_d) \\
&- \frac{3}{8\pi}\xi_0 \exp^2[-R_d(1-2\lambda)^2] \\
&- \frac{1}{4\pi}\xi_0 \exp(-R_d)\exp[-R_d(1-2\lambda)^2] \\
&+ \frac{\operatorname{erf}^2[\sqrt{R_d}(1-2\lambda)]}{16}\left\{4 - R_d\left[\frac{12\lambda^2}{\xi_0} + 20\lambda + 3\xi_0\right]\right\} \\
&+ \frac{\operatorname{erf}^2(\sqrt{R_d})}{16}\left\{-4 - R_d\left[\frac{12\lambda^2}{\xi_0} - 12\lambda + 3\xi_0\right]\right\} \\
&- \frac{\operatorname{erf}(\sqrt{R_d})\operatorname{erf}[\sqrt{R_d}(1-2\lambda)]}{8}R_d\left\{\frac{4\lambda^2}{\xi_0} - 4\lambda + \xi_0\right\} \\
&+ \frac{(\xi_0 - 2\lambda)}{8}\sqrt{\frac{R_d}{\pi}}\exp(-R_d)\operatorname{erf}(\sqrt{R_d}) \\
&- \frac{3(\xi_0 - 2\lambda)}{8}\sqrt{\frac{R_d}{\pi}}\exp(-R_d)\operatorname{erf}[\sqrt{R_d}(1-2\lambda)] \\
&- \frac{3\xi_0 - 2\lambda(1+2\xi_0)}{8}\sqrt{\frac{R_d}{\pi}}\exp[-R_d(1-2\lambda)^2]\operatorname{erf}[\sqrt{R_d}(1-2\lambda)] \\
&- \frac{3\xi_0 + 10\lambda}{8}\sqrt{\frac{R_d}{\pi}}\exp[-R_d(1-2\lambda)^2]\operatorname{erf}(\sqrt{R_d}) \qquad \text{for } \lambda \le \frac{\xi_0}{2}
\end{aligned}
\tag{9-41}
$$

and

$$
\begin{aligned}
h(\lambda) = 1 &+ 2\xi_0 R_d - \frac{3}{8\pi}\xi_0 \exp^2(-R_d) - \frac{3}{8\pi}\xi_0\exp^2[-R_d(1-2\lambda)^2] \\
&- \frac{1}{4\pi}\xi_0\exp(-R_d)\exp[-R_d(1-2\lambda)^2] \\
&+ \frac{\operatorname{erf}^2[\sqrt{R_d}(1-2\lambda)]}{4}(1 - 4\xi_0 R_s) \\
&- \frac{\operatorname{erf}^2(\sqrt{R_d})}{4} + \frac{\xi_0}{4}\sqrt{\frac{R_d}{\pi}}\exp(-R_d)\operatorname{erf}(\sqrt{R_d}) \\
&- \frac{\xi_0(1-2\lambda)}{4}\sqrt{\frac{R_d}{\pi}}\exp[-R_d(1-2\lambda)^2]\operatorname{erf}[\sqrt{R_d}(1-2\lambda)] \\
&- \xi_0\sqrt{\frac{R_d}{\pi}}\exp[-R_d(1-2\lambda)^2]\operatorname{erf}(\sqrt{R_d}) \qquad \text{for } \frac{\xi_0}{2} \le \lambda \le \frac{1}{2}
\end{aligned}
\tag{9-42}
$$

For large signal-to-noise ratios, the loop will operate in the vicinity of $\lambda = 0$. Hence, $h(\lambda)$ may be approximated by its value at the origin; that is,

$$h(0) = \frac{S(0, 0)}{(K_2)^2 \xi_0 N_0 T/4} = 1 + \frac{\xi_0 R_d}{2} - \frac{\xi_0}{2\pi} \exp^2(-R_d)$$
$$- \frac{\xi_0}{2\pi}\left[\frac{1}{\sqrt{\pi}} \exp(-R_d) + \sqrt{R_d}\,\text{erf}(\sqrt{R_d})\right]^2 \tag{9-43}$$

and

$$\lim_{R_d \to \infty} h(0) = 1 \tag{9-44}$$

Note also that we have normalized $K_1 = 1$ in all the above equations with no loss in generality since the results depend only on the ratio K_2/K_1 (see Fig. 9-16).

Figures 9-20 and 9-21 illustrate the normalized loop nonlinearity $g_n(\lambda)$

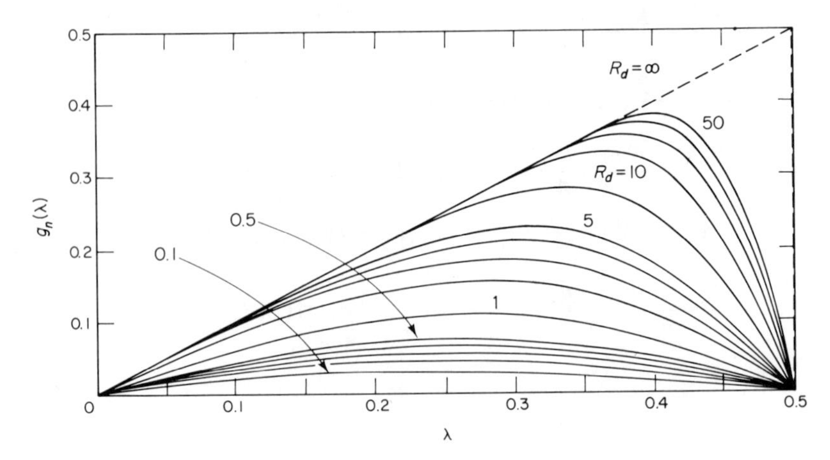

Fig. 9-20 Normalized phase detector characteristic as a function of signal-to-noise ratio—$\xi_0 = 1$

as a function of R_d for $\xi_0 = 1$ and $\xi_0 = 0.5$, respectively. The limiting curve ($R_d \longrightarrow \infty$) is indicated by dashed lines. Similarly, Figs. 9-22 and 9-23 illustrate the normalized spectrum $h(\lambda)$ versus λ for $\xi_0 = 1$ and $\xi_0 = 0.5$, respectively, and at the same values of R_d. An important observation at this point is that both the normalized loop nonlinearity and noise spectrum are highly dependent on signal-to-noise ratio and the "window" of integration in the quadrature branch. Such dependence is brought about by the way in which the error signal e_k depends on the probability of making correct decisions on the data transitions together with the measure of sync error obtained over an interval $\xi_0 T$ in the quadrature branch.

An application of the Fokker-Planck equation enables one to evaluate the steady-state p.d.f. of λ. For a first-order loop filter $F(s) = K_p$, the solu-

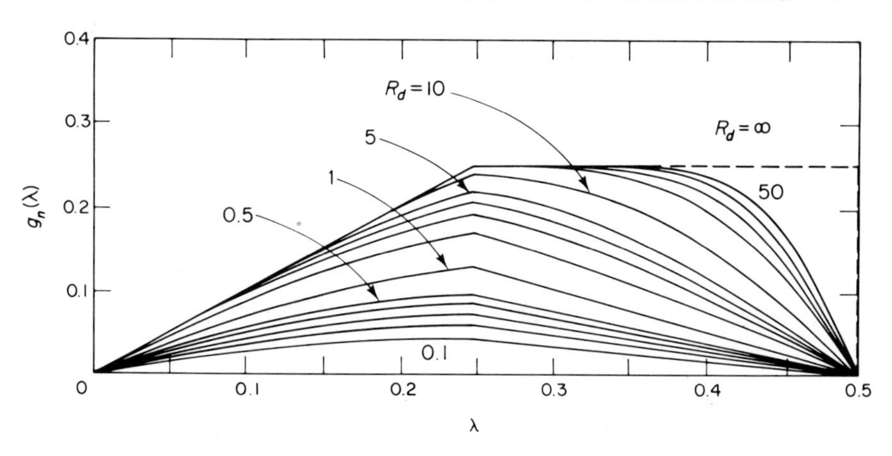

Fig. 9-21 Normalized phase detector characteristic as a function of signal-to-noise ratio—$\xi_0 = .5$

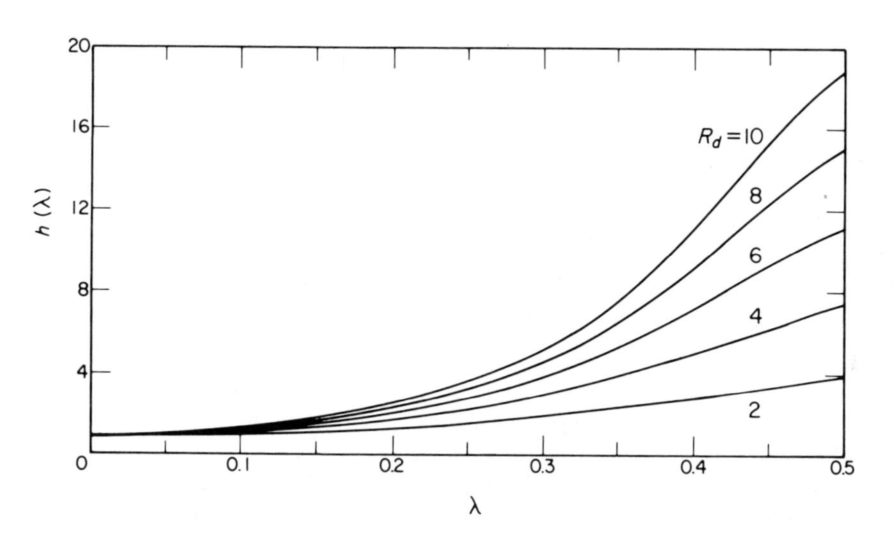

Fig. 9-22 Normalized additive noise spectrum as a function of signal-to-noise ratio—$\xi_0 = 1$

tion is of the form

$$p(\lambda) = C_1 \exp\left\{-\int_0^\lambda \frac{2R_d\delta_{s0}g_n(y) + \xi_0 K_2[dh(y)/dy]}{\xi_0 K_2 h(y)}\,dy\right\} \qquad |\lambda| \leq \tfrac{1}{2} \quad (9\text{-}45)$$

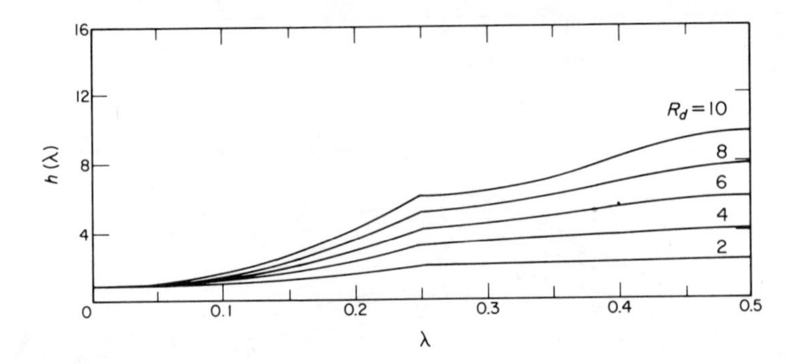

Fig. 9-23 Normalized additive noise spectrum as a function of signal-to-noise ratio—$\xi_0 = .5$

where
$$\delta_{s0} \triangleq \frac{2}{W_{L_{s0}}T} \qquad W_{L_{s0}} = \frac{AK_VK_p}{2} = \frac{W_{L_s}}{K_2} \tag{9-46}$$

and C_1 is chosen so that

$$2\int_0^{1/2} p(\lambda)\,d\lambda = 1 \tag{9-47}$$

In the above, the subscript s denotes behavior in the linear region ($R_d \longrightarrow \infty$) and the subscript 0 denotes normalization with respect to the condition $\xi_0 = 1$. The latter follows from the fact that K_2 is arbitrarily chosen equal to unity when $\xi_0 = 1$. The actual linear loop bandwidth W_L is given in terms of W_{L_s} by

$$W_L = K_g W_{L_s} = \frac{AK_VK_pK_2K_g}{2} \tag{9-48}$$

where K_g is the slope of $g(\lambda)$ at $\lambda = 0$; that is,

$$K_g = \operatorname{erf}(\sqrt{R_d}) - \frac{\xi_0}{2}\sqrt{\frac{R_d}{\pi}}\exp(-R_d) \tag{9-49}$$

Hence, the loop bandwidth W_L is a function of the signal-to-noise ratio R_d. The normalized mean-squared sync error σ_λ^2 is then given by

$$\sigma_\lambda^2 = 2\int_0^{1/2} \lambda^2 p(\lambda)\,d\lambda \tag{9-50}$$

If $S(0, \lambda)$ is approximated by $S(0, 0)$, then (9-45) reduces to

$$p(\lambda) = C_1 \exp\left[\frac{-2R_d\delta_{s0}}{\xi_0 K_2 h(0)}\int_0^{\lambda} g_n(y)\,dy\right] \qquad |\lambda| \leq \tfrac{1}{2} \tag{9-51}$$

Figure 9-24 is a plot (in dashed lines) of σ_λ^2 versus R_d for $\delta_{s0} = 20$ and 100. Included (in solid line) are the corresponding values obtained by optimizing the value of ξ_0 to be discussed shortly. We observe several interesting results, namely:

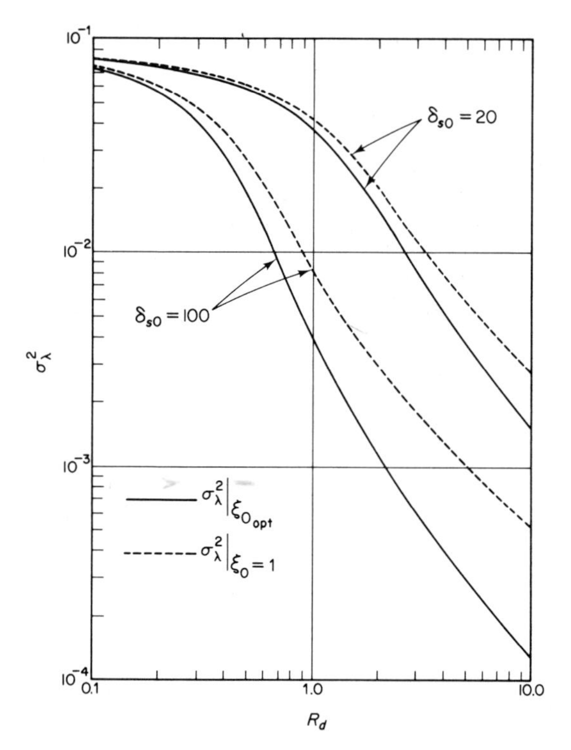

Fig. 9-24 Variance of the normalized symbol sync error versus signal-to-noise ratio for various values of δ_{s0}; $\xi_0 = 1$, and $\xi_0 = \xi_{opt}$

1. For small $R_d\delta_{s0}$, the p.d.f. $p(\lambda)$ becomes uniform in the interval $(-\frac{1}{2}, \frac{1}{2})$, and thus is independent of ξ_0 whether or not the variation of $S(0, \lambda)$ with λ is taken into account. In fact,

$$\lim_{R_d \to 0} \sigma_\lambda^2 = \frac{1}{12} \qquad \text{for all } \xi_0 \tag{9-52}$$

2. For large $R_d\delta_{s0}$, and $\xi_0 = 1$, $p(\lambda)$ has the form

$$p(\lambda) = C_1[1 + 4R_d\lambda^2]^{-(1+\delta_{s0}/4)} \qquad |\lambda| \leq \tfrac{1}{2} \tag{9-53}$$

with zero mean and variance

$$\sigma_\lambda^2 = \frac{1}{2R_d\delta_{s0}}\left[\frac{1}{1 - 2/\delta_{s0}}\right] \tag{9-54}$$

The factor $1 - 2/\delta_{s0}$ represents the increase in mean-squared sync error due to considering the variation of the noise spectrum with λ, which for large δ_{s0} becomes negligible.

As mentioned earlier, it is desired to optimize the performance of the symbol synchronizer by varying the integration interval $\xi_0 T$ in the quadrature branch along with the input signal-to-noise ratio R_d. In terms of the above model, we choose a system constraint whereby the *average* power of the equivalent feedback reference signal that is cross-correlated with the input signal plus noise will be maintained constant as ξ_0 is varied at fixed R_d. In particular, a VCO is proposed (Fig. 9-25) whose output $r(t + \hat{\epsilon})$, the cross-correlating reference signal, is a rectangular pulse train of random amplitude and variable pulse width. The polarity of the pulses is chosen according to the statistics of the in-phase branch output I_k in the data transition

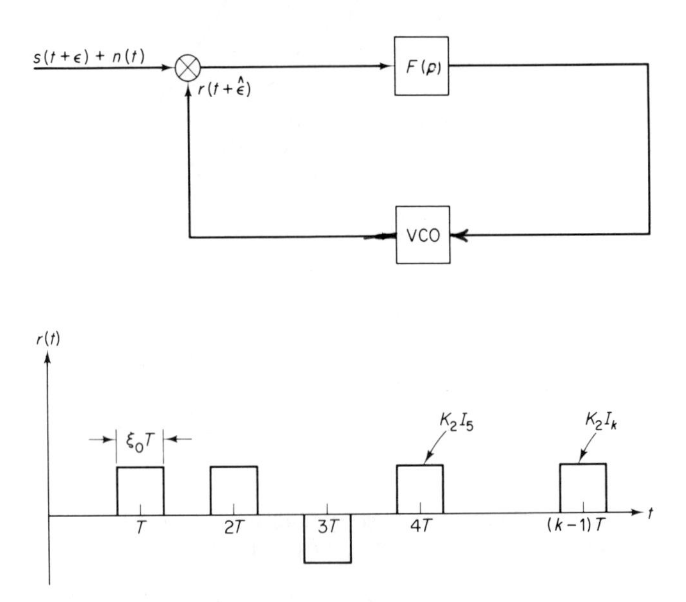

Fig. 9-25 An equivalent model for the DTTL

loop. The width of the pulses is chosen coincident with the integration interval in the quadrature branch and the amplitude K_2 is chosen to keep the average power of $r(t)$ constant at fixed R_d. Cross-correlating $r(t + \hat{\epsilon})$ with the input signal $s(t + \epsilon)$ plus noise $n(t)$ and then low-pass filtering results in the identical loop nonlinearity and additive noise source given in Fig. 9-11 and described by (9-40), (9-41), and (9-42).

Referring now to Fig. 9-25, the above statements can be expressed mathematically by

$$r(t) = \sum_{k=-\infty}^{\infty} K_2 I_k p_k(t, \xi_0) \tag{9-55}$$

where $p_k(t, \xi_0)$ is a unit rectangular pulse defined by

$$p_k(t, \xi_0) = \begin{cases} 1 & \left(k - \frac{\xi_0}{2}\right)T \leq t \leq \left(k + \frac{\xi_0}{2}\right)T \\ 0 & \text{otherwise} \end{cases} \tag{9-56}$$

The average power of $r(t)$ as defined in (9-55) is then

$$P_{AV} \triangleq E_{n,s,\lambda}\left\{\frac{1}{T}\int_0^T r^2(t)\,dt\right\} \tag{9-57}$$

where the symbol $E_{n,s,\lambda}$ denotes averaging over the noise, signal, and sync error p.d.f.'s. Substituting (9-55) and (9-56) in (9-57) gives

$$P_{AV} = K_2^2 \xi_0 E_{n,s,\lambda}\{I_k^2\} \tag{9-58}$$

Averaging over all possible, equally-likely sequences formed from the symbols a_{k-2}, a_{k-1}, and a_k, we obtain

$$E_{n,s}\{I_k^2\} = \tfrac{1}{2} + \tfrac{1}{8}\operatorname{erf}^2[\sqrt{R_d}(1-2\lambda)] - \tfrac{1}{8}\operatorname{erf}^2(\sqrt{R_d}) \tag{9-59}$$

Finally,

$$E_{n,s,\lambda}\{I_k^2\} = 2\int_0^{1/2}\{\tfrac{1}{2} + \tfrac{1}{8}\operatorname{erf}^2[\sqrt{R_d}(1-2\lambda)] - \tfrac{1}{8}\operatorname{erf}^2(\sqrt{R_d})\}p(\lambda)\,d\lambda \tag{9-60}$$

where $p(\lambda)$ is the steady-state p.d.f. of the sync error as given in (9-45). We wish to normalize K_2 such that $K_2 = 1$ when $\xi_0 = 1$ for every R_d. Then,

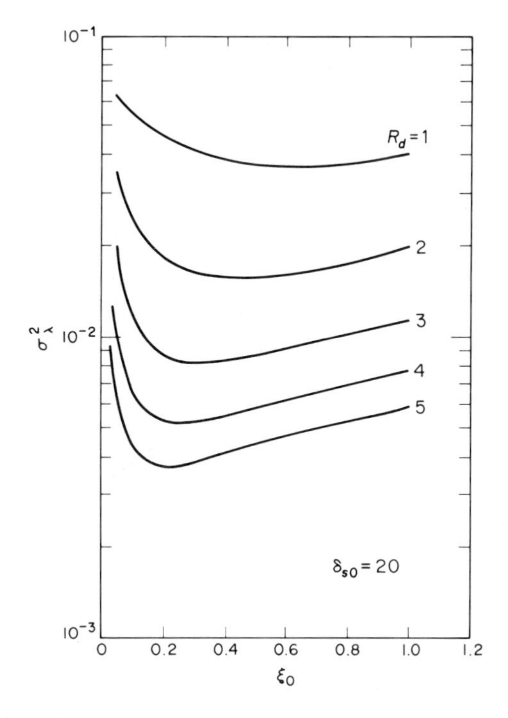

Fig. 9-26 Variance of the normalized symbol sync error versus normalized integration interval for the quadrature branch; $\delta_{s0} = 20$; signal-to-noise ratio is a parameter

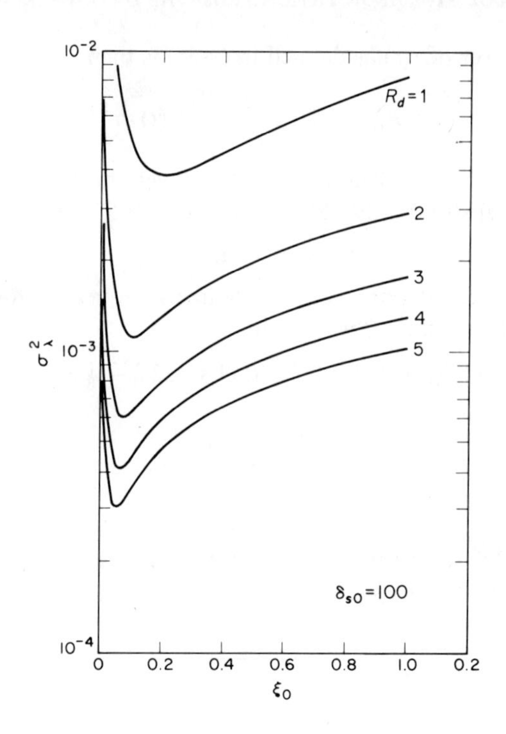

Fig. 9-27 Variance of the normalized symbol sync error versus normalized integration interval in the quadrature branch; $\delta_{s0} = 100$; signal-to-noise ratio is a parameter

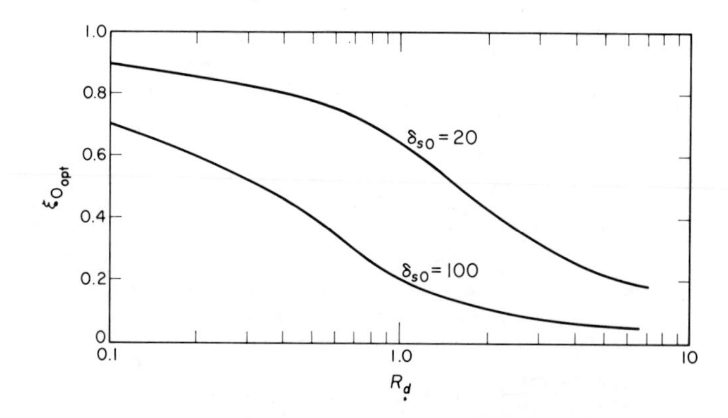

Fig. 9-28 Optimum (minimum σ_λ^2) normalized integration interval for the quadrature branch versus signal-to-noise ratio with δ_{s0} as a parameter

from (9-58) and (9-60) and the fixed average power constraint, K_2 must satisfy,

$$K_2^2\xi_0 \int_0^{1/2} \{\tfrac{1}{2} + \tfrac{1}{8}\operatorname{erf}^2[\sqrt{R_d}(1-2\lambda)] - \tfrac{1}{8}\operatorname{erf}^2(\sqrt{R_d})\}p(\lambda)\,d\lambda$$
$$= \int_0^{1/2} \{\tfrac{1}{2} + \tfrac{1}{8}\operatorname{erf}^2[\sqrt{R_d}(1-2\lambda)] - \tfrac{1}{8}\operatorname{erf}^2(\sqrt{R_d})\}p(\lambda)\Big|_{K_2=\xi_0=1} d\lambda \qquad (9\text{-}61)$$

The above equation can be solved numerically on a digital computer using iterative techniques.

With K_2 chosen as the solution of (9-61) and a given signal-to-noise ratio, the mean-squared symbol sync error can be optimized by choosing the value of ξ_0 that minimizes σ_λ^2 as defined in (9-54). Curves of σ_λ^2 versus ξ_0 with R_d and δ_{s0} as parameters are illustrated in Figs. 9-26 and 9-27. We observe that for each value of R_d, an optimum ξ_0, say $\xi_{0\,\text{opt}}$, exists along with a corresponding value of σ_λ^2. Figure 9-28 illustrates the variation of $\xi_{0\,\text{opt}}$ versus R_d for $\delta_{s0} = 20$ and $\delta_{s0} = 100$; the corresponding values of $\sigma_{\lambda\,\text{opt}}^2$ are plotted in Fig. 9-24.

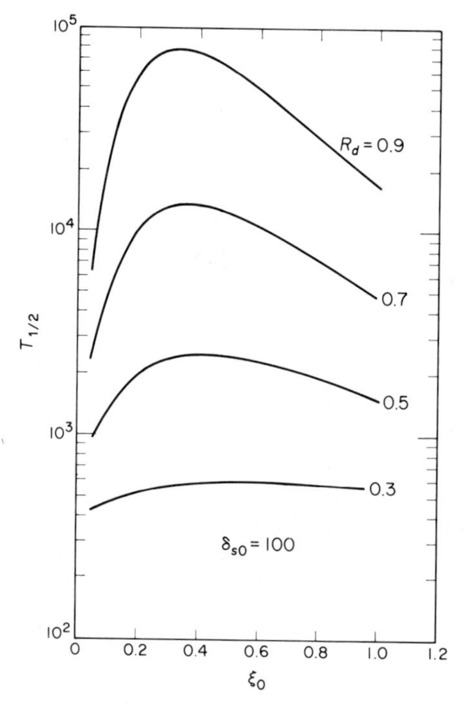

Fig. 9-29 Mean time to first half-cycle slip versus normalized integration interval for the quadrature branch; $\delta_{s0} = 100$; signal-to-noise ratio is a parameter

For large $R_d\delta_{s0}$, $p(\lambda)$ has the form

$$p(\lambda) = C_1\xi_0 K_2\left(1 + \frac{4R_d\lambda^2}{\xi_0}\right)^{-[1+(\delta_{s0}/4K_2)]} \qquad |\lambda| \leq \tfrac{1}{2} \tag{9-62}$$

with zero mean and variance

$$\sigma_\lambda^2 = \frac{\xi_0 K_2}{2R_d\delta_{s0}}\left(\frac{1}{1 - 2K_2/\delta_{s0}}\right) \tag{9-63}$$

where K_2 is again the solution to (9-61). This asymptote when plotted versus R_d at fixed δ_{s0} has a steeper slope for the optimum design than for the case of full symbol integration ($\xi_0 = 1$) in the quadrature integrate and dump. A similar result was observed previously in Fig. 9-15.

It is also possible to optimize (maximize) the mean time to first slip performance of the symbol synchronizer under consideration by again varying ξ_0 at fixed R_d. An expression for the mean time to first slip of half of a cycle (which incidentally is one-half the mean time to first slip of a full cycle) is given by[13,22]

$$\begin{aligned} T_{1/2} &= \frac{1}{\gamma}\int_0^{1/2} p(z)\,dz \int_0^{1/2} \frac{1}{h(x)p(x)}\,dx \\ &= \frac{1}{2\gamma}\int_0^{1/2} \frac{1}{h(x)p(x)}\,dx \end{aligned} \tag{9-64}$$

where $p(x)$ and $h(x)$ are given by (9-45), and (9-41) and (9-42), respectively, and

$$\gamma = \frac{4K_2}{\rho\delta_{s0}} \tag{9-65}$$

with

$$\rho = \frac{2R_d\delta_{s0}}{K_2\xi_0} \tag{9-66}$$

Figure 9-29 plots $T_{1/2}$ versus ξ_0 as a function of R_d. In this figure the normalized time-bandwidth product δ_{s0} is held constant. For each signal-to-noise ratio, there is an optimum (maximum) mean time to first slip. The value of ξ_0 at which the maximum occurs ($\xi_{0\,\mathrm{opt}}$) is a function of the signal-to-noise ratio R_d and δ_{s0}. This functional relationship is illustrated in Fig. 9-30. The optimum mean time to first half slip $T_{1/2\,\mathrm{opt}}$ is plotted against R_d in Fig. 9-31 for $\delta_{s0} = 100$ along with the values of $T_{1/2}|_{\xi_0=1}$, the latter corresponding to the case of no optimization. It is interesting to compare the values of ξ_0 that minimize σ_λ^2 (Fig. 9-28) with those that maximize $T_{1/2}$ (Fig. 9-30). In the actual design of a system, ξ_0 is chosen as some compromise between these two optimum values, depending on the relative importance of σ_λ^2 and $T_{1/2}$. Alternately, one can constrain $T_{1/2}$ to be above a certain threshold value and choose ξ_0 on the basis of σ_λ^2.

Although all of the above results have for simplicity been presented for a first-order loop, extension to the case of higher-order loops is straightforward.[13]

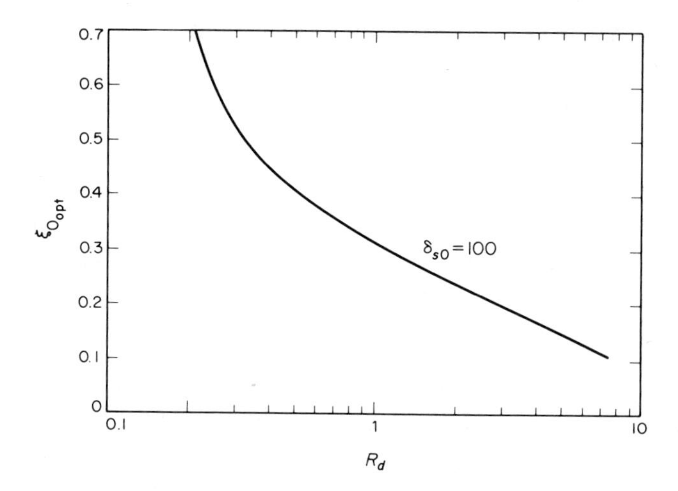

Fig. 9-30 Optimum (maximum $T_{1/2}$) normalized integration interval for the quadrature branch versus signal-to-noise ratio; $\delta_{s0} = 100$

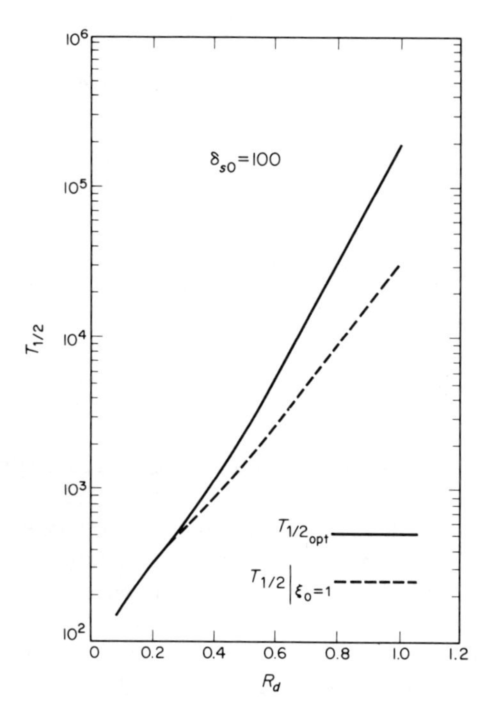

Fig. 9-31 Mean time to first half-cycle slip versus signal-to-noise ratio; $\delta_{s0} = 100$; $\xi_0 = 1$ and $\xi_0 = \xi_{0_{opt}}$

The Early-Late Gate Symbol Synchronizer and a Comparison of Several Synchronizer Configurations

In practice, there are several other classes of symbol synchronizer configurations that exploit the idea of shaping the equivalent loop nonlinearity. One such class contains synchronizers of the *early-late gate* integration type (Fig. 9-32).[23-25] Several possible circuit topologies (Fig. 9-33) have been suggested for implementing the associated phase detector characteristic. Of

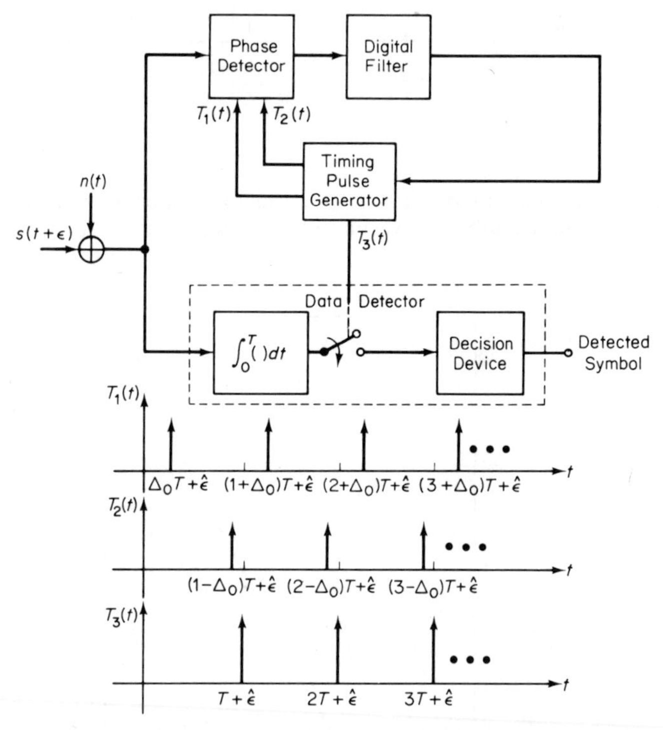

Fig. 9-32 Early-late gate symbol synchronizer and associated data detector

these, the one of most interest is the *absolute value type* (Fig. 9-33a). We shall also for comparison purposes present the results for the phase detector topology of Fig. 9-33b, which synthesizes a difference-of-squares type of loop error signal. The limiter approximation topology of Fig. 9-33c is merely a circuit simplification of Fig. 9-33b that allows one to use a chopper rather than an analog multiplier to form the loop error signal.

Absolute Value Type of Early-Late Gate Symbol Synchronizer (*AVTS*). Figure 9-34 is a functional diagram of the absolute value type symbol synchronizer (AVTS). An error signal e_k is generated by differencing the abso-

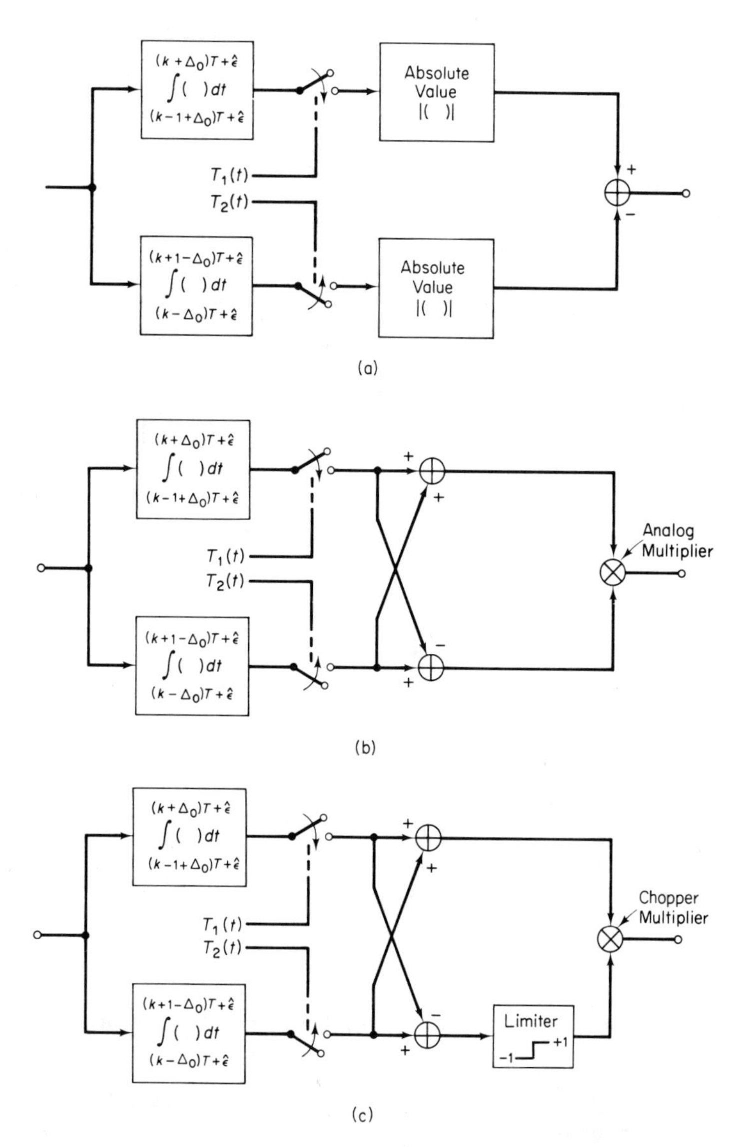

Fig. 9-33 Three different phase detector topologies

lute values of two integrals of the input signal plus noise, each taken over a full symbol interval T, starting $\Delta_0 T$ early and $\Delta_0 T$ late, respectively.* In the absence of noise, the ideal operation of the loop is as follows: For a given phase offset λT between the actual transition times $\{t_k\}$ and their local estimates

* We note that $\Delta_0 T$ is equivalent to $\Delta\epsilon/2$ of Fig. 9-10.

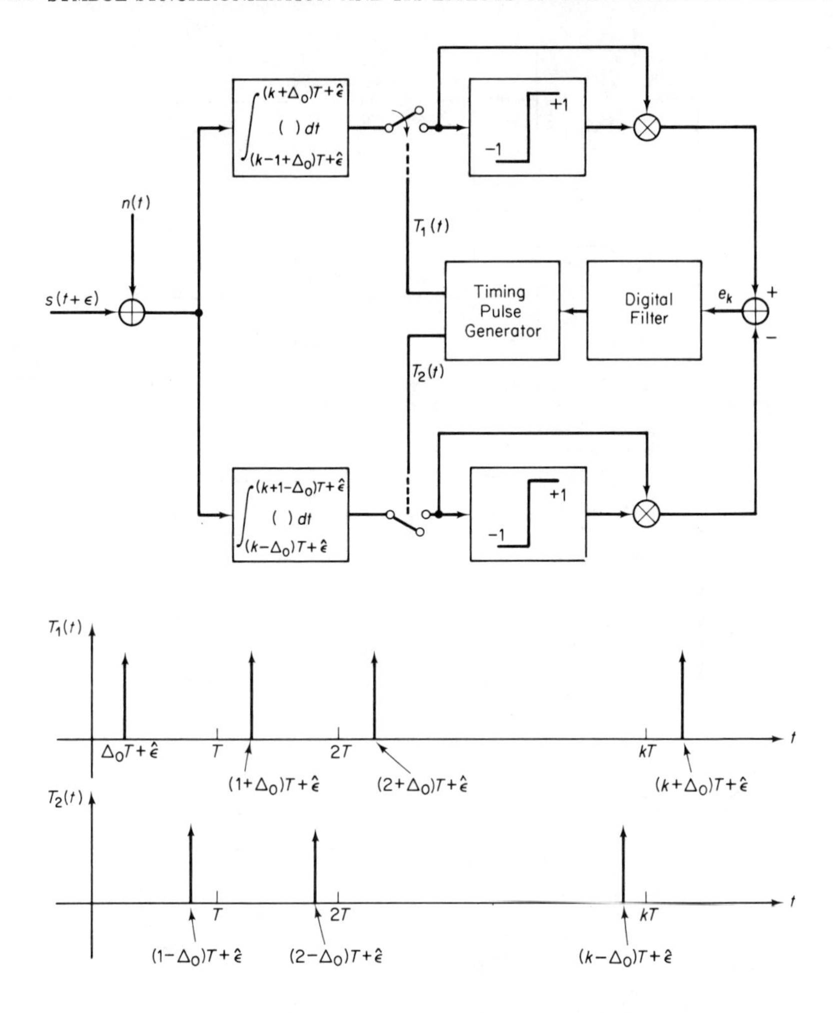

Fig. 9-34 An absolute value type of early-late gate symbol synchronizer (AVTS)

$\{\hat{t}_k\}$, the error signal e_k is: (1) zero when no transition occurs at t_k, and (2) linearly proportional to λ when a transition occurs at t_k independent of its polarity. The polarity independence in (2) is a direct consequence of taking absolute values before differencing. The filtered error signal is used to drive a timing pulse generator, which controls the charging and discharging instants of the matched filters. If the closed loop is designed to be narrowband relative to the symbol rate, then as before one can talk about an equivalent loop nonlinearity, $g(\lambda)$, and an equivalent additive noise source with spectrum $S(\omega, \lambda)$. Hence, the PLL model of Fig. 9-11 is once again applicable using the new definitions of $g(\lambda)$ and $S(0, \lambda)$ as given by

$$\frac{g(\lambda)}{2AT} = g_n(\lambda) \triangleq \begin{cases} \dfrac{\alpha_0}{4}\operatorname{erf}(\alpha_0\sqrt{R_d}) - \dfrac{\beta_0}{4}\operatorname{erf}(\beta_0\sqrt{R_d}) \\ \quad + \dfrac{1}{4\sqrt{R_d\pi}}[\exp(-\alpha_0^2 R_d) - \exp(-\beta_0^2 R_d)] \\ \qquad\qquad\qquad \text{for } \lambda \leq \Delta_0 \\ \dfrac{\alpha_1}{4}\operatorname{erf}(\alpha_1\sqrt{R_d}) - \dfrac{\beta_1}{4}\operatorname{erf}(\beta_1\sqrt{R_d}) \\ \quad + \dfrac{1}{4\sqrt{R_d\pi}}[\exp(-\alpha_1^2 R_d) - \exp(-\beta_1^2 R_d)] \\ \qquad\qquad\qquad \text{for } \Delta_0 \leq \lambda \leq \frac{1}{2} \end{cases} \tag{9-67}$$

and

$$\begin{aligned} h(\lambda) \triangleq \frac{S(0, \lambda)}{N_0 T} = 1 &+ R_d\left[1 + \frac{\alpha_0^2 + \beta_0^2}{2}\right] - f(1 - \Gamma_0, \sqrt{2R_d}, \sqrt{2R_d}) \\ &- f(1 - \Gamma_0, \alpha_0\sqrt{2R_d}, -\beta_0\sqrt{2R_d}) \\ &+ \frac{1}{2}\Bigg\{\Bigg[\sqrt{R_d}\operatorname{erf}(\sqrt{R_d}) + \alpha_0\sqrt{R_d}\operatorname{erf}(\alpha_0\sqrt{R_d}) \\ &+ \frac{1}{\sqrt{\pi}}\exp(-R_d) + \frac{1}{\sqrt{\pi}}\exp(-\alpha_0^2 R_d)\Bigg] \\ &\times \Bigg[\sqrt{R_d}\operatorname{erf}(\sqrt{R_d}) + \beta_0\sqrt{R_d}\operatorname{erf}(\beta_0\sqrt{R_d}) \\ &+ \frac{1}{\sqrt{\pi}}\exp(-R_d) + \frac{1}{\sqrt{\pi}}\exp(-\beta_0^2 R_d)\Bigg]\Bigg\} \\ &+ \frac{1}{2}\Bigg[\alpha_0\sqrt{R_d}\operatorname{erf}(\alpha_0\sqrt{R_d}) - \beta_0\sqrt{R_d}\operatorname{erf}(\beta_0\sqrt{R_d}) \\ &+ \frac{1}{\sqrt{\pi}}\exp(-\alpha_0^2 R_d) - \frac{1}{\sqrt{\pi}}\exp(-\beta_0^2 R_d)\Bigg]^2 \\ &- \frac{1}{2}[f(\Gamma_0, \sqrt{2R_d}, \sqrt{2R_d}) + f(\Gamma_0, \alpha_0\sqrt{2R_d}, \beta_0\sqrt{2R_d}) \\ &+ f(\Gamma_0, \sqrt{2R_d}, \alpha_0\sqrt{2R_d}) + f(\Gamma_0, \sqrt{2R_d}, \beta_0\sqrt{2R_d})] \\ &\qquad\qquad \text{for } 0 \leq \lambda \leq \Delta_0 \end{aligned} \tag{9-68}$$

and

$$\begin{aligned} h(\lambda) = 1 &+ R_d\left[1 + \frac{\alpha_1^2 + \beta_1^2}{2}\right] - \frac{1}{2}[f(1 - \Gamma_0, \sqrt{2R_d}, \sqrt{2R_d}) \\ &+ f(1 - \Gamma_0, \sqrt{R_d}, -\beta_1\sqrt{2R_d}) + f(1 - \Gamma_0, \sqrt{2R_d}, \alpha_1\sqrt{2R_d}) \\ &+ f(1 - \Gamma_0, \alpha_1\sqrt{2R_d}, -\beta_1\sqrt{2R_d})] \\ &+ \frac{1}{2}\Bigg\{\Bigg[\sqrt{R_d}\operatorname{erf}(\sqrt{R_d}) + \alpha_1\sqrt{R_d}\operatorname{erf}(\alpha_1\sqrt{R_d}) + \frac{1}{\sqrt{\pi}}\exp(-R_d) \\ &+ \frac{1}{\sqrt{\pi}}\exp(-\alpha_1^2 R_d)\Bigg]\Bigg[\sqrt{R_d}\operatorname{erf}(\sqrt{R_d}) + \beta_1\sqrt{R_d}\operatorname{erf}(\beta_1\sqrt{R_d}) \end{aligned}$$

$$+\frac{1}{\sqrt{\pi}}\exp(-R_d)+\frac{1}{\sqrt{\pi}}\exp(-\beta_1^2R_d)\Big]\Big\}$$
$$+\frac{1}{2}\Big[\alpha_1\sqrt{R_d}\,\mathrm{erf}(\alpha_1\sqrt{R_d})-\beta_1\sqrt{R_d}\,\mathrm{erf}(\beta_1\sqrt{R_d})$$
$$+\frac{1}{\sqrt{\pi}}\exp(-\alpha_1^2R_d)-\frac{1}{\sqrt{\pi}}\exp(-\beta_1^2R_d)\Big]$$
$$-f(\Gamma_0\sqrt{2R_d},\sqrt{2R_d})-f(\Gamma_0,\alpha_1\sqrt{2R_d},\beta_1\sqrt{2R_d})$$
$$\text{for } \Delta_0\le\lambda\le\tfrac{1}{2} \tag{9-69}$$

where
$$\alpha_0=1-2(\Delta_0-\lambda)$$
$$\beta_0=\beta_1=1-2(\Delta_0+\lambda)$$
$$\alpha_1=1-2(\lambda-\Delta_0)$$
$$\Gamma_0=1-2\Delta_0 \tag{9-70}$$

and $f(\xi, b, a)$ is defined by

$$f(\xi,b,a)=\frac{1}{\sqrt{2\pi}}\int_0^\infty r(\xi,b,a,x_n)\,dx_n \tag{9-71}$$

with

$$r(\xi,b,a,x_n)=\sqrt{1-\xi^2}x_n\exp\Big[-\Big(\frac{x_\pi^2+a^2}{2}\Big)\Big]$$
$$\times\Big(\frac{2}{\sqrt{\pi}}\exp\Big\{-\Big[\frac{(\xi x_n)^2+(b-\xi a)^2}{2(1-\xi^2)}\Big]\Big\}\cosh\Big[ax_n-\frac{\xi(b-\xi a)x_n}{1-\xi^2}\Big]$$
$$+\Big[\frac{\xi x_n+(b-\xi a)}{\sqrt{1-\xi^2}}\Big]\frac{e^{ax_n}}{2}\mathrm{erf}\Big[\frac{\xi x_n+(b-\xi a)}{\sqrt{2(1-\xi^2)}}\Big]$$
$$+\Big[\frac{\xi x_n-(b-\xi a)}{\sqrt{1-\xi^2}}\Big]\frac{e^{-ax_n}}{2}\mathrm{erf}\Big[\frac{\xi x_n-(b-\xi a)}{\sqrt{2(1-\xi^2)}}\Big]\Big) \tag{9-72}$$

Figure 9-35 illustrates the normalized loop nonlinearity $g_n(\lambda)$, defined above, as a function of R_d for $\Delta_0 = 1/4$. The limiting curve ($R_d \to \infty$) is also indicated by dashed lines, where the value of Δ_0 has been chosen to make this curve triangular shaped and hence produce the broadest linear region of operation. One could again conceive of optimizing this absolute value type of symbol synchronizer by varying Δ_0 at fixed R_d and an imposed average power constraint.

The steady-state performance of the AVTS can again be obtained by application of the Fokker-Planck equation. The p.d.f. of λ is given by an equation quite similar to (9-45); that is,

$$p(\lambda)=C_1\frac{1}{h(\lambda)}\exp\Big[-2R_d\delta_s\int_0^\lambda\frac{g_n(y)}{h(y)}\,dy\Big] \tag{9-73}$$

where C_1 is a normalization constant,

$$\delta_s\triangleq\frac{2}{W_{L_s}T}\qquad W_{L_s}=\frac{2AK_VK_p}{2}=\frac{W_L}{K_g} \tag{9-74}$$

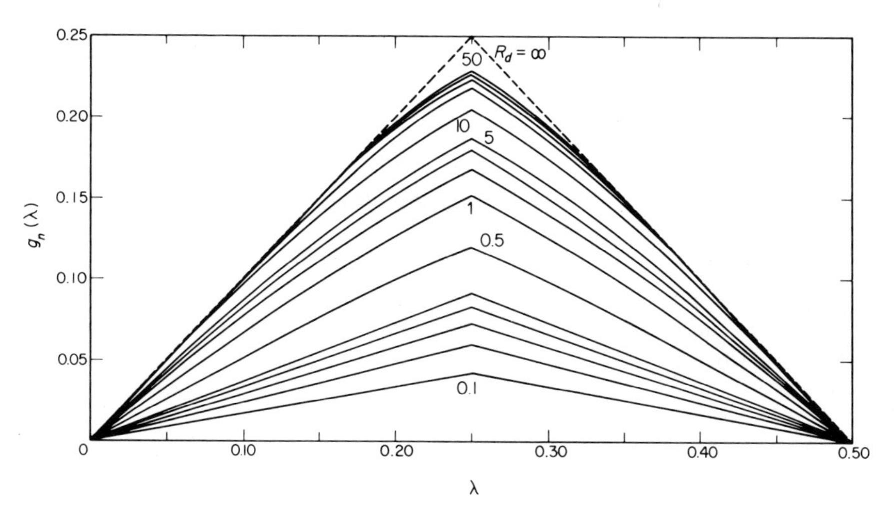

Fig. 9-35 Normalized phase detector characteristic as a function of signal-to-noise ratio; $\Delta_0 = \frac{1}{4}$

and K_g is the slope of $g_n(\lambda)$ at $\lambda = 0$:

$$K_g = \text{erf}\left(\frac{\sqrt{R_d}}{2}\right) \tag{9-75}$$

The subscript s has the same significance as before.

Difference of Squares Loop (DSL). For the difference of squares loop (DSL), which employs the phase detector topology of Figure 9-33b, the p.d.f. of the normalized sync error may be obtained by an extension of the work of Layland.[23] Briefly,

$$p(\lambda) = C_1 \exp\left[\frac{-4R_d^2\delta_s \int_0^\lambda g_n(y)\,dy}{1 + \frac{5}{4}R_d}\right] \tag{9-76}$$

with C_1 chosen to satisfy (9-47) and σ_λ^2 defined by (9-50). For this case,

$$\delta_s \triangleq \frac{2}{W_L T} \qquad W_L = \frac{A^2 T K_V K_p}{2} \tag{9-77}$$

$$\int_0^\lambda g_n(y)\,dy = \begin{cases} \dfrac{\lambda^2}{2} & \text{for } \lambda \leq \frac{1}{4} \\[2ex] \dfrac{1}{16} - \dfrac{[\frac{1}{2} - \lambda]^2}{2} & \text{for } \frac{1}{4} \leq \lambda \leq \frac{1}{2} \end{cases} \tag{9-78}$$

An important point to note for this case is that the normalized nonlinearity $g_n(\lambda)$ does not degrade with signal-to-noise ratio. Hence, the bandwidth W_L is not a function of R_d.

A Performance Comparison of Several Symbol Synchronizers. In comparing the performance of several different symbol synchronizer configurations, one must choose a fixed operating condition that is common to all. One possible basis of comparison is to require equal loop bandwidths for all configurations at every R_d. This comparison is meaningful when considering the actual design of a loop having a specified bandwidth operating over a given range of input signal-to-noise ratios.

Figure 9-36 plots σ_λ^2 versus R_d for $\delta_s = 20$ and $\delta_s = 100$. Included in this plot are the corresponding curves for the DTTL with full integration in the quadrature branch and the DSL. The asymptotic behaviors of σ_λ^2 for large R_d are described by

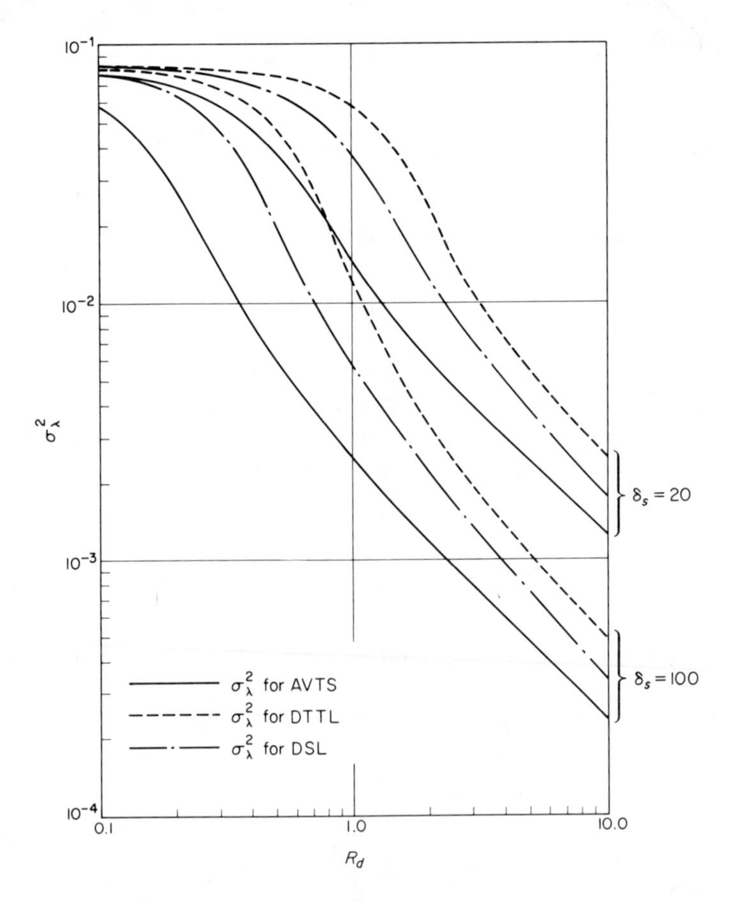

Fig. 9-36 A comparison of the performance of three symbol synchronizer configurations in terms of the variance of the normalized symbol sync error versus signal-to-noise ratio; $\delta_s = 20$, and $\delta_s = 100$

$$\begin{aligned} \text{AVTS:} \quad \sigma_\lambda^2 &= \frac{1}{4R_d\delta_s} \\ \text{DTTL:} \quad \sigma_\lambda^2 &= \frac{1}{2R_d\delta_s} \\ \text{DSL:} \quad \sigma_\lambda^2 &= \frac{5}{16R_d\delta_s} \end{aligned} \tag{9-79}$$

Hence, in the linear region, the AVTS synchronizer offers a 3-dB advantage over the DTTL and a 0.97-dB advantage over the DSL. Furthermore, the more, the AVTS appears to be uniformly better than the other two configurations at all values of signal-to-noise ratio. In the limit as $R_d \longrightarrow 0$, all three configurations give the variance of a uniform distribution, that is, $\sigma_\lambda^2 = 1/12$.

A comparison of performance on the basis of mean time to first slip can also been made for the three symbol synchronizer topologies. As one might expect, the relative rankings of the three configurations remains the same as in the case where σ_λ^2 is compared at each value of R_d and δ_s.

9-3. SYMBOL SYNCHRONIZATION OVER A SEPARATE CHANNEL

The methods used for providing symbol sync over a separate channel are a direct application of the ranging techniques discussed in Chap. 4. Let us start by considering how the double-loop tracking device could be used in an application where only symbol sync is required. As indicated in Chap. 4, the input to the double-loop device is the PN code signal $s_{\text{PN}}(t)$ multiplied by the clock signal $\text{Ck}_2(t)$ at a frequency $f_{Ck} = 1/t_0$ where t_0 is the code digit period. This composite signal will now be used as the synchronizing signal and referred to as the *sync subcarrier*. Then for a data rate of $\mathcal{R}_s$ symbols/sec, the PN code length (in digits) p and the subcarrier (clock) rate could be chosen such that

$$\mathcal{R}_s p = f_{\text{Ck}} \tag{9-80}$$

This implies that the length (in sec) of a PN code word is identical to the duration of a single data symbol, and hence a single state detector on the PN code generator is sufficient to provide a source of symbol sync pulses.

Alternately, to conserve transmission bandwidth (but unfortunately increase the PN acquisition time), the parameters $\mathcal{R}_s$, p, and $2f_{\text{Ck}}$ can be related by

$$\mathcal{R}_s p = f_{\text{Ck}} n \tag{9-81}$$

where $n = \log_2 (p + 1)$ is the number of shift register stages required to generate the code. Now, the length of one PN code word is equivalent to the duration of n data symbols, and thus n state detectors are required on the PN code generator. This same principle can be used to achieve simultaneous symbol and word sync in applications where this is necessary. Letting $\mathcal{R}_w$ denote the data word rate, the required relationship is

$$\mathcal{R}_s p = f_{\mathrm{Ck}} \frac{\mathcal{R}_s}{\mathcal{R}_w} \tag{9-82}$$

where the ratio $\mathcal{R}_s/\mathcal{R}_w$ is analogous to n above and represents the number of data symbols per word transmitted.

The question remains as to where to locate the data spectrum (the data-modulated subcarrier) relative to that of the transmitted PN synchronization signal. The envelope of the power spectrum of this signal in w/Hz is given by[26]

$$S(f) = \left(\frac{2}{f_{\mathrm{Ck}}}\right)\left[\frac{\sin^4(\pi f/2f_{\mathrm{Ck}})}{(\pi f/2f_{\mathrm{Ck}})^2}\right] \tag{9-83}$$

where the total power has been normalized to unity. This spectrum exhibits a broad null in the vicinity of the frequency $2f_{\mathrm{Ck}}$, which therefore presents an excellent choice for the location of the data signal spectrum. Thus, the data $d(t)$ are biphase modulated onto a square-wave data subcarrier $\mathbb{S}(t)$ at frequency $2f_{\mathrm{Ck}}$, and the composite signal at the output of the RF transmitter (assuming for simplicity a two-channel system) is of the form

$$x(t) = \sqrt{2P}\sin\left[\omega_0 t + (\cos^{-1} m_d)\, d(t)\mathbb{S}(t) + (\cos^{-1} m_s) s_{\mathrm{PN}}(t)\mathrm{Ck}_2(t)\right] \tag{9-84}$$

The design of the modulation factors m_d and m_s for precisely this system has been discussed in Chap. 7.

9-4. ERROR PROBABILITY PERFORMANCE

No discussion of symbol synchronization would be complete without considering the effect of symbol sync error on system error probability performance. Since the optimum detector for known signals is a cross-correlator, we are interested primarily in examining the degradation in performance of this device when the symbol synchronization reference is noisy. The first step in the presentation will be to derive the error probability of the correlation detector conditioned on a symbol sync error. This conditional error probability can then be averaged over the p.d.f. $p(\lambda)$ of the sync error (which depends on the symbol synchronizer configuration used) to yield the average error probability of the receiver. When compared with a source of perfect symbol sync, the degradation in performance due to symbol sync error can be assessed. The section concludes with a determination of the error probability performance degradation in the presence of both noisy symbol and carrier references.

Conditional Error Probability for a Fixed Symbol Sync Error

Consider the situation where the symbol sync is in error by $\epsilon = \lambda T$ sec. The cross-correlator statistic on which a decision is made for the nth transmitted symbol now depends on the received data in the time interval $(n-1)T + \hat{\epsilon} \le t \le nT + \hat{\epsilon}$ where $\hat{\epsilon}$ is the sync estimate obtained from the symbol synchronizer. Stated mathematically,

$$\hat{a}_n = \text{sgn}\left\{\int_{(n-1)T+\hat{\epsilon}}^{nT+\hat{\epsilon}} y(t)p_s[t-(n-1)T-\hat{\epsilon}]\,dt\right\} \tag{9-85}$$

where $\hat{a}_n$ is the nth detected symbol and $y(t)$, the correlation detector input, is defined by (9-1) together with (9-2). Making the change of variables $t' = t-(n-1)T-\hat{\epsilon}$ in (9-85), and separating the signal and noise correlation components, we get

$$\hat{a}_n = \text{sgn}\, Q$$

where

$$\begin{aligned} Q \triangleq \int_0^T & p_s(t')[a_{n-1}p_s(t'+T-\lambda T) + a_n p_s(t'-\lambda T) \\ & + a_{n+1}p_s(t'-T-\lambda T]\,dt' \\ & + \int_0^T p_s(t')n[t'+(n-1)T+\hat{\epsilon}]\,dt' \end{aligned} \tag{9-86}$$

We note that in general three input symbols are involved in estimating the nth transmitted symbol. Actually, if λ is known to be positive (or negative), then only two symbols are involved, namely the nth and one of its adjacent symbols. Denoting the signal and noise components of Q by Q_s and Q_n, respectively, then

$$Q_s = a_{n-1}R_s[-T(1-\lambda)] + a_nR_s(\lambda T) + a_{n+1}R_s[T(1+\lambda)] \tag{9-87}$$

where $R_s(\tau)$ is the signal pulse autocorrelation function defined by

$$\begin{aligned} R_s(\tau) &= \int_\tau^T p_s(t')p_s(t'-\tau)\,dt' \\ &= \int_0^{T-\tau} p_s(t'+\tau)p_s(t')\,dt' \end{aligned} \tag{9-88}$$

With regard to the input symbols affecting the calculation of Q_s, one of four situations can occur: either the nth symbol and its adjacent symbol are alike (both negative or both positive) or they are unlike (negative followed by positive, or vice versa). Hence, the overall error probability conditioned on knowing λ is given by

$$\begin{aligned} P_E(\lambda) = \sum_{k=1}^{2} & [\text{Prob}\{n\text{th and adjacent symbols} = (-1)^k\} \\ & \times \text{Prob}\{\text{error}\,|\,n\text{th and adjacent symbols} = (-1)^k\} \\ & + \text{Prob}\{n\text{th symbol} = (-1)^k \text{ and adjacent symbol} = (-1)^{k+1}\} \\ & \times \text{Prob}\{\text{error}\,|\,n\text{th and adjacent symbol} = (-1)^{k+1}\}] \end{aligned} \tag{9-89}$$

Since the random variable Q is conditionally Gaussian on a_{n-1}, a_n, a_{n+1} and fixed λ, with variance

$$\sigma_Q^2 = E\{Q_N^2\} = \frac{N_0 ST}{2} \tag{9-90}$$

and the input symbols occur with equal probability and are independent, then

$$P_E(\lambda) = \tfrac{1}{4}\,\text{erfc}\,\{\sqrt{R_d}\,[r_s(|\lambda|T) + r_s(T - |\lambda|T)]\} + \tfrac{1}{4}\,\text{erfc}\,\{\sqrt{R_d}\,[r_s(|\lambda|T) - r_s(T - |\lambda|T)]\} \tag{9-91}$$

where

$$r_s(\tau) \triangleq \frac{R_s(\tau)}{ST} \tag{9-92}$$

For NRZ data—that is, $p_s(t) = \sqrt{S};\ 0 \leq t \leq T$,

$$r_s(\tau) = \begin{cases} 1 - \dfrac{|\tau|}{T} & |\tau| \leq T \\ 0 & |\tau| \geq T \end{cases} \tag{9-93}$$

Substituting (9-93) in (9-91), we get

$$P_E(\lambda) = \tfrac{1}{4}\,\text{erfc}\,(\sqrt{R_d}) + \tfrac{1}{4}\,\text{erfc}\,[\sqrt{R_d}(1 - 2|\lambda|)] \qquad |\lambda| \leq \tfrac{1}{2} \tag{9-94}$$

If the data is Manchester coded, then the conditional error probability is given by (see Prob. 9-7)

$$P_E(\lambda) = \tfrac{1}{4}\,\text{erfc}\,[\sqrt{R_d}\,(1 - 2|\lambda|)] + \tfrac{1}{4}\,\text{erfc}\,[\sqrt{R_d}\,(1 - 4|\lambda|)] \qquad |\lambda| \leq \tfrac{1}{4} \tag{9-95}$$

When the data format is either RZ or Miller coding, then the expression for $P_E(\lambda)$ as given by (9-91) does not apply. The reasons for this are as follows. From Chap. 1, we recall that an RZ data sequence contains *two* basic pulse waveforms, for example, $p_{s_1}(t)$ and $p_{s_2}(t)$; thus, the decision rule of (9-85) must be altered to accommodate this fact (see Prob. 9-8). For Miller coding, there are *four* elementary pulse waveforms; furthermore, the occurrence of these waveforms in a data sequence is not independent from symbol interval to symbol interval (see Prob. 9-9). Despite these differences, one can take an approach similar to that leading up to (9-91) and produce expressions for the conditional error probability associated with RZ and Miller-coded data. The results are given by

$$P_E(\lambda) = \frac{1}{4}\,\text{erfc}\left(\sqrt{\frac{R_d}{2}}\right) + \text{erfc}\left[\sqrt{\frac{R_d}{2}}(1 - 4|\lambda|)\right] \qquad |\lambda| \leq \tfrac{1}{4} \quad \text{(RZ)} \tag{9-96}$$

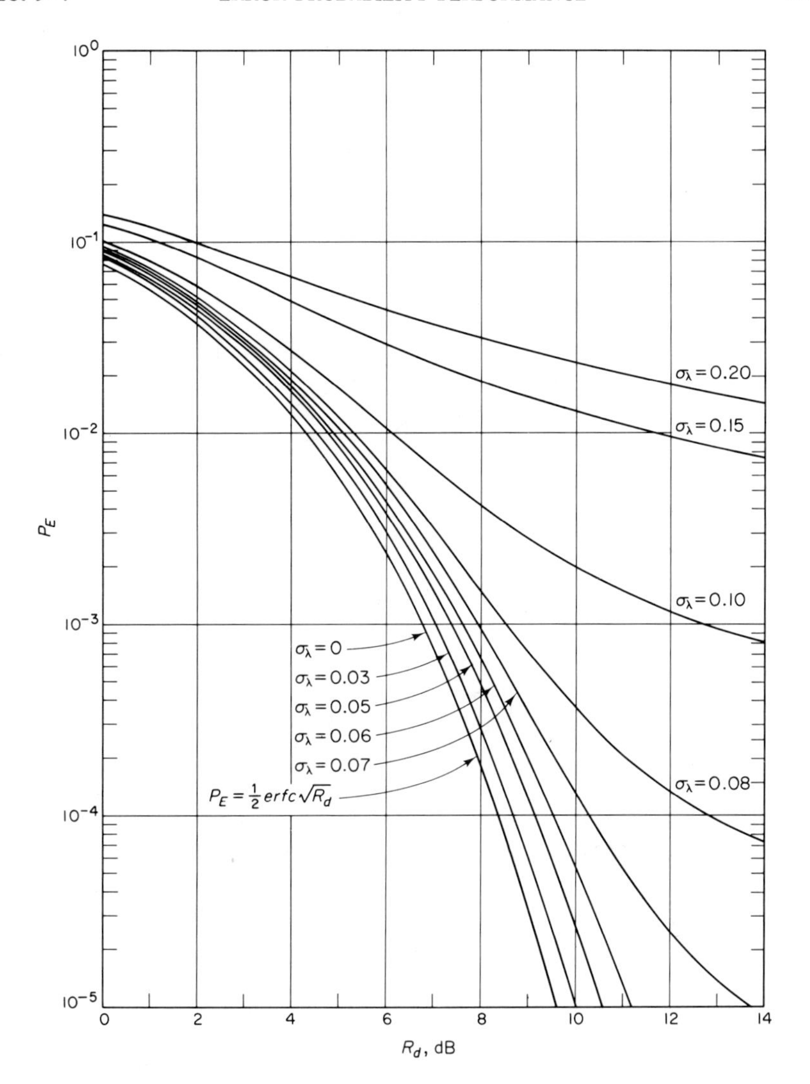

Fig. 9-37 Average probability of error versus signal-to-noise ratio with standard deviation of the symbol sync error as a parameter (NRZ)

and

$$P_E(\lambda) = \frac{1}{2} + \frac{3}{16}\operatorname{erf}^2(a) + \frac{1}{16}\operatorname{erf}^2(b) - \frac{3}{4\sqrt{\pi}}\int_0^{\sqrt{2}c} \operatorname{erf}(x)\exp[-(x-\sqrt{2}\,b)^2]\,dx - \frac{3}{4\sqrt{\pi}}\int_0^{\sqrt{2}c} \operatorname{erf}(x)\exp[-(x-\sqrt{2}\,a)^2]\,dx \qquad (9\text{-}97)$$

$$|\lambda| \leq \tfrac{1}{4}$$

(Miller code)

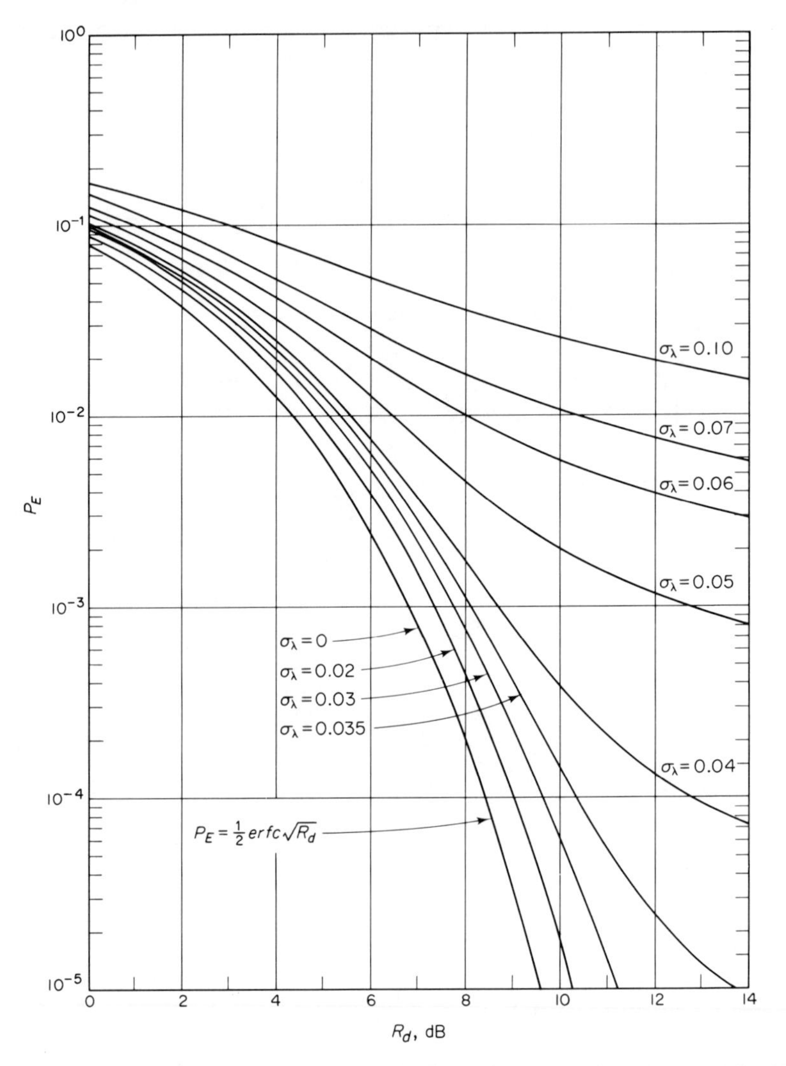

Fig. 9-38 Average probability of error versus signal-to-noise ratio with standard deviation of the symbol sync error as a parameter (Manchester code)

where $a \triangleq \sqrt{R_d/2}\,(1 - 4|\lambda|)$, $b \triangleq \sqrt{R_d/2}$, and $c \triangleq \sqrt{R_d/2}\,(1 - 2|\lambda|)$. The average error probability P_E is determined from

$$P_E = \int_{-\lambda_{\max}}^{\lambda_{\max}} p(\lambda) P_E(\lambda)\, d\lambda \tag{9-98}$$

where, as before, $p(\lambda)$ is the probability distribution of the normalized sync error, and $\lambda_{\max}$ is the maximum value of λ for which $P_E(\lambda)$ is defined; for example, $\lambda_{\max} = \frac{1}{2}$ for NRZ data and $\lambda_{\max} = \frac{1}{4}$ for Manchester, RZ and Miller-coded data. The p.d.f.—$p(\lambda)$—depends on the way in which symbol

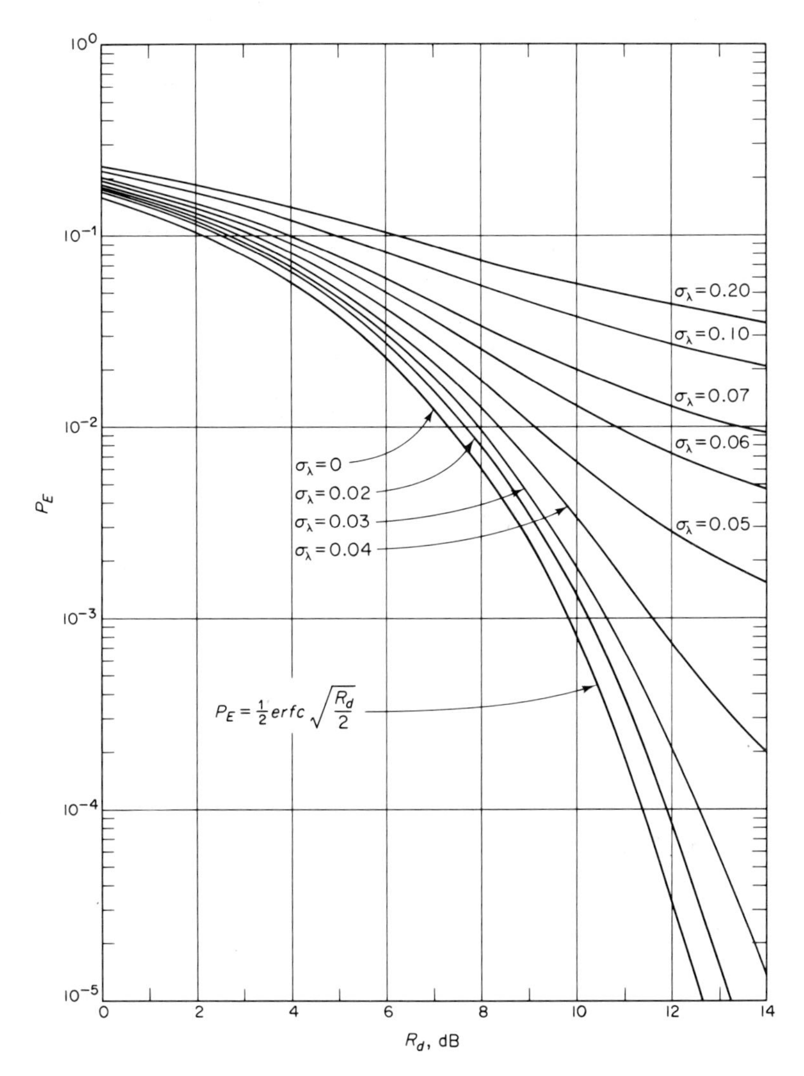

Fig. 9-39 Average probability of error versus signal-to-noise ratio with standard deviation of the symbol sync error as a parameter (RZ)

sync is derived at the receiver. Various solutions for $p(\lambda)$ corresponding to particular symbol synchronizer configurations have been found in earlier sections. For the purpose of comparing the average error probability performance of the various data formats, we postulate a Tikhonov p.d.f. for $p(\lambda)$ (see Chap.2), which is characterized entirely in terms of the variance σ_λ^2 of the sync error. Thus, for NRZ data,

$$p(\lambda) = \frac{\exp\left[\cos 2\pi\lambda/(2\pi\sigma_\lambda)^2\right]}{I_0[(1/2\pi\sigma_\lambda)^2]} \qquad |\lambda| \leq \tfrac{1}{2} \tag{9-99}$$

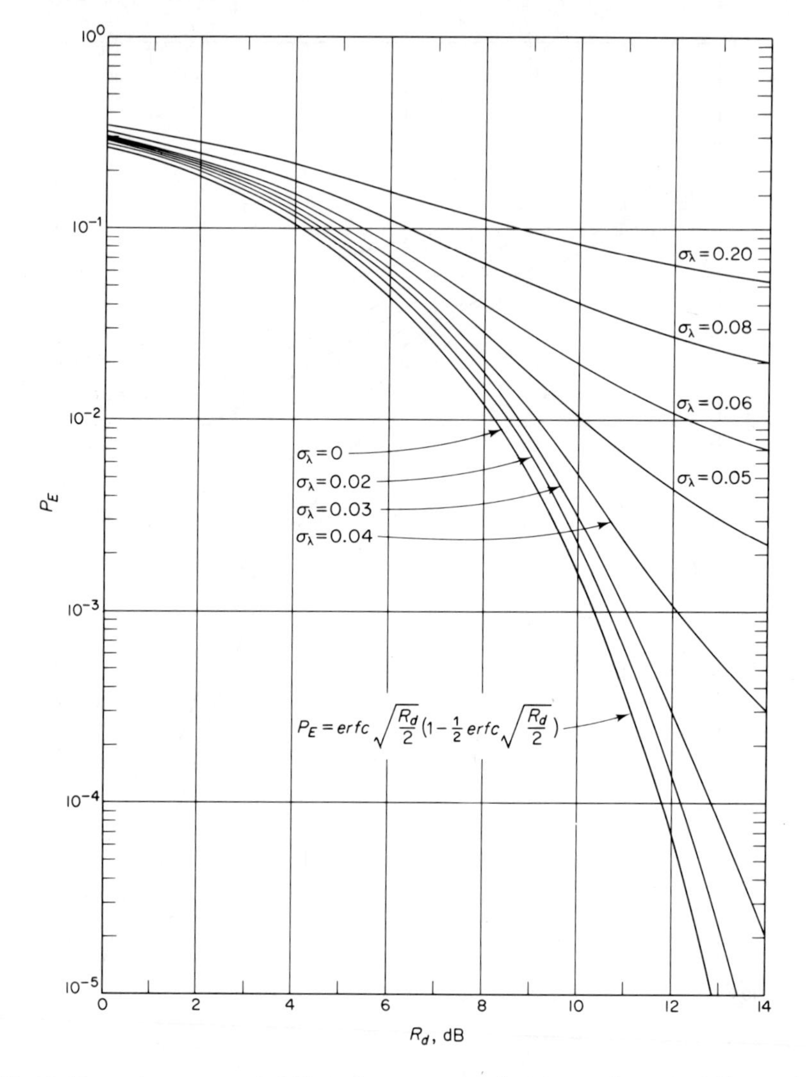

Fig. 9-40 Average probability of error versus signal-to-noise ratio with standard deviation of the symbol sync error as a parameter (Miller code)

whereas for baseband signaling techniques that employ transitions in the middle of the symbol interval, $p(\lambda)$ is characterized by

$$p(\lambda) = \frac{2 \exp\,[\cos 4\pi\lambda/(4\pi\sigma_\lambda)^2]}{I_0[(1/4\pi\sigma_\lambda)^2]} \qquad |\lambda| \leq \tfrac{1}{4} \tag{9-100}$$

Substituting (9-99) or (9-100) into (9-98) together with the appropriate expression for $P_E(\lambda)$ and performing numerical integration on a digital computer provides the graphical results of Figs. 9-37 through 9-40. In these curves the average error probability is plotted versus R_d with the standard deviation σ_λ of the normalized sync error λ as a parameter.

The degradation in system performance relative to a perfectly synchronized system can be expressed in terms of the amount by which the signal-to-noise ratio must be increased to achieve the same error probability. Then, for a given data signal-to-noise ratio R_d, and a data sequence formed from the set of symbols $\pm p_s(t)$, the degradation (in dB) is given by $\Delta R_d = 10 \log_{10}(R_{d\lambda} - R_d)$ where $R_{d\lambda}$ is the solution of

$$\frac{1}{2} \operatorname{erfc} \sqrt{R_d} = \frac{1}{4} \int_{-1/2}^{1/2} p(\lambda) \Big[\operatorname{erfc} \{\sqrt{R_{d\lambda}} [r_s(|\lambda|T) + r_s(T - |\lambda|T)]\} + \operatorname{erfc} \{\sqrt{R_{d\lambda}} [r_s(|\lambda|T) - r_s(T - |\lambda|T)]\} \Big] d\lambda \tag{9-101}$$

System Performance Due to Combined Noisy Reference and Symbol Sync Losses

In discussing the degradation in system error probability performance when both the symbol sync and the carrier reference signal are noisy, we consider two definite physical situations or system mechanizations. First, the two errors are assumed to be correlated—that is, the symbol and subcarrier (carrier) sync are derived from the same source. A typical example of this situation occurs when symbol and subcarrier (carrier) sync are both derived from the double-loop tracking device discussed earlier. Second, the two reference errors are statistically independent, as in a system where the carrier reference and symbol sync are derived from different sources. Regardless of the methods used for producing symbol and subcarrier references, the error probability conditioned on a fixed value of λ, and fixed subcarrier phase error ϕ_s and radian frequency ω_s, can be computed. For a data sequence whose symbols are chosen from the set $\pm p_s(t)$, this conditional error probability is given by (see Prob. 9-9)

$$\begin{aligned} P_E(\phi_s, \lambda) = \tfrac{1}{4} \operatorname{erfc} \{ & \sqrt{R_d'} \cos \phi_s [r_s(\lambda T) + r_s(T - \lambda T)] \\ & - \sqrt{R_d'} [\cos \phi_s r_{sc}(\lambda T) + \cos (2m\pi\lambda + \phi_s) r_{sc}(T - \lambda T)] \\ & - \sqrt{R_d'} [\sin \phi_s r_{ss}(\lambda T) + \sin (2m\pi\lambda + \phi_s) r_{ss}(T - \lambda T)]\} \\ + \tfrac{1}{4} \operatorname{erfc} \{ & \sqrt{R_d'} \cos \phi_s [r_s(\lambda T) - r_s(T - \lambda T)] \\ & - \sqrt{R_d'} [\cos \phi_s r_{sc}(\lambda T) - \cos (2m\pi\lambda + \phi_s) r_{sc}(T - \lambda T)] \\ & - \sqrt{R_d'} [\sin \phi_s r_{ss}(\lambda T) - \sin (2m\pi\lambda + \phi_s) r_{ss}(T - \lambda T)]\} \\ & \text{for} \quad 0 \leq \lambda \leq \tfrac{1}{2} \\ & \qquad |\phi_s| \leq \pi \end{aligned} \tag{9-102}$$

and

$$
\begin{aligned}
P_E(\phi_s, \lambda) = \tfrac{1}{4}\,\text{erfc}\,\{&\sqrt{R_d'}\cos\phi_s\,[r_s(|\lambda|T) + r_s(T - |\lambda|T)] \\
&-\sqrt{R_d'}\,[\cos(2m\pi|\lambda| - \phi_s)\,r_{sc}(|\lambda|T) + \cos\phi_s\,r_{sc}(T - |\lambda|T)] \\
&+\sqrt{R_d'}\,[\sin(2m\pi|\lambda| - \phi_s)\,r_{ss}(|\lambda|T) - \sin\phi_s\,r_{ss}(T - |\lambda|T)]\} \\
+ \tfrac{1}{4}\,\text{erfc}\,\{&\sqrt{R_d'}\cos\phi_s\,[r_s(|\lambda|T) - r_s(T - |\lambda|T)] \\
&-\sqrt{R_d'}\,[\cos(2m\pi|\lambda| - \phi_s)\,r_{sc}(|\lambda|T) - \cos\phi_s\,r_{sc}(T - |\lambda|T)] \\
&+\sqrt{R_d'}\,[\sin(2m\pi|\lambda| - \phi_s)\,r_{ss}(|\lambda|T) + \sin\phi_s\,r_{ss}(T - |\lambda|T)]\} \\
&\qquad\text{for} \quad -\tfrac{1}{2} \le \lambda \le 0 \\
&\qquad\qquad |\phi_s| \le \pi
\end{aligned}
\tag{9-103}
$$

where $r_s(\tau)$ is defined by (9-88) and (9-92),

$$
\begin{aligned}
r_{ss}(\tau) &\triangleq \frac{\int_0^{T-\tau} p_s(t' + \tau)p_s(t')\sin 2\omega_s t'\,dt'}{ST} \\
r_{sc}(\tau) &\triangleq \frac{\int_0^{T-\tau} p_s(t' + \tau)p_s(t')\cos 2\omega_s t'\,dt'}{ST}
\end{aligned}
\tag{9-104}
$$

$$
R_d' \triangleq \frac{R_d}{1 - \cos(2m\pi\lambda + \phi_s)r_{sc}(0) - \sin(2m\pi\lambda + \phi_s)r_{ss}(0)}
$$

and m is an integer defined by $m \triangleq \omega_s T/\pi$; that is, $m/2$ represents the number of subcarrier cycles per symbol period. Also, in the above, we have assumed that the reference subcarrier is sinusoidal and that ϕ_s is slowly varying relative to the symbol period. For NRZ data, (9-102) and (9-103) reduce to

$$
\begin{aligned}
P_E(\phi_s, \lambda) &= \tfrac{1}{4}\,\text{erfc}\,\{\sqrt{R_d}\cos\phi_s\} \\
&+ \tfrac{1}{4}\,\text{erfc}\left\{\sqrt{R_d}\left[(1 - 2|\lambda|)\cos\phi_s - \frac{\text{sgn}\,\lambda}{m\pi}[\sin\phi_s - \sin(2m\pi\lambda + \phi_s)]\right]\right\} \\
&\qquad |\lambda| \le \tfrac{1}{2} \\
&\qquad |\phi_s| \le \pi
\end{aligned}
\tag{9-105}
$$

Furthermore, for large m,

$$
\begin{aligned}
P_E(\phi_s, \lambda) &= \tfrac{1}{4}\,\text{erfc}\,(\sqrt{R_d}\cos\phi_s) \\
&+ \tfrac{1}{4}\,\text{erfc}\,[\sqrt{R_d}(1 - 2|\lambda|)\cos\phi_s]
\end{aligned}
\tag{9-106}
$$

Dependent Symbol and Subcarrier Synchronization References. Here we present an approximate theory that can be used to assess the performance of the data detector when the double-loop device is used to provide sources of symbol and subcarrier sync. Letting t_0 denote the digit period of the PN code and assuming that the subcarrier is a square wave, then the p.d.f. of the phase error in the inner loop conditioned on knowing λ is approximately given by

$$p(\phi_s \mid \lambda) = \frac{\exp[\alpha_s(1 - l\mid\lambda\mid)\cos\phi_s]}{2\pi I_0[\alpha_s(1 - l\mid\lambda\mid)]} \qquad |\phi_s| \leq \pi \tag{9-107}$$

where l is the number of PN code digits per symbol ($l = T/t_0$) and α_s is the effective signal-to-noise ratio in the inner loop tracking bandwidth. The statistics of the outer loop can be approximated by

$$p(\lambda) = l\frac{\exp[\alpha_s \cos 2\pi l\lambda]}{I_0(\alpha_s)} \qquad |\lambda| \leq \frac{1}{2l} \tag{9-108}$$

Thus, the joint p.d.f. $p(\phi_s, \lambda)$ becomes

$$p(\phi_s, \lambda) = l\frac{\exp\{\alpha_s[(1 - l\mid\lambda\mid)\cos\phi_s + \cos 2\pi l\lambda]\}}{2\pi I_0(\alpha_s)I_0[\alpha_s(1 - l\mid\lambda\mid)]} \qquad |\lambda| \leq \frac{1}{2l}, \quad |\phi_s| \leq \pi \tag{9-109}$$

Letting $z = l\lambda$, the variance of the subcarrier phase error is given by

$$\sigma^2_{\phi_s} = \int_0^\pi \int_0^{1/2} \frac{2\phi_s^2 \exp\{\alpha_s[(1 - z)\cos\phi_s + \cos 2\pi z]\}}{\pi I_0(\alpha_s)I_0[\alpha_s(1 - z)]}\, dz\, d\phi_s \tag{9-110}$$

while the variance of the misalignment between the reference code and the incoming code is

$$\sigma^2_\lambda = \frac{2}{l^2}\int_0^{1/2} \frac{z^2 \exp(\alpha_s \cos 2\pi z)}{I_0(\alpha_s)}\, dz \tag{9-111}$$

For the case of practical interest, $l \gg 1$; thus,

$$\sigma^2_\lambda \ll \sigma^2_{\phi_s} \tag{9-112}$$

Averaging the conditional error probability of (9-106) over the joint p.d.f. of ϕ_s and λ as given in (9-109), we get an expression for the average error probability, which can be written as

$$\begin{aligned} P_E &= \int_{-1/2l}^{1/2l} \int_{-\pi}^{\pi} p(\phi_s, \lambda) P_E(\phi_s, \lambda)\, d\phi_s\, d\lambda \\ &= \frac{1}{2}\int_0^{1/2}\int_0^\pi \frac{\exp\{\alpha_s[(1 - z)\cos\phi_s + \cos 2\pi z]\}}{\pi I_0(\alpha_s)I_0[\alpha_s(1 - z)]}\Big\{\operatorname{erfc}(\sqrt{R_d}\cos\phi_s) \\ &\quad + \operatorname{erfc}\Big[\sqrt{R_d}\Big(1 - \frac{2z}{l}\Big)\cos\phi_s\Big]\Big\}\, d\phi_s\, dz \end{aligned} \tag{9-113}$$

For large l, (9-113) reduces to

$$\begin{aligned} P_E &= \int_0^{1/2l}\int_0^\pi 2p(\phi_s, \lambda)\operatorname{erfc}(\sqrt{R_d}\cos\phi_s)\, d\phi_s\, d\lambda \\ &= \int_0^\pi p(\phi_s)\operatorname{erfc}(\sqrt{R_d}\cos\phi_s)\, d\phi_s \end{aligned} \tag{9-114}$$

From (9-114), the important conclusion is that in systems that obtain symbol and subcarrier sync by means of a double-loop tracking device, the average error probability is relatively independent of the symbol timing error; that is, the subcarrier phase error is the most significant.

Independent Symbol and Subcarrier Synchronization References. Here symbol sync is derived from a double-loop tracking device, and subcarrier sync is obtained by means of a squaring or Costas loop. Since the symbol and subcarrier sync are derived from independent sources, the average error probability for this case is given by

$$P_E = \int_{-1/2l}^{1/2l} \int_{-\pi/2}^{\pi/2} p_d(\phi_s)p(\lambda)P_E(\phi_s, \lambda)\, d\phi_s\, d\lambda \tag{9-115}$$

where $p(\lambda)$ and $p_d(\phi_s)$ are the p.d.f. of the normalized sync error and subcarrier phase error, respectively. The p.d.f. $p_d(\phi_s)$ is given in (6-36) with ϕ replaced by ϕ_s, while $p(\lambda)$ is obtained from (9-109) by integrating out the variable ϕ_s.

The practical implication of this method for obtaining symbol and subcarrier sync is that $\sigma^2_{\phi_s} \ll \sigma^2_\lambda$—that is, $p_d(\phi_s) \cong \delta(\phi_s)$—and consequently the error probability is largely determined by the statistics of the symbol sync error. Under this assumption, the average error probability becomes

$$\begin{aligned} P_E &= \int_{-1/2l}^{1/2l} p(\lambda)P_E(0, \lambda)\, d\lambda \\ &= \frac{1}{2}\int_0^{1/2}\int_0^{\pi} \frac{\exp\{\alpha[(1-z)\cos\phi + \cos 2\pi z]\}}{\pi I_0(\alpha)I_0[\alpha(1-z)]} \\ &\quad \times \left\{\operatorname{erfc}\sqrt{R_d} + \operatorname{erfc}\left[\sqrt{R_d}\left(1 - \frac{2z}{l}\right)\right]\right\} d\phi\, dz \end{aligned} \tag{9-116}$$

Note that we have omitted the s subscript on ϕ and α so as not to confuse them with the parameters of the squaring loop.

9-5. CONCLUSIONS

In this chapter, we have treated certain aspects of the symbol synchronizer design problem. One aspect that has been omitted is that pertaining to acquisition behavior because the theory that allows one to make trade-off studies between the acquisition and tracking performance of symbol synchronizers is not complete, particularly when the input is random data plus noise. Nevertheless, certain special cases have been investigated.[27,28] The question of what constitutes an *optimum* symbol synchronizer is largely open, that is, symbol synchronizer design is an art and not a science. For practical applications, an optimum symbol synchronizer should reflect the best compromise between acquisition and tracking performance. For example, if the allowable value of the mean-squared sync error or symbol slippage rate is specified for a given signal-to-noise ratio, then one would like to maximize the frequency detuning that can be tolerated and still provide for acceptable performance in the acquisition mode. Alternately, if the system data rates are known and the allowable acquisition time specified, then the design engineer concentrates his efforts on optimizing the tracking performance.

PROBLEMS

9-1. Show that for Manchester coded data, the MAP estimate of symbol sync is that value of ϵ which maximizes

$$f(y, \epsilon) = \prod_{k=0}^{K} \cosh \left\{\frac{2}{N_0} \int_{(k-1)T+\epsilon}^{(k-1/2)T+\epsilon} y(t)\, dt - \frac{2}{N_0} \int_{(k-1/2)T+\epsilon}^{kT+\epsilon} y(t)\, dt\right\}$$

9-2. Consider a data sequence formed by transmitting with equal probability one of two equal energy pulse waveforms, that is, $p_{s_1}(t)$ or $p_{s_2}(t)$, independently in each symbol interval.

a) Following a procedure similar to that given in Sec. 9-2, show that the MAP estimate of symbol sync is that value of ϵ which maximizes

$$f(y, \epsilon) = \prod_{k=0}^{K} \left\{\exp\left[-\frac{2}{N_0} \int_{T_k(\epsilon)} y(t)p_{s_1}[t - (k - 1)T - \epsilon]\, dt\right] + \exp\left[-\frac{2}{N_0} \int_{T_k(\epsilon)} y(t)p_{s_2}[t - (k - 1)T - \epsilon]\, dt\right]\right\}$$

b) From the result of part a), show that for RZ data the MAP estimate of symbol sync is that value of ϵ which maximizes

$$f(y, \epsilon) = \prod_{k=0}^{K} \left\{\exp\left[-\frac{2}{N_0} \int_{T_k(\epsilon)} y(t)\, dt\right] + \exp\left[-\frac{2}{N_0} \int_{(k-1)T+\epsilon}^{(k-1/2)T+\epsilon} y(t)\, dt + \frac{2}{N_0} \int_{(k-1/2)T+\epsilon}^{kT+\epsilon} y(t)\, dt\right]\right\}$$

c) Find $f(y, \epsilon)$ if the set of symbols corresponds to an FSK system, that is, $p_{s_1}(t) = \sqrt{2S} \sin \omega_1 t$ and $p_{s_2}(t) = \sqrt{2S} \sin \omega_2 t$.

9-3. Consider the case where the additive channel noise $n(t)$ is Gaussian but has a non-white spectrum. If $R_n(\tau)$ denotes the autocorrelation function of the noise, then show that the MAP estimate of symbol sync is that value of ϵ which maximizes

$$f(y, \epsilon) = \prod_{k=0}^{K} \cosh \left\{\frac{2}{N_0} \int_{T_k(\epsilon)} y(t)q[t - (k - 1)T - \epsilon]\, dt\right\}$$

where $q(t)$ is the solution of the integral equation

$$p_s[t - (k - 1)T - \epsilon] = \int_{T_k(\epsilon)} R_n(t - x)q[x - (k - 1)T - \epsilon]\, dx$$

9-4. Consider a data sequence where the kth transmitted symbol $s(t; a_k)$ is chosen from the N-ary set $a_k p_s(t)$; $a_k = \pm 1, \pm 2, \ldots, \pm N/2$. If this data sequence is transmitted over the additive white Gaussian noise channel, show that the MAP estimate of symbol sync is given by that value of ϵ which maximizes

$$f(y, \epsilon) =$$
$$\prod_{k=0}^{K} \left[\sum_{i=1}^{N/2} \exp(-i^2 E/N_0) \cosh \left\{ \frac{2i}{N_0} \int_{T_k(\epsilon)} y(t) p_s[t - (k-1)T - \epsilon] \, dt \right\} \right]$$

9-5. Evaluate Eq. (9-91) for the three pulse shapes in Fig. 9-4.

9-6. Consider a situation where

$$r_s(\tau) = \begin{cases} 1 & |\tau| \leq \dfrac{T}{2} \\ 0 & \dfrac{T}{2} \leq |\tau| \leq T \end{cases}$$

a) Show that for this case the conditional error probability as given by (9-91) is

$$P_E(\lambda) = \tfrac{1}{2} \operatorname{erfc}(\sqrt{R_d})$$

which is independent of λ and corresponds to the detection error probability for perfectly synchronized PSK signals.

b) Can the $r_s(\tau)$ given in part a) be realized with any physically realizable $p_s(t)$?

c) For physically realizable pulses, and $r_s(\tau) = 0$ for $\tau > T$, Boas and Kac[29] have shown that

$$r_s(\tau) \leq \cos\left(\frac{\pi}{1 + [T/\tau]}\right)$$

where $[T/\tau]$ denotes the greatest integer not exceeding T/τ. Find an expression for $P_E(\lambda)$ in terms of this upper bound on $r_s(\tau)$.

d) Does the expression for $P_E(\lambda)$ found in part c), necessarily represent a lower bound on this quantity?

9-7. Consider a data sequence which is Manchester coded.

a) Show that the correlation function $r_s(\tau)$ as defined by Eqs. (9-88) and (9-92) is given by

$$r_s(\tau) = \begin{cases} 1 - 3\dfrac{|\tau|}{T} & |\tau| \leq \dfrac{T}{2} \\ \dfrac{|\tau|}{T} - 1 & \dfrac{T}{2} \leq |\tau| \leq T \end{cases}$$

b) By direct substitution of the result of part a) into Eq. (9-91), verify that the conditional error probability for Manchester coding is given by (9-95).

9-8. Consider a data sequence formed by transmitting with equal probability one of two equal energy pulse waveforms, that is, $p_{s_1}(t)$ or $p_{s_2}(t)$ independently in each symbol interval. It is easily shown that the decision rule [analogous to (9-85)] for this type of sequence is given by

$$\text{choose } p_{s_1}(t) \quad \text{if} \quad Q > 0$$
$$\text{choose } p_{s_2}(t) \quad \text{if} \quad Q < 0$$

where

$$Q \triangleq \int_{(n-1)T+\hat{\epsilon}}^{nT+\hat{\epsilon}} y(t)\{p_{s_1}[t-(n-1)T-\hat{\epsilon}] - p_{s_2}[t-(n-1)T-\hat{\epsilon}]\}\,dt$$

a) Show that the conditional error probability for this type of sequence is given by

$$\begin{aligned}
P_E(\lambda) = {} & \tfrac{1}{8}\operatorname{erfc}\{\sqrt{R_d'}[r_{s_{11}}(\lambda T) + r_{s_{11}}(T-\lambda T) \\
& \qquad - r_{s_{21}}(\lambda T) - r_{s_{12}}(T-\lambda T)]\} \\
& + \tfrac{1}{8}\operatorname{erfc}\{\sqrt{R_d'}[r_{s_{11}}(\lambda T) + r_{s_{21}}(T-\lambda T) \\
& \qquad - r_{s_{21}}(\lambda T) - r_{s_{22}}(T-\lambda T)]\} \\
& + \tfrac{1}{8}\operatorname{erfc}\{\sqrt{R_d'}[r_{s_{22}}(\lambda T) + r_{s_{22}}(T-\lambda T) \\
& \qquad - r_{s_{12}}(\lambda T) - r_{s_{21}}(T-\lambda T)]\} \\
& + \tfrac{1}{8}\operatorname{erfc}\{\sqrt{R_d'}[r_{s_{22}}(\lambda T) + r_{s_{12}}(T-\lambda T) \\
& \qquad - r_{s_{12}}(\lambda T) - r_{s_{11}}(T-\lambda T)]\} \\
& \qquad \text{for} \quad \lambda > 0
\end{aligned}$$

and

$$\begin{aligned}
P_E(\lambda) = {} & \tfrac{1}{8}\operatorname{erfc}\{\sqrt{R_d'}[r_{s_{11}}(|\lambda|T) + r_{s_{11}}(T-|\lambda|T) \\
& \qquad - r_{s_{12}}(|\lambda|T) - r_{s_{21}}(T-|\lambda|T)]\} \\
& + \tfrac{1}{8}\operatorname{erfc}\{\sqrt{R_d'}[r_{s_{11}}(|\lambda|T) + r_{s_{12}}(T-|\lambda|T) \\
& \qquad - r_{s_{12}}(|\lambda|T) - r_{s_{22}}(T-|\lambda|T)]\} \\
& + \tfrac{1}{8}\operatorname{erfc}\{\sqrt{R_d'}[r_{s_{22}}(|\lambda|T) + r_{s_{22}}(T-|\lambda|T) \\
& \qquad - r_{s_{21}}(|\lambda|T) - r_{s_{12}}(T-|\lambda|T)]\} \\
& + \tfrac{1}{8}\operatorname{erfc}\{\sqrt{R_d'}[r_{s_{22}}(|\lambda|T) + r_{s_{21}}(T-|\lambda|T) \\
& \qquad - r_{s_{21}}(|\lambda|T) - r_{s_{11}}(T-|\lambda|T)]\} \\
& \qquad \text{for} \quad \lambda < 0
\end{aligned}$$

where

$$r_{s_{ij}}(\tau) \triangleq \frac{\int_0^{T-\tau} p_{s_i}(t'+\tau)p_{s_j}(t')\,dt'}{ST}$$

$$R_d' \triangleq \frac{R_d}{2[1 - r_{s_{12}}(0)]}$$

and we have made use of the fact that $r_{s_{ij}}(\tau) = r_{s_{ji}}(-\tau)$ for all $i, j = 1, 2$.

b) Show that if $p_{s_1}(t) = -p_{s_2}(t)$, the expression for conditional error probability reduces to Eq. (9-91).

c) For RZ data, evaluate $r_{s_{ij}}(\tau)$ for $i, j = 1, 2$, and, by direct substitution of these results into the expression for $P_E(\lambda)$ given in part a), verify Eq. (9-96).

9-9. For a Miller-coded data format, the appropriate decision rule, in the presence of symbol sync error, for the nth symbol is

$$\hat{a}_n = \text{sgn}\,[|Q_1| - |Q_2|]$$

where

$$Q_1 = \int_{(n-1)T+\hat{\epsilon}}^{nT+\hat{\epsilon}} y(t)p_{s_1}[t - (n-1)T - \hat{\epsilon}]\,dt$$

$$Q_2 = \int_{(n-1)T+\hat{\epsilon}}^{nT+\hat{\epsilon}} y(t)p_{s_2}[t - (n-1)T - \hat{\epsilon}]\,dt$$

and

$$p_{s_1}(t) = \begin{cases} \sqrt{S} & 0 \le t \le T \\ 0 & \text{elsewhere} \end{cases}$$

$$p_{s_2}(t) = \begin{cases} \sqrt{S} & 0 \le t \le T/2 \\ -\sqrt{S} & T/2 \le t \le T \\ 0 & \text{elsewhere} \end{cases}$$

Derive the expression for conditional error probability as given by (9-97).

9-10. Consider the situation where the data symbols are transmitted on a sinusoidal subcarrier. Then, analogous to Eq. (9-1), the received signal is given by

$$y(t) = \sum_{k=0}^{K} a_k p_s[t - (k-1)T - \epsilon]\{\sqrt{2}\sin[\omega_s(t-\epsilon)]\} + n(t)$$

and the decision rule for the nth data symbol is

$$\hat{a}_n = \text{sgn}\left\{\int_{(n-1)T+\hat{\epsilon}}^{nT+\hat{\epsilon}} y(t)p_s[t - (n-1)T - \hat{\epsilon}]\sin[\omega_s(t-\epsilon) - \phi_s]\,dt\right\}$$

Following an approach analogous to that leading up to Eq. (9-91), verify (9-102) and (9-103).

9-11. Using (9-102) and (9-103), derive an expression for $P_E(\phi_s, \lambda)$ if the data symbols are Manchester coded on the sinusoidal subcarrier.

REFERENCES

1. Golomb, S. W., et. al., *Digital Communications with Space Applications.* Prentice-Hall, Inc., Englewood Cliffs, N. J., 1964.

2. Stiffler, J. J., *Theory of Synchronous Communications*, Prentice-Hall, Inc., Englewood Cliffs, N. J., 1971.

3. VAN TREES, H. L., *Detection, Estimation, and Modulation Theory, Part I*, John Wiley & Sons, Inc., New York, N. Y., 1968.

4. STIFFLER, J. J., "Maximum Likelihood Symbol Synchronization," Jet Propulsion Laboratory, Pasadena, Calif., SPS 37–35, Vol. IV (October, 1965) 349–357.

5. SAGE, A. P., and MCBRIDE, A. L., "Optimum Estimation of Bit Synchronization," *IEEE Transactions on Aerospace and Electronic Systems*, Vol. AES-5, No. 3 (May, 1969) 525–536.

6. WINTZ, P. A., and LUECKE, E. J., "Performance of Optimum and Suboptimum Synchronizers," *IEEE Transactions on Gommunication Technology*, Vol. COM-17, No. 3 (June, 1969) 380–389.

7. MENGALI, U., "A Self Bit Synchronizer Matched to the Signal Shape," *IEEE Transactions on Aerospace and Electronic Systems*, Vol. AES-7, No. 4 (July, 1971) 686–693.

8. VAN HORN, J. H., "A Theoretical Synchronization System for use with Noisy Digital Signals," *IEEE Transactions on Communication Technology*, Vol. COM-12, No. 3 (September, 1964) 82–90.

9. WINTZ, P. A., and HANCOCK, J. C., "An Adaptive Receiver Approach to the Time Synchronization Problem," *IEEE Transactions on Communication Technology*, Vol. COM-13, No. 1 (March, 1965) 90–96.

10. PRICE, R., and GREEN, P. E., JR., "A Communication Technique for Multipath Channels," *Proceedings of the IRE*, Vol. 46 (March, 1958) 555–570.

11. MOSIER, R. F., and CLABAUGH, R. G., "Kineplex, a Bandwidth-efficient Binary Transmission System," *Transactions of the AIEE* (*Communications and Electronics*), Vol. 76 (January, 1958) 723–728.

12. DOELZ, M. L., HEALD, E. T., and MARTIN, D. L., "Binary Data Transmission Techniques for Linear Systems," *Proceedings of the IRE*, Vol. 45 (May, 1957) 656–661.

13. LINDSEY, W. C., *Synchronization Systems in Communication and Control*. Prentice-Hall, Inc., Englewood Cliffs, N.J., 1972.

14. SHAFT, P., "Optimum Design of the Nonlinearity in Signal Tracking Loops," Stanford Research Institute, Menlo Park, Calif., May, 1968.

15. STIFFLER, J. J., "On the Selection of Signals for Phase-Locked Loops," *IEEE Transactions on Communication Technology*, Vol. COM-16, No. 2 (April, 1968) 239–244.

16. LAYLAND, J. W., "On Optimal Signals for Phase-Locked Loops," *IEEE Transactions on Communication Technology*, Vol. COM-17, No. 5 (October, 1969) 526–531.

17. LINDSEY, W. C., and TAUSWORTHE, R. C., "Digital Data-Transition Tracking Loops," Jet Propulsion Laboratory, Pasadena, Calif., SPS 37–50, Vol. III (April, 1968) 272–276.

18. ———, and ANDERSON, T. O., "Digital Data-Transition Tracking Loops," *Proceedings of the International Telemetering Conference*, Los Angeles, Calif. (Oct., 1968) 259–271.

19. TAUSWORTHE, R. C., "Analysis and Design of the Symbol-Tracking Loop," Jet Propulsion Laboratory, Pasadena, Calif., SPS 37–51, Vol. II (May, 1968) 145–147.

20. SIMON, M. K., "An Analysis of the Steady-State Phase Noise Performance of a Digital Data-Transition Tracking Loop," Jet Propulsion Laboratory, Pasadena, Calif., SPS 37–55, Vol. III (February, 1969) 54–62.

21. ———, "Optimization of the Performance of a Digital Data Transition Tracking Loop," *IEEE Transactions on Communication Technology*, Vol. COM-18, No. 3 (October, 1970) 686–690.

22. VITERBI, A. J., *Principles of Coherent Communication.* McGraw-Hill Book Company Inc., New York, N.Y., 1966.

23. LAYLAND, J. W., "Telemetry Bit Synchronization Loop," Jet Propulsion Laboratory, Pasadena, Calif., SPS 34–46, Vol. III (July, 1967) 204–215.

24. CARL, C., "Relay Telemetry Modulation System Development," Jet Propulsion Laboratory, Pasadena, Calif., SPS 37–50, Vol. III (April, 1968) 326–331.

25. SIMON, M. K., "Nonlinear Analysis of an Absolute Value Type of Early-Late-Gate Bit Synchronizer," *IEEE Transactions on Communication Technology*, Vol. COM-18, No. 3 (October, 1970) 589–596.

26. SPRINGETT, J. C., "Telemetry and Command Techniques for Planetary Spacecraft," Jet Propulsion Laboratory, Pasadena, Calif., Technical Report 32–495 (January, 1965).

27. MCRAE, D. D. and SMITH, E. F., "An Acquisition Aid for Digital Phase-Locked Loops," *Proceedings of the International Conference on Communications*, Philadelphia, Pa. (June, 1972) 14–18—14–22.

28. ———, and SMITH, E. F., "Bit Synchronization,", *Proceedings of the International Telemetering Conference*, Los Angeles, Calif. (October, 1972) 539–552.

29. BOAS, R., and KAC, M., "Inequalities for Fourier Transforms of Positive Functions," *Duke Mathematical Journal*, Vol. 12 (1945) 189–206.

Chapter 10

NONCOHERENT COMMUNICATION OVER THE GAUSSIAN CHANNEL

10-1. INTRODUCTION

Frequently, a communications design engineer is faced with the problem of designing a communication system in which the economics are such that the transmitter power is limited, the transmitting oscillator cannot be completely stabilized, and the physics are such that channel multipath exists. In such situations, the phase coherent techniques discussed in earlier chapters may not apply. The reasons are several: first of all, a small received data signal-to-noise spectral density ratio S/N_0 requires one to choose a data transmission rate commensurately low so as to provide an acceptable error probability in the telemetry channel. At the same time, for coherent links, the bandwidth of the carrier-tracking loop must be decreased in order to provide a good coherent reference for data demodulation; however, the problem of *oscillator phase instability* does not permit lowering the loop bandwidth below a certain value necessary to maintain phase coherence. Stated another way, one cannot compensate for a limited P_c/N_0 simply by decreasing the loop bandwidth proportionately since the loop cannot efficiently track spectrally spread carriers of bandwidth wider than the loop bandwidth.

Secondly, system performance is extremely sensitive to the problem of optimum power allocation. If one suppresses the carrier to make efficient use of the total transmitter power for tracking as well as data detection, then, for a fixed transmitter power, one is still faced with the problems of oscillator instabilities and channel multipath. If the design specification cannot be met by using phase coherent detection of suppressed carrier signals, one must resort to considering data transmission and data detection techniques which are less sensitive to the problems of oscillator instabilities and channel multipath. Therefore, one turns to the method commonly referred to as *noncoherent communication*. Noncoherent systems have been proposed as a solution to the problems of telemetering from a certain class of small planetary capsules or probes and/or for deep space probes wherein available transmitter power is limited to a few watts. Such system concepts also appear to be quite attractive for outer planet entry probe to spacecraft or Earth links, or for use as a back-up to larger systems in the event that the primary telemetry link suffers catastrophic failure.

In this chapter, we explore five main problem areas that enter into the design of a practical noncoherent telemetry system. Each area is discussed from the viewpoint of existing literature that has evolved during the past decade relative to the subject of noncoherent communications. In particular we discuss the problems of transmitter design, receiver design, time and frequency synchronization of the receiver (acquisition and tracking), and system performance.

10-2. TRANSMITTER CHARACTERIZATION

In practice, the set of *transmitted signals* $\{x_k(t), k = 1, \ldots, N\}$ as well as the *reference signals* used in the demodulation and detection process cannot be generated as "pure" sinusoids. In any practical transmitter, or receiver, the phase process of the generated tones always undergoes a certain amount of diffusion with the passage of time. This diffusion leads to phase instabilities in the transmitter/receiver pair and causes *diffusional spreading* of the transmitted tones and the stored receiver tones. It is reasonable, however, to assume that the transmitter instabilities are of much greater significance than any receiver instabilities owing to the fact that ground-based reference tones can be generated from very stable oscillator circuits—that is, frequency synthesizers—whereas the transmitter frequencies are usually derived from a single crystal oscillator.

Considering then that the transmitter oscillator is inherently noisy, it is reasonable to characterize the set of transmitted tones by

$$x_k[t, \theta(t)] = \sqrt{2S_k(t)} \cos [\omega_k t + \theta(t)] \qquad \begin{matrix} 0 \leq t \leq T \\ k = 1, 2, \ldots, N \end{matrix} \tag{10-1}$$

where $S_k(t)$ are amplitude variations due to shot noise, spurious oscillations generated in feedback circuits, and the like. The process $\theta(t)$ in (10-1) accounts for the fact that diffusional spreading of the oscillator phase takes place at a rate determined by the physics of the oscillator mechanization. For most cases of interest, the amplitude variations can be neglected so that we can assume that $S_k(t) = S$ for $k = 1, \ldots, N$, and $0 \leq t \leq T$. Thus the transmitted signal set is conveniently approximated by

$$x_k[t, \theta(t)] = \sqrt{2S} \cos [\omega_k t + \theta(t)] \qquad \begin{matrix} 0 \le t \le T \\ k = 1, 2, \ldots, N \end{matrix} \tag{10-2}$$

In (10-2) the parameter S represents the total received power; $\omega_k \triangleq 2\pi f_k$ with $f_k = f_0 + k\Delta f$; $\Delta\omega$ is the frequency spacing of the tones; and f_0 is the carrier frequency.

It is convenient to separate the process $\theta(t)$ into three terms:[1,2]

$$\theta(t) = \omega_d t + \underbrace{\theta_s(t)}_{\substack{\text{rapidly} \\ \text{varying or} \\ \text{short-term} \\ \text{variations}}} + \underbrace{\theta_l(t)}_{\substack{\text{slowly} \\ \text{varying or} \\ \text{long-term} \\ \text{variations}}} \tag{10-3}$$

where $f_d = \omega_d/2\pi$ represents the uncertainty existing at the receiver relative to the carrier frequency, $\theta_s(t)$ is a random process whose *correlation time* is short relative to the duration T, and $\theta_l(t)$ is a random process whose *correlation time* is long relative to T. It is further assumed that the two random processes $\theta_s(t)$ and $\theta_l(t)$ are independent, although not necessarily separable in their effect on system performance. Before continuing with the system evaluation, we briefly discuss the sources of the three instability terms in (10-3). The fixed frequency offset term f_d is attributable to such factors as any shock that might be imparted to the transmitter crystal, fixed Doppler residual, and the like. The underlying source for $\theta_s(t)$ is the thermal and semiconductor noise associated with the transmitter oscillator circuits that cause a so-called "line width" in the transmitter signal spectrum. A number of factors contribute to the presence of $\theta_l(t)$, such as long-term temperature changes of the local environment, component aging, and power supply variations. The effects produced by the long-term variations $\theta_l(t)$ are probably minimal; however, they must be tracked if the demodulation of the RF carrier is to be shifted to baseband successfully. On the other hand, the diffusional spreading of the spectrum of the transmitted tones caused by $\theta_s(t)$ could be appreciable in that such smearing would also tend to produce a reduction in the effective signal-to-noise ratio at the receiver.

The effect of $\theta_s(t)$ on the power spectrum of a T-sec segment of signal is illustrated in Fig. 10-1. Measurements conducted on S-band transmitters have typically placed the loss due to $\theta_s(t)$—that is, L_{θ_s}—between 0.3 and 1.5 dB. Further experimental evidence reveals that the derivative of $\theta_s(t)$—that is, $\dot{\theta}_s(t)$—can be characterized as a finite power, low-pass, Gaussian process whose power spectrum behaves as $f^{2-\epsilon}$ ($\epsilon > 0$) in the neighborhood of zero frequency. If the effect of the process $\dot{\theta}_s(t)$ is then considered as a frequency modulation on the carrier, then because $\theta_s(t)$ has infinite variance, the power spectrum of the modulated signal does not possess a discrete component at the carrier frequency. It has further been observed experimentally that at

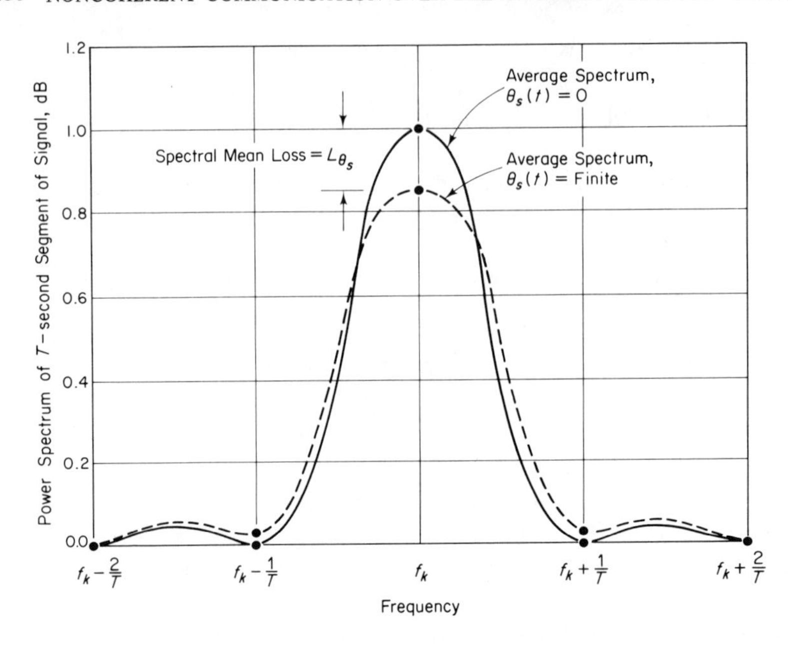

Fig. 10-1 Average power spectra of signal

frequencies far from the carrier the transmitter spectrum is highly dependent on the spectrum of $\dot{\theta}_s(t)$, while in the vicinity of $f_0 + f_d$ the transmitter's power spectrum is "bell shaped" and essentially independent of the spectrum of $\dot{\theta}_s(t)$. Unfortunately, little is known at present about the exact characteristics of $\theta_s(t)$ or $\dot{\theta}_s(t)$, and hence a system performance evaluation that includes its effect can at best be approximate. Later in this chapter we suggest several methods that can be used to compensate for the short-term instability, but, in view of the lack of knowledge about its characteristics, we cannot provide conclusive results.

A special case of (10-1) is the so-called "ideal" MFSK transmitter, which is assumed to have no phase instabilities, no frequency uncertainty, and no amplitude variations, and which transmits signals chosen from a set of N orthogonal tones. For such a transmitter, (10-1) simplifies to

$$x_k(t) = \sqrt{2S} \cos \omega_k t \qquad \begin{matrix} 0 \leq t \leq T \\ k = 1, 2, \ldots, N \end{matrix} \tag{10-4}$$

where $\omega_k = 2\pi(f_0 + k/T)$. The next section is concerned with optimum noncoherent detection of this signal set.

10-3. OPTIMUM NONCOHERENT DETECTION

The term *noncoherent* as it is used here implies a lack of knowledge of the received carrier phase. Hence, corresponding to the transmission of $x_i(t)$

as specified by (10-4), we observe at the receiver input

$$y(t) = \sqrt{2S}\cos(\omega_i t + \theta) + n(t) \qquad 0 \le t \le T \tag{10-5}$$

where θ is the random phase shift, introduced by the channel, which is assumed to be a uniformly distributed random variable between $-\pi$ and π, and independent of frequency. For the present we assume that the receiver is operating with perfect *time* and *frequency* synchronization; that is, it has a complete knowledge of the instant in time at which the modulation may change from one tone to another and also the exact location of the carrier frequency. Later this assumption is relaxed to the point where noisy estimates of this information are considered in the characterization of system performance.

Optimum Receiver Structures

Under the foregoing assumptions, it may be shown[3] that the optimum receiver in the sense of maximum *a posteriori* probability computes from the observed data $y(t)$ the following set of samples:

$$M_k^2 = X_k^2 + Y_k^2 \qquad k = 1, 2, \ldots, N \tag{10-6}$$

where

$$X_k = \sqrt{2S}\int_0^T y(t)\cos\omega_k t\, dt \quad \text{and} \quad Y_k = \sqrt{2S}\int_0^T y(t)\sin\omega_k t\, dt \tag{10-7}$$

and selects, on the basis of these samples, the signal that gives rise to the largest M_k^2. We note that (10-6) and (10-7) depend on the *envelope* of the cross-correlation function of the received signal $y(t)$ and the kth stored message waveform. By "stored" we mean that the waveforms are stored in time and in frequency (assumed perfect). Alternately, the probability computer contains a set of N matched filters (one matched to each signaling waveform), each followed by an envelope detector (usually a linear rectifier and low-pass filter is sufficient) and a device that samples the output envelopes. Figure 10-2 illustrates for one particular element the cross-correlation version of the receiver and its alternate matched filter realization.

An alternate implementation of the above receiver is its mechanization in the frequency domain as a *spectrum analyzer receiver*. This implementation is realized as follows. First we write M_k^2, the test statistic, in terms of X_k and Y_k; that is,

$$M_k^2 = 2S\left[\left(\int_0^T y(t)\cos\omega_k t\, dt\right)^2 + \left(\int_0^T y(t)\sin\omega_k t\, dt\right)^2\right] \tag{10-8}$$

This may be rewritten as

$$M_k^2 = 2S\left\{\int_0^T \int_{-t}^{T-t} y(t)y(t+\tau)[\cos\omega_k t\cos\omega_k(t+\tau) + \sin\omega_k t\sin\omega_k(t+\tau)]\, d\tau\, dt\right\} \tag{10-9}$$

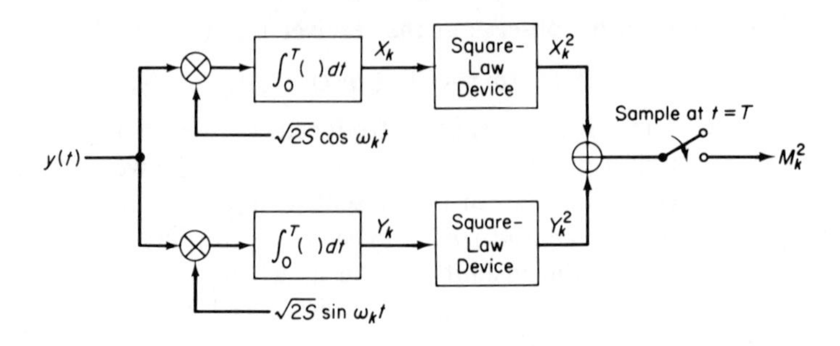

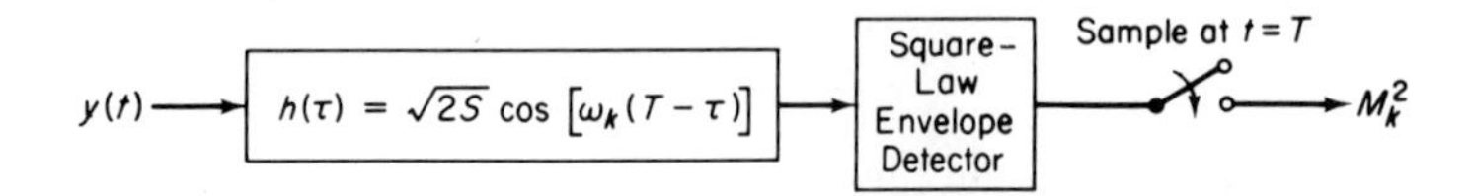

Fig. 10-2 Alternate ways to mechanize an element of the optimum noncoherent receiver

Using the well-known identity for the cosine of the difference of two angles, (10-9) simplifies to

$$M_k^2 = 2S\int_0^T \int_{-t}^{T-t} y(t)y(t+\tau)\cos\omega_k\tau\, d\tau\, dt \tag{10-10}$$

Using simple algebraic manipulations, it is relatively straightforward to show that the above is identically equivalent to

$$M_k^2 = 4S\int_0^T \int_0^{T-t} y(t)y(t+\tau)\cos\omega_k\tau\, d\tau\, dt \tag{10-11}$$

Defining the finite time autocorrelation function of $y(t)$ as

$$\phi_y(\tau) = \int_0^{T-t} y(t)y(t+\tau)\, dt \qquad 0 \leq \tau \leq T \tag{10-12}$$

we see that (10-11) may be written as

$$M_k^2 = 4S\int_0^T \phi_y(\tau)\cos\omega_k\tau\, d\tau \tag{10-13}$$

Since $\phi_y(\tau)$ is an even function, the quantity M_k^2 as defined by (10-13) may be recognized as the spectral density of the receiver input evaluated at the frequency f_k. Thus

$$M_k^2 = 4S\Phi_y(f_k) \qquad k = 1, 2, \ldots, N \tag{10-14}$$

where $\Phi_y(f)$ is the Fourier transform of the autocorrelation function $\phi_y(\tau)$. This interpretation of the decision process leads to the high-pass spectrum analyzer receiver of Fig. 10-3. This mechanization allows for the use of a digital computer to perform data processing and detection.

In practice, demodulation is the first operation performed on the incom-

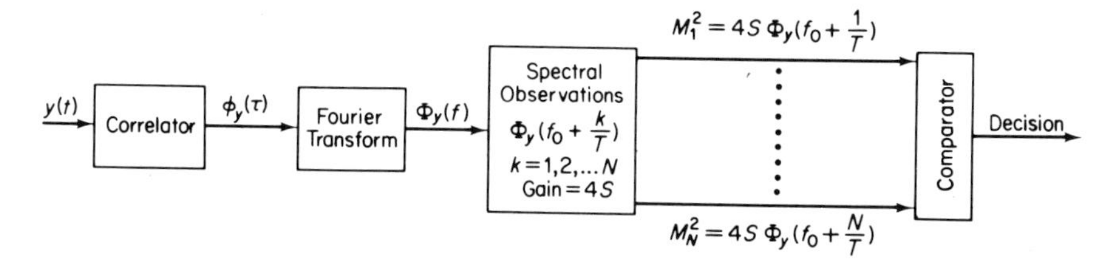

Fig. 10-3 High-pass spectrum analyzer receiver

ing signal. The high-pass waveform $y(t)$ is heterodyned to a low-pass equivalent $z(t)$ by means of a reference waveform at the carrier frequency. Thus any practical receiver must be provided with an estimate of f_0. This heterodyning operation reduces the high-pass spectrum analyzer receiver of Fig. 10-3 to the low-pass equivalent illustrated in Fig. 10-4 with the corresponding set of spectral observations $\{\Phi_z(k/T), k = 1, 2, \ldots, N\}$.

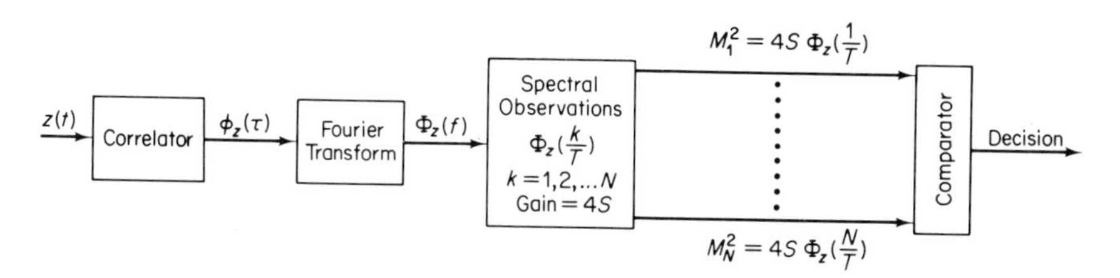

Fig. 10-4 Low-pass spectrum analyzer receiver

Error Probability Performance of the Optimum Receiver

The performance of the three receiver mechanizations indicated in Figures 10-2, 10-3, and 10-4 are identical. In fact, it has been shown[4-6] that the word error probability is given by

$$P_E(N) = 1 - \int_0^\infty \exp[-(x + R_d)] I_0[\sqrt{4R_d x}] G(x)\, dx \tag{10-15}$$

where $G(x) \triangleq [1 - \exp(-x)]^{N-1}$

and $R_d = ST/N_0$ denotes the data signal-to-noise ratio. Actually, this expression for error probability is due to Reiger[7] and Turin.[3] On expanding $G(x)$ by the binomial theorem and performing a term-by-term integration, we find the word error probability for a particular degree of coding to be given by

$$P_E(N) = \sum_{k=1}^{N-1} \frac{(-1)^{k+1}}{k+1} {}_{N-1}C_k \exp\left(-\frac{kR_d}{k+1}\right) \tag{10-16}$$

Since the signaling alphabet is orthogonal, the bit error probability $P_B(N)$ is

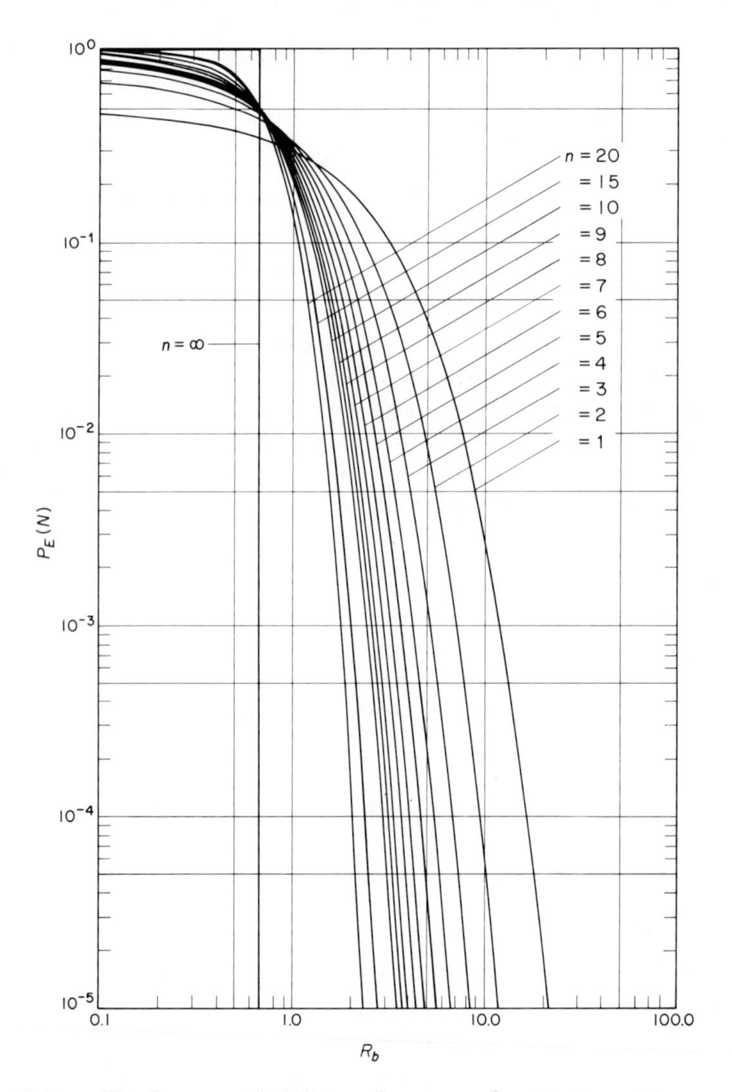

Fig. 10-5 Word error probability performance of an ideal noncoherent receiver (orthogonal signaling)

related to the word error probability by (5-54). The word error probability performance of a noncoherent system that transmits an $n = \log_2 N$ bit word by sending one bit at a time (so-called "uncoded transmission") is given by

$$P_E(N) = 1 - [1 - P_B(2)]^n \tag{10-17}$$

with

$$P_B(2) = \frac{1}{2} \exp\left(-\frac{R_b}{2}\right) \tag{10-18}$$

and $R_b = R_d/n$. Figures 10-5 and 10-6 are plots of (10-15) and (5-54), respec-

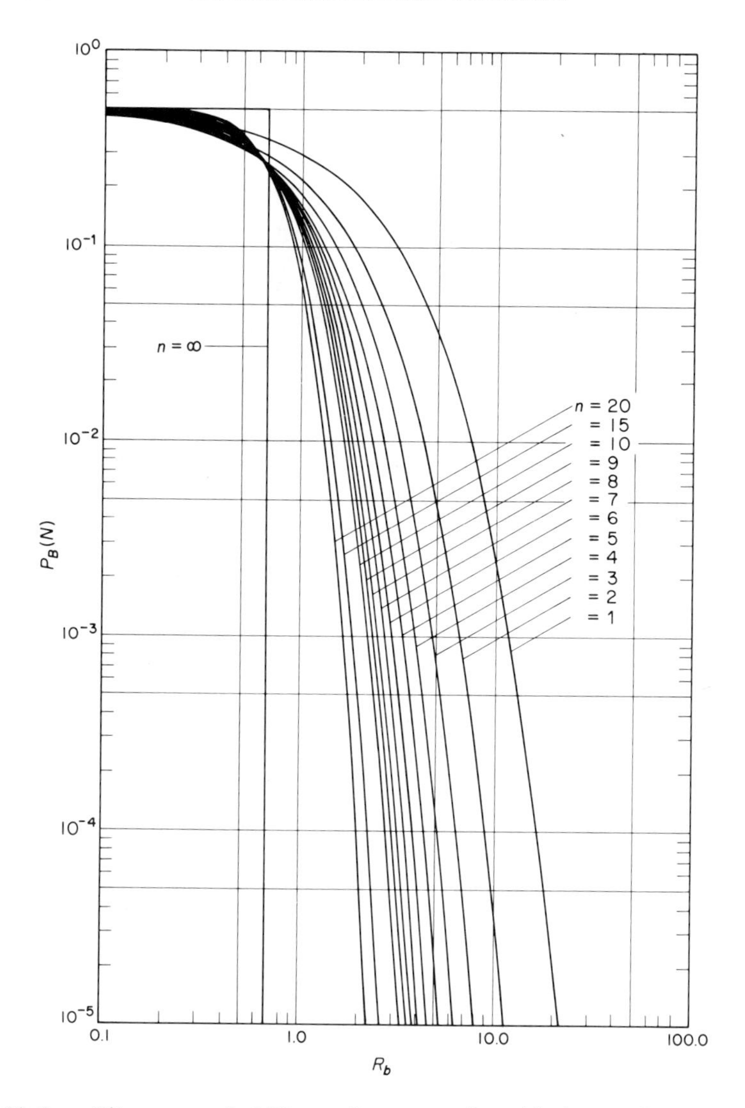

Fig. 10-6 Bit error probability performance of an ideal noncoherent receiver (orthogonal signaling)

tively, while Fig. 10-7 illustrates for various values of n the uncoded word error probability as specified by (10-17) and (10-18). The data corresponding to Figs. 10-5 and 10-6 are tabulated in Table 10-1. Coded and uncoded word error probabilities are compared in Figs. 10-8 and 10-9 for $n = 5$ and 10, respectively. Also shown in these two figures are results given in Chap. 5 for coherent communication employing orthogonal and bi-orthogonal codes. For orthogonal codes we observe that the type of detection performed—

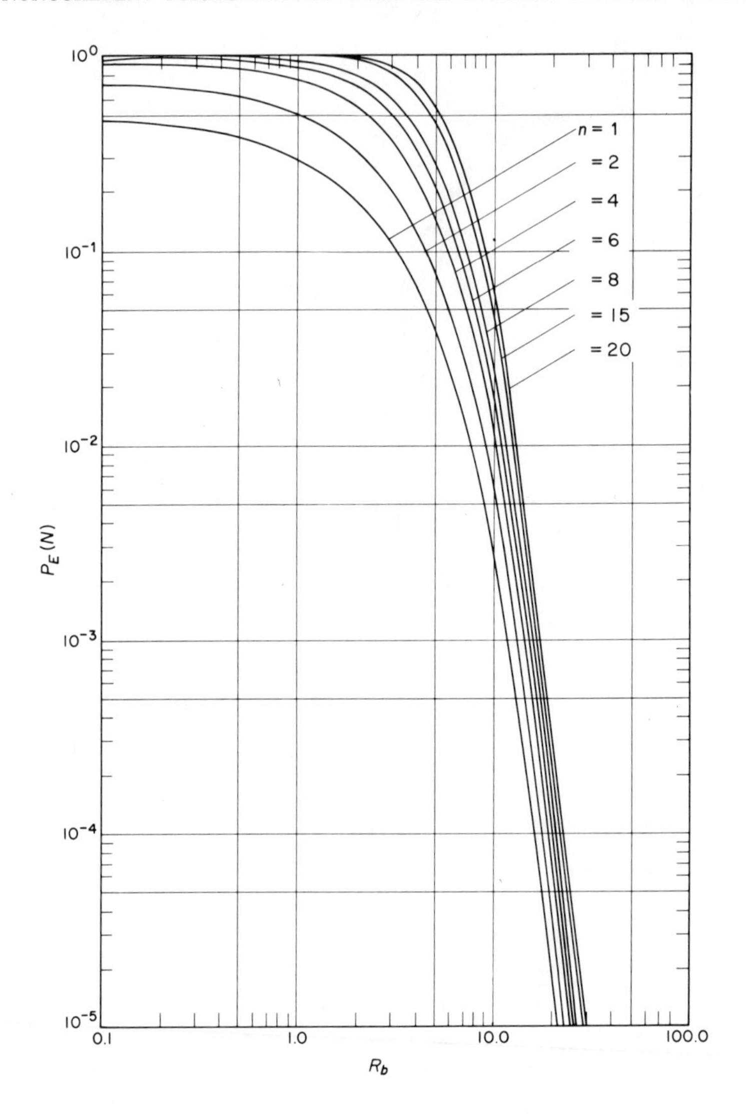

Fig. 10-7 Uncoded word error probability performance of an ideal noncoherent receiver

whether a coherent or noncoherent detector is mechanized—is not very critical for error probabilities less than 10^{-3}. As a matter of fact, the use of 5-bit bi-orthogonal codes and coherent reception is not appreciably more effective (given a certain energy-to-noise ratio) than the use of orthogonal codes and noncoherent reception. On the other hand, a comparison of coded versus uncoded transmission indicates that coding the transmitted signal into n-bit

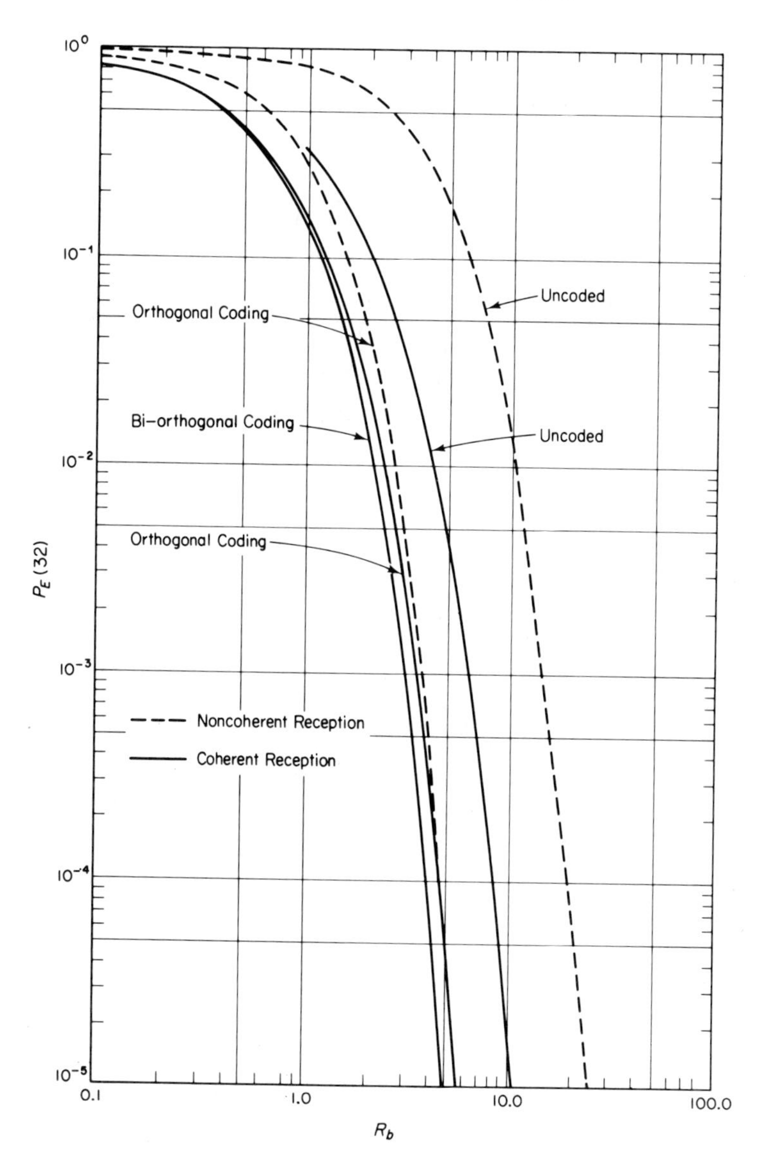

Fig. 10-8 A comparison of coded and uncoded, coherent and noncoherent error probabilities ($n = 5$)

orthogonal words (rather than transmitting an n-bit word one bit at a time) is more effective in noncoherent communications.

For example, if five bits of information are to be sent with a word error probability of 10^{-3} (typical for telemetry systems), the use of orthogonal codes and coherent detection will reduce the required signal-to-noise ratio

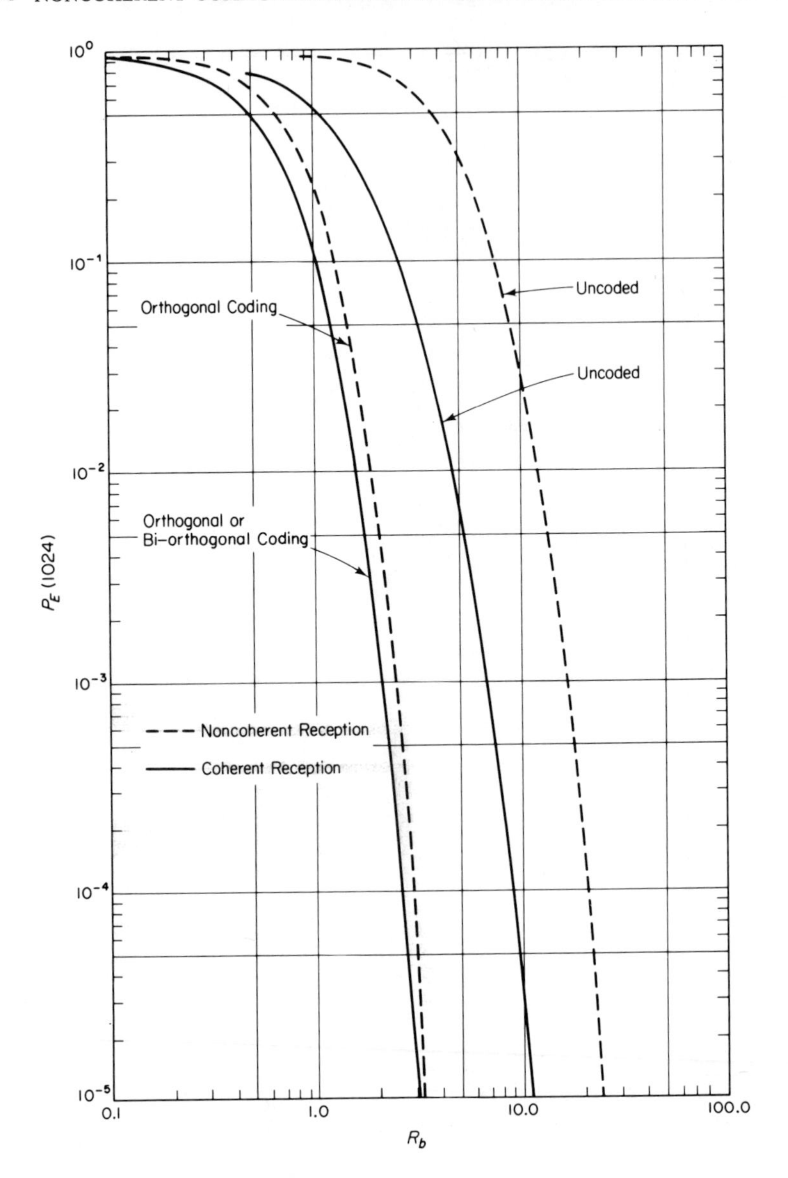

Fig. 10-9 A comparison of coded and uncoded, coherent and noncoherent error probabilities ($n = 10$)

per bit R_b by approximately 2.5 dB under that required for equivalent performance using bit-by-bit detection. Under the same word error probability conditions, the use of orthogonal codes and noncoherent detection will reduce the required R_b by approximately 6 dB under that required for equivalent performance using bit-by-bit detection. For ten bits and $P_E(1024) = 10^{-3}$, or-

TABLE 10-1 Table of Error Probabilities*

n	R_b	$P_E(N)$	$P_B(N)$
1	0.10	0.47566299	0.47566299
1	0.20	0.45245639	0.45245639
1	0.30	0.43038283	0.43038283
1	0.40	0.40938688	0.40938688
1	0.50	0.38941585	0.38941585
1	0.60	0.37041960	0.37041960
1	0.70	0.35235050	0.35235050
1	0.80	0.33516322	0.33516322
1	0.90	0.31881469	0.31881469
1	1.00	0.30326390	0.30326390
1	2.00	0.18393379	0.18393379
1	3.00	0.11156219	0.11156219
1	4.00	0.67666810 (−1)	0.67666810 (−1)
1	5.00	0.41042460 (−1)	0.41042460 (−1)
1	6.00	0.24893686 (−1)	0.24893686 (−1)
1	7.00	0.15098834 (−1)	0.15098834 (−1)
1	8.00	0.91579139 (−2)	0.91579139 (−2)
1	9.00	0.55545525 (−2)	0.55545525 (−2)
1	10.00	0.33690023 (−2)	0.33690023 (−2)
1	20.00	0.22702006 (−4)	0.22702006 (−4)
1	30.00	0.15491873 (−6)	0.15491873 (−6)
2	0.10	0.69729300	0.46486200
2	0.20	0.64740086	0.43160057
2	0.30	0.60033873	0.40022582
2	0.40	0.55605758	0.37070505
2	0.50	0.51448822	0.34299215
2	0.60	0.47554594	0.31703062
2	0.70	0.43913440	0.29275626
2	0.80	0.40514895	0.27009930
2	0.90	0.37347940	0.24898626
2	1.00	0.34401217	0.22934145
2	2.00	0.14596746	0.97311642 (−1)
2	3.00	0.59142382 (−1)	0.39428255 (−1)
2	4.00	0.23265217 (−1)	0.15510145 (−1)
2	5.00	0.89725611 (−2)	0.59817073 (−2)
2	6.00	0.34135198 (−2)	0.22756799 (−2)
2	7.00	0.12862816 (−2)	0.85752111 (−3)
2	8.00	0.48142263 (−3)	0.32094842 (−3)
2	9.00	0.17931499 (−3)	0.11954332 (−3)
2	10.00	0.66558807 (−4)	0.44372537 (−4)
2	20.00	0.52466877 (−8)	0.34977917 (−8)
3	0.10	0.81067010	0.46324006
3	0.20	0.74726009	0.42700577
3	0.30	0.68567334	0.39181334
3	0.40	0.62657519	0.35804297

* The negative integer in parentheses following each entry in the table represents the power of ten by which the entry should be multiplied.

Table of Error Probabilities (Cont'd.)

n	R_b	$P_E(N)$	$P_B(N)$
3	0.50	0.57043348	0.32596199
3	0.60	0.51755436	0.29574534
3	0.70	0.46811344	0.26749339
3	0.80	0.42218257	0.24124718
3	0.90	0.37975258	0.21700148
3	1.00	0.34075232	0.19471561
3	2.00	0.10403854	0.59450594 (−1)
3	3.00	0.28086227 (−1)	0.16049273 (−1)
3	4.00	0.70613553 (−2)	0.40350601 (−2)
3	5.00	0.16994870 (−2)	0.97113545 (−3)
3	6.00	0.39791019 (−3)	0.22737724 (−3)
3	7.00	0.91543186 (−4)	0.52310392 (−4)
3	8.00	0.20826269 (−4)	0.11900725 (−4)
3	9.00	0.47052477 (−5)	0.26887130 (−5)
4	0.10	0.87480467	0.46656249
4	0.20	0.80776387	0.43080740
4	0.30	0.73896835	0.39411645
4	0.40	0.67047250	0.35758533
4	0.50	0.60384502	0.32205068
4	0.60	0.54023067	0.28812303
4	0.70	0.48041543	0.25622156
4	0.80	0.42489003	0.22660802
4	0.90	0.37390866	0.19941795
4	1.00	0.32754132	0.17468870
4	2.00	0.72300648 (−1)	0.38560346 (−1)
4	3.00	0.12880714 (−1)	0.68697145 (−2)
4	4.00	0.20424174 (−2)	0.10892893 (−2)
4	5.00	0.30286457 (−3)	0.16152777 (−3)
4	6.00	0.43187463 (−4)	0.23033313 (−4)
4	7.00	0.60210688 (−5)	0.32112367 (−5)
5	0.10	0.91417912	0.47183438
5	0.20	0.84895959	0.43817269
5	0.30	0.77720004	0.40113550
5	0.40	0.70247324	0.36256683
5	0.50	0.62770140	0.32397491
5	0.60	0.55513959	0.28652366
5	0.70	0.48641986	0.25105541
5	0.80	0.42262875	0.21813096
5	0.90	0.36439869	0.18807674
5	1.00	0.31200098	0.16103277
5	2.00	0.49863067 (−1)	0.25735777 (−1)
5	3.00	0.58190925 (−2)	0.30034026 (−2)
5	4.00	0.57668620 (−3)	0.29764449 (−3)
5	5.00	0.52287368 (−4)	0.26987028 (−4)
5	6.00	0.45188212 (−5)	0.23322948 (−5)
6	0.10	0.93979762	0.47735752
6	0.20	0.87905609	0.44650468

Table of Error Probabilities (Cont'd.)

n	R_b	$P_E(N)$	$P_B(N)$
6	0.30	0.80685618	0.40983171
6	0.40	0.72794692	0.36975082
6	0.50	0.64658495	0.32842410
6	0.60	0.56626057	0.28762442
6	0.70	0.48961055	0.24869107
6	0.80	0.41845143	0.21254675
6	0.90	0.35388105	0.17974910
6	1.00	0.29641064	0.15055778
6	2.00	0.34335385 (−1)	0.17440195 (−1)
6	3.00	0.26108418 (−2)	0.13261419 (−2)
6	4.00	0.16071591 (−3)	0.81633480 (−4)
6	5.00	0.88698038 (−5)	0.45052972 (−5)
7	0.10	0.95711961	0.48232799
7	0.20	0.90197754	0.45453987
7	0.30	0.83091322	0.41872792
7	0.40	0.74927559	0.37758770
7	0.50	0.66247386	0.33384509
7	0.60	0.57526560	0.28989762
7	0.70	0.49140357	0.24763644
7	0.80	0.41354422	0.20840023
7	0.90	0.34332261	0.17301296
7	1.00	0.28151613	0.14186639
7	2.00	0.23662358 (−1)	0.11924338 (−1)
7	3.00	0.11677939 (−2)	0.58849460 (−3)
7	4.00	0.44457374 (−4)	0.22403716 (−4)
7	5.00	0.14906905 (−5)	0.75121413 (−6)
8	0.10	0.96913247	0.48646649
8	0.20	0.91988883	0.46174812
8	0.30	0.85096917	0.42715315
8	0.40	0.76769318	0.38535187
8	0.50	0.67635687	0.33950462
8	0.60	0.58295815	0.29262213
8	0.70	0.49244574	0.24718845
8	0.80	0.40841021	0.20500591
8	0.90	0.33308576	0.16719599
8	1.00	0.26753381	0.13429148
8	2.00	0.16334709 (−1)	0.81993833 (−2)
8	3.00	0.52168079 (−3)	0.26186330 (−3)
8	4.00	0.12243868 (−4)	0.61459419 (−5)
8	5.00	0.25043727 (−6)	0.12570968 (−6)
9	0.10	0.97760764	0.48976038
9	0.20	0.93412402	0.46797603
9	0.30	0.86799131	0.43484496
9	0.40	0.78391840	0.39272624
9	0.50	0.68878537	0.34506664
9	0.60	0.58976436	0.29545925
9	0.70	0.49305712	0.24701100

Table of Error Probabilities (Cont'd.)

n	R_b	$P_E(N)$	$P_B(N)$
9	0.80	0.40327123	0.20203020
9	0.90	0.32329471	0.16196369
9	1.00	0.25448456	0.12749128
9	2.00	0.11298195 (−1)	0.56601523 (−2)
9	3.00	0.23295512 (−3)	0.11670550 (−3)
9	4.00	0.33635154 (−5)	0.16850488 (−5)
10	0.10	0.98365984	0.49231069
10	0.20	0.94557257	0.47324844
10	0.30	0.88261934	0.44174106
10	0.40	0.79840960	0.39959503
10	0.50	0.70009521	0.35038979
10	0.60	0.59592947	0.29825599
10	0.70	0.49340657	0.24694444
10	0.80	0.39822520	0.19930723
10	0.90	0.31397626	0.15714159
10	1.00	0.24231875	0.12127781
10	2.00	0.78295896 (−2)	0.39186215 (−2)
10	3.00	0.10402556 (−3)	0.52063625 (−4)
10	4.00	0.92324475 (−6)	0.46207362 (−6)
15	0.10	0.99642596	0.49822818
15	0.20	0.97806794	0.48904889
15	0.30	0.93228243	0.46615544
15	0.40	0.85337610	0.42670107
15	0.50	0.74569228	0.37285751
15	0.60	0.62101884	0.31051889
15	0.70	0.49348616	0.24675061
15	0.80	0.37514593	0.18757869
15	0.90	0.27370068	0.13685451
15	1.00	0.19228964	0.96147759 (−1)
15	2.00	0.12810497 (−2)	0.64054440 (−3)
15	3.00	0.18542031 (−5)	0.92712985 (−6)
20	0.10	0.99917572	0.49958833
20	0.20	0.99075913	0.49538004
20	0.30	0.95947231	0.47973661
20	0.40	0.89028011	0.44514047
20	0.50	0.78007104	0.39003589
20	0.60	0.64073464	0.32036763
20	0.70	0.49265479	0.24632763
20	0.80	0.35536443	0.17768238
20	0.90	0.24143049	0.12071536
20	1.00	0.15520509	0.77602618 (−1)
20	2.00	0.21528282 (−3)	0.10764151 (−3)
20	3.00	0.35731771 (−7)	0.17865902 (−7)

thogonal or bi-orthogonal coding and coherent reception reduce the required R_b by approximately 5 dB under that required without coding. Finally, using noncoherent reception, 10-bit orthogonal codes, and $P_E(1024) = 10^{-3}$, one realizes an improvement of approximately 8.5 dB over the uncoded noncoherent procedure of bit-by-bit detection.

Finally by a limiting procedure similar to that employed in Chap. 5, it may be shown[6] that the asymptotic ($N \longrightarrow \infty$) error probability performance of the optimum noncoherent receiver is identical to that for coherent communications as described by (5-75).

10-4. SUBOPTIMUM NONCOHERENT DETECTION

Exact evaluation of the spectral observations is difficult when one elects to implement the spectrum analyzer receivers of Figs. 10-3 and 10-4. However, it is possible to approximate the spectral estimate $\Phi_z(f)$ which then results in a receiver performance which is suboptimum.

Techniques for Approximating the Evaluation of the Spectral Observations

There are various approaches to approximating the spectral observations $\{\Phi_z(k/T), k = 1, \ldots, N\}$. First, one can evaluate the autocorrelation function $\phi_z(\tau)$ directly and then evaluate the Fourier transform to obtain $\Phi_z(f)$.[8] This is referred to as the *autocorrelation function* method.

A practical realization of this method that is convenient for digital processing is discussed in Refs. 9 and 10 and summarized here. Since the Fourier transform of the random process $z(t)$ is essentially zero outside the frequency interval $|f| \leq N/T$, then in accordance with the sampling theorem[11] we have, to a good approximation,

$$z(t) \cong z_a(t) \triangleq \sum_{n=1}^{2BT} z\left(\frac{n}{2B}\right) \operatorname{sinc}\left[2\pi B\left(t - \frac{n}{2B}\right)\right] \qquad (10\text{-}19)$$

where $\operatorname{sinc} x \triangleq \sin x/x$, $B \triangleq N/T$, and $z(n/2B)$ is a sample of $z(t)$ at $t = n/2B$. In terms of the above, the autocorrelation function of $z_a(t)$ is given by

$$\phi_{z_a}(\tau) = \int_0^{T-\tau} z_a(t) z_a(t+\tau)\, dt = \sum_{n=1}^{2BT} \sum_{m=1}^{2BT} z_n z_m \operatorname{sinc}\left[2\pi B\left(\tau + \frac{n-m}{2B}\right)\right] \qquad (10\text{-}20)$$

which approximates that of $\phi_z(\tau)$. We observe that for $\tau = k/2B$, k an integer, the above expression simplifies to

$$\phi_{z_a}\left(\frac{k}{2B}\right) = \begin{cases} \dfrac{1}{2B} \displaystyle\sum_{n=1}^{2BT-k} z_n z_{n+k} & k = 0, \pm 1, \pm 2, \ldots, \pm(2BT-1) \\ 0 & \text{otherwise} \end{cases} \qquad (10\text{-}21)$$

Since the spectrum of $\phi_{z_a}(\tau)$ is limited to the frequency range $|f| \leq B$ one obtains by means of the sampling theorem

$$\phi_{z_a}(\tau) = \sum_{k=-(2BT-1)}^{2BT-1} \phi_{z_a}\left(\frac{k}{2B}\right) \operatorname{sinc}\left[2\pi B\left(\tau - \frac{k}{2B}\right)\right] \tag{10-22}$$

Taking Fourier transforms, the spectrum of $z(t)$ is approximated by

$$\Phi_z(f) \cong \Phi_{z_a}(f) = \frac{1}{2B}\phi_{z_a}(0) + \frac{1}{B}\sum_{k=1}^{2BT-1} \phi_{z_a}\left(\frac{k}{2B}\right) \cos\left(\frac{\pi k f}{B}\right) \qquad -B \leq f \leq B \tag{10-23}$$

These expressions for the correlation coefficients and for the approximate spectral observations $\Phi_{z_a}(k/T)$ form the basis for the computational procedure.

Second, one can calculate the spectrum by the *discrete Fourier transform* approach. This approach takes samples $z(k/2B)$, $k = 1, 2, \ldots, 2BT$; evaluates

$$X(f) = \sum_{k=1}^{N} z\left(\frac{k}{2B}\right) \cos\left(\frac{\pi k f}{B}\right) \quad \text{and} \quad Y(f) = \sum_{k=1}^{N} z\left(\frac{k}{2B}\right) \sin\left(\frac{\pi k f}{B}\right) \tag{10-24}$$

and then computes the spectrum

$$\Phi_z(f) \cong X^2(f) + Y^2(f) \tag{10-25}$$

which yields the spectral observations $\Phi_z(k/T)$, $k = 1, 2, \ldots, 2BT$. The discrete Fourier transform method permits greatly simplified analysis.[12]

Third, the receiver can be implemented with a small computer possessing an analog-to-digital converter programmed with the *Cooley-Tukey Fast Fourier Transform* algorithm.[8,13] The Fast Fourier Transform (FFT) approach produces results at frequency multiples of $1/T$ provided that the number of samples of the process is a power of two. Although the FFT method does not give better results than the discrete Fourier transform or the autocorrelation function method, it has the advantage that considerably less time is required to perform numerical operations.

Error Probability Performance in the Presence of Time Domain Truncation

Here we consider the error probability performance of the spectrum analyzer receiver when spectral estimates are evaluated by means of *time domain truncation* of the autocorrelation function. This approach reduces the amount of computation time required but increases the error probability.

The simplest form of time domain truncation of the autocorrelation function is to apply a rectangular window to the function; that is, compute only $\phi_{z_a}(j/2B)$ for $j = 0, 1, 2, \ldots, J$ where $J \leq 2BT - 1$. When the correlation function is truncated in this manner, the computation of the *approximate spectral observation* $\Phi_{z_a}(k/T)$ is based on the expression

$$\phi_{z_a}(\tau; J) \triangleq \phi_{z_a}(\tau)W(\tau) \tag{10-26}$$

where $W(\tau)$ is the *truncating window* defined by

$$W(\tau) \triangleq \begin{cases} 1 & |\tau| \leq \left(\frac{J+1}{2BT}\right)T \\ 0 & \text{all other } \tau \end{cases} \tag{10-27}$$

Hence, from (10-19),

$$\Phi_z(f) \cong \Phi_{z_a}(f; J) \triangleq \frac{\phi_{z_a}(0)}{2B} + \frac{1}{B}\sum_{j=1}^{J} \phi_{z_a}\left(\frac{j}{2B}\right)\cos\left(\frac{\pi j f}{B}\right) \qquad |f| \leq B \tag{10-28}$$

The quantity $1 - [(J+1)/2BT]$ is a measure of the fractional truncation. With truncation, the frequency separation Δf between tones must be adjusted so that they remain envelope orthogonal at the receiver. Thus,

$$\Delta f = \frac{1}{T}\left(\frac{2BT}{J+1}\right) \tag{10-29}$$

yields orthogonal signals when truncation is employed. Hence, the minimum total transmission bandwidth B_{tr} required for N tones must now satisfy the inequality

$$B_{tr} = \frac{N}{T}\left(\frac{2BT}{J+1}\right) \geq \frac{N}{T} \tag{10-30}$$

Using a result from the theory of sequences of j-dependent random variables,[14] Springett and Charles[10] are able to approximate the p.d.f. of the spectral estimates $\Phi_{z_a}(f_l; J)$, $l = 1, 2, \ldots, N$ and hence the error probability performance under the assumptions $S/N_0B \ll 1$ and $J + 1 \ll 2BT$. Under these conditions, and assuming that the signal corresponding to f_k was in fact transmitted, the p.d.f. of $\Phi_{z_a}(f; J)$ at $f = f_l$ is Gaussian with mean

$$\mu_l \triangleq E\{\Phi_{z_a}(f_l; J)\} = \frac{T}{2B}[N_0B + R_{s_k}(0)] + 2\sum_{j=1}^{J}\left[\frac{2BT - j}{(2B)^2}\right]R_{s_k}(j)\cos\left[2\pi f_l\left(\frac{j}{2B}\right)\right] \tag{10-31}$$

Here

$$R_{s_k}(j) \triangleq S\cos\left[2\pi f_k\left(\frac{j}{2B}\right)\right] \tag{10-32}$$

The variance of $\Phi_{z_a}(f_l; J)$—that is,

$$\sigma_l^2 \triangleq E\{\Phi_{z_a}^2(f_l; J)\} - [E\{\Phi_{z_a}(f_l; J)\}]^2 \tag{10-33}$$

is quite tedious to evaluate and hence we outline only the major steps. The first term in (10-33) can be written as a double sum, namely,

$$E\{\Phi_{z_a}^2(f_l; J)\} = \sum_{j=0}^{J}\sum_{h=0}^{J}\epsilon_j\epsilon_h\left(\frac{1}{2B}\right)^2 E\{\phi_{z_a}(j)\phi_{z_a}(h)\}\cos 2\pi f_l\left(\frac{j}{2B}\right)\cos 2\pi f_l\left(\frac{h}{2B}\right) \tag{10-34}$$

where $\epsilon_0 = 1$ and $\epsilon_n = 2$ $(n \neq 0)$. The elements $E\{\phi_{z_a}(j)\phi_{z_a}(h)\}$ are now considered for the following four cases: (1) $j = h = 0$; (2) $j = 0,\ h \neq 0$; (3) $j = h \neq 0$; and (4) $j \neq h \neq 0,\ j < h$. The results are:

$$
\begin{aligned}
E\{\phi_{z_a}(j)\phi_{z_a}(h)\} \\
&= \left(\frac{T}{2B}\right)(N_0B)^2[2BT(1 + R_d\delta)^2 + 2(1 + 2R_d\delta)] \qquad j = h = 0 \\
&= \frac{2BT - h}{(2B)^2}(N_0B)[2BT(1 + R_d\delta) + 4]R_{s_k}(h) \qquad j = 0,\ h \neq 0 \\
&= \frac{2BT - h}{(2B)^2}(N_0B)^2\left\{1 + [(2BT - h)(R_d\delta)^2 + 4R_d\delta]\frac{R^2_{s_k}(h)}{R^2_{s_k}(0)}\right. \\
&\qquad \left. - \left(\frac{R_d\delta}{2BT - h}\right)\left[2h\left(\frac{R_{s_k}(2h)}{R_{s_k}(0)} - \frac{R_d\delta}{2}\right)\right]\right\} \qquad j = h \neq 0
\end{aligned}
$$

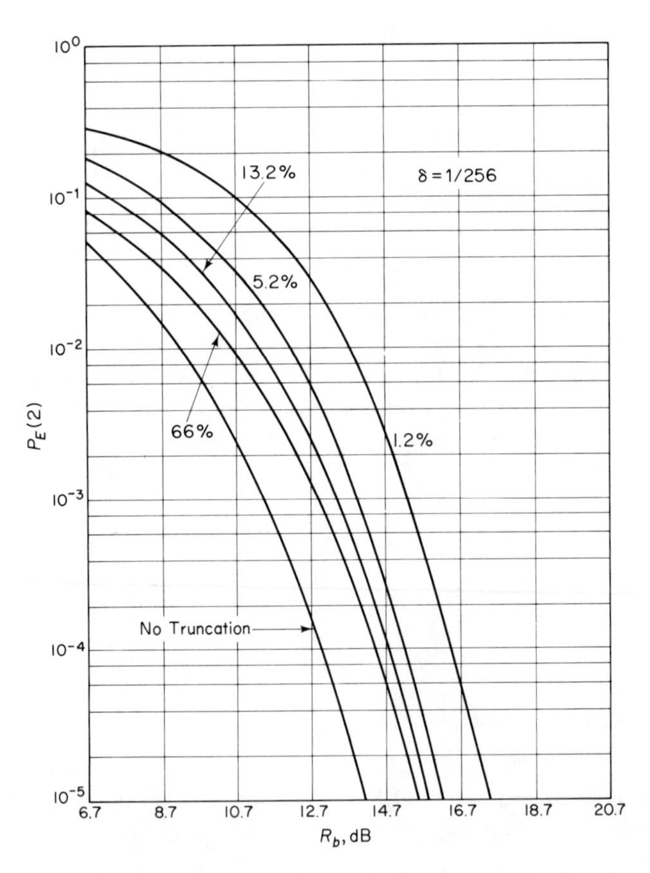

Fig. 10-10 Probability of error performance for the spectrum analyzer receiver with autocorrelation function truncation; $n = 1$ (Courtesy of F. J. Charles and J. C. Springett)

$$= \frac{2BT - h}{(2B)^2}(N_0 B)^2 \left\{[(2BT - j)(R_d\delta)^2 + 2R_d\delta(2 + R_d\delta)]\frac{R_{s_k}(j)R_{s_k}(h)}{R_{s_k}^2(0)} - \frac{jR_d\delta}{2BT - h}(2 + R_d\delta)\frac{R_{s_k}(j+h)}{R_{s_k}(0)}\right\} \qquad j \neq h \neq 0, j < h \qquad (10\text{-}35)$$

where

$$\delta \triangleq \frac{1}{BT} \qquad (10\text{-}36)$$

The variance of the spectral estimates can now be obtained by weighting each j, h element of the covariance matrix by $\epsilon_j \epsilon_h (1/2B)^2$ [$\cos 2\pi f_l(j/2B) \times \cos 2\pi f_l(h/2B)$], summing the resulting $(J + 1)^2$ elements, and subtracting the square of the mean. Although closed-form expressions can be obtained for both (10-31) and (10-33), it is simpler to evaluate them numerically on a digital computer. Assuming that the N Gaussian distributed spectral esti-

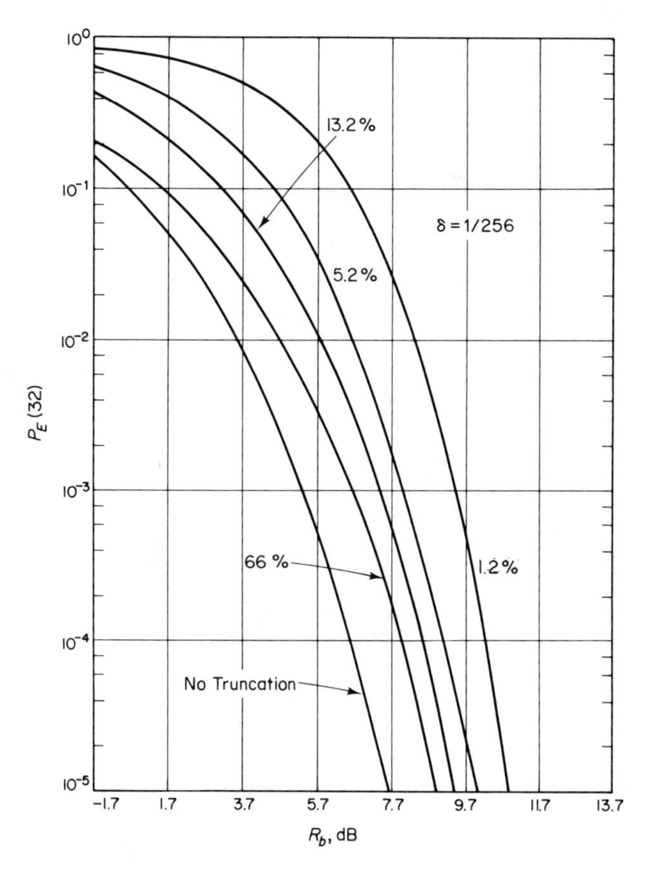

Fig. 10-11 Probability of error performance for the spectrum analyzer receiver with autocorrelation function truncation; $n = 5$ (Courtesy of F. J. Charles and J. C. Springett)

mates $\Phi_{z_a}(f_1; J), \Phi_{z_a}(f_2; J), \ldots, \Phi_{z_a}(f_N; J)$ are uncorrelated, an expression for the probability of correct reception of a word is given by

$$P_c(N) = \frac{1}{\sqrt{\pi}} \int_{-\infty}^{\infty} \exp(-x^2) \left\{ \frac{1}{2} \operatorname{erfc}\left[-\frac{\sigma_k}{\sigma_l} x - \frac{\mu_k - \mu_l}{\sqrt{2}\,\sigma_l} \right] \right\}^{N-1} dx \qquad (10\text{-}37)$$

where the μ's and σ are defined in (10-31) and (10-33) respectively. Also in (10-37), any value of $l \neq k$ is suitable since μ_l and σ_l are independent of l for $l \neq k$ (see Prob. 10-2). Numerical computation of $P_E(N) = 1 - P_c(N)$ was carried out in Ref. 10 for various values of the normalized window width $(J + 1)/2BT$ in percent. The results are illustrated in Figs. 10-10 and 10-11 where the word error probability $P_E(N)$ is plotted versus the signal-to-noise ratio per bit $R_b = R_d/n$ in dB for $n = 1$ and $n = 5$ bits, respectively. Although the results were plotted for small percent truncations (including no truncation), one should be suspect of applying these results to a practical system, since the assumptions regarding the independence and Gaussian nature of the spectral estimates break down when $J \longrightarrow 2BT$. Comparing the no-truncation curve of Figs. 10-10 and 10-11 with the ideal noncoherent results of Fig. 10-5, one observes that the approximation breaks down in such a way as to give a pessimistic estimate of the correct answer.

10-5. NONCOHERENT DETECTION IN THE PRESENCE OF SHORT-TERM OSCILLATOR INSTABILITY

When a short-term frequency instability $\dot{\theta}_s(t)$ frequency modulates the nominal carrier, then the methods and formulas developed here apply in the same approximate sense if the signal correlation function $R_{s_k}(j)$ is redefined to include the effects of the instability. Letting $R_{\dot{\theta}_s}(\tau)$ denote the autocorrelation function of the assumed stationary frequency modulation process,

$$R_{\dot{\theta}_s}(\tau) \triangleq E\{\dot{\theta}_s(t)\dot{\theta}_s(t + \tau)\} \qquad (10\text{-}38)$$

then the modulated signal has a correlation function[15]

$$R_{s_k}(j) = S \cos\left[2\pi f_k \left(\frac{j}{2B}\right)\right] e^{-\Omega(j/2B)} \qquad (10\text{-}39)$$

where

$$\Omega(\tau) \triangleq \int_0^{|\tau|} [|\tau| - z] R_{\dot{\theta}_s}(z)\, dz = 2\int_0^{\infty} S_{\dot{\theta}_s}(f) \frac{[1 - \cos 2\pi f\tau]}{(2\pi f)^2}\, df \qquad (10\text{-}40)$$

and $S_{\dot{\theta}_s}(f)$ is the Fourier transform of $R_{\dot{\theta}_s}(\tau)$. Substituting (10-39) and (10-40) into (10-31) through (10-37), one can evaluate in principle the effect of a short-term instability on system error probability performance.

Intuitively, one might expect that an optimum rectangular truncation window, in the sense of minimizing the error probability, might exist for a fixed S/N_0B and power in the $\dot{\theta}_s(t)$ process because of the tradeoff between the smoothing of the spectrum and the reduction of the peak spectral estimates when truncation is applied in the time domain. Indeed this is so and has been observed experimentally. Analytic support for this phenomenon (at least for large truncation) can be obtained by using the above development once the correlation function of the $\dot{\theta}_s(t)$ process can be characterized. Furthermore, it appears that the optimum "shape" of the window to be applied to the autocorrelation function of the $z_a(t)$ process should be related to the spectrum of $\dot{\theta}_s(t)$. These problems are still open to full investigation.

A second method of compensation for short-term frequency instability, which is somewhat related to the above discussion, would make use of a good estimate of the autocorrelation function of $\theta_s(t)$—that is, $\hat{R}_{\theta_s}(\tau)$—which in turn could be used as a "window" to produce "smoothed" spectral estimates leading to the detection process (Figs. 10-12 and 10-13). The estimate $\hat{R}_{\theta_s}(\tau)$

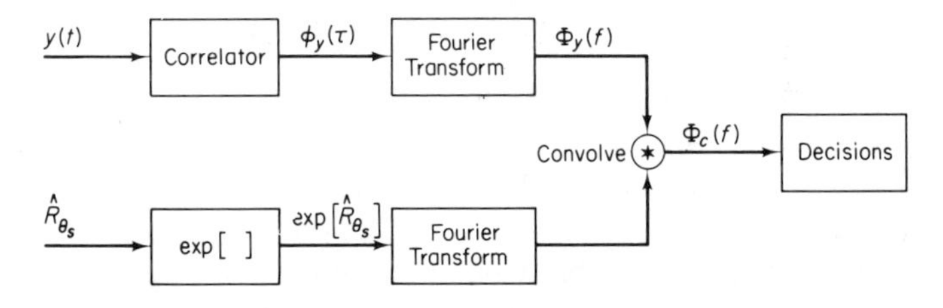

Fig. 10-12 A method of compensating for short-term instability by convolution

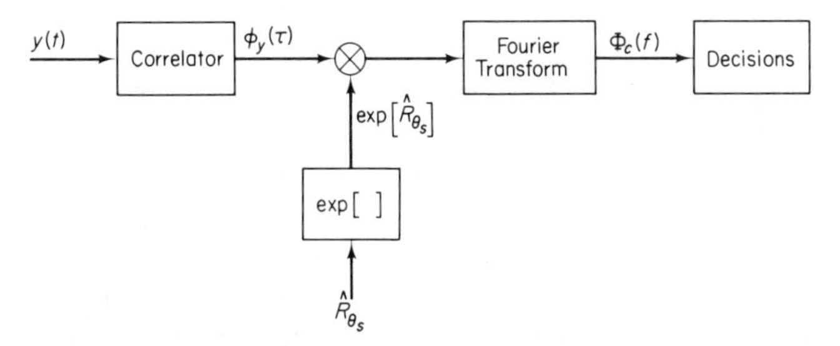

Fig. 10-13 A method of compensating for short-term instability by autocorrelation function weighting

for this so-called "semicoherent" type of detection would have to be obtained prior to actual use of the transmitter. This general idea of incorporating *a priori* knowledge about the statistical characterization of $\theta_s(t)$ obtained from direct measurement with the detection process should not be overlooked. In fact, it has been shown[16] for coherent reception that one should attempt to "match" the receiver with the channel output by passage of the stored waveforms through a channel-simulating filter (*estimator-correlator*) followed by cross-correlation of the output of this filter with the input to the receiver (Fig. 10-14).

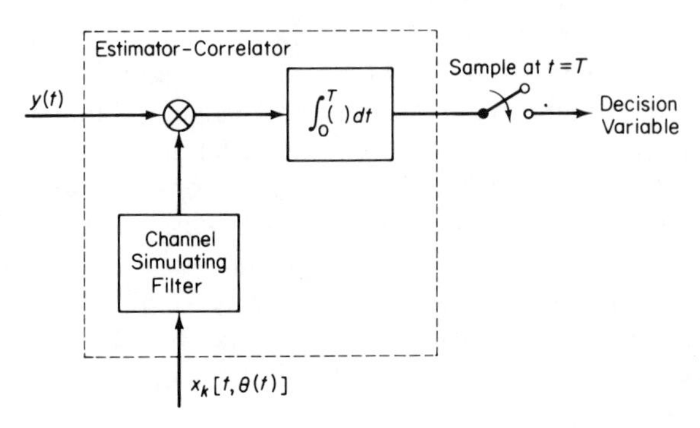

Fig. 10-14 Adapting or matching the receiver to the transmitted signal

To complete the picture, results from studies of the effects on receiver performance when $\theta(t)$ of (10-3) is a deterministic process and assumed to be a frequency ramp clearly show a reduction in effective signal-to-noise ratio (see Prob. 10-4).[17] Finally, it should be pointed out that $\theta(t)$ affects not only the system error probability but also the performance of any frequency-acquisition loop through f_d and any frequency-tracking loop through the processes $\theta_s(t)$ and $\theta_t(t)$. Thus, further study is warranted in these areas to effect a design.

10-6. TIME SYNCHRONIZATION OF THE OPTIMUM RECEIVER

In a noncoherent telemetry system one must be able to provide the receiver with an estimate of the instants in time at which the modulation may change from one tone to another. As is the usual case in estimating zero crossings of a random process, one must first acquire the probable instants when transitions take place and then track the instants by an appropriate tracking loop mechanization.

One method for "time synchronizing the receiver" under the assumption of perfect frequency sync is based on obtaining the maximum-likelihood estimate of the normalized (with respect to T) timing error λ_t between the transitions of the received signal and the present synchronization state of the receiver.[2,12] The synchronization procedure is as follows.

An initial acquisition sequence consisting of the highest and lowest frequencies of the signaling set is transmitted for a fixed interval of time T_{acq}. In each T-sec interval, the maximum-likelihood estimate of λ_t is given in terms of the discrete Fourier spectrum estimates evaluated for the known frequencies in the acquisition sequence by (see Appendix A)

$$\hat{\lambda}_t = \frac{M'_N - M'_1}{2(M'_1 + M'_N)} \tag{10-41}$$

where the prime on M_1 and M_N denotes the fact that the integrals in (10-7) are computed over a T-sec interval in quadrature with the primary detection interval (Fig. 10-15). The conditional p.d.f. of this estimator is given by

$$\begin{aligned} p(\hat{\lambda}_t \mid \lambda_t) &= \left(\frac{1}{4} - \hat{\lambda}_t^2\right) \exp\left[-R_d\left(\lambda_t^2 + \frac{1}{2}\right)\right] \\ &\times \int_0^\infty y^3 \exp\left[-\frac{y^2}{2}\left(\lambda_t^2 + \frac{1}{2}\right)\right] I_0\left[\sqrt{2R_d}\left(\frac{1}{2} - \lambda_t\right)\right. \\ &\times \left.\left(\frac{1}{2} - \hat{\lambda}_t\right)y\right] I_0\left[\sqrt{2R_d}\left(\frac{1}{2} + \lambda_t\right)\left(\frac{1}{2} + \hat{\lambda}_t\right)y\right] dy \end{aligned} \tag{10-42}$$

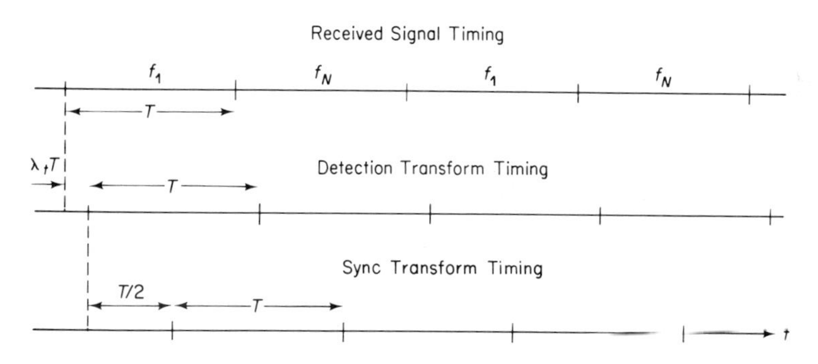

Fig. 10-15 Timing diagram

For $\lambda_t = 0$, the conditional p.d.f. of (10-42) is symmetrical in $\hat{\lambda}_t$; whereas, for $\lambda_t \neq 0$, the estimate becomes biased in the direction of λ_t. Based upon these properties, the conditional mean $E\{\hat{\lambda}_t \mid \lambda_t\}$ (Fig. 10-16) is suitable for use as a phase detector characteristic in a closed loop synchronizing scheme (Fig. 10-17). Also, since for a single T-sec interval the variance of $\hat{\lambda}_t$ is quite large, a linear filter must be incorporated in the loop for smoothing out this estimate. This filtered estimate is compared with the present timing error in the loop, and the difference of these two quantities is used to form the new estimate of epoch.

Time synchronization tracking can be performed by the same technique as initial acquisition. Since it is reasonable to assume that in an operating system any change in timing would be due to a slow drift between the transmitter and receiver timing oscillators, the loop of Fig. 10-17 can also be used for tracking if the loop filter time constant is switched to be long, relative to that used during the acquisition interval. The major difference between the two functions is that the signaling frequencies are known in advance during

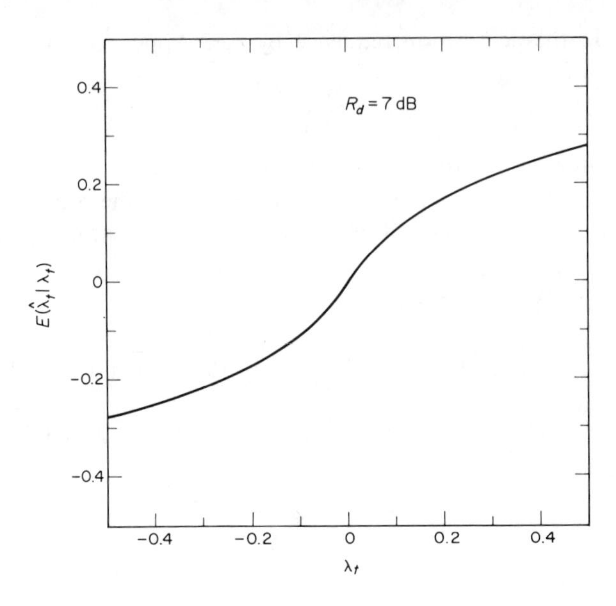

Fig. 10-16 Equivalent phase detector characteristic for time synchronization loop (Courtesy of H. D. Chadwick)

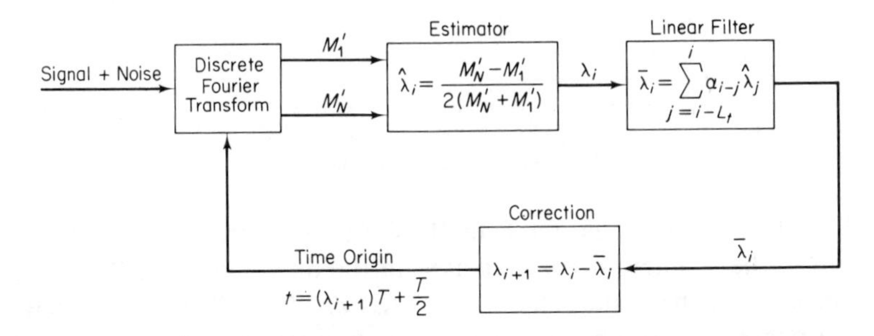

Fig. 10-17 Time synchronization loop

acquisition, whereas, for tracking, the detection function must be performed to determine which frequencies are present and hence which frequencies should be used to form the maximum-likelihood estimate of the tracking error.

10-7. ERROR PROBABILITY PERFORMANCE OF THE OPTIMUM RECEIVER IN THE PRESENCE OF TIMING UNCERTAINTY

Regardless of the method by which the timing estimate is obtained, it can be shown[2] that the conditional word error probability in the presence of a fixed timing error λ_t is given by

$$P_E(\lambda_t) = 1 - \int_0^\infty [1 - Q(\sqrt{2R_d}\lambda_t, y)]\left[1 - \exp\left(\frac{-y^2}{2}\right)\right]^{N-2}$$
$$\times\, y \exp\{-\tfrac{1}{2}[y^2 + 2R_d(1 - \lambda_t)]^2\} I_0[\sqrt{2R_d}(1 - \lambda_t)y]\, dy \quad (10\text{-}43)$$

where

$$Q(\alpha, \beta) \triangleq \int_\beta^\infty x \exp\left(-\frac{x^2 + \alpha^2}{2}\right) I_0(\alpha x)\, dx \quad (10\text{-}44)$$

is Marcum's Q function.[18] The expression for error probability as given in (10-43) has been evaluated on a digital computer and is plotted versus R_b in dB in Fig. 10-18 for $n = 5$ and various values of λ_t. We note from this figure the considerable increase in error probability caused by a small timing uncertainty.

If the synchronization loop of Fig. 10-17 is used for acquisition, then the error probability performance of the optimum receiver in the presence of

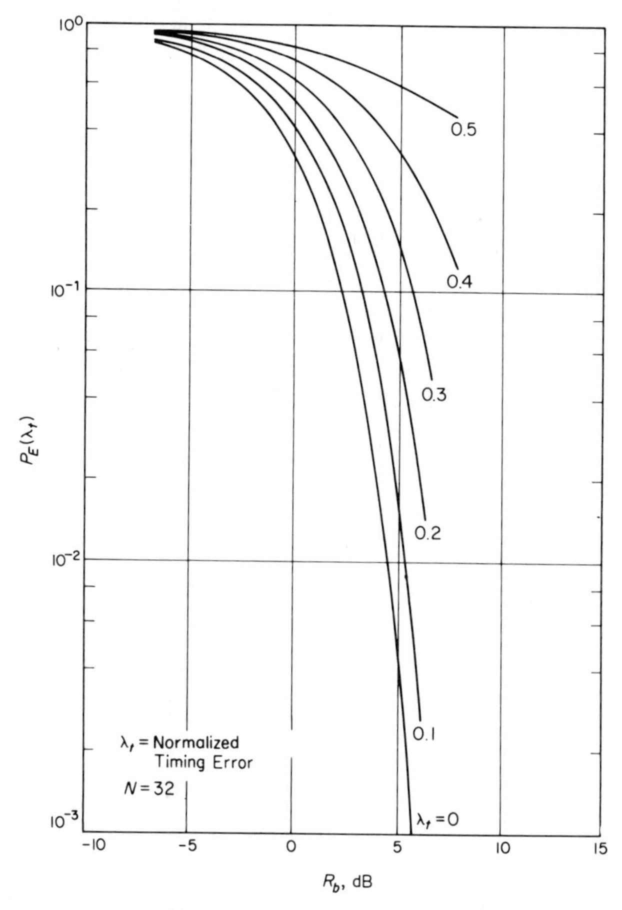

Fig. 10-18 Error probability performance in the presence of timing error; $N = 32$ (Courtesy of H. D. Chadwick and J. C. Springett)

the residual timing error remaining after an initial acquisition interval is plotted in Fig. 10-19 versus R_d for several values of L_t. Here, L_t is the number of $2T$-sec periods in one acquisition sequence (the total acquisition time $T_{acq} = 2L_tT$). The curves in this figure were obtained by experimentally comput-

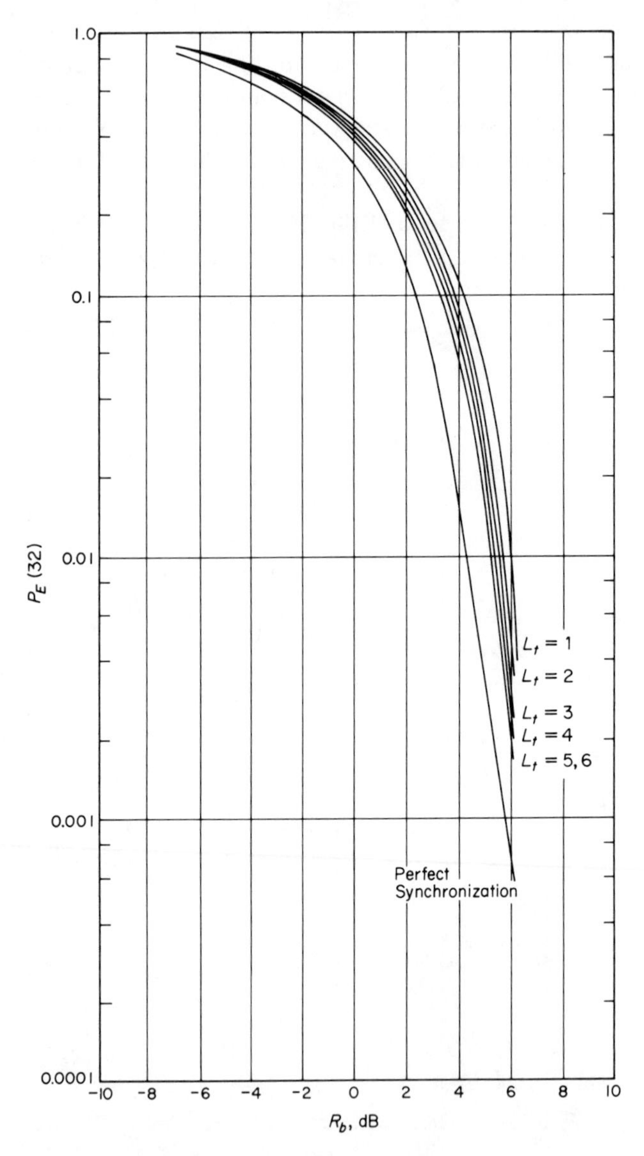

Fig. 10-19 Error probability performance of time synchronized receiver with acquisition time as a parameter; $N = 32$ (Courtesy of H. D. Chadwick and J. C. Springett)

ing a p.d.f. for the residual timing error, then averaging (10-43) over this p.d.f. Increasing L_t beyond the value six does not produce any significant increase in accuracy.

10-8. FREQUENCY SYNCHRONIZATION OF THE OPTIMUM RECEIVER

Frequency synchronization of an MFSK receiver refers to estimating at the receiver the correct location of the carrier frequency f_0 discussed earlier. It is clear that the expected frequency offset f_d and the long-term effects $\theta_l(t)$ affect the design and operation of any frequency-synchronization loop. Typically, the offset term f_d might be many orders of magnitude larger than the signal bandwidth B, and hence acquisition of the carrier requires searching over a wide frequency band to locate the signaling frequency spectrum (Fig. 10-20). In Sec. 10-6, time synchronization was achieved assuming perfect

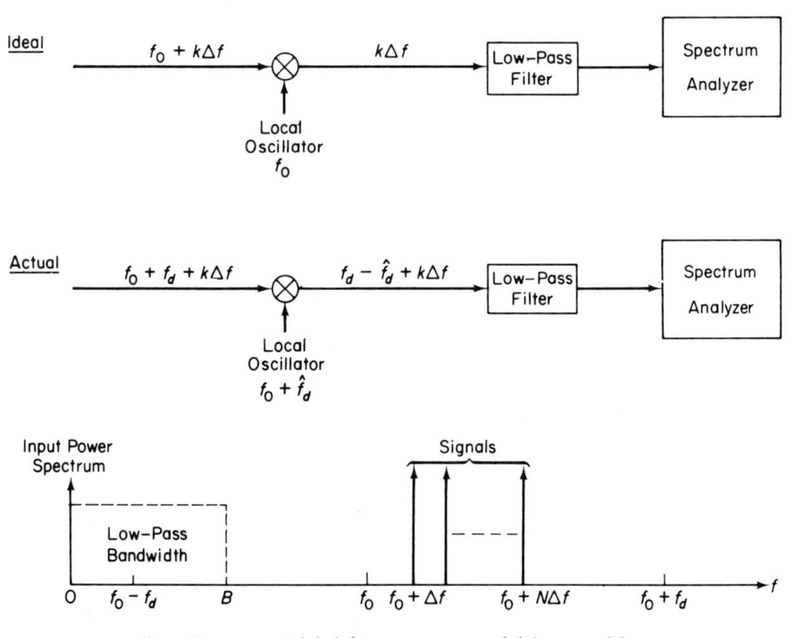

Fig. 10-20 Initial frequency acquisition problem

frequency synchronization had already been obtained. Hence, to be consistent here we must propose a method for frequency acquisition that does not depend on having time synchronization. Nevertheless, we assume that the synchronizing sequence available for time synchronization is also available for acquiring the carrier frequency.

A practical solution[2,19] to the acquisition problem is to step the receiver local oscillator in frequency from one end of the uncertainty band to the other, that is, $f_0 - f_d$ to $f_0 + f_d$, in steps of width B, each time computing the spectrum for a $2T$-sec observation interval; the width of the detection interval must be $2T$ to guarantee that both signal frequencies are present in equal energy. Because of the low signal-to-noise ratio in the receiver, one might actually average the spectral estimates in each band over several $2T$-sec segments. One can then either search the entire uncertainty band for the two largest spectral peaks or stop the search immediately after the first peak exceeds a given preset threshold μ_0'. The latter technique is preferable since it results in a smaller average acquisition time. The threshold is established based on the values of the spectral components due to noise only. For example, μ_0' can be set by requiring that the probability of any noise component exceeding it be less than some number P_n. Letting $f_{d_{\max}}$ denote the maximum value of the carrier frequency uncertainty, then there at most $2Nf_{d_{\max}}/B$ spectral estimates to be searched. Since the noise components in these spectral estimates are statistically independent, then the total probability P_n is approximately

$$P_n = 2N(f_{d_{\max}}/B)\{\exp[-\mu_0'^2/2\sigma^2]\} \tag{10-45}$$

where the exponential term in (10-45) is the probability of any *one* of the noise components exceeding the threshold and $\sigma^2 = N_0 ST$. Rearranging (10-45) gives

$$\frac{\mu_0'}{\sigma} = \sqrt{2 \ln \left(\frac{2N f_{d_{\max}}}{BP_n}\right)} \tag{10-46}$$

Having established the threshold, the probability of no acquisition is obtained by computing the probability P_s that the *lower* frequency signal peak does not exceed the preset threshold, that is,

$$P_s = 1 - Q\left(\gamma, \frac{\mu_0'}{\sigma}\right) \tag{10-47}$$

where

$$\gamma \triangleq \sqrt{m_t R_d}\left(\frac{\sin \pi\lambda_f}{\pi\lambda_f}\right) \tag{10-48}$$

In (10-48), m_t is the number of $2T$-sec intervals over which the spectrum is averaged in each band and λ_f is the normalized fractional uncertainty between two adjacent spectral lines separated by Δf—that is, $0 \leq |\lambda_f| < 0.5$. The minimum value of γ (which gives the largest value of P_s) occurs for $|\lambda_f| = 0.5$; that is,

$$\gamma_{\min} = \frac{2}{\pi}\sqrt{m_t R_d} \tag{10-49}$$

We note that the expression for probability of no acquisition as given in (10-47) includes the case where the lower frequency signal estimate fails to exceed the threshold but the upper frequency signal estimate does. With no prior knowledge of the distribution of the initial frequency error, the worst case assumption is that of a uniform distribution between $-f_{d\max}$ and $f_{d\max}$. Under this assumption, the average time to acquire would be one half of the time to examine the entire spectrum or

$$T_{\text{acq}} = 2m_t T f_{d\max}/B \tag{10-50}$$

Assuming now that the acquisition procedure just described has been successful in that the frequency offset has been corrected to within an error tolerance $\lambda_f \Delta_f$ with $|\lambda_f| \leq 0.5$, one can proceed with the frequency-tracking phase of the synchronization problem. Using a similar maximum-likelihood procedure to estimate λ_f as in the time synchronization case, one can design a frequency-tracking loop similar to Fig. 10-17. The details of such a design are given preliminary consideration in Ref. 20.

10-9. ERROR PROBABILITY PERFORMANCE OF THE OPTIMUM RECEIVER IN THE PRESENCE OF FREQUENCY UNCERTAINTY

The frequency acquisition scheme just discussed is designed to resolve the frequency error to within half of the spacing between signaling frequencies. From that point on the tracking loop should guarantee that the frequency error remains between these bounds. Hence, the probability of error for the detection process need only be calculated for values of normalized frequency error less than one-half. Assuming that the system is perfectly time-synchronized, the probability of error, conditioned on a fixed normalized frequency error λ_f, is given by[2,20]

$$P_E(\lambda_f) = 1 - \int_0^\infty y\left[1 - \exp\left(-\frac{y^2}{2}\right)\right]^{N-1} \times \exp\left\{-\frac{1}{2}\left[y^2 + 2R_d\left(\frac{\sin \pi\lambda_f}{\pi\lambda_f}\right)^2\right]\right\} I_0\left[\sqrt{2R_d}\left(\frac{\sin \pi\lambda_f}{\pi\lambda_f}\right)y\right] dy \tag{10-51}$$

10-10. ERROR PROBABILITY PERFORMANCE OF THE OPTIMUM RECEIVER IN THE PRESENCE OF COMBINED TIME AND FREQUENCY ERRORS

Thus far, we have considered *independently* the degradation in system performance due to time and frequency errors with the resulting error probability expressions given by (10-43) and (10-51), respectively. To a first approximation, the combined effect of both types of error can be found by cascading their effects when the errors are small. The resulting expression for the probability of error, conditioned on both time and frequency errors, is[21]

$$P_E(\lambda_t, \lambda_f) = 1 - \int_0^\infty [1 - Q(\gamma_2, y)]\left[1 - \exp\left(-\frac{y^2}{2}\right)\right]^{N-2} \times \exp\left\{-\frac{1}{2}[y^2 + \gamma_1^2]\right\} I_0[\gamma_1 y]dy \tag{10-52}$$

where

$$\gamma_1 \triangleq \sqrt{2R_d}(1 - \lambda_t)\left[\frac{\sin \pi\lambda_f}{\pi\lambda_f}\right]$$
$$\gamma_2 \triangleq \sqrt{2R_d}\lambda_t\left[\frac{\sin \pi\lambda_f}{\pi\lambda_f}\right] \tag{10-53}$$

Clearly, when λ_t and λ_f are independently set equal to zero, (10-52) reduces to (10-43) and (10-51), respectively.

10-11. FREQUENCY SYNCHRONIZATION AND ERROR PROBABILITY PERFORMANCE OF A SUBOPTIMUM RECEIVER

The frequency synchronization procedures of Sec. 10-8 can be applied to a practical suboptimum receiver whose spectral estimates are computed via the discrete Fourier transform approach. If $2BT$ samples are used for calculating the discrete Fourier transforms, and these transforms are evaluated at frequencies k/T; $k = 0, 1, \ldots, BT$, then the total number of spectral contributions to be searched during the acquisition procedure is now $2f_{d\max}T$. If, as in the optimum receiver case, the threshold technique is used for acquisition, then, analogous to (10-46), the equation that establishes the threshold for the suboptimum receiver is

$$\frac{\mu_0'}{\sigma} = \sqrt{2 \ln\left(\frac{2f_{d\max}T}{P_n}\right)} \tag{10-54}$$

The probability of no acquisition is still given by (10-47); however, the normalized relative frequency error λ_f is now normalized with respect to the frequency spacing $1/T$. Finally, the conditional error probability for the suboptimum receiver is given by[19]

$$P_E(\lambda_f) = 1 - \int_0^\infty y \prod_{\substack{i=-N/2 \\ i\neq 0}}^{N/2-1} \left\{1 - Q\left[\sqrt{2Rd}\left(\frac{\sin \pi(iBT/N - \lambda_f)}{\pi(iBT/N - \lambda_f)}\right), y\right]\right\} \times \exp\left\{-\frac{1}{2}\left[y^2 + 2R_d\left(\frac{\sin \pi\lambda_f}{\pi\lambda_f}\right)^2\right]\right\} I_0\left[\sqrt{2R_d}\left(\frac{\sin \pi\lambda_f}{\pi\lambda_f}\right)y\right]dy \tag{10-55}$$

and is plotted in Figs. 10-21 and 10-22 for $N = 32$ with $2BT = 64$ and $2BT = 2048$ respectively. We note from these results that the conditional error probability for a given frequency error is quite sensitive to the number of samples used in evaluating the discrete Fourier transform. Hence, the num-

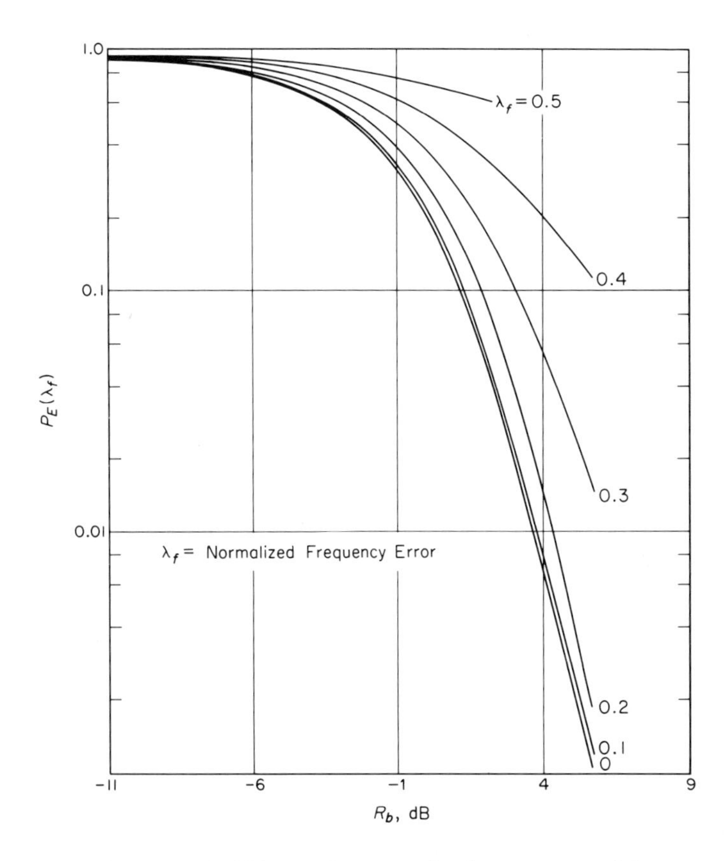

Fig. 10-21 Error probability performance in the presence of frequency error; $N = 32$; $2BT = 64$ (Courtesy of H. D. Chadwick)

ber of samples is an important parameter in the practical design of an MFSK receiver. Also, since

$$\lim_{\alpha \to 0} Q(\alpha, \beta) = \exp(-\beta^2/2) \tag{10-56}$$

then, as the ratio of the number of samples to the number of signals, that is, $2BT/N$, becomes large, (10-55) reduces to (10-51).

APPENDIX A

DERIVATION OF THE MAXIMUM-LIKELIHOOD ESTIMATOR $\hat{\lambda}_t$

For an input sequence alternating between the lowest and highest frequencies in the signal set, the received signal is written as

$$y(t) = \begin{cases} \sqrt{2S}\cos(\omega_1 t + \phi_1) + n(t) & -T/2 \le t \le 0 \\ \sqrt{2S}\cos(\omega_N t + \phi_N) + n(t) & 0 \le t \le T/2 \end{cases} \tag{A-1}$$

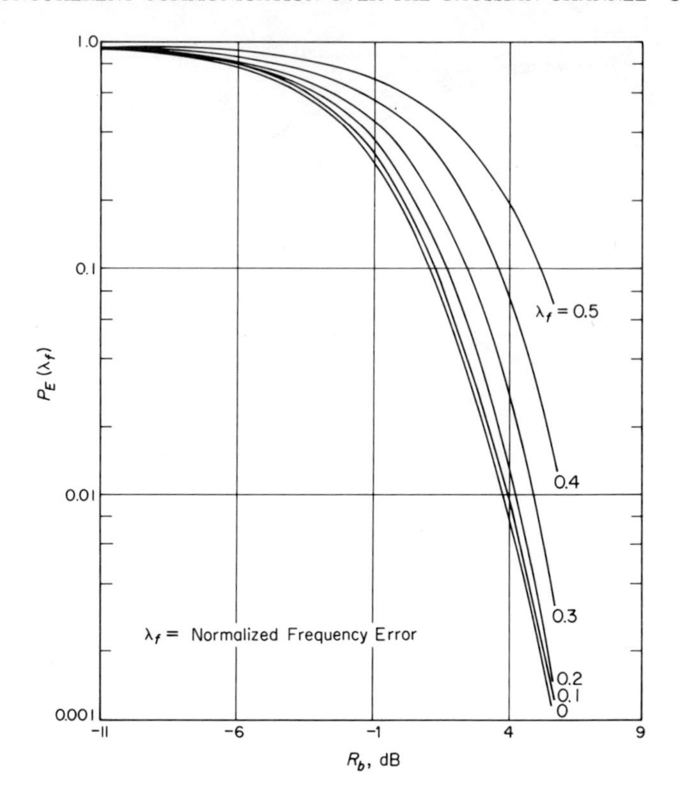

Fig. 10-22 Error probability performance in the presence of frequency error; $N = 32$; $2BT = 2048$ (Courtesy of H. D. Chadwick and J. C. Springett)

After demodulating $y(t)$ with a reference signal (assumed to have unit power) of radian frequency ω_0 and unknown phase θ_r, the equivalent low-pass signal becomes

$$z(t) = \begin{cases} \sqrt{2S}\cos(\Delta\omega t + \theta_1) + N(t, \theta_r) & -T/2 \leq t \leq 0 \\ \sqrt{2S}\cos(N\Delta\omega t + \theta_N) + N(t, \theta_r) & 0 \leq t \leq T/2 \end{cases} \tag{A-2}$$

where $N(t, \theta_r)$ is approximately low-pass white Gaussian noise of single-sided spectral density N_0 and

$$\begin{aligned} \theta_1 &= \phi_1 - \theta_r \\ \theta_N &= \phi_N - \theta_r \end{aligned} \tag{A-3}$$

are random phases uniformly distributed between $-\pi$ and π.

Using (10-7) as applied to the low-pass signal of above, the Fourier spectral components X'_1, X'_N, Y'_1, Y'_N computed over the quadrature symbol interval $(-T/2 + \lambda_t T, T/2 + \lambda_t T)$ are Gaussian random variables with conditional means given by:

$$\begin{aligned} E\{X'_1 \mid \theta_1, \lambda_t\} &= ST(\tfrac{1}{2} - \lambda_t)\cos\theta_1 \\ E\{X'_N \mid \theta_N, \lambda_t\} &= ST(\tfrac{1}{2} + \lambda_t)\cos\theta_N \\ E\{Y'_1 \mid \theta_1, \lambda_t\} &= -ST(\tfrac{1}{2} - \lambda_t)\sin\theta_1 \\ E\{Y'_N \mid \theta_N, \lambda_t\} &= -ST(\tfrac{1}{2} + \lambda_t)\sin\theta_N \end{aligned} \tag{A-4}$$

and variances

$$\sigma^2_{X'_1} = \sigma^2_{X'_N} = \sigma^2_{Y'_1} = \sigma^2_{Y'_N} \triangleq \sigma^2 = \frac{N_0 ST}{2} \tag{A-5}$$

In the above, λ_t is the normalized (with respect to T) relative timing error. Hence, the conditional p.d.f.'s of these spectral components become

$$\begin{aligned} p(X'_1 \mid \theta_1, \lambda_t) &= \frac{1}{\sqrt{2\pi}\,\sigma}\exp\left\{-\frac{1}{2\sigma^2}\left[X'_1 - ST\left(\frac{1}{2} - \lambda_t\right)\cos\theta_1\right]^2\right\} \\ p(X'_N \mid \theta_N, \lambda_t) &= \frac{1}{\sqrt{2\pi}\,\sigma}\exp\left\{-\frac{1}{2\sigma^2}\left[X'_N - ST\left(\frac{1}{2} + \lambda_t\right)\cos\theta_N\right]^2\right\} \\ p(Y'_1 \mid \theta_1, \lambda_t) &= \frac{1}{\sqrt{2\pi}\,\sigma}\exp\left\{-\frac{1}{2\sigma^2}\left[Y'_1 + ST\left(\frac{1}{2} - \lambda_t\right)\sin\theta_1\right]^2\right\} \\ p(Y'_N \mid \theta_N, \lambda_t) &= \frac{1}{\sqrt{2\pi}\,\sigma}\exp\left\{-\frac{1}{2\sigma^2}\left[Y'_N + ST\left(\frac{1}{2} + \lambda_t\right)\sin\theta_N\right]^2\right\} \end{aligned} \tag{A-6}$$

In arriving at (A-4) through (A-6), two simplifying assumptions have been made—namely, that the contributions to the expected value at the signaling frequencies Δf and $N\Delta f$ are due only to the in-phase component of the signal, and that the expected value at the nonsignaling frequencies are small. These assumptions are easily justified by the relative magnitude of the terms involved.

Since the Fourier spectral components are mutually independent, the logarithm of their joint conditional p.d.f. can be written as

$$\begin{aligned} \log p(X'_1, X'_N, Y'_1, Y'_N \mid \theta_1, \theta_N, \lambda_t) = K'_0 &- \frac{1}{2\sigma^2}\left\{\left[X'_1 - ST\left(\frac{1}{2} - \lambda_t\right)\cos\theta_1\right]^2\right. \\ &+ \left[X'_N - ST\left(\frac{1}{2} + \lambda_t\right)\cos\theta_N\right]^2 + \left[Y'_1 + ST\left(\frac{1}{2} - \lambda_t\right)\sin\theta_1\right]^2 \\ &\left.+ \left[Y'_N + ST\left(\frac{1}{2} + \lambda_t\right)\sin\theta_N\right]^2\right\} \end{aligned} \tag{A-7}$$

where K'_0 is a constant not involving the unknown parameters.

To solve for $\hat{\lambda}_t$, it is necessary to simultaneously solve for the maximum-likelihood estimates of the other unknown parameters—$\hat{\theta}_1$, $\hat{\theta}_N$. Differentiating (A-7) with respect to the parameter being estimated and equating the result to zero gives an equation whose solution is the maximum-likelihood estimate of the parameter.

Solving first for $\hat{\theta}_1$ gives

$$\frac{\partial \log p(X'_1, X'_N, Y'_1, Y'_N | \hat{\theta}_1, \hat{\theta}_N, \hat{\lambda}_t)}{\partial \hat{\theta}_1} = \frac{1}{2\sigma^2}\left\{2\left[X'_1 - ST\left(\frac{1}{2} - \lambda_t\right)\cos\hat{\theta}_1\right]\right.$$
$$\times\left[ST\left(\frac{1}{2} - \lambda_t\right)\sin\hat{\theta}_1\right] + 2\left[Y'_1 + ST\left(\frac{1}{2} - \lambda_t\right)\sin\hat{\theta}_1\right]$$
$$\left.\times\left[ST\left(\frac{1}{2} - \lambda_t\right)\cos\hat{\theta}_1\right]\right\} = 0 \tag{A-8}$$

which can be simplified to

$$X'_1 \sin\hat{\theta}_1 + Y'_1 \cos\hat{\theta}_1 = 0$$

or

$$\hat{\theta}_1 = \tan^{-1}\left(-\frac{Y'_1}{X'_1}\right) \tag{A-9}$$

Similarly, solving for $\hat{\theta}_2$ gives

$$\hat{\theta}_N = \tan^{-1}\left(-\frac{Y'_N}{X'_N}\right) \tag{A-10}$$

The final step is to write the equation for λ_t:

$$\frac{\partial \log p(X'_1, X'_N, Y'_1, Y'_N | \hat{\theta}_1, \hat{\theta}_N, \hat{\lambda}_t)}{\partial \hat{\lambda}_t} = 0$$

which yields

$$\hat{\lambda}_t = \frac{1}{2ST}[-X'_1\cos\hat{\theta}_1 + Y'_1\sin\hat{\theta}_1 + X'_N\cos\hat{\theta}_N - Y'_N\sin\hat{\theta}_N] \tag{A-11}$$

Substituting the solutions for $\hat{\theta}_1$ and $\hat{\theta}_N$ as given by (A-11) and (A-12), respectively, into (A-11) gives

$$\hat{\lambda}_t = \frac{M'_N - M'_1}{2ST} \tag{A-12}$$

where

$$M'_1 = \sqrt{(X'_1)^2 + (Y'_1)^2}$$
$$M'_N = \sqrt{(X'_N)^2 + (Y'_N)^2} \tag{A-13}$$

If, in addition, the parameter S is also unknown, then a maximum-likelihood estimate of S is given by

$$\hat{S} = \frac{1}{T}\left[\frac{M'_1(\frac{1}{2} - \hat{\lambda}_t) + M'_N(\frac{1}{2} + \hat{\lambda}_t)}{2(\frac{1}{4} + \hat{\lambda}_t^2)}\right] \tag{A-14}$$

Substituting (A-14) into (A-12) and solving for $\hat{\lambda}_t$ gives

$$\hat{\lambda}_t = \frac{M'_N - M'_1}{2(M'_N + M'_1)}$$

This is the required maximum-likelihood estimator of λ_t in terms of the Fourier spectral components M'_1 and M'_N.

PROBLEMS

10-1. Consider the more general problem of noncoherent reception of a set of N equal energy, equiprobable, non-orthogonal but equally cross-correlated signals. Mathematically speaking, the received waveform in the absence of noise corresponding to the transmission of the kth signal in the set can be written in the complex form

$$y_s(t) = \text{Re}\{\xi_k(t)e^{j(2\pi f_0 t + \theta)}\}$$

where Re denotes the real part of the quantity in braces, $\xi_k(t)$ is a complex process which has a spectrum centered at zero frequency, and θ is an unknown angle uniformly distributed between 0 and 2π. Assuming, as stated above, that the signal set has the properties

$$\int_0^T |\xi_k(t)|^2\, dt = 2ST \qquad k = 1, 2, \ldots, N$$

$$\int_0^T \xi_j(t)\xi_k^*(t)\, dt = 2\lambda ST \qquad j, k = 1, 2, \ldots, N$$
$$j \neq k$$
$$\lambda \text{ real and non-negative}$$

then the probability of word error corresponding to the optimum receiver is given by the double integral[4]

$$P_E(N) = 1 - (1 - \lambda)\exp(-R_d)\int_0^\infty \int_0^\infty rs \exp\left[-\frac{1}{2}(r^2 + s^2)\right]$$
$$\times I_0(\sqrt{2R_d(1-\lambda)}r)I_0(\sqrt{\lambda}\,rs)[1 - Q(\sqrt{\lambda}\,s, r)]^{N-1}\, dr\, ds$$

where $Q(\alpha, \beta)$ is Marcum's Q function defined in (10-44).

a) Show that when $\xi_k(t) = \sqrt{2S}\exp(j2\pi k\Delta f t)$; $k = 1, 2, \ldots, N$ with Δf an integer multiple of $1/T$ corresponding to the orthogonal MFSK signaling alphabet discussed in the chapter, the general error probability expression that has been given reduces to (10-22).

b) Using the identity

$$\int_0^\infty s \exp\left[-\frac{1}{2}(s^2 + c^2)\right] I_0(cs)Q(as, b)\, ds$$
$$= Q\left(\frac{ac}{\sqrt{1+a^2}}, \frac{b}{\sqrt{1+a^2}}\right)$$

show that for $N = 2$, the general error probability expression simplifies to

$$P_E(2) = Q\left(\frac{1}{2}\sqrt{2R_d(1 - \sqrt{1-\lambda^2})}, \frac{1}{2}\sqrt{2R_d(1 + \sqrt{1-\lambda^2})}\right)$$
$$- \frac{1}{2}\exp\left(\frac{-R_d}{2}\right)I_0\left(\frac{\lambda R_d}{2}\right)$$

c) Assuming $N = 2$, show that for small λ the preceding error probability expression has the approximate form

$$P_E(2) \cong \frac{1}{2}\exp\left(\frac{-R_d}{2}\right) + \exp\left(\frac{-R_d}{2}\right)\left[\frac{R_d}{4}(1 - \sqrt{1 - \lambda^2}) + \frac{R_d^2\lambda^2}{32}\right]$$

Hint: For small values of its argument, the zero order Bessel function can be approximated by the first two terms of its power series, that is,

$$I_0(x) \cong 1 + \frac{x^2}{4}$$

Also, for small α and $\alpha < \beta$,

$$Q(\alpha, \beta) \cong \exp\left(-\frac{\alpha^2 + \beta^2}{2}\right)\left[1 + \frac{\alpha^2}{2}\left(1 + \frac{\beta^2}{2}\right)\right]$$

10-2. Using the trigonometric identities

$$\sum_{k=0}^{n} \cos^2 k\theta = \frac{n+2}{2} + \frac{\cos[(n+1)\theta]\sin n\theta}{2\sin\theta}$$

$$\sum_{k=1}^{n-1} k\cos k\theta = \frac{n\sin\{[(2n-1)/2]\theta\}}{2\sin\theta/2} - \frac{1-\cos n\theta}{4\sin^2\theta/2}$$

show that the mean of the spectral estimate $\Phi_{z_a}(f_l; J)$ is given by

$$\mu_l \triangleq E\{\Phi_{z_a}(f_l; J)\} = \begin{cases} \dfrac{N_0 T}{2}\left\{1 + \dfrac{S}{N_0 B}\left[J - \dfrac{(J-1)(J+1)}{4BT}\right]\right\} & f_l = \dfrac{l}{T} \\ & l = k \\ \dfrac{N_0 T}{2}\left\{1 + \dfrac{S}{N_0 B}\left[\dfrac{J+1}{2BT} - 1\right]\right\} & f_l = \dfrac{l}{T} \\ & l \neq k \end{cases}$$

where k/T is the signaling frequency. A similar independence on l for $l \neq k$ can be shown for the variance of $\Phi_{z_a}(f_l; J)$, but the mathematics required is cumbersome.

10-3. Assume that a short-term frequency instability $\dot{\theta}_s(t)$ frequency modulates the nominal carrier and is characterized by an RC spectrum, that is,

$$R_{\dot{\theta}_s}(\tau) = S_s \exp[-2\pi B_s \tau]$$

$$S_{\dot{\theta}_s}(f) = \frac{S_s}{\pi B_s}\left[\frac{1}{1 + (f/B_s)^2}\right]$$

Using (10-39) and (10-40), show that the frequency-modulated carrier has a correlation function given by

$$R_{s_k}(j) = S\exp\left\{-\frac{S_s}{(2\pi B_s)^2}\left[e^{-j(\pi B_s/B)} + j\left(\frac{\pi B_s}{B}\right) - 1\right]\right\}\cos 2\pi f_k\left(\frac{j}{2B}\right)$$

10-4. As a first step in evaluating the effect of frequency drift on a system design, it is useful to study the performance of an MFSK receiver in the presence of a *linear* frequency drift. Corresponding then to the transmission of $x_i[t, \theta(t)]$ of (10-2), the received signal is characterized over a T-second interval by

$$y(t) = \sqrt{2S}\cos\left[\omega_i t + \tfrac{1}{2}\alpha_\omega t^2 + \theta\right] + n(t)$$

with $\omega_i = \omega_0 + i\Delta\omega$, and ω_0 the carrier radian frequency. For the optimum receiver of Fig. 10-2,

a) Show that the mean and variance of the Fourier coefficients X_k and Y_k as defined in (10-7) are given by

$$E\{X_k\} = \frac{S}{\sqrt{\alpha_\omega/\pi}}[\cos\theta_k\,\Delta C_k + \sin\theta_k\,\Delta S_k]$$

$$E\{Y_k\} = \frac{S}{\sqrt{\alpha_\omega/\pi}}[\cos\theta_k\,\Delta S_k - \sin\theta_k\,\Delta C_k] \qquad k = 1, 2, \ldots, N$$

$$\sigma_{X_k}^2 = \sigma_{Y_k}^2 = \frac{N_0 ST}{2}$$

where $$\theta_k \triangleq \frac{[(i-k)\Delta\omega]^2}{2\alpha_\omega} - \theta$$

$$\Delta C_k \triangleq C\left[\frac{(i-k)\Delta\omega + \alpha_\omega T}{\sqrt{\pi\alpha_\omega}}\right] - C\left[\frac{(i-k)\Delta\omega}{\sqrt{\pi\alpha_\omega}}\right]$$

$$\Delta S_k \triangleq S\left[\frac{(i-k)\Delta\omega + \alpha_\omega T}{\sqrt{\pi\alpha_\omega}}\right] - S\left[\frac{(i-k)\Delta\omega}{\sqrt{\pi\alpha_\omega}}\right]$$

and $C(z)$, $S(x)$ are the Fresnel integrals defined by

$$C(z) = \int_0^z \cos\left(\frac{\pi}{2}t^2\right)dt$$

$$S(z) = \int_0^z \sin\left(\frac{\pi}{2}t^2\right)dt$$

b) Since X_k and Y_k have a joint Gaussian p.d.f. conditioned on θ, and θ is uniformly distributed between 0 and 2π, show that the marginal joint p.d.f. of X_k and Y_k is given by

$$p(X_k, Y_k) = \frac{1}{\pi N_0 ST}\exp\{-R_{dk}\}\exp\left\{-\frac{X_k^2 + Y_k^2}{N_0 ST}\right\}$$
$$\times I_0\left\{2\sqrt{R_{dk}\left(\frac{X_k^2 + Y_k^2}{N_0 ST}\right)}\right\} \qquad k = 1, 2, \ldots, N$$

where

$$R_{dk} \triangleq R_d\left[\frac{\Delta C_k^2 + \Delta S_k^2}{\alpha_\omega T^2/\pi}\right]$$

c) Using the result of part b), show that the p.d.f. of the spectral estimate $M_k \triangleq \sqrt{X_k^2 + Y_k^2}$ is given by

$$q(M_k) = \frac{2M_k}{N_0ST} \exp\{-R_{dk}\} \exp\left\{-\frac{M_k^2}{N_0ST}\right\} I_0\left\{2\sqrt{R_{dk}\left(\frac{M_k^2}{N_0ST}\right)}\right\} \qquad k = 1, 2, \ldots, N$$

d) Since the probability of correctly detecting the transmitted signal for the optimum receiver corresponds to the event $[0 \leq M_i \leq \infty, M_k \leq M_i$ for all $k \neq i]$, write an expression for this probability of correct reception using the set of spectral estimate p.d.f.'s given in part c).

10-5. Suppose that in Prob. 10-4 a constant radian frequency error ω_d were present in addition to the linear frequency drift α_ω. Discuss how the results of parts a) through d) are changed by this additional frequency instability.

REFERENCES

1. *Short-Term Frequency Stability*, IEEE-NASA Symposium, Goddard Space Flight Center (November, 1964) 23–24.
2. CHADWICK, H. D. and SPRINGETT, J. C., "The Design of a Low Data Rate MFSK Communication System," *IEEE Transactions on Communication Technology*, Vol. COM-18, No. 6 (December, 1970) 740–750.
3. TURIN, G. L., "Communication Through Noisy-Random Multipath Channels," *IRE Convention Record*, Part 4 (1956) 154–166.
4. NUTTALL, A. H., "Error Probabilities for Equicorrelated M-ary Signals under Phase Coherent and Phase Incoherent Reception," *IRE Transactions on Information Theory*, Vol. IT-8, No. 3 (July, 1962) 305–314.
5. ARTHURS, E., and DYM, H. "On the Optimum Detection of Digital Signals in the Presence of White Gaussian Noise," *IRE Transactions on Communications Systems*, Vol. CS-10, No. 4 (December, 1962) 336–372.
6. LINDSEY, W. C., "Coded Noncoherent Communications," *IEEE Transactions on Space Electronics and Telemetry*, Vol. SET-11, No. 1 (March, 1965) 6–13.
7. REIGER, S., "Error Rates in Data Transmission," *Proceedings of the IRE (Correspondence)*, Vol. 46, No. 5 (May, 1958) 919–920.
8. FERGUSON, M., "Communication at Low Data Rates: Spectral Analysis Receivers," *IEEE Transactions on Communication Technology* Vol. COM-16, No. 5 (October, 1968) 657–668.
9. CHARLES, F. J. and SHEIN, N. P., *A Preliminary Study of the Application of Noncoherent Techniques to Low Power Telemetry*, Jet Propulsion Laboratory, Pasadena, Calif., Technical Memorandum 3341–65–14 (Reorder No. 65–815) (November, 1965).
10. ——— and SPRINGETT, J. C., "The Statistical Properties of the Spectral Estimates Used in the Decision Process by a Spectrum Analyzer Receiver," *Proceedings of the National Telemetering Conference*, San Francisco, Calif. (May, 1967).

11. SHANNON, C. E., "Communication in the Presence of Noise," *Proceedings of the IRE*, Vol. 37, No. 1 (January, 1949) 10–21.

12. CHADWICK, H. D., "Time Synchronization in an MFSK Receiver," Jet Propulsion Laboratory, Pasadena, Calif., SPS 37–48, Vol. III (1967) 252–264, also presented at the Canadian Symposium on Communications, Montreal, Quebec (November, 1968).

13. COOLEY, J. W. and TUKEY, J. W., "An Algorithm for the Machine Calculation of Complex Fourier Series," *Math. of Computation*, Vol. 19 (1965) 297–301.

14. FRASER, D.A.S., *Nonparametric Methods in Statistics*. John Wiley & Sons, Inc., New York, N.Y. (1957) 215.

15. MIDDLETON, D., *An Introduction to Statistical Communication Theory*. McGraw-Hill Book Company, New York, N.Y. (1960) 604.

16. KAILATH, T., "Correlation Detection of Signals Perturbed by a Random Channel," *IRE Transactions on Information Theory*, Vol. IT-6, No. 3 (June, 1960) 361–366.

17. JEAN, F. L., "Performance of a Conventional MFSK Receiver in the Presence of Frequency Drift," Jet Propulsion Laboratory, Pasadena, Calif., SPS 37–50, Vol. III (1968) 338–342.

18. MARCUM, J., "Table of Q Functions," RAND Corporation Report M-339 (January, 1950).

19. CHADWICK, H. D., "Frequency Acquisition in an MFSK Receiver," Jet Propulsion Laboratory, Pasadena, Calif., SPS 37–52, Vol. III (1968) 239–247.

20. ———, "Frequency Tracking in an MFSK Receiver," Jet Propulsion Laboratory, Pasadena, Calif., SPS 37–57, Vol. III (1969) 47–54.

21. ———, "The Error Probability of a Spectrum Analysis Receiver with Incorrect Time and Frequency Synchronization," presented at the International Symposium on Information Theory, Ellenville, N.Y. (January 30, 1969).

Chapter 11

TRACKING LOOPS WITH IMPROVED PERFORMANCE

11-1. INTRODUCTION

In practice, one may be confronted with a set of system requirements for which the use of a PLL as a tracking receiver would not provide adequate performance. To avoid making radical changes in resident system equipment, however, the system designer seeks a solution to his problem that uses the PLL as a subelement of a more sophisticated configuration. Two such tracking loop configurations are discussed in this chapter. The first uses the principle of decision-directed feedback and hence is referred to as the *data-aided carrier-tracking loop*. The second scheme, the *hybrid carrier-tracking loop*, bridges the advantages of the PLL and the Costas loop.

From a theoretical point of view, we are investigating only a special case of a more general problem, which has received widespread attention in the literature,[1–8] namely, the establishment of the optimum demodulator for estimating random messages transmitted by a nonlinear modulation over a random channel. For the analog communication system problem, Van Trees[1-4] and Thomas and Wong,[5] among many others, have applied the principles of *maximum a posteriori* (MAP) probability and Bayes' criterion in seeking a solution. The MAP approach leads to an integral equation for the message estimate, and the solution to this equation corresponds to a physically unrealizable demodulator. Van Trees[3] suggests making an approximation to the unrealizable demodulator for the purpose of implementation. Snyder,[6] on the other hand, takes the state variable approach to the optimum demodulation problem based upon the theory of Markov processes. It turns out that the state variable approach leads directly to a physically realizable demodulator, which is "equivalent" to the physically realizable portion of the Van Trees[3] MAP estimator. Snyder[6] establishes quasi-optimum (optimum for large signal-to-noise ratios) phase and frequency demodulators of a stationary Gaussian message corrupted by additive white Gaussian noise. In fact, the quasi-optimum demodulators reduce to a PLL for large signal-to-noise ratios.

In the digital domain one can use a MAP procedure similar to that presented in Sec. 9-2 to arrive at a MAP estimator of the random phase for a single-channel communication system.[8] This estimate can then be used to suggest certain closed-loop tracking configurations. The motivation behind this approach follows along the lines of the discussions in Chap. 9.

11-2. THE MAP ESTIMATOR OF PHASE FOR A SINGLE-CHANNEL SYSTEM

The problem of estimating the phase and frequency in a phase-coherent communication system where a single data channel phase modulates a transmitted RF carrier has been considered in previous chapters. Thus far, we have assumed the use of a PLL for tracking the RF phase and frequency, and no attempt has been made to use the power in the data channel for improving performance. A good deal of insight into the efficient use of the data power to improve tracking capability can be gained by considering the receiver structure which forms the MAP estimate of the RF phase (frequency assumed known) given a finite past record of the received signal plus noise. The method used follows the well-known theory of MAP estimation of an unknown parameter in Gaussian noise. It is relatively straightforward to apply this theory to the problem of RF carrier phase estimation in a single-channel system and hence we leave the details to an appendix and present only the significant results in the text.

In a single-channel system an RF transmitter is phase modulated by a single data channel to provide an output signal of the form

$$x(t) = \sqrt{2P}\sin\left[\omega_0 t + (\cos^{-1} m_1)m(t)\right] \tag{11-1}$$

where ω_0 is the nominal carrier frequency and m_1 is the modulation factor. The modulation $m(t)$ is assumed digital and may be placed directly on the RF carrier or alternately on a separate subcarrier. Thus, two cases exist: either $m(t) = d(t)$, or $m(t) = d(t)\mathbb{S}(t)$, where $d(t)$ is a binary waveform representing data with transitions occurring at T-sec intervals and $\mathbb{S}(t)$ is a square-wave subcarrier of period T/N. Here, N represents the number of subcarrier cycles in the interval T. For simplicity, we will pursue the first case and later indicate the necessary modification of the receiver structure when the modulation is placed on a square-wave subcarrier.

The transmitted signal $x(t)$ is sent over an additive white Gaussian noise

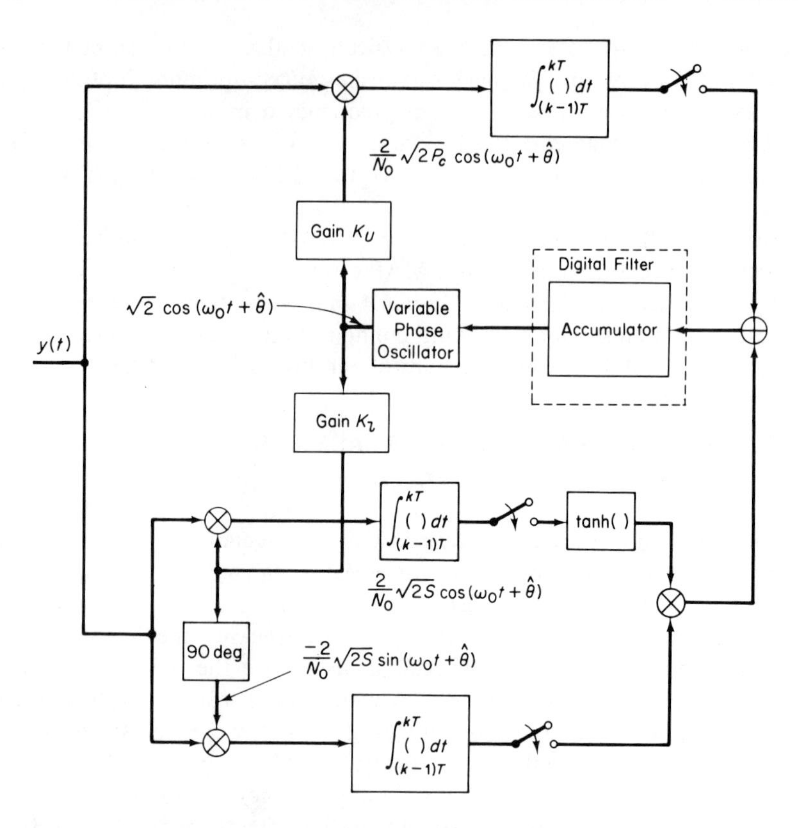

Fig. 11-1 A discrete carrier-tracking loop motivated by the MAP approach (data modulated directly on carrier)

channel, which also introduces an arbitrary phase shift $\theta(t)$. Thus, the received signal $y(t)$ is of the form

$$y(t) = \sqrt{2P}\sin\,[\omega_0 t + (\cos^{-1} m_1)m(t) + \theta(t)] + n_i(t) \tag{11-2}$$

Then, using the theory of MAP estimation and assuming $\theta(t) = \theta =$ constant for K symbol intervals, the best estimate of θ in the MAP sense, based on a KT observation of the received signal $y(t)$, is the value of θ that maximizes the function (see Appendix A)

$$f(\theta) = \prod_{k=1}^{K} \exp\left[\frac{2}{N_0}\int_{(k-1)T}^{kT} \sqrt{2P_c}\,y(t)\sin(\omega_0 t + \theta)\,dt\right] \times \cosh\left[\frac{2}{N_0}\int_{(k-1)T}^{kT} \sqrt{2S}\,y(t)\cos(\omega_0 t + \theta)\,dt\right] \tag{11-3}$$

or

$$\ln f(\theta) = \sum_{k=1}^{K} \left\{\frac{2}{N_0}\int_{(k-1)T}^{kT} \sqrt{2P_c}\,y(t)\sin(\omega_0 t + \theta)dt + \ln\cosh\left[\frac{2}{N_0}\int_{(k-1)T}^{kT} \sqrt{2S}\,y(t)\cos(\omega_0 t + \theta)\,dt\right]\right\} \tag{11-4}$$

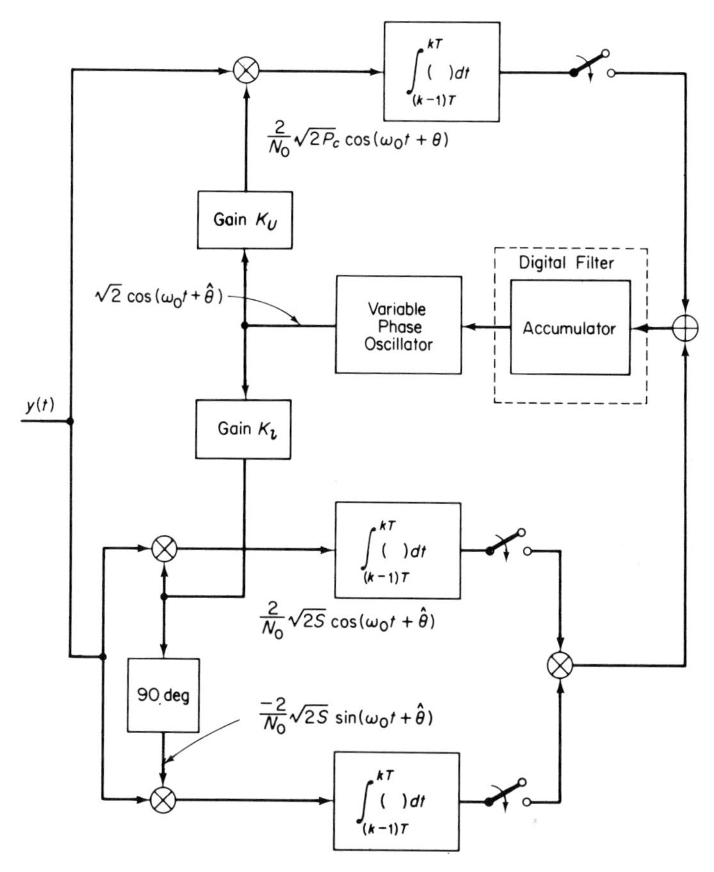

Fig. 11-2 The limiting case of Fig. 11-1 for small signal-to-noise ratio

where $P_c \triangleq m_1^2 P$ and $S \triangleq (1 - m_1^2)P$. We interpret (11-4) as follows. For each value of θ, one correlates the received signal $y(t)$ with the stored replicas $(2/N_0)[\sqrt{2P_c}\sin(\omega_0 t + \theta)]$ and $(2/N_0)[\sqrt{2S}\cos(\omega_0 t + \theta)]$ over each symbol interval, adding the result of the first correlation to the log hyperbolic cosine of the second. This statistic is then accumulated over K successive symbol intervals and stored. The value of θ yielding the largest accumulated value is then declared as the best estimate $\hat{\theta}$. Actually, θ takes on a continuum of values between $-\pi$ and π, and hence to apply the above technique practically, in each KT observation interval, one must quantize θ into a number of levels compatible with the size of buffer storage one is willing to provide and the estimation accuracy desired.

At this point, one might propose a closed-loop system, incorporating the above ideas, that would continuously update the estimate of the input phase θ. To see this, we differentiate $f(\theta)$, or equivalently $\ln f(\theta)$, with respect to θ and equate to zero. The solution of this equation is then $\hat{\theta}$, the MAP

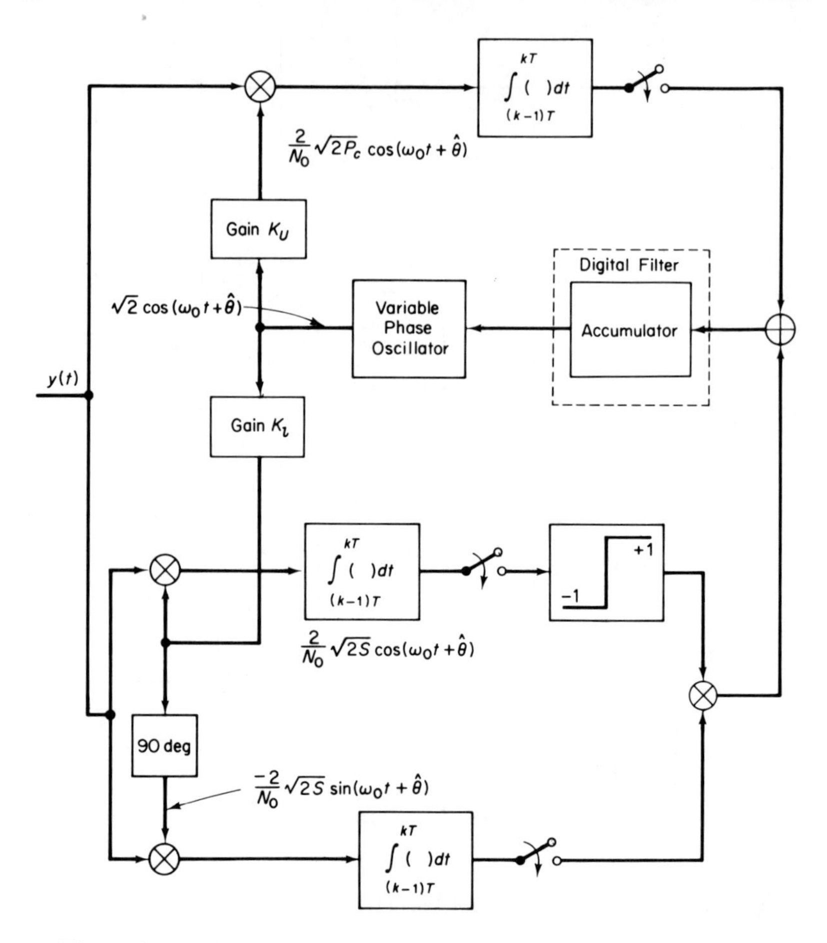

Fig. 11-3 The limiting case of Fig. 11-1 for large signal-to-noise ratio

estimate. Hence,

$$\begin{aligned} g(\hat{\theta}) = \frac{d}{d\theta} \ln f(\theta)\,|_{\theta=\hat{\theta}} = \sum_{k=1}^{K} \Bigg\{ &\frac{2}{N_0} \int_{(k-1)T}^{kT} \sqrt{2P_c}\, y(t) \cos(\omega_0 t + \hat{\theta})\, dt \\ &- \tanh\left[\frac{2}{N_0} \int_{(k-1)T}^{kT} \sqrt{2S}\, y(t) \cos(\omega_0 t + \hat{\theta})\, dt\right] \\ &\times \frac{2}{N_0} \int_{(k-1)T}^{kT} \sqrt{2S}\, y(t) \sin(\omega_0 t + \hat{\theta})\, dt \Bigg\} = 0 \end{aligned} \qquad (11\text{-}5)$$

By using $g(\hat{\theta})$ as an error signal in a tracking loop, one obtains the configuration of Fig. 11-1. In practice, the receiver structure shown is difficult to construct so that one is interested in examining its limiting forms for small and large signal-to-noise ratios. Specifically, one makes the following approximations to the hyperbolic tangent function:

$$\begin{aligned} \tanh x &\cong x \qquad |x| \ll 1 \\ \tanh x &\cong \operatorname{sgn} x \quad |x| \gg 1 \end{aligned} \tag{11-6}$$

which lead to the circuit configurations of Figs. 11–2 and 11–3.

Most interesting, perhaps is the way in which the power splits between the upper and lower branches of each configuration as a function of the modulation factor m_1. Specifically, from (11-5) and Fig. 11-1, the signal power in the upper branch is proportional to m_1^2, whereas that in the lower branch varies as $1 - m_1^2$. As we shall see for the data-aided loop, splitting the power between upper and lower branches as m_1^2 and $1 - m_1^2$, respectively, gives the maximum improvement in loop signal-to-noise ratio relative to that of a PLL when both are operated in the linear region.

Finally, for the case where $d(t)$ is placed on a square-wave subcarrier $\mathbb{S}(t)$, the configuration of Fig. 11-1 is modified as in Fig. 11-4. The subcarrier is assumed to be a perfect reference (no phase jitter) or to have a random phase independent of the RF phase being estimated. In practice, the subcarrier reference signal is derived from a suppressed carrier-tracking loop whose input

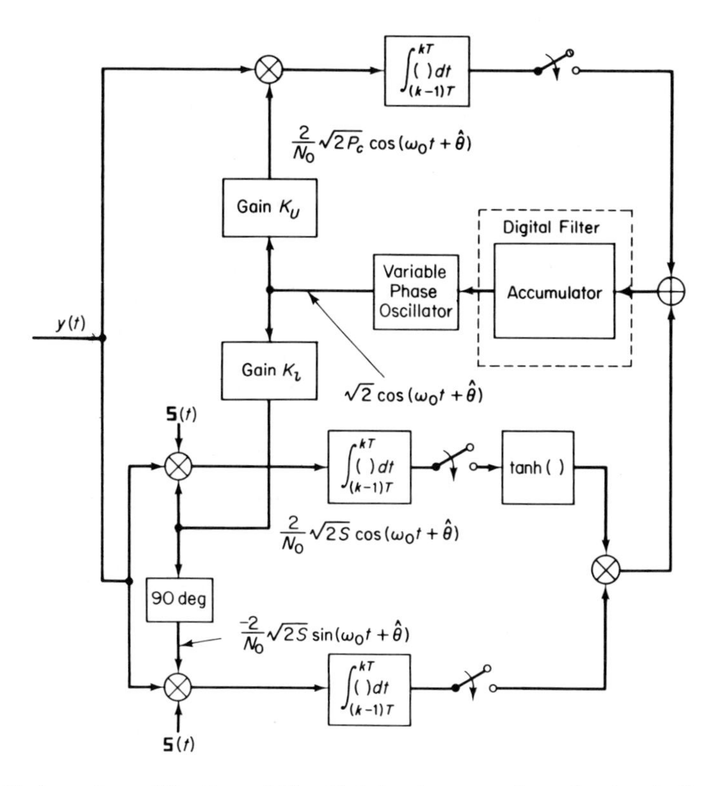

Fig. 11-4 A modification of Fig. 11-1 for the case where the data is first modulated on a subcarrier

amplitude is degraded by the cosine of the RF phase error and hence depends on the input RF phase. However, the subcarrier phase error is usually negligibly small relative to the RF phase noise component, and hence to a first-order approximation it can be ignored.

In the next two sections we study two specific types of tracking loops which, when digital modulation is present on the RF carrier, yield improved performance over the PLL. This improvement comes about because the PLL tracks only the carrier component and ignores the power in the modulation sidebands.

11-3. DATA-AIDED CARRIER-TRACKING LOOPS

Here we introduce a carrier-tracking loop that uses the power in the modulation sidebands as well as the power in the carrier for purposes of establishing a coherent reference signal.[9] Based on the principle of decision-directed feedback,[10] the motivation behind the data-aided loop (DAL) centers around the use of the power in the composite signal sidebands to enhance the effective loop signal-to-noise ratio thereby reducing the noisy reference loss and ultimately the probability of error of the receiver. We first consider the case of a received signal in the form of a carrier phase modulated by a biphase-modulated data subcarrier. An estimate of the biphase-modulated data subcarrier term is formed which, when fed back to the carrier-tracking loop, can be used to recover the power in the sideband components for carrier-tracking purposes. The net effect is cumulative in that any reduction in the loss due to the noisy reference results in a further improvement in the performance of the data detector, the subcarrier-tracking loop, and so on. Ordinarily, the sideband power, being centered around the subcarrier frequency, is filtered out by the carrier-tracking loop (a PLL) and hence is not available for improving carrier-tracking performance.

Consider the block diagram illustrated in Fig. 11-5, where the upper half

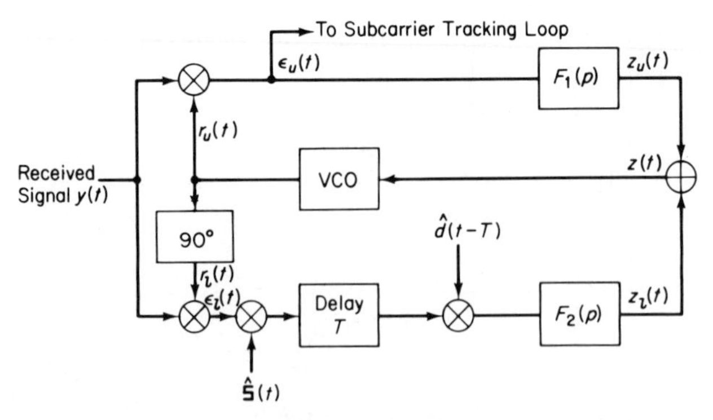

Fig. 11-5 The data-aided carrier-tracking loop (DAL)

of the configuration constitutes a PLL. The modification proposed here makes use of a quadrature channel to insert a signal component into the VCO whose power is proportional to that of the biphase-modulated data subcarrier. Note that the shape of the equivalent phase detector characteristic is affected by the power in the signal's sidebands, the phase error in the subcarrier-tracking loop, the symbol sync error, and the conditional probability of error of the data detector. However, cursory examination of these effects indicates that the *symbol sync error* and *subcarrier sync error* are small relative to the loss caused by the carrier-tracking loop phase error; consequently, essentially all the sideband power can be recovered and used for improving the principal source of system degradation—the noisy reference loss.

The Stochastic Integro-differential Equation of Operation

For a single-channel communication system, the received signal $y(t)$ is of the form given in Eq. (11-2). It is convenient to characterize reference signals $r_u(t)$ and $r_l(t)$ in Fig. 11-4 by

$$r_u(t) = \sqrt{2}K_1 \cos[\omega_0 t + \hat{\theta}(t)] \quad \text{and} \quad r_l(t) = -\sqrt{2}K_2 \sin[\omega_0 t + \hat{\theta}(t)] \tag{11-7}$$

Assuming an upper loop multiplier gain K_{1m} and neglecting the double frequency terms it is easy to show that in operator form

$$z_u(t) = K_1 K_{1m} F_1(p)\{\sqrt{P_c} \sin \varphi(t) + N_u[t, \varphi(t)]\} \tag{11-8}$$

where $N_u[t, \varphi(t)]$ is modeled exactly the same way as $N[t, \varphi(t)]$ of (2-7).

Similarly, assuming a lower loop multiplier gain of K_{2m} and neglecting double-frequency components, the output signal $z_l(t)$ is given by

$$\begin{aligned} z_l(t) = {} & K_2 K_{2m} F_2(p) \exp(-pT) \\ & \times \{\sqrt{S}\, d(t)\hat{d}(t)\mathsf{S}(t)\hat{\mathsf{S}}(t) \sin \varphi(t) + \hat{d}(t) N_l[t, \varphi(t)]\} \end{aligned} \tag{11-9}$$

where $\hat{d}(t)$ is the estimate of $d(t)$ and $\hat{\mathsf{S}}(t)$ is the estimate of $\mathsf{S}(t)$. In arriving at (11-9), we have also assumed that a band-pass filter isolates the biphase-modulated data subcarrier from the dynamic phase error $\epsilon_l(t)$. Furthermore, since the correlation time of the additive noise $n_i(t)$ is small in comparison with that of the data-modulated subcarrier, it is easy to show that the process $N_l[t, \varphi(t)]$ can be modeled in the same way as $N_u[t, \varphi(t)]$, and that due to the orthogonal nature of the reference signals $r_l(t)$ and $r_u(t)$, the random processes $N_l[t, \varphi(t)]$ and $N_u[t, \varphi(t)]$ are statistically independent at time t.

In order to simplify (11-9) any further, it is necessary to make certain assumptions, which in practice are often readily satisfied. When the loop bandwidth is small relative to the signaling data rate $1/T$, the phase error is essentially constant over many symbol intervals, and the product signal

$d(t)\hat{d}(t)\mathsf{S}(t)\hat{\mathsf{S}}(t)$ can be replaced by its statistical average. Assuming for simplicity perfect subcarrier sync ($\hat{\mathsf{S}} = \mathsf{S}$), then*

$$\begin{aligned} E\{d\hat{d}\mathsf{S}\hat{\mathsf{S}}\} &= (+1) \times \text{Prob}\{d = \hat{d}\} + (-1) \times \text{Prob}\{d \neq \hat{d}\} \\ &= (+1)[1 - P_E(\varphi)] + (-1)P_E(\varphi) \\ &= 1 - 2P_E(\varphi) \end{aligned} \tag{11-10}$$

where $P_E(\varphi)$ is the symbol error probability of the data detector conditioned on the loop phase error. At this point, it is not necessary to restrict $P_E(\varphi)$ to any particular type of data detection algorithm.

From the assumption on the relation between loop bandwidth and signaling rate, we also can deduce that the factor $\exp(-j\omega T)$ is approximately unity for all ω within this loop bandwidth and hence has negligible effect from a steady-state performance standpoint. Mechanistically, however, the delay in the lower loop is important in assuring that the $d(t)\mathsf{S}(t)$ signal produced in the lower loop is multiplied by the $\hat{d}(t)\hat{\mathsf{S}}(t)$ signal corresponding to the same symbol interval. The delay is also important from a signal acquisition standpoint.

Under the above assumptions, (11-9) simplifies to

$$z_l = K_2 K_{2m} F_2(p)\{\sqrt{S}[1 - 2P_E(\varphi)] \sin\varphi + \hat{d}N_l(t, \varphi)\} \tag{11-11}$$

Finally, as noted in Chap. 2, the instantaneous VCO phase $\hat{\theta}$ is related to its input $z \triangleq z_l + z_u$ through

$$\hat{\theta} = \frac{K_V}{p} z = \frac{K_V}{p}[F_1(p)(yr_u) + F_2(p)(yr_l\hat{d}\hat{\mathsf{S}})] \tag{11-12}$$

where K_V is the gain of the VCO. Thus, from (11-8), (11-11), and (11-12)

$$\begin{aligned} \dot{\varphi} = \dot{\theta} &- K_u F_1(p)[\sqrt{P_c} \sin\varphi + N_u(t, \varphi)] \\ &- K_l F_2(p)\{\sqrt{S}[1 - 2P_E(\varphi)] \sin\varphi + \hat{d}N_l(t, \varphi)\} \end{aligned} \tag{11-13}$$

where $K_u \triangleq K_1 K_{1m} K_V$ and $K_l \triangleq K_2 K_{2m} K_V$ represent the upper and lower open-loop gains, respectively. Equation (11-13) represents the stochastic integro-differential equation of loop operation for the configuration illustrated in Fig. 11-5.

Nonlinear Analysis for Second-Order DALs with Identical Loop Filters

For the case where the loop filters $F_1(s)$ and $F_2(s)$ are identical and equal to $F(s)$, (11-13) reduces to

$$\dot{\varphi} = \Omega_0 - F(p)[\{K_u\sqrt{P_c} + K_l\sqrt{S}[1 - 2P_E(\varphi)]\} \sin\varphi + N_e(t, \varphi)] \tag{11-14}$$

---

* Hereafter, for simplicity of notation in the equations, we suppress the dependence on t in all variables that are functions of time.

where we have assumed that $\theta(t) = \Omega_0 t + \theta_0$ and $N_e(t, \varphi) \triangleq K_u N_u(t, \varphi) + K_l \hat{d}(t) N_l(t, \varphi)$ is approximately represented by a low-pass Gaussian noise process with spectral density

$$S_{N_e}(\omega) = \frac{N_0 K_u^2}{2}(1 + G^2) \tag{11-15}$$

where $G \triangleq K_l/K_u$. The case of interest here is whėre the loop filter $F(s)$ is second-order and given by (2-9). Then, noticing the resemblance between (11-14) and (2-96), it is easily shown that the p.d.f. of ϕ is given by (2-99) where the potential function $U_0(\phi)$ is now defined by

$$U_0(\phi) = -\beta\phi - \alpha\left\{\frac{\cos\phi - GM\int \sin\phi[1 - 2P_E(\phi)]d\phi}{1 + GM[1 - 2P_E(0)]}\right\} \tag{11-16}$$

with $M \triangleq \sqrt{(1 - m_1^2)/m_1^2}$. The parameters α and β are defined by (2-17) with the following replacements:

$$\begin{aligned} P_c &\longrightarrow P_c\left\{\frac{1 + GM[1 - 2P_E(0)]}{\sqrt{1 + G^2}}\right\}^2 \\ K &\longrightarrow K_u\sqrt{1 + G^2} \\ \sin\phi &\longrightarrow \sin\phi\left\{\frac{1 + GM[1 - 2P_E(\phi)]}{1 + GM[1 - 2P_E(0)]}\right\} \end{aligned} \tag{11-17}$$

The Selection of the Upper and Lower Loop Gains

To optimize the design, we need to minimize the variance of the phase error for a given m_1 by proper choice of the loop gains K_u and K_l. This optimization procedure is extremely tedious and must be carried out by means of numerical integration on a digital computer. Consequently, at this point we propose to optimize the loop as follows. Let $\sqrt{P_c}$ and K be taken as the parameters of a PLL. Then the signal-to-noise ratio in the loop bandwidth ρ is given by (2-17). For the DAL in its linear region of operation there is an analogous parameter

$$\rho_e = \frac{4\sqrt{P_{ce}}}{N_0 K_e} = \frac{2\sqrt{P_{ce}}}{N_0 W_{Le}} \tag{11-18}$$

where, in terms of the parameters of a PLL,

$$\sqrt{P_{ce}} = \sqrt{P_c}\left\{\frac{1 + GM[1 - 2P_E(0)]}{\sqrt{1 + G^2}}\right\} \tag{11-19}$$

and

$$K_e = K\left(\frac{K_u}{K}\right)(\sqrt{1 + G^2}) \tag{11-20}$$

Using (11-18) through (11-20) and the definition for ρ, we find that the improvement in loop signal-to-noise ratio is given by

$$I \triangleq \frac{\rho_e}{\rho} = \left(\frac{K}{K_u}\right)\left\{\frac{1 + GM[1 - 2P_E(0)]}{1 + G^2}\right\} \tag{11-21}$$

where for a first-order loop

$$\frac{W_{Le}}{W_L} = \left(\frac{K_u}{K}\right)\{1 + GM[1 - 2P_E(0)]\} \tag{11-22}$$

The optimization procedure is carried out as follows: Choose G such that I is maximized subject to the constraint that $W_{Le}/W_L = 1$. Differentiating I with respect to G and equating to zero, yields an optimum value of G, say G_0, given by

$$G_0 = M[1 - 2P_E(0)] \tag{11-23}$$

Also since $W_{Le}/W_L = 1$,

$$\frac{K}{K_u} = 1 + G_0 M[1 - 2P_E(0)] \tag{11-24}$$

so that the maximum value of I, say $I_{\max}$, is given by

$$I_{\max} \triangleq \left(\frac{\rho_e}{\rho}\right)_{\max} = \frac{\{1 + G_0 M[1 - 2P_E(0)]\}^2}{1 + G_0^2} \tag{11-25}$$

Substitution for G_0 into (11-25) gives

$$I_{\max} = \frac{1}{m_1^2}\left\{[1 - 2P_E(0)]^2 + m_1^2\{1 - [1 - 2P_E(0)]^2\}\right\} \tag{11-26}$$

and

$$\begin{aligned} K_u &= \left\{\frac{m_1^2}{[1 - 2P_E(0)]^2 + m_1^2\{1 - [1 - 2P_E(0)]^2\}}\right\}K \\ K_l &= \sqrt{\frac{1 - m_1^2}{m_1^2}}\left\{\frac{m_1^2[1 - 2P_E(0)]}{[1 - 2P_E(0)]^2 + m_1^2\{1 - [1 - 2P_E(0)]^2\}}\right\}K \end{aligned} \tag{11-27}$$

In the limit as the data detector error probability $P_E(0)$ goes to zero, one obtains the largest relative improvement—that is, $I_{\max} = 1/m_1^2$.

For the optimum value of G as in (11-23), the potential function of Eq. (11-16) becomes

$$U_0(\phi) = \left[-\beta\phi - \alpha\left\{\frac{\cos\phi - M^2[1 - 2P_E(0)]\int \sin\phi[1 - 2P_E(\phi)]d\phi}{1 + M^2[1 - 2P_E(0)]^2}\right\}\right] \tag{11-28}$$

and the replacements of (11–17) simplify to

$$\begin{aligned} P_c &\longrightarrow P_c\{1 + M^2[1 - 2P_E(0)]^2\} \\ K &\longrightarrow K_u\sqrt{1 + M^2[1 + 2P_E(0)]^2} \\ \sin\phi &\longrightarrow \sin\phi\left\{\frac{1 + M^2[1 - 2P_E(0)][1 - 2P_E(\phi)]}{1 + M^2[1 - 2P_E(0)]^2}\right\} \end{aligned} \tag{11-29}$$

Tracking Performance for the Case of Perfect Ambiguity Resolution

The variance of the phase error is given by

$$\sigma_{\phi e}^2 = \int_{-\pi}^{\pi} \phi^2 p(\phi) d\phi \tag{11-30}$$

and must be evaluated numerically on a digital computer using Eqs. (2-99), (11-28), and (11-29). For values of m_1^2 where $p(\phi)$ becomes trimodal within the interval $(-\pi, \pi)$, it is desirable to apply an alternate definition for loop performance. The system now possesses two additional unstable singularities (ϕ_1, ϕ_2) in the interval $(-\pi, \pi)$, and the phase values $-\pi$, π therefore act as stable nodes. If the stable points occurring between 0 and π (or $-\pi$) can be resolved, the variance of the phase error should be defined as

$$\sigma_{\phi e}^2 = \int_{-\pi}^{-\phi_1} (\pi + \phi)^2 p(\phi) d\phi + \int_{-\phi_1}^{\phi_2} \phi^2 p(\phi) d\phi + \int_{\phi_2}^{\pi} (\pi - \phi)^2 p(\phi) d\phi \tag{11-31}$$

where ϕ_1 and ϕ_2 are values of phase error for which the potential function $U_0(\phi)$ has its minimum in an interval of width 2π. It is a simple matter to show that for the case of zero detuning, $p(\phi)$ is symmetric in ϕ, and $\phi_1 = \phi_2$, thus allowing simplification of (11-31).

Mean Time to First Slip or First Loss of Phase Synchronization

The next performance parameter we investigate is the mean time to first loss of phase synchronization. For simplicity of presentation, a first-order PLL with zero detuning is assumed and the improvement in first-slip time relative to that of the PLL is computed.

Based on the theory presented in Chap. 2, it is easy to show that under the above assumptions together with a zero initial condition on the phase error, the equivalent bandwidth–mean time to first slip product can be expressed as

$$W_{Le}\tau_e(2\pi) = \frac{\rho_e}{2} \int_0^{\pi} \frac{d\phi}{p(\phi)} \tag{11-32}$$

where $p(\phi)$ is given by (2-99) together with (11-16) and (11-17). When $G = G_0$ and $W_{Le}/W_L = 1$, (11-32) reduces to

$$W_{Le}\tau_e(2\pi) = \frac{\rho}{2m_1^2} \left\{[1 - 2P_E(0)]^2 + m_1^2\{1 - [1 - 2P_E(0)]^2\}\right\} \int_0^{\pi} \frac{d\phi}{p(\phi)} \tag{11-33}$$

For the PLL with zero frequency offset and zero initial phase error,

$$W_L\tau(2\pi) = \pi^2 \rho I_0^2(\rho) \tag{11-34}$$

so that

$$\frac{\tau_e(2\pi)}{\tau(2\pi)} = \left\{\frac{[1 - 2P_E(0)]^2 + m_1^2\{1 - [1 - 2P_E(0)]^2\}}{2\pi^2 m_1^2 I_0^2(\rho)}\right\} \int_0^\pi \frac{d\phi}{p(\phi)} \tag{11-35}$$

For large ρ the relative improvement in equivalent bandwidth–mean time to first slip product becomes

$$\frac{W_{Le}\tau_e(2\pi)}{W_L\tau(2\pi)} \cong \exp\,[2(\rho_e - \rho)] \tag{11-36}$$

while for $G = G_0$ and $W_{Le}/W_L = 1$,

$$\frac{\tau_e(2\pi)}{\tau(2\pi)} \cong \exp\left\{2\rho\left[\frac{(1 - m_1^2)[1 - 2P_E(0)]^2}{m_1^2}\right]\right\} \tag{11-37}$$

Finally, for small ρ, $I_0(\rho) \cong 1$, $p(\phi) \cong 1/2\pi$, and (11–36) reduces to

$$\frac{\tau_e(2\pi)}{\tau(2\pi)} \cong \frac{1}{m_1^2}\left\{[1 - 2P_E(0)]^2 + m_1^2\{1 - [1 - 2P_E(0)]^2\}\right\} \tag{11-38}$$

DAL and PLL Performance Comparisons (PSK Signals)

Before we consider detailed performance of the DAL, it is of interest to study its equivalent phase detector characteristic with $\Omega_0 = 0$ as given by

$$\mathcal{S}(\phi) = \frac{\{1 + GM[1 - 2P_E(\phi)]\}}{\sqrt{1 + G^2}} \sin\phi \tag{11-39}$$

With $G = G_0$, the above equation becomes

$$\mathcal{S}(\phi) = \frac{1 + M^2[1 - 2P_E(\phi)][1 - 2P_E(0)]}{\sqrt{1 + M^2[1 - 2P_E(0)]^2}} \sin\phi \tag{11-40}$$

The function $\mathcal{S}(\phi)$ is illustrated in Figs. 11-6 through 11-8 with m_1^2 and $R_d = ST/N_0$ as parameters, and $P_E(\phi)$ as given by (6-21)—that is, the case of PSK signals. [For the remainder of this section we assume that $P_E(\phi)$ is characterized as above.] For comparison, the phase dector characteristic of a PLL, obtained from (11-39) by letting $G = 0$ (opening the lower loop), also is shown.

For $\Omega_0 = 0$ and a first-order loop ($r \longrightarrow \infty$), the p.d.f. $p(\phi)$, with $G = G_0$ and $W_{Le}/W_L = 1$, is illustrated in Figs. 11-9 through 11-13 for various values of ρ with δ and m_1^2 as parameters. Figures 11-9 and 11-12 show how $p(\phi)$ is affected by a change in δ. It is interesting to note from Fig. 11-13 that for $\rho < 1$ the p.d.f. becomes trimodal; however, such values of ρ are merely of academic interest.

The mean value of the first-slip time $\tau_e(2\pi)$ in the DAL as computed from Eqs. (2-99), (11-28), (11-29), and (11-33), is plotted in Fig. 11-14 versus ρ for fixed δ and several values of m_1^2. Note the increase in slope of these

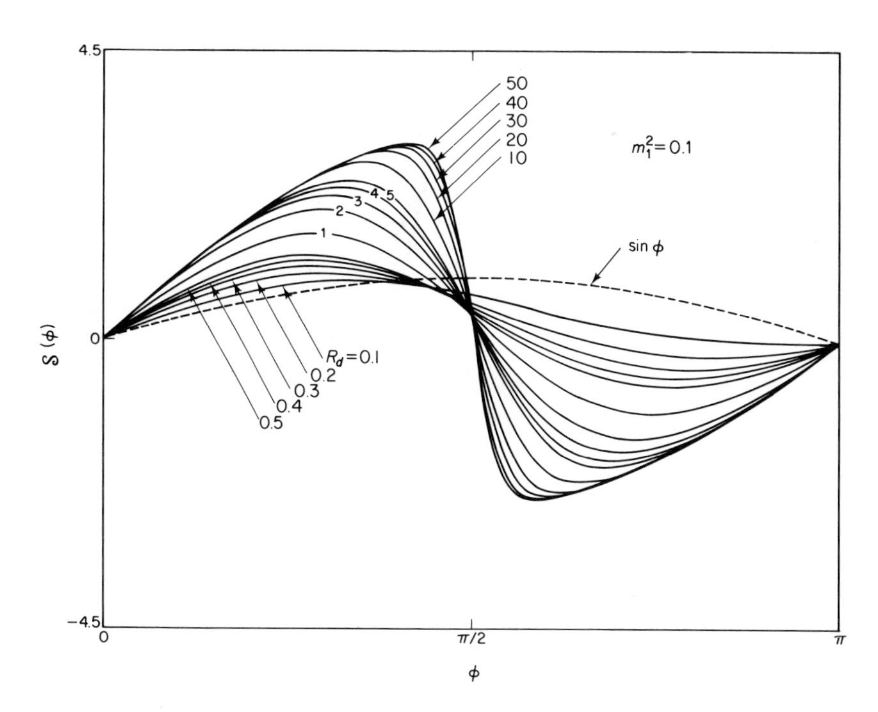

Fig. 11-6 Equivalent loop phase detector characteristic with R_d as a parameter: $m_1^2 = 0.1$

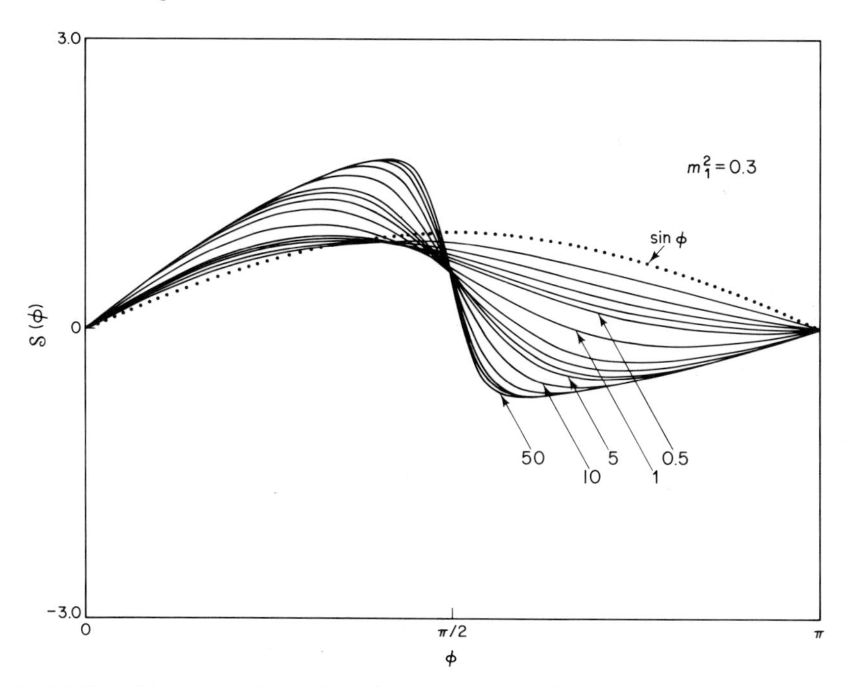

Fig. 11-7 Equivalent loop phase detector characteristic with R_d as a parameter: $m_1^2 = 0.3$

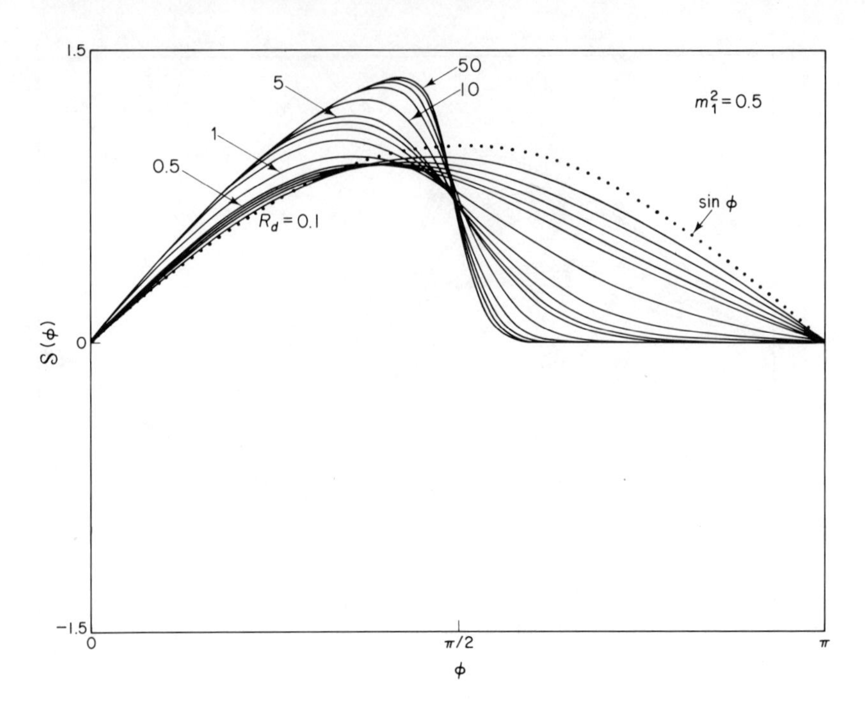

Fig. 11-8 Equivalent loop phase detector characteristic with R_d as a parameter: $m_1^2 = 0.5$

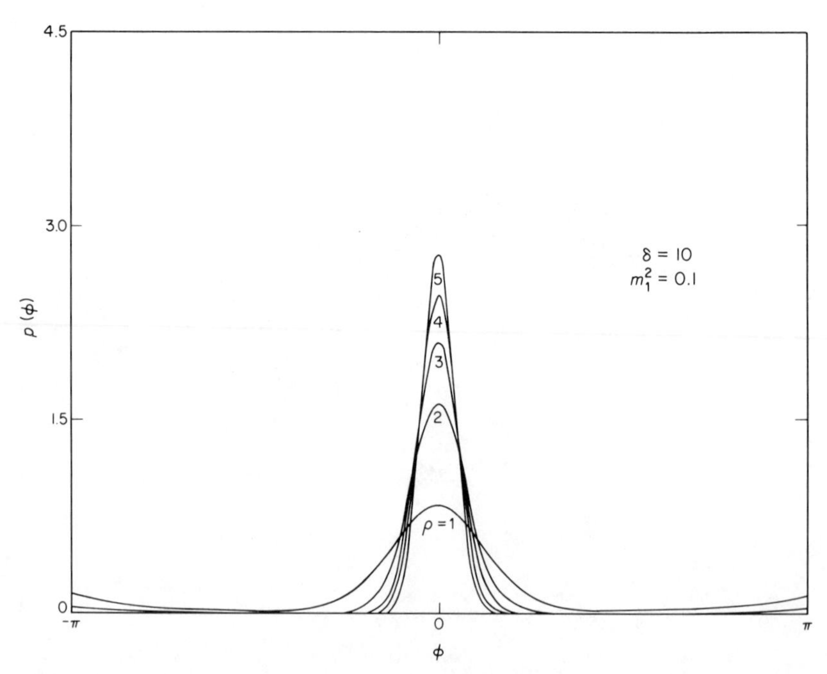

Fig. 11-9 Phase error p.d.f. with loop signal-to-noise ratio of PLL as a parameter: $\delta = 10$, $m_1^2 = 0.1$

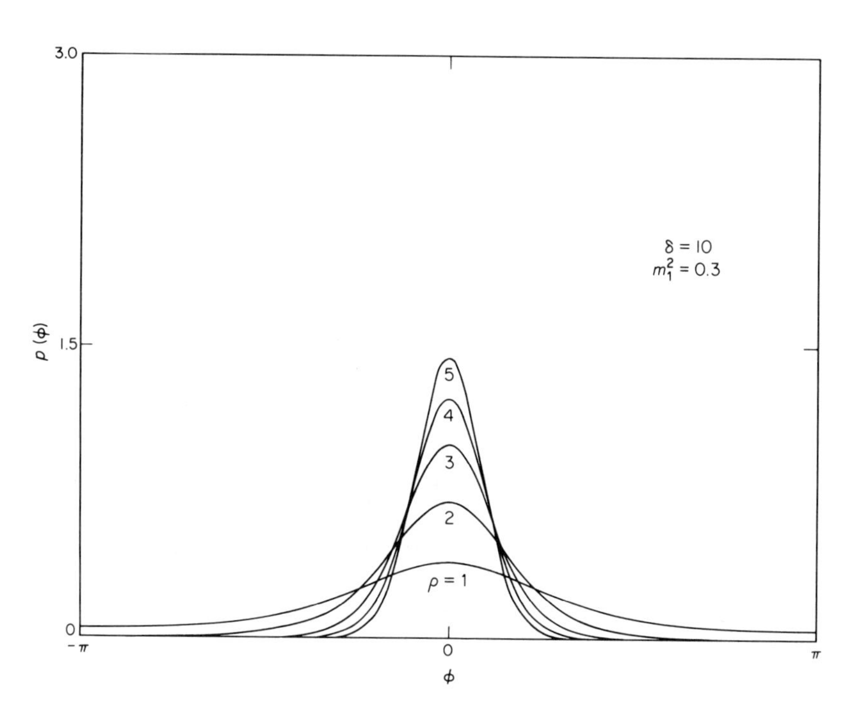

Fig. 11-10 Phase error p.d.f. with loop signal-to-noise ratio of PLL as a parameter: $\delta = 10,\ m_1^2 = 0.3$

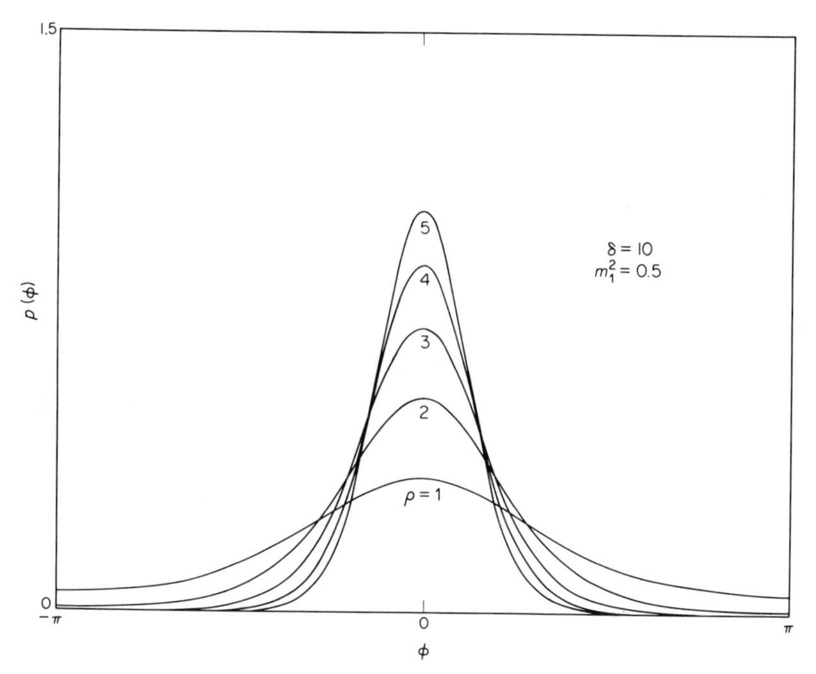

Fig. 11-11 Phase error p.d.f. with loop signal-to-noise ratio of PLL as a parameter: $\delta = 10,\ m_1^2 = 0.5$

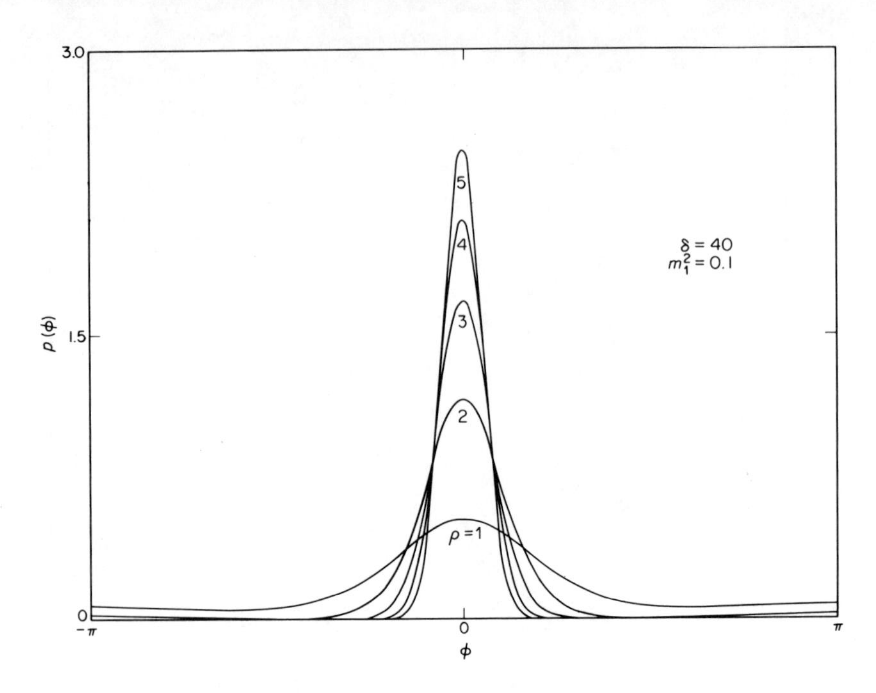

Fig. 11-12 Phase error p.d.f. with loop signal-to-noise ratio of PLL as a parameter: $\delta = 40$, $m_1^2 = 0.1$

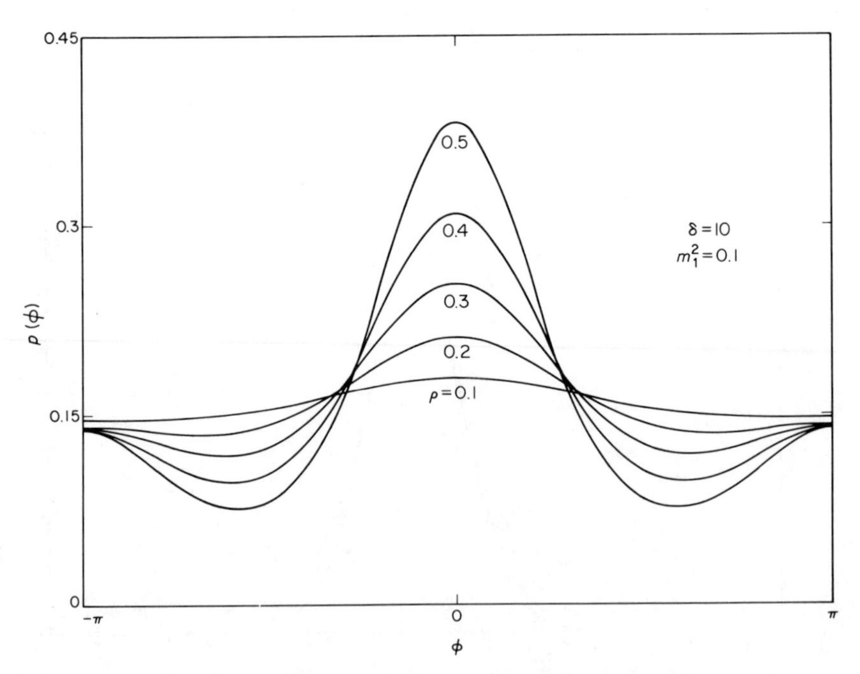

Fig. 11-13 Phase error p.d.f. with loop signal-to-noise ratio of PLL as a parameter: $\delta = 10$, $m_1^2 = 0.1$

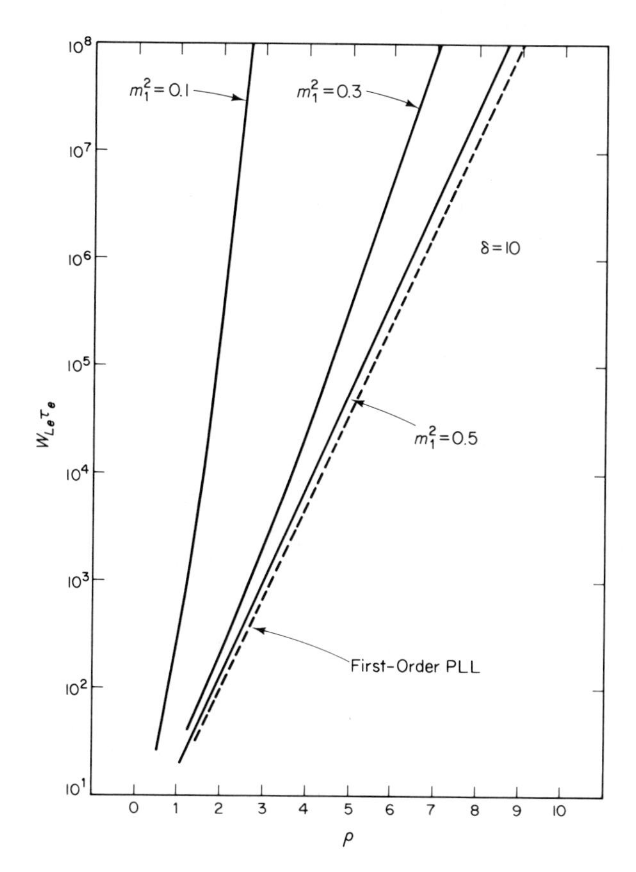

Fig. 11-14 Mean time to first slip of a cycle versus loop signal-to-noise ratio of PLL with m_1^2 as a parameter: $\delta = 10$

curves with decreasing m_1^2 in accordance with (11-34). Figure 11-15 shows the behavior of the variance of the phase error for a first-order loop with zero detuning as a function of ρ for various values of m_1^2 and δ fixed. For the case $m_1^2 = 0.5$, the results, as given by (11-30) and (11-31), are identical; hence, only one curve appears. Figure 11-16 compares the performance of the DAL to that of a PLL. This curve depicts the ratio of the mean-squared phase error in the DAL to that of a PLL (in dB) versus ρ for various values of m_1^2 with $\delta = 5$. Note the limiting asymptotes, indicated by the dashed lines, as predicted by (11-26) with $P_E(0) \longrightarrow 0$. This is so since in the limit as $\rho \longrightarrow \infty$, $\sigma_\phi^2/\sigma_{\phi e}^2 \cong \rho_e/\rho$. In principle, one can obtain a relative improvement in mean-squared phase error in the nonlinear region of operation greater than that predicted for linear behavior. This property is undoubtedly inherent in the nonlinear nature of the loop phase detector characteristic.

To illustrate the equivalent performance of a coherent data detector, we

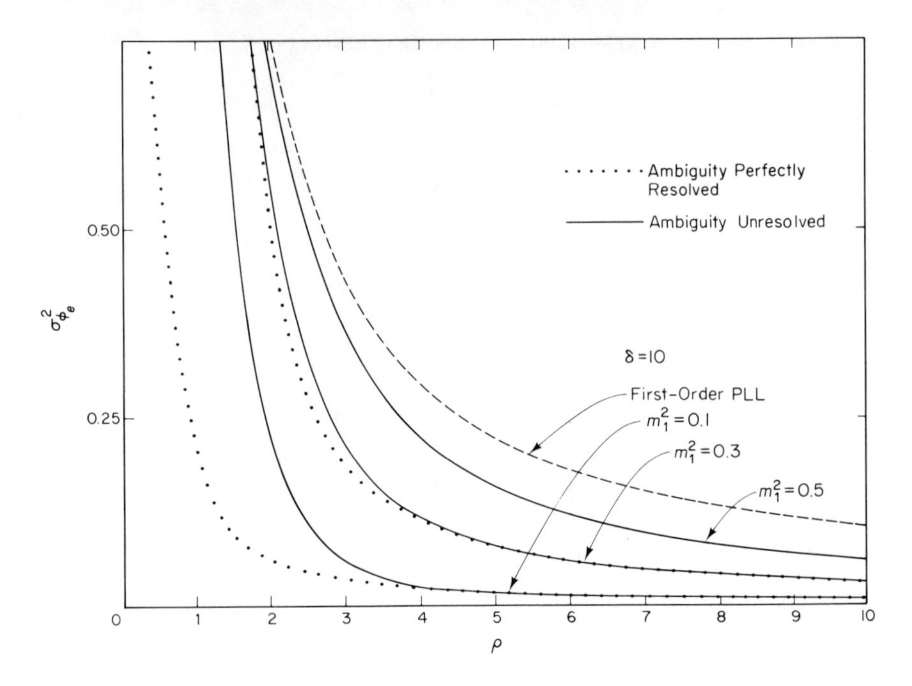

Fig. 11-15 Variance of the phase error versus loop signal-to-noise ratio of PLL with m_1^2 as a parameter: $\delta = 10$

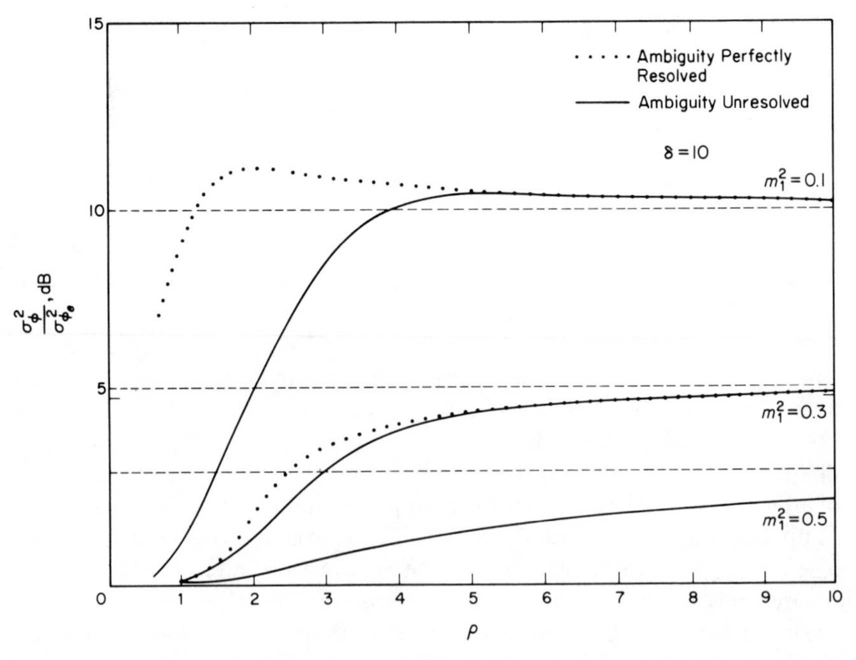

Fig. 11-16 Relative improvement (in dB) of the variance of the phase error of the DAL over that of the PLL with m_1^2 as a parameter: $\delta = 10$

make use of the expression for average error probability as given by

$$P_E = \int_{-\pi}^{\pi} p(\phi)P_E(\phi)d\phi \tag{11-41}$$

or when a "genie" is present to resolve the phase ambiguities

$$P_E = \int_{-\pi}^{-\phi_1} p(\phi)P_E(\pi + \phi)d\phi + \int_{-\phi_1}^{\phi_2} p(\phi)P_E(\phi)d\phi + \int_{\phi_2}^{\pi} p(\phi)P_E(\pi - \phi)d\phi \tag{11-42}$$

Performing the numerical integration necessary for evaluation of (11-41) and (11-42) with $P_E(\phi)$ of (6-21) produces the results given in Fig. 11-17. For comparison purposes we superimpose: (1) the result from Chap. 7 for the case where the power between the carrier and data sidebands is divided optimally, and (2) the ideal PSK performance curve.

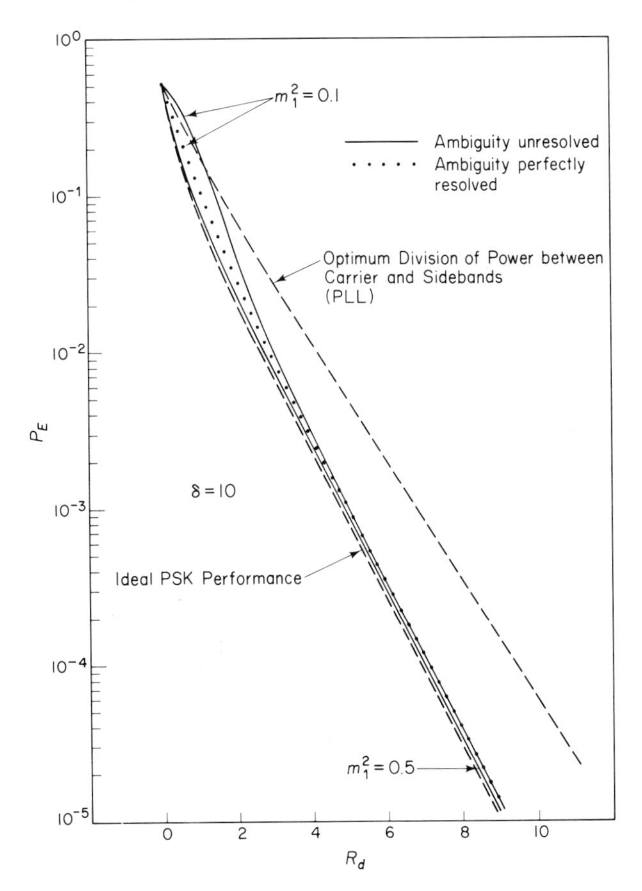

Fig. 11-17 Error probability performance of a PSK system using a DAL with m_1^2 as a parameter: $\delta = 10$

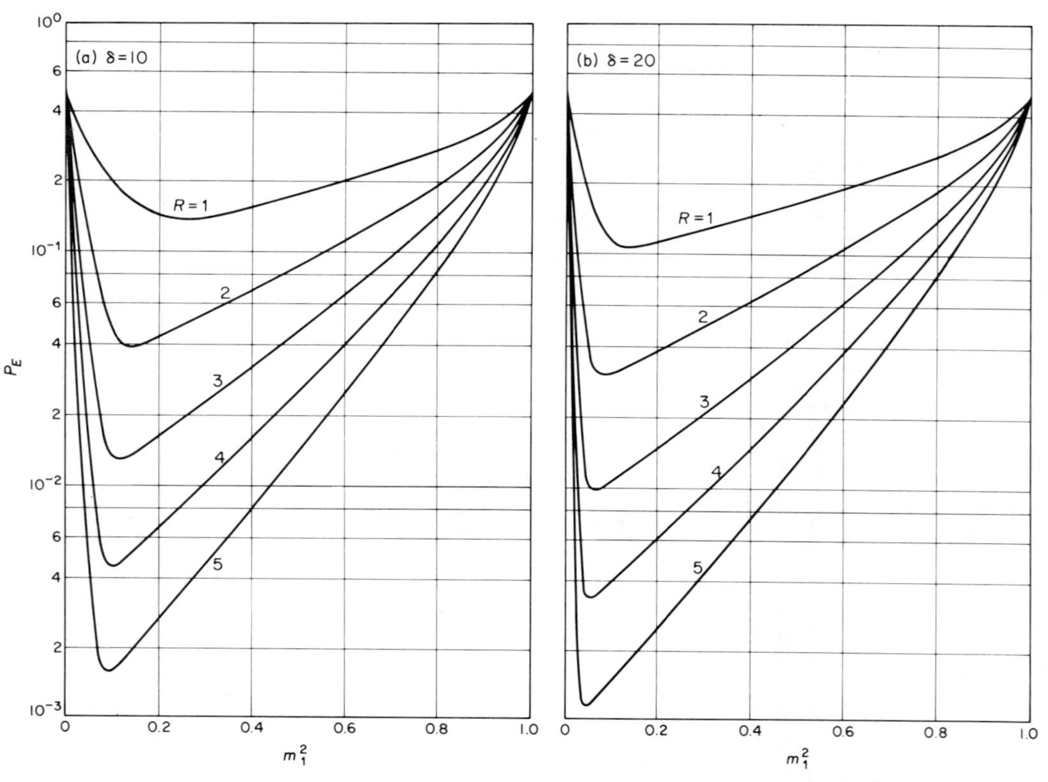

Fig. 11-18 Error probability vs modulation factor squared for fixed total signal-to-noise ratio (ambiguity unresolved)

We now consider the optimum choice of a modulation factor in a phase-coherent communication system using a DAL to track the RF carrier. The criterion of optimization used in Sec. 7-2 for a single-channel system with a PLL can also be applied here since it does not depend on the RF tracking loop configuration. Hence we wish to choose the value of modulation factor $m_{1\text{opt}}$ that minimizes the system error probability for a fixed total energy-to-noise ratio. For the case where no attempt is made to resolve the phase ambiguities and PSK signals are transmitted, Fig. 11-18 plots P_E versus m_1^2 for fixed δ and $R = PT/N_0$ as a parameter. The curves in this figure clearly indicate a minimum P_E and corresponding optimum m_1^2; not surprisingly, they have the same general shape as the equivalent illustrations in Chap. 7. For the situation where the phase ambiguities are perfectly resolved, Fig. 11-19

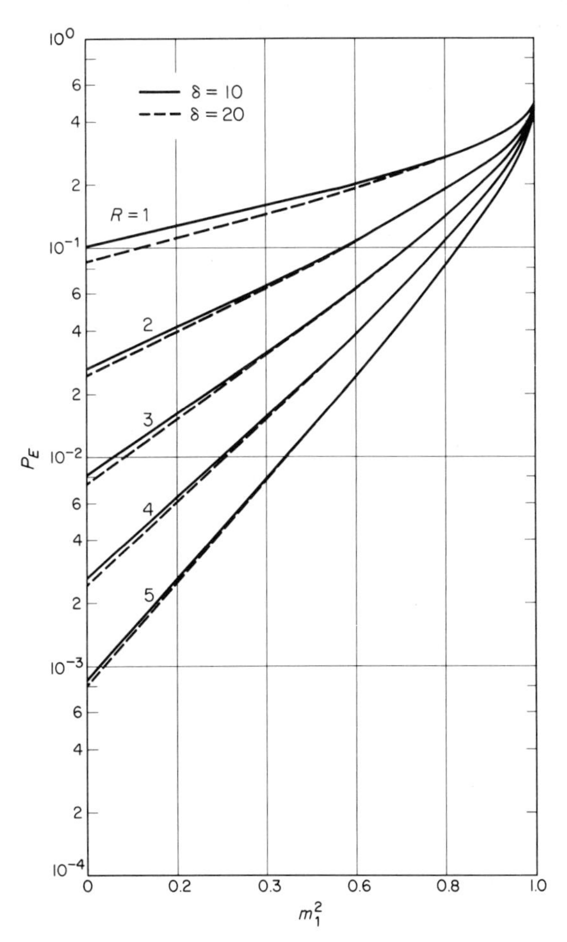

Fig. 11-19 Error probability vs modulation factor squared for fixed total signal-to-noise ratio (ambiguity perfectly resolved)

plots P_E versus m_1^2 for the same values of δ and R as in Fig. 11-18. We note here that the best error probability performance is achieved for $m_1^2 = 0$ (a suppressed-carrier system), regardless of the values of δ and R.

11-4. HYBRID CARRIER-TRACKING LOOPS

We have considered a carrier-tracking loop configuration in which results of the data detector are fed back and used to increase the effective loop signal-to-noise ratio bandwidth. Two factors of prime importance in implementing this configuration are the necessity of symbol timing and the delay element in the lower loop, which accounts for the fact that data decisions are not available until the end of each symbol interval. Figure 11-20 illustrates another scheme for improving the performance of the carrier-tracking loop, which does not make use of the decision-feedback principle. This loop uses feed-forward techniques to accomplish its ultimate goal.

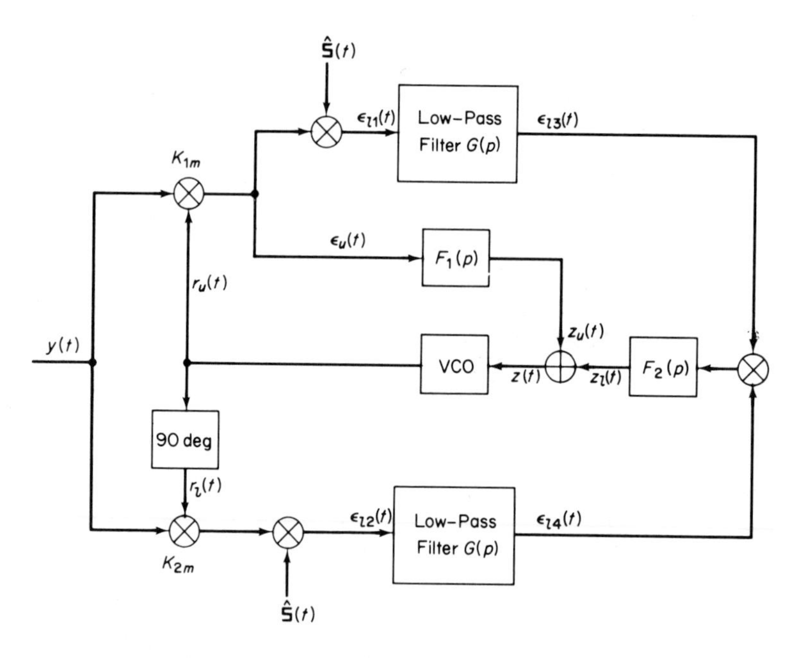

Fig. 11-20 The hybrid carrier-tracking loop (HTL)

From the standpoint of phase-coherent communication systems, it should once again be intuitively clear that any efficient receiver, if it is to track the carrier coherently and extract the modulation from the received signal, should make effective use of the power in both components; that is, coherence of the sideband component at the carrier frequency should be exploited in the carrier-tracking loop. In Chap. 2, we discussed how one can track the carrier frequency and phase of the sideband components by means of a squaring or Costas loop. For quite some time the PLL has been used successfully to track the frequency and phase of the carrier component. The hybrid tracking loop (HTL)[11] illustrated in Fig. 11-20 combines these two approaches and can be used to estimate the carrier frequency and phase. With the appropriate gains, the HTL operates as a PLL when $m_1 = 1$, and its operation is somewhat analogous to the Costas loop when $m_1 = 0$.

The Stochastic Integro-differential Equation of Operation

The received signal $y(t)$ of Eq. (11-2) is applied to three phase detectors, two of which have gain K_{1m} and the other K_{2m}. The reference signals driving these phase detectors are denoted by $r_u(t)$ and $r_l(t)$ (Fig. 11-20). The output $\epsilon_u(t)$ of the upper loop phase detector is applied to the loop filter $F_1(s)$, which removes the modulation sidebands and forms the signal $z_u(t)$. Thus, in operator form,

$$z_u(t) = K_{1m}F_1(p)\epsilon_u(t) = K_{1m}F_1(p)y(t)r_u(t) \tag{11-43}$$

The lower loop makes use of the power in the sidebands to form the estimate of $\theta(t)$ in the following way. Using the estimate $\hat{\mathsf{S}}(t)$ to demodulate the square-wave subcarrier $\mathsf{S}(t)$, the upper phase detector output $\epsilon_{l1}(t) = K_{1m}\hat{\mathsf{S}}(t)y(t)r_u(t)$ and the lower phase detector output $\epsilon_{l2}(t) = K_{2m}\hat{\mathsf{S}}(t)y(t)r_l(t)$ are then applied to low-pass filters $G(s)$, which remove high-frequency noise and signal components. This produces low-pass signals $\epsilon_{l3}(t)$ and $\epsilon_{l4}(t)$, which are then multiplied and applied to the lower loop filter $F_2(s)$ to form $z_l(t)$:

$$z_l(t) = F_2(p)[G(p)\epsilon_{l1}(t)G(p)\epsilon_{l2}(t)] \tag{11-44}$$

Assuming as before that the VCO acts as an ideal integrator with gain K_V, the phase estimate $\hat{\theta}$ of the VCO output is related to z_u and z_l through

$$\begin{aligned}\hat{\theta} &= \frac{K_V}{p}(z_u + z_l)\\ &= \frac{K_V}{p}\{K_{1m}F_1(p)(yr_u) + K_{1m}K_{2m}F_2(p)[G(p)(\hat{\mathsf{S}}yr_u)G(p)(\hat{\mathsf{S}}yr_l)]\}\end{aligned} \tag{11-45}$$

where we have again dropped the dependence on t in our notation. The first term in Eq. (11-45) is due to the PLL nature of the HTL while the second term is attributable to the squaring operation familiar in Costas loops. Both terms combined yield the hybrid system. Combining (11-43) through (11-45), the stochastic integro-differential equation of loop operation becomes

$$\dot{\varphi} = \Omega_0 - K_V\{K_{1m}F_1(p)(yr_u) + K_{1m}K_{2m}F_2(p)[G(p)(\hat{\mathsf{S}}yr_u)G(p)(\hat{\mathsf{S}}yr_l)]\} \tag{11-46}$$

Defining the reference signals as in (11-7) and assuming a perfect square-wave subcarrier reference $\hat{\mathfrak{S}}(t)$, the loop equation reduces to the form

$$\dot{\varphi} = \Omega_0 - K_u F_1(p)[\sqrt{P_c} \sin \varphi + N_u(t, \varphi)] - K_l F_2(p)\left[\frac{S}{2} \sin 2\varphi + N_l(t, \varphi)\right] \quad (11\text{-}47)$$

In the above, $K_l \triangleq K_2 K_{2m} K_u$, $K_u \triangleq K_1 K_{1m} K_V$, $N_u(t, \varphi)$ is the usual band-limited noise process (single-sided spectral density of N_0 w/Hz) affecting the performance of the upper loop and

$$N_l(t, \varphi) = N_1(t)N_2(t) + d(t)\sqrt{S}[N_1(t) \sin \varphi + N_2(t) \cos \varphi] \quad (11\text{-}48)$$

The noise processes $N_1(t)$ and $N_2(t)$ are orthogonal, band-limited Gaussian with two-sided low-pass spectral densities

$$S_{N_1}(\omega) = S_{N_2}(\omega) = \frac{N_0}{2} |G(j\omega)|^2 \quad (11\text{-}49)$$

We note that $N_l(t, \varphi)$, although spectrally equivalent, is not the same as that arising in the Costas or squaring loop. Furthermore, the processes $N_u(t, \varphi)$ and $N_l(t, \varphi)$ are statistically independent since they correspond to different frequency bands of the input wideband noise spectrum.

If we now assume that $F_1(p) = F_2(p) = F(p)$, then (11-47) reduces to

$$\dot{\varphi} = \Omega_0 - F(p)\left[K_u\left\{\sqrt{P_c} \sin \varphi + \frac{GS}{2} \sin 2\varphi\right\} + N_e(t, \varphi)\right] \quad (11\text{-}50)$$

where $G \triangleq K_l/K_u$ again denotes the ratio of gain in the lower loop to that in the upper loop, and

$$N_e(t, \varphi) = K_u[N_u(t, \varphi) + GN_l(t, \varphi)] \quad (11\text{-}51)$$

Note that the shape of the equivalent phase detector characteristic (the term in braces), relative to that of a PLL, is modified by the addition of a double harmonic term, which serves to make it more more "rectangular." It was pointed out in Chap. 9 that the optimum phase detector characteristic for zero loop detuning, is "rectangular" when the additive noise is white and Gaussian.

Nonlinear Analysis for Second-Order HTLs with Identical Loop Filters

For narrowband loops, an analysis based upon the "fluctuation equation" approach[12,13] can, to a good approximation, be applied. Thus, we can make use of the results given in Chap. 2 to solve for the statistical dynamics of interest. For the loop filter of Eq. (2-9), the potential function $U_0(\phi)$, is given by

$$U_0(\phi) = -\beta\phi - \alpha\left\{\frac{\cos \phi + (GM\sqrt{S}/4) \cos 2\phi}{1 + GM\sqrt{S}}\right\} \quad (11\text{-}52)$$

where α and β are defined by (2-17) with the following replacements

$$P_c \longrightarrow P_c\left[\frac{1+GM\sqrt{S}}{\sqrt{1+G^2S\mathcal{S}_L^{-1}}}\right]^2$$
$$K \longrightarrow K_u\sqrt{1+G^2S\mathcal{S}_L^{-1}}$$
$$\sin\phi \longrightarrow \sin\phi\left[\frac{1+(GM\sqrt{S}/2)(\sin 2\phi/\sin\phi)}{1+GM\sqrt{S}}\right] \tag{11-53}$$

The parameter $\mathcal{S}_L$ is the squaring loss defined in Chap. 2 by (2-79), with A^2 replaced by S and $M = \sqrt{S/P_c}$.

The Selection of the Upper and Lower Loop Gains

The procedure for selecting K_l and K_u will again be based on the desire to optimize system performance as in Sec. 11-3. The effective loop signal-to-noise ratio of the HTL may still be defined as in (11-18), but with $\sqrt{P_{ce}}$ and K_e now given by

$$\sqrt{P_{ce}} = \frac{\sqrt{P_c}[1+GM\sqrt{S}]}{\sqrt{1+G^2S\mathcal{S}_L^{-1}}} \quad \text{and} \quad K_e = K\left(\frac{K_u}{K}\right)(\sqrt{1+G^2S\mathcal{S}_L^{-1}}) \tag{11-54}$$

Hence, the improvement in loop signal-to-noise ratio relative to a PLL is

$$I \triangleq \frac{\rho_e}{\rho} = \left(\frac{K}{K_u}\right)\left[\frac{1+GM\sqrt{S}}{1+G^2S\mathcal{S}_L^{-1}}\right] \tag{11-55}$$

The ratio of the effective bandwidth of a first-order HTL to that of a PLL may also be determined using (11-54); that is,

$$\frac{W_{Le}}{W_L} = \left(\frac{K_u}{K}\right)[1+GM\sqrt{S}] \tag{11-56}$$

Constraining the bandwidth ratio W_{Le}/W_L equal to unity, differentiating I of (11-55) with respect to G, and equating to zero produces an optimum value of G, say G_0, which is given by

$$G_0 = \frac{M}{\sqrt{S\mathcal{S}_L^{-1}}} \tag{11-57}$$

Substituting Eqs. (11-56) and (11-57) into (11-55) together with the constraint $W_{Le}/W_L = 1$ gives

$$I_{\max} = 1 + M^2\mathcal{S}_L = \frac{1}{m_1^2}[\mathcal{S}_L + m_1^2(1-\mathcal{S}_L)] \tag{11-58}$$

Recalling from the definition of $\mathcal{S}_L$ that $\lim_{\rho\to\infty}\mathcal{S}_L = 1$ for any $G(s)$ filter characteristic, we observe that

$$\lim_{\rho\to\infty} I_{\max} = \frac{1}{m_1^2} \tag{11-59}$$

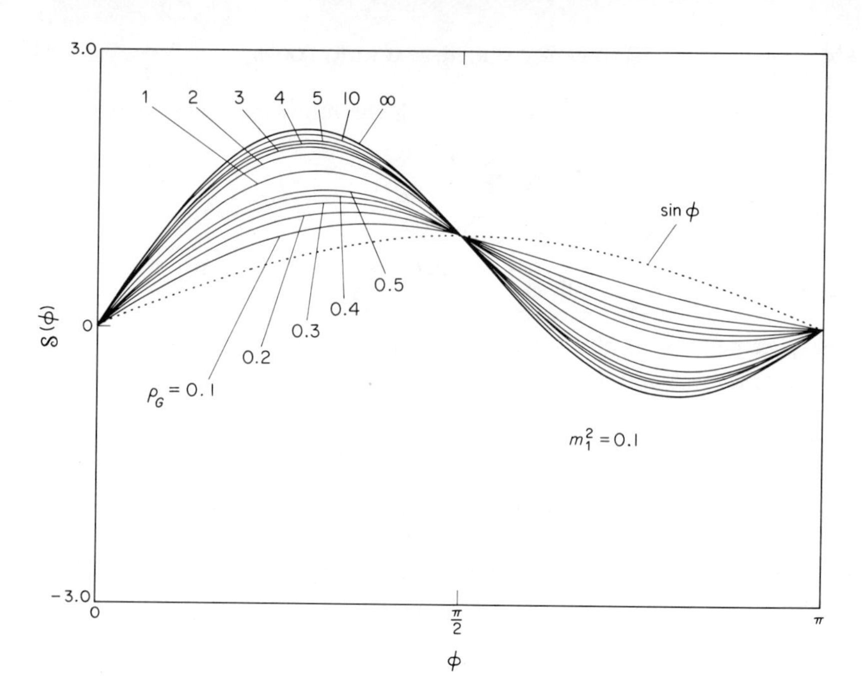

Fig. 11-21 Equivalent loop phase detector characteristic with ρ_G as a parameter: $m_1^2 = 0.1$

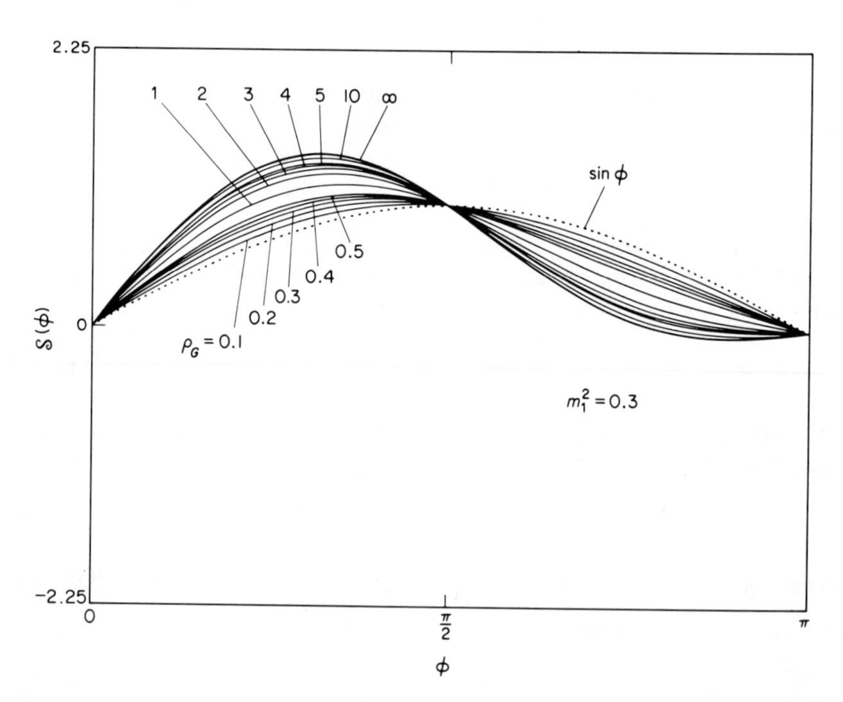

Fig. 11-22 Equivalent loop phase detector characteristic with ρ_G as a parameter: $m_1^2 = 0.3$

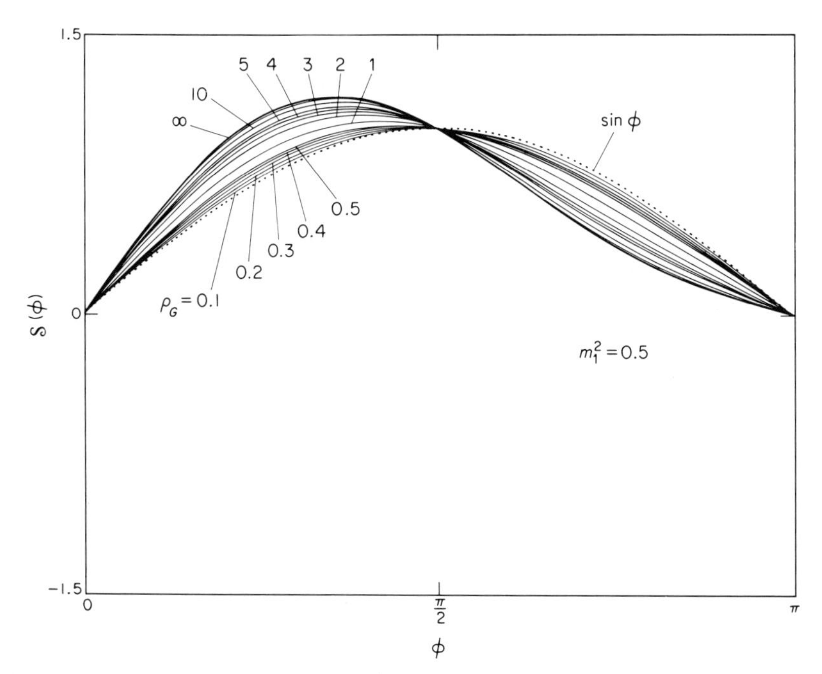

Fig. 11-23 Equivalent loop phase detector characteristic with ρ_G as a parameter: $m_1^2 = 0.5$

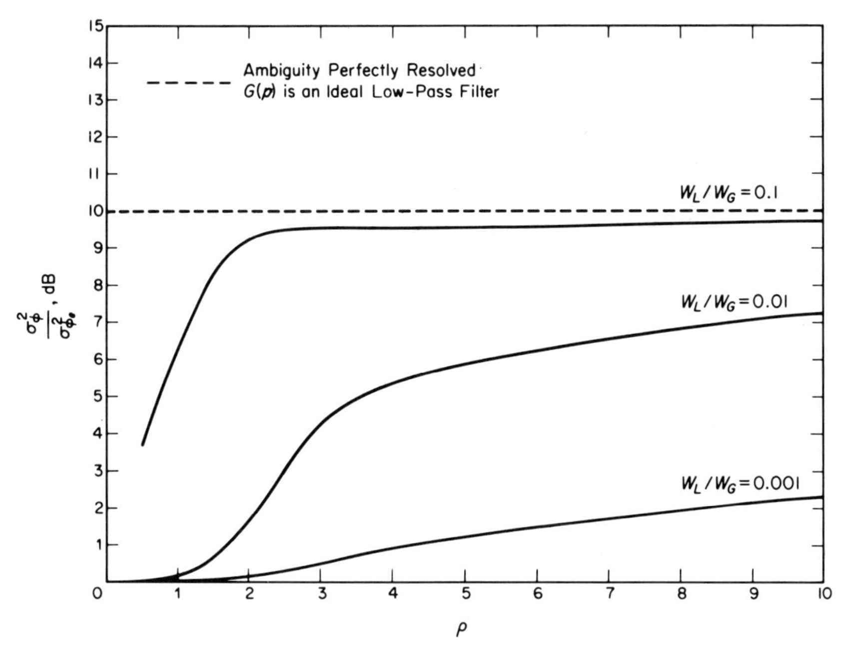

Fig. 11-24 The effect of squaring loss on the relative improvement (in dB) of the variance of the phase error of the HTL over that of the PLL, $m_1^2 = 0.1$

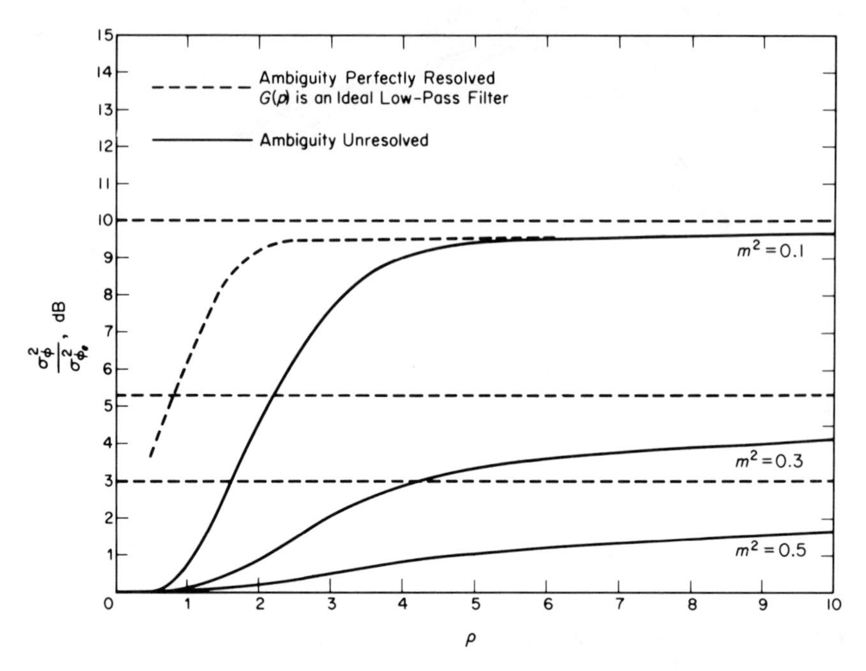

Fig. 11-25 Relative improvement (in dB) of the variance of the phase error of the HTL over that of the PLL with m_1^2 as a parameter: $W_L/W_G = 0.1$

which is the same result as that obtained for the DAL. Finally, for the optimum G of (11-57) and $W_{Le}/W_L = 1$, the upper and lower loop gains, respectively, are given by

$$K_u = \left[\frac{m_1^2}{\mathcal{S}_L + m_1^2(1 - \mathcal{S}_L)}\right]K \quad \text{and} \quad K_l = \sqrt{\frac{1 - m_1^2}{m_1^2 S}}\left[\frac{m_1^2\mathcal{S}_L}{\mathcal{S}_L + m_1^2(1 - \mathcal{S}_L)}\right]K \tag{11-60}$$

The equations for phase error variance as given by Eqs. (11-30) and (11-31) are directly applicable here if one uses $U_0(\phi)$ of (11-52) to characterize the phase error p.d.f. $p(\phi)$. Once again, additional unstable singularities may be created at ϕ_1, ϕ_2, and one must consider the possibility of resolving them.

With respect to mean time to first slip, the expression given in (11-32) for a first-order loop is again appropriate here, using the $p(\phi)$ described above and $\rho_e = I\rho$, with I defined by (11-55). Using (11-34), one can again calculate the ratio of equivalent mean time to first slip of the HTL to that of the PLL; that is,

$$\frac{\tau_e(2\pi)}{\tau(2\pi)} = \left[\frac{\mathcal{S}_L + m_1^2(1 - \mathcal{S}_L)}{2\pi^2 m_1^2 I_0^2(\rho)}\right]\int_0^\pi \frac{d\phi}{p(\phi)} \tag{11-61}$$

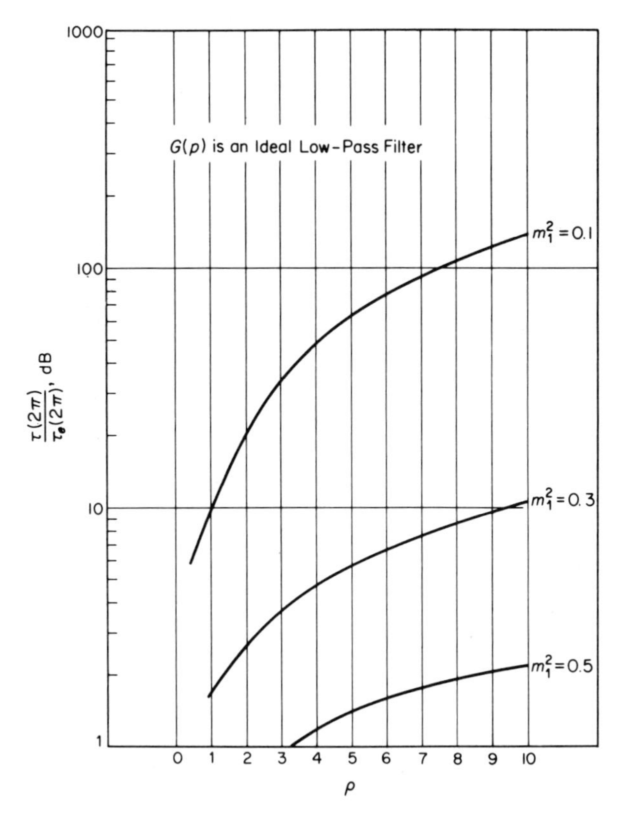

Fig. 11-26 Relative improvement (in dB) of the mean time to first slip of a cycle of the HTL over that of the PLL with m_1^2 as a parameter: $W_L/W_G = 0.1$

HTL and PLL Performance Comparisons

A comparison of the performance of an HTL and a PLL differs primarily in two respects from the performance comparison in Sec. 11-3. First, since the hybrid scheme does not depend upon decisions made on the data, its performance is independent of the symbol error probability. Second, because of the dependence of the squaring loss on the transfer function of the low-pass filter $G(s)$, it is necessary to assume a functional form for this characteristic in order to obtain specific numerical results.

To begin the performance comparison, we look at the equivalent phase detector characteristic of the HTL with $\Omega_0 = 0$ and $G = G_0$, which is given by

$$\mathcal{S}(\phi) = \frac{\sin \phi + (M^2 \mathcal{S}_L/2) \sin 2\phi}{\sqrt{1 + M^2 \mathcal{S}_L}} \tag{11-62}$$

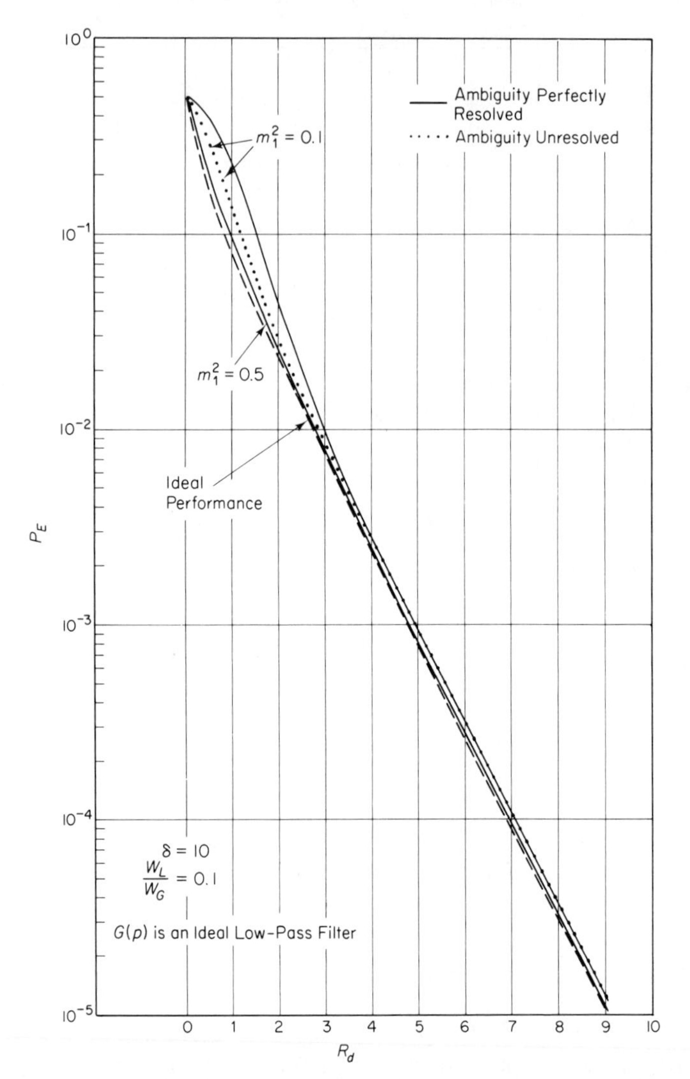

Fig. 11-27 Error probability performance of a PSK system using an HTL with m_1^2 as a parameter: $\delta = 10$, $W_L/W_G = 0.1$

Plots of $\mathcal{S}(\phi)$ versus ϕ with m_1^2 and $\rho_G \triangleq 2P/(N_0 W_G)$ as parameters, and $G(s)$ characterized by an ideal filter characteristic are given in Figs. 11-21, 11-22, and 11-23; W_G is the two-sided bandwidth of the filter $G(s)$, and $\mathcal{S}_L$ is given in terms of ρ_G by

$$\mathcal{S}_L = \frac{1}{1 + \{1/[(1 - m_1^2)\rho_G]\}} = \frac{1}{1 + \{1/[M^2 \rho(W_L/W_G)]\}} \tag{11-63}$$

Fixing ρ_G in the above is equivalent to simultaneously holding the total power-to-noise ratio in the loop bandwidth and the ratio of W_L to W_G constant. The effect of the squaring loss on the tracking improvement obtainable with an HTL relative to the use of a PLL is illustrated in Fig. 11-24 where $\sigma_\phi^2/\sigma_{\phi e}^2$ is plotted in dB versus ρ for fixed m_1^2, an ideal low-pass filter for $G(s)$, and three values of W_L/W_G. Also assumed in the figure is that the ambiguities caused by the trimodal nature of the phase error p.d.f. have been perfectly resolved. We observe that at $\rho = 10$, and $W_L/W_G = 0.01$ (or below), the relative improvement in mean-squared tracking error is far from its asymptotic value of $1/m_1^2$. A plot of $\sigma_\phi^2/\sigma_{\phi e}^2$ in dB versus ρ for $W_L/W_G = 0.1$ and three values of m_1^2 is given in Fig. 11-25. Again an ideal low-pass filter is assumed for $G(s)$, and results are given for the case where the ambiguities are perfectly resolved, and the case where they are not. For $m_1^2 = 0.3$ and $m_1^2 = 0.5$, these two cases are indistinguishable. Figure 11-26 illustrates the relative improvement in the mean time to first slip performance obtained by using the HTL [see Eq. (11-61)]. The same $G(s)$ and bandwidth ratio W_L/W_G as above are assumed. Finally, Fig. 11-27 illustrates the error probability performance of a binary PSK system using an HTL for purposes of carrier tracking. Again as in the DAL situation (Fig. 11-17) we observe near-ideal performance for values of P_E less than 10^{-2}.

It is worthwhile considering other circuit implementations of the basic HTL illustrated in Fig. 11-20. A band-pass equivalent of the HTL having performance identical to that of the low-pass version just discussed is shown in Fig. 11-28, where the band-pass filter $H(s)$ is assumed to have an equivalent transfer characteristic to $G(s)$ of Fig. 11-20. We present this

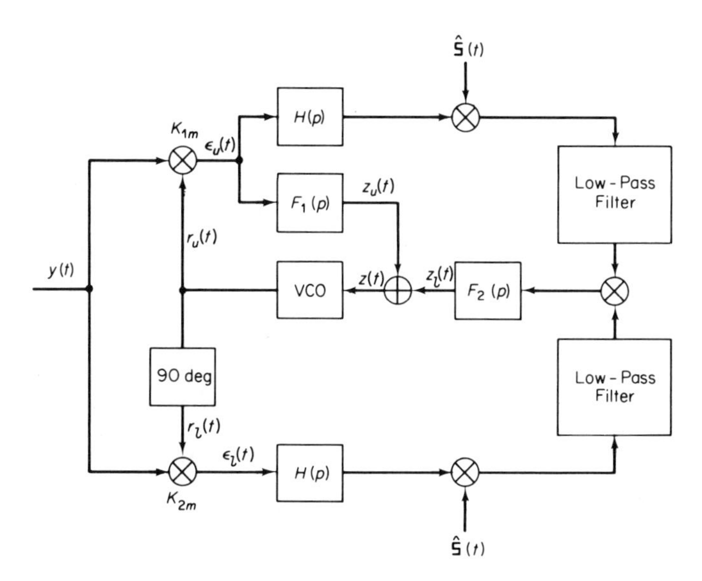

Fig. 11-28 A band-pass version of the HTL

circuit configuration principally as a stepping stone to the next implementation to be discussed. Development of Eq. (11-47) assumed a perfect subcarrier reference in the receiver. The subcarrier reference jitter can easily be included to demonstrate its ultimate effect on loop performance. More important is the possibility of eliminating the provision for a subcarrier reference, thus simplifying the loop mechanization as well as the signal acquisition procedure—that is, the signal can be acquired without opening the lower loop until the RF carrier is acquired, then acquiring the subcarrier, and then closing the lower loop.

The simplified loop with subcarrier references omitted is shown in Fig. 11-29. The loop equation is now given by (11-47) with $N_l(t, \varphi)$ redefined as

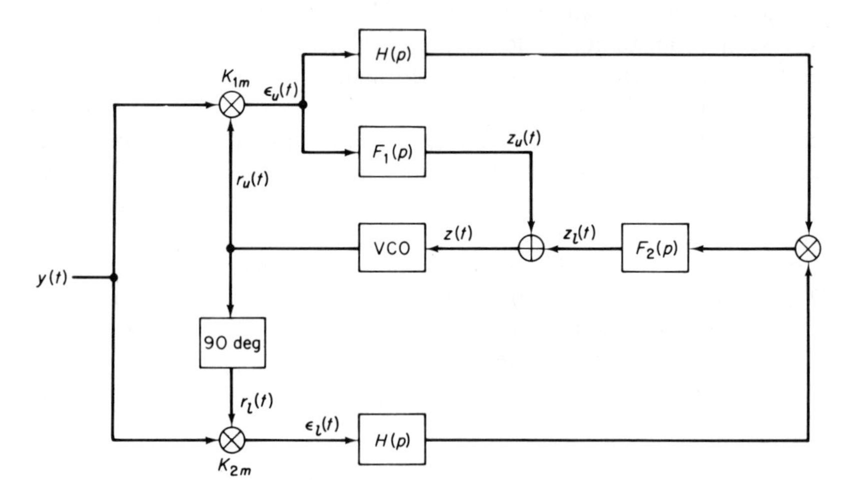

Fig. 11-29 The HTL with subcarriers eliminated

$$N_l(t, \varphi) \triangleq \sqrt{2}\,N_1(t)N_2(t) + d(t)\sqrt{S}[N_1(t)\sin\varphi + N_2(t)\cos\varphi] \qquad (11\text{-}64)$$

and loop performance is obviously inferior, since we get twice as much noise variance from the the noise × noise term. For this circuit, loop performance is found from (11-52) through (11-61) with $\mathcal{S}_L^{-1}$ defined by

$$\mathcal{S}_L^{-1} \triangleq 1 + \frac{4}{N_0 S}\int_{-\infty}^{\infty} R_{N_1}^2(\tau)d\tau \qquad (11\text{-}65)$$

The coherent receiver analyzed here is essentially a linear combination of a PLL and a modified Costas loop. It should be pointed out that this receiver is easier to implement than the DAL, proposed in the previous section, which employs decision-directed feedback. The HTL does not require implementation of a delay in the loop nor does it require the use of the data estimate–subcarrier estimate product. It is well known that extraneous delays in a PLL reduce the loop acquisition range, etc.

11-5. APPLICATIONS TO MULTI-CHANNEL SYSTEMS

The principles of the DAL and HTL are clearly not limited to single-channel communication system applications. In particular, when the phase modulation $\theta(t)$ consists of several data-modulated subcarriers, coherency of the multiple sideband components can be exploited at the carrier frequency by means of multiple loops. Consider first as an example the two-channel system discussed in Chap. 7. A DAL configuration that makes use of the power in both sideband components and the cross-modulation component to improve RF carrier-tracking performance is illustrated in Fig. 11-30. Assuming identical loop filters in each of the four branches—$F_c(s) = F_1(s) = F_2(s) = F_{12}(s) = F(s)$—and the loop phase error essentially constant over the longer of the two symbol intervals, then by an approach similar to that given in Sec. 11-2, the stochastic integro-differential equation describing the loop operation can be expressed in the form

$$\begin{aligned}\dot{\varphi} = \Omega_0 - F(p)\Big[\{&K_c\sqrt{m_1^2 m_2^2 P} \\ &+ K_{12}\sqrt{(1-m_1^2)(1-m_2^2)P}[1-2P_{E1}(\varphi)][1-2P_{E2}(\varphi)] \\ &+ K_{d1}\sqrt{m_2^2(1-m_1^2)P}[1-2P_{E1}(\varphi)] \\ &+ K_{d2}\sqrt{m_1^2(1-m_2^2)P}[1-2P_{E2}(\varphi)]\}\sin\varphi + N_e(t,\varphi)\Big]\end{aligned} \tag{11-66}$$

where

$$\begin{aligned}K_c &= K_1 K_{cm} K_V \qquad & K_{12} &= K_1 K_{12m} K_V \\ K_{d1} &= K_2 K_{1m} K_V \qquad & K_{d2} &= K_2 K_{2m} K_V\end{aligned} \tag{11-67}$$

and

$$\begin{aligned}N_e(t,\varphi) = K_c N_c(t,\varphi) &+ K_{12}\hat{d}_1(t)\hat{d}_2(t)N_{12}(t,\varphi) + K_1\hat{d}_1(t)N_1(t,\varphi) \\ &+ K_2\hat{d}_2(t)N_2(t,\varphi)\end{aligned} \tag{11-68}$$

The noise processes $N_c(t, \varphi)$, $N_{12}(t, \varphi)$, $N_1(t, \varphi)$ and $N_2(t, \varphi)$ are all approximately low-pass Gaussian and mutually independent since they result from different frequency bands of the input wideband noise spectrum. Hence, $N_e(t, \varphi)$ has a spectral density

$$S_{N_e}(\omega) = \frac{N_0}{2}[K_c^2 + K_{12}^2 + K_1^2 + K_2^2] \tag{11-69}$$

Again the assumption is required that the bandwidth of the noise components is wide relative to the higher of the two data rates.

The p.d.f. of ϕ still has the canonical form of (2-99) with the potential function $U_0(\phi)$ now given by

$$U_0(\phi) = -\beta\phi + \alpha\frac{\int f(\phi)\sin\phi\,d\phi}{f(0)} \tag{11-70}$$

with

$$\begin{aligned}f(\phi) \triangleq 1 &+ G_{12}M_1M_2[(1-2P_{E1}(\phi)][1-2P_{E2}(\phi)] \\ &+ G_1M_1[1-2P_{E1}(\phi)] + G_2M_2[1-2P_{E2}(\phi)]\end{aligned} \tag{11-71}$$

and
$$G_{12} \triangleq \frac{K_{12}}{K_c} \qquad G_1 \triangleq \frac{K_{d1}}{K_c} \qquad G_2 \triangleq \frac{K_{d2}}{K_c}$$
$$M_1 \triangleq \sqrt{\frac{1-m_1^2}{m_1^2}} \qquad M_2 \triangleq \sqrt{\frac{1-m_2^2}{m_2^2}} \tag{11-72}$$

Definitions of the parameters α and β as given by (2-17) once again apply here with the following replacements

$$P_c \longrightarrow P_c\left[\frac{f(0)}{\sqrt{1+G_{12}^2+G_1^2+G_2^2}}\right]^2$$
$$K \longrightarrow K_c\sqrt{1+G_{12}^2+G_1^2+G_2^2} \tag{11-73}$$
$$\sin\phi \longrightarrow \sin\phi\left[\frac{f(\phi)}{f(0)}\right]$$

The p.d.f. $p(\phi)$ of (2-99), together with its associated $U_0(\phi)$ as described by Eqs. (11-70) through (11-73), is sufficient to determine all steady-state tracking and error probability performance characteristics of a two-channel system using the DAL of Fig. 11-30.

In a similar manner, one can extend the notion of an HTL to two-channel system applications. The loop is now designed so that a feed-forward path is provided for each of the two data channels and one for the cross-

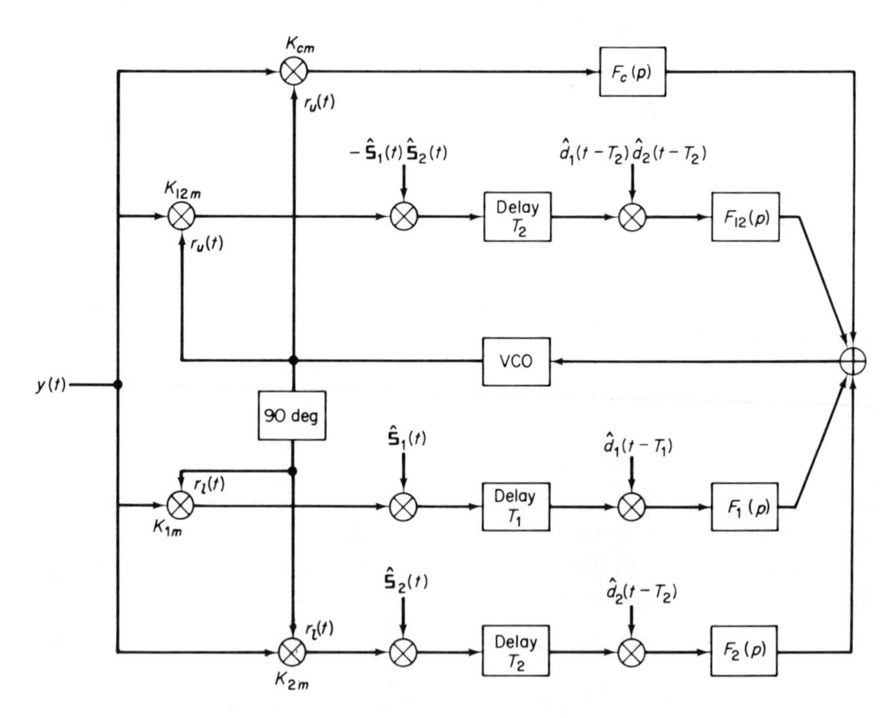

Fig. 11-30 A DAL for a two-channel system

modulation product (Fig. 11-31). In general, low-pass filters $G_1(s)$, $G_2(s)$, and $G_{12}(s)$ will be designed differently (perhaps only in bandwidth) since the modulation bandwidths of the data channels and the cross-modulation term are not necessarily restricted to be equal. However, for simplicity one may assume that loop filters $F_c(s)$, $F_1(s)$, $F_2(s)$, and $F_{12}(s)$ are all equal to $F(s)$.

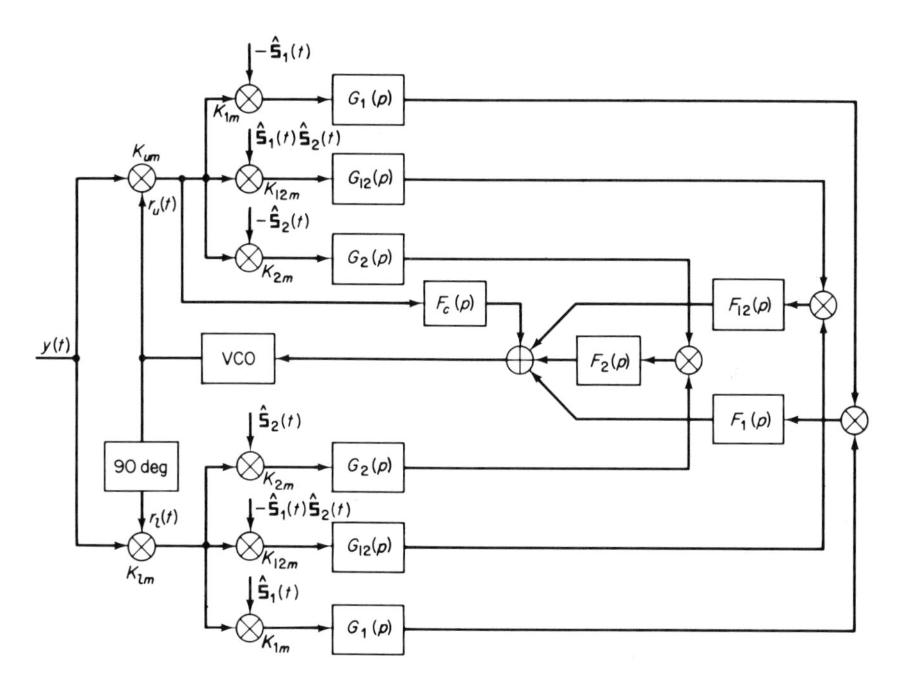

Fig. 11-31 An HTL for a two-channel system

Under these conditions, the stochastic integro-differential equation describing loop operation is given by

$$\dot{\varphi} = \Omega_0 - F(p)\Big[K_c\sqrt{m_1^2 m_2^2 P}\sin\varphi + \{K_{12}(1 - m_1^2)(1 - m_2^2)P$$

$$+ K_{d1}m_2^2(1 - m_1^2)P + K_{d2}m_1^2(1 - m_2^2)P\}\frac{\sin 2\phi}{2} + N_e(t, \varphi)\Big] \tag{11-74}$$

where

$$K_c = K_1 K_{um} K_V \qquad K_{12} = K_{12m}^2 K_2 K_{lm} K_c$$
$$K_{d1} = K_{1m}^2 K_2 K_{lm} K_c \qquad K_{d2} = K_{2m}^2 K_2 K_{lm} K_c \tag{11-75}$$

and

$$N_e(t, \varphi) = K_c N_c(t, \varphi) + K_{12}N_{12}(t, \varphi) + K_1 N_1(t, \varphi) + K_2 N_2(t, \varphi) \tag{11-76}$$

$N_c(t, \varphi)$ is the usual low-pass noise process (single-sided spectral density N_0 w/Hz) that affects the PLL part of Fig. 11-31 and

$$N_1(t, \varphi) = N_{u1}(t)N_{l1}(t) + d_1(t)\sqrt{m_2^2(1 - m_1^2)P}[N_{u1}(t) \sin \varphi + N_{l1}(t) \cos \varphi]$$
$$N_2(t, \varphi) = N_{u2}(t)N_{l2}(t) + d_2(t)\sqrt{m_1^2(1 - m_2^2)P}[N_{u2}(t) \sin \varphi + N_{l2}(t) \cos \varphi]$$
$$N_{12}(t, \varphi) = N_{u12}(t)N_{l12}(t) + d_1(t)d_2(t)\sqrt{(1 - m_1^2)(1 - m_2^2)P} \times [N_{u12}(t) \sin \varphi + N_{l12}(t) \cos \varphi] \quad (11\text{-}77)$$

The noise components $N_{u1}(t)$, $N_{l1}(t)$, $N_{u2}(t)$, $N_{l2}(t)$, $N_{u12}(t)$, and $N_{l12}(t)$, are pairwise independent, band-limited Gaussian noise processes with two-sided, low-pass spectral densities

$$S_{N_{u1}}(\omega) = S_{N_{l1}}(\omega) = \frac{N_0}{2}|G_1(j\omega)|^2$$
$$S_{N_{u2}}(\omega) = S_{N_{l2}}(\omega) = \frac{N_0}{2}|G_2(j\omega)|^2 \quad (11\text{-}78)$$
$$S_{N_{u12}}(\omega) = S_{N_{l12}}(\omega) = \frac{N_0}{2}|G_{12}(j\omega)|^2$$

Again, without going into any detailed discussion, we shall present the potential function associated with the phase error p.d.f. from which various performance measures that characterize the system can be determined. Defining the squaring loss coefficients $\mathcal{S}_{L1}$, $\mathcal{S}_{L2}$, and $\mathcal{S}_{L12}$ by

$$\mathcal{S}_{L1} = \left[1 + \frac{2}{N_0 P_1}\int_{-\infty}^{\infty} R_{N_{u1}}^2(\tau)\, d\tau\right]^{-1}$$
$$\mathcal{S}_{L2} = \left[1 + \frac{2}{N_0 P_2}\int_{-\infty}^{\infty} R_{N_{u2}}^2(\tau)\, d\tau\right]^{-1} \quad (11\text{-}79)$$
$$\mathcal{S}_{L12} = \left[1 + \frac{2}{N_0 P_{12}}\int_{-\infty}^{\infty} R_{N_{u12}}^2(\tau)\, d\tau\right]^{-1}$$

and recalling the definitions of Eq. (11-72), the potential function $U_0(\phi)$ is given by

$$U_0(\phi) = -\beta\phi - \alpha\left\{\frac{\cos \phi + (C_m/4) \cos 2\phi}{1 + C_m}\right\} \quad (11\text{-}80)$$

with
$$C_m \triangleq G_{12}M_{12}P_{12} + G_1M_1P_1 + G_2M_2P_2 \quad (11\text{-}81)$$

and
$$P_{12} \triangleq (1 - m_1^2)(1 - m_2^2)P$$
$$P_1 \triangleq m_2^2(1 - m_1^2)P \quad (11\text{-}82)$$
$$P_2 \triangleq m_1^2(1 - m_2^2)P$$

The necessary replacements in the definitions of α and β as given by (2-17) are

$$P_c \longrightarrow P_c\left[\frac{1 + C_m}{\sqrt{1 + G_{12}^2P_{12}\mathcal{S}_{L12}^{-1} + G_1^2P_1\mathcal{S}_{L1}^{-1} + G_2^2P_2\mathcal{S}_{L2}^{-1}}}\right]^2$$
$$K \longrightarrow K_c\sqrt{1 + G_{12}^2P_{12}\mathcal{S}_{L12}^{-1} + G_1^2P_1\mathcal{S}_{L1}^{-1} + G_2^2P_2\mathcal{S}_{L2}^{-1}} \quad (11\text{-}83)$$
$$\sin \phi \longrightarrow \sin \phi\left[\frac{1 + (C_m/2)(\sin 2\phi/\sin \phi)}{1 + C_m}\right]$$

Although in this section we have not specifically discussed the choice of the system loop gains (e.g., G_1, G_2, and G_{12}) to optimize the two-channel DAL or HTL, it should be clear from the foregoing how one should proceed. Typically, one could find the effective loop signal-to-noise ratio and maximize with respect to G_1, G_2, and G_{12} the ratio of this quantity to its equivalent for a PLL.

Generalization of previous results to a multi-channel system wherein a carrier is phase modulated by L sinusoidal data subcarriers and M binary-valued data subcarriers follows directly from the results of Sec. 7-4. In carrying through the analysis, one point arises that has not been discussed previously in this chapter. Demodulation of a transmitted square-wave subcarrier with a square-wave reference (assumed ideal in phase) produces a product signal that is unity for all time. When the transmitted data is biphase modulated onto a *sinusoidal* subcarrier, then demodulation by either a square-wave or sinusoidal reference results in a product signal $S(t)\,\hat{S}(t)$ whose average value is other than unity. For the case of a perfect square-wave reference, this average value equals $2/\pi$, whereas for a sinusoidal reference we must insert the factor 1/2. Hence, in describing the loop dynamics of a system with transmitted sinusoidal subcarriers, one of these two factors must be included in its appropriate place in the loop equation.

APPENDIX A
DERIVATION OF THE MAP ESTIMATE OF $\boldsymbol{\theta}$

We begin by specifying the vector space which appropriately characterizes the received signal, $y(t)$, in each symbol interval.[14] Expanding the signal and noise processes in Eq. (11-2) in terms of the basis functions $\sin m\omega_0 t$ and $\cos m\omega_0 t$, we get, for the kth symbol interval, $k = 1, 2, \ldots K$,

$$\begin{aligned} y(t) &= [\sqrt{2P_c}\cos\theta - \sqrt{2S}d_k\sin\theta]\sin\omega_0 t \\ &\quad + [\sqrt{2P_c}\sin\theta + \sqrt{2S}d_k\cos\theta]\cos\omega_0 t \\ &\quad + \sum_{m=1}^{\infty}\eta_{mk}\cos m\omega_0 t + \sum_{m=1}^{\infty}\xi_{mk}\sin m\omega_0 t \\ &= \sum_{m=1}^{\infty} a_{mk}(\theta, d_k)\cos m\omega_0 t \\ &\quad + \sum_{m=1}^{\infty} b_{mk}(\theta, d_k)\sin m\omega_0 t \qquad (k-1)T \le t \le kT \end{aligned} \tag{A-1}$$

where d_k is the polarity of the kth transmitted symbol, that is, $d_k = m(t)$, $(k-1) \le t \le kT$; η_{mk} and ξ_{mk} are zero mean, uncorrelated Gaussian variables all with variance N_0/T; and $a_{mk}(\theta, d_k)$, $b_{mk}(\theta, d_k)$ are conditionally Gaussian on θ and d_k. Letting $\boldsymbol{\eta}_k$, $\boldsymbol{\xi}_k$, $\mathbf{a}_k$, and $\mathbf{b}_k$ denote vectors with components η_{mk}, ξ_{mk}, $a_{mk}(\theta, d_k)$, and $b_{mk}(\theta, d_k)$ respectively ($m = 1, 2, \ldots$); then the following properties are noted:

1. $\boldsymbol{\eta}_k$ and $\boldsymbol{\xi}_l$ are independent for all $k, l = 1, 2, \ldots K$.

2. $E\{\boldsymbol{\eta}_k, \boldsymbol{\eta}_l\} = E\{\boldsymbol{\xi}_k, \boldsymbol{\xi}_l\} = \sigma^2 I$, where I is the identity matrix.

It is convenient to assume finite demensionality for the vectors, e.g., M, and later let M approach infinity. In this case, $\sigma^2 = (N_0/T)^M$.

3. The vectors $\mathbf{a}_k$ and $\mathbf{b}_k$ are Gaussian conditioned on fixed θ and d_k and are independent from symbol interval to symbol interval. Furthermore, all components of these vectors are zero mean except for the first components in each which represent the signal in each symbol interval and hence contain the entire dependence on θ and d_k.

From this we see that vectors $\mathbf{a}_k$ and $\mathbf{b}_k$ are sufficient to represent the received signal, $y(t)$, in the kth symbol interval. Furthermore, the received signal may be specified over the entire observation interval $0 \leq t \leq KT$ by the partitioned vector $\mathbf{c}_k = (\mathbf{a}_1, \mathbf{a}_2, \ldots \mathbf{a}_K, \ \mathbf{b}_1, \mathbf{b}_2, \ldots \mathbf{b}_K)$, thus the *a posteriori* p.d.f. $p[\theta \mid y(t)]$, $0 \leq t \leq KT$, is stochastically equivalent to the p.d.f. $p(\theta \mid \mathbf{c})$. Letting $\mathbf{d}$ be the data vector whose components are $d_1, d_2, \ldots d_k$ and noting that the random phase and symbol sequence are independent, one obtains from Bayes' rule:

$$p(\theta \mid \mathbf{c}) = \frac{p(\theta)}{p(\mathbf{c})} \int_{\mathbf{d}} p(\mathbf{c} \mid \theta, \mathbf{d}) p(\mathbf{d}) \, d\mathbf{d} \tag{A-2}$$

Since the transmitted symbols are assumed to be independent, their joint p.d.f. $p(\mathbf{d})$ can be expressed as:

$$p(\mathbf{d}) = \prod_{k=1}^{K} p(d_k) \tag{A-3}$$

where $p(d_k)$ for equiprobable symbols is specified by

$$p(d_k) = \tfrac{1}{2}; \; d_k = \pm 1 \qquad k = 1, 2, \ldots K \tag{A-4}$$

Furthermore, from properties 1, 2 and 3 following (A–1),

$$p(\mathbf{c} \mid \theta, \mathbf{d}) = \prod_{k=1}^{K} p(\mathbf{a}_k, \mathbf{b}_k \mid \theta, d_k) \tag{A-5}$$

Hence, combining (A-2), (A-3), and (A-4),

$$p(\theta \mid \mathbf{c}) = \frac{p(\theta)}{p(\mathbf{c})} \int_{\mathbf{d}} \prod_{k=1}^{K} p(\mathbf{a}_k, \mathbf{b}_k \mid \theta, d_k) p(d_k) \, d\mathbf{d} \tag{A-6}$$

Since $p(\theta)$ is assumed uniformly distributed between $-\pi$ and π and $p(\mathbf{c})$ is independent of θ, maximizing $p(\theta \mid \mathbf{c})$ with respect to θ is equivalent to maximizing

$$\int_{\mathbf{d}} \prod_{k=1}^{K} p(\mathbf{a}_k, \mathbf{b}_k \mid \theta, d_k) p(d_k) \, d\mathbf{d} = \prod_{k=1}^{K} \int_{d_k} p(\mathbf{a}_k, \mathbf{b}_k \mid \theta, d_k) p(d_k) d\, d_k \tag{A-7}$$

The conditional p.d.f. $p(\mathbf{a}_k, \mathbf{b}_k \mid \theta, d_k)$ is $2M$-dimensional Gaussian with a diagonal covariance matrix $(N_0/T)I$; that is,

$$p(\mathbf{a}_k, \mathbf{b}_k \mid \theta, d_k) = \frac{1}{(2\pi N_0/T)^M} \exp\left\{-\frac{\sum_{m=1}^{M} [a_{mk}(\theta, d_k) - E\{a_{mk}(\theta, d_k)\}]^2}{(2N_0/T)} - \frac{\sum_{m=1}^{M} [b_{mk}(\theta, d_k) - E\{b_{mk}(\theta, d_k)\}]^2}{(2N_0/T)}\right\} \tag{A-8}$$

Recalling that only $a_{1k}(\theta, d_k)$ and $b_{1k}(\theta, d_k)$ have nonzero mean, we can rewrite (A-8) as

$$p(\mathbf{a}_k, \mathbf{b}_k \mid \theta, d_k) = \frac{1}{(2\pi N_0/T)^M} \exp\left\{-\sum_{m=2}^{M} \frac{(a_{mk}^2 + b_{mk}^2)}{(2N_0/T)}\right\} \times \exp\left\{-\frac{(a_{1k} - \sqrt{2P_c}\sin\theta - \sqrt{2S}d_k\cos\theta)^2}{(2N_0/T)} - \frac{(b_{1k} - \sqrt{2P_c}\cos\theta + \sqrt{2S}d_k\sin\theta)^2}{(2N_0/T)}\right\} \tag{A-9}$$

Averaging over the distribution of d_k as given in (A-4), one obtains:

$$\int_{d_k} p(\mathbf{a}_k, \mathbf{b}_k \mid \theta, d_k)p(d_k)d\, d_k =$$

$$\frac{1}{2(2\pi N_0/T)^M} \exp\left\{-\frac{\sum_{m=1}^{M} (a_{mk}^2 + b_{mk}^2)}{(2N_0/T)} - \frac{(2S + 2P_c)}{(2N_0/T)}\right\} \times \left[\exp\left\{\frac{T}{N_0}a_{1k}[\sqrt{2P_c}\sin\theta + \sqrt{2S}\cos\theta] + \frac{T}{N_0}b_{1k}[\sqrt{2P_c}\cos\theta - \sqrt{2S}\sin\theta]\right\} + \exp\left\{\frac{T}{N_0}a_{1k}[\sqrt{2P_c}\sin\theta - \sqrt{2S}\cos\theta] + \frac{T}{N_0}b_{1k}[\sqrt{2P_c}\cos\theta + \sqrt{2S}\sin\theta]\right\}\right] \tag{A-10}$$

From the orthogonality property of the basis functions, the weighting coefficients can be expressed as

$$a_{1k} = \frac{2}{T}\int_{(k-1)T}^{kT} y(t)\cos\omega_0 t\, dt$$

$$b_{1k} = \frac{2}{T}\int_{(k-1)T}^{kT} y(t)\sin\omega_0 t\, dt \tag{A-11}$$

We have tacitly assumed in the above that the carrier frequency is an integer multiple of the data rate (i.e. the symbols are synchronous with the carrier) and that symbol sync is known exactly. Substituting (A-11) in (A-10), combining terms, and taking the limit as $M \longrightarrow \infty$, we get

$$\int_{d_k} p(\mathbf{a}_k, \mathbf{b}_k \mid \theta, d_k) p(d_k) d\, d_k = C_k \exp\left\{\frac{2}{N_0}\int_{(k-1)T}^{kT} \sqrt{2P_c}\, y(t) \sin(\omega_0 t + \theta)\, dt\right\} \times \cosh\left\{\frac{2}{N_0}\int_{(k-1)T}^{kT} \sqrt{2S}\, y(t) \cos(\omega_0 t + \theta)\, dt\right\} \tag{A-12}$$

where C_k is independent of θ. Thus, from (A-7) and (A-12), the best estimate of θ in the maximum *a posteriori* sense, based on a KT observation of the received signal, $y(t)$, is that value of θ, that is, $\hat{\theta}$, which maximizes the function

$$f(\theta) = \prod_{k=1}^{K} \exp\left\{\frac{2}{N_0}\int_{(k-1)T}^{kT} \sqrt{2P_c}\, y(t) \sin(\omega_0 t + \theta)\, dt\right\} \times \cosh\left\{\frac{2}{N_0}\int_{(k-1)T}^{kT} \sqrt{2S}\, y(t) \cos(\omega_0 t + \theta)\, dt\right\} \tag{A-13}$$

or alternately:

$$\ln f(\theta) = \sum_{k=1}^{K} \left\{\frac{2}{N_0}\int_{(k-1)T}^{kT} \sqrt{2P_c}\, y(t) \sin(\omega_0 t + \theta)\, dt + \ln\cosh\left[\frac{2}{N_0}\int_{(k-1)T}^{kT} \sqrt{2S}\, y(t) \cos(\omega_0 t + \theta)\, dt\right]\right\} \tag{A-14}$$

REFERENCES

1. Van Trees, H. L., *Detection, Estimation and Modulation Theory*, John Wiley & Sons, Inc., New York, N.Y., 1967.
2. ———, "Analog Communication Over Randomly Time-Varying Channels," *1964 WESCON Record*, Part 4, 1964.
3. ———, "Analog Communication Over Randomly Time-Varying Channels," *IEEE Transactions on Information Theory*, Vol. IT-12, No. 1 (January, 1966) 51–63.
4. ———, "The Structure of Efficient Demodulators for Multi-Dimensional Phase Modulated Signals," *IEEE Transactions on Communication Systems*, Vol. CS-11, No. 3 (September, 1963) 261–271.
5. Thomas, J. B., and Wong, E., "On the Statistical Theory of Optimum Demodulation," *IRE Transactions on Information Theory*, Vol. IT-6, No. 4 (September, 1960) 420–425.
6. Snyder, D. L., "The State-Variable Approach to Analog Communication Theory," *IEEE Transactions on Information Theory*, Vol. IT-14, No. 1 (January, 1968) 94.
7. Riter, S., "An Optimum Phase Reference Detector for Fully Modulated Phase Shift Keyed Signals," *IEEE Transactions on Aerospace and Electronic Systems*, Vol. AES-5, No. 4 (July, 1969) 627–631.

8. BUTMAN, S., and SIMON, M. K., "On the Receiver Structure for a Single-Channel Phase-Coherent Communication System," Jet Propulsion Laboratory, Pasadena, Calif. SPS 37–62, Vol. III (April, 1970) 103–108.

9. LINDSEY, W. C., and SIMON, M. K., "Data-Aided Carrier Tracking Loops," *IEEE Transactions on Communication Technology*, Vol. COM-19, No. 2 (April, 1971) 157–168.

10. PROAKIS, J. G., DROUILHET, P. R., JR., and PRICE, R., "Performance of Coherent Detection Systems using Decision-Directed Channel Measurement," *IEEE Transactions on Communication Systems*, Vol. CS-12, No. 1 (March, 1964) 54–63.

11. LINDSEY, W. C., "Hybrid Carrier and Modulation Tracking Loops," *IEEE Transactions on Communication Technology*, Vol. COM-20, No. 1 (February, 1972) 24–26.

12. STRATONOVICH, R. L., *Topics in the Theory of Random Noise*, Gordon and Breach, New York, N.Y., 1967.

13. LINDSEY, W. C., *Synchronization Systems in Communication and Control*, Prentice-Hall, Inc., Englewood Cliffs, N. J., 1972.

14. DAVENPORT, W. B., JR., and ROOT, W. L., *An Introduction to the Theory of Random Signals and Noise*, McGraw-Hill Book Co., Inc., New York, N.Y., 1958.

INDEX

D

E

F

G

H

I

T

U

V

W

Y